The Cauchy Problem

GIAN-CARLO ROTA, *Editor*
ENCYCLOPEDIA OF MATHEMATICS AND ITS APPLICATIONS

GIAN-CARLO ROTA, *Editor*
ENCYCLOPEDIA OF MATHEMATICS AND ITS APPLICATIONS

GIAN-CARLO ROTA, *Editor*
ENCYCLOPEDIA OF MATHEMATICS AND ITS APPLICATIONS

ENCYCLOPEDIA OF MATHEMATICS
and Its Applications

GIAN-CARLO ROTA, Editor
Department of Mathematics
Massachusetts Institute of Technology
Cambridge, Massachusetts

Editorial Board

GIAN-CARLO ROTA, *Editor*

ENCYCLOPEDIA OF MATHEMATICS AND ITS APPLICATIONS

Volume 18

Section: Analysis
Felix E. Browder, *Section Editor*

The Cauchy Problem

H. O. Fattorini

Departments of Mathematics
and System Science
University of California, Los Angeles

Foreword by
Felix Browder
University of Chicago

1983

Addison-Wesley Publishing Company
Advanced Book Program/World Science Division
Reading, Massachusetts

London · Amsterdam · Don Mills, Ontario · Sydney · Tokyo

Library of Congress Cataloging in Publication Data

Fattorini, H. O. (Hector O.), 1938-
   The Cauchy problem.

   (Encyclopedia of mathematics and its applications;
v. 18)
   Bibliography: p.
   Includes index.
   1. Cauchy problem.   2. Initial value problems.
I. Title, II. Series.
QA378.F37   1983   515.3'54   82-22669
ISBN 0-201-13517-5

Manufactured in the United States of America

ABCDEFGHIJ-HA-89876543

Contents

Editor's Statement

A large body of mathematics consists of facts that can be presented and described much like any other natural phenomenon. These facts, at times explicitly brought out as theorems, at other times concealed within a proof, make up most of the applications of mathematics, and are the most likely to survive change of style and of interest.

This ENCYCLOPEDIA will attempt to present the factual body of all mathematics. Clarity of exposition, accessibility to the non-specialist, and a thorough bibliography are required of each author. Volumes will appear in no particular order, but will be organized into sections, each one comprising a recognizable branch of present-day mathematics. Numbers of volumes and sections will be reconsidered as times and needs change.

It is hoped that this enterprise will make mathematics more widely used where it is needed, and more accessible in fields in which it can be applied but where it has not yet penetrated because of insufficient information.

GIAN - CARLO ROTA

Foreword

The Cauchy problem (whose name was coined by Jacques Hadamard in his classical treatise *Lectures on Cauchy's Problem in Linear Partial Differential Equations* published in the Silliman Lecture Series by Yale University Press in 1921) is one of the major problems of the theory of partial differential equations, both in its classical form as it arose in the late nineteenth and early twentieth centuries and in the modern theory, which has seen such a meteoric development since the Second World War. In the classical period, it appeared in two significantly different forms: first as the basic formulation for the most fundamental result in the theory of partial differential equations in the analytic domain—the Cauchy–Kowalewski theorem—as well as the classical boundary value problem, which was relevant to the study of both the wave equation and the more general class of second-order equations of hyperbolic type. In the Cauchy–Kowalewski theorem, the basic local existence theorem for a general (or in the classical case, general second-order) partial differential equation in analytic form with the highest normal derivative near a point on a surface written in terms of derivatives of lower normal-order is given in terms of the Cauchy data—that is, the prescription of the lower normal derivatives on the surface. The Cauchy problem for the wave equation is solvable globally (i.e., for the whole space, or at the very least a non-microscopic region) in terms of Cauchy data on a non-characteristic surface.

The present volume is devoted to an extensive development and exposition of the application of the concept of the *abstract Cauchy problem* to partial differential equations. It might therefore be of value for the relatively uninitiated reader to have stated in a simple and nontechnical form, the reasons that have motivated this concept and that make it worthwhile to study the *abstract Cauchy problem* even in the domain of partial differential equations in which the *concrete Cauchy problem* still plays a major role.

The reasons appear in the discussions of Hadamard's book (and his papers that it summarizes) where for the first time a clear view was developed of the major thrust of the modern theory of boundary value problems for partial differential equations. During the last four decades of the nineteenth century both mathematicians and mathematical physicists (the two fields not being fundamentally distinct at that time) devoted a great deal of effort of the highest degree of originality and technical power to the study of a number of classical boundary value problems for partial differential equations of the greatest significance in mathematical physics, complex function theory, and differential geometry. These include in partic- ular the Dirichlet problem for the Laplace equation, the initial-boundary value problem for the heat equation, and the Cauchy problem for the wave equation. Yet it seems justified to say that the significance of these results for a general theory of partial differential equations and a corresponding theory of boundary value problems was first clarified by Hadamard's very simple criterion for the well-posedness of boundary value problems: that solutions should exist for general classes of reasonable non-analytic data, that they should be uniquely determined by data in such classes, and continuous in appropriate norms on the data and the solutions. Hadamard pointed out that many non-classical problems could be easily shown to be *ill-posed*. It is to this criterion and the corresponding analysis of partial differential operators in terms of their characteristic forms that we owe the basic principles of classification of partial differential equations, a principle clearly absent in the fundamental result in the analytic case—the theorem of Cauchy–Kowalewski.

We must look beyond these remarks (which of course appear in every contemporary textbook on partial differential equations) to a conceptual consequence of considerable historical interest. Hadamard's remark was essentially the application to the concrete analytic area of partial differential equations of a fundamental principle from another developing area of mathematical study: functional analysis. His criterion could be most clearly stated in the general language: Examine the mapping or operator T, which assigns to each datum element f the corresponding solution u of the boundary value problem. Then the problem is well-posed if $u = T(f)$ is well-defined for each f, and continuous from the space of data (ap- propriately normed) to the space of solutions.

The abstract Cauchy problem is both a refinement and a more specific form for the direction of thought implied by Hadamard's criterion. Whatever the formal dates may be for its detailed formulation, it seems to have arisen between the two World Wars as a response to developments in two originally distinct mathematical domains. Most explicitly, it appears in the study of the initial-value problem for the Schrödinger equation, which is central in the Schrödinger formulation of the quantum theory, an ordinary differential equation of first order for an unknown function u from the real line to an infinite-dimensional Hilbert space. The initial-value problem for this equation,

$$\frac{du}{dt} = \frac{i}{\hbar} H(u),$$

gives a paradigm for the study of such initial-value problems in infinite-dimensional spaces. The corresponding development in the 1930s and beyond of the theory of Markoff processes and more general stationary random processes with such basic principles as the semi-group law embodied in the Chapman–Kolmogoroff equations gave rise to strong motives for the post-war study of the theory of one-parameter semi-groups—that is, maps from the non-negative real numbers t in R^+ to operators $U(t)$ in such infinite-dimensional spaces as $C(\Omega)$ or $L^1(d\mu)$ satisfying the semi-group law $U(t)U(s) = U(t + s)$. The central theme of this study was the determination of the infinitesimal generator of the semi-group—that is, an operator A such that $U(t)f = u(t)$ could be characterised as the solution (at least for suitably nice f) of the differential equation $du/dt(t) = Au(t)$, with $u(0) = f$. Conversely, the theory of analytic semi-groups [and its extension to the corresponding theory of evolution equations that generalize the semi-group law to the corresponding composition conditions that are valid for ordinary differential equations

$$\frac{du}{dt} = A(t)u(t)]$$

developed a sophisticated apparatus for the generation of semi-groups by linear operators A (usually closed and densely defined but not continuous), which is a significant branch of the modern theory of linear functional analysis. More recently still, there has been a very flourishing development of a corresponding theory for nonlinear operators A and nonlinear mappings $U(t)$, the latter being non-expansive or Lipschitzian mappings, which has extended a significant part of the linear theory to the nonlinear domain.

The present volume by Professor Fattorini appears at an interesting juncture in this process of development. The paradigm of the theory of one-parameter semi-groups of linear mappings and their relation to their infinitesimal generators has reached and passed its apogee. Even the nonlinear theory has probably achieved its principal goals as far as very general results are concerned. Yet the application of this now-classical machinery to the basic core of problems concerning the classical equations of mathematical physics has not been thoroughly exposed and digested in the expository literature available to nonspecialists and the mathematically interested in various domains of science and engineering. One needs a form of exposition to use in the analysis of applied problems rather than on the frontier of technical research in this area of functional analysis.

Professor Fattorini's book makes an important contribution to this process. Its central concern is with the equations of mathematical physics. It spends relatively little space on the other broad theme of application—the theory of stochastic processes. This is perfectly natural, however, since the nature of the terrain of these two major domains of application have such significant conceptual and technical differences. One may hope with such treatments as this book presents that the relatively completed paradigm of the abstract Cauchy problem will become an easily usable and well-understood instrument in the domains for which it was designed to apply.

FELIX E. BROWDER

Preface

Consider an initial value or initial-boundary value problem

$$u_t = Au, \quad u = u_0 \quad \text{for} \quad t = 0, \tag{1}$$

where A is, say, a partial differential operator in the "space variables" $x_1, \ldots, x_m$. It was discovered independently by E. Hille and K. Yosida about forty years ago that (1) can be studied to advantage under the form of an "ordinary differential" initial value problem

$$u'(t) = Au(t), \quad u(0) = u_0, \tag{2}$$

where now A is thought of as an operator in a suitable function space E, u is a function of t taking values in E, and the boundary conditions (if any) are included in the definition of the space or of the domain of A. The formal similarity of (2) with a system of ordinary differential equations provides us with heuristic insight on the original problem; for instance, we may expect to be able to write the solution of (2) (thus of (1)) in the form $u(t) = \exp(tA)u_0$ (this is in fact true if the exponential is correctly interpreted). Moreover, results on (2) may apply to many different types of partial differential equations or even to more general equations, an expectation that is borne out in practice.

A large body of theory along these lines was developed during and after the forties and now permeates most advanced treatments of hyperbolic and parabolic initial-boundary value problems; it applies equally well to equations not purely differential, such as the neutron transport equation. Highlights of this development have been the introduction of dissipative operators by R. S. Phillips and the extension of many of the basic results to operators A depending on t by T. Kato and H. Tanabe. Finally, the theory of the initial value problem (2) has been extended in the last twenty years to include nonlinear equations; this has proved to be a deep and fruitful field where important research is taking place even today. Along the way, the theory of abstract differential equations and its equivalent formulation—semigroup theory—have found significant applications in many areas; among the most recent have been singular perturbations and control theory. We may also mention that semigroup theory has been an essential language in such computational developments as finite difference methods for partial differential equations.

Nowadays, many volumes devoted wholly or partly to the treatment of semigroup theory exist, foremost among them the encyclopedic treatise of Hille and Phillips, still the standard reference in many areas. In contrast, accounts of the applications to particular partial differential and other equations are scarcer, usually being part of treatises on partial differential equations. Other basic applications are only found within the research literature and are for this reason not readily accessible to nonspecialists.

I have attempted to bridge this gap, at least partly, in the present volume by collecting some basic results on the equation (2) and on its time-dependent version that can be readily applied to a variety of equations and are (or may be suspected to be) in a reasonably definitive form. Most of the material presented is on applications of these results. Anything resembling completeness in so vast a field is of course out of the question, but I hope the wide range of examples presented will provide the reader with fairly general and useful ideas on how to fit an equation like (1) into the mold of abstract differential equations, and on what the general results mean when applied to particular equations. A specialist may find here and there (perhaps in the large bibliography) some new facts; however, the intended audience for this book is scientists, engineers, and applied mathematicians looking for efficient ways to handle particular problems.

The prospective reader is expected to have some familiarity with ordinary differential equations and a good knowledge of real variable theory, in particular the Lebesgue integral and Lebesgue spaces; an acquaintance with complex variable at the undergraduate level is sufficient. Also, a knowledge of elementary functional analysis is necessary; most of what is needed is included, partly without proofs, in the introductory Chapter 0, although the only indispensable information there is that on resolvents of unbounded operators. No familiarity with the theory of partial differential equations is assumed (except in some parts of Chapters 1 and 4); however, some information on the classical equations (Laplace, wave, heat) may help to put results in perspective. Within these requirements this book is essentially self-contained with the exception of numerous results on Sobolev spaces and two on singular integrals, which are stated without proof in Chapter 4. Some distribution theory is used through the text; the few facts on vector-valued distributions needed in Chapter 8 are presented there with complete proofs.

The contents of this book can be described as follows. Chapter 1 consists almost exclusively of examples drawn from problems of mathematical physics; the resulting equations are treated by ad hoc methods (Fourier series and transforms) and the results provide motivation for the definition of well-posed Cauchy problem. The resulting theory is examined in Chapter 2. Chapter 3 discusses the particular case corresponding to dissipative operators and some related facts (such as semigroups in Banach lattices) with applications to second order ordinary differential operators and symmetric hyperbolic equations. Chapter 4 is on abstract parabolic equations (chiefly the analytic case) and on applications to second order parabolic equations. Chapter 5 deals with perturbation theory; the applications include the neutron transport equation and the Schrödinger and Dirac equations with potentials. Other topics include continuous and discrete approximations to abstract differential equations, among them finite difference methods, which are illustrated with a parabolic initial-boundary value problem. Further considerations on the idea of well posed problem are found in Chapter 6, where formulations different from that of Cauchy problem are introduced in several examples. The theory of the equation (2) with A depending on t is the subject of Chapter 7. Finally, we present in Chapter 8 a brief account of the theory of the Cauchy problem in the sense of distributions. Here the restriction to purely differential equations of the type of (2) is unreasonable and we consider instead hereditary equations; relations with system theory are pointed out. In the case of the equation (2) this formulation is seen to be equivalent to the so-called mildly well-posed Cauchy problem, brought into existence as a tool for the treatment of hyperbolic equations with multiple characteristics.

Some shortcuts through the text will be evident to the reader. For instance, Chapter 1 may be passed over except for the definition of well-posed Cauchy problem and for some results on self-adjointness of the unperturbed Schrödinger and Dirac operators to be used in Chapter 5. Chapters 2, 3, and the first two sections of Chapter 4 are basic for the understanding of most of the subsequent material, as are Sections 1 and 3 of Chapter 5; the rest of these chapters and the remaining ones are fairly independent of each other.

Numerous paragraphs labelled "Example" can be found throughout the text. Some of these are worked out in detail; others are not and should be understood as exercises. The most difficult ones are starred and references are given.

Sections and paragraphs in small type can be omitted without detriment to the comprehension of subsequent material.

Each chapter ends with "Miscellaneous" comments; these give historical information, discuss parts of the theory not treated in detail, and provide bibliographical indications.

The selection of topics reflects of course the tastes and limitations of the author. Without a doubt, the most important subject we have not covered is that of nonlinear equations (although some references to quasilinear equations are found in Chapter 2). Even within the field of linear equations much has been omitted or is barely mentioned in passing. Some of these topics are: applications of semigroup theory to probability, random evolutions and differential operators of order greater than two, as well as the theory of the equation (2) not directly associated with the well posed Cauchy problem.

I am happy to acknowledge my indebtedness to numerous individuals and institutions in relation to the writing of this book. A set of lecture notes that became part of Chapters 2, 3, 4, and 5 was written during 1972 and 1973 at the Universidad de Buenos Aires, in preparation for a projected course on applications of functional analysis. My thanks go to my colleagues there for making possible a fruitful stay, which would not have been viable without the support of the Consejo Nacional de Investigaciones Científicas y Técnicas of Argentina. Likewise, I am grateful to my colleagues at the Università di Firenze and to the Consiglio Nazionale delle Richerche of Italy for their arrangements for my visit during the spring of 1975, at which time the actual writing began.

Finally, I wish to point out that the completion of this book would have been impossible without the effective and understanding support of the National Science Foundation given during the entire period of its preparation.

To my wife Natalia go my sincere thanks for her patience and understanding, as well as for her constant encouragement.

H. O. FATTORINI

The Cauchy Problem

Chapter 0

Elements of Functional Analysis

1. BANACH AND HILBERT SPACES. BOUNDED OPERATORS. INTEGRATION OF VECTOR-VALUED FUNCTIONS

We shall denote by $E, F, X, Y, \dots$ etc. real or complex Banach spaces; a Hilbert space will be named H. Unless otherwise stated all spaces will be assumed *complex*. We denote by $\|\cdot\|_E$ the norm in a Banach space E, by $(\cdot, \cdot)_H$ the scalar product in a Hilbert space H, and by $(E; F)$ the space of all linear bounded operators from E into F endowed with the norm $\|A\| = \sup\{\|Au\|_F; \|u\|_E \le 1\}$ (the "uniform operator norm"). Clearly we have $\|Au\|_F \le \|A\|_{(E; F)}\|u\|_E$; also, if $A \in (F; X)$ and $B \in (E; F)$, then $AB \in (E; X)$ and $\|AB\|_{(E; X)} \le \|A\|_{(F; X)}\|B\|_{(E; F)}$. When confusion is not likely (that is, nearly always), we shall drop the subindices from norms and scalar products; also $(E; E)$ sometimes will be abbreviated to (E).

The following two results are basic.

1.1 Banach-Steinhaus Theorem. *Let $\{A_n\}$ be a sequence in $(E; F)$ such that (a) $\|A_n\| \le C$ for all n. (b) $\lim A_n u$ exists for u in a dense subset of E. Then $Au = \lim A_n u$ exists for all $u \in E$, $A \in (E; F)$, and $\|A\| \le \liminf \|A_n\|$.*

1.2 Uniform Boundedness Theorem. *Let $\mathcal{B}$ be a subset of $(E; F)$ such that $\{Bu; B \in \mathcal{B}\}$ is bounded in F for each $u \in E$. Then $\|B\| \le C$ $(B \in \mathcal{B})$; that is, $\mathcal{B}$ is bounded in $(E; F)$.*

For proofs see Hille-Phillips [1957: 1, pp. 41 and 26].

We recall that invertible operators are an open subset of $(E; F)$. In fact, if $A \in (E; F)$ possesses an inverse $A^{-1} \in (F; E)$, then so does every operator $B \in (E; F)$ with $\|B - A\| < \|A^{-1}\|^{-1}$ and

$$B^{-1} = A^{-1} \sum_{n=0}^{\infty} \left((A - B) A^{-1} \right)^n \in (F; E), \tag{1.1}$$

the series convergent in the norm of $(F; E)$ in the range of B indicated above.

Given $A \in (E)$, the *resolvent set* of A (written ρ or $\rho(A)$) is the set of all complex λ such that $\lambda I - A$ (I the identity operator in E) has an inverse $R(\lambda) = (\lambda I - A)^{-1} \in (E)$, which is called the *resolvent* or *resolvent operator* of A. The complement $\sigma = \sigma(A)$ is the *spectrum* of A. As a consequence of our previous observations on inverses, we deduce that $\rho(A)$ is always open; precisely, if $\lambda \in \rho(A)$, then every μ with $|\mu - \lambda| \le \|R(\lambda)\|^{-1}$ belongs to $\rho(A)$ and

$$R(\mu) = \sum_{n=0}^{\infty} (\lambda - \mu)^n R(\lambda)^{n+1}, \tag{1.2}$$

the series convergent in the indicated range; moreover, $\rho(A)$ contains the set $\langle \lambda; |\lambda| > \|A\| \rangle$ and

$$R(\lambda) = \sum_{n=0}^{\infty} \lambda^{-(n+1)} A^n. \tag{1.3}$$

In particular, $\rho(A)$ is never empty and $\sigma(A)$ is compact. It can be shown (Hille-Phillips [1957: 1, p. 125]) that $\sigma(A)$ is nonempty as well.

If E is a real Banach space, all the previous statements have an immediate counterpart, except that $\sigma(A)$ may be empty (this can be exemplified by a real matrix without real eigenvalues).

Back in the general case, let f be a function defined in a compact interval $a \le t \le b$ and taking values in E. If f is continuous (or only piecewise continuous), the Riemann integral of f can be defined and shown to exist in the same way as in the scalar case as a limit of Riemann sums,

$$\int_a^b f(t) \, dt = \lim \sum_{k=1}^{n} (t_k - t_{k-1}) f(\xi_k),$$

the limit understood in the norm of E. Integrals over multidimensional sets are similarly defined, and improper Riemann integrals (that is, integrals over noncompact sets) are obtained in the usual way as limits of integrals on a convenient sequence of compact subsets. All the properties of ordinary Riemann integrals that make sense in the vector-valued setting can be proved by rather trivial modifications of the classical proofs. An important existence criterion for improper integrals (which, to fix ideas, we state only for the real line) is the following: if J is a (finite or infinite) interval in $\mathbb{R} = (-\infty, \infty)$ and f is an E-valued piecewise continuous function defined in J and such that $\|f(\cdot)\|$ is integrable in J, then f is integrable over J and

$$\left\| \int_J f(t)\, dt \right\| \le \int_J \|f(t)\|\, dt. \qquad (1.4)$$

This result holds as well, of course, in multidimensional regions. We note in passing that the fundamental theorem of calculus holds: if f is an E-valued continuous function defined in $a \le t \le b$ and

$$g(t) = \int_a^t f(s)\, ds, \qquad (1.5)$$

then

$$g'(t) = f(t) \quad (a \le t \le b), \qquad (1.6)$$

the derivative understood as the limit in E of the corresponding quotient of increments. Conversely, if g is continuously differentiable in $a \le t \le b$ with derivative f, then g is given by (1.5) modulo a constant (that is, modulo an element of E that does not depend on t).

Let $A(\cdot)$ be a function defined in an interval J of the real line with values in $(E; F)$. If A is continuous, the previously outlined integration theory can be automatically applied to it. However, we shall have to contend most of the time with functions that are merely *strongly continuous*, that is, such that $t \to A(t)u$ is continuous in F for each $u \in E$. The integral can then be defined "elementwise"; that is, although $\int A(t)\, dt$ may not make sense, $\int A(t)u\, dt$ does for every $u \in E$. We state the following useful result, restricted for definiteness to the real line.

1.3 Lemma. *Let $A(\cdot)$ be strongly continuous in an interval J. Assume that $\|A(\cdot)\|$ is integrable in J. Then the operator defined by*

$$Au = \int_J A(t)u\, dt \qquad (1.7)$$

belongs to $(E; F)$; precisely,

$$\|A\| \le \int_J \|A(t)\|\, dt. \qquad (1.8)$$

The proof is an immediate consequence of (1.4) and the preceding comments.

We note in passing that the integrand in (1.8) is not necessarily continuous under the present assumptions. However, $\|A(t)\| = \sup\{\|A(t)u\|; \|u\| \le 1\}$, where each $t \to \|A(t)u\|$ is continuous, so that $\|A(\cdot)\|$ is lower semicontinuous. On the other hand, if J' is a compact subinterval of J, $\|A(\cdot)u\|$ is continuous, hence bounded in J' for each $u \in E$. It follows from the uniform boundedness theorem (1.2) that $\|A(\cdot)\|$ is bounded in J'. This gives sense to the integrability assumption in Lemma 1.3.

A rather complete exposition of the theory of Riemann (or, more generally, Riemann-Stieltjes) integration of vector-valued functions can be found in Hille-Phillips [1957: 1, Sec. 3.2] and in Hille [1972: 1, Sec. 7.4]. We shall also use a few times the theory of Lebesgue integrals (as generalized by Bochner) of vector-valued functions, which is in Hille-Phillips [1957: 1, Sec. 3.7] and Hille [1972: 1, Sec. 7.5]. We note that both integration theories can be made to function equally well in a real or in a complex space.

2. LINEAR FUNCTIONALS: THE DUAL SPACE. VECTOR-VALUED ANALYTIC FUNCTIONS

Using the notation in the previous section we define E^*, the *dual* of E, as $E^* = (E; \mathbb{C})$ ($\mathbb{C}$ the complex numbers). The elements of E^* are the bounded linear functionals in E. Application of a functional $u^* \in E^*$ to an element u of E will be denoted $\langle u^*, u \rangle$, $u^*(u)$, or $\langle u, u^* \rangle$. It is clear from the definition of the norm in E^* that the "Cauchy-Schwarz inequality"

$$|\langle u^*, u \rangle| \leq \|u^*\| \|u\| \tag{2.1}$$

holds for all $u \in E$, $u^* \in E^*$.

The existence of abundant linear functionals in E is assured by the

2.1   Hahn-Banach Theorem.   *Let N be a subspace of E, $f: N \to \mathbb{C}$ a linear functional such that $|f(u)| \leq C\|u\|$ ($u \in N$). Then there exists a $u^* \in E^*$ such that* (a) $u^*(u) = f(u)$ ($u \in N$). (b) $\|u^*\| \leq C$.

Theorem 2.1 is actually a very particular case of the general Hahn-Banach theorem whose proof can be found in Banach [1932: 1, p. 27] or Hille-Phillips [1957: 1, p. 29]. We note that the Hahn-Banach theorem holds also in real spaces (actually, this was the case originally proved by Banach).

2.2   Corollary.   *Let N be a closed subspace of E, $u \notin E$. Then there exists a functional $u^* \in E^*$ such that $u^*(N) = \{0\}$, $u^*(u) = 1$, $\|u^*\| = 1/d$, where $d = \text{dist}(u, N) > 0$.*

The proof is immediately obtained by defining a functional f in the subspace $\{w = v + \lambda u; v \in N, \lambda \in \mathbb{C}\}$ generated by N and u by $f(v + \lambda u) = \lambda$ and applying Theorem 2.1 (see Hille-Phillips [1957: 1, p. 30]). Taking N itself equal to $\{0\}$, we obtain

2.3   Corollary.   *Let $u \in E$. Then there exists a $u^* \in E^*$ with*

$$\langle u^*, u \rangle = \|u\|^2 = \|u^*\|^2. \tag{2.2}$$

2.5 Lemma. *An arbitrary power series $\Sigma a_n(\zeta - z)^n$ with coefficients n E converges absolutely and uniformly in $|\zeta - z| \leq d'$ if $d' < r$,*

$$1/r = \lim \sup \|a_n\|^{1/n} \tag{2.6}$$

and diverges if $|\zeta - z| > r$. The function $f(\zeta) = \Sigma a_n(\zeta - z)^n$ is analytic in $|\zeta - z| < r$ and the successive derivatives are obtained by differentiation of (2.5) term by term.

For the proofs of these and other results the reader may consult Hille-Phillips [1957: 1, Sec. 3.11]. We note the following interesting consequence of Lemma 2.5. Let $A \in (E)$. Define $r = r(A)$, the *spectral radius* of A, by $r = \sup\{|\lambda|; \lambda \in \sigma(A)\}$. It follows from (1.2) that $R(\lambda)$ is analytic in $\rho(A)$. Applying then Lemma 2.4 to $R(1/\lambda)$, we deduce that (1.3) must converge for $|\lambda| > r$; thus it results from Lemma 2.5 that

$$r(A) = \lim \sup \|A^n\|^{1/n}. \tag{2.7}$$

We note finally that the limsup on the right-hand side of (2.7) is actually a limit; this follows from Pólya-Szego [1954: 1, p. 17, Nr. 98] and from the obvious inequality $\|A^{m+n}\| \leq \|A\|^m \|A\|^n$.

A thorough treatment of the theory of vector-valued analytic functions can be found in Hille-Phillips [1957: 1, Ch. 3] and also in Hille [1972: 1, Ch. 8]. We note in particular that the facts concerning Laurent series and development in power series about singular points extend to the present case.

3. UNBOUNDED OPERATORS; THE RESOLVENT. CLOSED OPERATORS

We shall deal most of the time with operators (e.g., differential operators) that are neither bounded nor everywhere defined. We extend then our previous definition of linear operator by allowing A to be defined in a subspace $D(A)$ of a Banach space E and to take values in another Banach space F. Only the case $E = F$ will be used. The subspace $D(A)$ is called the *domain* of A; in most instances $D(A)$ will be dense in E (we say then that A is *densely defined*) but not always. The *sum* of two operators is defined by $(A + B)u = Au + Bu$ with $D(A + B) = D(A) \cap D(B)$ while the *product* or *composition* is $(AB)u = A(Bu)$, where $D(AB)$ consists of all $u \in D(B)$ with $Bu \in D(A)$. (Note that, even if A and B are densely defined, $A + B$ and AB may not be; in fact, both $D(A + B)$ and $D(AB)$ may reduce to $\{0\}$.) If A is bounded $(\|Au\| \leq C\|u\|, u \in D(A))$ and densely defined, we can extend A by continuity to all of E; we shall then usually assume that all bounded operators are everywhere defined.

An intermediate notion between that of bounded operator and the general definition introduced in this section is that of closed operator. An

By virtue of the obvious bilinearity of the function $\langle$ i
an element $u \in E$ gives rise to a $u^{**} \in E^{**} = (E^*)^*$ througl

$$\langle u^{**}, u^* \rangle = \langle u, u^* \rangle.$$

It is clear that the map $u \to u^{**}$ from E into E^{**} thus obt
moreover, it follows from (2.1) that $\|u^{**}\|_{E^{**}} \leq \|u\|_E$, and fro
actually have $\|u^{**}\|_{E^{**}} = \|u\|$. We thus see that E can be ider
and metrically) with a subspace of E^{**}; in a less precise fashi
that "E is a closed subspace of E^{**}." If $E = E^{**}$—tha
continuous linear functional in E^* can be expressed by the for
some $u \in E$—we say that the space E is *reflexive*.

Let Ω a domain (open connected set) in the complex pl
a function defined in Ω with values in E. We say that f is
holomorphic, in Ω if the derivative $f'(\zeta)$ exists (as the limit in th
of the corresponding quotient of increments) for all $\zeta \in \Omega$.
define analyticity by means of linear functionals as follows: f is
if and only if the scalar function $\langle u^*, f(\cdot) \rangle$ is analytic in Ω for
Clearly the first definition implies the second. Remarkably enou
are actually equivalent. If $A(\cdot)$ is an $(E; F)$-valued function de
the first definition of analyticity (existence of the limit of the
increments in the norm of $(E; F)$) is likewise equivalent to the
that $\langle u^*, A(\cdot)u \rangle$ is an ordinary analytic function for any $u \in$
$u^* \in F^*$ (for a proof see Hille-Phillips [1957: 1, p. 93]). Vec
analytic functions share most properties of ordinary analytic
Roughly speaking, the proofs are carried out by imitating the pro
scalar case (replacing when appropriate modulus by norm) or by
functionals and then using the scalar theory. We state a few facts
be of use later.

2.4   Lemma.   *Let f be a E-valued analytic function in Ω. The*
is a simple closed contour such that Γ and its interior are contained in

$$f^{(n)}(z) = \frac{n!}{2\pi i} \int_\Gamma \frac{f(\zeta)}{(\zeta - z)^{n+1}} d\zeta$$

if z belongs to the interior of Γ and Γ is oriented counterclockwise. (b) $\frac{1}{2}$
f can be developed in power series about z,

$$f(\zeta) = \sum_{n=0}^{\infty} a_n (\zeta - z)^n$$

where $a_n = f^{(n)}(z)/n!$ and the series (2.5) converges absolutely and unif.
in $|\zeta - z| \leq d'$ if $d' < d = \text{dist}(z, \text{ boundary of } \Omega)$ (that is, $\sum \|a_n\| \|\zeta$
converges uniformly in $|\zeta - z| \leq d'$).

operator A is *closed* if and only if whenever $\{u_n\}$ is a sequence in $D(A)$ such that $u_n \to u$ and $Au_n \to v$ for some $u, v \in E$, it follows that $u \in D(A)$ and $Au = v$. Clearly a bounded, everywhere defined operator is closed. An equivalent definition of closedness is the following. Consider the Cartesian product $E \times E$ endowed with any norm that imparts to it the product topology (for instance, $\|(u, v)\|_{E \times E} = \|u\| + \|v\|$). Then $E \times E$ is a Banach space. The *graph* of the operator A is

$$\Gamma(A) = \{\{u, Au\}; u \in D(A)\}.$$

Clearly $\Gamma(A)$ is always a subspace of $E \times E$ and it is closed if and only if A is closed according to the previous definition.

A fundamental result is:

3.1 Closed Graph Theorem. *Let A be closed and everywhere defined. Then A is bounded.*

This theorem is equivalent to the following result, which is of great importance in its own right:

3.2 Open Map Theorem. *Let $B \in (E; F)$. Assume that $B(E) = F$. Then if U is an open set in E, $B(U)$ is an open set in F.*

3.3 Corollary. *Let $B \in (E; F)$. Assume that B is one-to-one and $B(E) = F$; that is, assume that B has an inverse B^{-1}. Then $B^{-1} \in (F; E)$.*

For proofs of these and more general results, see Banach [1932: 1, p. 38], Hille-Phillips [1957: 1, pp. 46–47], or Dunford-Schwartz [1958: 1, pp. 55–58].

3.4 Lemma. *Let A be a closed operator in E. Assume that $u(\cdot)$ is defined in a (bounded or unbounded) interval J and takes values in $D(A)$; suppose, moreover, that $u(\cdot)$, $Au(\cdot)$ are continuous (or only piecewise continuous) and that $\|u(\cdot)\|$, $\|Au(\cdot)\|$ are integrable in J. Then $\int_J u(t)\, dt \in D(A)$ and*

$$A \int_J u(t)\, dt = \int_J Au(t)\, dt. \tag{3.1}$$

The proof is immediate for compact intervals; in fact, we only have to approximate $\int u\, dt$ by a sequence of Riemann sums $u_n = \sum_k (t_{k,n} - t_{k-1,n}) u(\xi_{k,n})$ and observe that Au_n is a sequence of Riemann sums approximating $\int Au\, dt$. If J is not compact, we approximate $\int_J u\, dt$ by $\int_{J_n} u\, dt$, where J_n is a sequence of compact subintervals of J, and apply the previous observation.

Naturally, this result can be greatly generalized; the integral may be multidimensional or understood in the sense of Lebesgue–Bochner. For a very general statement the reader may consult Hille-Phillips [1957: 1, p. 83].

Given an operator A (not necessarily densely defined) in E, we define, as in the bounded case, the *resolvent set* of A (denoted again $\rho(A)$)

as the set of all complex λ such that $\lambda I - A$ has a bounded inverse $R(\lambda) = R(\lambda; A)$. In the present situation this means

$$(\lambda I - A)R(\lambda) = I \tag{3.2}$$

and

$$R(\lambda)(\lambda I - A)u = u \quad (u \in D(A)). \tag{3.3}$$

Note that, in view of the definition of the domain of the composition of two operators, (3.2) includes the requirement that $R(\lambda)E \subseteq D(\lambda I - A) = D(A)$. The spectrum of A, $\sigma(A)$, is again defined as the complement of $\rho(A)$. Unlike as in the case where A is bounded, there are examples where $\rho(A)$ or $\sigma(A)$ are empty, but it is still true that $\rho(A)$ is open and that $R(\cdot)$ is an analytic function in $\rho(A)$. Its Taylor development about $\lambda \in \rho(A)$ is given as before by

$$R(\mu) = \sum_{n=0}^{\infty} (\lambda - \mu)^n R(\lambda)^{n+1}, \tag{3.4}$$

the series convergent in (E) for $|\mu - \lambda| \le \|R(\lambda)\|^{-1}$ (in particular, every such μ belongs to the resolvent set $\rho(A)$). For a proof see Dunford-Schwartz [1958; 1, p. 566]. An important ingredient in the proof is the following observation: if $B \in (E; E)$ is one-to-one, then its inverse B^{-1} (which may not be everywhere or even densely defined) is closed. This shows that $\lambda I - A = R(\lambda)^{-1}$ is closed if $\lambda \in \rho(A)$, a fortiori A itself is closed. The converse is not true: a closed operator may very well have an empty resolvent set. It follows immediately from (3.4) that powers and derivatives of the resolvent are related by the following formula:

$$R(\lambda)^{(n)} = (-1)^n n! R(\lambda)^{n+1} \quad (\lambda \in \rho(A), n \ge 1). \tag{3.5}$$

An immediate consequence of formula (3.4) and following comments is:

3.5 Lemma. *Let the sequence* $\{\lambda_n\}$ *belong to* $\rho(A)$; *assume that* $\lambda_n \to \lambda$ *and that* $\|R(\lambda_n)\|$ *remains bounded* (*or, more generally, that* $\|R(\lambda_n)\|$ $= o(|\lambda - \lambda_n|)$. *Then* $\lambda \in \rho(A)$.

In fact, the disk $|\mu - \lambda_n| \le \|R(\lambda_n)\|^{-1}$ must be contained in $\rho(A)$ for all n; for sufficiently large n the disk contains λ.

In intuitive terms, Lemma 3.5 states that "the resolvent must blow up (rapidly enough) when we approach the spectrum."

The next result, equally elementary, establishes that the resolvent may be constructed by analytic continuation.

3.6 Lemma. *Let* Ω *be an open connected set in* $\mathbb{C}$ *such that* $\Omega \cap \rho(A) \ne \varnothing$, *and let* $Q(\lambda)$ *be a* (E)-*valued analytic function in* Ω *such that* $Q(\lambda) = R(\lambda)$ *in* $\Omega \cap \rho(A)$. *Then* $\Omega \subseteq \rho(A)$ *and* $Q(\lambda) = R(\lambda)$ *in* Ω.

To prove Lemma 3.6 use Lemma 3.5 as follows. Take $\lambda_0 \in \Omega \cap \rho(A)$, $\lambda \in \Omega$ and join them by a curve Γ in Ω. Let $\Gamma' = \Gamma \cap \rho(A)$. Then Γ' is open in Γ; but $R(\lambda) = Q(\lambda)$ remains bounded in Γ' so that by Lemma 3.5 Γ' must as well be closed in $\rho(A)$. By connectedness $\Gamma' = \Gamma$.

Equalities (3.2) and (3.3) are sometimes called the *first resolvent equation(s)*. *The second resolvent equation*, which results rather easily from the definition, is

$$R(\lambda) - R(\mu) = (\mu - \lambda)R(\lambda)R(\mu) \quad (\lambda, \mu \in \rho(A)). \qquad (3.6)$$

Another useful equation involving the resolvent can be obtained taking $u \in D(A^m)$, writing (3.3) in the form $R(\lambda)u = \lambda^{-1}u + \lambda^{-1}AR(\lambda)u$, replacing the left-hand side into the right-hand side, and iterating. The result is

$$R(\lambda)u = \frac{1}{\lambda}u + \frac{1}{\lambda^2}Au + \cdots + \frac{1}{\lambda^m}A^{m-1}u$$

$$+ \frac{1}{\lambda^m}R(\lambda)A^m u \quad (\lambda \in \rho(A), \lambda \neq 0). \qquad (3.7)$$

We have observed in the previous section that if A is bounded, then $R(\lambda)$ exists for $|\lambda|$ large enough. This property is shared by the resolvents of some unbounded operators. However, we have:

3.7   Lemma.   *Assume $R(\lambda)$ exists for $|\lambda| > a$ and has a pole at infinity. Then A is bounded.*

In fact, we must have

$$R(\lambda) = \sum_{n=-\infty}^{m} \lambda^n B_n \quad (|\lambda| > a), \qquad (3.8)$$

where $B_n \in (E)$ $(-\infty < n \leq m)$. The coefficients are given by the usual formulas

$$B_n = \frac{1}{2\pi i}\int_{|\lambda| = a + 1} \lambda^{-(n+1)}R(\lambda)\, d\lambda,$$

and it follows then from Lemma 3.4 that $B_n(E) \subseteq D(A)$. Making use of (3.2) in (3.8) and equating coefficients in the power series thus obtained we arrive at the relations $B_m = 0$, $B_n = AB_{n+1}$ $(n \neq -1)$, $B_{-1} = AB_0 + I$, which show that $B_0 = B_1 = \cdots = B_m = 0$, $B_{-1} = I$, $B_{-2} = A$, so that $A \in (E)$ as claimed.

3.8   Example.   (a) Let $E = L^2(0, 1)$ (see Section 7), A the operator

$$Au(x) = u'(x), \qquad (3.9)$$

$D(A)$ defined as the set of all $u \in L^2(0, 1)$ that are absolutely continuous in $0 < t < 1$ and such that $u' \in L^2(0, 1)$. Then A is densely defined and closed but $\rho(A) = \varnothing$ (the equation $(\lambda I - A)u = v$ has multiple solutions.) (b) Let B_0 be the restriction of A defined by $u(0) = 0$. Then B_0 is densely defined and $\rho(B_0) = \mathbb{C}$. $R(\lambda)$ has an essential singularity at infinity. (c) Let B be the restriction of A defined by $u(0) = u(1) = 0$. Then $\rho(B) = \varnothing$ (the equation $(\lambda I - B)u = v$ has no solution except for certain right-hand sides.)

It is sometimes useful to classify complex numbers in the spectrum of an operator. We shall only single out *eigenvalues*—that is, complex numbers λ such that there exists $u \in D(A)$, $u \neq 0$ such that $(\lambda I - A)u = 0$, or

$$Au = \lambda u. \tag{3.10}$$

Solutions $u \neq 0$ of (3.10) are called *eigenvectors* (corresponding to the eigenvalue λ); when λ is an eigenvalue, the set of all solutions of (3.10) is a nontrivial subspace $E(\lambda)$ of E called the *eigenspace corresponding to λ*. If A is closed, its eigenspaces (if any) are closed. Given an eigenvalue λ, a vector $u \in D(A^m)$ is called a *generalized eigenvector* if there exists an integer $m \geq 1$ such that

$$(\lambda I - A)^m u = 0 \tag{3.11}$$

for some $u \in D(A^m)$, $u \neq 0$; *generalized eigenspaces* $E_g(\lambda)$ are defined in the obvious way. A generalized eigenvector has *degree m* if it satisfies (3.11), but $(\lambda I - A)^{m'} u \neq 0$ for $m' < m$; the *degree* of an eigenvalue λ is the maximum of the degree of all its generalized eigenvectors (which may be infinite). Equally infinite may be $m(\lambda)$, the *multiplicity* of λ, defined as dim $E(\lambda)$ and dim $E_g(\lambda)$, the *generalized multiplicity* of λ, denoted $m_g(\lambda)$. Obviously $E(\lambda) \subseteq E_g(\lambda)$, so that $m(\lambda) \leq m_g(\lambda)$.

3.9 Example. *Spectral theory of finite-dimensional operators.* Let $E = \mathbb{C}^m$ ($\mathbb{C}^m$ is m-dimensional unitary space), A an operator from E into E. Then (a) $\sigma(A)$ consists of a finite number of eigenvalues $\lambda_1, \ldots, \lambda_n$ ($n \leq m$), (b) $E = E_g(\lambda_1) \oplus \cdots \oplus E_g(\lambda_n)$, where $\oplus$ indicates direct sum (so that $m_g(\lambda_1) + \cdots + m_g(\lambda_n) = m$).

*3.10 Example. *Spectral theory of compact operators.* An everywhere defined operator A in a Banach space E is *compact* if and only if $\{Au_n\}$ contains a convergent subsequence whenever $\{u_n\}$ is bounded. Then (a) $\sigma(A)$ consists of a (finite or countable) sequence $\{\lambda_1, \lambda_2, \ldots\}$ of complex numbers having an accumulation point at zero when the sequence is infinite, and (b) Each nonzero $\lambda \in \sigma(A)$ is an eigenvalue with $m_g(\lambda) < \infty$ (hence with finite multiplicity). However, unlike in the finite-dimensional case, the space E may not be spanned by the generalized eigenvectors, even in an approximate sense. Consider the *Volterra operator*

$$Vu(x) = \int_0^x u(t)\, dt \tag{3.12}$$

in $L^2(0,1)$; A is compact and $\sigma(A) = \{0\}$. Since 0 is not an eigenvalue of V, V has no generalized eigenvectors at all (see Dunford-Schwartz [1958: 1, Sec. 7.4] or Kato [1976: 1, Sec. 3.7] for complete treatments of spectral properties of compact operators).

3.11 Example. Let A be an operator in a Banach space with $\rho(A) \neq \varnothing$ and such that $R(\lambda)$ is compact for some $\lambda \in \rho(A)$. Then: (a) $R(\lambda)$ is compact for all $\lambda \in \rho(A)$ (use (3.6) and the fact that the sum of two compact operators and the composition of a compact operator and a bounded operator is compact). (b) $\sigma(A)$ is empty or consists of a (finite or infinite) sequence $\lambda_1, \lambda_2, \dots$ of eigenvalues with $m_g(\lambda_k) < \infty$. If the sequence is infinite, then $|\lambda_k| \to \infty$. The case $\sigma(A) = \varnothing$ is exemplified by the operator B_0 of Example 3.8(b); here $B_0^{-1} = V$ (V the Volterra operator defined by (3.12)).

*3.12 Example. *A functional calculus for bounded operators.* Let $A \in (E)$, $\mathcal{K}(A)$ the class of all (scalar-valued) functions f holomorphic in some open set $\mathcal{O}_f \supset \sigma(A)$ (depending in general on f). For every $f \in \mathcal{K}(A)$ we define

$$f(A) = \frac{1}{2\pi i} \int_\Gamma f(\lambda) R(\lambda) \, d\lambda, \tag{3.13}$$

where Γ is a finite union of simple closed curves contained in $\mathcal{O}_f$ and containing $\sigma(A)$ in their interiors. Then: (a) $f(A) \in (E)$. (b) If $f, g \in \mathcal{K}(A)$, then $f + g$ and fg belong to $\mathcal{K}(A)$ and $(f + g)(A) = f(A) + g(A)$, $(fg)(A) = f(A)g(A)$. (c) If $f(\lambda) \equiv 1$, $g(\lambda) \equiv \lambda$, then $f(A) = I$, $g(A) = A$. (d) If $f(\lambda) = \Sigma a_n \lambda^n$, the series convergent in a disk $|\lambda| < a$ containing $\sigma(A)$, then $f(A) = \Sigma a_n A^n$, the series convergent in (E). (e) *Spectral mapping theorem*: if $f \in \mathcal{K}(A)$, then $\sigma(f(A)) = f(\sigma(A))$ (see Dunford-Schwartz [1958: 1, p. 566]). The functional calculus can be extended to unbounded operators A with "sufficiently small spectrum"; the algebra $\mathcal{K}(A)$ consists now of all functions f holomorphic in some open set $\mathcal{O}_f \supset \sigma(A)$ and at infinity. Formula (3.13) becomes

$$f(A) = f(\infty) I + \frac{1}{2\pi i} \int_\Gamma f(\lambda) R(\lambda) \, d\lambda, \tag{3.14}$$

with Γ described as before. The properties are rather similar to those for the case $A \in (E)$; for additional details see Taylor [1958: 1] or Dunford-Schwartz [1958: 1].

4. ADJOINTS

If $B \in (E)$, its adjoint is defined by the well-known formula $\langle B^*u^*, u \rangle = \langle u^*, Bu \rangle$ ($u \in E$, $u^* \in E^*$). Then $B^* \in (E^*)$ (moreover $\|B^*\|_{(E^*)} = \|B\|_{(E)}$) and

$$(A + B)^* = A^* + B^*, \quad (AB)^* = B^*A^*, \tag{4.1}$$

if A belongs as well to (E) (see Hille-Phillips [1957: 1, p. 43]). The definition of the adjoint is more complicated when we consider unbounded operators. Let A be a densely defined operator in E. Then $u^* \in E^*$ belongs to the domain of the adjoint A^* if and only if $u \to \langle u^*, Au \rangle$ is a continuous linear functional *of* u—that is, if there exists $v^* \in E^*$ such that

$$\langle v^*, u \rangle = \langle u^*, Au \rangle, \tag{4.2}$$

and we set

$$A^*u = v^*. \tag{4.3}$$

Since we assume $D(A)$ dense in E, (4.2) is sufficient to identify v^*, thus A^* is well defined. It may well be the case, however, that $D(A^*)$ reduces to $\{0\}$. As we shall see below, this rather unfortunate situation can be avoided if we impose some mild assumptions on A. To clarify this we begin by identifying the dual of $E \times E$ endowed with the norm $\|(u, v)\| = \|u\| + \|v\|$.

4.1 Lemma. *The dual space $(E \times E)^*$ can be linearly and isometrically identified with the space $E^* \times E^*$ normed with $\|(u^*, v^*)\|_{E^* \times E^*} = \max(\|u^*\|, \|v^*\|)$, an element of $E^* \times E^*$ acting as a functional on $E \times E$ through the formula*

$$\langle \{u^*, v^*\}, \{u, v\} \rangle = \langle u^*, v \rangle - \langle v^*, u \rangle. \tag{4.4}$$

The proof is simple. It is clear that (4.4) defines a linear functional in $E \times E$ whose norm does not surpass $\max(\|u^*\|, \|v^*\|)$. These two numbers must in fact coincide; if, say, $\|u^*\| \geq \|v^*\|$, we can insert in (4.4) elements of the form $\{0, v\}$ with $\|v\| = 1$ and $\|u^*(u)\|$ arbitrarily close to $\|u^*\|$. On the other hand, let Φ be a continuous linear functional in $E \times E$. Then $\Phi(\{u, v\}) = \langle u^*, v \rangle - \langle v^*, u \rangle$, where u^*, v^* are defined by $\langle u^*, v \rangle = \Phi(\{0, v\})$ $(v \in E)$ and $\langle v^*, u \rangle = -\Phi(\{u, 0\})$ $(u \in E)$.

The reasons for choosing the identification (4.4) over the far more symmetric $\langle \{u^*, v^*\}, \{u, v\} \rangle = \langle u^*, u \rangle + \langle v^*, v \rangle$ (which works equally well) will be clear in a few lines. We digress now briefly. If K is a subset of a Banach space E, we define $K^\perp \subseteq E^*$, the *orthogonal* of K, as follows:

$$K^\perp = \{u^* \in E^*; \langle u^*, u \rangle = 0, u \in K\}.$$

It is easy to check that $K^\perp$ is always a closed subspace of E^*. If E is a reflexive space, then $K^{\perp\perp} = (K^\perp)^\perp$ may be considered as a subspace of E. We have:

4.2 Lemma. *Let E be reflexive, K a subspace of E. Then $K^{\perp\perp} = \overline{K}$ = closure of K; in particular, if K is closed, then $K^{\perp\perp} = K$.*

To prove this we observe that $K^{\perp\perp} \supseteq K$ even if K is not a subspace. On the other hand, if $u \notin \overline{K}$, we may construct through Corollary 2.2 a functional u^* such that $u^*(\overline{K}) = \{0\}$, $u^*(u) = 1$, which proves that $u \notin K^{\perp\perp}$.

The relation with adjoint theory is established by:

4.3   Lemma.   *Under the identification established in Lemma* 4.1 *we have*

$$\Gamma(A^*) = \Gamma(A)^{\perp}. \tag{4.5}$$

The proof is an immediate consequence of (4.2) and the definition of orthogonal, and is left to the reader.

If E is a reflexive space, it is easy to see that the product $E \times E$ is as well reflexive; that is, $E \times E$ can be identified with the dual of $(E \times E)^* = E^* \times E^*$ using formula (4.4). We obtain combining the previous results:

4.4   Theorem.   *Let A be densely defined and closed and E reflexive. Then $D(A^*)$ is dense in E^*.*

In fact, if this were not true, we could find a nonzero $u \in E$ with $\langle u^*, u \rangle = 0$ for all $u^* \in D(A^*)$. But then $\langle 0, u \rangle \in \Gamma(A^*)^{\perp}$, and because of Lemma 4.2 $\langle 0, u \rangle \in \Gamma(A)$, which is impossible ($A0 = 0 \neq u$).

If E is not reflexive, $D(A^*)$ may not be dense in E^*; we note, however, that A^* is always closed. This follows directly from the definition or from (4.5).

The operator A is called *closable* if $\overline{\Gamma(A)}$ is the graph of an operator $\bar{A}$, which is then called the *closure* of A. Clearly A is closable if and only if whenever a sequence $\{u_n\}$ in $D(A)$ is such that $u_n \to 0$ and $Au_n \to v$, then $v = 0$ (observe than an arbitrary subspace of $E \times E$ is the graph of some linear operator if and only if it does *not* contain any element of the form $\langle 0, v \rangle$, $v \neq 0$).

Since an arbitrary set and its closure have the same orthogonal, we obtain from Lemma 4.3 that

$$\Gamma(\bar{A}^*) = \Gamma(\bar{A})^{\perp} = \overline{\Gamma(A)}^{\perp} = \Gamma(A)^{\perp} = \Gamma(A^*)$$

that is, $\bar{A}^* = A^*$; the adjoint of a closable operator coincides with the adjoint of its closure (so that it is densely defined if E is reflexive).

4.5   Lemma.   *Let A be densely defined and closable and E reflexive. Then*

$$A^{**} = \bar{A}. \tag{4.6}$$

In fact, $\Gamma(A^{**}) = \Gamma(A^*)^{\perp} = \Gamma(A)^{\perp \perp} = \overline{\Gamma(A)} = \Gamma(\bar{A})$ (we have used Lemma 4.2 in the third equality). Note that $A^{**} = (A^*)^*$ is well defined since A^* is densely defined.

Computations with adjoints of unbounded operators are considerably more complicated than those involving bounded operators. For instance, instead of (4.1) we have

$$A^* + B^* \subseteq (A + B)^*, \quad A^*B^* \subseteq (BA)^*, \tag{4.7}$$

where $A \subseteq B$ means $\Gamma(A) \subseteq \Gamma(B)$ (that is, B is an extension of A). To prove (4.7) (which is immediate), we assume of course that both $D(A)$ and $D(B)$ are densely defined; also in the first inclusion $D(A + B)$, and in the second $D(BA)$ must be dense so that every adjoint can be calculated.

Better results can be obtained if B is assumed to be bounded and everywhere defined. In that case we have

$$(A^* + B^*) = (A + B)^*, \quad A^*B^* = (BA)^*, \tag{4.8}$$

under the sole assumption that A is densely defined (note that $D(A + B) = D(BA) = D(A)$). The proofs are again a direct consequence of the definition of adjoint and are omitted.

4.6   Lemma.   *Let A be densely defined. Then*

$$\rho(A^*) \supseteq \rho(A) \tag{4.9}$$

and

$$R(\lambda; A^*) = R(\lambda; A)^*. \tag{4.10}$$

Proof.   Let $\lambda \in \rho(A)$. By virtue of the first equality (4.8), $(\lambda I - A)^* = \lambda I - A^*$; we apply then the second inclusion relation (4.7) obtaining

$$R(\lambda; A)^*(\lambda I - A^*) \subseteq I, \tag{4.11}$$

which is (3.3) for A^*. On the other hand, making use of the rather evident fact that

$$A \subseteq B \quad \text{implies} \quad B^* \subseteq A^*, \tag{4.12}$$

we apply the second equality (4.8) to $R(\lambda; A)(\lambda I - A) \subseteq I$ obtaining

$$(\lambda I - A^*)R(\lambda; A)^* = I. \tag{4.13}$$

This shows that $\lambda \in \rho(A^*)$ and that (4.10) holds.

4.7   Corollary.   *Let A be densely defined and closed and E reflexive. Then*

$$\rho(A^*) = \rho(A) \tag{4.14}$$

and (4.10) holds.

Proof.   According to (4.6), $A^{**} = A$. The results then follow from two applications of (4.9).

4.8   Example.   Let A, B_0, B be the operators in Example 3.8. Then $A^* = -B$, $B^* = -A$, $B_0^* = -B_1$, $B_1^* = -B_0$, where B_1 is the restriction of A defined by $u(1) = 0$.

4.9   Example.   Let $A \in (E), f \in \mathcal{H}(A)$ (see Example 3.12). Then

$$f(A^*) = f(A)^*.$$

5.   GENERALIZED SEQUENCES. WEAK CONVERGENCE

A set $\mathcal{Q} = \{\alpha, \beta, \ldots\}$ with a partial ordering relation $\leq$ is called a *directed set* if given $\alpha, \beta \in \mathcal{Q}$ there exists $\gamma \in \mathcal{Q}$ with $\alpha \leq \gamma$, $\beta \leq \gamma$. A *generalized sequence* or *net* in E is a function $\alpha \to u_\alpha$ from a directed set into E. We say that u_α *converges to* $u \in E$ (in symbols, $u_\alpha \to u$) if, given $\varepsilon > 0$ there exists $\alpha_0 \in \mathcal{Q}$ with

$$\|u - u_\alpha\| \leq \varepsilon \quad \text{whenever} \quad \alpha \geq \alpha_0.$$

A net $\{v_\beta; \beta \in \mathcal{B}\}$ is a *subnet* of $\{u_\alpha; \alpha \in \mathcal{Q}\}$ if and only if there exists a map $j: \mathcal{B} \to \mathcal{Q}$ with the following properties: $u_{j(\beta)} = v_\beta$ and for every $\alpha \in \mathcal{Q}$ there exists $\beta \in \mathcal{B}$ such that if $\gamma \geq \beta$ in $\mathcal{B}$, then $j(\gamma) \geq \alpha$ in $\mathcal{Q}$ (note that, although we use the same symbols for elements and order relations in the two directed sets $\mathcal{Q}$ and $\mathcal{B}$, these sets may be quite different).

A generalized sequence in E is said to *converge E^*-weakly* to $u \in E$ if and only if

$$\langle u_\alpha, u^* \rangle \to \langle u, u^* \rangle$$

for all $u^* \in E^*$, and a generalized sequence $\{u_\alpha^*\}$ in E^* *converges E-weakly* to $u^* \in E^*$ if and only if

$$\langle u_\alpha^*, u \rangle \to \langle u^*, u \rangle$$

for all $u \in E$. Clearly both notions are weaker than ordinary convergence; note that in the space E^* we have two different definitions, that of E-weak convergence and E^{**}-weak convergence, the second one being in general more demanding than the first since $E \subseteq E^{**}$. Both notions coincide if E is reflexive, of course.

It is easy to see that if $u_\alpha \to u$ weakly (in any of the two possible ways if the net is in E^*), then

$$\|u\| \leq \liminf \|u_\alpha\|,$$

where the $\liminf$ of a real-valued generalized sequence is defined in the same way as for a sequence.

5.1   Theorem of Alaoglu.   *Let $\{u_\alpha^*\}$ be a bounded generalized sequence in E^*. Then there exists a subnet $\{v_\beta^*\}$ of $\{u_\alpha^*\}$ that converges E-weakly to some $u^* \in E^*$.*

A proof can be found in Dunford-Schwartz [1958: 1, p. 424].

A space E is *separable* if it contains a countable dense set. The following sequential version of Theorem 5.1 holds:

5.2   Theorem.   *Let E be separable, and $\{u_n^*\}$ a bounded sequence in E^*. Then there exists a subsequence $\{u_{n(k)}^*\}$ of $\{u_n^*\}$ that converges weakly to some $u^* \in E^*$.*

We return for a moment to the matter of adjoints. In case E is not reflexive, it is important to have some indication that $D(A^*)$ is "large enough." This is provided by the following weak analogue of Theorem 4.4.

5.3   Theorem. *Let A be a densely defined, closed operator in the Banach space E. Then $D(A^*)$ is E-weakly dense in E^* (that is, for every $u^* \in E^*$ there exists a generalized sequence $\{u_\alpha^*\} \subset D(A^*)$ with $u_\alpha^* \to u^*$ E-weakly).*

For a proof, see Hille-Phillips [1957: 1, p. 43].

The role of generalized sequences in topology is discussed with more detail in Dunford-Schwartz [1958: 1, I.7.1] or in Kelley [1955: 1, Ch. 2].

6.   NORMAL, SYMMETRIC, SELF-ADJOINT, AND UNITARY OPERATORS IN HILBERT SPACE

Let H be a complex Hilbert space. Then, as is well known, if $u^* \in H^*$, there exists an element $\gamma u^* \in H$ such that

$$\langle u^*, u \rangle = (\gamma u^*, u) \quad (u \in H). \tag{6.1}$$

It is immediate that the map $\gamma: E^* \to E$ is additive, isometric, and transforms E^* onto E; on the other hand, since $\langle u^*, u \rangle$ is linear in u^* while (v, u) is conjugate linear in v, we have

$$\gamma(\lambda u^*) = \bar{\lambda}\gamma u^* \quad (u^* \in E^*, \lambda \in \mathbb{C}). \tag{6.2}$$

Hitherto we have defined adjoints with respect to the bilinear form $\langle u^*, u \rangle$; in the present case we can also use the scalar product (u, v) and the operators thus obtained will be called *Hilbert space adjoints* (we use the notation A^+ for them). Clearly A^+ is an operator in H (instead of H^*), and A^* and A^+ are related by

$$A^* = \gamma^{-1}A^+\gamma, \quad A^+ = \gamma A^*\gamma^{-1}. \tag{6.3}$$

Most of the time we will ignore this distinction, however, and write A^* also for A^+. This will originate no confusion. A closed, densely defined operator A is called *normal* if

$$A^*A = AA^*. \tag{6.4}$$

(This equality of course includes the requirement that $D(A^*A) = D(AA^*)$.) A densely defined operator A is *symmetric* if

$$(Au, v) = (u, Av) \quad (u, v \in D(A)).$$

Clearly this is equivalent to

$$A \subseteq A^*. \tag{6.5}$$

A is called *self-adjoint* if

$$A = A^*. \tag{6.6}$$

It is evident that if A is everywhere defined and symmetric, then it is self-adjoint (note also that, since $A = A^*$, A must also be closed, thus bounded by the closed graph theorem). Finally, a bounded, everywhere defined operator B is called *unitary* if $B^* = B^{-1}$.

A *projection* is a bounded symmetric operator P with

$$P^2 = P. \tag{6.7}$$

A *spectral measure* (denoted $P(d\lambda)$) is a map from the Borel sets in the plane to (H) whose values are projections and which satisfies $P(\varnothing) = 0$, $P(\mathbb{C}) = I$,

$$P(\cup e_j)u = \sum P(e_j)u \quad (u \in E) \tag{6.8}$$

for every sequence $e_1, e_2, \cdots$ of disjoint Borel sets, and

$$P(e_1 \cap e_2) = P(e_1)P(e_2) \tag{6.9}$$

for any two Borel sets e_1, e_2.

6.1 Theorem. *Let A be normal. Then there exists a spectral measure $P(d\lambda)$ such that $P(e) = 0$ if $e \cap \sigma(A) = \varnothing$, $P(e)H \subseteq D(A)$, and $AP(e) \supseteq P(e)A$ for any bounded Borel set e (moreover, $P(e)$ commutes with all bounded operators that commute with A), the domain of A consists of all u with*

$$\int_{\sigma(A)} |\lambda|^2 \|P(d\lambda)u\|^2 < \infty \tag{6.10}$$

and

$$(Au, v) = \int_{\sigma(A)} \lambda (P(d\lambda)u, v) \quad (u \in D(A), v \in H). \tag{6.11}$$

Note that $\sum \|P(e_j)u\|^2 = \sum (P(e_j)u, P(e_j)u) = \sum (P(e_j)u, u) = (P(e)u, u) = (P(e)u, P(e)u) = \|P(e)\|^2$, where the e_j are disjoint Borel sets and $e = \cup e_j$ so that $\|P(d\lambda)u\|^2$ is a measure; the same is true of $(P(d\lambda)u, v)$. Moreover, it follows from the definition of $P(d\lambda)$ that

$$\int_{\sigma(A)} \|P(d\lambda)u\|^2 = \|u\|^2 \tag{6.12}$$

for every $u \in E$. Observe next that, since $(P(e)u, u) = \|P(e)u\|^2 \geq 0$, $(P(u)u, v)$ is a (semidefinite) scalar product so that $(P(e)u, v) \leq \|P(e)u\|^2 \|P(e)v\|^2 \leq \|P(e)u\|^2 \|v\|^2$. By the Cauchy-Schwartz inequality, $|\int \lambda (P(d\lambda)u, v)| \leq \|v\|^2 \int |\lambda|^2 \|P(d\lambda)u\|^2$ so that the right-hand side of (6.11) exists and defines an element of H if u satisfies (6.10).

Theorem 6.1 particularizes to self-adjoint and to unitary operators as follows: a normal operator A is self-adjoint if $\sigma(A)$ is contained in the real axis and, conversely, if A is self-adjoint, then A is normal and every element in the spectrum is real. It follows that the spectral measure provided by Theorem 6.1 vanishes for any e that does not intersect the real axis. Therefore, condition (6.10) takes the form

$$\int_{-\infty}^{\infty} \lambda^2 \|P(d\lambda)u\|^2 < \infty \qquad (6.13)$$

and the integral defining A is

$$(Au, v) = \int_{-\infty}^{\infty} \lambda(P(d\lambda)u, v) \quad (u \in D(A), v \in H). \qquad (6.14)$$

We note that $\sigma(A)$ is contained in $\lambda \geq \omega$ if and only if $(Au, u) \geq \omega \|u\|^2$ for $u \in D(A)$. Finally, in the case where $A = U$ is unitary, $\sigma(A)$ is contained in the unit circle $|\lambda| = 1$ and (6.11) becomes

$$(Uu, v) = \int_{|\lambda|=1} \lambda(P(d\lambda)u, v); \qquad (6.15)$$

the spectral measure $P(d\lambda)$ vanishes outside of $|\lambda| = 1$. Observe that when A is bounded (which is automatically true in the unitary case) condition (6.10) is always satisfied since P must vanish in the complement of $\sigma(A)$.

The spectral measure in Theorem 6.1 (which is uniquely determined by A) is called the *resolution of the identity* associated with A.

Theorem 6.1 is the basis of the following *functional calculus* for A. Let $\mathcal{F}(A)$ be the set of all complex-valued Borel measurable functions defined in $\sigma(A)$ and essentially bounded with respect to the resolution of the identity $P(d\lambda)$ ($|f(\lambda)| \leq C$ except in a set e with $P(e) = 0$). Define

$$\Phi_f(u, v) = \int_{\sigma(A)} f(\lambda)(P(d\lambda)u, v) \quad (u, v \in H). \qquad (6.16)$$

Then Φ_f is linear in v, conjugate linear in u, and bounded ($|\Phi_f(u, v)| \leq C\|u\| \|v\|$). Therefore, there exists a unique operator $B \in (H)$ with $(Bu, v) = \Phi_f(u, v)$ ($u, v \in H$). We define then

$$f(A) = B. \qquad (6.17)$$

The functional calculus enjoys the following properties:

$$(f + g)(A) = f(A) + g(A), \quad (fg)(A) = f(A)g(A), \qquad (6.18)$$

$$\|f(A)\| = \|f\|, \qquad (6.19)$$

where $\|f\|$ indicates the essential supremum of f with respect to $P(d\lambda)$. Moreover, if $f(\lambda) \equiv 1$, $f(A) = I$. We also have

$$f(A)^* = \bar{f}(A). \qquad (6.20)$$

The functional calculus can be defined as well when f is unbounded, in which case the operator $f(A)$ will be in general unbounded. We shall not find occasion to make use of this general version. For a proof of Theorem 6.1 and for further details see Dunford-Schwartz [1963: 1].

A problem of considerable interest is that of extending a symmetric operator A to a self-adjoint operator, if such an extension exists. To this end we define the *deficiency indices* of a symmetric operator A as follows: the *positive deficiency index* d_+ of A is the dimension of the orthogonal complement of $(A + iI)H$ (which, of course, may be infinite), and the *negative deficiency index* d_- is similarly defined with respect to $(A - iI)H$. Then A can be extended to a self-adjoint operator if $d_+ = d_-$ and, in particular, A is self-adjoint if and only if $d_+ = d_- = 0$. If d_- and d_+ are different, this extension is not possible, although we can always extend A to a symmetric operator such that one or the other of the deficiency indices are zero; these operators are *maximal symmetric* (that is, they are not properly contained in any other symmetric operator).

We note, finally, an important particular case of Theorem 6.1 and of the functional calculus. When the spectrum of A consists of isolated points $\lambda_1, \lambda_2, \ldots$, then each λ_j is an eigenvalue of A with eigenspace $H_j = P_j H$, $P_j = P(\{\lambda_j\})$. Here (6.12) becomes

$$\sum_j \|P_j u\|^2 = \|u\|^2 \quad (u \in E) \tag{6.21}$$

while, by virtue of (6.9), $(P_i u, P_j u) = (P_i P_j u, u) = 0$ if $i \neq j$. Condition (6.10) describing the domain of A becomes

$$\sum_j |\lambda_j|^2 \|P_j u\|^2 < \infty$$

and

$$Au = \sum_j \lambda_j P_j u.$$

In particular, it results from (6.19) and the comments following that u can be developed in a series of eigenvectors of A, any two terms in the development being orthogonal (in fact, eigenvectors corresponding to different eigenvalues are always orthogonal).

6.2   Example.   When A is a bounded self-adjoint operator, both the "holomorphic" functional calculus in Example 3.12 and the "measurable" calculus described above apply. The latter contains the former: if $f \in \mathcal{K}(H)$, the operator $f(A)$ can be defined by (6.16) and (6.17). A similar observation holds in relation to the functional calculus for unbounded operators in Example 3.12.

6.3   Example.   Let A be a symmetric operator. Then its deficiency index d_+ (resp. d_-) is the dimension of the subspace $H_\lambda = \{u \in D(A^*); A^*u = \lambda u\}$ for any complex λ with $\operatorname{Im} \lambda > 0$ (for any complex λ with $\operatorname{Im} \lambda < 0$). See Dunford-Schwartz [1963: 1, p. 1232].

6.4 Example. From the operators in Example 3.8, none is self-adjoint or even symmetric. However, iB is symmetric and both deficiency indices equal 1. For α complex, $|\alpha| = 1$, let A_α be the restriction of A defined by the boundary condition $u(1) = \alpha u(0)$. Then each iA_α is a self-adjoint extension of iB. If $\alpha = e^{i\varphi}$, then $\sigma(iA_\alpha)$ consists of the sequence of eigenvalues $-(\varphi + 2k\pi)$, $k = \ldots, -1, 0, 1, \ldots$.

7. SOME SPECIAL SPACES AND THEIR DUALS

We denote as customary by $\mathbb{R}^m$ m-dimensional Euclidean space, the elements of $\mathbb{R}^m$ indicated by $x = (x_1, \ldots, x_m)$, $y = (y_1, \ldots, y_m)$, and so on. Let Ω be a Borel subset of $\mathbb{R}^m$ and μ a positive Borel measure in Ω. Given p, $1 \le p < \infty$, $L^p(\Omega; \mu)$ is the space of all μ-measurable functions u in Ω with

$$\|u\| = \|u\|_p = \|u\|_{L^p(\Omega, \mu)} = \left(\int_\Omega |u(x)|^p \mu(dx) \right)^{1/p} < \infty. \qquad (7.1)$$

Here $\|\cdot\|_p$ is a norm and $L^p(\Omega; \mu)$ is a Banach space with respect to it. (Strictly speaking, the elements of $L^p(\Omega; \mu)$ are not functions but equivalence classes of functions, the equivalence relation being equality almost everywhere with respect to μ.) The space $L^\infty(\Omega; \mu)$ consists of all μ-measurable, μ-essentially bounded functions u with the norm

$$\|u\| = \mu\text{-ess.}\sup\{|u(x)|; \; x \in \Omega\} \qquad (7.2)$$

and is likewise a Banach space; the same observation on equivalence classes applies. When μ is the Lebesgue measure, we write simply $L^p(\Omega; \mu) = L^p(\Omega)$.

7.1 F. Riesz Representation Theorem. *Let* $1 \le p < \infty$. *Then the dual space* $L^p(\Omega; \mu)^*$ *can be linearly and metrically identified with* $L^{p'}(\Omega; \mu)$, $p'^{-1} + p^{-1} = 1$, *an element* u^* *of* $L^{p'}$ *acting on* L^p *as follows:*

$$\langle u^*, u \rangle = \int_\Omega u^*(x) u(x) \mu(dx). \qquad (7.3)$$

For a proof see Dunford-Schwartz [1958: 1, p. 286] (note that inequality (2.1) becomes the Hölder inequality $\int u^* u \, d\mu \le \|u^*\|_{p'} \|u\|_p$).

The spaces $L^p(\Omega; \mu)$ are always separable if $1 \le p < \infty$. The space $L^2(\Omega; \mu)$ is a Hilbert space; the scalar product is

$$(u, v) = \int_\Omega \bar{u}(x) v(x) \mu(dx) \qquad (7.4)$$

(compare with (7.3)). We shall make occasional use of L^p spaces of vector-valued functions; we limit ourselves here to the finite-dimensional case. The space $L^p(\Omega; \mu)^\nu$ consists of all vector-valued functions $u(x) =$

$(u_1(x), \ldots, u_\nu(x))$, where each of the components is μ-measurable and

$$\|u\|_p = \left(\int_\Omega \sum_{j=1}^\nu |u_j(x)|^p \mu(dx) \right)^{1/p} < \infty$$

in the case $1 \leq p < \infty$; again $L^p(\Omega; \mu)^\nu$ is a separable Banach space, and the duals can be identified as in the scalar case. The space $L^2(\Omega; \mu)^\nu$ is a Hilbert space, the scalar product being

$$(u, v) = \int_\Omega (u(x), v(x)) \mu(dx)$$

$$= \int_\Omega \left(\sum_{j=1}^n \bar{u}_j(x) v_j(x) \right) \mu(dx). \tag{7.5}$$

The other spaces where subsequent developments take place are spaces of continuous functions. Let K be a compact subset of $\mathbb{R}^m$. The space $C(K)$ consists of all continuous functions u in K endowed with the supremum norm

$$\|u\| = \sup\{|u(x)|; \ x \in K\}.$$

Then $C(K)$ is a Banach space and its dual can be identified as follows. Let $\Sigma(K)$ be the space of all finite Borel measures defined in K. For $\mu \in \Sigma(K)$ define the measure $|\mu|$, the *total variation* of μ by the formula

$$|\mu|(e) = \sup \Sigma |\mu(e_j)|, \tag{7.6}$$

where e is any μ-measurable set and the supremum is taken over all possible decompositions of e into a disjoint union of a finite number of μ-measurable sets $\{e_j\}$. Then $|\mu| \in \Sigma(K)$ and we can define a norm in $\Sigma(K)$ that makes it a Banach space thus:

$$\|\mu\| = |\mu|(K).$$

7.2 F. Riesz Representation Theorem. *The dual space $C(K)^*$ can be linearly and metrically identified with $\Sigma(K)$, an element μ of $\Sigma(K)$ acting on $C(K)$ as follows*:

$$\langle \mu, u \rangle = \int_K u(x) \mu(dx). \tag{7.7}$$

See Dunford-Schwartz [1958: 1, pp. 97, 127 and 265] for the proof of a more general result where K may be a compact Hausdorff space.

Subspaces and extensions of $C(K)$ will also be used. If J is a closed subset of K, we denote by $C_J(K)$ the subspace of $C(K)$ consisting of all u that vanish in J; the dual $C_J(K)^*$ can then be identified with the quotient of $\Sigma(K)$ by the equivalence relation: $\mu_1 \sim \mu_2$ if μ_1 and μ_2 coincide outside of J—that is, if $\mu_1(e) = \mu_2(e)$ for every Borel subset e of K such that

$e \cap J = \emptyset$. This space can in turn be linearly and metrically identified with the subspace $\Sigma_J(K)$ of $\Sigma(K)$ consisting of all measures μ with $\mu(J) = 0$. To see this it suffices to notice that for every $\mu \in \Sigma(K)$, there exists another $\mu_J \in \Sigma(K)$ such that $\mu \sim \mu_J$ and $\mu_J(J) = 0$, the measure μ_J being defined by $\mu_J(e) = \mu(e \cap (K \setminus J))$; note that μ_J is the element of least norm in its equivalence class. In applications, however, we will most of the time ignore the equivalence relation and represent linear functionals by (7.7), the value of μ on J being irrelevant. The set J will usually be the boundary of K or a subset thereof.

If K is closed but not compact, define $C_0(K)$ to be the space of all continuous functions in K with

$$\lim_{|x| \to \infty} u(x) = 0.$$

The space $\Sigma(K)$ is defined and normed in the same way as when K is compact. The identification of the dual of $C_0(K)$ is provided by

7.3   Theorem.   *The dual space $C_0(K)^*$ can be linearly and metrically identified with $\Sigma(K)$, an element $\mu \in \Sigma$ acting on $C_0(K)$ according to (7.7).*

Theorem 7.3 is an immediate consequence of Theorem 7.2. In fact, let $K \cup \{\infty\}$ be the Alexandroff or one-point compactification of K (Royden [1968: 1, p. 168]). Then $C_0(K)$ coincides with the subspace $C_{\{\infty\}}(K \cup \{\infty\})$ consisting of all functions in $C(K \cup \{\infty\})$ that vanish at ∞; accordingly, $C_0(K)^*$ can be linearly and metrically identified with $\Sigma_{\{\infty\}}(K \cup \{\infty\})$, which is linearly and metrically isomorphic to $\Sigma(K)$.

The same observations made in connection with $C_J(K)$ apply to the space $C_{0,J}(K)$ consisting of all $u \in C_0(K)$ that vanish in a closed subset J of K: the dual space $C_{0,J}(K)^*$ can be identified with $\Sigma_J(K)$ linearly and metrically.

Although we will work most of the time with complex L^p, C, C_0, and Σ spaces (i.e., the functions and measures take complex values), we shall also find occasion to use spaces of real-valued functions. When necessary to distinguish between the two cases we shall do so by means of the subindices $\mathbb{C}$ and $\mathbb{R}$. (For instance, we shall write $L^p(K; \mu)_{\mathbb{C}}$, $C(K)_{\mathbb{R}}$, etc.)

We note finally that the spaces $C(K)$ and $C_0(K)$ (and of course any of their subspaces) are separable.

Let now Ω be a domain (open connected set) in $\mathbb{R}^m$. The space $\mathcal{D}(\Omega)$ (of Schwartz test functions in Ω) consists of all infinitely differentiable functions φ whose support $\operatorname{supp} \varphi = \overline{\{x; \varphi(x) \neq 0\}}$ is compact and contained in Ω. When $\Omega = \mathbb{R}^m$, we write simply $\mathcal{D}(\mathbb{R}^m) = \mathcal{D}$ and we introduce as well the space $\mathcal{S}(\mathbb{R}^m) = \mathcal{S}$ (whose elements are also called Schwartz test functions), consisting of all infinitely differentiable functions ψ dying down at infinity faster than any power of $|x|$ together with all their derivatives. $\mathcal{D}^{(k)}(\Omega)$ consists of all functions φ, k times continuously differentiable and having compact support contained in Ω. Clearly $\mathcal{D}(\Omega) = \cap_k \mathcal{D}^{(k)}(\Omega)$. On the other hand, $\mathcal{D}^{(k)}(\overline{\Omega})$ consists of all functions φ defined and k times continuously differentiable in Ω, vanishing outside a bounded set and such that all partial derivatives of order $\leq k$ are continuous in $\overline{\Omega}$. The space $C^{(k)}(\overline{\Omega})$ is defined in the same way but omitting the requirement of bounded support; obviously, if Ω is bounded $\mathcal{D}^{(k)}(\overline{\Omega})$ and $C^{(k)}(\overline{\Omega})$ coincide. We also define $\mathcal{D}(\overline{\Omega}) = \cap_k \mathcal{D}^{(k)}(\overline{\Omega})$ and $C^{(\infty)}(\overline{\Omega}) = \cap_k C^{(k)}(\overline{\Omega})$ and write simply $\mathcal{D}^{(k)}$, $C^{(k)}$, and so on when $\Omega = \mathbb{R}^m$. More often than not these spaces will be given no topology or notion of convergence, with the partial exception of $\mathcal{D}$ and other spaces of test functions in Chapter 8; also the functions in them may be real or complex, the distinction indicated with a subindex $\mathbb{C}$ or $\mathbb{R}$ as in L^p or C spaces.

Some liberties with the notations just introduced (and with others) will be taken in subsequent material. For instance, $C(\Omega)$ will be written $C[a, b]$ (not $C([a, b])$) when $\Omega = [a, b]$, $C(\Omega)$ will be abbreviated to C whenever Ω has been previously identified, and so on.

8.   CONVOLUTION AND MOLLIFIERS. SOBOLEV SPACES

We consider the spaces $L^p(\mathbb{R}^m) = L^p(\mathbb{R}^m; \mu)$, where $\mu(dx) = dx$ is the ordinary Lebesgue measure and $1 \leq p \leq \infty$. If $f \in L^p(\mathbb{R}^m)$, $g \in L^q(\mathbb{R}^m)$, we define their *convolution* by

$$(f * g)(x) = \int f(x - y) g(y) \, dy. \tag{8.1}$$

(All the integrals in this section are over $\mathbb{R}^m$.)

8.1   Theorem of Young.   *If* $f \in L^p$ *and* $g \in L^q$, $1 \leq p, q \leq \infty$, $p^{-1} + q^{-1} \geq 1$, *then the integral in* (8.1) *exists for almost all* $x \in \mathbb{R}^m$, $f * g$ *belongs to* L^r $(r^{-1} = p^{-1} + q^{-1} - 1)$ *and*

$$\|f * g\|_r \leq \|f\|_p \|g\|_q.$$

For a proof see Stein-Weiss [1971: 1, p. 178]. As a particular case of this result we obtain that if β is a function in L^1, the operator $Ju = \beta * u$ is bounded and $\|J\| \leq \|\beta\|_1$.

An important instance of this kind of operator is the following. Let β be a nonnegative function in $\mathcal{D}$ with support in (say) $|x| \leq 1$ and $\int \beta \, dx = 1$. For $n \geq 1$ define $\beta_n(x) = n^m \beta(nx)$, and

$$J_n u = \beta_n * u. \tag{8.2}$$

8.2   Theorem.   (a) *The operator* J_n *is a bounded operator in* $L^p(\mathbb{R}^m)$, $1 \leq p \leq \infty$, *with norm* ≤ 1 $(n \geq 1)$ (b) *if* $1 \leq p < \infty$,

$$\|J_n u - u\|_p \to 0 \quad \text{as} \quad n \to \infty$$

for any $u \in L^p$.

Part (a) results from the preceding comments; see Stein-Weiss [1971: 1] for a proof of part (b).

The interest of this result lies in that each $J_n u$ is infinitely differentiable; in fact, if $D^\alpha = (D^1)^{\alpha_1} \cdots (D^m)^{\alpha_m}$ $(\alpha = (\alpha_1, \ldots, \alpha_m)$, $D^j = \partial/\partial x_j)$ is an arbitrary differentiation monomial, then

$$D^\alpha(J_n u) = D^\alpha(\beta_n * u) = D^\alpha \beta_n * u$$

so that we can approximate an arbitrary $u \in L^p$ by infinitely differentiable functions in the L^p norm. Moreover, if the support of u is bounded, $\text{supp}(J_n u) \subseteq \{x;\ \text{dist}(x, \text{supp}\, u) \le 1/n\}$ is also bounded. Since any function u in L^p can be approximated by a function with bounded support (say, its restriction to $|x| \le n$ for n large enough), we obtain in particular that $\mathcal{D}$ is dense in $L^p(\mathbb{R}^m)$, that is,

$$\overline{\mathcal{D}} = L^p(\mathbb{R}^m) \quad (1 \le p < \infty).$$

We shall call the sequence $\{\beta_n\}$ a *mollifier*, or *mollifying sequence*, or a δ-*sequence*.

Given an arbitrary domain $\Omega \subseteq \mathbb{R}^m$, an integer $j \ge 0$ and a real number p, $1 \le p < \infty$, we define $W^{j,p}(\Omega)$ as the space of all functions $u \in L^p(\Omega)$ such that

$$D^\alpha u \in L^p(\Omega)$$

for $|\alpha| = \alpha_1 + \cdots + \alpha_n \le j$ (the derivatives understood in the sense of distributions). Then the space $W^{j,p}(\Omega)$ is a Banach space with the norm

$$\|u\| = \|u\|_{j,p} = \|u\|_{W^{j,p}(\Omega)}$$

$$= \left(\sum_{|\alpha| \le j} \int_\Omega |D^\alpha u(x)|^p\, dx \right)^{1/p}. \tag{8.3}$$

Clearly $W^{j,2}(\Omega)$ is a Hilbert space for any j, the scalar product defined by

$$(u, v) = (u, v)_{j,2} = (u, v)_{W^{j,2}(\Omega)}$$

$$= \sum_{|\alpha| \le j} \int_\Omega D^\alpha \bar{u}(x) D^\alpha v(x)\, dx. \tag{8.4}$$

For proofs of these facts and those that follow, as well as for additional details, the reader may consult Dunford-Schwartz [1963: 1, Ch. 16], Lions-Magenes [1968: 1, Ch. 1], or Friedman [1969: 1, Part 1]. When $p = 2$ the notation $H^j(\Omega)$ is often used instead of $W^{j,2}(\Omega)$.

In case $\Omega = \mathbb{R}^m$, the space $H^j(\mathbb{R}^m) = W^{j,2}(\mathbb{R}^m)$ can be easily characterized using the Fourier-Plancherel transform in $L^2(\mathbb{R}^m)$,

$$\mathcal{F}u(\sigma) = \tilde{u}(\sigma) = \lim_{a \to \infty} \frac{1}{(2\pi)^{m/2}} \int_{|x| \le a} e^{i(\sigma, x)} u(x)\, dx \tag{8.5}$$

where $\sigma = (\sigma_1, \ldots, \sigma_m)$, $(\sigma, x) = \sigma_1 x_1 + \cdots + \sigma_m x_m$, and the limit in (8.5) is understood in the sense of the norm of $L^2(\mathbb{R}^m)$.[1] The operator $\mathcal{F}$ is an isometric isomorphism from $L_x^2(\mathbb{R}^m)$ onto $L_\sigma^2(\mathbb{R}^m)$ (subindices are self explanatory). The inverse of $\mathcal{F}$ is given by

$$u(x) = \mathcal{F}^{-1}\tilde{u}(x) = \lim_{a \to \infty} \frac{1}{(2\pi)^{m/2}} \int_{|\sigma| \le a} e^{-i(\sigma, x)} \tilde{u}(\sigma)\, d\sigma, \qquad (8.6)$$

the limit understood as in (8.5). A function $u \in L^2$ satisfies $D^\alpha u \in L_x^2$ if and only if $\sigma^\alpha \mathcal{F} u \in L_\sigma^2$, where we have set $\sigma^\alpha = \sigma_1^{\alpha_1} \sigma_2^{\alpha_2} \cdots \sigma_m^{\alpha_m}$; moreover,

$$\mathcal{F}(D^\alpha u)(\sigma) = (-i\sigma)^\alpha \mathcal{F} u(\sigma).$$

It follows that $u \in H^j(\mathbb{R}^m)$ if and only if $(1 + |\sigma|^2)^{j/2} \tilde{u} \in L^2$ (here $|\sigma|^2 = \sigma_1^2 + \cdots + \sigma_n^2$) and the norm $\|u\|'_{j,2} = \|(1 + |\sigma|^2)^{j/2} \tilde{u}\|_2$ is equivalent to the norm of $H^j(\mathbb{R}^m)$. These observations make natural the introduction of the spaces $H^s(\mathbb{R}^m)$ (here s is a nonnegative parameter) that consist of all $u \in L_x^2$ such that $(1 + |\sigma|^2)^{s/2} \tilde{u}$ belongs to L_σ^2. These are Hilbert spaces under the scalar product

$$(u, v) = (u, v)_{s,2} = (u, v)_{H^s(\mathbb{R}^m)}$$

$$= \int_{\mathbb{R}^m} (1 + |\sigma|^2)^s \bar{\tilde{u}}(\sigma) \tilde{v}(\sigma)\, d\sigma \qquad (8.7)$$

corresponding to the norm

$$\|u\| = \|u\|_{s,2} = \|u\|_{H^s(\mathbb{R}^m)} = \|(1 + |\sigma|^2)^{s/2} \tilde{u}\|_{L^2(\mathbb{R}^m)}.$$

We shall also make use of H^s spaces of vector-valued functions $u = (u_1, \ldots, u_\nu)$. These spaces are denoted by $H^s(\mathbb{R}^m)^\nu$; the scalar product is defined by (8.7) with $(u(\sigma), v(\sigma))$ in the integrand.

Sometimes, functions that belong to L^p of $W^{j,p}$ only *locally* will appear. For instance, let Ω be a domain in $\mathbb{R}^m$. We write $u \in L^p_{\text{loc}}(\Omega)$ or $u \in L^p_{\text{loc}}$ in Ω to indicate that the function u, defined in Ω, belongs to $L^p(K)$ for every compact set $K \subseteq \Omega$. Similarly, $u \in W^{j,p}_{\text{loc}}(\Omega)$ means that $u \in W^{j,p}(\Omega')$ for every bounded domain Ω' with $\bar{\Omega}' \subset \Omega$.

---

[1]Note that, due to the factor $(2\pi)^{-m/2}$ in our definition of the Fourier transform, the convolution formula takes the form $\mathcal{F}(u * v) = (2\pi)^{m/2} \mathcal{F} u \mathcal{F} v$.

Chapter 1

The Cauchy Problem for Some Equations of Mathematical Physics: The Abstract Cauchy Problem

The purpose of this chapter is to introduce the notions of properly posed Cauchy problem in $t \geqslant 0$ (Section 1.2) and in $-\infty < t < \infty$ (Section 1.5) for the equation $u'(t) = Au(t)$ in an arbitrary Banach space E. These definitions will be fundamental in the rest of this work. In the case A is a differential operator in a function space E, there are relations with Cauchy problems that are well posed in the sense of Hadamard; these relations are explored in Section 1.7. The rest of the chapter is understood as motivation for the central idea of well posed Cauchy problem: several equations of mathematical physics are examined by ad hoc Fourier series and Fourier integral methods with the aim of discovering properties of solutions which will be generalized later to wide classes of equations.

Sections 1.1 and 1.3 deal with the heat-diffusion equation in a two-dimensional square. We find that (a) the equation produces a properly posed Cauchy problem in the spaces L^p ($1 \leqslant p < \infty$) and in spaces of continuous functions; (b) nonnegative initial data give rise to nonnegative solutions; (c) the L^1 norm of nonnegative solutions is preserved in time; (d) solutions become extremely smooth in arbitrarily small time. All of these properties are instances of general theorems on parabolic equations to be found in Chapter 4.

Section 1.4 treats the Schrödinger equation. Its properties are in a sense opposite to those of the heat-diffusion equation; for instance, the Cauchy problem is well posed in the space L^2 (but not in

L^p spaces, $p \neq 2$ or in spaces of continuous functions), the L^2 norm of arbitrary solutions remains constant and, finally, solutions do not become smoother than their initial data. Conclusions of the same type are obtained for the Maxwell and Dirac equations in Section 1.6; these are the prototype of results on symmetric hyperbolic systems to be found in Chapters 3 and 5.

1.1.   THE HEAT EQUATION IN A SQUARE

Consider the equation

$$\frac{\partial u}{\partial t} = \kappa \Delta u = \kappa \left(\frac{\partial^2 u}{\partial x^2} + \frac{\partial^2 u}{\partial y^2} \right) \quad (t \geqslant 0) \tag{1.1.1}$$

(κ a positive constant) in the square $\Omega = \{(x, y); \ 0 < x, y < \pi\}$ with initial condition

$$u(x, y, 0) = u_0(x, y) \quad (0 \leqslant x, y \leqslant \pi) \tag{1.1.2}$$

and boundary condition

$$u(x, y, t) = 0 \quad ((x, y) \in \Gamma, t \geqslant 0), \tag{1.1.3}$$

where Γ is the boundary of Ω. It is customary to call $u = u(x, y, t)$ a *classical solution* of (1.1.1), (1.1.2), and (1.1.3) if u is continuously differentiable up to the degree required by the equation and the initial and boundary conditions and satisfies identically all of them. We make precise these somewhat vague requirements by assuming that: u is continuous in $\bar{\Omega} \times [0, \infty)$, u_t exists and is continuous in the same region, $u_x, u_y, u_{xx}, u_{xy}, u_{yy}$ exist in $\Omega \times [0, \infty)$ and are continuous in $\bar{\Omega} \times [0, \infty)$, the equation (1.1.1) holds in $\Omega \times [0, \infty)$, and the initial and boundary conditions are satisfied.

We can write the initial-boundary value problem (1.1.1), (1.1.2), (1.1.3) as a pure initial value problem for an ordinary differential equation in a Banach space as follows. Using the notation in Section 7 we take $E = C_\Gamma(\bar{\Omega})$, the space of all continuous functions in $\bar{\Omega}$ that vanish at the boundary Γ, endowed with the supremum norm, and define an operator A in E as follows:

$$Au = \kappa(u_{xx} + u_{yy}) = \kappa \Delta u \tag{1.1.4}$$

where $D(A)$ consists of all $u \in E$ such that $u_x, u_y, u_{xx}, u_{xy}, u_{yy}$ exist in Ω and are continuous in $\bar{\Omega}$, and $\Delta u \in E$ (that is, $\Delta u = 0$ on Γ). Let u be a classical solution of (1.1.1), (1.1.2), and (1.1.3) in the sense made precise above. Define a function $t \to u(t) \in E$ in $t \geqslant 0$ as follows:

$$u(t)(x, y) = u(x, y, t) \quad (t \geqslant 0, (x, y) \in \bar{\Omega}). \tag{1.1.5}$$

Clearly $u(t) \in D(A)$ for all $t \geqslant 0$. On the other hand, if we define $v(t) \in E$ by

$$v(t)(x, y) = u_t(x, y, t) \quad (t \geqslant 0, (x, y) \in \overline{\Omega})$$

(note that $u_t = \kappa \Delta u \in E$ by hypothesis), we see, on account of the uniform continuity of u_t on compact subsets of $\overline{\Omega} \times [0, \infty)$, that $v(\cdot)$ is continuous in $t \geqslant 0$. Moreover, if $t > 0$, $|h| < t$, we have

$$\|h^{-1}(u(t + h) - u(t)) - v(t)\|_E$$

$$= \sup_{(x, y) \in \Omega} |h^{-1}(u(x, y, t + h) - u(x, y, t)) - u_t(x, y, t)|$$

$$\to 0 \quad \text{as} \quad h \to 0$$

by virtue of an elementary application of the mean value theorem and of the continuity of u_t in $\overline{\Omega} \times [0, \infty)$; the same can be proved when $t = 0$, but we have to take $h > 0$. It turns out that $u(\cdot)$ is continuously differentiable in the sense of the norm of E in $t \geqslant 0$ and

$$u'(t) = Au(t) \quad (t \geqslant 0). \tag{1.1.6}$$

On the other hand,

$$u(0) = u_0, \tag{1.1.7}$$

where u_0 is the initial function in (1.1.2). Conversely, let $t \to u(t) \in E$ be a function defined and continuously differentiable in $t \geqslant 0$, and such that $u(t) \in D(A)$ and (1.1.6) is satisfied for $t \geqslant 0$. Then it is rather easy to show that

$$u(x, y, t) = u(t)(x, y) \quad (t \geqslant 0, (x, y) \in \Omega) \tag{1.1.8}$$

is a classical solution of (1.1.1), (1.1.2) (with $u_0(x, y) = u(0)(x, y)$), and (1.1.3). This shows the complete equivalence of our original initial-boundary value problem and the initial value or Cauchy problem (1.1.6), (1.1.7). Making use of this equivalence we try to obtain additional information on the latter, taking advantage of the fact that (1.1.1) can be explicitly solved by separation of variables. Let

$$u_0(x, y) = \sum_{m=1}^{\infty} \sum_{n=1}^{\infty} a_{mn} \sin mx \sin ny \tag{1.1.9}$$

with

$$\sum_{m=1}^{\infty} \sum_{n=1}^{\infty} (m^2 + n^2)|a_{mn}| < \infty. \tag{1.1.10}$$

Then we obtain a classical solution of (1.1.1), (1.1.2), and (1.1.3) setting

$$u(x, y, t) = \sum_{m=1}^{\infty} \sum_{n=1}^{\infty} e^{-\kappa(n^2 + m^2)t} a_{mn} \sin mx \sin ny. \tag{1.1.11}$$

(this follows from elementary theorems on differentiation of multiple Fourier series, which can be found, for instance, in Zygmund [1959: 1]) and a fortiori a solution of (1.1.6) and (1.1.7). On the other hand, let $u(\cdot)$ be an arbitrary solution of (1.1.6) and (1.1.7), and let $u(x, y, t)$ be the function defined by (1.1.8). Then it follows from the maximum principle for the heat equation (Bers-John-Schechter [1964: 1, p. 96]) that

$$|u(x, y, t)| \leq \sup_{(\xi, \eta) \in \bar{\Omega}} |u_0(\xi, \eta)| \quad ((x, y) \in \bar{\Omega}, t \geq 0)$$

or

$$\|u(t)\|_E \leq \|u_0\|_E \quad (t \geq 0). \tag{1.1.12}$$

We have then proved that solutions of the Cauchy problem (1.1.6), (1.1.7) exist for u_0 satisfying (1.1.10) (note that the subspace of all such u_0 is dense in E) and that arbitrary solutions are continuously dependent on their initial value (as inequality (1.1.12) and linearity of the equation (1.1.6) show). Clearly this last property implies in particular that solutions are unique; that is, if two solutions coincide for $t = 0$, they coincide forever.

1.2. THE ABSTRACT CAUCHY PROBLEM

Taking as motivation the observations in the previous section we examine the equation

$$u'(t) = Au(t) \quad (t \geq 0) \tag{1.2.1}$$

where A is a densely defined operator in an arbitrary (real or complex) Banach space E. A *solution* of (1.2.1) is a function $t \to u(t)$ continuously differentiable for $t \geq 0$ (i.e., such that $u'(t) = \lim h^{-1}(u(t+h) - u(t))$ exists and is continuous in the norm of E in $t \geq 0$, the limit being one-sided for $t = 0$) and such that $u(t) \in D(A)$ and (1.2.1) is satisfied for $t \geq 0$. We may of course define solutions in intervals other than $[0, \infty)$, say $(-\infty, \infty)$ or $[0, T]$.

We say that the *Cauchy problem* (or *initial value problem*) for (1.2.1) is *well posed* (or *properly posed*) in $t \geq 0$) if the following two assumptions are satisfied:

(a) Existence of solutions for sufficiently many initial data. *There exists a dense subspace D of E such that, for any $u_0 \in D$, there exists a solution $u(\cdot)$ of (1.2.1) in $t \geq 0$ with*

$$u(0) = u_0 \tag{1.2.2}$$

(b) Continuous dependence of solutions on their initial data. *There exists a nondecreasing, nonnegative function $C(t)$ defined in $t \geq 0$ such that*

$$\|u(t)\| \leq C(t)\|u(0)\| \quad (t \geq 0) \tag{1.2.3}$$

for any solution of (1.2.1).

Note that (b) refers to *any* solution of (1.2.1), that is, not only to the solutions in (a), where $u(0) \in D$. Note also that, since our definition of solution requires that $u(t) \in D(A)$ in $t \geq 0$, we must have

$$D \subseteq D(A). \tag{1.2.4}$$

Hypothesis (b) can be given the following equivalent but slightly more natural form:

(b′) *Let $\{u_n(\cdot)\}$ be a sequence of solutions of (1.2.1) with $u_n(0) \to 0$. Then $u_n(t) \to 0$ uniformly on compacts of $t \geq 0$.*

We examine some immediate implications of assumptions (a) and (b). Let $u \in D$. Define, for $t \geq 0$,

$$S(t)u = u(t), \tag{1.2.5}$$

where $u(\cdot)$ is the (only) solution of (1.2.1) with $u(0) = u$. In view of (1.2.3), $S(t)$ is a bounded operator in D (with norm $\leq C(t)$). Since D is dense in E we can extend $S(t)$ to a bounded operator in E, which we denote by the same symbol, and

$$\|S(t)\| \leq C(t) \quad (t \geq 0). \tag{1.2.6}$$

The operator-valued function $S(\cdot)$ is called the *propagator* (or *solution operator* or *evolution operator*) of the equation (1.2.1). If $u(\cdot)$ is an arbitrary solution of (1.2.1), then

$$u(t) = S(t)u(0) \quad (t \geq 0), \tag{1.2.7}$$

where $u_0 = u(0)$. This is just the definition of $S(t)$ when $u(0) \in D$; if $u(0)$ does not belong to D, let $\{u_n\}$ be a sequence of elements of D such that $u_n \to u(0)$. Then it results from (1.2.3) that $S(t)u_n \to u(t)$, whereas, by definition of S outside of D, $S(t)u(0) = \lim S(t)u_n$, thus proving (1.2.7) in general.

Let u be an arbitrary element of E, and let $\{u_n\}$ be a sequence in D with $u_n \to u$. It follows from (1.2.6) that the (continuously differentiable) functions $S(\cdot)u_n$ converge uniformly to $S(\cdot)u$ on compacts of $t \geq 0$, thus $S(\cdot)u$ is continuous in $t \geq 0$. The function $u(t) = S(t)u$ will be called a *generalized solution* of (1.2.1). Clearly, $u(\cdot)$ need not be a genuine solution of (1.2.1) if u does not belong to D.

Generalized solutions of (1.2.1) can be defined in a way more akin to the general distribution-theoretical idea of solution of a differential equation. Let A be closed and let $u(t)$ be a continuous (or just locally integrable) function in $t \geq 0$. Then we say that $u(\cdot)$ is a *weak solution* of (1.2.1) and (1.2.2) if and only if, for every $u^* \in D(A^*)$ and every test function $\varphi \in \mathcal{D}$ we have

$$\int_0^\infty \langle A^*u^*, u(t) \rangle \varphi(t) \, dt = -\int_0^\infty \langle u^*, u(t) \rangle \varphi'(t) \, dt - \langle u^*, u_0 \rangle \varphi(0).$$

$$\tag{1.2.8}$$

As the following result shows, both definitions are equivalent.

1.2.1 Lemma. *Assume the Cauchy problem for* (1.2.1) *is properly posed, and let* $u(\cdot)$ *be a continuous (or only locally integrable) weak solution of* (1.2.1) *and* (1.2.2). *Then*

$$u(t) = S(t)u_0 \quad (t \geqslant 0) \tag{1.2.9}$$

(almost everywhere if $u(\cdot)$ *is only assumed to be locally integrable). Conversely, every function of the form* (1.2.9) *is a weak solution of* (1.2.1) *and* (1.2.2).

Lemma 1.2.1 shows that generalized and weak solutions of (1.2.1) and (1.2.2) coincide. We postpone the proof until Section 2.4, where a more general result will be shown (Theorem 2.4.6).

The case where the operator A in (1.2.1) is everywhere defined and bounded is of no special interest in applications since nontrivial differential operators are never bounded in the spaces considered in this volume or in other usual Banach spaces of functions. However, the theory becomes utterly simple here and provides some heuristic suggestions for the general case.

1.2.2 Example. Let $A \in (E)$. Then the Cauchy problem for (1.2.1) is properly posed in $t \geqslant 0$ with $D = E$. To see this we define $S(\zeta)$ for any ζ complex by

$$S(\zeta) = \sum_{n=0}^{\infty} \frac{\zeta^n}{n!} A^n \quad (t \geqslant 0), \tag{1.2.10}$$

the series being convergent in the norm of (E) uniformly on compacts of $\mathbb{C}$ since $\|\zeta^n A^n\| \leqslant |\zeta|^n \|A\|^n$; moreover,

$$\|S(\zeta)\| \leqslant \sum_{n=0}^{\infty} \frac{|\zeta|^n}{n!} \|A\|^n = e^{|\zeta|\,\|A\|}. \tag{1.2.11}$$

It follows from the elementary theory of vector-valued analytic functions (see Section 2, especially Lemma 2.5) that S is analytic in $\mathbb{C}$ and that

$$S'(\zeta) = AS(\zeta) = S(\zeta)A.$$

Hence $u(\cdot) = S(\cdot)u_0$ is a solution of the initial value problem (1.2.1), (1.2.2) for any $u_0 \in E$ thus proving (a). In view of (1.2.11) we only have to show that solutions of (1.2.1) with the same initial data coincide or, equivalently, that if $u(\cdot)$ is a solution with $u(0) = 0$, then $u(t) = 0$ $(t \geqslant 0)$. To see this, we integrate $u'(s) = Au(s)$ in $0 \leqslant s \leqslant t$, apply A on both sides, use (1.2.1), integrate again, and so on. We obtain

$$u(t) = \frac{1}{(n-1)!} \int_0^t (t-s)^{n-1} A^n u(s)\, ds, \tag{1.2.12}$$

whence $\|u(t)\| \leqslant C a^n \|A\|^n / n!$ in $0 \leqslant t \leqslant a$, where C is a bound for $\|u(t)\|$ there. Letting $n \to \infty$, it follows that u vanishes identically in $t \geqslant 0$.

There exists an interesting Laplace transform relation among the propagator $S(\cdot)$ and the resolvent operator $R(\lambda) = (\lambda I - A)^{-1}$ corresponding to the scalar relation $\mathfrak{L}(e^{at}) = 1/(\lambda - a)$. ($\mathfrak{L}$ indicates the Laplace transform.) Noting that the partial sums of (1.2.10) are bounded in the norm of (E) by $\exp(|\zeta|\,\|A\|)$, we take $\lambda \in \mathbb{C}$ such that $\operatorname{Re}\lambda > \|A\|$ and integrate $\exp(-\lambda t)S(t)$ in $t \geqslant 0$; term-by-term integration is easily justified and yields

$$\int_0^{\infty} e^{-\lambda t} S(t)\, dt = \sum_{n=0}^{\infty} \lambda^{-(n+1)} A^n = R(\lambda). \tag{1.2.13}$$

(See (1.3).) This formula, generalized to unbounded A in Chapter 2, will be the basis of our treatment of the abstract Cauchy problem. Of some importance also is the representation of the propagator as the inverse Laplace transform of $R(\lambda)$; in our case, we can verify by means of another term-by-term integration that

$$S(\zeta) = \frac{1}{2\pi i} \int_{\Gamma} e^{\lambda \zeta} R(\lambda) \, d\lambda \qquad (1.2.14)$$

where Γ is a simple closed contour, oriented counterclockwise and enclosing $\sigma(A)$ in its interior. Formula (1.2.14) has counterparts for unbounded A (see Example 2.1.9). Note also that (1.2.14) is nothing but the prescription to compute $e^{\zeta A}$ according to the functional calculus sketched in Example 3.12.

1.2.3 Example. (a) Using (1.2.14) show that if $r_M = \sup\{\mathrm{Re}\,\lambda; \ \lambda \in \sigma(A)\}$ and if $\omega > r_M$, then

$$\|S(t)\| \leqslant Ce^{\omega t} \quad (t \geqslant 0) \qquad (1.2.15)$$

for a suitable constant C. (b) Show that (1.2.15) does not necessarily hold if $\omega = r_M$. (c) Using (1.2.13) show that (1.2.15) cannot hold if $\omega < r_M$.

1.3.   THE DIFFUSION EQUATION IN A SQUARE

Equation (1.1.1) describes the evolution in time of the temperature of an homogeneous plate occupying the square $0 \leqslant x, y \leqslant \pi$ (κ is the ratio of the conductivity to the specific heat; see Bergman-Schiffer [1953: 1, Ch. 1]), the boundary condition (1.1.3) expressing the fact that the temperature at the boundary is kept equal to zero. In this case, the choice of the space E in Section 1.1 is natural enough. However, the equation is also a model for diffusion processes; in that case $u(x, y, t)$ is the concentration at (x, y) at time t of the diffusing substance while the boundary condition (1.1.3) expresses that particles reaching the boundary Γ are absorbed. (See Bharucha-Reid [1960: 1, Ch. 3], where the interpretation of κ is also discussed.) In the diffusion case the supremum norm has no obvious physical meaning; on the other hand, $\int u(x, y; t) \, dx \, dy$ is the total amount of matter present at time t and this suggests that the L^1 norm is the natural choice here. (Observe, incidentally, that only nonnegative solutions should be admitted since densities cannot be negative; the same is true for heat processes if we measure temperatures in the absolute scale. We shall comment on this later in this section). Since no additional complication is involved, we take $E = L^p(\Omega)$, where $1 \leqslant p < \infty$. The main difference with the case $E = C_\Gamma(\bar{\Omega})$ considered in Section 1 is that now functions in E are only defined modulo a null set, and thus it makes no sense trying to impose the boundary condition on every function in E. This difficulty, however, may be readily circumvented by including the boundary condition in the definition of the domain of A. In fact, we define here $D(A)$ to be the set of all u such that $u_x, u_y, u_{xx}, u_{xy}, u_{yy}$ exist in Ω, are continuous in $\bar{\Omega}$, and such that $u = 0$ on Γ. Condition (a) of the definition of well-posed Cauchy problem with $E = L^p(\Omega)$ is a consequence of the same condition in $E = C_\Gamma(\bar{\Omega})$; in fact, since $C_\Gamma(\bar{\Omega}) \subseteq L^p(\Omega)$ and convergence in $C_\Gamma(\bar{\Omega})$ implies convergence in $L^p(\Omega)$, a solution of (1.1.6) and (1.1.7) in $C_\Gamma(\bar{\Omega})$ is as well a solution in $L^p(\Omega)$. (We take D again to be the set of all u_0 whose Fourier development (1.1.9) satisfies (1.1.10).)

We check now condition (b). Let $u(\cdot)$ be an arbitrary solution (in $E = L^p(\Omega)$) of

$$u'(t) = Au(t) \quad (t \geq 0), \tag{1.3.1}$$

$$u(0) = u_0. \tag{1.3.2}$$

For each $t \geq 0$ we develop $u(x, y, t) = u(t)(x, y)$ in Fourier series,

$$u(x, y, t) \sim \sum_{n=1}^{\infty} \sum_{m=1}^{\infty} a_{mn}(t) \sin mx \sin ny. \tag{1.3.3}$$

We have

$$a'_{mn}(t) = \lim_{h \to 0} \frac{a_{mn}(t+h) - a_{mn}(t)}{h}$$

$$= \lim_{h \to 0} \frac{4}{\pi^2} \int_\Omega \frac{u(x, y, t+h) - u(x, y, t)}{h} \sin mx \sin ny \, dx \, dy.$$

$$\tag{1.3.4}$$

The quotient of increments in the integrand converges in the norm of $L^p(\Omega)$ to $Au(t)(x, y) = \kappa \Delta u(x, y, t)$. An application of Hölder's inequality shows that we can take limits under the integral sign, obtaining

$$a'_{mn}(t) = \frac{4\kappa}{\pi^2} \int_\Omega \Delta u(x, y, t) \sin mx \sin ny \, dx \, dy$$

$$= -\frac{4\kappa(m^2 + n^2)}{\pi^2} \int_\Omega u(x, y, t) \sin mx \sin ny \, dx \, dy$$

$$= -\kappa(m^2 + n^2) a_{mn}(t) \quad (t \geq 0).$$

Since $a_{mn}(0) = a_{mn}$, where $\{a_{mn}\}$ are the Fourier coefficients of u_0, it follows that $a_{mn}(t) = a_{mn} \exp(-\kappa(m^2 + n^2)t)$ and thus that $u(x, y, t)$ equals the sum of its Fourier series in $t > 0$:

$$u(x, y, t)$$

$$= \sum_{m=1}^{\infty} \sum_{n=1}^{\infty} e^{-\kappa(m^2 + n^2)t} a_{mn} \sin mx \sin ny$$

$$= \int_\Omega \left(\frac{4}{\pi^2} \sum_{m=1}^{\infty} \sum_{n=1}^{\infty} e^{-\kappa(m^2 + n^2)t} \sin mx \sin m\xi \sin ny \sin n\eta \right) u_0(\xi, \eta) \, d\xi \, d\eta$$

$$= \int_\Omega G(x, y, \xi, \eta, t) u_0(\xi, \eta) \, d\xi \, d\eta \quad (t > 0, (x, y) \in \Omega), \tag{1.3.5}$$

where the interchange of summation and integration is easily justified and G is well defined and infinitely differentiable for $t > 0$. We deduce next some immediate properties of G. Observe first that, as pointed out in Section 1.1, if $u_0 \in D$, then the function $u(x, y, t)$ defined by (1.3.5) is a classical solution of (1.1.1), (1.1.2), and (1.1.3) to which we can apply the maximum principle; it follows in particular that if u_0 is nonnegative, the same is true of $u(x, y, t)$. Then, if $\{\varphi_n\}$ is a mollifying sequence in $\mathcal{D}(\mathbb{R}^2)$ (see Section 8),

$$\int_\Omega G(x, y, \xi, \eta, t) \varphi_n(x' - \xi, y' - \eta) \, d\xi \, d\eta \geq 0 \qquad (1.3.6)$$

for (x, y), $(x', y') \in \Omega$, and n large enough. But the integral (1.3.6) approaches $G(x, y, x', y', t)$ when $n \to \infty$, so that

$$G(x, y, \xi, \eta, t) \geq 0 \quad (t > 0, (x, y), (\xi, \eta) \in \overline{\Omega}). \qquad (1.3.7)$$

On the other hand, if n is a positive integer, we can choose $\varphi_n \in \mathcal{D}(\mathbb{R}^2)$ with support in Ω such that $0 \leq \varphi_n(\xi, \eta) \leq 1$ and

$$\int_\Omega (1 - \varphi_n(\xi, \eta)) \, d\xi \, d\eta \leq \frac{1}{n}. \qquad (1.3.8)$$

Again by the maximum principle,

$$\int_\Omega G(x, y, \xi, \eta, t) \varphi_n(\xi, \eta) \, d\xi \, d\eta \leq 1 \qquad (1.3.9)$$

for $(x, y) \in \Omega$, whereas, by virtue of (1.3.8), the integral on the left-hand side of (1.3.9) converges to $\int G(x, y, \xi, \eta, t) \, d\xi \, d\eta$ as $n \to \infty$. We deduce then that

$$\int_\Omega G(x, y, \xi, \eta, t) \, d\xi \, d\eta \leq 1 \quad (t > 0, (x, y) \in \Omega). \qquad (1.3.10)$$

We go back now to (1.3.5). Let $1 \leq p < \infty$, $p'^{-1} = 1 - p^{-1}$. Applying Hölder's inequality we obtain[1]

$|u(x, y, t)|$

$$= \left| \int_\Omega G(x, y, \xi, \eta, t) u_0(\xi, \eta) \, d\xi \, d\eta \right|$$

$$\leq \int_\Omega G(x, y, \xi, \eta, t)^{1/p} |u_0(\xi, \eta)| G(x, y, \xi, \eta, t)^{1/p'} \, d\xi \, d\eta$$

$$\leq \left(\int_\Omega G(x, y, \xi, \eta, t) |u_0(\xi, \eta)|^p \, d\xi \, d\eta \right)^{1/p} \left(\int_\Omega G(x, y, \xi, \eta, t) \, d\xi, d\eta \right)^{1/p'},$$

$$\qquad (1.3.11)$$

where we have used (1.3.7). It follows from (1.3.10) and from the immediate fact that G is symmetric in (x, ξ) and in (y, η) that

$$\int_\Omega G(x, y, \xi, \eta, t) \, dx \, dy \leq 1$$

---

[1] The argument is modified in an obvious way when $p = 1$.

for $(\xi, \eta) \in \Omega$. Taking then the p-th power of (1.3.11) and integrating in Ω we obtain

$$\|u(t)\|_p \leqslant \|u_0\|_p \quad (t \geqslant 0) \tag{1.3.12}$$

for $1 \leqslant p < \infty$, where $\|\cdot\|_p$ indicates the norm of $L^p(\Omega)$ and $u(\cdot)$ is the L^p-valued function $u(t)(x, y) = u(x, y, t)$. Since $u(\cdot)$ was an arbitrary solution of (1.3.1) and (1.3.2), inequality (1.3.12) completes the proof that the Cauchy problem for (1.3.1) is well posed in $L^p(\Omega)$.

We note that (1.3.7) together with the representation (1.3.5) (which holds for an arbitrary solution of (1.3.1)) establish the important fact that a solution whose initial value is nonnegative remains nonnegative forever, a fact that is essential if the diffusion interpretation of (1.3.1) is to make sense. The same property holds of course for the solutions in $C_\Gamma(\bar{\Omega})$ treated in Section 1.2 since, as we have pointed out before, a solution of (1.1.6) in $C_\Gamma(\bar{\Omega})$ is automatically a solution of (1.3.1) in $L^p(\Omega)$.

We have seen in Section 1.1 that in the $C_\Gamma(\bar{\Omega})$ setting there was a complete correspondence between classical solutions of the original equation (1.1.1) and solutions of the abstract differential equation (1.1.6). In the $L^p(\Omega)$ case, however, one wonders what kind of solution of (1.1.1) is provided by solutions of (1.3.1). Differentiability of u in the space variables of course depends on how the domain of the operator A is defined. It turns out that differentiability with respect to t in the L^p sense involves no more than existence of the partial derivative in the sense of distributions and "mean continuity" conditions on it. We make this more precise below.

1.3.1 Lemma. *Let Ω be an arbitrary domain in Euclidean space $\mathbb{R}^m$, and let $t \to u(t)$ be a function with values in $L^p(\Omega)$ $(1 \leqslant p < \infty)$ defined and continuously differentiable in $0 \leqslant t \leqslant T$. Then the function $u(x, t) = u(t)(x)$ satisfies $D_t u = v$ in the sense of distributions, where $v(x, t) = u'(t)(x)$. Conversely, let $u(x, t)$, $v(x, t)$ be functions in $L^p(\Omega \times (0, T))$ such that $D_t u = v$ in the sense of distributions and assume that $v(t)(x) = v(x, t)$ is continuous in $0 \leqslant t \leqslant T$ as a $L^p(\Omega)$-valued function. Then we can modify $u(x, t)$ in a null set in $\Omega \times (0, T)$ in such a way that $u(t)(x) = u(x, t)$ is continuously differentiable as a $L^p(\Omega)$-valued function and $u'(t) = v(t)$ $(0 \leqslant t \leqslant T)$.*

Proof. Let $\varphi \in \mathcal{D}$ with support in $\Omega \times (0, T)$. Then we have

$$\int_{\Omega \times (0,T)} u(x, t) D_t \varphi(x, t) \, dx \, dt$$

$$= \lim_{h \to 0} \int_{\Omega \times (0,T)} u(x, t) \frac{1}{h} \{\varphi(x, t + h) - \varphi(x)\} \, dx \, dt$$

$$= \lim_{h \to 0} \int_{\Omega \times (0,T)} \frac{1}{h} \{u(x, t - h) - u(x, t)\} \varphi(x, t) \, dx \, dt$$

$$= \int_{\Omega \times (0,T)} v(x, t) \varphi(x, t) \, dx \, dt.$$

(Here we take $|h| < d$, where d is the distance from the support of φ to the top and bottom of the cylinder $\Omega \times (0, T)$.) This proves the first half of Lemma 1.3.1. To prove the converse, we consider the function

$$w(x, t) = \int_0^t v(x, s)\, ds \quad (x \in \Omega, 0 \leqslant t \leqslant T).$$

This function is defined almost everywhere in $\Omega \times (0, T)$, and an integration by parts shows that, always in the sense of distributions,

$$D_t w = v$$

in $\Omega \times (0, T)$ (note that $w \in L^p(\Omega \times (0, T))$ if v does). It then follows that $D_t(u - w) = 0$, thus $u - w$ must equal almost everywhere a function of x alone; this implies that

$$\frac{1}{h}\{u(x, t + h) - u(x, t)\} = \frac{1}{h}\int_t^{t+h} v(x, s)\, ds \tag{1.3.13}$$

almost everywhere (of course we take h sufficiently small; if $t = 0$, then $h > 0$, if $t = T$, then $h < 0$). We obtain making use of Hölder's inequality that

$$\left\| \frac{1}{h}\{u(x, t + h) - u(x, t)\} - v(x, t)\} \right\|_p$$

$$= \left(\int_\Omega \left| \frac{1}{h}\int_t^{t+h}(v(x, s) - v(x, t))\, ds \right|^p dx \right)^{1/p}$$

$$\leqslant \left(\frac{1}{h}\int_{\Omega \times (t, t+h)} |v(x, s) - v(x, t)|^p\, dx\, ds \right)^{1/p} \to 0 \quad \text{as} \quad h \to 0$$

(with the adequate modifications if $h < 0$). This ends the proof.

Lemma 1.3.1 shows, roughly speaking, that solutions of an equation like (1.1.1) considered as an equation in a L^p space will only be weak solutions in the sense of distribution theory, at least as regards t-dependence. This need not concern us unduly since classical solutions are in no way "required by nature" (see Truesdell-Toupin [1960: 1, p. 232]) and weak solutions of one kind or another are perfectly acceptable to model physical phenomena. On the other hand, weak solutions are, from a mathematical point of view, usually conducive to simpler and more powerful theories, as made evident by the work of Sobolev, Friedrichs, and Bochner in the thirties and forties and by the widespread use of distribution theory in the modern treatment of partial differential equations.

It should be pointed out, however, that Lemma 1.3.1 is somewhat irrelevant for the heat-diffusion equation and will only come fully into its own in the next two sections. The reason for this observation is the following result, where Ω is again the square $\{(x, y); 0 < x, y < \pi\}$.

1.3.2 Lemma. Let $u(\cdot)$ be an arbitrary solution of (1.3.1) in $L^p(\Omega)$, $1 \leqslant p < \infty$ or in $C_\Gamma(\overline{\Omega})$. Then $u(x, y, t) = u(t)(x, y) = (S(t)u)(x, y)$ can be extended to a function $u(z_1, z_2, \zeta)$ holomorphic for $z_1 \in \mathbb{C}$, $z_2 \in \mathbb{C}$, and $\text{Re}\,\zeta > 0$. *Any of the partial derivatives can be obtained by term-by-term differentiation of* (1.1.11).

For the proof, it suffices to observe that if $|\text{Im}\,z_1|$, $|\text{Im}\,z_2| \leqslant a$, and $\text{Re}\,\zeta \geqslant \varepsilon$, where a and ε are positive, the general term of the series in (1.3.5) is bounded in absolute value by

$$|a_{mn}| e^{(m+n)a - \kappa(m^2 + n^2)\varepsilon} \tag{1.3.14}$$

where the $|a_{mn}|$ are uniformly bounded.

As a consequence of this result we see that the solution $u(\cdot)$ will in fact be a classical solution (and more) for $t > 0$ (although the derivative on the right at $t = 0$ must be understood in the L^p sense).

Another notable property of the heat-diffusion equation is that, due to the rapidly decreasing multipliers $\exp(-\kappa(n^2 + m^2)t)$ in the Fourier series (1.3.5) of a solution, the values of a family of solutions at any fixed $t > 0$ tend to "bunch up" even if the family of initial values is widely dispersed. A precise statement of this phenomenon is:

1.3.3 Lemma. *Let* $\{u_k\}$ *be a sequence in* $L^1(\Omega)$ *with* $\|u_k\|_1 \leqslant C$ ($n \geqslant 1$) *and let* $t > 0$. *Then the sequence* $\{S(t)u_k\} \subseteq C_\Gamma(\bar\Omega)$, ($S(\cdot)$ *the evolution operator of* (1.3.1)) *contains a subsequence convergent in the norm of* $C_\Gamma(\bar\Omega)$.

Proof. Using Lemma 1.3.2 for the computation of the derivatives we have, noting that $|a_{mn}| \leqslant 4\pi^{-2}\|u(0)\|_1$ for any L^1 solution of (1.3.1):

$$\|S(t)u_k\|_{C_\Gamma(\bar\Omega)}, \|D^j S(t)u_k\|_{C_\Gamma(\bar\Omega)} \leqslant C$$

for $k \geqslant 1$ and some constant $C > 0$. The second set of bounds and the mean value theorem are easily seen to imply that the family $\{u_k(t)\}$ is equicontinuous (more precisely, equi-Lipschitz continuous) in $\bar\Omega$, thus the result follows from the Arzelà-Ascoli theorem (Dunford-Schwartz [1958: 1, p. 266]).

Considering the relations between the different L^p norms among them and with the norm of C_Γ, we obtain as a consequence of Lemma 1.3.3 that the propagators $S(t)$ of (1.3.1) in any of the spaces L^p or C_Γ are *compact* operators for any $t > 0$ (see Example 3.10 for definitions and properties).

A somewhat similar treatment of the heat-diffusion equation can be given for the boundary condition

$$D^\nu u(x, y, t) = 0 \quad (t > 0, (x, y) \in \Gamma), \tag{1.3.15}$$

where ν denotes the outer normal vector at the boundary (which is well defined except at the four corners of Ω) and D^ν is the derivative in the direction of ν. The role of the function G is now played by

$$H(x, y, \xi, \eta, t) = \frac{1}{\pi^2} + \frac{2}{\pi^2} \sum_{m=1}^{\infty} e^{-\kappa m^2 t} \cos mx \cos m\xi$$

$$+ \frac{2}{\pi^2} \sum_{n=1}^{\infty} e^{-\kappa n^2 t} \cos ny \cos n\eta$$

$$+ \frac{4}{\pi^2} \sum_{m=1}^{\infty} \sum_{n=1}^{\infty} e^{-\kappa(m^2+n^2)t} \cos mx \cos m\xi \cos ny \cos n\eta.$$

$$(1.3.16)$$

The following computation is easily justified by the analogue of Lemma 1.3.2 for the present boundary conditions and the fact that $u(x, y, t)$ is nonnegative for all t:

$$\frac{d}{dt} \|u(t)\|_1 = \frac{d}{dt} \int_\Omega |u(x, y, t)| \, dx \, dy = \frac{d}{dt} \int_\Omega u(x, y, t) \, dx \, dy$$

$$= \int_\Omega D_t u(x, y, t) \, dx \, dy = \kappa \int_\Omega \Delta u(x, y, t) \, dt$$

$$= \kappa \int_\Gamma D^\nu u \, d\sigma = 0 \quad (t > 0) \tag{1.3.17}$$

where $d\sigma$ is the differential of length on Γ. It follows that, in Banach space notation,

$$\|u(t)\|_1 = \|u_0\|_1 \quad (t \geq 0). \tag{1.3.18}$$

This is not surprising since the diffusion interpretation of the boundary condition (1.3.14) is that no diffusing matter enters or leaves Ω through the boundary Γ (Bharucha-Reid [1960: 1]). On the other hand, there is no physical reason for the conservation of norm property (1.3.18) to hold in L^p norms ($p > 1$) or in the supremum norm; even in L^1, we cannot expect the norm of solutions of arbitrary sign to remain constant. Elementary examples confirm this negative insight.

1.3.4 Example. (a) Exhibit a nonnegative classical solution u of (1.1.1) with boundary condition (1.3.15) such that

$$\|u(t)\|_p < \|u(0)\|_p \quad (1 < p < \infty, t > 0),$$

and similarly for the supremum norm. (b) Exhibit a classical solution with

$$\|u(t)\|_1 < \|u(0)\|_1 \quad (t > 0).$$

1.3.5 Example. Complete the analysis of the boundary condition (1.3.15) by showing that

$$H(x, y, \xi, \eta, t) \geq 0 \quad (t > 0, (x, y), (\xi, \eta) \in \bar{\Omega})$$

and

$$\int_\Omega H(x, y, \xi, \eta, t)\, d\xi\, d\eta = 1 \quad (t > 0, (x, y) \in \Omega).$$

State and prove the analogues of Lemmas 1.3.2 and 1.3.3.

1.4. THE SCHRÖDINGER EQUATION

In quantum mechanics, the state of a particle of mass $\mathfrak{m}$ (say, in three-dimensional space) under the influence of an external potential U is described by the equation

$$-\frac{h}{i}\frac{\partial \psi}{\partial t} = -\frac{h^2}{2\mathfrak{m}}\Delta\psi + U\psi \qquad (1.4.1)$$

where h is Planck's constant and the function $|\psi|^2$ is interpreted as the probability density of the particle (the probability of finding the particle in a set e at time t is $\int_e |\psi(x, t)|^2\, dx$). Clearly this interpretation imposes that

$$\int_{\mathbb{R}^3} |\psi(x, t)|^2\, dx = 1$$

for all t and suggests that $L^2(\mathbb{R}^3)_\mathbb{C}$ is the proper space for the study of (1.4.1). We shall only consider here the free particle case ($U = 0$) and carry out our study in $\mathbb{R}^m$ for any $m \geqslant 1$ as no additional difficulties are involved. (For complete details about the physical situation modelled by (1.4.1) see Schiff [1955: 1] or Fock [1976: 1].)

We study then the equation

$$u'(t) = \mathcal{A}u(t) \qquad (1.4.2)$$

where $\mathcal{A}$ is the operator in $E = L^2(\mathbb{R}^m)$ defined by

$$\mathcal{A}u = i\kappa\Delta u = i\kappa\big((D^1)^2 u + \cdots + (D^m)^2 u\big), \qquad (1.4.3)$$

κ an arbitrary real number. The domain of $\mathcal{A}$ consists of all $u \in L^2(\mathbb{R}^m)$ such that Δu (understood in the sense of distributions) belongs to L^2. The simplest way to study (1.4.2) is by means of the Fourier-Plancherel transform $\mathcal{F}$ (see Section 8). The operator $\mathcal{F}$ is an isometric isomorphism from $L^2(\mathbb{R}^m_x)$ onto $L^2(\mathbb{R}^m_\sigma)$ that transforms the operator $\mathcal{A}$ into the multiplication operator

$$\mathfrak{M}\tilde{u}(\sigma) = -i\kappa|\sigma|^2\tilde{u}(\sigma),$$

$D(\mathfrak{M})$ defined as the set of all $\tilde{u} \in L^2(\mathbb{R}^m_\sigma)$ such that $|\sigma|^2\tilde{u} \in L^2(\mathbb{R}^m_\sigma)$; precisely, $u \in D(\mathcal{A})$ if and only if $\mathcal{F}u = \tilde{u} \in D(\mathfrak{M})$ and

$$\mathcal{F}(\mathcal{A}u) = \mathfrak{M}(\mathcal{F}u).$$

Hence, if $u_0 \in D(\mathcal{A})$, we obtain a solution of (1.4.2) setting

$$u(x, t) = \mathcal{F}^{-1}\big(\exp(-i\kappa|\sigma|^2 t)\mathcal{F}u_0(\sigma)\big) \qquad (1.4.4)$$

where $\mathscr{F}^{-1}$ is the inverse Fourier transform (verification of this formula is based on the isometric character of $\mathscr{F}$ and $\mathscr{F}^{-1}$ and is left to the reader). Since $D(\mathcal{C})$ is dense in E part (a) of the definition in Section 1.2 is verified. To check condition (b) we observe that if $u \in D(\mathcal{C})$, then $(\mathcal{C}u, u) = (\mathscr{F}\mathcal{C}u, \mathscr{F}u) = i\kappa \int |\sigma|^2 |\tilde{u}|^2 \, d\sigma$ so that

$$\text{Re}(\mathcal{C}u, u) = 0. \tag{1.4.5}$$

Accordingly, if $u(\cdot)$ is a solution of (1.4.2) we have

$$D_t \|u(t)\|^2 = 2\,\text{Re}(u'(t), u(t))$$
$$= 2\,\text{Re}(\mathcal{C}u(t), u(t)) = 0$$

so that $\|u(t)\|$ is constant:

$$\|u(t)\| = \|u(0)\|.$$

We note that the preceding arguments can be justified just as well for t negative as for t positive; in other words, the Cauchy problem for (1.4.2) is "well posed for all t." This idea will be formalized in the next section; in the rest of this one we examine the equation (1.4.2) in the space $E = L^p(\mathbb{R}^m)$, where $1 \leqslant p < \infty$. The operator $\mathcal{C}$ is still defined by (1.4.3) but now we take $D(\mathcal{C}) = \mathcal{S}(\mathbb{R}^m) = \mathcal{S}$, which is dense in L^p. If $u \in \mathcal{S}$, then (1.4.4) provides a solution of (1.4.2) in L^p; this follows easily from the facts that the Fourier transform $\mathscr{F}$ and its inverse $\mathscr{F}^{-1}$ are continuous isomorphisms of $\mathcal{S}$ onto itself and that convergence in $\mathcal{S}$ implies convergence in any L^p space (see L. Schwartz [1966: 1]). Because of the convolution theorem for Fourier transforms of distributions (L. Schwartz [1966: 1, p. 268]), we can write (1.4.4) in the form $u = \mathscr{F}^{-1}(\exp(-i\kappa|\sigma|^2 t)) * u_0$ for $u_0 \in \mathcal{S}$, or

$$u(x, t) = \frac{1}{(4\pi i \kappa t)^{m/2}} \int_{\mathbb{R}^m} e^{i|x-y|^2/4\kappa t} u_0(y) \, dy. \tag{1.4.6}$$

We examine in detail the case $E = L^1(\mathbb{R}^m)$. If the Cauchy problem for (1.4.2) were properly posed in L^1, then, given $t > 0$ there would exist a constant C such that

$$\|u(t)\|_1 \leqslant C\|u_0\|_1 \tag{1.4.7}$$

for any function of the form (1.4.6). Assume this is the case, and let $\langle \varphi_n \rangle$ be a δ-sequence (Section 8). Setting $u_0 = \varphi_n$ in (1.4.6) and calling u_n the function so obtained, we see that

$$u_n(x, t) \to k(x, t) = (4\pi i \kappa t)^{-m/2} \exp(i|x - y|^2/4\kappa t)$$

for all x. Since $\|\varphi_n\|_1 = 1$ for all n, it follows from Fatou's theorem that $k(\cdot, t)$ must be in L^1, which is false. This shows that (1.4.7) cannot hold, and then that the Cauchy problem for (1.4.2) is not properly posed in L^1. Much more is true. In fact, we have

*1.4.1 Example (Hörmander [1960: 1, p. 109]). Let κ be a real number, $\kappa \neq 0$. Then the operator

$$Bu = \mathcal{F}^{-1}\left(e^{i\kappa|\sigma|^2}\mathcal{F}u\right)$$

from $\mathcal{S}$ into $\mathcal{S}$ satisfies

$$\|Bu\|_p \leqslant C\|u\|_p \quad (u \in \mathcal{S})$$

if and only if $p = 2$.

However, this sweeping negative result does not mean that L^p results for the Schrödinger equation are totally nonexistent. We shall reexamine the subject in the next section.

An important difference between the Cauchy problem for the Schrödinger equation and for the heat-diffusion equation is the following (another one is pointed out in the next section). While solutions of the heat equation either in $C_\Gamma(\overline{\Omega})$ or in $L^p(\Omega)$ become extremely regular with the passage of time (Lemma 1.3.2), a solution of the Schrödinger equation has no tendency to become smoother than its initial value. A concrete formulation of this statement is:

1.4.2 Lemma. *Let* $u(\cdot)$ *be a solution of* (1.4.2) *in* $L^2(\mathbb{R}^m)$, *and assume that* $u(t_0) \in H^s(\mathbb{R}^m)$ *for some* t_0. *Then* $u(t) \in H^s(\mathbb{R}^m)$ *for* $-\infty < t < \infty$; *in particular,* $u(0) \in H^s(\mathbb{R}^m)$.

To prove this result we only have to observe that (1.4.4) can be obviously generalized to

$$u(x, t) = \mathcal{F}^{-1}\left(\exp\left(-i\kappa|\sigma|^2(t - t_0)\right)\mathcal{F}u_0(\sigma, t_0)\right)$$

and take a look at the definition of the spaces H^s (see Section 8); we note that the H^s norm of a solution is constant for all t.

1.4.3 Remark. The domain of the operator $\mathcal{C}$ in (1.4.3) coincides with the space $H^2(\mathbb{R}^m)$ if $\kappa \neq 0$ (see Section 8); moreover, the norm $(\|u\| + \kappa^{-2}\|\mathcal{C}u\|^2)^{1/2}$ (obviously equivalent to the graph norm $\|u\| + \|\mathcal{C}u\|$) is nothing but the norm $\|\cdot\|'_{2,2}$ in $H^2(\mathbb{R}^m)$ (see Section 8).

1.4.4 Remark. The operator $\mathcal{C}$ is *closed*. This can be seen as follows. Assume $\{u_n\} \subset D(\mathcal{C})$, $u_n \to u$, $\mathcal{C}u_n \to v$ in $L^2(\mathbb{R}^m_x)$. Then $\mathcal{F}u_n \to \mathcal{F}u$ and $|\sigma|^2\mathcal{F}u_n \to \mathcal{F}v$ in $L^2(\mathbb{R}^m_\sigma)$. Passing if necessary to a subsequence we may assume that $\mathcal{F}u_n \to \mathcal{F}u$ and $|\sigma|^2\mathcal{F}u_n \to \mathcal{F}v$ a.e., thus $\mathcal{F}v = |\sigma|^2\mathcal{F}u$, and it follows that $u \in D(\mathcal{C})$ and $\mathcal{C}u = v$. As the following example shows, this result holds in much greater generality, although the proof cannot use the Fourier transform, since $p \neq 2$.

1.4.5 Example. Let

$$\mathcal{C}u = \sum_{|\alpha| \leqslant r} A_\alpha D^\alpha u \tag{1.4.8}$$

be a differential operator of order r in $L^p(\mathbb{R}^m)^\nu$ (see Section 7) with

$m, \nu \geqslant 1$, $1 \leqslant p < \infty$, where the A_α are $\nu \times \nu$ complex-valued matrices and $D(\mathcal{Q})$ is the set of all $u \in L^p(\mathbb{R}^m)^\nu$ such that $\mathcal{Q}u \in L^p(\mathbb{R}^m)^\nu$ (derivatives understood in the sense of distributions). Then $\mathcal{Q}$ is a closed operator.

1.4.6 Example. Let $\mathcal{Q}$ be as in Example 1.4.5, $p = 2$, $\mathcal{P}(\lambda) = \mathcal{P}(\lambda_1, \ldots, \lambda_m)$ the *characteristic polynomial* of $\mathcal{Q}$,

$$\mathcal{P}(\lambda) = \sum_{|\alpha| \leqslant r} \lambda^\alpha A_\alpha \qquad (1.4.9)$$

where $\lambda^\alpha = \lambda_1^{\alpha_1} \cdots \lambda_m^{\alpha_m}$. Assume that $\mathcal{P}(\lambda)$ commutes with its complex matrix adjoint

$$\mathcal{P}(\lambda)' = \sum_{|\alpha| \leqslant r} \bar{\lambda}^\alpha A'_\alpha. \qquad (1.4.10)$$

(Equivalently, assume that $A_\alpha A'_\beta = A'_\beta A_\alpha$ for $|\alpha|, |\beta| \leqslant r$.) Then $\mathcal{Q}$ is *normal*. To see this we use the Fourier-Plancherel transform $\mathcal{F}$, still defined by (8.5) as a vector integral in $L^2(\mathbb{R}^m)^\nu$, and having properties similar to those in the case $\nu = 1$. To compute $\mathcal{Q}^*$, take $u \in D(\mathcal{Q}^*)$. Then $v \to (u, \mathcal{Q}v)$ is a continuous linear functional of v in the L^2 norm for $v \in D(\mathcal{Q})$; in other words,

$$\int (\mathcal{P}(-i\sigma)'\mathcal{F}u(\sigma), \tilde{v}(\sigma)) \, d\sigma \leqslant C \left(\int |\tilde{v}(\sigma)|^2 \, d\sigma \right)^{1/2}, \qquad (1.4.11)$$

where $\tilde{v}$ indicates a generic element of $\mathcal{F}D(\mathcal{Q})$. Since this subspace is dense in $L^2(\mathbb{R}_\sigma^m)^\nu$, it follows that $\mathcal{P}(-i\sigma)'\mathcal{F}u(\sigma)$ belongs to $L^2(\mathbb{R}_\sigma^m)^\nu$ and

$$\mathcal{Q}^*u = \mathcal{F}^{-1}(\mathcal{P}(-i\sigma)'\mathcal{F}u(\sigma)). \qquad (1.4.12)$$

Conversely, if $u \in L^2(\mathbb{R}_x^m)^\nu$ is such that $\mathcal{P}(-i\sigma)'\mathcal{F}u(\sigma) \in L^2(\mathbb{R}_\sigma^m)^\nu$, then $u \in D(\mathcal{Q}^*)$ and (1.4.12) holds. Since $\mathcal{Q}$ itself can be expressed by

$$\mathcal{Q}u = \mathcal{F}^{-1}(\mathcal{P}(-i\sigma)\mathcal{F}u(\sigma)), \qquad (1.4.13)$$

the domain of $\mathcal{Q}$ consisting of all $u \in L^2(\mathbb{R}_x^m)^\nu$ such that $\mathcal{P}(-i\sigma)\mathcal{F}u(\sigma) \in L^2(\mathbb{R}_\sigma^m)^\nu$, it follows easily from the equality $\mathcal{P}(\lambda)\mathcal{P}(\lambda)' = \mathcal{P}(\lambda)'\mathcal{P}(\lambda)$ that

$$\mathcal{Q}^*\mathcal{Q} = \mathcal{Q}\mathcal{Q}^*$$

so that $\mathcal{Q}$ is indeed normal. Obviously, the commutation condition for $\mathcal{P}(\lambda)$ is automatically satisfied in the scalar case—that is, when $\nu = 1$. In view of the expression (1.4.10) for $\mathcal{P}(\lambda)'$, we see that $\mathcal{Q}^*$ can also be defined, in the same fashion as $\mathcal{Q}$, as the operator whose domain consists of all $u \in L^2(\mathbb{R}^m)^\nu$ such that $\mathcal{Q}'u \in L^2(\mathbb{R}^m)^\nu$, where $\mathcal{Q}'$ is the formal adjoint of $\mathcal{Q}$,

$$\mathcal{Q}' = \sum_{|\alpha| \leqslant r} (-1)^{|\alpha|} A'_\alpha D^\alpha u \qquad (1.4.14)$$

with $\mathcal{Q}^*u = \mathcal{Q}'u$. In particular, if $\mathcal{Q}' = \mathcal{Q}$, the operator $\mathcal{Q}$ is *self-adjoint*.

    It follows from the preceding remarks that if $\mathcal{Q}$ is the Schrödinger operator (1.4.3), then $i\mathcal{Q}$ is self-adjoint.

1.5.   THE CAUCHY PROBLEM IN $(-\infty, \infty)$

We present here an abstract formulation of the situation encountered in Section 1.4 in relation to the Schrödinger equation. As in Section 1.2, A is a densely defined operator in an arbitrary Banach space E.

We say that the Cauchy problem for

$$u'(t) = Au(t) \quad (-\infty < t < \infty) \tag{1.5.1}$$

is *well posed* (or *properly posed*) *in* $-\infty < t < \infty$ if and only if (a) and (b) of Section 1.2 hold with the following modifications: in (a) we replace "there exists a solution of (1.2.1) in $t \geqslant 0$" by "there exists a solution of (1.5.1) in $-\infty < t < \infty$" and in (b) we assume the existence of a function $C(t)$ such that $C(t)$ and $C(-t)$ are nondecreasing in $t \geqslant 0$ and

$$\|u(t)\| \leqslant C(t)\|u(0)\| \quad (-\infty < t < \infty). \tag{1.5.2}$$

In this language, the results of the previous section can be expressed by stating that the Cauchy problem for the Schrödinger equation is properly posed in $(-\infty, \infty)$ with $E = L^2(\mathbb{R}^m)$. We note that this is far from true for the heat equation considered in Section 1.1; in fact, consider the initial-value problem (1.1.6), (1.1.7) in the space $E = C_\Gamma(\overline{\Omega})$ with $u_0(x, y) = u_n(x, y) = \sin nx \sin ny$. A solution in $(-\infty, \infty)$ is given by

$$u(t)(x, y) = e^{-2\kappa n^2 t} u_n(x, y). \tag{1.5.3}$$

However, if $t < 0$, $\|u(t)\| = \exp(2\kappa n^2 |t|)$ whereas $\|u_0\| = 1$, which shows that an inequality of the type of (1.5.2) will never hold for t negative. The same counterexample works of course in $L^p(\Omega)$ for any $p \geqslant 1$. We can thus say that the problem of solving the heat (or diffusion) equation backwards in time is "improperly posed." We shall study some problems of this type in Chapter 6.

1.5.1   Example. If $A \in (E)$, the Cauchy problem for (1.5.1) is properly posed in $-\infty < t < \infty$; it suffices to notice that the manipulations in Example 1.2.2 make just as much sense for $t \leqslant 0$ as for $t \geqslant 0$. We note the following analogue of (1.2.13) (which is similarly proved), expressing $R(\lambda)$ by means of the propagator $S(t)$ for $t \leqslant 0$: if $\operatorname{Re} \lambda < -\|A\|$, then

$$\int_0^\infty e^{\lambda t} S(-t) \, dt = -R(\lambda). \tag{1.5.4}$$

1.6.   THE MAXWELL EQUATIONS

Let $\mathcal{E}$, $\mathcal{H}$ be three-dimensional vector functions of $x = (x_1, x_2, x_3)$ and t. We consider the system

$$\frac{\partial \mathcal{E}}{\partial t} = c \operatorname{rot} \mathcal{H} \tag{1.6.1}$$

$$\frac{\partial \mathcal{H}}{\partial t} = - c \operatorname{rot} \mathcal{E} \tag{1.6.2}$$

where c is a positive constant and $\mathcal{E}$, $\mathcal{H}$ are supposed to satisfy in addition

$$\operatorname{div} \mathcal{E} = \operatorname{div} \mathcal{H} = 0 \tag{1.6.3}$$

for all x, t. The system (1.6.1), (1.6.2), (1.6.3) describes the evolution in time of the electromagnetic field $(\mathcal{E}, \mathcal{H})$ in a homogeneous, isotropic medium occupying the whole space, in the absence of charges and currents (see Kline-Kay [1965: 1] for a thorough description of the physical setting and of the constant c). We write (1.6.1), (1.6.2) as a vector differential equation $u'(t) = \mathfrak{A} u(t)$ as follows. The space E is $L^2(\mathbb{R}^3)^6$ (see Section 7), consisting of all six-dimensional complex vector functions $(\mathcal{E}, \mathcal{H}) = u = (u_1, u_2, \ldots, u_6)$ with components in $L^2(\mathbb{R}^3)$ with scalar product $(u, v) = \Sigma(u_j, v_j)$, where $v = (v_1, v_1, \ldots, v_6)$ and $(\cdot, \cdot)$ in the sum indicates the scalar product in $L^2(\mathbb{R}^3)$. This choice is of course motivated by physics, since the function

$$\frac{1}{8\pi} \| u(t) \|^2 = \frac{1}{8\pi} \int_{\mathbb{R}^3} \left(|\mathcal{E}(x, t)|^2 + |\mathcal{H}(x, t)|^2 \right) dx$$

is interpreted as the energy of the electromagnetic field, which should remain constant in time in the absence of external excitation. The operator $\mathfrak{A}$ is symbolically expressed by

$$\mathfrak{A} = \begin{bmatrix} 0 & 0 & 0 & 0 & -cD^3 & cD^2 \\ 0 & 0 & 0 & cD^3 & 0 & -cD^1 \\ 0 & 0 & 0 & -cD^2 & cD^1 & 0 \\ 0 & cD^3 & -cD^2 & 0 & 0 & 0 \\ -cD^3 & 0 & cD^1 & 0 & 0 & 0 \\ cD^2 & -cD^1 & 0 & 0 & 0 & 0 \end{bmatrix}$$

$$= A_1 D^1 + A_2 D^2 + A_3 D^3, \tag{1.6.4}$$

where $D^j = \partial / \partial x_j$ and A_1, A_2, A_3 are 6×6 symmetric matrices, the domain of $\mathfrak{A}$ consisting of all those $u \in L^2(\mathbb{R}^3)^6$ such that $\mathfrak{A} u$ (taken in the sense of distributions) belongs to $L^2(\mathbb{R}^3)^6$.

A Fourier analysis of the equation

$$u'(t) = \mathfrak{A} u(t) \tag{1.6.5}$$

is carried out much in the same way as for the Schrödinger equation. The Fourier-Plancherel transform of functions in $L^2(\mathbb{R}^3)^6$ is defined again by (8.5) (which is now a vector integral) and defines an isometric isomorphism from $L^2(\mathbb{R}^3_x)^6$ into $L^2(\mathbb{R}^3_\sigma)^6$ that transforms the operator $\mathfrak{A}$ into the multiplication operator $\tilde{u}(\sigma) \to (-i \Sigma \sigma_j A_j) \tilde{u}(\sigma)$. If $u_0 \in D(\mathfrak{A})$, we obtain a solution of (1.6.5) thus:

$$u(x, t) = \mathcal{F}^{-1} \left(\exp \left(-it \sum_{j=1}^{3} \sigma_j A_j \right) \mathcal{F} u_0 \right). \tag{1.6.6}$$

If $u \in D(\mathfrak{A})$, then $(\mathfrak{A} u, u) = (\mathcal{F} \mathfrak{A} u, \mathcal{F} u) = i \int \Sigma \sigma_j (A_j u(\sigma), u(\sigma)) d\sigma$ so that again we have

$$\mathrm{Re}(\mathfrak{A}\mathfrak{u},\mathfrak{u}) = 0, \tag{1.6.7}$$

which implies (as in the previous section) that if $\mathfrak{u}(\cdot)$ is a solution of (1.6.5), then

$$\|\mathfrak{u}(t)\| = \|\mathfrak{u}(0)\| \quad (-\infty < t < \infty). \tag{1.6.8}$$

We have then proved that the Cauchy problem for (1.6.5) is well posed in $-\infty < t < \infty$ in the space $L^2(\mathbb{R}^3)^6$.

We have, however, neglected the two equations (1.6.3). To take care of them we simply note that

$$\frac{\partial}{\partial t}(\mathrm{div}\,\mathfrak{E}(x,t)) = \mathrm{div}\,\frac{\partial\mathfrak{E}}{\partial t}(x,t)$$

$$= c\,\mathrm{div}\,\mathrm{rot}\,\mathfrak{H}(x,t) = 0 \tag{1.6.9}$$

(the differentiations understood in the sense of distributions; see Lemma 1.3.1 in relation to the t-derivative), and we prove similarly that

$$\frac{\partial}{\partial t}(\mathrm{div}\,\mathfrak{H}(x,t)) = 0 \tag{1.6.10}$$

so that conditions (1.6.3) will be satisfied for all t if they are satisfied by the initial conditions $\mathfrak{E}(x,0)$, $\mathfrak{H}(x,0)$.[2]

We consider now different spaces. Denote by $C_0(\mathbb{R}^3)^6$ the Banach space of all vector functions $\mathfrak{u} = (u_1, u_2, \ldots, u_6)$, where all the components belong to $C_0(\mathbb{R}^3)$, with norm

$$\|\mathfrak{u}\| = \sup_{1 \leqslant j \leqslant 6} \|u_j\|_{C_0},$$

and by $L^p(\mathbb{R}^3)^6$ $(1 \leqslant p < \infty)$ the space consisting of all 6-vectors with components in $L^p(\mathbb{R}^3)$ (see Section 7 for details). The operator $\mathfrak{A}$ is defined by the expression (1.6.4), but we now take as $D(\mathfrak{A})$ the space $\mathfrak{S}^6$ of all 6-vectors with components in $\mathfrak{S}$. As in the case of the Schrödinger equation, it is easy to show that the function given by (1.6.6) provides a solution of

$$\mathfrak{u}'(t) = \mathfrak{A}\mathfrak{u}(t) \quad (-\infty < t < \infty) \tag{1.6.11}$$

in $C_0(\mathbb{R}^3)^6$, and in $L^p(\mathbb{R}^3)^6$ for $1 \leqslant p < \infty$. But there is no continuous dependence on initial data, as the following result shows:

*1.6.1 Example (Brenner [1966: 1]). Let $A_1, A_2, \ldots, A_m$ be $\nu \times \nu$ symmetric matrices, and let $\mathfrak{S}^\nu$ be the set of all ν-dimensional vectors with

---

[2] The argument runs as follows. The function $t \to \mathfrak{E}(\cdot, t)$ with values in $L^2(\mathbb{R}^m)$ is continuously differentiable. Since $L^2(\mathbb{R}^m)$ is contained in the space $\mathfrak{D}'(\mathbb{R}^m)$ (of all distributions defined in $\mathbb{R}^m$) with continuous inclusion, $t \to \mathfrak{E}(\cdot, t) \in \mathfrak{D}'(\mathbb{R}^m)$ is continuously differentiable as well, and so is $t \to \mathrm{div}\,\mathfrak{E}(\cdot, t)$ since div is a continuous operator in $\mathfrak{D}'(\mathbb{R}^m)$. Now, $D_t\,\mathrm{div}\,\mathfrak{E}(\cdot, t) = \mathrm{div}\,D_t\mathfrak{E}(\cdot, t) = 0$; hence $\mathrm{div}\,\mathfrak{E}(\cdot, t)$ is constant. We are making use here of the fact that functions with values in a locally convex space that have derivative zero everywhere must be constant; this can be easily proved applying linear functionals and using the "scalar" theorem. A similar reasoning takes care of $\mathfrak{H}$.

components in $\mathfrak{S}$. The operator

$$Bu = \mathfrak{F}^{-1}\left(\exp\left(i\sum_{j=1}^{m}\sigma_j A_j\right)\mathfrak{F}u\right)$$

satisfies

$$\|Bu\|_p \leqslant C\|u\|_p$$

($\|\cdot\|_p$ is the norm in $L^p(\mathbb{R}^m)^r$) for $p \neq 2$, $1 \leqslant p \leqslant \infty$ if and only if the matrices $A_1, \ldots, A_m$ commute.

However, we cannot discard offhand the possibility that the restrictions (1.6.3) on the space may improve the situation. This is in fact not the case, but a little more work is necessary to clarify the point.

1.6.2   Lemma.   *Let* $\varphi = (\varphi_1, \varphi_2, \varphi_3)$ *be a three-dimensional vector function with components in* $\mathfrak{S}$. *Then we can write*

$$\varphi = \psi + \eta \tag{1.6.12}$$

where the components of ψ, η *are infinitely differentiable and*

$$\operatorname{div}\psi = 0, \quad \operatorname{rot}\eta = 0. \tag{1.6.13}$$

Moreover, ψ *and* η *belong to* $L^p(\mathbb{R}^3)^3$ *for every* p, $1 < p < \infty$, *and there exists a constant* $C = C_p$ *depending only on* p *such that*

$$\|\psi\|_p \leqslant C\|\varphi\|_p, \quad \|\eta\|_p \leqslant C\|\varphi\|_p \tag{1.6.14}$$

Proof.   Let $\mathfrak{S}$ be the (closed) subspace of $L^2(\mathbb{R}^3)^3$ consisting of all solenoidal vectors ψ ($\operatorname{div}\psi = 0$ in the sense of distributions).[3] Applying the projection theorem we obtain a decomposition of the type of (1.6.12) with $\psi \in \mathfrak{S}$, $\eta \in \mathfrak{S}^\perp$. Now, if ξ is an arbitrary vector with components in $\mathfrak{S}$, $\operatorname{rot}\xi \in \mathfrak{S}$; it follows that

$$\int (\eta, \operatorname{rot}\xi)\, dx = 0$$

for all such ξ, hence $\operatorname{rot}\eta = 0$ in the sense of distributions. Observe next that $\Delta\eta = \operatorname{grad}\operatorname{div}(\varphi - \psi) - \operatorname{rot}\operatorname{rot}\eta = \operatorname{grad}\operatorname{div}\varphi$. Define a function $\tilde{\eta}$ by

$$\tilde{\eta}(x) = -\frac{1}{4\pi}\int \frac{1}{|y|}\operatorname{grad}\operatorname{div}\varphi(x - y)\, dy. \tag{1.6.15}$$

Then $\tilde{\eta}$ is infinitely differentiable; moreover, $\Delta\tilde{\eta} = \operatorname{grad}\operatorname{div}\varphi$ (Courant-Hilbert [1962: 1, p. 246]) so that $\Delta(\eta - \tilde{\eta}) = 0$ in the sense of distributions.

Let $\tilde{\eta}_1$ be the first component of $\tilde{\eta}$. We write the integral in (1.6.15) as the limit of the integral $\tilde{\eta}_{1,\varepsilon}$ in $|y| \geqslant \varepsilon$ and apply the divergence theorem twice. The

---

[3]Vectors in $\mathfrak{S}$ need not have first derivatives in L^2.

details of the computation follow:

$$\tilde{\eta}_{1,\varepsilon}(x) = -\frac{1}{4\pi}\int_{|y|\geqslant\varepsilon}\frac{1}{|y|}D_x^1\operatorname{div}\varphi(x-y)\,dy$$

$$= -\frac{1}{4\pi}\int_{|y|\geqslant\varepsilon}\frac{1}{|y|}D_y^1\operatorname{div}_y\varphi(x-y)\,dy$$

$$= \frac{1}{4\pi}\int_{|y|=\varepsilon}\frac{y_1}{|y|^2}\operatorname{div}_y\varphi(x-y)\,d\sigma$$

$$-\frac{1}{4\pi}\int_{|y|\geqslant\varepsilon}\frac{y_1}{|y|^3}\operatorname{div}_y\varphi(x-y)\,dy$$

$$= \frac{1}{4\pi}\int_{|y|=\varepsilon}\frac{y_1}{|y|^2}\operatorname{div}_y\varphi(x-y)\,d\sigma$$

$$+\frac{1}{4\pi}\int_{|y|=\varepsilon}\frac{1}{|y|^4}\left(y_1^2\varphi_1(x-y)+y_1y_2\varphi_2(x-y)\right.$$

$$\left.+y_1y_3\varphi_3(x-y)\right)d\sigma$$

$$+\frac{1}{4\pi}\int_{|y|\geqslant\varepsilon}\frac{1}{|y|^5}\left((|y|^2-3y_1^2)\varphi_1(x-y)-3y_1y_2\varphi_2(x-y)\right.$$

$$\left.-3y_1y_3\varphi_3(x-y)\right)dy$$

where $d\sigma$ indicates the area differential on $|y|=\varepsilon$. We take now limits as $\varepsilon\to 0$. The first surface integral is easily seen to tend to zero on account of the fact that $y_1|y|^{-2}$ is homogeneous of degree -1. In the second integral we note that $y_1y_j|y|^{-4}$ is homogeneous of degree -2 with

$$\int_{|y|=\varepsilon}\frac{y_1y_j}{|y|^4}\,d\sigma = \begin{cases}\dfrac{4\pi}{3} & \text{if } j=1 \\ 0 & \text{if } j=2,3\end{cases}$$

and write $\varphi_j(x-y)=\varphi_j(x)+(\varphi_j(x-y)-\varphi(x))$. The final result is

$$\tilde{\eta}_1(x) = \tfrac{1}{3}\varphi_1(x)+\sum_{j=1}^{3}\lim_{\varepsilon\to 0}\int_{|y|\geqslant\varepsilon}\frac{\Omega_j(y)}{|y|^3}\varphi_j(x-y)\,dy \tag{1.6.16}$$

where the Ω_j are smooth off the origin, homogeneous of degree zero, and satisfy

$$\int_{|y|=1}\Omega_j(y)\,d\sigma = 0 \quad (j=1,2,3).$$

Accordingly, the second inequality (1.6.14) for $\tilde{\eta}_1$ follows from the Calderón-Zygmund theorem for singular integrals (Stein-Weiss [1971: 1, p. 255]); $\tilde{\eta}_2$ and $\tilde{\eta}_3$ are similarly handled. It remains to be shown that $\eta=\tilde{\eta}$ a.e. which reduces to proving that a function $f\in L^2(\mathbb{R}^3)$, which satisfies $\Delta f=0$ in the sense of distributions, must be identically zero. To do this we note that if $\varphi\in\mathcal{D}$, then $\varphi*f$ is an ordinary harmonic function tending to zero at infinity, thus by Liouville's theorem (Stein-Weiss [1971: 1, p. 40]) $\varphi*f$ is identically zero; since φ is arbitrary, $f=0$ a.e. as desired.

We observe finally that each of the inequalities (1.6.14) implies the other, so that the proof of Lemma 1.6.2 is finished.

We can now deal with the complete system (1.6.1), (1.6.2), (1.6.3) in $L^p(\mathbb{R}^3)^6$ for $1 < p < \infty$. Let $\mathfrak{E}_s^p$ be the closed subspace of $L^p(\mathbb{R}^3)^6$ consisting of all vectors u with

$$\text{div}(u_1, u_2, u_3) = \text{div}(u_4, u_5, u_6) = 0. \tag{1.6.17}$$

Consider $\mathfrak{A}_s$, the restriction of $\mathfrak{A}$ to $D(\mathfrak{A}_s) = D(\mathfrak{A}) \cap \mathfrak{E}_s^p$, the space of all 6-vectors with components in $\mathfrak{S}$ satisfying (1.6.17). Assuming that the Cauchy problem for

$$u'(t) = \mathfrak{A}_s u(t) \tag{1.6.18}$$

is well posed in $\mathfrak{E}_s^p$, we shall obtain a contradiction below.

Let $u \in D(\mathfrak{A}) = \mathfrak{S}^6$. Write

$$u = \mathfrak{v} + \mathfrak{w}$$

where the vectors $\mathfrak{v}$, $\mathfrak{w}$ correspond to the decomposition (1.6.12) in the first and last three coordinates of u, so that $\mathfrak{v} \in \mathfrak{E}_s^p$ and

$$\text{rot}(w_1, w_2, w_3) = \text{rot}(w_4, w_5, w_6) = 0. \tag{1.6.19}$$

By virtue of Lemma 1.6.2, $\mathfrak{w} \in L^2(\mathbb{R}^3)^6$ and $\mathfrak{A}\mathfrak{w} = 0$, where $\mathfrak{A}$ is the operator (1.6.4) as defined in $L^2(\mathbb{R}^3)^6$. Accordingly $(\sigma_1 A_1 + \sigma_2 A_2 + \sigma_3 A_3)\mathcal{F}\mathfrak{w} = 0$ and it follows that

$$\mathcal{F}^{-1}\left(\exp\left(-it \sum_{j=1}^{3} \sigma_j A_j\right)\mathcal{F}\mathfrak{w}\right) = \mathfrak{w}. \tag{1.6.20}$$

Now, if $u(\cdot)$ is the solution of (1.6.11) with $u(0) = u$, we have, by virtue of (1.6.6) and (1.6.20),

$$u(t) = \mathcal{F}^{-1}\left(\exp\left(-it \sum_{j=1}^{3} \sigma_j A_j\right)\mathcal{F}u\right)$$

$$= \mathcal{F}\left(\exp\left(-it \sum_{j=1}^{3} \sigma_j A_j\right)\mathcal{F}\mathfrak{v}\right) + \mathfrak{w}$$

$$= \mathfrak{v}(t) + \mathfrak{w}.$$

If $S_s(\cdot)$ is the propagator of (1.6.18), an approximation argument shows that $\mathfrak{v}(t) = S_s(t)\mathfrak{v}$, thus for t fixed we must have

$$\|u(t)\|_p \leqslant \|\mathfrak{v}(t)\|_p + \|\mathfrak{w}\|_p \leqslant C'\|\mathfrak{v}\|_p + \|\mathfrak{w}\|_p \leqslant C\|u\|_p \quad (u \in \mathfrak{S})$$

(where we have used both inequalities (1.6.14) at the end). However, this estimate is forbidden by Example 1.6.1. This is a contradiction and shows that the Cauchy problem for (1.6.18) cannot be well posed in $\mathfrak{E}_s^p$ for any p in the range $1 < p < \infty$ except for $p \neq 2$.

Let $\mathfrak{E}_s$ be the subspace of $C_0(\mathbb{R}^3)^6$ consisting of all vectors u satisfying (1.6.17). That the Cauchy problem for (1.6.18) is not well posed in the space $\mathfrak{E}_s$ (that is, that the Cauchy problem for Maxwell's equations (1.6.1), (1.6.2), (1.6.3) is not well posed in the supremum norm) is of course not news for the physicist; in fact, an electromagnetic wave that is initially small everywhere may become enormously large (even infinite) in certain regions at a later time through the phenomenon of "focusing of waves." Examples of these regions are the focuses and caustic surfaces of geometrical optics. (See Kline-Kay [1965: 1], especially p. 326 for a discussion of caustics; for explicit examples of focusing of waves for the wave equation, to which the system (1.6.1), (1.6.2), (1.6.3) can be reduced, see Bers-John-Schechter [1964: 1, p. 13].)

The failure of the present treatment in the L^p ($p \neq 2$) and the C_0 cases for the Schrödinger and Maxwell equations confronts us with the following problem. Assume, say, that we want to predict the behavior of an electromagnetic field at some fixed point x_0 in space. Then it is obvious that mean square estimates like (1.6.8) are of small comfort and we really need bounds in the supremum norm. Although existence of this type of estimate seems to have been ruled out by the previous counterexamples, we can try to remedy the situation by measuring the initial state of the field in a different norm. To make the result more widely applicable, we work with the *symmetric hyperbolic system*

$$D_t u = \sum_{j=1}^{m} A_j D^j u + Bu, \qquad (1.6.21)$$

where $u = (u_1, \ldots, u_\nu)$ and $A_1, \ldots, A_m$, B are complex constant matrices, $A_1, \ldots, A_m$ self-adjoint, and B skew-adjoint. The basic space for the treatment will be $H^s(\mathbb{R}^m)^\nu$, consisting of all vector functions $u = (u_1, \ldots, u_\nu)$ with

$$\left(1 + |\sigma|^2\right)^{s/2} \mathfrak{F} u(\sigma) \in L^2(\mathbb{R}^m)^\nu$$

endowed with the L^2-norm of $\left(1 + |\sigma|^2\right)^{s/2} \mathfrak{F} u(\sigma)$ (see Section 8). It is exceedingly simple to extend the Fourier analysis carried out for (1.6.5) to the space $L^2(\mathbb{R}^m)^\nu$; the operator $\mathfrak{A}$ is

$$\mathfrak{A} u = \sum_{j=1}^{m} A_j D^j u + Bu \qquad (1.6.22)$$

where $D(\mathfrak{A})$ is the space of all $u \in H^s(\mathbb{R}^m)^\nu$ such that $\mathfrak{A} u$ (always understood in the sense of distributions) belongs to $H^s(\mathbb{R}^m)^\nu$; equivalently, $u \in H^s$ belongs to $D(\mathfrak{A})$ if and only if

$$\left(1 + |\sigma|^2\right)^{s/2} \left(-i \sum_{j=1}^{m} \sigma_j A_j + B\right) \mathfrak{F} u(\sigma) \in L^2(\mathbb{R}^m)^\nu$$

(where the factor $-i$ and the summand B may of course be omitted). If $u_0 \in D(\mathfrak{A})$, the function

$$u(t) = \mathfrak{F}^{-1}\left(\exp\left(-it \sum_{j=1}^{m} \sigma_j A_j + tB\right)\right) \mathfrak{F} u \qquad (1.6.23)$$

is a solution of

$$u'(t) = \mathcal{Q}u(t) \quad (-\infty < t < \infty). \tag{1.6.24}$$

Moreover, if $u \in D(\mathcal{Q})$, we have

$$(\mathcal{Q}u, u)_{H^s} = \int (1 + |\sigma|^2)^s \left\{ i \sum_{j=1}^m \sigma_j (A_j \tilde{u}(\sigma), \tilde{u}(\sigma)) + (B\tilde{u}(\sigma), \tilde{u}(\sigma)) \right\} d\sigma.$$

$$\tag{1.6.25}$$

Since $A'_j = A_j$, $(A_j \tilde{u}(\sigma), \tilde{u}(\sigma))$ is real; on the other hand, $B' = -B$ so that $(B\tilde{u}(\sigma), \tilde{u}(\sigma))$ is imaginary, and

$$\operatorname{Re}(\mathcal{Q}u, u)_{H^s} = 0. \tag{1.6.26}$$

This implies again that an arbitrary solution of (1.6.24) has constant H^s norm, hence the Cauchy problem for (1.6.24) is properly posed. As we shall see below, this will yield estimates in the L^p and C_0 norms for the solution in terms of the H^s norm of $u(0)$. Let u be an arbitrary element of $H^s(\mathbb{R}^m)'$. Since $(1 + |\sigma|^2)^{s/2} \tilde{u} \in L^2$ and $(1 + |\sigma|^2)^{-s/2} \in L^2$ if $s > m/2$, we obtain from the Schwarz inequality that $\tilde{u}(\sigma) = (1 + |\sigma|^2)^{s/2} \tilde{u}(\sigma)(1 + |\sigma|^2)^{-s/2}$ belongs to $L^1(\mathbb{R}^m)$ with norm

$$\|\tilde{u}\|_1 \leqslant K_s^{1/2} \|\tilde{u}\|_{H^s},$$

where $K_s = \int (1 + |\sigma|^2)^{-s} d\sigma$. It follows from well known properties of the Fourier transform that u (if necessary modified in a null set) belongs to $C_0(\mathbb{R}^m)$ with norm

$$\|u\|_{C_0} \leqslant (2\pi)^{-m/2} \|\tilde{u}\|_1 \leqslant (2\pi)^{-m/2} K_s^{1/2} \|\tilde{u}\|_{H^s}.$$

If $s \leqslant m/2$, estimates in the supremum norm are no longer possible; we use L^p norms instead. Let

$$2 < p < \frac{2m}{m - 2s} \leqslant \infty \tag{1.6.27}$$

and p' defined by $p'^{-1} + p^{-1} = 1$ so that

$$1 \leqslant \frac{2m}{m + 2s} < p' < 2.$$

Define next $q = 2/p' \geqslant 1$ and q' by $q'^{-1} + q^{-1} = 1$; we have

$$q' = \frac{2}{2 - p'} > \frac{m + 2s}{2s},$$

hence

$$p'q's = \frac{2p's}{2 - p'} > m,$$

and we obtain from Hölder's inequality for the exponents q, q' applied to the product $|\tilde{u}|^{p'} = |(1 + |\sigma|^2)^{s/2}\tilde{u}|^{p'}(1 + |\sigma|^2)^{-p's/2}$ that $\tilde{u} \in L^{p'}$ with norm

$$\|\tilde{u}\|_{p'} \leqslant K(p)\|\tilde{u}\|_{H^s}$$

where $K(p) = K_r^h$, $h = 1/p'q' = (2 - p')/2p' = (p - 2)/2p$, $r = p'q's/2 = p's/(2 - p') = ps/(p - 2)$. We make now use of the Hausdorff-Young theorem for Fourier integrals (Stein-Weiss [1971: 1, p. 178]) to deduce that $u \in L^p$ with norm

$$\|u\|_p \leqslant (2\pi)^{-d}\|\tilde{u}\|_{p'},$$

where $d = m(p - 2)/2p$.

We collect all these observations.

1.6.3   Theorem.   *The Cauchy problem for* (1.6.24) *is well posed in* $-\infty < t < \infty$ *in all spaces* $E = H^s(\mathbb{R}^m)^\nu$, $s \geqslant 0$, *in particular in* $E = L^2(\mathbb{R}^m)^\nu$; *if u is a solution in* L^2 *such that* $u(0) \in H^s$, *then* $u(t) \in H^s$ *for all t and*

$$\|u(t)\|_{H^s} = \|u(0)\|_{H^s} \quad (-\infty < t < \infty). \tag{1.6.28}$$

For $s > m/2, u(t)$ *(after eventual modification is a null set) belongs to* $C_0(\mathbb{R}^m)$ *and*

$$\|u(t)\|_{C_0} \leqslant C\|u(0)\|_{H^s} \quad (-\infty < t < \infty) \tag{1.6.29}$$

with $C = (2\pi)^{-m/2}K_s^{1/2}$. *If* $s \leqslant m/2$ *and p satisfies* (1.6.27), *then* $u(t) \in L^p$ *and*

$$\|u(t)\|_{L^p} \leqslant C(p)\|u(0)\|_{H^s} \quad (-\infty < t < \infty) \tag{1.6.30}$$

with $C(p) = (2\pi)^{-d}K(p)$. *In particular, if* $s = m/2$, $u(t) \in L^p$ *and the estimates* (1.6.30) *hold for any* $p \geqslant 1$.

Observe that estimates for the derivatives of $u(t)$ can be obtained in the same way. For instance, if $s > m/2 + k$ and $\alpha = (\alpha_1, \ldots, \alpha_m)$ is a multi-index of nonnegative integers with $|\alpha| \leqslant k$, the Fourier transform of $D^\alpha u$ is $(-i\sigma)^\alpha \tilde{u} \in H^{s-k}$; hence estimates of the form

$$\|D^\alpha u(t)\|_{C_0} \leqslant C\|u(0)\|_{H^s} \quad (-\infty < t < \infty) \tag{1.6.31}$$

hold. The same remark applies of course to L^p norms; details are left to the reader.

We note finally that results of precisely the same form hold for the Schrödinger equation. The proofs are identical: once (1.6.28) is established by direct Fourier analysis, inequalities (1.6.29) if $s > m/2$ or (1.6.30) if $s \leqslant m/2$ apply.

1.6.4   Example.   Prove in detail the results about the Schrödinger equation mentioned above (see Lemma 1.4.2).

1.6.5   Example.   The theory of the symmetric hyperbolic system (1.6.21) can be easily modified to embrace the case where B is an arbitrary $\nu \times \nu$

complex matrix. Write

$$B = B_1 + B_2,$$

where $B_1 = \frac{1}{2}(B + B')$ is self-adjoint and $B_2 = \frac{1}{2}(B - B')$ is skew-adjoint. Using (1.6.25) we obtain

$$\left| \mathrm{Re}(\mathcal{C}u, u)_{H^s} \right| \leqslant \int \left(1 + |\sigma|^2\right)^s \left| (B_1 \tilde{u}(\sigma), \tilde{u}(\sigma)) \right| d\sigma$$

$$\leqslant \omega \|u\|_{H^s} \qquad\qquad (1.6.32)$$

instead of the stronger condition (1.6.26), where ω is the maximum of the absolute value of the eigenvalues of B_1. Accordingly, if $u(\cdot)$ is an arbitrary solution of (1.6.24), we must have

$$\left| D_t \|u(t)\|^2 \right| \leqslant 2\omega \|u(t)\|^2$$

which is easily seen to imply that

$$\|u(t)\| \leqslant e^{\omega |t|} \|u(0)\| \quad (-\infty < t < \infty). \qquad (1.6.33)$$

This estimate guarantees continuous dependence on the initial datum. Existence of solutions is proved in the same way.

1.6.6 Example. *Relativistic quantum mechanics: the Dirac equation.* A relativistic description of the motion of a particle of mass m with spin $\frac{1}{2}$ (electron, proton, etc.) is provided by the *Dirac equation*

$$ihD_t\psi = ihc \sum_{j=1}^{3} \alpha_j D^j\psi - \beta\, \mathrm{m}\, c^2\psi + U\psi. \qquad (1.6.34)$$

Here h is again Planck's constant, c is the speed of light, $\alpha_1, \alpha_2, \alpha_3, \beta$ are the 4×4 matrices

$$\alpha_1 = \begin{bmatrix} 0 & 0 & 0 & 1 \\ 0 & 0 & 1 & 0 \\ 0 & 1 & 0 & 0 \\ 1 & 0 & 0 & 0 \end{bmatrix}, \quad \alpha_2 = \begin{bmatrix} 0 & 0 & 0 & -i \\ 0 & 0 & i & 0 \\ 0 & -i & 0 & 0 \\ i & 0 & 0 & 0 \end{bmatrix},$$

$$\alpha_3 = \begin{bmatrix} 0 & 0 & 1 & 0 \\ 0 & 0 & 0 & -1 \\ 1 & 0 & 0 & 0 \\ 0 & -1 & 0 & 0 \end{bmatrix}, \quad \beta = \begin{bmatrix} 1 & 0 & 0 & 0 \\ 0 & 1 & 0 & 0 \\ 0 & 0 & -1 & 0 \\ 0 & 0 & 0 & -1 \end{bmatrix},$$

and $\psi(x, t)$ is defined in $\mathbb{R}^3$ for each t and takes values in $\mathbb{C}^4$, that is, $\psi = (\psi_1, \psi_2, \psi_3, \psi_4)$. For each t, $|\psi(x, t)|^2 = \Sigma |\psi_k(x, t)|^2$ is interpreted as a probability density as for the Schrödinger equation. (See Schiff [1955: 1, p. 323] for additional details.) In the case of a free particle ($U = 0$), the equation (1.6.34) can be written as a symmetric hyperbolic system of the form (1.6.21), where

$$\mathcal{Q} = c\alpha_1 D^1 + c\alpha_2 D^2 + c\alpha_3 D^3 + \frac{imc^2}{h}\beta. \tag{1.6.35}$$

Thus Theorem 1.6.3 applies; in particular, the Cauchy problem for the Dirac equation is well posed in $E = L^2(\mathbb{R}^3)^4$ and the conservation of norm property (1.6.28) holds.

In perturbation results in Chapter 5, a better identification of the domain of $\mathcal{Q}$ will be necessary. The following result holds for the general symmetric hyperbolic equation (1.6.22):

1.6.7 Example. Assume that

$$\det\left(\sum_{j=1}^{m} \sigma_j A_j \right) \neq 0 \quad (|\sigma| = 1). \tag{1.6.36}$$

Then we have

$$D(\mathcal{Q}) = H^1(\mathbb{R}^m)^\nu, \tag{1.6.37}$$

the norm of $H^1(\mathbb{R}^m)^\nu$ equivalent to the graph norm $\|u\| + \|\mathcal{Q}u\|$ in $D(\mathcal{Q})$. Note first that, independently of Assumption (1.6.36), $H^1 \subseteq D(\mathcal{Q})$ and $\|\mathcal{Q}u\| \leqslant C\|u\|_{H^1}$ for $u \in H^1$. On the other hand, if (1.6.36) is satisfied, we use the facts that each $\Sigma\sigma_j A_j$ is nonsingular for any σ, $|\sigma| = 1$ and that the function $\sigma \to \Sigma\sigma_j A_j$ is homogeneous of degree 1 to conclude the existence of a constant $\theta > 0$ such that if $u \in \mathbb{C}^\nu$,

$$\left| \left(\sum_{j=1}^{m} \sigma_j A_j \right) u \right| \geqslant \theta |\sigma| \, |u| \quad (\sigma \in \mathbb{R}^m) \tag{1.6.38}$$

wherefrom the opposite inclusion and inequality among norms follow. Condition (1.6.36) is satisfied for the Dirac equations (where $\det(\Sigma\sigma_j\alpha_j) = -c^4|\sigma|^4$), but not for the Maxwell equations since in that case $\det(\Sigma\sigma_j A_j) = 0$. We note (see the next Example) that condition (1.6.36) is necessary for the coincidence of $D(\mathcal{Q})$ with H^1.

In the applications of perturbation theory to the Dirac equation it is important to compute the best constant θ in (1.6.38). This is immediate since $\sigma_1\alpha_1 + \sigma_2\alpha_2 + \sigma_3\alpha_3$ is an orthogonal matrix, the length of each row (or column) being $c|\sigma|$. Accordingly, $|(\sigma_1\alpha_1 + \sigma_2\alpha_2 + \sigma_3\alpha_3)u| = c|\sigma| \, |u|$ so that

$$\theta = c. \tag{1.6.39}$$

1.6.8 Example. Assume $\det(\Sigma\sigma_j A_j) = 0$ for some $\sigma \neq 0$. Then

$$D(\mathcal{Q}) \neq H^1(\mathbb{R}^m)^\nu.$$

1.6.9 Example. The apparent lack of a more precise characterization of the domain of the Maxwell operator $\mathfrak{A}$ in (1.6.4) is in fact due to the neglect of conditions (1.6.3). To see this we consider the restriction $\mathfrak{A}_s$ of $\mathfrak{A}$ to the subspace $\mathfrak{E}_s^2 = \mathfrak{E} \times \mathfrak{E} \subseteq L^2(\mathbb{R}^3)^6$ of all vectors $\mathfrak{u} = (\mathcal{E}, \mathcal{H})$ satisfying (1.6.17). We have

$$\|\mathfrak{A}_s(\mathcal{E}, \mathcal{H})\|^2 = c^2(\|\text{rot } \mathcal{E}\|^2 + \|\text{rot } \mathcal{H}\|^2). \tag{1.6.40}$$

Let $\zeta = (\zeta_1, \zeta_2, \zeta_3)$ be a vector in $\mathbb{C}^3$ such that

$$\sigma_1 \zeta_1 + \sigma_2 \zeta_2 + \sigma_3 \zeta_3 = 0$$

for some $\sigma = (\sigma_1, \sigma_2, \sigma_3) \in \mathbb{R}^3$. Then we check easily that

$$|\sigma_2 \zeta_3 - \sigma_3 \zeta_2|^2 + |\sigma_3 \zeta_1 - \sigma_1 \zeta_3|^2 + |\sigma_1 \zeta_2 - \sigma_2 \zeta_1|^2 = |\sigma|^2 |\zeta|^2.$$

This identity, applied pointwise to the Fourier transform of a vector $\mathcal{E} \in \mathfrak{E}$, shows that rot $\mathcal{E} \in L^2(\mathbb{R}^3)^3$ if and only if $\mathcal{E} \in H^1(\mathbb{R}^3)^3$; applying this and (1.6.40) to both components $\mathcal{E}$ and $\mathcal{H}$ of an element of $D(\mathfrak{A}_s)$ we conclude that

$$D(\mathfrak{A}_s) = H^1(\mathbb{R}^3)^6 \cap \mathfrak{E}_s^2. \tag{1.6.40}$$

1.6.10   Example.   The operators $i\mathfrak{A}$ ($\mathfrak{A}$ the Maxwell operator (1.6.4)) and $i\mathcal{Q}$ ($\mathcal{Q}$ the Dirac operator (1.6.33)) are self-adjoint, the first in $L^2(\mathbb{R}^3)^6$, the second in $L^2(\mathbb{R}^3)^4$. More generally, $i\mathcal{Q}$ is self-adjoint in $L^2(\mathbb{R}^m)^\nu$, where $\mathcal{Q}$ is the symmetric hyperbolic operator (1.6.22). It suffices to apply the relevant portion of Example 1.4.6; the formal adjoint of $\mathcal{Q}$ is

$$\mathcal{Q}' = -\sum_{j=1}^m A_j D^j + B'$$

1.7.   MISCELLANEOUS NOTES

The classical *Cauchy problem* is that of solving a general partial differential equation or system $P(u) = 0$, the *Cauchy data* of the unknown function u (its value and the value of its normal derivatives up to order $r - 1$, r the order of the equation) prescribed on a hypersurface $\mathfrak{H}$ of m-dimensional Euclidean space $\mathbb{R}^m$. Extending early work of Cauchy, Sophia Kowalewska succeeded in proving in 1875 that the Cauchy problem for a wide class of analytic partial differential equations can always be solved (if only locally) if the surface $\mathfrak{H}$ and the Cauchy data of u are analytic and $\mathfrak{H}$ is nowhere characteristic. This result, known as the Cauchy-Kowalewska theorem, is usually seen nowadays in the simplified version of Goursat (see Bers-John-Schechter [1964: 1]; a different proof has been given by Hörmander [1969: 1]). A misunderstanding seems to have subsequently arisen in relation to the possibility of solving the Cauchy problem above for nonanalytic Cauchy data along the lines of the following argument: approximate the Cauchy data of u by analytic Cauchy data; solve the equation; take limits. The fallacy involved in overlooking continuous dependence of the solution on its Cauchy data was pointed out by Hadamard in [1923: 1, p. 33] (although he was aware of it no less than twenty years before): "I have often maintained, against different geometers, the importance of this distinction. Some of them indeed argued that you may always consider any functions as analytic, as, in the contrary case, they could be approximated with any required precision by analytic ones. But, in my opinion, this objection would not apply, the question not being whether such an approximation would alter the data very little but whether it would alter the solution very little."[1] Hadamard presents the classical counterexample to the approximation argument—namely, the Cauchy problem for the Laplace equation (which will be found in its essential features in Chapter 6)—and singles out those Cauchy problems for which continuous dependence holds; *correctly set* problems in his terminology. Other names (well set, well posed, properly posed) are in current use now.

The concept of a properly posed problem in the sense of Hadamard does not fit into the abstract framework introduced in Section 1.2. To clarify this we restrict ourselves to the particular case where the equation is linear and of first order and where the hypersurface $\mathfrak{H}$ is the hyperplane $x_m = 0$. The requirement that $\mathfrak{H}$ be nowhere characteristic implies that the equation or system can be written in the form

$$D^m u = \sum_{j=1}^{m-1} A_j(x) D^j u + B(x) u + f(x), \qquad (1.7.1)$$

where u, f are vector functions of $(x_1,\ldots,x_m)$ and the A_j, B are matrix functions of appropriate dimensions; we assume that A_j, B, f have derivatives of all orders. The Cauchy data of u reduce to its value for $x_m = 0$,

$$u(x_1,\ldots,x_{m-1},0) = \varphi(x_1,\ldots,x_{m-1}). \qquad (1.7.2)$$

The Cauchy problem is *properly posed in the sense of Hadamard* if solutions exist for, say, φ infinitely differentiable and depend continuously on φ; according to Hadamard, continuous dependence is understood in the topology of uniform convergence (on compact sets) of derivatives up to a certain order, not necessarily the same for the data and the solution. It is remarkable that continuous dependence in this sense is a consequence of existence and uniqueness, as the following result (where we assume that $f = 0$) shows:

1.7.1   Theorem. *Let $X_m > 0$. Assume a unique infinitely differentiable solution u of (1.7.1), (1.7.2) exists in $|x_m| \leqslant X_m$ for every $\varphi \in \mathfrak{D}(\mathbb{R}^{m-1})$. Then, given $a, b > 0$ and an integer $M \geqslant 0$ there exists an integer $N \geqslant 0$ such that if $\langle \varphi_n \rangle \subset \mathfrak{D}(\mathbb{R}^{m-1})$ with the support of each φ_n contained in $x_1^2 + \cdots + x_{m-1}^2 \leqslant b^2$ and*

$$\lim_{n \to \infty} |D^\beta \varphi_n(x)| = 0$$

uniformly in $\mathbb{R}^{m-1}$ for $\beta = (\beta_1,\ldots,\beta_{m-1})$ with $|\beta| \leqslant N$, then

$$\lim_{n \to \infty} |D^\alpha u_n(x)| = 0$$

[1]From J. Hadamard, *Lectures on Cauchy's Problem in Linear Partial Differential Equations*, (New Haven: Yale University Press, 1923).

uniformly in $x_1^2 + \cdots + x_m^2 \leqslant a$, $|x_m| \leqslant X_m$ *for all* $|\alpha| \leqslant M$.

Proof. Let $\mathcal{D}_b$ the subspace of $\mathcal{D}(\mathbb{R}^{m-1})$ consisting of all φ with support in $x_1^2 + \cdots + x_{m-1}^2 \leqslant b^2$. The space $\mathcal{D}_b$ is a Fréchet space equipped with the translation invariant metric

$$\rho(\varphi,\psi) = \sum_{k=0}^{\infty} 2^{-k} \|\varphi - \psi\|_k / (1 + \|\varphi - \psi\|_k), \qquad (1.7.3)$$

where $\|\cdot\|_k$ denotes the maximum of all partial derivatives of order $\leqslant k$. Similarly, the space $C_{X_m}^{(\infty)}$ of all infinitely differentiable functions of $x_1, \ldots, x_m$ defined in $|x_m| \leqslant X_m$ is a Fréchet space endowed with the translation invariant metric

$$d(u,v) = \sum_{k=0}^{\infty} \sum_{l=0}^{\infty} 2^{-(k+l)} \|u - v\|_{k,l} / (1 + \|u - v\|_{k,l}), \qquad (1.7.4)$$

where $\|\cdot\|_{k,l}$ denotes the maximum of all partial derivatives of order $\leqslant k$ in $x_1^2 + \cdots + x_{m-1}^2 \leqslant l^2$, $|x_m| \leqslant X_m$. The linear operator $\mathfrak{S}\varphi = u$ (from $\mathcal{D}_b$ into $C_{X_m}^{(\infty)}$) from Cauchy datum φ to solution u provided by the assumptions of Theorem 1.7.1 is easily seen to be closed, thus it must be continuous by virtue of the closed graph theorem for Fréchet spaces (Banach [1932: 1, p. 41]). It follows that there exists $\delta > 0$ with

$$d(u,0) \leqslant 2^{-(M+l+1)} \text{ if } \rho(\varphi,0) \leqslant \delta$$

with $l \geqslant a$. Accordingly, if N is so large that

$$2^{-(N+1)} + 2^{-(N+2)} + \cdots \leqslant \delta/2$$

and φ is such that

$$\|\varphi\|_N \leqslant \delta/4$$

(note that $\|\cdot\|_k$ increases with K), we shall have $\rho(\varphi,0) \leqslant \delta$ so that $\|u\|_{M,l} / (1 + \|u\|_{M,l}) \leqslant \frac{1}{2}$, thus $\|u\|_{M,l}$ must remain bounded. Hence there exists a constant C (depending on a, b, M) such that

$$\|u\|_{M,a} \leqslant C \|\varphi\|_N \qquad (1.7.5)$$

This concludes the proof.

We note that a rather obvious approximation argument based on (1.7.5) shows that (1.7.1), (1.7.2) will actually have a solution for every initial datum φ with support in $x_1^2 + \cdots + x_{m-1}^2 \leqslant c^2 < b^2$ having continuous derivatives of order $\leqslant N$, these solutions having continuous derivatives of order $\leqslant M$ in $x_1^2 + \cdots + x_{m-1}^2 \leqslant a^2$, $|x_m| < X_m$; it suffices to approximate the φ in question in the norm $\|\cdot\|_N$ by a sequence $\{\varphi_n\}$ in $\mathcal{D}$ with support in $x_1^2 + \cdots + x_{m-1}^2 \leqslant b^2$ and let $n \to \infty$; of course, we assume that $M \geqslant 1$ so that limits can be taken in the equation (1.7.1).

There is evidence in [1923: 1] that Hadamard was aware of the relation between existence of solutions for sufficiently differentiable initial data and continuous dependence. Contrasting properly and improperly posed problems, he states on p. 32: "On the contrary, none of the physical problems connected with $\nabla^2 u = 0$ is formulated analytically in Cauchy's way. All of them lead to statements such as Dirichlet's, i.e. with only one numerical datum at every point of the boundary. Such is the case with the equation of heat. All this agrees with the fact that Cauchy's data, if not analytic, do not determine any solution of any one of these two equations.

"This remarkable agreement between the two points of view appears to me as an evidence that the attitude which we adopted above—that is, making a rule not to assume analyticity of data—agrees better with the true and inner nature of things that Cauchy's and his successors' previous conception."[2]

The first result of the type of Theorem 1.7.1 was given by Banach [1932: 1 p. 44] as an application of his inverse function-closed graph theorem; the result there is not restricted to Cauchy problems, however. Many variants have been proved using the same basic idea: we note the following result, due to Lax [1957: 2], where the phenomenon of finite domains of dependence is deduced from mere existence and uniqueness of solutions of (1.7.1), (1.7.2) for φ having arbitrary growth at infinity.

1.7.2   Theorem.   *Let $X_m > 0$. Assume a unique infinitely differentiable solution u of (1.7.1), (1.7.2) exists in $|x_m| \leqslant X_m$ for every φ infinitely differentiable in $\mathbb{R}^{m-1}$. Then, given $a > 0$ and an integer $M \geqslant 0$, there exists $b, C > 0$ and an integer $N \geqslant 0$ (all three depending on a, M) such that*

$$\|u\|_{M,a} \leqslant C \|\varphi\|_{N,b}. \qquad (1.7.6)$$

The proof is essentially the same as that of Theorem 1.7.1; this time the space of Cauchy data is $C^{(\infty)}(\mathbb{R}^{m-1})$ consisting of all φ infinitely differentiable in $\mathbb{R}^{m-1}$ endowed with a metric of the type of (1.7.4). Inequality (1.7.6) implies that the solution u in $x_1^2 + \cdots + x_{m-1}^2 \leqslant a^2$, $|x_m| \leqslant X_m$ depends only on the restriction of the initial datum to $x_1^2 + \cdots + x_{m-1}^2 \leqslant b^2$.

An abstract version of the Cauchy problem incorporating Hadamard's basic ideas could be formulated along the lines of Section 1.2 by looking at the differential operator on the right-hand side of (1.7.1) as a x_m-dependent linear operator in $C^{(\infty)}(\mathbb{R}^{m-1})$ or $\mathcal{D}(\mathbb{R}^{m-1})$ with values in $C^{(\infty)}(\mathbb{R}^{m-1})$; setting $x_m = t$, $F = C^{(\infty)}(\mathbb{R}^{m-1})$, and $E = \mathcal{D}(\mathbb{R}^{m-1})$ or $C^{(\infty)}(\mathbb{R}^{m-1})$, depending on whether the setting is that of Theorems 1.7.1 or 1.7.2, we are led to the following *abstract Cauchy problem in the sense of Hadamard*: given two linear topological spaces E, F with $E \subseteq F$ (but where the topology of E is not necessarily that inherited from F) and a family of operators $\{A(t); 0 \leqslant t \leqslant T\}$ with domains $D(A(t))$, the Cauchy problem for

$$u'(t) = A(t)u(t) \quad (0 \leqslant t \leqslant T) \qquad (1.7.7)$$

is *properly posed in the sense of Hadamard* if solutions $u(\cdot)$ (defined in an

---

[2]From J. Hadamard, *Lectures on Cauchy's Problem in Linear Partial Differential Equations*, (New Haven: Yale University Press, 1923).

obvious way) exist for initial data $u(0)$ in a dense set $D \subseteq E$ and depend continuously on them, in the sense that $\lim u_\gamma(t) = 0$ in F, uniformly in $0 \leqslant t \leqslant T$ whenever (the generalized sequence) $\langle u_\gamma(0) \rangle$ converges to zero in E (note that in the examples considered above $D(A(t)) \equiv E$ and $A(t): E \to F$ is continuous). Different abstract schemes of this type, corresponding to different definitions of solution could be devised. A setting where E and F are Banach spaces is the following: E is the Banach space $BC^{(N)}(\mathbb{R}^{m-1})$ of functions with bounded and continuous partial derivatives of order $\leqslant N$ (equipped with the norm $\|\cdot\|_N$, the supremum of all these derivatives in $\mathbb{R}^{m-1}$) and $F = BC^{(M)}(\mathbb{R}^{m-1})$. This last version, however, is probably not very faithful to Hadamard's formulation since it incorporates a priori restrictions on the behavior of the solutions at infinity (implicit in their membership in $BC^{(M)}(\mathbb{R}^{m-1})$ for all t), which are wholly absent from his original definition of correctly set problem; however, these restrictions at infinity allow us to treat as well characteristic Cauchy problems (such as the usual initial value problem for the heat equation) for which the local schemes in Theorems (1.7.1) and (1.7.2) are inadequate, since solutions may not exist or uniqueness may fail (see Hörmander [1961: 1, p. 121]).

    The abstract Cauchy problem in the sense described above does not seem to have been studied in full generality; for the case where E and F are Banach spaces, see Kreĭn-Laptev-Cvetkova [1970: 1], where $T = \infty$ and the problem is understood in a somewhat different sense (see also Sova [1977: 1]). On the other hand, there exists a vast literature on properly posed "concrete" Cauchy problems for partial differential equations, undoubtedly one of the central questions in the theory: see, for instance, Petrovskiĭ [1938: 1], A. Lax [1956: 1], Lax [1957: 2], Hörmander [1955: 1] and [1969: 1] (additional references can be found there), Strang [1966: 1], [1967: 1], [1969: 1], Flaschka-Strang [1971: 1], Gelfand-Šilov [1958: 3], and Friedman [1963: 1].

    The abstract Cauchy problem for differential equations in Banach spaces was introduced by Hille [1952: 3] (see also [1953: 2], [1953: 3], [1954: 1], [1954: 2], [1957: 1], Phillips [1954: 2]). It should be pointed out that Hille's prescription for a well-set Cauchy problem is somewhat different from that in Section 1.2 not only in the minor technical matter of definition of solution but in that continuous dependence is not assumed; the matter of interest is that solutions (that satisfy a certain growth condition at infinity) should be unique. That continuous dependence follows from existence and uniqueness in certain cases can be seen much in the same way as in Theorems 1.7.1 and 1.7.2, as the following result (which is due to Phillips [1954: 2] in a slightly more general formulation) shows.

    1.7.3   Theorem. *Let A be a densely defined operator in the Banach space E such that $\rho(A) \neq \varnothing$. Assume that for every $u \in D(A)$ there exists a unique solution of*

$$u'(t) = Au(t) \quad (t \geqslant 0) \tag{1.7.8}$$

in the sense of Section 1.2 *with* $u(0) = u$. *Then the Cauchy problem for* (1.7.8) *is properly posed in the sense of Section* 1.2.

Proof. Let $T > 0$, $C^{(1)} = C^{(1)}(0, T; E)$ the Banach space of all E-valued continuously differentiable $u(\cdot)$ defined in $0 \leqslant t \leqslant T$ endowed with the norm $\|u(\cdot)\|_1 = \sup\|u(t)\| + \sup\|u'(t)\|$ (the suprema taken in $0 \leqslant t \leqslant T$). Consider $D(A)$ equipped with (the norm equivalent to) its graph norm $\|(\lambda I - A)u\|$ for some $\lambda \in \rho(A)$ and the operator $\mathfrak{S} : D(A) \to C^{(1)}$ defined by $\mathfrak{S}u = u(\cdot)$, $u(\cdot)$ the solution of (1.7.8) with $u(0) = u$. If $u_n \to u$ in $D(A)$ and $\mathfrak{S}u_n(\cdot) \to v(\cdot)$ in $C^{(1)}$, it is easy to see taking limits in (1.7.8), and using the fact that A is closed that $v(\cdot)$ is a (then the only) solution of (1.7.8) with $v(0) = u$; it follows that $\mathfrak{S}$ is closed, hence bounded by the closed graph theorem. Given $u \in D(A)$, it is plain that $v(\cdot) = R(\lambda)\mathfrak{S}u$ is a (then the only) solution of $v'(t) = Av(t)$ with $v(0) = R(\lambda)u$, so that $\mathfrak{S}R(\lambda)u = R(\lambda)\mathfrak{S}u$. Hence we have, for $u(\cdot) = \mathfrak{S}u$,

$$\sup\|u(t)\| = \sup\|(\lambda I - A)\mathfrak{S}R(\lambda)u\|$$
$$\leqslant C'\|v(\cdot)\|_1 \leqslant C\|(\lambda I - A)R(\lambda)u\| = C\|u\|.$$

Since T is arbitrary, the result follows.

The definition of properly posed abstract Cauchy problem given in Section 1.2 was introduced by Lax in a 1954 New York University seminar (where A was allowed to depend on time). The time-invariant version was published in Lax-Richtmyer [1956: 1]. Although also inspired by Hadamard's basic idea, this formulation is less general since solutions and initial data are measured in the same norm. Yet, it seems to be ample enough (together with its time-dependent counterpart, to be examined in Chapter 7) to model linear physical systems where a "measure of the state" such as energy, amount of diffusing matter, or temperature, remains bounded or at least does not become infinite in finite time; ideally, the norm of the space E should be this measure (although the choice of norm is many times dictated by mere mathematical convenience, as is the L^2 norm in the treatment of the heat equation). However, not all physical systems fit into this scheme; see Birkhoff [1964: 1, p. 312]. The term propagator was used to denote the operator $S(t)$ by Segal [1963: 1].

An interesting problem in this connection is: given a system (or class of systems) of partial differential equations, we seek functional norms that allow the Cauchy problem to fit in the scheme of Section 1.2, or, alternatively, we look for conditions on the system that permit the use of a given functional norm like the L^2 or the C norms. For results in both directions see Birkhoff-Mullikin [1958: 1], Birkhoff [1964: 1], Lax-Richtmyer [1956: 1] Richtmyer-Morton [1967: 1], Kreiss [1959: 1], [1962: 1], [1963: 1], and Strang [1969: 1], [1969: 2], [1969: 3], [1969: 4]. See also L. Schwartz [1950: 1], where the "state spaces" are spaces of distributions. We note that the surge of interest in the theory of the abstract Cauchy problem during and after the fifties was motivated in part by the rapid development of finite-difference methods (to approximate solutions of differential equations) made practical by the availability of high-speed computers.

The notion of properly posed problem in Section 1.2 may of course be modified in many ways: solutions may be defined in suitably weak senses, the function $C(\cdot)$ in the a priori estimate (1.2.3) can be assumed only finite, and so on. Different properties of the propagator $S(\cdot)$ result. Some of these variants will be examined in Section 2.5.

In relation to the examples in Sections 1.4 and 1.6, we note that the Schrödinger, Maxwell, and Dirac operators are skew-adjoint (that is, i times a self-adjoint operator). This was implicitly known when the corresponding equations were introduced in physics, since the conservation of norm property for the solutions (required on physical grounds) is a consequence of skew-adjointness, as we have seen in the corresponding examples. An abstract formulation of this argument is:

1.7.4 Example. Let A be a self-adjoint operator in the Hilbert space H. Then the Cauchy problem for
$$u'(t) = iAu(t) \qquad (1.7.9)$$
is well posed in $-\infty < t < \infty$ and, for any solution $u(\cdot)$ we have
$$\|u(t)\| = \|u(0)\|. \qquad (1.7.10)$$

The proof is essentially the same as that for, say, the Schrödinger operator, the role of the Fourier transform being taken over by the spectral integral and the functional calculus.

We should mention, however, that the rigorous proof of skew-adjointness for the Maxwell, Schrödinger, and Dirac operators is rather recent: for the last two see Kato [1951: 1], [1951: 2].

In the matter of Example 1.7.4, we point out that the converse is as well true (see Theorem 3.6.7) with a partial converse holding in the case where the Cauchy problem for (1.7.9) is well posed in $t \geqslant 0$ and (1.7.10) holds there (Theorem 3.6.6). Equation (1.7.9) is sometimes called the *abstract Schrödinger equation* (see J. L. B. Cooper [1948: 1]).

Example 1.6.1 generalizes an earlier result of Littman [1963: 1], where the nonexistence of L^p estimates for the wave equation ($p \neq 2$) is deduced from direct estimation of the explicit solution. For L^p estimates for symmetric hyperbolic systems, not necessarily with constant coefficients (as well as for estimates in other functional norms), see Brenner [1972: 1], [1973: 1], [1975: 1], [1977: 1], and Brenner-Thomée-Wahlbin [1975: 1]; estimates for the Schrödinger and other equations are in Da Prato-Giusti [1967: 1], [1967: 2], Lanconelli [1968: 1], [1968: 2], [1970: 1], and Marshall-Strauss-Wainger [1980: 1]. Most lie deeper than Theorem 1.6.3.

Example 1.2.2 deserves some comment. As pointed out there, it does not apply to any nontrivial differential operator A in the usual Banach spaces of classical analysis. However, many differential operators are continuous in certain linear topological spaces (such as the spaces of Schwartz test functions $\mathcal{D}$, S or the distribution spaces $\mathcal{D}'$, S') hence we may ask whether (1.2.10) makes sense in these conditions. The answer is in the negative, as the series is in general divergent.

We note finally that it is possible in some cases to deduce that a Cauchy problem is properly posed in the sense of Hadamard from the fact that it is properly posed in the sense of Section 1.2; this is usually done using "comparison-of-norms" results like the Sobolev embedding theorems. An example of this sort of reasoning is that following Theorem 1.6.3, although estimates in the H^s norms involve of course growth conditions at infinity. Estimates in local norms could be obtained, however, by using domain of dependence arguments.

Chapter 2

Properly Posed Cauchy Problems: General Theory

We continue in this chapter our examination of the Cauchy problem for the equation $u'(t) = Au(t)$ initiated in Sections 1.2 and 1.5. The main result is Theorem 2.1.1, where necessary and sufficient conditions are given in order that the Cauchy problem be well posed in $t \geq 0$. These conditions involve restrictions on the location of the spectrum of A and inequalities for the norm of the powers of the resolvent of A. The proof presented here, perhaps not the shortest or the simplest, puts in evidence nicely the fundamental Laplace transform relation between the propagator $S(t)$ and the resolvent of A; in fact, $S(t)$ is obtained from the resolvent using (a slight variant of) the classical contour integral for computation of inverse Laplace transforms. Section 2.2 covers the corresponding result for the Cauchy problem in the whole real line, and the adjoint equation. The adjoint theory is especially interesting in nonreflexive spaces and will find diverse applications in Chapters 3 and 4.

One of the first results in this chapter is the proof of the *semigroup* or *exponential equations* $S(0) = I$, $S(s + t) = S(s)S(t)$ for the propagator $S(t)$. We show in Section 2.3 that any strongly continuous operator-valued function satisfying the exponential equations must be the propagator of an equation $u'(t) = Au(t)$. Finally, Section 2.4 deals with several results for the inhomogeneous equation $u'(t) = Au(t) + f(t)$.

2.1. THE CAUCHY PROBLEM IN $t \geqslant 0$

We examine in this section the question of finding necessary and sufficient conditions on the densely defined operator A in order that the Cauchy problem for

$$u'(t) = Au(t) \quad (t \geqslant 0) \tag{2.1.1}$$

be well posed in $t \geqslant 0$. An answer is given in Theorem 2.1.1; the conditions on A bear on its resolvent set $\rho(A)$ and on the behavior of the resolvent $R(\lambda) = (\lambda I - A)^{-1}$ in $\rho(A)$.

We assume that the Cauchy problem for (2.1.1) is well posed in the sense of Chapter 1. Let $S(\cdot)$ be the propagator of (2.1.1). Then

$$S(0) = I, \quad S(s+t) = S(s)S(t), \quad s, t \geqslant 0. \tag{2.1.2}$$

The first equality (2.1.2) is evident from the definition of S. To prove the other one, let $u \in D$ and $t \geqslant 0$ fixed, and consider the function $u(s) = S(s+t)u$, $s \geqslant 0$. Due to the fact that A does not depend on t, $u(\cdot)$ is a solution of (2.1.1); hence it follows from (1.2.7) that $u(s) = S(s)u(0) = S(s)S(t)u$. The second equality in (2.1.2) results then from denseness of D.

An important consequence of (2.1.2) is the following. Let $t \geqslant 0$, n the largest integer not surpassing t. Then $S(t) = S(t-n)S(1)^n$; accordingly, $\|S(t)\| \leqslant \|S(t-n)\| \, \|S(1)\|^n \leqslant C \cdot C^n = C \exp(n \log C) \leqslant C \exp(t \log C)$, where $C = \sup\{C(t); \ 0 \leqslant t \leqslant 1\}$ (here $C(\cdot)$ is the function in our definition of well-posed Cauchy problem in Section 1.2). It follows that *there exist constants C, ω such that*

$$\|S(t)\| \leqslant Ce^{\omega t} \quad (t \geqslant 0). \tag{2.1.3}$$

This implies that every solution (and also every generalized solution) of (2.1.1) grows at most exponentially at infinity.

We can deduce from (2.1.2) more precise information on the behavior of $S(\cdot)$ at infinity. In fact, let $\omega(t) = \log\|S(t)\|$ $(t \geqslant 0)$. Clearly $\omega(\cdot)$ (which may take the value $-\infty$) is *subadditive*—that is, $\omega(s+t) \leqslant \omega(s) + \omega(t)$ for all $s, t \geqslant 0$. It follows then essentially from a result in Pólya and Szego [1954: 1, p. 17, Nr. 98] that

$$\omega_0 = \inf_{t \geqslant 0} \frac{\omega(t)}{t} = \lim_{t \to \infty} \frac{\omega(t)}{t} < \infty,$$

where we allow the value $-\infty$ for ω_0. Clearly (2.1.3) holds for any $\omega > \omega_0$ (but for no $\omega < \omega_0$); it may not necessarily hold for ω_0 itself.

Let λ be a complex number such that $\text{Re } \lambda > \omega$ (ω the constant in (2.1.2)). Define

$$Q(\lambda)u = \int_0^\infty e^{-\lambda t}S(t)u \, dt \quad (u \in E). \tag{2.1.4}$$

The norm of the integrand is bounded by $C\|u\|\exp(\omega - \text{Re }\lambda)t$, thus by Lemma 1.3, (2.1.4) defines a bounded operator in E. We have already seen in Example 1.2.2 that in the case where A is bounded, $Q(\lambda) = R(\lambda)$, which suggests that the same equality may hold in general. This is in fact so, at least if we assume A *closed*. In fact, let $u \in D$, $T > 0$. Making use of Lemma 3.4, we have

$$A\int_0^T e^{-\lambda t}S(t)u\, dt = \int_0^T e^{-\lambda t}AS(t)u\, dt$$

$$= \int_0^T e^{-\lambda t}(S(t)u)'\, dt$$

$$= e^{-\lambda T}S(T)u - u + \lambda\int_0^T e^{-\lambda t}S(t)u\, dt.$$

Letting $T \to \infty$ and making use again of the fact that A is closed, we see that $Q(\lambda)u \in D(A)$ and $AQ(\lambda)u = \lambda Q(\lambda)u - u$. Let now $u \in E$, $\{u_n\}$ a sequence in D with $u_n \to u$. Then $AQ(\lambda)u_n = \lambda Q(\lambda)u_n - u_n \to \lambda Q(\lambda)u - u$ while evidently $Q(\lambda)u_n \to Q(\lambda)u$. Accordingly $Q(\lambda)u \in D(A)$ and $AQ(\lambda)u = \lambda Q(\lambda)u - u$; in other words, $Q(\lambda)E \subseteq D(A)$ and

$$(\lambda I - A)Q(\lambda) = I. \tag{2.1.5}$$

It is plain that (2.1.5) shows that $(\lambda I - A): D(A) \to E$ is onto. We show next that $(\lambda I - A)$ is one-to-one. In fact, assume that there exists $u \in D(A)$ with $Au = \lambda u$. Then $u(t) = e^{\lambda t}u$ is a solution of (2.1.1); since $\|u(t)\| = e^{(\mathrm{Re}\,\lambda)t}\|u\|$, this contradicts (2.1.3) unless $u = 0$. We see then that $(\lambda I - A)$ is invertible and that $(\lambda I - A)^{-1} = Q(\lambda)$ is bounded. (Compare with (1.2.13).)

The fact that $\rho(A)$ contains the half plane $\mathrm{Re}\,\lambda > \omega$ (in particular, that it is nonempty) has some interesting consequences. In fact, let $u \in D$, $\lambda \in \rho(A)$. Clearly $u(t) = R(\lambda)S(t)u$ is a solution of (2.1.1), so that in view of (1.2.7),

$$R(\lambda)S(t)u = S(t)R(\lambda)u \quad (t \geq 0). \tag{2.1.6}$$

Since D is dense in E, (2.1.6) must hold for all $u \in E$. Noting that $D(A) = R(\lambda)E$, we see that

$$S(t)D(A) \subseteq D(A) \quad (t \geq 0)$$

and, making use of (2.1.6) for $(\lambda I - A)u$ instead of u and then applying $(\lambda I - A)$ to both sides, we obtain

$$AS(t)u = S(t)Au \quad (u \in D(A), t \geq 0). \tag{2.1.7}$$

Observe next that if we integrate the differential equation (2.1.1) and make use of (2.1.7), it follows that

$$S(t)u - u = \int_0^t (S(s)u)'\, ds = \int_0^t S(s)Au\, ds$$

for $u \in D$. Applying $R(\lambda)$ to both sides of this equality and using (2.1.6), we obtain

$$R(\lambda)(S(t)u - u) = R(\lambda)\int_0^t S(s)Au\, ds$$

$$= \int_0^t S(s)AR(\lambda)u\, ds.$$

Both sides of this equality are bounded operators of u, thus it must hold for all $u \in D(A)$. Making use of the fact that $R(\lambda)$ is one-to-one, we conclude that

$$S(t)u - u = \int_0^t S(s)Au\,ds = \int_0^t AS(s)u\,ds \qquad (2.1.8)$$

for all $u \in D(A)$. But then $S(\cdot)u$ must be a solution of (2.1.1) for any $u \in D(A)$; that is, the assumption that the Cauchy problem for (2.1.1) is well posed implies that we may take

$$D = D(A). \qquad (2.1.9)$$

All elements of $D(A)$ are then initial data of solutions of (2.1.1). We try now to obtain more information about $R(\lambda) = Q(\lambda)$. Estimates for the norm of arbitrary powers of $R(\lambda)$ can be obtained from (2.1.4) by differentiating both sides with respect to λ. On the left-hand side we use formula (3.5), whereas on the right-hand side we differentiate under the integral sign. (That this is permissible, say, for the first differentiation can be justified as follows on the basis of the dominated convergence theorem: the integrand corresponding to the quotient of increments is $h^{-1}(e^{-(\lambda+h)t} - e^{-\lambda t})S(t)u$, which is bounded in norm, uniformly with respect to small h by $Cte^{-(\operatorname{Re}\lambda - |h| - \omega)t}$. The higher differentiations can be handled in the same way.) We obtain the formula

$$R(\lambda)^n u = \frac{1}{(n-1)!}\int_0^\infty t^{n-1}e^{-\lambda t}S(t)u\,dt \quad (\operatorname{Re}\lambda > \omega, n \geq 1). \qquad (2.1.10)$$

Noting that the integrand is bounded in norm by $Ct^{n-1}e^{-(\operatorname{Re}\lambda - \omega)t}$, we deduce from (2.1.10) that

$$\|R(\lambda)^n\| \leq C(\operatorname{Re}\lambda - \omega)^{-n} \quad (\operatorname{Re}\lambda > \omega, n \geq 1). \qquad (2.1.11)$$

It is remarkable (but not unexpected, in view of a result in Widder [1931; 1] that identifies Laplace transforms of bounded functions) that inequalities (2.1.11) are as well *sufficient* for the Cauchy problem to be well posed. In fact, we have

2.1.1 Theorem. *Let A be closed. The Cauchy problem for (2.1.1) is well posed and its propagator satisfies (2.1.3) if and only if $\sigma(A)$ is contained in the half-plane $\operatorname{Re}\lambda \leq \omega$ and $R(\lambda)$ satisfies inequalities (2.1.11) (C and ω the same constants in (2.1.3)).*

We have already proved half of Theorem 2.1.1. To show the other half we begin by constructing certain solutions of (2.1.1). Let $u \in D(A^3)$, $\omega' > \omega, 0$. Define

$$u(t) = u + tAu + \tfrac{1}{2}t^2A^2u$$

$$+ \frac{1}{2\pi i}\int_{\omega'-i\infty}^{\omega'+i\infty}\frac{e^{\lambda t}}{\lambda^3}R(\lambda)A^3u\,d\lambda \quad (t \geq 0). \qquad (2.1.12)$$

A simple deformation-of-contour argument shows that the integral on the right-hand side vanishes for $t \leq 0$, so that

$$u(0) = u. \tag{2.1.13}$$

It is also plain that

$$\|u(t)\| \leq C(u) e^{\omega' t} \quad (t \geq 0). \tag{2.1.14}$$

Differentiating under the integral sign (justification of this runs along the same lines as that of (2.1.10)) and making use of another deformation-of-contour argument and of Lemma 2.1, we obtain that $u(\cdot)$ is continuously differentiable, $u(t) \in D(A)$ for all t and

$$u'(t) - Au(t) = -\tfrac{1}{2} t^2 A^3 u$$

$$+ \frac{1}{2\pi i} \int_{\omega' - i\infty}^{\omega' + i\infty} \frac{e^{\lambda t}}{\lambda^3} (\lambda I - A) R(\lambda) A^3 u \, d\lambda = 0 \quad (t \geq 0).$$

We see then that $u(\cdot)$ is a solution of (2.1.1) satisfying the initial condition (2.1.13). We try now to improve the estimate (2.1.14) for $u(\cdot)$. To this end we observe that, making use of the equality $u'(t) - Au(t) = 0$, of an argument similar to the one leading to (2.1.5), and of the fact that $(\lambda I - A)$ is one-to-one, we obtain

$$\int_0^\infty e^{-\lambda t} u(t) \, dt = R(\lambda) u \quad (\operatorname{Re} \lambda \geq \omega'). \tag{2.1.15}$$

(Note that the integral in (2.1.15) makes sense in view of (2.1.14); a more direct, but less tidy proof of (2.1.15) can be achieved computing its left-hand side by replacing $u(\cdot)$, given by (2.1.12), into it.) Differentiating now (2.1.15) repeatedly with respect to λ and arguing once again as in the justification of (2.1.10), we obtain

$$R(\lambda)^n u = \frac{1}{(n-1)!} \int_0^\infty s^{n-1} e^{-\lambda s} u(s) \, ds \quad (\operatorname{Re} \lambda > \omega').$$

Hence, if $t > 0$,

$$\left(\frac{n}{t}\right)^n R\left(\frac{n}{t}\right)^n u = \int_0^\infty \gamma_n(t, s) u(s) \, ds \quad (n > \omega' t), \tag{2.1.16}$$

where

$$\gamma_n(t, s) = \frac{1}{(n-1)!} \left(\frac{n}{t}\right)^n s^{n-1} e^{-ns/t} \quad (s \geq 0, t > 0).$$

It is immediate that each γ_n is positive and infinitely differentiable. It follows from (2.1.16) applied to $A = 0$, or directly, that

$$\int_0^\infty \gamma_n(t, s) \, ds = 1 \quad (t > 0, n \geq 1). \tag{2.1.17}$$

Clearly $\gamma_n(t, 0) = \gamma_n(t, \infty) = 0$. Moreover, γ_n, as a function of s, increases in

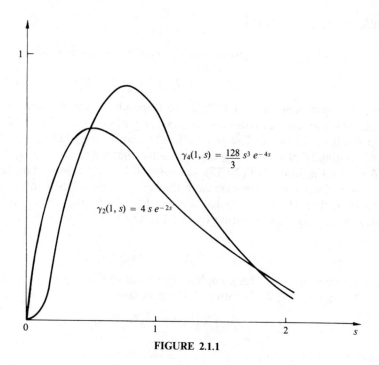

$\gamma_4(1, s) = \dfrac{128}{3} s^3 e^{-4s}$

$\gamma_2(1, s) = 4 s e^{-2s}$

FIGURE 2.1.1

$0 \leqslant s \leqslant (n-1)t/n$, decreases in $(n-1)t/n \leqslant s < \infty$, and we obtain from Stirling's formula that

$$\gamma_n(t, s) = \frac{1}{s} \sqrt{\frac{n}{2\pi}} \left(\frac{s}{t} e^{1-s/t} \right)^n (1 + o(1)) \text{ as } n \to \infty. \qquad (2.1.18)$$

In view of this asymptotic relation, of the preceding comments and of the inequality $\rho e^{1-\rho} < 1$ which holds for $\rho > 0$, $\rho \neq 1$, we obtain: if $0 < t' < t$ and $(n-1)/n \geqslant t'/t$, then

$$\gamma_n(t, s) \leqslant \gamma_n(t, t') \to 0 \text{ as } n \to \infty \quad (0 \leqslant s \leqslant t') \qquad (2.1.19)$$

Let now k be an arbitrary positive constant. An elementary analysis shows that the function $e^{ks}\gamma_n(t, s)$ $(n > kt)$ is decreasing in $s \geqslant (n-1)t/(n-kt)$ $\to t$ as $n \to \infty$, thus if we select $t'' > t$ and make use again of the asymptotic estimation (2.1.18), we obtain

$$e^{kt}\gamma_n(t, s) \to 0 \text{ as } n \to \infty \qquad (2.1.20)$$

uniformly in $s \geqslant t''$.

We prove next that

$$u(t) = \lim_{n \to \infty} \int_0^\infty \gamma_n(t, s) u(s) \, ds \qquad (2.1.21)$$

using the previous estimates for γ_n in a way familiar in the theory of mollifiers. Let $n > \omega' t$. Then

$$\left\| u(t) - \int_0^\infty \gamma_n(t,s) u(s)\, ds \right\| \leqslant \int_0^\infty \gamma_n(t,s) \| u(t) - u(s) \|\, dt$$

$$= I_n(\delta) + J_n(\delta) + K_n(\delta), \qquad (2.1.22)$$

where the right-hand side of (2.1.22) corresponds to the division of the domain of integration into the subdomains $(0, t - \delta), (t - \delta, t + \delta), (t + \delta, \infty)$, where $0 < \delta < t$. Let $\varepsilon > 0$. Taking δ sufficiently small and availing ourselves of the continuity of $u(\cdot)$ at t, we may assume that $\| u(t) - u(s) \| \leqslant \varepsilon/3$ for $t - \delta < s < t + \delta$; in view of (2.1.17), this assures that $\| J_n(\delta) \| \leqslant \varepsilon/3$ independently of n. Once δ has been chosen in this way, we use (2.1.19) and choose n_0 so large that $\| I_n(\delta) \| \leqslant \varepsilon/3$ for $n \geqslant n_0$. The third integral is similarly treated, observing that by virtue of (2.1.14) and (2.1.20) with $k = 2\omega'$ we have

$$\| \gamma_n(t,s) u(s) \| \leqslant C' \rho_n e^{-\omega' s} \quad (s \geqslant t + \delta),$$

where $\rho_n \to 0$ as $n \to \infty$. This completes the proof of (2.1.21).

We combine (2.1.21) with (2.1.16) to obtain

$$u(t) = \lim_{n \to \infty} \left(\frac{n}{t} \right)^n R\left(\frac{n}{t} \right)^n u \quad (t > 0) \qquad (2.1.23)$$

Making use of inequalities (2.1.11), we deduce that

$$\| u(t) \| \leqslant \lim_{n \to \infty} C\left(\frac{n}{t} \right)^n \left(\frac{n}{t} - \omega \right)^{-n} \| u \| = C e^{\omega t} \| u \| \quad (t \geqslant 0). \quad (2.1.24)$$

We define

$$\tilde{S}(t) u = u(t) \quad \left(u \in D(A^3), t \geqslant 0 \right)$$

where u is the function in (2.1.12). Making use of arguments similar to those employed in the treatment of $S(\cdot)$ (see Section 1.2), we show that $\tilde{S}$ can be extended to a (E)-valued strongly continuous function defined in $t \geqslant 0$, satisfying $\tilde{S}(0) = I$ and

$$\| \tilde{S}(t) \| \leqslant C e^{\omega t} \quad (t \geqslant 0), \qquad (2.1.25)$$

C, ω the constants in (2.1.24) (and in the hypotheses of Theorem 2.1.1). Clearly, the proof will be complete if we can establish that

$$u(t) = \tilde{S}(t) u(0) \qquad (2.1.26)$$

for *all* solutions of (2.1.1) (and not only for those originating in $D(A^3)$ defined by (2.1.12)). In order to show (2.1.26), we observe first that it follows immediately from the definition of $\tilde{S}(t)$ that $\tilde{S}(t) R(\lambda) u = R(\lambda) \tilde{S}(t) u$ if $u \in D(A^3)$; by the usual continuity argument, this equality must then hold for all $u \in E$. This implies that $\tilde{S}(t) D(A) \subseteq D(A)$ and

$$\tilde{S}(t) A u = A \tilde{S}(t) u \quad (u \in D(A), t \geqslant 0).$$

On the other hand, it is a simple consequence of (2.1.12) and the definition of $\tilde{S}$ that $M(t) = \tilde{S}(t)R(\lambda)^3$ is a continuously differentiable (E)-valued function in $t \geq 0$ and that $M'(t) = AM(t)$. Accordingly, if $u(\cdot)$ is *any* solution of (2.1.1) we have

$$\frac{d}{dt} M(t - s)u(s) = 0 \quad (0 \leq s \leq t).$$

Then $R(\lambda)^3 u(t) = M(0)u(t) = M(t)u(0) = R(\lambda)^3 \tilde{S}(t)u(0)$, from which (2.1.26) results since $R(\lambda)^3$ is one-to-one. This ends the proof of Theorem 2.1.1.

Note, incidentally, that although solutions of (2.1.1) are initially defined only for initial data in $D(A^3)$, this is enough to establish that the Cauchy problem for (2.1.1) is well posed and thus, by virtue of (2.1.9) and preceding comments, solutions actually exist for arbitrary initial data in $D(A)$.

2.1.2 Remark. Formula (2.1.23) can be written in the following manner:

$$S(t)u = \lim_{n \to \infty} \left(I - \frac{t}{n}A \right)^{-n} u \quad (t > 0). \tag{2.1.27}$$

Strictly speaking we have proved this only for $u \in D(A^3)$. However, it follows from inequalities (2.1.11) that $\|(I - (t/n)A)^{-n}\|$ is bounded for $n > \omega t$. Therefore (2.1.27) actually holds for all $u \in E$ (Theorem 1.1) and can be used to justify writing

$$S(t) = e^{tA} \quad (t \geq 0). \tag{2.1.28}$$

More results in this direction will be found later (see Section 2.5).

2.1.3 Example. Using the convergence arguments leading to (2.1.21) show that the limit in (2.1.27) is uniform on compacts of $t \geq 0$.

2.1.4 Remark. In the sufficiency part of Theorem 2.1.1, inequalities (2.1.11) need only be assumed for μ *real*, $\mu > \omega$. In fact, assume that $R(\mu)$ exists for $\mu > \omega$ and

$$\|R(\mu)^n\| \leq C(\mu - \omega)^{-n} \quad (\mu > \omega, n \geq 1). \tag{2.1.29}$$

Let $\mu > \omega$. We know (Section 3) that if the series

$$\sum_{j=0}^{\infty} (\mu - \lambda)^j R(\mu)^{j+1} \tag{2.1.30}$$

converges in (E) for some complex λ, then $\lambda \in \rho(A)$ and (2.1.30) equals $R(\lambda)$. But, in view of (2.1.29) the series converges in (E) whenever $|\lambda - \mu| < |\mu - \omega|$. Differentiating term by term and making use of formula (3.5), we obtain

$$R(\lambda)^n = \frac{(-1)^{n-1}}{(n-1)!} R(\lambda)^{(n-1)}$$

$$= \frac{1}{(n-1)!} \sum_{j=n-1}^{\infty} j(j-1)\cdots(j-n+2)(\mu-\lambda)^{j-n+1} R(\mu)^{j+1}$$

(2.1.31)

in $|\lambda - \mu| < |\mu - \omega|$. Replacing inequalities (2.1.29) into (2.1.31) we obtain the following estimates:

$$\|R(\lambda)^n\| \leq \frac{C}{(n-1)!} \sum_{j=n-1}^{\infty} j(j-1)\cdots(j-n+2)\frac{|\mu-\lambda|^{j-n+1}}{(\mu-\omega)^{j+1}}.$$

(2.1.32)

Let next λ be such that $\operatorname{Re}\lambda > \omega$. Take μ real and so large that the disk $|\zeta - \mu| \leq |\lambda - \mu|$ entirely belongs to $\operatorname{Re}\zeta > \omega$ and let $\mu_0 = \mu - |\lambda - \mu|$. Since $|\lambda - \mu| = \mu - \mu_0$, we obtain, making use of (2.1.31) for $A = \omega$, that

$$\|R(\lambda)^n\| \leq C(\mu_0 - \omega)^{-n} \quad (n \geq 1).$$ (2.1.33)

We let now $\mu \to +\infty$ and note that $\mu_0 \to \operatorname{Re}\lambda$, thus obtaining inequalities (2.1.11). Of course inequalities (2.1.29) need only be assumed for a sequence $\mu_n \to \infty$.

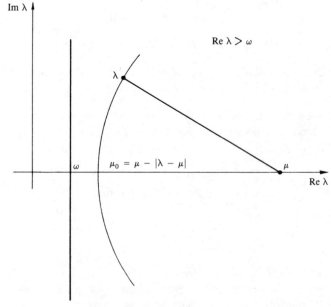

FIGURE 2.1.2

*2.1.5 Example (Phillips [1954: 1]). Obviously, the first inequality (2.1.29) implies all the others if $C = 1$, but if $C > 1$, not even the first inequality (2.1.11) implies necessarily the others (see also Hille-Phillips [1957: 1, p. 371]).

2.1.6 Example. Let A be a (not necessarily densely defined) operator such that $R(\lambda)$ exists for large real λ and satisfies the first inequality (2.1.29). Then, if $u \in \overline{D(A)}$,

$$\lim_{\lambda \to \infty} \lambda R(\lambda) u = u.$$

(*Hint*: Prove this first for $u \in D(A)$ and use Theorem 1.1.)

In all the following three examples, A is an operator satisfying the assumptions of Theorem 2.1.1.

2.1.7 Example. $D(A^\infty) = \cap_{n=1}^{\infty} D(A^n)$ is dense in E. (*Hint*: Consider the set of all elements of E of the form $\int \varphi(t) S(t) u \, dt$, where $u \in E$ and φ is a test function with support in $t > 0$.)

2.1.8 Example. If $u \in D(A^\infty)$, then

$$S(t) u \sim \sum_{n=0}^{\infty} \frac{t^n}{n!} A^n u, \tag{2.1.34}$$

where " $\sim$ " means "asymptotic in the sense of Poincaré as $t \to 0$"; precisely, we have

$$\lim_{t \to 0} \frac{1}{t^n} \left\{ S(t) u - \sum_{j=0}^{n-1} \frac{t^j}{j!} A^j u \right\} = \frac{1}{n!} A^n u. \tag{2.1.35}$$

2.1.9 Example. (a) If $u \in D(A)$, $\omega' > \omega, 0$ (ω the constant in (2.1.3)), $u \in D(A)$, then

$$S(t) u = \lim_{r \to \infty} \frac{1}{2\pi i} \int_{\omega' - ir}^{\omega' + ir} e^{\lambda t} R(\lambda) u \, d\lambda \tag{2.1.36}$$

for $t > 0$, the limit being uniform on compact subsets of $t > 0$. For $t = 0$ the limit is $\frac{1}{2} u$.

(b) If $u \in E$, $\omega' > \omega, 0$ and $u \in E$, then

$$S(t) u = \lim_{r \to \infty} \frac{1}{2\pi i} \int_{\omega' - ir}^{\omega' + ir} \left(1 - \frac{|\operatorname{Im} \lambda|}{r} \right) e^{\lambda t} R(\lambda) u \, d\lambda \tag{2.1.37}$$

in $t > 0$ with the same provisos as in (a). Note that this last integral can be considered as the Cesàro limit of the integral in (2.1.36).

The two formulas should be compared with (1.2.14).

2.2. THE CAUCHY PROBLEM IN $-\infty < t < \infty$. THE ADJOINT EQUATION.

We attempt in this section the characterization of those densely defined operators A that make the Cauchy problem for the equation

$$u'(t) = Au(t) \tag{2.2.1}$$

well posed in $-\infty < t < \infty$ in the sense of Section 1.5. This characterization turns out to be entirely reducible to the results in the previous section. In fact, we have

2.2.1   Theorem.   *Let A be closed. The Cauchy problem for* (2.2.1) *is well posed in* $(-\infty, \infty)$ *and its propagator S satisfies*

$$\|S(t)\| \leqslant Ce^{\omega|t|} \quad (-\infty < t < \infty) \tag{2.2.2}$$

if and only if $\sigma(A)$ is contained in the strip $|\operatorname{Re}\lambda| \leqslant \omega$ *and*

$$\|R(\lambda)^n\| \leqslant C(|\operatorname{Re}\lambda| - \omega)^{-n} \quad (|\operatorname{Re}\lambda| > \omega, n \geqslant 1). \tag{2.2.3}$$

Proof.   Assume the Cauchy problem for (2.2.1) is well posed in $-\infty < t < \infty$. Reasoning as in Theorem 2.1.1 we deduce that $R(\lambda)$ exists in $\operatorname{Re}\lambda > \omega$ and $\operatorname{Re}\lambda < -\omega$ and (compare with (1.5.4))

$$R(\lambda)u = \int_0^\infty e^{-\lambda t}S(t)u\,dt \quad (u \in E, \operatorname{Re}\lambda > \omega), \tag{2.2.4}$$

$$R(\lambda)u = -\int_0^\infty e^{\lambda t}S(-t)u\,dt \quad (u \in E, \operatorname{Re}\lambda < -\omega), \tag{2.2.5}$$

whence inequalities (2.2.3) follow from repeated differentiation as inequalities (2.1.11) follow from (2.1.4). Conversely, assume that the inequalities hold. Then, since $R(\lambda; -A) = -R(-\lambda; A)$ it follows from Theorem 2.1.1 that the Cauchy problem for (2.2.1) and for

$$u'(t) = -Au(t) \tag{2.2.6}$$

are both well posed in $t \geqslant 0$. Let $u \in D(A), u_+(\cdot)$ (resp. $u_-(\cdot)$) the solution of (2.2.1) (resp. (2.2.6)) with $u_+(0) = u$ (resp. $u_-(0) = u$). Then

$$u(t) = \begin{cases} u_+(t) & (t \geqslant 0) \\ u_-(-t) & (t \leqslant 0) \end{cases}$$

is a solution of (2.2.1) in $-\infty < t < \infty$. On the other hand, if $\{u_n(\cdot)\}$ is a sequence of solutions of (2.2.1) in $-\infty < t < \infty$ with $u_n(0) \to 0$, we see by applying the second part of the definition of well posed problem to $u_+(t) = u(t)$ and $u_-(t) = u(-t)$ in $t \geqslant 0$ that $u_n(t) \to 0$ uniformly on bounded subsets of $(-\infty, \infty)$. This ends the proof of Theorem 2.2.1.

We introduce some notations. If A is a densely defined operator in E, we say that $A \in \mathcal{C}_+(C, \omega)$ if the Cauchy problem for (2.1.1) is well posed in $t \geqslant 0$ and (2.1.3) holds. When knowledge of one or both constants is not essential, we simply write $A \in \mathcal{C}_+(\omega)$ or $A \in \mathcal{C}_+$. It is plain from the comments preceding inequality (2.1.3) that $\mathcal{C}_+(\omega) = \cup\{\mathcal{C}_+(C, \omega); 1 \leqslant C < \infty\}$, $\mathcal{C}_+ = \cup\{\mathcal{C}_+(\omega); -\infty < \omega < \infty\}$. In the same fashion we say that $A \in \mathcal{C}(C, \omega)$ when the Cauchy problem for (2.2.1) is well posed in $-\infty < t < \infty$ and

$$\|S(t)\| \leqslant Ce^{\omega|t|} \quad (-\infty < t < \infty). \tag{2.2.7}$$

It is clear from (2.1.3) applied to $t \to S(t)$ and to $t \to S(-t)$ that whenever the Cauchy problem for (2.2.1) is well posed, there exist constants C, ω such that (2.2.7) holds (we do not attempt here to distinguish between the exponential growth of $S(\cdot)$ for $t \geq 0$ and for $t \leq 0$). The classes $\mathcal{C}(\omega), \mathcal{C}$ are defined in the same way as the classes $\mathcal{C}_+(\omega), \mathcal{C}$, that is, $\mathcal{C}(\omega) = \cup\{\mathcal{C}(C, \omega); 1 \leq C < \infty\}$, $\mathcal{C} = \cup\{\mathcal{C}(\omega); 0 \leq \omega < \infty\}$.

2.2.2 Example. Show that $\mathcal{C}(\omega) = \varnothing$ if $\omega < 0$.

The problem of deciding whether a given operator A belongs to $\mathcal{C}$ or to $\mathcal{C}_+$ is in general not too easy to solve by direct application of Theorems 2.1.1 or 2.2.1 since their hypotheses involve the verification of infinite sets of inequalities for the resolvent operator $R(\lambda)$. The following result is useful when the form of the solutions of

$$u'(t) = Au(t) \tag{2.2.8}$$

can be guessed in advance.

2.2.3 Lemma. (a) *Let $S(\cdot)$ be a weakly measurable (E)-valued function defined in $t \geq 0$ such that*

$$\|S(t)\| \leq Ce^{\omega t} \quad (t \geq 0).$$

Let A be a closed, densely defined operator such that $R(\mu)$ exists for $\mu > \omega'$ and assume that for every $u \in E, u^* \in E^*$

$$\int_0^\infty e^{-\mu t}\langle u^*, S(t)u\rangle \, dt = \langle u^*, R(\mu)u\rangle \quad (\mu > \omega, \omega'). \tag{2.2.9}$$

Then $A \in \mathcal{C}_+(C, \omega)$ and the propagator of (2.2.8) coincides with $S(\cdot)$

(b) *Let $S(\cdot)$ be a weakly measurable (E)-valued function defined in $-\infty < t < \infty$ and such that*

$$\|S(t)\| \leq Ce^{\omega|t|} \quad (-\infty < t < \infty),$$

and let A be a closed, densely defined operator such that $R(\mu)$ exists for μ real, $|\mu| > \omega'$ and the equalities (2.2.9) and

$$\int_0^\infty e^{-\mu t}\langle u^*, S(-t)u\rangle \, dt = -\langle u^*, R(-\mu)u\rangle \quad (\mu > \omega, \omega') \tag{2.2.10}$$

hold. Then $A \in \mathcal{C}(C, \omega)$ and the propagator of (2.2.8) equals $S(\cdot)$

Proof. It follows from (2.2.9) that

$$\langle u^*, R(\mu)^n u\rangle = \frac{1}{(n-1)!} \int_0^\infty t^{n-1} e^{-\mu t}\langle u^*, S(t)u\rangle \, dt \quad (\mu > \omega, \omega', n \geq 1)$$

whence we obtain the inequalities $|\langle u^*, R(\mu)^n u\rangle| \leq C\|u^*\|\|u\|(\mu - \omega)^{-n}$ for $\mu > \omega, \omega'$ and $n \geq 1$. In view of the arbitrariness of u and u^*, inequalities (2.1.29) result and Theorem 2.1.1 (as modified by Remark 2.1.4) applies to show that $A \in \mathcal{C}_+$. If $\tilde{S}$ is the propagator of (2.2.8), then comparing (2.1.10) (for $n = 1$) with (2.2.9) we find out that the functions $\langle u^*, S(\cdot)u\rangle$ and

$\langle u^*, \tilde{S}(\cdot)u \rangle$ $(u^* \in E, u \in E)$ have the same Laplace transform, thus by a well known uniqueness theorem they must coincide. This ends the proof of (a). To prove (b) we only need to observe that it was a consequence of the proof of Theorem 2.2.1 that $A \in \mathcal{C}$ if and only if both A and $- A$ belong to $\mathcal{C}_+$.

We examine some artificial—but illustrative—examples where Lemma 2.2.3 can be used. Other applications of the theory will be seen in the following chapters.

2.2.4   Example.   Let $E = C_0(- \infty, \infty)$. Define

$$Au(x) = u'(x), \tag{2.2.11}$$

where $D(A)$ consists of all $u \in E$ whose derivative exists and belongs to E. Clearly $D(A)$ is dense in E (any test function belongs to $D(A)$). To compute $R(\lambda)$ we examine the differential equation

$$u'(x) - \lambda u(x) = - v(x), \tag{2.2.12}$$

where $v \in E$. It is an elementary exercise to show that if $\mathrm{Re}\,\lambda < 0$, the only solution of (2.2.12) that belongs to E is

$$R(\lambda)v(x) = u(x) = - \int_{-\infty}^{x} e^{\lambda(x-y)}v(y)\,dy \quad (-\infty < x < \infty), \tag{2.2.13}$$

and since it follows from (2.2.12) that $u' \in E$, it is clear that $u \in D(A)$. For $\mathrm{Re}\,\lambda > 0$, the formula is

$$R(\lambda)v(x) = u(x) = \int_{x}^{\infty} e^{\lambda(x-y)}v(y)\,dy. \tag{2.2.14}$$

In both cases u depends continuously on v in the norm of E.

Let now $S(\cdot)$ be the family of operators in (E) defined by

$$S(t)u(x) = u(x+t) \quad (-\infty < x, t < \infty). \tag{2.2.15}$$

It is plain that $S(\cdot)$ is strongly continuous and also that $\|S(t)\| = 1$ for $- \infty < t < \infty$. Moreover, if $\mu > 0$,

$$\left(\int_{0}^{\infty} e^{-\mu t} S(t)v\,dt \right)(x) = \int_{0}^{\infty} e^{-\mu t}v(x+t)\,dt$$

$$= \int_{x}^{\infty} e^{\mu(x-y)}v(y)\,dy$$

and a similar computation takes care of the case $\mu < 0$. It follows then from Lemma 2.2.3 that $A \in \mathcal{C}(1,0)$ and that $S(\cdot)$ is the propagator of (2.2.8).

2.2.5   Example.   The same analysis applies in the spaces $E = L^p(- \infty, \infty)$, $1 \leq p < \infty$. The operator A is defined by (2.2.11), but $D(A)$ is now the set of all $u \in E$ such that u' exists (in the sense of distributions) and belongs to E; equivalently, $D(A)$ consists of all absolutely continuous functions in E whose derivative (which must exist almost everywhere) belongs to E. Formulas (2.2.13) and (2.2.14) make sense in this context (verification that $u \in E$ involves Young's theorem 8.1). It is again true that $A \in \mathcal{C}(1,0)$ and that the propagator of (2.2.8) is given by the formula (2.2.15). In each case (as well as in the one examined in Example 2.2.4), $\sigma(A)$ coincides with the imaginary axis. The group $S(\cdot)$ is *isometric* in $C_0(-\infty,\infty)$ and in $L^p(-\infty,\infty)$:

$$\|S(t)u\| = \|u\| \quad (-\infty < t < \infty). \tag{2.2.16}$$

2.2.6 Example. Let $E = L^p(-\infty,0)$, $1 \leqslant p < \infty$, and let A be again defined by (2.2.11); now $D(A)$ consists of all absolutely continuous u in E such that $u' \in E$ and $u(0) = 0$. We examine the equation (2.2.12), extending v to $-\infty < x < \infty$ by setting $v(x) = 0$ for $x \geqslant 0$. If $\operatorname{Re}\lambda > 0$, formula (2.2.14) provides a solution satisfying $u(x) = 0$ and this solution is unique; clearly, $u \in D(A)$. Moreover, u depends continuously on v in the L^p norm so that $u = R(\lambda)v$. Consider now the operator valued function

$$S(t)u(x) = \begin{cases} u(x+t), & x \leqslant -t \\ 0, & -t \leqslant x \leqslant 0 \end{cases} \quad (x \leqslant 0, t \geqslant 0). \tag{2.2.17}$$

For $\mu > 0$, we have

$$\left(\int_0^\infty e^{-\mu t} S(t)v \, dt\right)(x) = \int_0^{-x} e^{-\mu t} v(x+t) \, dt$$
$$= \int_x^0 e^{\mu(x-y)} v(y) \, dy$$

almost everywhere in x (recall that integrals are computed in the L^p norm), thus A belongs to $\mathcal{C}_+(1,0)$ and $S(\cdot)$ is the propagator of (2.2.8). Note that each $S(t)$ is *isometric*:

$$\|S(t)u\| = \|u\| \quad (t \geqslant 0) \tag{2.2.18}$$

although, unlike in Examples 2.2.4 and 2.2.5, $S(t)$ is not invertible except for $t = 0$. The same argument works in spaces of continuous functions, but we must incorporate the boundary condition $u(0) = 0$ in the definition of the space to achieve denseness of the domain of A. Accordingly, we define E as the space of all continuous functions in $-\infty \leqslant x \leqslant 0$ such that $u(0) = 0$ and $u(x) \to 0$ as $x \to -\infty$ and $Au = u'$ with domain $D(A)$ consisting of all u in E with u' in E. Again each $S(t)$ is an isometry.

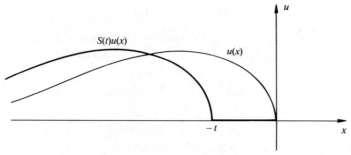

FIGURE 2.2.1

2.2.7 Example. In the previous example replace E by $L^p(0, \infty)$ $(1 \leqslant p < \infty)$ or $C_0[0, \infty)$ and define A in a similar way (although without boundary conditions at zero). In each case $A \in \mathcal{C}_+(1, 0)$ and the propagator is given by (2.2.15) (where we take the restriction of $u(x + t)$ to $x \geqslant 0$).

The last four examples provide results for the partial differential equation

$$\frac{\partial u}{\partial t} = \frac{\partial u}{\partial x} \qquad (2.2.19)$$

as an abstract differential equation.

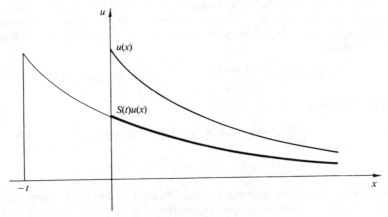

FIGURE 2.2.2

2.2.8 Example. Let H be a Hilbert space, A a normal operator, $P(d\lambda)$ its resolution of the identity (see Section 6). Then:

(a) $A \in \mathcal{C}_+$ if and only if $\{\operatorname{Re} \lambda \colon \lambda \in \sigma(A)\}$ is bounded above. If this condition is satisfied, then $A \in \mathcal{C}_+(1, \omega)$, where $\omega = \sup\{\operatorname{Re} \lambda; \lambda \in \sigma(A)\}$. The propagator of (2.2.8) is given by the formula

$$S(t) = \exp(tA) \quad (t \geqslant 0), \qquad (2.2.20)$$

and we have

$$\|S(t)\| = e^{\omega t} \quad (t \geqslant 0)$$

where the right-hand side of (2.2.20) can be computed using the functional calculus for normal operators (see Section 6); we may write

$$S(t)u = \int_{\sigma(A)} e^{\lambda t} P(d\lambda) u. \qquad (2.2.21)$$

(b) $A \in \mathcal{C}$ if and only if $\sigma(A)$ is contained in a strip parallel to the imaginary axis. If this condition is satisfied, then $A \in \mathcal{C}(1, \omega)$, $\omega = \sup(\omega_1, \omega_2)$, where $\omega_1 = \sup\{\operatorname{Re} \lambda; \lambda \in \sigma(A)\}$ and $-\omega_2 = \inf\{\operatorname{Re} \lambda; \lambda \in \sigma(A)\}$

(see Section 6 and references there.) In particular, if $\sigma(A)$ is contained in the imaginary axis, then $A \in \mathcal{C}(1,0)$. (Recall that if the spectrum of a normal operator is contained in the imaginary axis, then $A = iB$, where B is self-adjoint.) The propagator $S(t)$ can be computed by formula (2.2.20) and we have

$$\|S(t)\| = e^{\omega_1 t}, \quad \|S(-t)\| = e^{\omega_2 t} \quad (t \geqslant 0).$$

Let A be an operator in $\mathcal{C}_+(C,\omega)$. It is of great interest to know whether the adjoint A^* belongs to $\mathcal{C}_+$ as well. The question has a very simple answer when E is reflexive; in fact, in this case $D(A^*)$ is dense in E^*, $\rho(A^*) = \rho(A)$, and $R(\lambda; A^*) = R(\lambda; A)^*$ (Section 4). It follows then that A^* satisfies as well (and with the same constants) inequalities (2.1.11) and thus belongs to $\mathcal{C}_+$. The same is true, of course, in relation to the class $\mathcal{C}$. To identify $S^*(\cdot)$, the propagator of

$$u'(t) = A^* u(t), \tag{2.2.22}$$

we observe that if $u \in E$ and $u^* \in E^*$,

$$\int_0^\infty e^{-\mu t} \langle S^*(t) u^*, u \rangle \, dt = \langle R(\mu; A^*) u^*, u \rangle = \langle u^*, R(\mu; A) u \rangle$$

$$= \int_0^\infty e^{-\mu t} \langle u^*, S(t) u \rangle \, dt$$

$$= \int_0^\infty e^{-\mu t} \langle S(t)^* u^*, u \rangle \, dt \quad (\mu > \omega).$$

It follows then from uniqueness of Laplace transforms that $S^*(t) = S(t)^*$ for $t \geqslant 0$. If A belongs to $\mathcal{C}$, a similar argument takes care of $S^*(t)$ for t negative. We formalize these comments in the following

2.2.9 Theorem. *Let $A \in \mathcal{C}_+(C,\omega)$, and assume E reflexive. Then $A^* \in \mathcal{C}_+(C,\omega)$ as well. The solution operator of (2.2.22) is $S(\cdot)^*$, where $S(\cdot)$ is the solution operator of 2.2.8. The same conclusion holds for the class $\mathcal{C}(C,\omega)$.*

If the space E is not reflexive, the situation is considerably more complicated. In fact, it may be no longer true that A^* is densely defined; worse yet, there might be elements u^* of $D(A^*)$ for which no solution of (2.2.22) with u^* as initial datum exists.

2.2.10 Example. We use the result in Example 2.2.5 with $p = 1$. In this case $E^* = L^\infty(-\infty, \infty)$. To compute the adjoint of A, we observe that if $u \in D(A^*)$, then there exists $v(= A^* u)$ in E^* such that

$$\int u(x) w'(x) \, dx = \int v(x) w(x) \, dx \tag{2.2.23}$$

for all $w \in D(A)$. Since this equality must hold in particular when w is a test function, it follows that $u' = -v$ (in the sense of distributions), so that u, if necessary modified in a null set, is absolutely continuous. Conversely, if u is

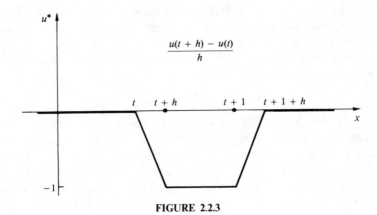

FIGURE 2.2.3

bounded, absolutely continuous and $u' \in L^\infty$, then (2.2.23) follows from integration by parts in the style of Lebesgue. In conclusion, $A^* = -d/dx$, the domain of A^* described by the previous comments. It is plain that $D(A^*)$ is not dense in E^*, for if $u \in E^*$ is a non-null function taking only the values 0 and 1, its distance to a continuous function must be at least $\frac{1}{2}$.

Take now a function u_0^* in $D(A^*)$ such that $A^* u_0^* \notin \overline{D(A^*)}$; for instance, $u_0(x) = 0 \ (x \leqslant 0), u(x) = x \ (0 \leqslant x \leqslant 1), u(x) = 1 \ (x \geqslant 1)$. Assume that (2.2.22) has a solution $u(t)(x) = u(t, x)$ with initial condition

$$u(0) = u_0^*. \tag{2.2.24}$$

It is not difficult to see that this solution must be of the form

$$u(t, x) = u_0^*(x - t).$$

But $u'(t)$ does not exist in the L^∞ norm for any t. Then there is actually no solution (in the sense of the definition in Chapter 1) of (2.2.22),(2.2.24) for our choice of u_0^*.

A way out of this difficulty is to replace the dual space E^* by the space $E^\# = \overline{D(A^*)}$ and restrict A^* to this subspace in the following sense: define $A^\# u^* = A^* u^*$, where $D(A^\#)$ is the set of all $u^* \in D(A^*)$ with $A^* u^* \in E^\# = \overline{D(A^*)}$. In the example at hand, these entities can be easily identified. In fact, $E^\#$ consists of all uniformly continuous bounded functions in $(-\infty, \infty)$ (we leave to the reader the proof of this simple fact) and $D(A^\#)$ is then the set of all u such that u, u' are uniformly continuous and bounded. It is not difficult to see that $D(A^\#)$ is dense in $E^\#$, that $S(t)^*$ maps $E^\#$ into $E^\#$ for all t, and that $S^\#(\cdot) = $ restriction to $E^\#$ of $S(\cdot)^*$ is strongly continuous. It follows then from Lemma 2.2.3 much in the same way as in the case of a reflexive space that $A^\# \in \mathcal{C}(1,0)$ and that $S^\#(\cdot)$ is the propagator of

$$u'(t) = A^\# u(t). \tag{2.2.25}$$

It is also immediate that $E^\#$ can also be characterized as the set of all $u^* \in E^*$ such that $S(\cdot)^* u^*$ is continuous in the L^∞ norm.

It is remarkable that all the previous observations have a counterpart in the general theory.

2.2.11   Theorem.   *Let $A \in \mathcal{C}_+(C, \omega)$ with E^* not necessarily reflexive; let $E^\#$ be the closure of $D(A^*)$ in E^*, $A^\# u^* = A^* u^*$, the domain of $A^\#$ consisting of all $u^* \in D(A^*)$ such that $A^* u^* \in E^\#$. Then $D(A^\#)$ is dense in $E^\#$ and $A^\# \in \mathcal{C}_+(C, \omega)$ in $E^\#$; the propagator of (2.2.25) is $S^\#(\cdot)$, the restriction of $S(\cdot)^*$ to $E^\#$. The subspace $E^\#$ can also be characterized as the set of all $u^* \in E^*$ for which $S(\cdot)^* u^*$ is continuous in $t \geqslant 0$. The same conclusions hold for the class $\mathcal{C}$.*

Proof.   We have

$$A^* R(\lambda; A^*) u^* = \lambda R(\lambda; A^*) u^* - u^* \quad (u^* \in E^*)$$

so that if $u^* \in E^\#$, then $A^* R(\lambda; A^*) u^*$ belongs as well to $E^\#$. In other words,

$$R(\lambda; A^*) E^\# \subseteq D(A^\#). \tag{2.2.26}$$

Since $\|R(\lambda: A^*)\| = \|R(\lambda; A)\| = O(1/\lambda)$ as $\lambda \to \infty$, we obtain from Example 2.1.6 that

$$\lim_{\lambda \to \infty} \lambda R(\lambda; A^*) u^* = u^* \tag{2.2.27}$$

for all $u^* \in E^\#$. This, combined with (2.2.26), clearly shows that $D(A^\#)$ is dense in $E^\#$. We have obtained as a consequence the fact that $R(\lambda; A^*) E^\# \subseteq E^\#$, so that equalities (4.11) and (4.13) imply

$$R(\lambda; A^\#) = R(\lambda; A^*)|_{E^\#}, \tag{2.2.28}$$

where the right-hand side of (2.2.28) is the restriction of $R(\lambda; A^*)$ to $E^\#$. It is then obvious that $\|R(\lambda; A^\#)\| \leqslant \|R(\lambda; A^*)\| = \|R(\lambda; A)\|$ and that a similar inequality holds for the powers $R(\lambda; A^\#)^n$, hence Theorem 2.1.1 implies that $A^\# \in \mathcal{C}_+(C, \omega)$. Let $S^\#(\cdot)$ be the propagator of $u'(t) = A^\# u(t)$. Then if $u^* \in E^\#$, $u \in E$,

$$\int_0^\infty e^{-\lambda t} \langle S^\#(t) u^*, u \rangle \, dt = \langle R(\lambda; A^\#) u^*, u \rangle = \langle R(\lambda; A^*) u^*, u \rangle$$

$$= \langle u^*, R(\lambda; A) u \rangle$$

$$= \int_0^\infty e^{-\lambda t} \langle u^*, S(t) u \rangle \, dt \quad (\lambda > \omega).$$

By uniqueness of Laplace transforms and arbitrariness of u, we obtain that $S^\#(t) u^* = S(t)^* u^*$, hence

$$S^\#(t) = S(t)^*|_{E^\#}$$

It is clear from this that $S(\cdot)^* u^*$ is continuous in $t \geqslant 0$ if $u^* \in E^\#$. Conversely, let u^* be an arbitrary element of E^* such that $S(\cdot)^* u^*$ is continuous at $t = 0$. If $\varepsilon > 0$ and δ is so small that $\|S^*(t) u^* - u^*\| \leqslant \varepsilon$ for

$0 \leqslant t \leqslant \delta$, we obtain

$$\|\lambda R(\lambda; A^*)u^* - u^*\| \leqslant \lambda \int_0^\infty e^{-\lambda t}\|S(t)u^* - u^*\|dt$$

$$\leqslant \varepsilon + o(1) \quad \text{as } \lambda \to \infty$$

dividing the interval of integration in $(0, \delta)$ and (δ, ∞). It follows that $\lambda R(\lambda; A^*)u^* \to u^*$ as $\lambda \to \infty$. But $R(\lambda; A^*)u^* \in D(A^*)$, thus $u^* \in \overline{D(A^*)} = E^\#$. The proof of Theorem 2.2.11 is thus complete.

The result just proved would be of course of little use unless we can show $E^\#$ is "large enough" in some sense. Since $D(A^*)$ is weakly dense in E^* (Section 5), the same is true of $E^\#$ and of $D(A^\#)$; in particular, if $\langle u^*, u \rangle = 0$ for all $u^* \in D(A^\#)$, it follows that $u = 0$. More than this can be proved; in fact, we have the following result.

2.2.12 Example. Define a new norm in E by the formula

$$\|u\|' = \sup\{|\langle u^*, u \rangle|; u^* \in E^\#, \|u^*\| \leqslant 1\}.$$

Then

$$\|u\|' \leqslant \|u\| \leqslant C\|u\|',$$

where $C = \liminf_{\lambda \to \infty} \|\lambda R(\lambda; A)\|$ (Hille-Phillips [1957; 1, p. 422]).

The operator $A^\# \in \mathcal{C}_+$ associated with an $A \in \mathcal{C}_+$ will be called the *Phillips adjoint of A*. Likewise, $S^\#(\cdot)$ is the *Phillips adjoint of* $S(\cdot)$.

2.3. SEMIGROUP THEORY

Let $t \to S(t)$ be a function defined in $t \geqslant 0$ whose values are $n \times n$ matrices. Assume that

$$S(0) = I, \quad S(s+t) = S(s)S(t) \quad (s, t \geqslant 0) \qquad (2.3.1)$$

and that S is continuous. Then (see Lemma 2.3.5 and Example 2.3.10)

$$S(t) = e^{tA} \qquad (2.3.2)$$

for some $n \times n$ matrix A or, equivalently, S is the unique solution of the initial value problem

$$S'(t) = AS(t), \quad S(0) = I. \qquad (2.3.3)$$

This result is due to Cauchy [1821: 1] for the case $n = 1$ and to Pólya [1928: 1] in the general case. Roughly speaking, it can be said that the core of semigroup theory consists of generalizations of the Cauchy-Pólya theorem to functions $S(\cdot)$ taking values in an infinite dimensional space E. What makes the theory so rich is the fact that "continuity," which has only one reasonable meaning in the finite-dimensional case, can be understood in many ways when dim $E = \infty$; for instance, we may assume S continuous in the norm of (E), strongly continuous, or weakly continuous. Another fundamental difference is that, while the Cauchy-Pólya theorem guarantees extension of S to $t < 0$ when dim $E < \infty$, this extension may not exist at all in the general case. However, it can be said that the close connection between the initial-value problem (2.3.3), its solution (2.3.2), and the functional equation (2.3.1) is maintained in infinite-dimensional spaces if the correct definitions are adopted.

We call a *semigroup* any function $S(\cdot)$ with values in (E) defined in $t \geqslant 0$ and satisfying both equations (2.3.1) there. Note that we have proved in Section 2.1 that the propagator of an equation

$$u'(t) = Au(t) \tag{2.3.4}$$

for which the Cauchy problem is well posed is a strongly continuous semigroup. The converse is as well true. In fact, we have

2.3.1 Theorem. *Let $S(\cdot)$ be a semigroup strongly continuous in $t \geqslant 0$. Then there exists a (unique) closed, densely defined operator $A \in \mathcal{C}_{+}$ such that S is the evolution operator of (2.3.4).*

Proof. We define A, the *infinitesimal generator* of S, by means of the expression

$$Au = \lim_{h \to 0+} \frac{1}{h}(S(h) - I)u, \tag{2.3.5}$$

the domain of A consisting precisely of those u for which the limit (2.3.5) exists. We show first that $D(A)$ is dense in E. To this end, take $\alpha > 0$, $u \in E$ and define

$$u^{\alpha} = \frac{1}{\alpha} \int_0^{\alpha} S(s)u\, ds.$$

Some manipulations with the second equation (2.3.1) show that

$$\frac{1}{h}(S(h) - I)u^{\alpha} = \frac{1}{\alpha}\left\{ \frac{1}{h}\int_{\alpha}^{\alpha+h} S(s)u\, ds - \frac{1}{h}\int_0^{h} S(s)u\, ds \right\}. \tag{2.3.6}$$

In view of the continuity of S, the limit as $h \to 0$ of the right-hand side of (2.3.6) exists and equals $\alpha^{-1}(S(\alpha)u - u)$. Accordingly, $u^{\alpha} \in D(A)$ and

$$Au^{\alpha} = \frac{1}{\alpha}(S(\alpha)u - u). \tag{2.3.7}$$

But a similar argument shows that $u^{\alpha} \to u$ as $\alpha \to 0$, which establishes the denseness of $D(A)$.

We prove next that A is closed. To this end, observe that, if $u \in E$ and $t, h > 0$,

$$\frac{1}{h}(S(h) - I)S(t)u = S(t)\frac{1}{h}(S(h) - I)u,$$

hence if we take $u \in D(A)$ and let $h \to 0+$, $S(t)u \in D(A)$ and

$$AS(t)u = S(t)Au. \tag{2.3.8}$$

This clearly implies that

$$(Au)^\alpha = Au^\alpha = \frac{1}{\alpha}(S(\alpha)u - u) \quad (u \in D(A)). \qquad (2.3.9)$$

Let $\{u_n\}$ be a sequence in $D(A)$ with $u_n \to u$, $Au_n \to v$ for some $u, v \in E$. Then

$$\frac{1}{h}(S(h) - I)u = \lim_{n \to \infty} \frac{1}{h}(S(h) - I)u_n$$

$$= \lim_{n \to \infty} (Au_n)^h = v^h \quad (h > 0).$$

Taking limits as $h \to 0+$, we see that $u \in D(A)$ and that $Au = v$, which shows the closedness of A.

To prove that $A \in \mathcal{C}_+$, we begin by verifying the first half of the definition of well-posed Cauchy problem (Section 1.2). Take $u \in D(A)$ and integrate (2.3.8) in $0 \leqslant s \leqslant t$: we obtain

$$\int_0^t S(s) Au \, ds = \int_0^t AS(s)u \, ds = A \int_0^t S(s)u \, ds$$

$$= S(t)u - u,$$

where we have made use of (2.3.7) in the last equality. This clearly shows that $S(\cdot)u$ is differentiable in $t \geqslant 0$ and

$$S'(t)u = S(s)Au = AS(s)u,$$

thus settling the existence question. To show continuous dependence on initial data, we proceed in a way not unlike the closing lines of Theorem 2.1.1. Observe first that, since $S(\cdot)$ is strongly continuous, $\|S(\cdot)u\|$ must be bounded on bounded subsets of $t \geqslant 0$. By the uniform boundedness theorem (Section 1), there exist constants $C(t)$ such that

$$\|S(s)\| \leqslant C(t) \quad (0 \leqslant s \leqslant t) \qquad (2.3.10)$$

for all $t \geqslant 0$. Making use of this it is not difficult to show that if $u(\cdot)$ is an arbitrary solution of (2.3.4), then $v(s) = S(t - s)u(s)$ is continuously differentiable in $0 \leqslant s \leqslant t$ and $v'(s) = S(t - s)Au(s) - AS(t - s)u(s) = 0$. Hence

$$u(t) = S(t)u(0), \qquad (2.3.11)$$

and the fact that the continuous dependence assumption is satisfied is an immediate consequence of (2.3.10) and (2.3.11). This ends the proof of Theorem 2.3.1.

Considering semigroups (rather than differential equations) as our primary objects of interest, which seems to be traditional in the literature, we can combine Theorems 2.1.1 and 2.3.1 in the single

2.3.2   Theorem.   *The operator A is the infinitesimal generator of a strongly continuous semigroup S satisfying*

$$\|S(t)\| \leqslant Ce^{\omega t} \quad (t \geqslant 0) \qquad (2.3.12)$$

if and only if A is closed and densely defined, $R(\lambda)$ exists for $\text{Re}\,\lambda > \omega$, and

$$\|R(\lambda)^n\| \leqslant C(\text{Re}\,\lambda - \omega)^{-n} \quad (\text{Re}\,\lambda > \omega, n \geqslant 1). \quad (2.3.13)$$

The treatment of the case $-\infty < t < \infty$ is entirely similar. A function with values in (E), defined in $-\infty < t < \infty$ and satisfying (2.3.1) there is called a *group*. The analogues of Theorems 2.3.1 and 2.3.2 in this case are

2.3.3 Theorem. *Let $S(\cdot)$ be a strongly continuous group. Then there exists a (unique) closed, densely defined operator $A \in \mathcal{C}$ such that S is the evolution operator of (2.3.4).*

2.3.4 Theorem. *The operator A is the infinitesimal generator of a strongly continuous group S satisfying*

$$\|S(t)\| \leqslant Ce^{\omega|t|} \quad (-\infty < t < \infty) \quad (2.3.14)$$

if and only if A is closed and densely defined, $R(\lambda)$ exists for $|\text{Re}\,\lambda| > \omega$, and

$$\|R(\lambda)^n\| \leqslant C(|\text{Re}\,\lambda| - \omega)^{-n} \quad (\text{Re}\,\lambda > \omega, n \geqslant 1). \quad (2.3.15)$$

These results, together with the representation for S outlined in Remark 2.1.2, constitute an extension of the Cauchy-Pólya theorem to the infinite-dimensional case. If we replace strong continuity by uniform continuity, the corresponding extension is much simpler (Lemma 2.3.5) although of scarce interest for the study of the Cauchy problem (see Example 1.2.2).

It was already observed (Example 1.2.2) that every bounded operator A belongs to $\mathcal{C}$ and the solution of (2.3.3) is continuous (in fact analytic) in the norm of (E). Using Theorem 2.3.1, or directly, it is easy to show that S is the semigroup generated by A; in conclusion, a bounded operator always generates an uniformly continuous semigroup. We give a proof of the converse.

2.3.5 Lemma. *Let the semigroup $S(\cdot)$ be continuous at $t = 0$ in the norm of (E). Then its infinitesimal generator A is bounded.*

Proof. Let $C, \omega > 0$ be such that (2.3.12) holds. Then, if $\lambda > \omega$,

$$\left(R(\lambda) - \frac{1}{\lambda}I\right)u = \int_0^\infty e^{-\lambda t}(S(t) - I)u\,dt \quad (u \in E)$$

so that

$$\left\|R(\lambda) - \frac{1}{\lambda}I\right\| \leqslant \int_0^\infty e^{-\lambda t}\eta(t)\,dt, \quad (2.3.16)$$

where $\eta(t) = \|S(t) - I\|$ is continuous for $t = 0$, $\eta(0) = 0$, and is bounded by $Ce^{\omega t} + 1$ in $t \geqslant 0$. Let $\varepsilon > 0$ and $\delta > 0$ such that $\eta(t) \leqslant \varepsilon$ if $0 \leqslant t \leqslant \delta$. Splitting the interval of integration at $t = \delta$ we obtain

$$\int_0^\infty e^{-\lambda t}\eta(t)\,dt \leqslant \frac{\varepsilon}{\lambda} + o\left(\frac{1}{\lambda}\right)$$

as $\lambda \to \infty$. This shows, in combination with (2.3.16), that $\|R(\lambda) - \lambda^{-1}I\| = o(\lambda^{-1})$ as $\lambda \to \infty$. Accordingly, if λ is large enough, $\|\lambda R(\lambda) - I\| < I$, which implies that $\lambda R(\lambda)$—hence $R(\lambda)$—must have a bounded inverse. But $R(\lambda)^{-1} = \lambda I - A$, thus A must be itself bounded. This ends the proof of Lemma 2.3.5.

We note in passing that the result in Lemma 2.3.5 can be obtained with weaker hypotheses; in fact, we only need

$$\liminf_{\lambda \to \infty} \lambda \int_0^\delta e^{-\lambda t} \eta(t)\, dt < 1 \tag{2.3.17}$$

for some (then for any) $\delta > 0$.

2.3.6 Example. Prove that (2.3.17) is a consequence of

$$\limsup_{t \to 0+} \|S(t) - I\| < 1 \tag{2.3.18}$$

so that (2.3.18) implies boundedness of the infinitesimal generator.

2.3.7 Example. The condition

$$\liminf_{t \to 0+} \|S(t) - I\| = 0$$

does not imply boundedness of the infinitesimal generator, even if $S(\cdot)$ is a group. (*Hint*: Let $\{e_j; j \geq 0\}$ be a complete orthonormal sequence in a Hilbert space H, and let $\{\theta(j)\}$ be an increasing sequence of positive integers with $\theta(j+1) - \theta(j) \to \infty$ as $j \to \infty$. Define $S(t)\Sigma c_j e_j = \Sigma c_j \exp(i2^{\theta(j)}t)e_j$. Then $S(\cdot)$ is a strongly continuous group with $\|S(t)\| = 1$ for all t. If $t_n = 2^{-\theta(n)+1}\pi$ for n large enough,

$$\|S(t_n) - I\| \leq \sup_{j \geq 0} |\exp(i2^{\theta(j)-\theta(n)+1}\pi) - 1|$$

$$= \sup_{0 \leq j \leq n-1} |\exp(i2^{\theta(j)-\theta(n)+1}\pi) - 1|$$

$$\leq \sup_{0 \leq j \leq n-1} 2^{\theta(j)-\theta(n)+1}\pi$$

$$\leq 2^{-(\theta(n)-\theta(n-1))+1}\pi.$$

*2.3.8 Example. (Neuberger [1970: 1]). If S is a group and

$$\limsup_{t \to 0} \|S(t) - I\| < 2,$$

then A must be bounded. For generalizations see Kato [1970: 2] and Pazy [1971: 2].

As an application of the results in this section, we prove

2.3.9 Lemma. *Let A be densely defined (but not necessarily closed) and assume that $A \in \mathcal{C}_+$ and that $D = D(A)$ (that is, every element of $D(A)$ is the initial value of some solution of (2.3.4). Then A is closable and its closure $\overline{A}$ belongs as well to $\mathcal{C}_+$, thus $\overline{A}$ is characterized by Theorem 2.1.1.*

Proof. Let $S(\cdot)$ be the solution operator of (2.3.4) and B the infinitesimal generator of (the strongly continuous semigroup) S. Clearly

$A \subseteq B$, so that A is closable and $\bar{A} \subseteq B$. By virtue of Theorem 2.3.1, $B \in \mathcal{C}_+$, and it is not difficult to see that the inclusion $A \subseteq \bar{A} \subseteq B$ and the fact that $A, B \in \mathcal{C}_+$ imply that $\bar{A} \in \mathcal{C}_+$. But then, in view of Theorem 2.1.1 both $R(\lambda; \bar{A})$ and $R(\lambda; B)$ exist for λ large enough. If $\bar{A} \neq B$, let $\lambda \in \rho(\bar{A}) \cap \rho(B)$, $u \in D(B)$, $u \notin D(\bar{A})$, $v \in D(\bar{A})$ such that $(\lambda I - \bar{A})v = (\lambda I - B)u$. Then if $w = u - v$, $w \neq 0$ and $(\lambda I - B)w = 0$, which is a contradiction.

2.3.10 Example. Lemma 2.3.5 can be proved in a fairly direct way, without recourse to semigroup theory. In fact, let $S(\cdot)$ be a semigroup continuous at $t = 0$ in the norm of (E). Then S is (E)-continuous in $t \geqslant 0$ (this follows from (2.3.1)) and

$$\left\| \frac{1}{t} \int_0^t S(s)\, ds - I \right\| < 1$$

for t sufficiently small so that $\int_0^t S(s)u\, ds$ is nonsingular. Let now $t = nh$. Then

$$\int_0^t S(t)\, ds = \lim_{n \to \infty} h \sum_{j=1}^n S((j-1)h),$$

hence it follows that $h \Sigma S((j-1)h)$ is invertible for t, h small enough. Now,

$$\frac{1}{h}(S(h) - I) h \sum_{j=1}^n S((j-1)h) = S(t) - I,$$

therefore $h^{-1}(S(h) - I)$ has a limit $A \in (E)$ in the norm of (E) and

$$A \int_0^t S(s)\, ds = S(t) - I$$

for $t \leqslant \delta$ small enough, so that $S(\cdot)$ is differentiable in the norm of (E) and $S'(t) = AS(t)$ in $0 \leqslant t \leqslant \delta$. It follows from the remarks on uniqueness in Example 1.2.2 that $S(t) = \tilde{S}(t) = \exp(tA)$ for $t \leqslant \delta$, hence for all t since $S(t) = S(t/n)^n$, and a similar equation holds for $\tilde{S}$.

In the next four examples $S(\cdot)$ is a strongly continuous semigroup and A its infinitesimal generator (or, equivalently, A is a closed densely defined operator such that the Cauchy problem for (2.3.4) is properly posed and $S(\cdot)$ is the propagator of (2.3.4).)

*2.3.11 Example (see Hille-Phillips [1957: 1, p. 312]). Let $h > 0$, $u \in E$. Then

$$\exp\big(th^{-1}(S(h) - I)\big)u \to S(t)u \text{ as } h \to 0 \qquad (2.3.19)$$

uniformly on compacts of $t \geqslant 0$.

2.3.12 Example. Using the previous example, it is possible to prove the *Weierstrass approximation theorem* for continuous functions: if f is continuous in $a \leqslant x \leqslant b$, there exists a sequence of polynomials p_n such that $p_n \to f$ uniformly in $a \leqslant t \leqslant b$ (place the interval conveniently, extend f to a function in $C_0(-\infty, \infty)$ and apply (2.3.19) to the semigroup in Example 2.2.4).

2.3.13 Example (see Dunford-Schwartz [1958: 1, p. 624]). For $u \in E$,

$$\lim_{\lambda \to \infty} \exp(t\lambda AR(\lambda))u = S(t)u \qquad (2.3.20)$$

uniformly on compacts of $t \geqslant 0$.

2.3.14 Example. Let $u, v \in E$ such that $h^{-1}(S(h) - I)u \to v$ E^*-weakly. Then $h^{-1}(S(h) - I)u \to v$ in the norm of E (so that $u \in D(A)$ and $Au = v$).

*2.3.15 Example. Let $\{S(t); \ t \geqslant 0\}$ be a (E)-valued function satisfying both equations 2.3.1 and such that $S(\cdot)u$ is strongly measurable in E for $t \geqslant 0$ for all $u \in E$. Then $t \to S(t)u$ is continuous in $t > 0$ for all $u \in E$ (see Hille-Phillips [1957: 1, p. 305]). Continuity at $t = 0$ does not follow. By the uniform boundedness principle, $\|S(t)\|$ is bounded on compacts of $t > 0$ and essentially as in the strongly continuous case it can be proved that $\|S(t)\|$ grows exponentially at infinity. On the other hand, *weak measurability* of $S(\cdot)$ (measurability of $\langle u^*, S(t)u \rangle$ in $t \geqslant 0$ for all $u \in E$, $u^* \in E^*$) does not imply any continuity properties of $S(\cdot)$ in $t > 0$ or even boundedness of $\|S(t)\|$ in any subinterval of $(0, \infty)$ (loc. cit. p. 305 and references there).

2.3.16 Example. Let $\{S(t); \ -\infty < t < \infty\}$ be a strongly measurable group in the sense defined in the previous example. Then $S(\cdot)$ is strongly continuous in $-\infty < t < \infty$.

*2.3.17 Example. Given two real numbers $\omega_1 < \omega_2$, show that there exists a strongly continuous semigroup $S(\cdot)$ with infinitesimal generator A such that (a) $\sigma(A)$ is contained in the half plane $\operatorname{Re} \lambda \leqslant \omega_1$, and (b) $S(\cdot)$ does *not* satisfy

$$\|S(t)\| \leqslant Ce^{\omega_2 t} \quad (t \geqslant 0)$$

for any constant C; compare with Example 1.2.3 and Example 2.2.8 (see Hille-Phillips [1957: 1, p. 665], where $\sigma(A)$ is in fact empty; for another example see Zabczyk [1975: 1]).

2.4. THE INHOMOGENEOUS EQUATION

Let f be a continuous function defined in $t \geqslant 0$ with values in E, and let $A \in \mathcal{C}_+$. Consider the equation

$$u'(t) = Au(t) + f(t). \qquad (2.4.1)$$

Solutions of (2.4.1) are defined exactly as solutions of the homogeneous equation ($f = 0$) were defined in Section 1.2. Since the difference of two solutions of (2.4.1) is a solution of the homogeneous equation, it is clear that there is (if any) only one solution of (2.4.1) satisfying the initial condition

$$u(0) = u_0. \qquad (2.4.2)$$

Let $u(\cdot)$ be a solution of (2.4.1), (2.4.2), and let S be the propagator of the homogeneous equation. It is easy to show (see the last lines of the proof of Theorem 2.3.1) that if $t > 0$, the function $s \to S(t-s)u(s)$ is continuously differentiable in $0 \leqslant s \leqslant t$ with derivative $-S'(t-s)u(s) + S(t-s)u'(s) = S(t-s)(u'(s) - Au(s)) = S(t-s)f(s)$. Therefore, we obtain after integration the familiar formula of Lagrange:

$$u(t) = S(t)u_0 + \int_0^t S(t-s)f(s)\,ds. \tag{2.4.3}$$

We turn now to the question of which conditions on u_0, f will guarantee that $u(\cdot)$, defined by (2.4.3), is indeed a solution of (2.4.1). Clearly we need $u_0 \in D(A)$, as can be seen taking $f = 0$. On the other hand, if A is bounded, (2.4.3) is a solution of (2.4.1) for any f continuous in $t \geqslant 0$ (the proof is immediate). This is not true in general when A is unbounded. (See, however, Example 2.4.7.)

2.4.1 Example. Prove the statement above about bounded operators. Give an example of unbounded $A \in \mathcal{C}_+$ and a continuous f such that (2.4.3) is not a solution of (2.4.2).

Necessary and sufficient conditions on f in order that (2.4.3) be a solution of (2.4.1) are not known. Two simple sufficient conditions are given in the following result.

2.4.2 Lemma. Assume that, either (a) $f(t) \in D(A)$ and f, Af are continuous in $0 \leqslant t \leqslant T$, or (b) f is continuously differentiable. Moreover, let $u_0 \in D(A)$. Then (2.4.3) is a solution of the initial value problem (2.4.1), (2.4.2).

Proof. Obviously we may assume that $u_0 = 0$. To prove that (2.4.3) is a solution of (2.4.1), it is sufficient to prove that $u(t) \in D(A)$, and that $Au(\cdot)$ is continuous and

$$u(t) = \int_0^t (Au(s) + f(s))\,ds \quad (0 \leqslant t \leqslant T). \tag{2.4.4}$$

We work at first under assumption (a). In this case, the fact that $Au(t) \in D(A)$ follows directly from Lemma 3.4 and from (2.1.7); moreover,

$$Au(s) = \int_0^s S(s-r)Af(r)\,dr \quad (0 \leqslant s \leqslant T), \tag{2.4.5}$$

from which continuity can be readily verified. (See the following Lemma 2.4.5.) To obtain (2.4.4), it suffices to integrate (2.4.5) in $0 \leqslant s \leqslant t$, reverse the order of integration, and make use of the fact that $S(s)Au = (S(s)u)'$, which holds for any $u \in D(A)$ (see again (2.1.7) and following comments). In case assumption (b) is satisfied instead, we observe, arguing as in the beginning of this section, that the function

$$v(s) = \int_0^{t-s} S(r)f(s)\,dr \quad (0 \leqslant s \leqslant t \leqslant T)$$

is continuously differentiable in $0 \leqslant s \leqslant t$ with derivative

$$v'(s) = -S(t-s)f(s) + \int_0^{t-s} S(r)f'(s)\,dr.$$

Integrating in $0 \leqslant s \leqslant t$, we obtain

$$u(t) = \int_0^t S(s)f(0)\,ds$$

$$+ \int_0^t \left(\int_0^{t-s} S(r)f'(s)\,dr \right) ds.$$

Using (2.3.7) we deduce that

$$Au(t) = S(t)f(0) - f(t) + \int_0^t S(t-s)f'(s)\,ds$$

$$= S(t)f(0) - f(t) + \int_0^t S(s)f'(t-s)\,ds. \qquad (2.4.6)$$

Integrating (2.4.6) and reversing the order of integration, we obtain (2.4.4). This ends the proof.

It is natural to expect that if additional conditions are imposed on the operator A (or, equivalently, on the propagator $S(\cdot)$ of the homogeneous equation (2.4.1)), it should be possible to weaken the assumptions on f. Some instances of this heuristic principle will be examined in Section 2.5(j).

Even in the general case, the assumptions can be relaxed if one is willing to accept weaker definitions of solutions of (2.4.1). This was discussed in Section 1.2 in relation to the homogeneous equation, and it is possible to extend these ideas to the nonhomogeneous equation (2.4.1).

We call a function $u(\cdot)$ a *mild solution* of (2.4.1), (2.4.2) in $0 \leqslant t \leqslant T$ if $u(t) \in D(A)$ for almost all t, $Au(\cdot)$ is integrable in the sense of Lebesgue-Bochner in $0 \leqslant t \leqslant T$, and

$$u(t) = \int_0^t (Au(s) + f(s))\,ds + u_0 \quad (0 \leqslant t \leqslant T). \qquad (2.4.7)$$

Naturally, in order for (2.4.7) to make sense, we have to assume that f itself is integrable. Lemma 2.4.2 admits an immediate generalization in this context.

2.4.3 Example. Let assumptions (a) and (b) in Lemma 2.4.2 be replaced by (a') $f(t) \in D(A)$ for almost all t and $f(\cdot), Af(\cdot)$ are integrable in $0 \leqslant t \leqslant T$, or (b') $f(t) = \int_0^t g(s)\,ds$, where g is integrable in $0 \leqslant t \leqslant T$. Then $u(t)$, defined by (2.4.3), is a mild solution of (2.4.1) if $u_0 \in D(A)$. (*Hint:* the proof of Lemma 2.4.2 can be imitated word by word if we substitute the Riemann by the Lebesgue-Bochner integral.)

However, the assumptions in Example 2.4.3 are still excessive for some applications. Copying then the definition made in Section 1.2 for the homogeneous equation, we call an integrable function $u(\cdot)$ a *weak solution* of (2.4.1), (2.4.2) in $0 \leqslant t < T$ with f itself integrable if for every u^* in the domain of the adjoint operator A^*, and for every test function φ with support in $(-\infty, T]$ we have

$$\int_0^T (\langle A^*u^*, u(t)\rangle + \langle u^*, f(t)\rangle)\varphi(t)\, dt$$

$$= -\int_0^T \langle u^*, u(t)\rangle \varphi'(t)\, dt - \langle u^*, u_0\rangle \varphi(0). \qquad (2.4.8)$$

It is not difficult to see that weak solutions are represented as well by the Lagrange formula. To prove this we need the following two results.

2.4.4 Example. Let $C_T^1(A^*)$ be the space of all $D(A^*)$-valued functions $u^*(\cdot)$ defined in $-\infty < t < \infty$, with compact support contained in $(-\infty, T]$, continuously differentiable (in the norm of E^*) and such that $A^*u^*(\cdot)$ is continuous. Let $\mathscr{D}_T \otimes D(A^*)$ be the space of all functions of the form $\Sigma \varphi_k(t)u_k^*$, where the sum is finite and $\varphi_1, \varphi_2, \dots$ are test functions with support in $(-\infty, T]$, $u_1^*, u_2^*, \dots$ are elements of $D(A^*)$. Then $\mathscr{D}_T \otimes D(A^*)$ is dense in $C_T^1(A^*)$ in the following norm:

$$\||u^*(\cdot)\|| = \sup\|u^*(t)\| + \sup\|u^{*\prime}(t)\| + \sup\|A^*u^*(t)\|,$$

the suprema taken in $-\infty < t \leqslant T$.

2.4.5 Example. Let f be integrable in $0 \leqslant t \leqslant T$. Then

$$u(t) = \int_0^t S(t-s)f(s)\, ds \qquad (2.4.9)$$

is continuous in $t \geqslant 0$.

We only have to note that, if $t < t'$,

$$\|u(t') - u(t)\| \leqslant \int_0^t \|(S(t'-s) - S(t-s))f(s)\|\, ds$$

$$+ \int_t^{t'} \|S(t'-s)f(s)\|\, ds.$$

The two integrals can be handled by means of the dominated convergence theorem. In the first we observe that the integrand tends to zero as $t' - t \to 0$ while being bounded by a constant times $\|f(\cdot)\|$; the same estimate applies to the second integral while the shrinking of the domain of integration takes care of the convergence to zero.

2.4.6 Theorem. *Let f be integrable in $0 \leqslant t \leqslant T$, u_0 an arbitrary element of E. Then (2.4.3) is a weak solution of (2.4.1), (2.4.2), that is, (2.4.8) holds for every $u^* \in D(A^*)$ and every $\varphi \in \mathscr{D}$ with support in $-\infty < t \leqslant T$. Conversely, let $u(\cdot)$ be a weak solution of (2.4.1), (2.4.2) with f integrable and $u_0 \in E$. Then $u(\cdot)$ obeys the Lagrange formula (2.4.3) almost everywhere (and can be modified in a null set in such a way that it becomes continuous and satisfies (2.4.3) everywhere).*

Proof. Let f be integrable in $0 \leqslant t \leqslant T$. Then there exists a sequence $\{f_n\}$ of smooth (say, infinitely differentiable) E-valued functions such that

$$\int_0^T \| f_n(t) - f(t) \| \, dt \to 0 \text{ as } n \to \infty.$$

(This can be proved with the aid of mollifiers; see Section 8.) Let $\{u_n\}$ be a sequence in $D(A)$ such that $u_n \to u_0$. It follows easily that if $u_n(\cdot)$ is the function obtained from u_0 and $f_n(\cdot)$ in formula (2.4.3), then $\| u_n(t) - u(t) \| \to 0$ uniformly in $0 \leqslant t \leqslant T$. Since each $u_n(\cdot)$ is a genuine solution of (2.4.1), (2.4.2) (as was proved in Lemma 2.4.2), equality (2.4.8) holds for $u_n(\cdot)$. Taking limits, we obtain the same equality for $u(\cdot)$, which proves that $u(\cdot)$ is in fact a weak solution as claimed.

Conversely, let $u(\cdot)$ be an arbitrary weak solution. Making use of Example 2.4.4, we observe that (2.4.8) can be generalized to

$$\int_0^T (\langle A^* u^*(t), u(t) \rangle + \langle u^*(t), f(t) \rangle) \, dt$$

$$= -\int_0^T \langle u^{*\prime}(t), u(t) \rangle \, dt - \langle u^*(0), u_0 \rangle \qquad (2.4.10)$$

for all $u^*(\cdot)$ in $C_T^1(A^*)$ (in fact, (2.4.8) obviously implies (2.4.10) for $u^*(\cdot)$ in $\mathcal{D}_T \otimes D(A^*)$ and the density result takes care of the rest). Let now φ be a fixed positive test function with integral 1 and define $\varphi_n(s) = n\varphi(ns)$; let $A^\#$ be the Phillips adjoint of A, and $S^*(\cdot)$ the Phillips adjoint of $S(\cdot)$ (see Section 2.3). Choose t_0 in the interval $(0, T)$ and u^* in $D(A^\#)$ and define

$$v_n^*(t) = -\int_t^T S^\#(s - t) \varphi_n(s - t_0) u^* \, ds \quad (t \leqslant T).$$

Slight variations on the argument used in Lemma 2.4.2 (a) easily show that $v_n^*(t)$ is a solution of $v_n^{*\prime}(t) + A^\# v_n^*(t) = \varphi_n(t - t_0) u^*$ in $t < T$ with "final condition" $v_n^*(T) = 0$. Let finally $\chi(\cdot)$ be a test function that equals 1 in the interval $0 \leqslant t \leqslant T$. Define $u_n^*(t) = \chi(t) v_n^*(t)$. Clearly $u_n^* \in C_T^1(A^*)$. Writing (2.4.10) for u_n^* and reversing the order of integration in the double integral, we obtain

$$\int_0^T \langle u^*, u(t) \rangle \varphi_n(t - t_0) \, dt = \int_0^T \langle u^*, \int_0^t S(t - s) f(s) \, ds \rangle \varphi_n(t - t_0) \, dt$$

$$+ \int_0^T \langle u^*, S(t) u_0 \rangle \varphi_n(t - t_0) \, dt \qquad (2.4.11)$$

if n is large enough.

We invoke now a well-known result in approximation theory (Hille-Phillips [1957: 1, p. 91]) according to which if g is an integrable scalar function, then $\int g(t) \varphi_n(t - t_0) \, dt \to g(t_0)$ whenever t_0 is a Lebesgue point of g. But the set e of Lebesgue points of $u(\cdot)$ has full measure in $0 \leqslant t \leqslant T$ (Hille-Phillips [1957: 1, p. 88]), whereas every Lebesgue point of $u(\cdot)$ is a Lebesgue point of $\langle u^*, u(\cdot) \rangle$. Making use of this and of the fact (proved in

Example 2.4.5) that $t \to \int_0^t S(t-s)f(s)\,ds$ is continuous, we let $n \to \infty$ in (2.4.9), and conclude that

$$\langle u^*, u(t) \rangle = \left\langle u^*, \int_0^t S(t-s)f(s)\,ds + S(t)u_0 \right\rangle$$

for $t \in e$. Since $u^* \in D(A^\#)$ is arbitrary and $D(A^\#)$ is weakly dense in E^* (see the remarks after Theorem 2.2.11), we see that the Lagrange formula holds in e: then in view of Example 2.4.5, $u(\cdot)$ can be modified outside of e in such a way that it becomes continuous in $0 \leqslant t \leqslant T$ and (2.4.3) holds everywhere.

2.4.7 Example.[1] Let c_0 be the space of all sequences $\xi = \{\xi_n\}$ $(n \geqslant 0)$ satisfying $\lim \xi_n = 0$, endowed with the norm $\|\xi\| = \sup |\xi_n|$, A the operator defined by

$$A\xi = \{-n\xi_n\}, \tag{2.4.12}$$

the domain of A consisting of all $\xi \in c_0$ such that the right-hand side of (2.4.12) belongs to c_0. Then $A \in \mathcal{C}_+(1,0)$ (the propagator of $u' = Au$ is $S(t)\xi = \{\exp(-nt)\xi_n\}$) and (2.4.3) is a solution of (2.4.1), (2.4.2) in $0 \leqslant t \leqslant T$ for any $u_0 \in D(A)$ and any c_0-valued f continuous in $0 \leqslant t \leqslant T$.

2.5. MISCELLANEOUS COMMENTS

Strongly continuous groups of unitary operators in Hilbert space were characterized by Stone [1930: 1], [1932: 1]. His celebrated result belongs to the subject matter of Chapter 3 and will be commented on there. Uniformly bounded groups $U(\cdot)$ of bounded linear operators in Banach spaces were studied by Gelfand [1939: 1], who defined the infinitesimal generator A and proved that $U(t)u = \exp(tA)u$, where $\exp(tA)u$ is defined by the exponential series, convergent in a certain dense subset of E. Related results are due to Fukamiya [1940: 1]. Some special semigroups of linear transformations in function spaces were considered by Hille [1936: 1]. Two years later Hille [1938: 1] and Sz.-Nagy [1938: 1] discovered independently a representation theorem for semigroups of bounded self-adjoint operators in Hilbert space: this result also belongs to the class spawned by Stone's theorem and will be referred to in Chapter 3. Other early results in semigroup theory can be found in Hille [1938: 2], [1939: 1], [1942: 1], [1942: 2], and [1947: 1]; there is also related material in Romanoff [1947: 1]. The ancestors of Theorem 2.3.2 due to Cauchy and Pólya were already pointed out in Section 2.3. This theorem (now called the *Hille-Yosida theorem*) was discovered independently by Hille [1948: 1] and Yosida [1948: 1] in the particular case where

---

[1]Personal communication of A. Pazy (who credits the result to T. Kato).

$C = 1$. Hille's proof culminates in the construction of S by means of formula (2.1.27), the *Hille approximation* of $S(\cdot)$; in contrast, Yosida uses formula (2.3.20), the *Yosida approximation* of $S(\cdot)$. The proof for the general case ($C \geqslant 1$) was discovered, also independently, by Feller [1953: 2], Miyadera [1952: 1], and Phillips [1953: 1]; all three papers use the Yosida approximation. Groups $S(\cdot)$, which are continuous in the (E)-topology, were shown to have the form $\exp(tA)$ for some $A \in (E)$ by Nathan [1935: 1] (his method is different than those in Lemma 2.3.5 and Example 2.3.10). A generalization of this result (where $S(\cdot)$ takes values in a Banach algebra) was proved independently by Nagumo [1936: 1] and Yosida [1936: 1].

The proof of Theorem 2.3.2 presented here is due to Hille [1952: 2]. Its basis is the following formula to recover a function f from its Laplace transform $\hat{f}$,

$$f(t) = \lim_{n \to \infty} \frac{(-1)^n}{n!} \left(\frac{n}{t} \right)^{n+1} \hat{f}^{(n)} \left(\frac{n}{t} \right), \tag{2.5.1}$$

due to Post [1930: 1] in the scalar case; see also Widder [1946: 1, Chapter 7]. The obvious relation between the characterization of strongly continuous semigroups by inequalities (2.1.11) and Widder's identification in [1931: 1] of Laplace transforms of bounded functions was already pointed out in Section 2.1 (see also Widder [1971: 1]).

The first result on "measurability implies continuity" is due to von Neumann [1932: 1], who considered groups of unitary operators. The particular case of Example 2.3.15, where S is strongly measurable in the norm of (E), is due to Hille [1948: 1], while the full result was proved by Miyadera [1951: 1] and Phillips [1951: 1], which showed that a crucial hypothesis in earlier proofs of Dunford [1938: 1] and Hille [1948: 1] was actually unnecessary.

Example 2.3.11 is due to Hille [1942: 1] (see also [1948: 1]). Formula (2.3.19) is one of the *exponential formulas* that justify writing $S(t) = \exp(tA)$; other such formulas are the Hille and Yosida approximations of S. An ample collection of these can be found in Hille-Phillips [1957: 1, Sec. 10.4] together with some general principles on their obtention. Example 2.3.12 is due to Dunford and Segal [1946: 1] where it can also be found a short proof of (2.3.19) for semigroups with $\|S(t)\| \leqslant 1$. Example 2.3.8 is a particular case of a result on smooth semigroups (see the bibliography in Section 4.11). Lemma 2.3.9 is due to Kreĭn and Sobolevskiĭ [1958: 1].

It is worth pointing out that, although the relation between semigroups and abstract differential equations was in the forefront from the very beginning (see Yosida [1949: 1], [1951: 1], [1951: 2], [1952: 1], and Hille [1952: 1]) and was one of the main incentives for development of the theory, most of the results in this chapter have appeared for the first time in the "semigroup-generator" formulation rather than in "equation-propagator" language.

In a more general form, Lemma 2.2.3 was proved by Phillips [1955: 1]. Lemma 2.4.2 on sufficient conditions for the solvability of the nonhomogeneous equation (2.4.1) is also due to Phillips [1953: 1] as well as the entire #-adjoint theory ([1955: 1], [1955: 2]).

In the rest of this section we comment on several extensions and generalizations of the results in this chapter.

(a) *Discontinuous Semigroups.* The Cauchy problem for an abstract differential equation $u' = Au$ has been studied under assumptions weaker than the ones in Section 1.2 by Kreĭn ([1967: 1] and references there). On the one hand, solutions are only assumed to be continuously differentiable and to belong to the domain of A for $t > 0$ (although continuity in $t \geqslant 0$ is retained). On the other hand, the function $C(\cdot)$ in (1.2.3) is merely assumed to be finite for all $t > 0$; this means that a sequence of solutions whose initial values tend to zero will only tend to zero *pointwise* (compare with (b') in Section 1.2). A Cauchy problem for which these weaker assumptions hold is called *correct* or *well posed* by Kreĭn; in his terminology, the setting used in the present work is that of *uniformly correct* or *uniformly well posed* problems.

Assume the Cauchy problem for $u' = Au$ is correct in the sense of Kreĭn. Then we can define the propagator $S(t)$ in the same way used in Section 1.2 and show that it satisfies the semigroup equations (2.1.2). If u is an arbitrary element of E, we can approximate it by a sequence $\{u_n\}$ in the subspace D of possible initial data. Since $S(t)u_n \to S(t)u$ pointwise in $t \geqslant 0$, it follows that $S(\cdot)$ is strongly measurable there, thus strongly continuous in $t > 0$ by Example 2.3.15. However, it is not in general true that $S(\cdot)$ is strongly continuous at $t = 0$ or even bounded in norm near the origin; all we know is that solutions starting at D assume their initial data—that is, $S(t)u \to u$ as $t \to 0+$ for $u \in D$. (We note, incidentally that, by virtue of Example 2.3.16, a Cauchy problem that is well posed in the sense of Kreĭn in $(-\infty, \infty)$ must be uniformly well posed there, at least if solutions are defined as in Section 1.5.)

This, and many other applications, motivate the study of semigroups that are strongly continuous for $t > 0$ but not at $t = 0$, or even semigroups of a more general nature. An extremely ample theory has been developed by Feller [1952: 2], [1953: 1], and [1953: 2] with probabilistic applications in view. In its more general version, even less than weak measurability is assumed. A generation theorem for semigroups strongly continuous in $t > 0$ is given by Feller [1953: 2]; the notion of infinitesimal generator is understood in a sense somewhat different from that in Section 2.3.

Different classes of semigroups strongly continuous for $t > 0$ have been staked out by Hille, Phillips, Miyadera, and others on the basis of their properties of convergence to the identity as $t \to 0$. Ignoring certain additional assumptions, some of these classes can be described as follows.

Assume that S satisfies

$$\int_0^1 \|S(t)u\| \, dt < \infty \tag{2.5.2}$$

for all $u \in E$. In the terminology of Hille, S is of class C_0 if

$$\lim_{t \to 0+} S(t)u \to u \quad (u \in E), \tag{2.5.3}$$

of class $(1, C_1)$ if (2.5.3) holds in Cèsaro mean, that is, if

$$\lim_{t \to 0+} \frac{1}{t} \int_0^t S(s)u \, ds = u \quad (u \in E) \tag{2.5.4}$$

and of class $(1, A)$ if (2.5.3) holds in Abel mean; in other words,

$$\lim_{\lambda \to \infty} \lambda \int_0^\infty e^{-\lambda t} S(t)u \, dt = u \quad (u \in E). \tag{2.5.5}$$

(We note that strong continuity of S for $t > 0$ can be seen to imply that $\|S(t)\| = 0(e^{\omega t})$ as $t \to \infty$, hence the integral in (2.5.5) makes sense for sufficiently large λ.) Obviously, the class C_0 coincides with $\mathcal{C}$. Other classes are $(0, C_1)$ and $(0, A)$, which are obtained respectively from the classes $(1, C_1)$ and $(1, A)$ by reinforcing condition (2.5.2) to

$$\int_0^1 \|S(t)\| \, dt < \infty \tag{2.5.6}$$

(where the integral makes sense due to the fact that

$$\|S(t)\| = \sup\{\|S(t)u\|; \|u\| \leqslant 1\}$$

is lower semicontinuous, hence measurable in $t > 0$). For general references on these and other types of semigroups see Hille-Phillips [1957: 1]; other papers on the subject are Abdulkerimov [1966: 1], Mamedov-Sobolevskiĭ [1966: 1], Miyadera [1954: 1], [1954: 2], [1954: 3], [1954: 4], [1955: 1], [1972: 1], Miyadera-Oharu-Okazawa [1972/73: 1], Oharu [1971/72: 1], Okazawa, [1974: 1], Orlov [1973: 1], and Phillips [1954: 1]. Other classes of semigroups with discontinuities at $t = 0$ are the "semigroups of growth n" of Da Prato [1966: 1], [1966: 2], characterized by polinomial growth at zero ($t^n \|S(t)\| \leqslant C$ for $0 \leqslant t \leqslant 1$). For other results on these semigroups and related classes see Okazawa [1973: 1], [1974: 1], Ponomarev [1972: 2], Sobolevskiĭ [1971: 2], Wild [1975: 1], [1976: 1], [1976: 2], [1977: 1], Zabreĭko-Zafievskiĭ [1969: 1], and Zafievskiĭ [1970: 1].

(b) *Systems, Huygens' Principle, Ergodic Theory, Markov Processes.* The state at time t of many physical systems can be described (or modelled) by a point $x(t)$ in a *state space* X. Some of these systems obey (in the terminology of Hadamard [1923: 1], [1924: 1], [1924: 2]) *the minor premise of Huygens' principle*. This can be formulated as follows; *given two times $s < t$, the state of the system at time t is totally determined by the state of the* *system at time s.* In other words, there exists a *flow* or *evolution function* $\Phi(t, s): X \to X$ such that

$$x(t) = \Phi(t, s; x(s)). \tag{2.5.7}$$

We note that other systems, such as those described in Chapter 8 of this work, do not obey the major premise of Huygens' principle since the state $x(t)$ may depend on all (or several) of the past states, that is, on the *history* $\{x(s); \ s < t\}$, or subsets thereof. This can be remedied by taking the histories as new states. We restrict ourselves, however, to systems satisfying (2.5.7).

It follows immediately from the definition of $\Phi(t, s)$ that

$$\Phi(t, t) = i = \text{identity map}, \tag{2.5.8}$$

$$\Phi(t, s) \circ \Phi(s, r) = \Phi(t, r) \quad (r \leqslant s \leqslant t). \tag{2.5.9}$$

The system is *reversible* if Φ is defined for all s, t (not just for $s < t$) and (2.5.9) holds for r, s, t in arbitrary position; here not only the future is determined by the present but the past itself can be known from the present. A system (reversible or not) is *time-invariant* if Φ only depends on $t - s$; we can then write $\Phi(t, s) = \Psi(t - s)$ and equations (2.5.8) and (2.5.9) become

$$\Psi(0) = i, \tag{2.5.10}$$

$$\Psi(r) \circ \Psi(s) = \Psi(r + s). \tag{2.5.11}$$

It should be pointed out that the above framework is not limited to deterministic systems; for instance, $x(t)$ may be a probability distribution (as in diffusion processes or in quantum mechanics) or some space average in statistical mechanics. In all these cases a "particle-by-particle" description of the system is either impracticable or senseless.

In many cases of interest, the state space X can be given a structure of normed linear space in which the functions $\Phi(t, s)$ are continuous operators depending on s, t smoothly in one way or another; if the system is time-invariant, the operators $\Psi(t)$ linear, and the time dependence strongly continuous, the systems will be amenable to the theory in Section 2.3; in particular, we deduce that the evolution of the system will be governed by an abstract differential equation

$$x'(t) = Ax(t). \tag{2.5.12}$$

An example which fits into this kind of framework is that of a *Markov process*. Here the states $x(t)$ of the system are countably additive probability measures $x(t; d\xi)$ defined in a set Ω (for instance, $x(t; e)$ may be the probability of finding in the subset $e \subset \Omega$ at time t a particle performing Brownian motion), which we assume defined in a common σ-algebra $\mathfrak{S}$ of subsets of Ω. The law (2.5.7) governing the evolution of the system takes the form

$$x(t; e) = \Phi(t, s; x(s))$$

$$= \int_\Omega P(t, s, \xi; e) x(s; d\xi) \quad (e \in \mathfrak{S}), \tag{2.5.13}$$

where $P(t, s, \xi; e)$ is called the *transition probability* of the process. (In the Brownian motion example, $P(t, s, \xi; e)$ is the probability that the particle falls in e at time t starting at ξ at time s.) To give sense to (2.5.13) in general, we make the following assumptions on the transition probabilities:

(i) *If $0 \leqslant s \leqslant t$ and $\xi \in \Omega$, then $P(t, s, \xi; d\eta)$ is a probability measure in Ω defined in $\mathfrak{S}$; moreover,*

$$P(t, t, \xi; d\eta) = \delta_\xi(d\eta) \quad (t \geqslant 0), \tag{2.5.14}$$

where δ_ξ is the Dirac measure centered at ξ.

(ii) *For each $e \in \mathfrak{S}$, the function $P(t, s, \cdot; e)$ is $\mathfrak{S}$-measurable in Ω.*

(iii) *If $0 \leqslant r \leqslant s \leqslant t, \xi \in \Omega$ and $e \in \mathfrak{S}$, then*

$$P(t, r, \xi; e) = \int_\Omega P(t, s, \eta; e) P(s, r, \xi; d\eta). \tag{2.5.15}$$

Condition (i) guarantees that the function Φ maps probability measures into probability measures and that (2.5.8) holds. Condition (ii) is an obvious technical requirement while (2.5.15), called the *Chapman-Kolmogorov* equation of the process, implies (2.5.9). The Markov process in question is *temporally homogeneous* if P depends on $t - s$, in which case we can use transition probabilities of the form $P(t, \xi; e)$ and the Chapman-Kolmogorov equation becomes

$$P(s + t, \xi; e) = \int_\Omega P(s, \xi; d\eta) P(t, \eta; e). \tag{2.5.16}$$

We go back for a moment to the general description of a system sketched above. It is often the case that the states of a system are unobservable in practice, the only information available being that given by a measuring instrument. What one can read off the instrument will in general be of the form

$$f(x(t)) \tag{2.5.17}$$

with f a real-valued function (for instance, the probability densities $x(t, d\xi)$ of a Markov process may only be known through space averages

$$f_u(x(t)) = \int_\Omega u(\xi) x(t, d\xi)). \tag{2.5.18}$$

It follows that, instead of taking the states themselves as elements of a linear space, we may obtain a mathematical model better adapted to reality by taking E to be a space of functions in X and defining an evolution operator in E by

$$(S(t, s)f)(x) = f(\Phi(t, s; x)),$$

which, in the time-invariant case, will be given by

$$(S(t)f)(x) = f(\Psi(t; x)). \tag{2.5.19}$$

An obvious bonus is that S will always be a linear operator; naturally S satisfies the semigroup equations (2.1.2) in the time-invariant case (and the equations (7.1.7) an (7.1.8) in general). For a temporally homogeneous Markov process we have

$$f_u(\Psi(t, x)) = \int_\Omega \left\{ \int_\Omega u(\xi) P(t, \eta, d\xi) \right\} x(d\eta), \tag{2.5.20}$$

which suggests the idea of "cancelling out" the measure and studying directly the semigroup

$$(S(t)u)(\xi) = \int_\Omega u(\xi) P(t, \eta; d\xi) \tag{2.5.21}$$

in a space of $\mathfrak{S}$-measurable functions. Continuity of $S(\cdot)$ is assured by additional conditions on the transition probability (see Yosida [1978: 1, p. 398]).

In actual practice, not even $f(x(t))$ may be an observable quantity; what an instrument reads is some time average like

$$\frac{1}{T} \int_0^T f(\Phi(t; x)) \, dt,$$

which, if the system "tends to a steady state" sufficiently fast will be a good approximation to its limit as $T \to \infty$. This is one among the motivations for the study of diverse "limit averages" of a semigroup $S(\cdot)$ of linear bounded operators, for instance, the Cèsaro limit

$$\lim_{T \to \infty} \frac{1}{T} \int_0^T S(t) u \, dt$$

or the Abel limit

$$\lim_{\lambda \to \infty} \lambda \int_0^\infty e^{-\lambda t} S(t) u \, dt$$

in various topologies. This sort of problems comprise what is called *ergodic theory*. In its early stages it dealt with "concrete" semigroups of the form (2.5.19) (see Hopf [1937: 1]) and with discrete rather than continuous averages. The first works on the abstract theory (Von Neumann [1932: 2], Riesz [1938: 1], Visser [1938: 1], Yosida [1938: 1], and Kakutani [1938: 1]) also refer to the limit of the discrete average

$$\lim_{n \to \infty} \frac{1}{n} \sum_{j=0}^{n-1} S^j u,$$

where S is an operator whose powers S^j are uniformly bounded in the norm of (E). The theory for the continuous case was initiated by Wiener [1939: 1]. Some results in ergodic theory deal with general semigroups in Banach spaces: others (for instance, those involving pointwise convergence) with semigroups in certain function spaces. For accounts of ergodic theory and bibliography see Yosida [1978: 1]; other treatments are found in Hille-Phillips [1957: 1] and Dunford-Schwartz [1958: 1].

The study of Markov processes from the point of view of semigroup theory was initiated by Hille [1950: 1] and Yosida [1949: 1], [1949: 2]. A systematic investigation of the subject was conducted shortly after by Feller ([1952: 1], [1953: 1], [1954: 1], and subsequent papers). Short and very readable expositions can be found in Feller [1971: 1] and Yosida [1978: 1]. Other references are Dynkin [1965: 1], [1965: 2], Gihman-Skorohod [1975: 1], [1975: 2], [1979: 1], Itô-McKean [1974: 1], Stroock-Varadhan [1979: 1], and Mandl [1968: 1]. We limit ourselves to note that one of the central problems is the computation of the operator A in the equation (2.5.12), (or, rather, in the equation corresponding to the semigroup (2.5.21)), called the *Fokker-Planck* equation of the process, and the reconstruction of the transition probabilities from A. Under adequate assumptions on Ω (Ω a subset of $\mathbb{R}^m$ or more generally a manifold) and on the transition probabilities, the Fokker-Planck equation is a second-order parabolic differential equation, and the actual trajectories of the process are selected on the basis of "boundary conditions," sometimes of nonstandard type.

For results on the relation between Ψ and the semigroup defined by (2.5.19) see Neuberger [1971/72: 1], [1973: 2], and Durrett [1977: 1].

(c) *Higher Order Equations.* The considerations on systems in the previous subsection would seem to indicate that every conceivable physical process that satisfies the major premise of Huygens' principle must necessarily be described by a first-order equation of the form (2.5.12), at least if the necessary conditions of time invariance, linearity, and continuity are satisfied. Thus, even if the process stems from an equation of higher order in time like

$$u^{(n)}(t) + A_1 u^{(n-1)}(t) + \cdots + A_n u(t) = 0 \qquad (2.5.22)$$

(where $A_1, \ldots, A_n$ are, say, closed and densely defined operators in a Banach space E), a reduction of (2.5.22) to a first-order equation (conserving whatever existence, uniqueness, and continuous dependence properties it may have) should be possible. This is, in fact, so in examples like the wave and vibrating plate equation, where energy considerations dictate the reduction. An abstract treatment of these ideas was given by Weiss [1967: 1].

A study of the equation (2.5.22) in a different vein was carried out by the author [1970: 2]. The Cauchy problem for (2.5.22) is declared to be *well posed* in an interval $0 \leqslant t < T$ if solutions exist in the interval for initial data

$u(0), u'(0), \ldots, u^{(n-1)}(0)$ in a dense subspace D of E and depend continuously on them uniformly on compact intervals in the sense that

$$\|u(s)\| \leqslant C(t) \left(\sum_{j=0}^{n-1} \|u^{(j)}(0)\| \right) \quad (0 \leqslant s \leqslant t < T)$$

with $C(\cdot)$ as Section 1.2. This point of view leads to some curious properties: for instance, the Cauchy problem may be well posed in a finite interval without being well posed in $t \geqslant 0$ or, even if it is well posed in $t \geqslant 0$, the solutions of (2.5.22) may increase faster than any exponential (in fact, arbitrarily fast) at infinity. Admittedly, this may only indicate that the above definition of well-posed problem is too inclusive. A more restricted definition was attempted in the author's [1981: 1] for the case $n = 2$; it implies the existence of a strongly continuous semigroup $\mathfrak{S}(\cdot)$ in a suitable product space that *propagates* the solutions of (2.5.22) in the sense that $\mathfrak{S}(t)(u(0), u'(0)) = (u(t), u'(t))$, and generalizes an earlier result of Kisyński [1970: 1] where the equation is $u''(t) = Au(t)$.

Necessary and sufficient conditions in order that the Cauchy problem for (2.5.22) be well posed seem to be unknown, except in the particular case

$$u^{(n)}(t) = Au(t). \tag{2.5.23}$$

If $n \geqslant 3$, the Cauchy problem for (2.5.23) is well posed if and only if A is everywhere defined and bounded. This was proved by the author [1969: 2] (where the case in which E is a linear topological space is also dealt with) and independently by Chazarain [1968: 1], [1971: 1] with a different proof that works for more general equations. The case $n = 2$ is of more interest: here the equation is

$$u''(t) = Au(t), \tag{2.5.24}$$

and the substitution $u(t) \to u(-t)$ maps solutions into solutions, thus the Cauchy problem is well posed in $(-\infty, \infty)$ if it is well posed in $t \geqslant 0$. In the spirit of Section 1.2 we can define two propagators $C(\cdot)$ and $S(\cdot)$ by $C(t)u = u_0(t)$, $S(t)u = u_1(t)$, where $u_0(\cdot)$ (resp. $u_1(\cdot)$) is the solution of (2.5.24) satisfying $u(0) = u$, $u'(0) = 0$ (resp. $u(0) = 0$, $u'(0) = u$), and extend them to all of E by continuity. $C(\cdot)$ is strongly continuous and satisfies the *cosine functional equation(s)*

$$C(0) = I, \quad 2C(s)C(t) = C(s+t) + C(s-t) \tag{2.5.25}$$

for all s, t, whereas S obeys the formula

$$S(t)u = \int_0^t C(s)u \, ds, \tag{2.5.26}$$

hence is (E)-continuous and satisfies with $C(\cdot)$ the *sine functional equation(s)*

$$S(0) = 0, \quad 2S(s)C(t) = S(s+t) + S(s-t).$$

Functions $C(\cdot)$ satisfying (2.5.25) with values in a Banach algebra $\mathfrak{B}$ were considered by Kurepa [1962: 1]; among other results, it is shown that measurability implies continuity and that $C(t) = \cos(tA)$ (the cosine defined by its power series) for some $A \in \mathfrak{B}$. The theory of strongly continuous *cosine functions* or *cosine operator functions* was initiated by Sova [1966: 2], who among other things proved that there exist K, ω such that

$$\|C(t)\| \leqslant Ke^{\omega|t|} \quad (-\infty < t < \infty), \tag{2.5.27}$$

defined the infinitesimal generator of $C(\cdot)$ (roughly, the second derivative at the origin) and proved a generation theorem of Hille-Yosida type (which was independently discovered by Da Prato-Giusti [1967: 3]). Other results are due to the author [1969: 2] and [1969: 3], where the setting is (in part) that of linear topological spaces; in particular, a "measurability implies continuity" theorem of the type of Example 2.3.16 (which implies that the concepts of well posed and uniformly well posed problem coincide for the second-order equation) and an analogue of Lemma 2.2.3 for second order equations. We also point out a relation proved in [1969: 2] between the second order equation (2.5.24) and the first-order equation (2.1.1) that turns out to be extremely useful in singular perturbation problems: if the Cauchy problem for (2.5.24) is properly posed in the sense outlined above, then the Cauchy problem for (2.1.1) is as well properly posed; the propagator $V(t)$ of (2.1.1) is explicitly given by the "abstract Poisson formula"

$$V(t)u = \frac{1}{\sqrt{\pi t}} \int_0^\infty e^{-s^2/4t} C(s)u\,ds, \tag{2.5.28}$$

which incidentally shows that V can be extended to a (E)-valued analytic function $V(\zeta)$ in the half plane $\operatorname{Re}\zeta > 0$ (in fact, $A \in \mathfrak{C}(\pi/2-)$; see the forthcoming Chapter 4). The fact that well posedness of (2.5.24) implies the corresponding property of (2.1.1) was proved independently by Goldstein [1970: 4] in a different way.

Given a strongly continuous group $U(\cdot)$, it is immediate that

$$C(t) = \tfrac{1}{2}(U(t) + U(-t)) \tag{2.5.29}$$

is a strongly continuous cosine function whose infinitesimal generator is $A = B^2$, B the infinitesimal generator of $U(\cdot)$. The central problem in the theory of cosine functions is whether a representation of the form (2.5.29) (analogous to the equality $\cos at = \tfrac{1}{2}(e^{iat} + e^{-iat})$) exists for *any* cosine function. The answer is in the negative; it suffices to take $C(t) = \cos(tA)$, where A is a bounded operator having no (bounded or unbounded) square root.[2] Other counterexamples were given by Kisyński [1971: 1], [1972: 1], where C is uniformly bounded in norm: see the author [1981: 1] for

---

[2]In fact, it can be easily shown that if (2.5.29) holds then $A = B^2$, where A is the infinitesimal generator of $C(\cdot)$ and B is the infinitesimal generator of $U(\cdot)$.

additional information. It was proved by the author [1969: 3] that in certain spaces the representation (2.5.29) must hold if $C(\cdot)$ is uniformly bounded; the construction of U is based upon the Hilbert transform formula

$$U(t)u = C(t) + \frac{2it}{\pi} \int_0^\infty \frac{C(s)u}{t^2 - s^2} ds, \qquad (2.5.30)$$

the (singular) integral understood as the limit for $\varepsilon \to 0$ of the integrals in $|t - s| \geqslant \varepsilon$. The (E)-valued function $U(\cdot)$ can be shown to be a strongly continuous group in spaces where the Hilbert transform operator, defined for E-valued functions, has desirable continuity properties, such as $E = L^p$, $1 < p < \infty$ (in particular in Hilbert spaces). In the same spaces, a cosine function that is not uniformly bounded but merely satisfies (2.5.27) may not be decomposable in the form (2.5.29), but it can be proved (see again the author [1969: 1]) that if ω is the constant in (2.5.27), $b \geqslant \omega$ and $C_b(\cdot)$ the cosine function generated by $A_b = A - b^2 I$, then the decomposition holds for $C_b(\cdot)$; in other words, A_b posseses a square root $A_b^{1/2}$ that generates a group $U_b(\cdot)$. It follows that the second-order equation (2.5.23) can be reduced to the first-order system

$$u_1'(t) = \left(A_b^{1/2} + ibI \right) u_2(t), \quad u_2'(t) = \left(A_b^{1/2} - ibI \right) u_1(t)$$

in the product space $E \times E$, thus in a way the theory circles back into the reduction-of-order idea outlined at the beginning of this subsection. We indicate at the end additional bibliography on cosine functions and on the equation (2.5.24), pointing out by the way that some of the ideas used in the treatment of the equations (2.5.23) and (2.5.24) originate in early work of Hille [1954: 1], [1957: 1] (see also Hille-Phillips [1957: 1, Ch. 23]), where it is assumed at the outset that $A = B^n$ and the Cauchy problem is understood in a different sense.

The general equation (2.5.22) has received a great deal of attention during the last three decades, especially in the second-order case

$$u''(t) + Bu'(t) + Au(t) = 0.$$

Most of the literature deals with sufficient conditions on A, B that allow reduction to a first-order system and application of semigroup theory. In the references that follow, some works treat as well the case where the coefficients A, B depend on t; we include as well material on the equation (2.5.24) and on cosine functions. See Atahodžaev [1976: 1], Balaev [1976: 1], Buche [1971: 1], [1975: 1], Carasso [1971: 1], Carroll [1969: 2], [1979: 2] (which contain additional bibliography), Daleckiĭ-Fadeeva [1972: 1], Dubinskiĭ [1968: 1], [1969: 1], [1971: 1], Efendieva [1965: 1], Gašymov [1971: 1], [1972: 1], Giusti [1967: 1], Goldstein [1969: 1], [1969: 3], [1969: 4], [1970: 1], [1970: 2], [1970: 4], [1971: 1], [1972: 1], [1974: 2], [1975/76: 2], Goldstein-Sandefur [1976: 1], Golicev [1974: 1], Hersh [1970: 1], Jakubov [1964: 2], [1966: 1], [1966: 2], [1966: 4], [1966: 5], [1967: 3], [1970: 1], [1973: 1], Jurčuk

[1974: 1], [1976: 1], [1976: 2], Kolupanova [1972: 1], Kononenko [1974: 1], Krasnosel'skiĭ-Kreĭn-Sobolevskiĭ [1956: 1], [1957: 1], Kreĭn [1967: 1] (with more references), Kurepa [1973: 1], [1976: 1], Ladyženskaya [1955: 1], [1956: 1], [1958: 1], Lions [1957: 1], Lions-Raviart [1966: 1], Lomovcev-Jurčuk [1976: 1], Mamedov [1960: 1], [1964: 1], [1964: 2], [1965: 1], [1966: 1], Mamiĭ [1965: 1], [1965: 2], [1966: 1], [1967: 1], [1967: 2], Mamiĭ-Mirzov [1971: 1], Masuda [1967: 1], Maz'ja-Plamenevskiĭ [1971: 1], Melamed [1964: 1], [1965: 1], [1969: 1], [1971: 1], Nagy [1974: 1], [1976: 1], Obrecht [1975: 1], Pogorelenko-Sobolevskiĭ [1967: 1], [1967: 2], [1967: 3], [1970: 1], [1972: 1], Radyno-Jurčuk [1976: 1], Raskin [1973: 1], [1976: 1], Raskin-Sobolevskiĭ [1967: 1], [1968: 1], [1968: 2], [1969: 1], Russell [1975: 1], Sandefur [1977: 1], Sobolevskiĭ [1962: 1], Sova [1968: 1], [1969: 1], [1970: 1], Straughan [1976: 1], Travis [1976: 1], Tsutsumi [1971: 1], [1972: 1], Veliev [1972: 1], Veliev-Mamedov [1973: 1], [1974: 1], Yosida [1956: 1], [1957: 2], and the author [1971: 2]. In some of these works the equations in question are time-dependent and/or semilinear (see (k)) or the setting is, at least in part, more general than that of properly posed Cauchy problems (see (e) and also Chapter 6).

(d) *Semigroup and Cosine Function Theory in Linear Topological Spaces.* Consider the translation group

$$S(t)u(x) = u(x + t)$$

in the linear topological space E consisting of all functions u continuous in $-\infty < t < \infty$, endowed with the family of seminorms

$$\|u\|_n = \sup_{|x| \leqslant n} |u(x)|,$$

$n = 1, 2, \ldots$. Obviously, $S(\cdot)$ is strongly continuous in E, but it is easily seen that, in general, the integral

$$\int_0^\infty e^{-\lambda t} S(t) u \, dt \tag{2.5.31}$$

fails to exist in any reasonable sense whatever the value of λ, thus blocking any generalization of formula (2.1.10). That this generalization fails is not surprising, since the infinitesimal generator of S (in an obvious sense) is $A = d/dx$, with domain consisting of all u continuously differentiable in $-\infty < t < \infty$ and the resolvent set of A is empty (the equation $\lambda u - u' = v$ always has infinitely many solutions in E). This example makes clear that the main difficulty in generalizing semigroup theory to linear topological spaces is the loss of (2.1.3) (which, in the present setting must be translated into the requirement that $\{e^{-\omega t}S(t); t \geqslant 0\}$ be equicontinuous for some ω). If this is *assumed*, the Banach space results generalize readily to a locally convex quasicomplete space. This extension was suggested by Schwartz [1958: 2] and independently carried out by Miyadera [1959: 2] for Fréchet spaces; the general version is due to Komatsu [1964: 1] and Yosida in the first edition (appeared in 1965) of his treatise [1978: 1]. We refer the reader to that book for details.

The general case is tougher. The only assumption is that S is strongly continuous and that $\langle S(t); \ 0 \leqslant t \leqslant 1 \rangle$ is equicontinuous; this is a consequence of the strong continuity if the space E is barreled, as the uniform boundedness theorem (Köthe [1966: 1]) shows. Generation theorems have been found by T. Komura [1968: 1], Ouchi [1973: 1], Dembart [1974: 1], and Babalola [1974: 1], and can be roughly described as follows. In the approach of Komura the integral (2.5.31) is given a sense through the theory of Fourier-Laplace transforms due to Gelfand and Šilov [1958: 1] where (generalized) functions of arbitrary growth at infinity can be transformed: the *generalized resolvent* $\Re(\lambda; A)$ then appears as a vector-valued analytic functional rather than a function, and is gotten at through "approximate resolvents" $\langle \Re_n(\lambda) \rangle$ obtained cutting off the domain of integration in (2.5.31). The methods of Dembart, Ouchi, and Babalola are somewhat related; the generalized resolvent is not used and its place is taken by a sequence of *approximate resolvents* constructed in the way indicated above. These ideas are related to the *asymptotic resolvent* of Walbroeck [1964: 1].

The theory of cosine functions admits a similar extension to linear topological spaces. When $\langle e^{-\omega|t|}C(t); \ -\infty < t < \infty \rangle$ is assumed equicontinuous for ω sufficiently large, the treatment (due to the author [1969: 2] for barreled, quasi-complete E) uses the same ideas as in the Banach space case. The "reduction to first order" outlined in (c) can be carried out if an additional assumption is satisfied; one of its forms is the requirement that the function

$$G(t) = \int_0^1 \log s \big(S(s + t) + S(s - t) \big) \, ds$$

(S defined by (2.5.26)) satisfy (a) $G(t)E \subseteq D(A)$, and (b) $AG(\cdot)$ is a strongly continuous function of t for all t (as pointed out by Nagy [1976: 1], however, this assumption fails to hold even in some Banach spaces). The case where only $\langle S(t); \ 0 \leqslant t \leqslant 1 \rangle$ is assumed to be equicontinuous has been examined by Konishi [1971/72: 1] who obtained generation theorems (in the style of those of T. Komura for semigroups) and many additional results.

We include below several references on semigroups, cosine functions, and linear differential equations in linear topological spaces; some of these belong to the subjects treated in Chapters 4 and 5 and will be also found there. See Chevalier [1969: 1], [1970: 1], [1975: 1], Dembart [1973: 1], V. V. Ivanov [1973: 1], [1974: 1], [1974: 2], Kononenko [1974: 1], Lovelady [1974/75: 1], Máté [1962: 2], Milštein [1975: 1], Ouchi [1973: 1], Povolockiĭ-Masjagin, [1971: 1], Sarymsakov-Murtazaev [1970: 1], Singbal-Vedak [1965: 1], [1972: 1], Terkelsen [1969: 1], [1973: 1], Tillmann

[1960: 1], Ujishima [1971: 1], [1972: 1], Vainerman-Vuvunikjan [1974: 1], Vuvunikjan [1971: 1], Watanabe [1972: 1], [1973: 1], and Yosida [1963: 1].

(e) The equation

$$u'(t) = Au(t) \tag{2.5.32}$$

(as well as higher-order and time-dependent versions) has been extensively studied under hypotheses sometimes considerably weaker than (or not comparable to) those for a properly posed Cauchy problem. In some cases the objective is to prove existence and/or uniqueness results for solutions in $t \geqslant 0$ or other intervals, as in Ljubič [1966: 3]. We include a small sample of results.

2.5.1 Example. Let A be a densely defined operator such that $R(\lambda)$ exists in a *logarithmic region*

$$\Lambda = \Lambda(\alpha, \beta) = \{\lambda : \operatorname{Re}\lambda \geqslant \beta + \alpha \log|\lambda|\}$$

with $\alpha \geqslant 0$ (see Section 8.4, especially Figure 8.4.1) and assume that

$$\|R(\lambda)\| \leqslant C(1 + |\lambda|)^m \quad (\lambda \in \Lambda)$$

for some $m \geqslant 0$. Then (a) for every $u_0 \in D(A^\infty) = \cap D(A^n)$ there exists a genuine solution of (2.5.32) in $t \geqslant 0$ with $u(0) = u_0$; (b) genuine solutions of the initial value problem are unique.

Uniqueness holds in fact in a strong version: every solution of (2.5.32) in a finite interval $0 \leqslant t \leqslant T$ with $u(0) = 0$ is identically zero in $0 \leqslant t \leqslant T$. This is best proved with the "convolution inverse" methods of Chapter 8 (or using the next example). Solutions are constructed by inverse Laplace transform of $R(\lambda)$ in the style of formula (2.1.12); however, integration must be performed over the boundary Γ of Λ and further terms in formula (3.7) are necessary:

$$u(t) = \sum_{j=0}^{p-1} \frac{t^j}{j!} A^j u_0 + \frac{1}{2\pi i} \int_\Gamma \frac{e^{\lambda t}}{\lambda^p} R(\lambda) A^p u_0 \, d\lambda.$$

This expression defines a solution in $0 \leqslant t < t_p = (p - m - 2)/\alpha$ with initial value $u(0) = u_0$. Thus (making use of uniqueness at each step) we can define a solution 'by pieces" in $t \geqslant 0$.

We note that under the hypotheses above $D(A^\infty)$ is dense in E, thus the set of initial vectors is dense in E. See Ljubič [1966: 3] and Beals [1972: 2] for far more general statements in this direction. Uniqueness can be also established in much greater generality:

2.5.2 Example. Assume that $R(\lambda)$ exists for $\lambda > \omega$ and that

$$\|R(\lambda)\| = O(e^{r\lambda}) \text{ as } \lambda \to \infty$$

for some $r \geqslant 0$. Then if $u(\cdot)$ is a solution of (2.5.32) in $0 \leqslant t \leqslant T$ with $u(0) = 0$, we have $u(t) = 0$ in $0 \leqslant t \leqslant T - r$.

To prove this result we note that

$$(\lambda I - A)\int_0^T e^{-\lambda t}u(t)\,dt = -e^{-\lambda T}u(T),$$

hence

$$\left\|\int_0^T e^{-\lambda t}u(t)\,dt\right\| = O(e^{-\lambda(T-r)}) \text{ as } \lambda \to \infty,$$

which implies that $u(t) = 0$ in $0 \leqslant t \leqslant T - r$ by (a slight variant of) the Paley-Wiener theorem.

In other lines of treatment (for instance Agmon-Nirenberg [1963: 1], [1967: 1]) existence is not in question and only properties of solutions such as asymptotic behavior at infinity of theorems of Phragmén-Lindelöf type are of interest. Many of the results have connections with the abstract Cauchy problem, however. We shall not treat in this work the equation (2.5.32) in this vein and we limit ourselves to indicate some bibliography. Besides the papers above, information can be found in the treatises of Carroll [1969: 2], [1979: 2], Ladas-Lakshmikantham [1972: 2], and Zaidman [1979: 1], together with references on the subject. Other works are Aliev [1974: 1], Arminjon [1970: 1], [1970: 2], Beals [1972: 1], [1972: 2], Djedour [1972: 1], Goldstein-Lubin [1974: 1], Ladas-Lakshmikantham [1971: 1], [1972: 1], Levine [1970: 1], [1970: 2], [1972: 1], [1973: 1], [1973: 2], [1975: 1], Levine-Payne [1975: 1], Malik [1967: 1], [1972: 1], [1973: 1], [1975: 1], and Pazy [1967: 1].

An important area of study concerns *almost periodic* solutions of (2.5.32) and of time-dependent analogues; most of the results are generalizations to the infinite-dimensional setting of theorems on almost periodicity of solutions of ordinary differential equations due to Bohl, Bohr and Neugebauer, Favard and others (see Favard [1933: 1] for results and references). The setting is sometimes that of well-posed Cauchy problems, other times a more general one. See the treatises of Amerio-Prouse [1971: 1] and Zaidman [1978: 1] for expositions of the theory and bibliographical references. Additional works are Žikov [1965: 1], [1965: 2], [1966: 1], [1967: 1], [1970: 1], [1970: 2], [1970: 3], [1971: 1], [1972: 1], [1975: 1], Žikov-Levitan [1977: 1], Levitan [1966: 1], Zaki [1974: 1], Rao [1973/74: 1], [1974/75: 1], [1975: 1], [1975: 2], [1977: 1], Rao-Hengartner [1974: 1], [1974: 2], Aliev [1975: 1], Bolis Basit-Zend [1972: 1], Biroli [1971: 1], [1972: 1], [1974: 1], [1974: 2], Lovicar [1975: 1], Mišnaevskiĭ [1971: 1], [1972: 1], Moseenkov [1977: 1], Welch [1971: 1], and the author [1970: 4]. For almost periodic functions in Banach spaces see Corduneanu [1961: 1].

(f) *Transmutation.* In a somewhat vague description, a *transmutation formula* transforms a solution of an abstract differential equation into a solution of another abstract differential equation. Two rather trivial transmutation formulas are $u(t) \to e^{at}u(t)$ (transforming a solution of $u'(t) =$

$Au(t)$ into a solution of $u'(t) = (A + aI)u(t))$ and the "identity formula" $u(t) \to u(t)$, which transforms a twice differentiable solution of $u'(t) = Au(t)$ such that $u(t) \in D(A^2)$ into a solution of the second order equation $u''(t) = A^2 u(t)$. We list below a number of transmutation formulas. In all the following examples, A is a closed operator in the Banach space E and solutions are understood in the strong sense as in Section 1.2.

2.5.3   Example.   Let $u(\cdot)$ be a solution of

$$u''(t) = Au(t) \quad (t \geq 0) \tag{2.5.33}$$

such that $u(0) = u$, $u'(0) = 0$, and

$$\|u(t)\| \leq Ce^{\omega t} \quad (t \geq 0). \tag{2.5.34}$$

Then

$$v(t) = \frac{1}{\sqrt{\pi t}} \int_0^\infty e^{-s^2/4t} u(s) \, ds$$

is a solution of

$$v'(t) = Av(t) \quad (t \geq 0) \tag{2.5.35}$$

with $v(0) = u$, satisfying an estimate of the type of (2.5.34) with ω^2 instead of ω. This formula is an abstract version of the classical Poisson formula for solution of the heat equation $u_t = u_{xx}$, where E is a space of functions in $(-\infty, \infty)$ (say, L^2) and $u(t)(x) = u(x + t)$. In its abstract version it can be traced back (in a different context) to Romanoff [1947: 1] and it was used by the author in [1969: 2] to relate the Cauchy problem for (2.5.34) and (2.5.35) (see (2.5.28) and surrounding comments). See also the author [1983: 3] for an extension to linear topological spaces.

2.5.4   Example.   Let $u(\cdot)$ be an arbitrary solution of (2.5.33), $\alpha > -1$. Define

$$v(t) = \frac{\Gamma(\alpha + 3/2)}{\sqrt{\pi} \, \Gamma(\alpha + 1)} \int_{-1}^1 u(t\eta)(1 - \eta^2)^\alpha d\eta. \tag{2.5.36}$$

Then $v(\cdot)$ is a solution of

$$v''(t) + \frac{2(\alpha + 1)}{t} v'(t) = Av(t) \quad (t > 0) \tag{2.5.37}$$

with initial conditions $v(0) = u(0)$, $v'(0) = 0$. In case A is the m-dimensional Laplacian, formula (2.5.36) transforms the wave equation (2.5.33) into the *Darboux equation* (2.5.37) and can be used to compute an explicit solution of the wave equation (see Courant-Hilbert [1962: 1, p. 700]).

2.5.5   Example.   Let $u(\cdot)$ be a solution of (2.5.33) with $u'(0) = 0$. Define

$$v(t) = \int_0^t J_0 \left(c(t^2 - s^2)^{1/2} \right) u(s) \, ds, \tag{2.5.38}$$

where J_0 is the Bessel function of order zero. Then $v(\cdot)$ is a solution of the equation

$$v''(t) = (A - c^2 I) v(t) \quad (t \geqslant 0) \tag{2.5.39}$$

with $v(0) = 0$, $v'(0) = u(0)$. Formula (2.5.38) is an abstract version of a well known formula for solution of the equation $u_{tt} = u_{xx} - c^2 u$. For additional details and other formulas of the same type see Sova [1970: 4].

Transmutation formulas may involve equations in several variables, as is the case with the following.

2.5.6 Example. Let $v(t_1, t_2)$ be a solution of

$$D^1 D^2 v(t_1, t_2) = A v(t_1, t_2) \quad (t_1, t_2 \geqslant 0) \tag{2.5.40}$$

with $v(t, 0) = f(t)$, $v(0, t) = g(t)$, $t \geqslant 0$ (we assume that $f(0) = g(0)$). Then the "self-convolution"

$$u(t) = \int_0^t v(t - s, s) \, ds \tag{2.5.41}$$

satisfies

$$u''(t) = Au(t) + f'(t) + g'(t) \quad (t \geqslant 0) \tag{2.5.42}$$

with $u(0) = 0$, $u'(0) = f(0) = g(0)$. Conversely, let $u(\cdot)$ be a solution of (2.5.33). Then

$$v(t_1, t_2) = \frac{1}{\pi} \int_{-1}^{1} (1 - \eta^2)^{-1/2} u\left(2(t_1 t_2)^{1/2} \eta\right) d\eta \tag{2.5.43}$$

is a solution of (2.5.40) with $v(t_1, 0) = v(0, t_2) = u(0)$, $D^2 u(t_1, 0) = D^1 u(0, t_2) = 0$. These formulas are due to the author [1971: 1], where applications to the abstract Goursat problem can be found (see (g))

One of the principal applications of transmutation formulas is that of showing that the Cauchy problem for certain equations is properly posed starting from similar properties of other equations.

There are at least two heuristic guides for discovery of transmutations formulas (their justification is usually simple). One is that of looking first at the case where the operator A "is a number a" (strictly speaking, coincides with the operator of multiplication by a in one-dimensional Banach space). In this particular case transmutation formulas will be found in a table of integrals. Once we get hold of the formula in question, its extension to operational equations is usually immediate. For instance, the scalar ancestor of both (2.5.36) and (2.5.43) (the latter in the case $\nu = 0$) is

$$J_\nu(t) = \frac{t^\nu}{2^\nu \Gamma(\nu + 1/2) \Gamma(1/2)} \int_{-1}^{1} (1 - \eta^2)^{\nu - 1/2} \cos t\eta \, d\eta$$

(see Gradstein-Ryžik [1963: 1, formula 8.411]. Another, more systematic procedure is to essay formulas of the type

$$v(t) = \int_0^\infty \kappa(t,s)u(s)\,ds. \tag{2.5.44}$$

For instance, if u is the function in Example 2.5.3, the requirement that v be a solution of (2.5.35) with $v(0) = u(0)$ leads (after two integrations by parts) to the distributional initial-boundary value problem

$$\kappa_t(t,s) = \kappa_{ss}(t,s) \quad (t,s \geqslant 0), \qquad \kappa_s(t,0) = 0 \quad (t \geqslant 0),$$

$$\kappa(0,s) = \delta(s) \quad (s \geqslant 0),$$

where δ is the Dirac delta; from these conditions we obtain that $v(t) = (\pi t)^{-1/2}\exp(-s^2/4t)$ and we obtain anew the result in Example 2.5.3. This and other related arguments were systematically examined (in the context of partial differential equations) by Delsarte and Lions (see Delsarte [1938: 1], Delsarte-Lions [1957: 1], and Lions [1956: 1]). An interesting sideline on formula (2.5.44) is that in certain cases κ may be a distribution rather than a function. For instance, assume we have a solution $u(\cdot)$ of (2.5.33) with $u(0) = u$, $u'(0) = 0$ and we wish to obtain a solution $v(\cdot)$ of (2.5.39) with $v'(0) = 0$. Instead of deducing differential equations and boundary conditions for κ (a simple exercise left to the reader) we simply differentiate (2.5.38) obtaining the new transmutation formula

$$v(t) = u(t) - ct \int_0^t (t^2 - s^2)^{-1/2} J_1\big(c(t^2 - s^2)^{1/2}\big) u(s)\,ds,$$

which can be easily justified directly; here

$$\kappa(s,t) = \delta(t-s) - ct(t^2 - s^2)^{-1/2} J_1\big(c(t^2 - s^2)^{1/2}\big).$$

An extensive treatment of transmutation formulas and their applications as well as a large bibliography can be found in Carroll [1979: 2]. Other references are Bragg [1969: 1], [1974: 1], [1974/75: 1], [1974/75: 2], Bragg-Dettman [1968: 1], [1968: 2], [1968: 3], Butcher-Donaldson [1975: 1], Dettman [1969: 1], [1973: 1], Donaldson [1971: 1], [1972: 1], [1975: 1], [1977: 1], Carroll [1963: 1], [1966: 1], [1966: 2], [1969: 1], [1969: 2] (where additional references can be found), [1976: 1], [1978: 1], [1979: 1], Carroll-Showalter [1976: 1], Dunninger-Levine [1976: 1], [1976: 2], Bobisud-Hersh [1972: 1], and Hersh [1970: 1].

(g) *Equations in Several Variables.* There exists some literature on abstract differential equations involving partial derivatives of functions of several variables and related "initial value" (or, rather, boundary value) problems. One of these is the *Goursat* (or *characteristic initial value*) problem: find a solution of

$$D^1 D^2 u(t_1, t_2) = Au(t_1, t_2) \quad (t_1, t_2 \geqslant 0) \tag{2.5.45}$$

satisfying

$$u(t_1, 0) = f(t_1) \quad (t_1 \geqslant 0), \qquad u(0, t_2) = g(t_2) \quad (t_2 \geqslant 0) \tag{2.5.46}$$

(where of course $f(0) = g(0)$). This type of problem (together with higher order versions) has been considered by the author in [1971: 1]; one of the techniques employed is that of transmutation (see Section 2.5.(f), especially Example 2.5.6), but other devices must be used to identify completely the operators A that make the Goursat problem (2.5.45), (2.5.46) well posed in a suitable sense. See also Briš-Jurčuk [1971: 1], W. A. Roth [1972/73: 1]. Other works on problems in several variables are Daleckiĭ-Fadeeva [1972: 1] and Kolupanova [1972: 1]

(h) *Scattering.* Consider the wave equation

$$u_{tt} = c^2 \Delta u \qquad (2.5.47)$$

in m-dimensional Euclidean space $\mathbb{R}^m$ with m odd. It is known (and easily proved) that the energy of a solution $u(t) = u(t, \cdot)$,

$$E(t) = \tfrac{1}{2} \int \left(u_t^2 + c^2 |\text{grad } u|^2 \right) dx, \qquad (2.5.48)$$

remains constant with t, thus we can define a group $U_0(\cdot)$ of unitary operators in $H_0(\mathbb{R}^m) \oplus L^2(\mathbb{R}^m)$ by the rule

$$U_0(t)(u, v) = (u(t), u_t(t)). \qquad (2.5.49)$$

Here $H_0(\mathbb{R}^m)$ consists of all $u \in L^2_{\text{loc}}(\mathbb{R}^m)$ such that $|\text{grad } u| \in L^2(\mathbb{R}^m)$, two such functions declared equivalent when they differ a.e. by a constant: the norm in $H_0(\mathbb{R}^m)$ is

$$\|u\|_1 = c^2 \int |\text{grad } u|^2 \, dx.$$

Assume an obstacle (a compact set) K is planted in $\mathbb{R}^m$; now we look for solutions of (2.5.47) in $\Omega = \mathbb{R}^m \setminus K$ that vanish on the boundary Γ of Ω. The energy of these solutions is also constant, thus we can define another isometric semigroup $U(\cdot)$ by (2.5.49), this time in the space $H_\Gamma(\Omega) \oplus L^2(\mathbb{R}^m)$; here $H_\Gamma(\Omega)$ consists of all $u \in H_0(\mathbb{R}^m)$ that vanish on Γ (strictly speaking, such that some member of its equivalence class vanishes on Γ). It is plausible to conjecture that if $u \in H_\Gamma(\Omega) \oplus L^2(\mathbb{R}^m)$, the free solution $U_0(t)u$ and the perturbed solution $U(t)u$ (at least if the obstacle does not "trap energy") will differ by little when $t \to \infty$ (hence for $t \to -\infty$, since the equation (2.5.47) is invariant through the change $t \to -t$). A measure of the deviation of the perturbed group from the free one are the *wave operators.*

$$W_+ u = \lim_{t \to \infty} U(-t)U_0(t)u,$$
$$W_- u = \lim_{t \to \infty} U(t)U_0(-t)u. \qquad (2.5.50)$$

(Note that, in view of the explicit solution of (2.5.47) in $\mathbb{R}^m$ with m odd, if u has compact support then $U_0(t)$, $U_0(-t)$ will be zero on K for t sufficiently large, hence $U(-t)U_0(t)u$ and $U(t)U_0(-t)u$ are both defined.) Leaving aside technical details, the *scattering operator* is defined by

$$S = W_+^{-1} W_-$$

and describes, roughly speaking, the effect on an incoming wave $U_0(t)u$, t large, produced by the obstacle; the inverse problem of (acoustic) scattering theory is that of reconstructing the obstacle K from the the scattering operator S. Similar descriptions can be given of the scattering problems associated with the Maxwell, Schrödinger, and neutron transport equations. These physical situations have been the motivation for the creation of an *abstract scattering theory*, where one of the problems is that of showing existence and suitable properties of the wave operators W_+, W_-, with U and U_0 groups of unitary operators in a Hilbert space or more general groups. For expositions of scattering theory, both in the abstract version or in concrete problems such as that of acoustic or Schrödinger scattering, see the treatises of Kato [1976: 1] and Lax-Phillips [1967: 1]. See also Bondy [1976: 1], Foias [1975: 1], Helton [1974: 1], Kato-Kuroda [1970: 1], [1971: 1], and Lax-Phillips [1966: 1], [1971: 1], [1972: 1], [1972: 2], [1973: 1], [1976: 1].

(i) *Singular Differential Equations.*   Diverse applications lead to abstract differential equations where the time derivative of highest order is not explicitly given, for instance,

$$Mu'(t) + Lu(t) = 0. \tag{2.5.51}$$

One such example is the *pseudoparabolic equation*

$$u_t - \kappa u_{xxt} = u_{xx}, \tag{2.5.52}$$

where $\kappa > 0$. Assume, to fix ideas, that a solution of (2.5.52) is sought in the interval $0 \leqslant x \leqslant \pi$ with boundary conditions

$$u(0, t) = u(\pi, t) = 0. \tag{2.5.53}$$

Since the equation $\kappa u'' - u$ has no nontrivial solution satisfying $u(0) = u(\pi) = 0$, the Green function $G_\kappa(t, s)$ of the boundary value problem exists and we may transform the equation (2.5.52) into the partial integro-differential equation

$$u_t = -\int_0^\pi G_\kappa(t, s) u_{xx}(x, s) \, ds, \tag{2.5.54}$$

which is amenable to semigroup theory methods (although it would be obviously simpler to examine (2.5.52) directly by separation of variables). A similar reduction is available when $\kappa < 0$ as long as $-\kappa \neq 1/m^2$ ($m = 1, 2, \dots$). The abstract counterpart for (2.5.51) is the case where M is invertible; here the equation can be reduced to (2.1.1) with $A = M^{-1}L$. In other cases, however, M^{-1} does not possess a (bounded or unbounded) inverse and the reduction fails; the behavior of (2.5.51) can be expected to be quite different from that of (2.1.1). To gain some insight into the problem we note that the separation-of-variables solution of (2.5.52) is

$$u(x,t) = \sum a_n(t)\sin nx,$$

where $(1 + \kappa n^2)a_n'(t) = - n^2 a_n(t)$, hence if $\kappa = 1/m^2$, any possible initial condition must have zero mth Fourier coefficient.

Other examples of singular differential equations are

$$a(x)u_t = u_{xx} \tag{2.5.55}$$

(where $a(x)$ may vanish somewhere) and the fourth order equation

$$u_{xxtt} = u_{xxxx}. \tag{2.5.56}$$

Finally, in some time-dependent problems, the leading coefficient vanishes at the very instant where initial conditions are supposed to be given: an example is the *abstract Darboux equation* (2.5.37).

An attractive treatment of a class of singular equations can be found in the treatise of Showalter [1977: 1]. Other works on the subject are Showalter [1969: 1], [1970: 1], [1970: 2], [1972: 1], [1972: 2], [1973: 1], [1973/74: 1], [1975: 1], Carroll-Showalter [1976: 1], Baiocchi-Baouendi [1977: 1], Bragg [1974/75: 1], [1974/75: 2], Butcher-Donaldson [1975: 1], Bykov-Fomina [1973: 1], [1976: 1], Donaldson [1970: 1], [1971: 1], Donaldson-Goldstein [1976: 1], Dunninger-Levine [1976: 1], [1976: 2], Favini [1974: 1], [1974: 2], [1974: 3], [1975: 1], [1977: 1], Knops-Payne [1971: 1], Lagnese [1972: 2], [1973: 1], [1973: 2], [1974: 1], Levine [1972: 1], [1973: 3] [1974: 1], Schuss [1972: 1], Travis [1976: 1], and C. L. Wang [1975: 1].

(j) *The Inhomogeneous Equation*

$$u'(t) = Au(t) + f(t). \tag{2.5.57}$$

Using the explicit formula (2.4.3) for the solution of (2.5.57), numerous results on the correspondence $f(\cdot) \to u(\cdot)$ have been proved; in most of them the operator A is assumed to belong to subclasses of $\mathcal{C}_+$ (such as the class $\mathcal{C}$ introduced in Chapter 4). We quote below two results of this sort; in both of them we assume that $S(t)E \subseteq D(A)$, that the operator $AS(t)$ is bounded for each t, and

$$\|AS(t)\| \leqslant C/t \quad (t > 0) \tag{2.5.58}$$

for some C (see Sections 4.1 and 4.2).

2.5.7 Example (Pazy [1974: 1, p. 113]). Let f be an E-valued function defined in $0 \leqslant t \leqslant T$. Assume there exists a nonnegative function σ, also defined in $0 \leqslant t \leqslant T$, and such that

$$\|f(t) - f(s)\| \leqslant \sigma(t - s) \quad (0 \leqslant s \leqslant t \leqslant T),$$

$$\int_0^h \frac{\sigma(t)}{t}\, dt < \infty \text{ for some } h > 0.$$

Then

$$u(t) = \int_0^t S(t-s)f(s)\,ds \tag{2.5.59}$$

is a solution of (2.5.57) in the strong sense: $u(\cdot)$ is continuously differentiable and belongs to $D(A)$ in $0 \leqslant t \leqslant T$ and satisfies (2.5.57) there.

The assumptions on the function f are obviously much weaker than those in Lemma 2.4.2: the present result is an illustration of the heuristic principle that *stronger assumptions on S will allow weaker assumptions on f*. In the next result these requirements (as well as the definition of solution) are weakened further: we still assume (2.5.58).

*2.5.8 Example (De Simon [1964: 1]). Let E be a Hilbert space, f a strongly measurable E-valued function defined in $0 \leqslant t \leqslant T$ such that

$$\|f\|_p = \left(\int_0^T \|f(t)\|^p\,dt \right)^{1/p} < \infty,$$

where $1 < p < \infty$. Then, if $u(\cdot)$ is the function in (2.5.59), $u'(t)$ exists a.e., $u(t) \in D(A)$ a.e. and (2.5.57) is satisfied almost everywhere in $0 \leqslant t \leqslant T$. Morever, we have

$$\|u'\|_p \leqslant C\|f\|_p$$

for C independent of f. Finally, $u(t)$ is the integral of $u'(\cdot)$ in $0 \leqslant s \leqslant t$.

For references on the equation (2.5.57) (some of which deal with the case where A depends on t) see Anosov-Sobolevskiĭ [1971: 1], [1972: 1], Ball [1977: 1], Baras-Hassan-Veron [1977: 1], Dyment-Sobolevskiĭ [1970: 1], [1970: 2], [1971: 1], Gerštein-Sobolevskiĭ [1974: 1], [1975: 1], Crandall-Pazy [1968/69: 1], Polička-Sobolevskiĭ [1976: 1], Raskin-Jasakov [1970: 1], Sobolevskiĭ [1964: 3], [1964: 4], [1965: 1], [1966: 1], [1967: 2], [1971: 3], and De Graaf [1971: 1], [1972: 1].

(k) *Semilinear Equations.* Roughly speaking, a *quasilinear* or *semilinear* equation is a nonlinear perturbation of a linear equation, the perturbation terms being "less unbounded" than the linear part (in contrast with the genuinely nonlinear equations to be discussed in Section 3.8(d).) A first-order semilinear equation is of the form

$$u'(t) = Au(t) + B(t, u(t)), \tag{2.5.60}$$

where B is a nonlinear operator. To solve (2.5.60) with initial condition $u(0) = u_0$, assume that u is a genuine solution, that B is continuous and use formula (2.4.3); the Volterra integral equation

$$u(t) = S(t)u_0 + \int_0^t S(t-s)B(s, u(s))\,ds \tag{2.5.61}$$

results. On the other hand, even if $u_0 \in D(A)$, a continuous solution of

(2.5.61) may not be a solution of (2.5.60) ($u(\cdot)$ may not be differentiable); as in the linear case, we declare $u(\cdot)$ to be a *generalized solution* of the differential equation.

Equation (2.5.61) lends itself to solution by successive approximations or by use of fixed point theorems. In the following result we only assume that $A \in \mathcal{C}_+$.

2.5.9 Example (Segal [1963: 1]). Let B be defined and continuous in $[0, T] \times E$ and Lipschitz continuous in u uniformly with respect to t,

$$\|B(t, u) - B(t, v)\| \leqslant M \|u - v\| \quad (0 \leqslant t \leqslant T, u, v \in E). \quad (2.5.62)$$

Then, given $u_0 \in E$, (2.5.61) has a unique solution $u(\cdot)$ continuous in $0 \leqslant t \leqslant T$.

To prove this, we define a sequence of succesive approximations taking as $u_0(\cdot)$ any continuous E-valued function with $u_0(0) = u_0$ and setting

$$u_{n+1}(t) = S(t)u_0 + \int_0^t S(t - s) B(s, u_n(s)) \, ds \quad (n = 1, 2, \dots).$$

Using (2.5.62) the estimate

$$\|u_{n+1}(t) - u_n(t)\| \leqslant CM \int_0^t \|u_n(s) - u_{n-1}(s)\| \, ds$$

results, where C is a bound for $\|S(s)\|$ in $0 \leqslant s \leqslant T$. This leads to

$$\|u_{n+1}(t) - u_n(t)\| \leqslant K(CM)^n t^n / n! \quad (n \geqslant 1, 0 \leqslant t \leqslant T)$$

for some $K > 0$. The rest of the proof is routine. It is interesting to note that under stronger assumptions on S and B the solution obtained is genuine.

2.5.10 Example. Let S satisfy the assumptions in Example 2.5.7, and assume B is Lipschitz continuous jointly in s and u,

$$\|B(s, u) - B(t, v)\| \leqslant M(|s - t| + \|u - v\|) \quad (0 \leqslant s, t \leqslant T, u, v \in E).$$

Then the solution $u(\cdot)$ of (2.5.61) with $u_0 \in D(A)$ obtained in Example 2.5.9 is continuously differentiable in $0 \leqslant t \leqslant T$, $u(t) \in D(A)$, and (2.5.60) holds there.

We begin by proving that the generalized solution $u(\cdot)$ of (2.5.61) is Lipschitz continuous in $0 \leqslant t \leqslant T$. Let $T' < T$ be a number to be determined later, $0 \leqslant t \leqslant t' \leqslant T'$. Then

$$\|u(t') - u(t)\| \leqslant \|(S(t') - S(t))u_0\| + \int_t^{t'} \|S(s) B(t' - s, u(t' - s))\| \, ds$$

$$+ \int_0^t \|S(s)\| \|B(t' - s, u(t' - s)) - B(t - s, u(t - s))\| \, ds.$$

Let $\rho(t, t')$ be the maximum of $\|u(t' - s) - u(t - s)\|$ in $0 \leqslant s \leqslant t$. Then we obtain from the previous inequality that

$$\|u(t') - u(t)\| \leqslant C(\|Au_0\| + N)|t' - t| + CMT'(|t' - t| + \rho(t, t')),$$

where C (resp. N) is a bound for $\|S(s)\|$ (resp. $\|B(s, u(s))\|$) in $0 \leqslant s \leqslant T$. Write this inequality replacing t' by $t' - s$ and t by $t - s$ for $0 \leqslant s \leqslant t$. Noting that $\rho(t - s, t' - s) \leqslant \rho(t, t')$, we obtain

$$\rho(t, t') \leqslant C(\|Au_0\| + N)|t' - t| + CMT'(|t' - t| + \rho(t, t')).$$

If $T' \leqslant 1/2CM$, we deduce that

$$\rho(t, t') \leqslant C'|t' - t|$$

in $0 \leqslant t \leqslant t' \leqslant T'$. This shows that $u(\cdot)$ is Lipschitz continuous in $0 \leqslant t \leqslant T'$. Hence $B(\cdot, u(\cdot))$ is as well Lipschitz continuous and it follows from Example 2.5.7 that $u(\cdot)$ is a genuine solution in $0 \leqslant t \leqslant T'$. In particular,

$$u_1 = S(T')u_0 + \int_0^{T'} S(T' - s)B(s, u(s))\, ds \in D(A).$$

Now, if $T' \leqslant t \leqslant T$, we have

$$u(t) = S(t - T')u_1 + \int_{T'}^t S(t - s)B(s, u(s))\, ds$$

and we can argue exactly as before to deduce that u is Lipschitz continuous in $T' \leqslant t \leqslant 2T'$, thus a genuine solution in $T' \leqslant t \leqslant 2T'$. A finite number of steps then cover the interval $0 \leqslant t \leqslant T$.

In the following result, a rather strong assumption on S is traded off by a weakening of the requirements on B: however, only a local solution is obtained and uniqueness is lost.

2.5.11   Example (Pazy [1974: 1, p. 124]).   Assume $S(t)$ is compact for $t > 0$, and let B be continuous in $[0, T] \times E$. Then there exists $T' > 0$ (depending on u_0) such that (2.5.60) has a continuous solution in $0 \leqslant t \leqslant T'$.

The proof is an application of the Schauder-Tikhonov fixed point theorem to the operator

$$(\mathfrak{S}u(\cdot))(t) = S(t)u_0 + \int_0^t S(t - s)B(s, u(s))\, ds. \qquad (2.5.63)$$

We point out that the solution may not be unique (take E one-dimensional, $A = 0$, $B(s, u) = \sqrt{u}$). On the other hand, if only continuity of B is postulated, the mere assumption that $A \in \mathcal{C}_+$ may fail to produce solutions of (2.5.63); to see this it suffices to take $A = 0$ and make use of a celebrated counterexample of Dieudonné [1950: 1]. In a substantially improved version due to Godunov, it can be formulated as follows:

*2.5.12   Example (Godunov [1974: 1]).   There exists a continuous map $B: (-\infty, \infty) \times H \to H$ (H a separable Hilbert space) with the following property: for every real t_0 and $u \in H$, the initial-value problem

$$u'(t) = B(t, u(t)), \quad u(t_0) = u \qquad (2.5.64)$$

(or, equivalently, the integral equation (2.5.61) with $S(t) \equiv I$) has no solution in any interval containing t_0.

On the other hand, even continuity of B in the topology of E may be an excessively strong assumption, especially in view of the applications where A is a linear differential operator and B a (lower order) perturbation of A. Several results exist where (Hölder) continuity properties of B are postulated in the graph norm of fractional powers of A (see Sobolevskiĭ [1960: 1], [1960: 2], [1961: 1], [1964: 1], Kato-Fujita [1962: 1], Henry [1978: 1], Pecher [1975: 1], and Webb [1977: 2]. Other type of results where the assumptions are on $S(t)B$ instead than on B itself have been proved by Weissler [1979: 1]. A detailed treatment of the equation (2.5.60) can be found in Martin [1976: 1]. Other references on different types of semilinear equations are Aširov [1972: 1], Ahundov [1972: 1], Ahundov-Jakubov [1969: 1], [1970: 1], Ardito-Ricciardi [1974: 1], Akinyele [1976: 1], Bartak [1976: 1], Biroli [1972: 1], [1974: 1], Brézis-Strauss [1973: 1], Calvert [1976: 1], Chafee [1977: 1], Daleckiĭ-Kreĭn [1970: 1], Datko [1968: 1], [1970: 1], [1970: 2], [1972: 1], De Blasi-Myjak [1977: 1], Domšlak [1962: 1], [1965: 1], [1975: 1], Enikeeva [1968: 1], [1969: 1], [1969: 2], [1972: 1], Gel'man-Gerst [1968: 1], Gajewski [1971: 1], [1972: 1], Heinz-Von Wahl [1975: 1], Hughes-Kato-Marsden [1976: 1], Iannelli [1976: 1], Iooss [1972: 1], Jakubov [1966: 3], [1967: 2], [1967: 3], [1970: 3], [1970: 4], Jakubov-Aliev [1976: 1], Jakubov-Ismailova [1974: 1], Kakita [1974/75: 1], Kartsatos [1969: 1], Kielhöfer [1973: 1], [1975: 1], Kirjanen [1974: 1], Kluge-Bruckner [1974: 1], Konishi [1976: 2], Kreĭn [1967: 1], Lightbourne [1976: 1], Lunin [1973: 1], [1974: 1], Mamedov [1964: 2], [1965: 1], Mamiĭ [1966: 1], [1967: 1], [1967: 2], Mamiĭ-Mirzov [1971: 1], Martin [1975: 1], [1977: 1], Massey [1976: 1], [1977: 1], Monari [1971: 1], Nikloenko [1973: 1], Pavel [1972: 1], [1974: 1], [1974: 2], Pazy [1972: 1], [1975: 1], Pao [1972: 1], [1973: 1], [1974/75: 1], Pogorelenko-Sobolevskiĭ [1967: 2], [1967: 3], [1970: 1], [1972: 1], Raskin [1973: 1], Raskin-Sobolevskiĭ [1968: 1], [1968: 2], Rautmann [1973: 1], Reichelt [1975: 1], Ricciardi-Tubaro [1973: 1], Simonenko [1970: 1], [1973: 1], Sobolevskiĭ [1958: 3], [1968: 5], Tsutsumi [1971: 1], [1972: 1], Veliev-Mamedov [1973: 1], [1974: 1], Vilella-Bressan [1974: 1], Ward [1976: 1], Zarubin [1970: 1], [1970: 2], and Lovelady [1973/74: 1].

(l) *Approximation.* Consider the space $E = L^2(0, 2\pi)$. Given $u \in E$, write

$$(\mathfrak{A}(t)u)(x) = \sum e^{-|n|t} a_n e^{inx},$$

where the $\{a_n; -\infty < n < \infty\}$ are the Fourier coefficients of u. We check that $\{\mathfrak{A}(t); t \geq 0\}$ is a strongly continuous semigroup in E (called the *Abel-Poisson semigroup*); the fact that

$$\mathfrak{A}(t)u \to u \text{ as } t \to 0$$

simply expresses that the Fourier series of an element u of E converges in the sense of Abel to u in the L^2 norm. A natural question is how closely does $\mathfrak{A}(t)u$ approximate u for t small; for instance, which are the $u \in E$

such that

$$\|\mathfrak{A}(t)u - u\| = 0(t^{\alpha}) \text{ as } t \to 0+?$$

Similar questions may be formulated in the spaces L^p or C or in relation to different semigroups such as the *Weierstrass semigroup*

$$(\mathfrak{B}(t)u)(x) = \sum e^{-n^2 t} a_n e^{inx}$$

in $(0, 2\pi)$, and

$$(\mathfrak{D}(t)u)(x) = \frac{1}{2\sqrt{\pi t}} \int_{-\infty}^{\infty} e^{-(x-\xi)^2/4t} u(\xi) \, d\xi$$

in $(-\infty, \infty)$ and the *Cauchy-Poisson semigroup*[3]

$$(\mathfrak{X}(t)u)(x) = \frac{t}{\pi} \int_{-\infty}^{\infty} \frac{u(x-\xi)}{t^2 + \xi^2} \, d\xi$$

also in $(-\infty, \infty)$. Other problems are, say, the rate of convergence of exponential formulas (see the beginning of this section). These questions can be formulated for an arbitrary semigroup in a general Banach (or linear topological) space. This point of view originates in Hille [1936: 1]; several results of this sort can be found in Hille [1948: 1]. An exhaustive study is made in the treatise of Butzer and Berens [1967: 1] with a more detailed history of the subject and a long list of references. See also Butzer-Nessel [1971: 1]. Some of the more recent works are Butzer-Pawelke [1968: 1], Butzer-Westphal [1972: 1], Trebels-Westphal [1972: 1], Westphal [1968: 1], [1970: 1], [1970: 2], [1974: 1], [1974: 2], Berens-Butzer-Westphal [1968: 1], Berens-Westphal [1968: 1], Ditzian [1969: 1], [1970: 1], [1971: 1], [1971: 2], Ditzian-May [1973: 1], Köhnen [1972: 1], Terehin [1974: 1], [1975: 1], and Golovkin [1969: 1].

---

[3] Both Weierstrass semigroups express the solution of the heat equation in function of the initial condition, the first in $(0, 2\pi)$, the second in $(-\infty, \infty)$. The Cauchy-Poisson semigroup produces the only harmonic function $u(x, t)$ bounded in mean in the half plane $t > 0$ which takes the boundary value $u(x)$ for $t = 0$.

Chapter 3

Dissipative Operators and Applications

If we restrict our attention to equations $u'(t) = Au(t)$ having solutions whose norm does not increase with time, the theory of the Cauchy problem in Chapter 2 becomes simpler and more incisive. The resulting theory of dissipative operators is treated in Sections 3.1 and 3.6, with applications to second order ordinary differential operators and to symmetric hyperbolic operators in the rest of the chapter.

Section 3.7 develops a theory whose main application is to study partial differential (and more general) equations that possess the property of conserving nonnegative solutions.

3.1. DISSIPATIVE OPERATORS

As pointed out in Section 2.1, direct application of Theorem 2.1.1 is in general difficult, since it involves verification of the infinite set of inequalities (2.1.11), or at least of their "real" counterparts (2.1.29). When $C = 1$, however, only the first inequality needs to be checked. Since $A - \omega I \in \mathcal{C}(1,0)$ if and only if $A \in \mathcal{C}(1, \omega)$ (this can be seen observing that solutions of $u' = Au$ are mapped into solutions of $v' = (A - \omega I)v$ through $v(t) = e^{-\omega t}u(t)$), we may restrict ourselves to the case $\omega = 0$. We state the resulting particular case of Theorem 2.1.1.

117

3.1.1 Theorem. *The closed, densely defined operator A belongs to* $\mathcal{C}_+(1,0)$ *if and only if* $(0,\infty) \subseteq \rho(A)$ *and*

$$\|R(\lambda)\| \leqslant 1/\lambda \quad (\lambda > 0). \tag{3.1.1}$$

We note that the solution operator $S(\cdot)$ of

$$u'(t) = Au(t) \tag{3.1.2}$$

satisfies

$$\|S(t)\| \leqslant 1 \quad (t \geqslant 0). \tag{3.1.3}$$

The semigroup $S(\cdot)$ is called a *contraction semigroup*. If (3.1.2) is a model for a physical system and $\|\cdot\|$ measures, say, the energy of the states, then (3.1.3) means that as a state evolutions in time, its energy decreases (or at least, does not increase). This situation is of course common in practice.

The result corresponding to the Cauchy problem in $(-\infty,\infty)$ is

3.1.2 Theorem. *The closed, densely defined operator A belongs to* $\mathcal{C}(1,0)$ *if and only if* $(-\infty,0)\cup(0,\infty) \subseteq \rho(A)$, *and, for* λ *real,*

$$\|R(\lambda)\| \leqslant 1/|\lambda| \quad (\lambda \neq 0). \tag{3.1.4}$$

Note that in the case covered by Theorem 3.1.1, $\sigma(A)$ must be contained in the left half plane $\operatorname{Re}\lambda \leqslant 0$, whereas if A satisfies the hypotheses of Theorem 3.1.2, $\sigma(A)$ is a subset of the imaginary axis.

Theorem 3.1.1 can be cast in a considerably more convenient (and intuitive) form by using the notion of dissipative operator, which we define below after some preparatory work.

For $u \in E$ define the *duality set* of u as

$$\Theta(u) = \{u^* \in E^*; \|u^*\|^2 = \|u\|^2 = \langle u^*, u \rangle\}.$$

Clearly $\Theta(0) = \{0\}$. It follows from the material in Section 2 (Corollary 2.3) that $\Theta(u)$ is nonempty for all u. It is also true that each $\Theta(u)$ is a closed convex set. Closedness is immediate. On the other hand, let $u_1^*, u_2^* \in \Theta(u)$ and $u^* = \alpha u_1^* + (1-\alpha)u_2^*$ with $0 < \alpha < 1$. Then $\langle u^*, u \rangle = \|u\|^2$, while $\|u^*\| \leqslant \|u\|$. It follows that $\|u^*\| = \|u\|$ and thus that $u^* \in \Theta(u)$.

3.1.3 Example. Let E^* be *strictly convex* ($\|u^* + v^*\| < 2$ if $\|u^*\| = \|v^*\| = 1, u^* \neq v^*$). Then Θ is single-valued.

3.1.4 Example. Assume in addition that E^* is *uniformly convex* (if $\{u_n^*\}$, $\{v_n^*\}$ are such that $\|u_n^*\| = \|v_n^*\| = 1$, $\|u_n^* + v_n^*\| \to 2$ implies $\|u_n^* - v_n^*\| \to 0$). Then Θ is continuous.

3.1.5 Example. Assume there exists a map $\theta: E \to E^*$ such that $\theta(u) \in \Theta(u)$ and θ is conjugate linear ($\theta(\lambda u^* + \mu v^*) = \bar{\lambda}\theta(u^*) + \bar{\mu}\theta(v^*)$ for $u^*, v^* \in E^*$ and complex λ, μ). Then E is a Hilbert space. (*Hint:* Define $2(u,v) = \langle \overline{\theta(v)}, u \rangle + \langle \theta(u), v \rangle$.) The converse is of course true.

3.1.6 Example. Let Ω be a Borel subset of Euclidean space, and let $E = L^p(\Omega)$ $(1 \leqslant p < \infty)$. Then:

(a) If $1 < p < \infty$, Θ is single-valued and, if $u \neq 0$,

$$\Theta(u)(x) = \|u\|_p^{2-p} |u(x)|^{p-2} \bar{u}(x)$$

(where we make the convention that $|u(x)|^{p-2}\bar{u}(x)$ if $u(x) = 0$).

(b) If $p = 1$, $u \neq 0$, $\Theta(u)$ consists of all $u^* \in L^\infty(\Omega)$ with

$$u^*(x) = \|u\|_1 |u(x)|^{-1} \bar{u}(x) \quad (u(x) \neq 0),$$

$$|u^*(x)| \leqslant \|u\|_1 \text{ a.e.} \quad (u(x) = 0).$$

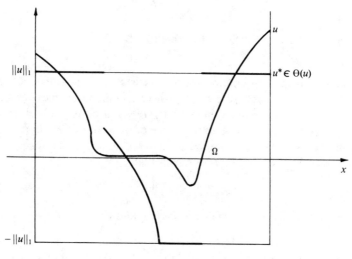

FIGURE 3.1.1

3.1.7 Example. (a) Let K be a compact subset of Euclidean space $\mathbb{R}^m$, and let $E = C(K)$ (Section 7). If $u \in C(K)$, let

$$m(u) = \{x \in K; |u(x)| = \|u\|\}.$$

Then $m(u)$ is nonempty and $\Theta(u)$ consists of all measures $\mu \in \Sigma(K)$ supported by $m(u)$ and such that

$$\int_e u(x)\mu(dx) \geqslant 0$$

for every Borel set $e \in m(u)$ (i.e., such that $u(x)\mu(dx)$ is a positive measure) and

$$\int_K |\mu|(dx) = \|u\|.$$

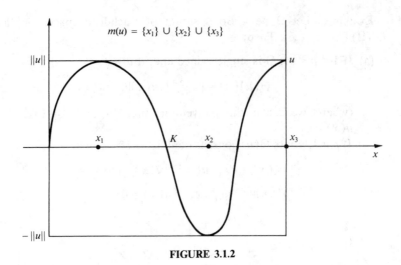

FIGURE 3.1.2

For instance, in the figure above, $\Theta(u)$ consists of all measures of the form $\frac{1}{2}\|u\|(\alpha\delta_1 - \delta_2 + (1-\alpha)\delta_3)$ $(0 \leqslant \alpha \leqslant 1)$, where δ_j is the Dirac measure at x_j.

(b) The same result applies to $C_J(K)$, J a closed subset of K (see Section 7), where $C_J(K)$ denotes the space of those $u \in C(K)$ that vanish in J, or, in case K is an unbounded closed set, to $C_{0,J}$, the space of all u continuous in K vanishing in J and such that

$$\lim_{|x| \to \infty} |u(x)| = 0.$$

A map $\theta: E \to E^*$ is called a *duality map* if

$$\theta(u) \in \Theta(u) \quad (u \in E).$$

An operator A (not necessarily closed or densely defined) is said to be *dissipative* if

$$\operatorname{Re}\langle u^*, Au \rangle \leqslant 0 \quad (u \in D(A), u^* \in \Theta(u)) \tag{3.1.5}$$

and *dissipative with respect to θ* (θ a given duality map) if

$$\operatorname{Re}\langle \theta(u), Au \rangle \leqslant 0 \quad (u \in D(A)). \tag{3.1.6}$$

It is plain that "dissipative with respect to θ" is in general a weaker notion than "dissipative," although the two notions coincide in a Hilbert space (where the only duality map is the identity) or in a space with a strictly convex dual (Example 3.1.3). Observe, incidentally, that in a Hilbert space inequalities (3.1.5) and (3.1.6) coalesce into

$$\operatorname{Re}(Au, u) \leqslant 0 \quad (u \in D(A)). \tag{3.1.7}$$

Note also that, in the general case, dissipativity of an operator with respect to a duality map is totally unaffected by the values of the duality map outside of $D(A)$.

Before considering examples of dissipative operators we justify their introduction in the context of the Cauchy problem.

3.1.8   Theorem.   *Let $A \in \mathcal{C}_+(1,0)$. Then A is dissipative and*

$$(\lambda I - A) D(A) = E \quad (\lambda > 0). \tag{3.1.8}$$

Conversely, let A be densely defined, dissipative with respect to some duality map θ, and let (3.1.8) be satisfied for some $\lambda_0 > 0$. Then $A \in \mathcal{C}_+(1,0)$.

Proof.   Let $A \in \mathcal{C}_+(1,0)$, S the evolution operator of (3.1.2), $u \in D(A)$, $u^* \in \Theta(u)$. We have

$$\operatorname{Re}\langle u^*, S(h)u \rangle \leqslant \|u^*\| \, \|S(h)\| \, \|u\| \leqslant \|u\|^2 \quad (h > 0),$$

hence

$$\operatorname{Re}\left\langle u^*, \frac{1}{h}(S(h)u - u)\right\rangle = \operatorname{Re}\frac{1}{h}\langle u^*, S(h)u \rangle - \frac{1}{h}\|u\|^2 \leqslant 0.$$

Letting $h \to 0$, inequality (3.1.5) follows. That (3.1.8) holds is evident, since every positive λ must belong to $\rho(A)$.

To prove the converse, let $\lambda > 0$ and $u \in D(A)$. Write

$$v = \lambda u - Au.$$

Then

$$\langle \theta(u), v \rangle = \lambda \langle \theta(u), u \rangle - \langle \theta(u), Au \rangle$$

so that in view of (3.1.6),

$$\lambda \|u\|^2 = \lambda \langle \theta(u), u \rangle = \operatorname{Re}\lambda \langle \theta(u), u \rangle$$
$$\leqslant \operatorname{Re}\langle \theta(u), v \rangle \leqslant \|\theta(u)\| \, \|v\| = \|u\| \, \|v\|.$$

If $u \neq 0$, it follows that $\lambda \|u\| \leqslant \|v\|$ (the inequality is obviously true if $u = 0$). Hence

$$\|u\| \leqslant \frac{1}{\lambda} \|(\lambda I - A)u\| \quad (u \in D(A)). \tag{3.1.9}$$

Inequality (3.1.9) plainly implies that $(\lambda I - A)^{-1}$ exists and is a bounded operator, although it does not follow that $(\lambda I - A)^{-1}$ is everywhere defined. (We do not yet know that $(\lambda I - A)D(A) = E$.) However, since (3.1.8) is assumed to hold for $\lambda = \lambda_0$, it results that $(\lambda_0 I - A)^{-1}$ is everywhere defined, then $\lambda_0 \in \rho(A)$ and $\|R(\lambda_0)\| < 1/\lambda_0$. By virtue of the results in Section 3, $R(\lambda)$ must then exist in $|\lambda - \lambda_0| < \lambda_0$, and, again because of (3.1.9), $\|R(\lambda)\| < 1/\lambda$ there; making use of the same argument for $\lambda_1 = 3\lambda_0/2, \lambda_2 = 5\lambda_0/2, \ldots$, we see that $R(\lambda)$ exists for every $\lambda > 0$ and satisfies

$$\|R(\lambda)\| < 1/\lambda \quad (\lambda > 0),$$

which places A under the hypotheses of Theorem 3.1.1. This completes the proof of Theorem 3.1.8.

The result corresponding to the Cauchy problem in $(-\infty, \infty)$ is

3.1.9   Theorem.   *Let $A \in \mathcal{C}(1,0)$. Then*

$$\operatorname{Re}\langle u^*, Au \rangle = 0 \quad (u \in D(A), u^* \in \Theta(u)) \tag{3.1.10}$$

and (3.1.8) *holds for all real* λ, $\lambda \neq 0$. *Conversely, assume that A is densely defined, that* (3.1.10) *is satisfied, and that* (3.1.8) *holds for some* $\lambda_0 > 0$ *as well as for some* $\lambda_1 < 0$. *Then* $A \in \mathcal{C}(1, 0)$.

The proof follows immediately from Theorem 3.1.8 and from the fact (already used in the previous chapter) that $A \in \mathcal{C}(C, \omega)$ if and only if $A, - A \in \mathcal{C}_+(C, \omega)$. We note that the fact that $\|S(t)\| \leqslant 1$ in $-\infty < t < \infty$ actually implies that each $S(t)$ is an *isometry*:

$$\|S(t)u\| = \|u\| \quad (-\infty < t < \infty). \tag{3.1.11}$$

In fact, if (3.1.11) is false for some $u \in E$ and some t_0, we would have $\|u\| = \|S(-t_0)S(t_0)u\| \leqslant \|S(t_0)u\| < \|u\|$, which is a contradiction.

3.1.10 Example. Let $A \in \mathcal{C}_+(C, \omega)$. Then the space E can be renormed (with a norm equivalent to the original one) in such a way that $A \in \mathcal{C}_+(1, \omega)$. (*Hint:* If $\omega = 0$, use $\|u\|' = \sup\{\|S(t)u\|; \ t \geqslant 0\}$). The same result holds in $\mathcal{C}(C, \omega)$.

In many of the future applications of Theorem 3.1.8, the operator A will be at first defined in a set of smooth functions, where verification of dissipativity usually reduces to integration by parts in one way of another. However, equality 3.1.8 may not hold, and in order to remedy this it will be necessary to extend the domain of definition of A; the crucial step is to show that dissipativity is not lost in the extension. The following result will be useful in this connection.

3.1.11 Lemma. *Let* A_0 *be densely defined and dissipative with respect to a duality map* $\theta_0 \colon D(A_0) \to E^*$. *Then* (a) A_0 *is closable.* (b) *Let* $A = \bar{A}_0$. *Then there exists a duality map* $\theta \colon D(A) \to E^*$ *whose restriction to* $D(A_0)$ *coincides with* θ_0 *such that* A *is dissipative with respect to* θ.

Proof. We begin by observing that if $\{u_n\}$ is a sequence in $D(A_0)$ such that $u_n \to u \in E$ and such that $\theta(u_n)$ is E-weakly convergent to some $u^* \in E^*$ (which can always be achieved by passing to a subsequence; see Theorem 5.3),[1] then

$$\|u^*\| \leqslant \lim\|\theta_0(u_n)\| = \lim\|u_n\| = \|u\|.$$

On the other hand,

$$\langle u^*, u \rangle = \lim\langle \theta_0(u_n), u_n \rangle$$
$$= \lim\|u_n\|^2 = \|u\|^2.$$

These two equalities imply that $\|u^*\| = \|u\|$ and that $u^* \in \Theta(u)$.

---

[1]Strictly speaking a subnet, unless E is separable (see Corollary 5.3). The same observation applies to subsequent arguments.

We prove first that A_0 is closable. Assume this fails to be the case. Then there exists a sequence $\{u_n\}$ such that $u_n \to 0$, $A_0 u_n \to v \neq 0$. Choose $u \in D(A_0)$ such that

$$\|u - v\| < \|u\| \tag{3.1.12}$$

and let ζ be an arbitrary complex number. Making use of the considerations at the beginning of the proof, we may assume that $\theta_0(u + \zeta u_n)$ converges weakly to some $u^* \in \Theta(u)$. Then

$$\mathrm{Re}\langle u^*, A_0 u \rangle + \mathrm{Re}\,\zeta \langle u^*, v \rangle = \lim \mathrm{Re}\langle \theta(u + \zeta u_n), A_0(u + \zeta u_n) \rangle \leqslant 0.$$

Since ζ is arbitrary, we must have

$$\langle u^*, v \rangle = 0. \tag{3.1.13}$$

However $\langle u^*, v \rangle = \langle u^*, u \rangle + \langle u^*, v - u \rangle$, which contradicts the previous equality since, in view of (3.1.12), $|\langle u^*, v - u \rangle| < \|u\|^2$.

Let A be the closure of A_0. We extend θ_0 to $D(A)$ as follows: if $u \in D(A_0)$, set $\theta(u) = \theta_0(u)$; if $u \in D(A) \setminus D(A_0)$, select a sequence $\{u_n\}$ in $D(A_0)$ such that $u_n \to u$, $A_0 u_n \to Au$ and $\{\theta_0(u_n)\}$ is weakly convergent to some u^* in E^*, and define $\theta(u) = u^*$ (we have already noted that u^* must belong to $\Theta(u)$). Then

$$\langle \theta(u), Au \rangle = \lim \langle \theta_0(u_n), A_0 u_n \rangle,$$

which plainly shows that A is dissipative with respect to θ. (The extension of θ_0 outside of $D(A_0)$ is of course far from unique in general.) This ends the proof of Lemma 3.1.11.

For future reference, densely defined dissipative operators satisfying (3.1.8) for some $\lambda > 0$ (or, equivalently, operators in $\mathcal{C}_+(1,0)$) will be called *m-dissipative*.

3.1.12 Remark. Let A be a *closed*, densely defined dissipative operator. It follows from the fundamental inequality (3.1.9) that $(\lambda I - A)D(A)$ is closed. This shows that, in order to establish (3.1.8) we only have to prove that $(\lambda I - A)D(A)$ is dense in E, that is,

$$\overline{(\lambda I - A)D(A)} = E. \tag{3.1.14}$$

Condition (3.1.14) can be given yet another equivalent form. Clearly $(\lambda I - A)D(A)$ fails to be dense in E if and only if there exists a $u^* \in E^*$, $u^* \neq 0$ such that

$$\langle u^*, \lambda u - Au \rangle = 0 \quad (u \in D(A)). \tag{3.1.15}$$

But (3.1.15) is in turn equivalent to: $u^* \in D(A^*)$ and

$$(\lambda I - A^*)u^* = 0 \tag{3.1.16}$$

so that (3.1.8) holds if and only if the equation (3.1.16) has no nontrivial solutions. A sufficient condition for this to happen is that A^* itself be dissipative with respect to some duality map $\theta^*: D(A^*) \to E^{**}$, for in this case inequality (3.1.9) for A^* shows that (3.1.16) implies $u^* = 0$. Note that it is not necessary here to require that $D(A^*)$ be dense in E^*.

If A is not closed, the preceding remarks can be applied to its closure (which exists because of Lemma 3.1.10). It follows that, if A_0 is densely defined and dissipative and $(\lambda I - A_0)D(A)$ is dense in E, then $A = \bar{A}_0$ is m-dissipative, that is, $A \in \mathcal{C}_+(1,0)$.

3.1.13 Remark. In many applications (say, in the theory of the heat equation) it is natural to consider spaces of *real-valued* rather than complex-valued functions. Since the theory heretofore developed works in complex spaces, we indicate how to fit real spaces in the complex theory.

Let E be a real Banach space. The *complexification* of E is the set E_C ("$E \oplus iE$") of all pairs (u, v) ("$u + iv$") of elements of E. Addition and multiplication by complex scalars $\lambda = \xi + i\eta$ are defined by $(u_1, v_1) + (u_2, v_2) = (u_1 + u_2, v_1 + v_2)$ and $(\xi, \eta)(u, v) = (\xi u - \eta v, \eta u + \xi v)$, whereas the norm is any of the usual product norms. The following modifications to the various results must be made.

Theorem 2.1.1. The statement is the same, except that inequalities (2.1.11) are replaced by their real counterparts. Necessity is shown in the same way (except that $\lambda > \omega$ is real). To prove sufficiency, we introduce an operator A_C in E_C by

$$A_C(u, v) = (Au, Av), \tag{3.1.17}$$

with $D(A_C) = D(A) \times D(A)$; clearly A_C is densely defined in E_C since A is densely defined in E. If I (resp. I_C) is the identity operator in E (resp. E_C) and λ is real, then $(\lambda I_C - A_C)(u, v) = ((\lambda I - A)u, (\lambda I - A)v)$, hence it follows that $\rho(A) = \rho(A_C) \cap \mathbb{R}$, $R(\lambda; A_C)(u, v) = (R(\lambda; A)u, R(\lambda; A)v)$ and the real inequalities (2.1.29) for A_C follow from the corresponding inequalities for A (with the same constants C, ω). It follows then that $A_C \in \mathcal{C}_+(C, \omega)$ in E_C. Since $u_C(\cdot) = (u(\cdot), v(\cdot))$ is a solution of $u_C'(t) = A_C u_C(t)$ if and only if $u(\cdot), v(\cdot)$ are solutions of $u'(t) = Au(t)$, we deduce that $A \in \mathcal{C}_+(C, \omega)$ as well.

Lemma 2.3.5, Lemma 2.3.9, Theorem 3.1.1, Theorem 3.1.2. Same statements and proofs.

Theorem 2.2.1. The statement and proof are the same, except that inequalities (2.2.3) are replaced by their real counterparts

$$\|R(\mu)^n\| \leq C(|\mu| - \omega)^{-n} \quad (|\mu| > \omega, n \geq 1). \tag{3.1.18}$$

Lemma 2.2.3, Theorem 2.2.9, Theorem 2.2.11, Theorem 2.3.1, Theorem 2.3.3. No change in statements or proofs.

Theorems 2.3.2, 2.3.4. Statements and proofs are the same but inequalities (2.3.13) and (2.3.15) are replaced by their real counterparts.

In a real Banach space, an operator A is *dissipative* if

$$\langle u^*, Au \rangle \leqslant 0 \quad (u \in D(A), u^* \in \Theta(u)) \tag{3.1.19}$$

and *dissipative with respect to θ* (θ a given duality map) if

$$\langle \theta(u), Au \rangle \leqslant 0 \quad (u \in D(A)). \tag{3.1.20}$$

Theorems 3.1.8, 3.1.9, Lemma 3.1.11. Same statements and proofs.

In the future, we shall mostly (but not always) consider complex spaces and leave it to the reader the formulation of the real counterparts of subsequent results (whenever applicable); in most instances the complex setting yields additional information.

We close this section with an observation about adjoints that is an immediate consequence of Theorem 3.1.8 and of the duality theory in Section 2.2, especially Theorems 2.2.9 and 2.2.11; the formulation and proofs are identical in the real or the complex case.

3.1.14 Theorem. *Let A be m-dissipative in E. Then the Phillips adjoint $A^{\#}$ defined in Theorem 2.2.11 is m-dissipative in $E^{\#}$. If E is reflexive, $A^* = A^{\#}$ is m-dissipative in $E^* = E^{\#}$.*

3.1.15 Example (Bishop-Phelps [1961: 1]; see also Ekeland [1979: 1]). If E is an arbitrary Banach space, the set $\Theta = \cup\{\Theta(u); u \in E\}$ is dense in E^*.

3.1.16 Example (James [1963: 1]). The Banach space E is reflexive if and only if

$$\Theta = E^*.$$

3.1.17 Example. Check Theorem 3.1.8 for the operators in Example 2.2.7. Check Theorem 3.1.9 for the operators in Examples 2.2.4 and 2.2.5.

3.2. ORDINARY DIFFERENTIAL OPERATORS IN THE WHOLE LINE

As an illustration of the theory in the preceding section, we examine the ordinary differential operator

$$A_0 u(x) = a(x)u''(x) + b(x)u'(x) + c(x)u(x). \tag{3.2.1}$$

The coefficients a, b, c are real-valued. We shall consider A_0 in the *complex* spaces $L_{\mathbb{C}}^p$, $C_{0,\mathbb{C}}$; although this is not always called for in applications, it will yield important additional information about A_0. Since all spaces under consideration are complex, we eliminate the subscript $\mathbb{C}$ throughout.

Our standing assumptions on the coefficients a, b, c are: a is twice continuously differentiable, b is continuously differentiable, c is continuous in $-\infty < x < \infty$.

We begin by considering the space $E = C_0(-\infty, \infty)$ and define $D(A_0)$ there as the set of all twice continuously differentiable u such that $A_0 u \in C_0$.

3.2.1 Lemma. *The operator A_0 is dissipative (dissipative with respect to a duality map θ) in C_0 if and only if*

$$a(x) \geqslant 0, \quad c(x) \leqslant 0 \quad (-\infty < x < \infty). \qquad (3.2.2)$$

Proof. Let $u \in D(A_0)$, $u \neq 0$,

$$m(u) = \{x \in \mathbb{R}; \ |u(x)| = \|u\|\}. \qquad (3.2.3)$$

We have already seen (Example 3.1.7) that $\Theta(u)$ consists of all measures $\mu \in \Sigma$ supported by $m(u)$, such that $u(x)\mu(dx)$ is a positive measure in $m(u)$ and $\int |\mu|(dx) = \|u\|$. We have

$$\langle \mu, A_0 u \rangle$$

$$= \int_{m(u)} A_0 u(x)\mu(dx)$$

$$= \int_{m(u)} u(x)^{-1}(a(x)u''(x) + b(x)u'(x) + c(x)u(x))u(x)\mu(dx). \qquad (3.2.4)$$

Write $u = u_1 + iu_2$ with u_1, u_2 real. Since $|u|^2 = u_1^2 + u_2^2$ reaches its maximum in $m(u)$, we must have there $\frac{1}{2}(|u|^2)' = u_1 u_1' + u_2 u_2' = 0$, $\frac{1}{2}(|u|^2)'' = u_1 u_1'' + u_2 u_2'' + u_1'^2 + u_2'^2 \leqslant 0$. Accordingly, if $x \in m(u)$,

$$u^{-1}u' = i\|u\|^{-2}(u_1 u_2' - u_1' u_2), \qquad (3.2.5)$$

$$u^{-1}u'' = \|u\|^{-2}(u_1 u_1'' + u_2 u_2'') + i\|u\|^{-2}(u_1 u_2'' - u_1'' u_2), \qquad (3.2.6)$$

where

$$u_1 u_1'' + u_2 u_2'' \leqslant -\left(u_1'^2 + u_2'^2\right). \qquad (3.2.7)$$

Replacing (3.2.5) and (3.2.6) into (3.2.4) and making use of inequality (3.2.7), we obtain

$$\mathrm{Re}\langle \mu, A_0 u \rangle = \leqslant -\int_{m(u)} (\|u\|^{-2}a(x)|u'(x)|^2 - c(x))u(x)\mu(dx) \leqslant 0$$

if inequalities (3.2.2) hold. To prove the converse let x_0, α, β be three arbitrary real numbers. If $\alpha > 0$, $\beta \leqslant 0$ it is immediate that there exists a nonnegative function $u \in D(A_0)$ having a single maximum at x_0 and such that $u(x_0) = \alpha$, $u''(x_0) = \beta$. Then $\Theta(u) = \{\alpha\delta_0\}$, where δ_0 is the Dirac measure at x_0. We have

$$\langle \delta_0, A_0 u \rangle = a(x_0)\beta + c(x_0)\alpha.$$

If one of the conditions (3.2.2) fails to hold, it is clear that we can find α, β making $\langle \delta_0, A_0 u \rangle$ positive. This ends the proof of Lemma 3.2.1.

We consider next the operator A_0 in the spaces $E = L^p(-\infty, \infty)$, $1 \leqslant p < \infty$. In this case the definition of $D(A_0)$ is not critical since A_0 will have to be extended later. We take $D(A_0) = \mathcal{D}_\mathbb{C}$, the space of complex-valued Schwartz test functions.

3.2.2 Lemma. *The operator A_0 is dissipative in L^p $(1 \leqslant p < \infty)$ if*

$$a(x) \geqslant 0, \quad c(x) + \frac{1}{p}(a''(x) - b'(x)) \leqslant 0 \quad (-\infty < x < \infty). \quad (3.2.8)$$

The first inequality (3.2.8) is necessary in all cases; the second is necessary when $p = 1$.

Proof. Recall that the only duality map $\theta : L^p \to L^{p'} (p'^{-1} + p^{-1} = 1)$ is $\theta(u) = \|u\|^{2-p} |u(x)|^{p-2} \bar{u}(x)$ $(u \neq 0)$. It is easy to see that if $p \geqslant 2$ and $u \in \mathcal{D}$, then $\theta(u)$ is continuously differentiable and

$$\|u\|^{p-2} \theta(u)'(x) = (p-2)|u|^{p-4} (u_1 u_1' + u_2 u_2') \bar{u} + |u|^{p-2} \bar{u}'$$
$$= (p-2)|u|^{p-4} \operatorname{Re}(\bar{u}u') \bar{u} + |u|^{p-2} \bar{u}', \quad (3.2.9)$$

while, on the other hand,

$$(|u|^p)' = p|u|^{p-2}(u_1 u_1' + u_2 u_2')$$
$$= p|u|^{p-2} \operatorname{Re}(\bar{u}u'). \quad (3.2.10)$$

Performing two integrations by parts we obtain

$$\|u\|^{p-2} \operatorname{Re}\langle \theta(u), A_0 u \rangle = \operatorname{Re} \int ((au')' + (b-a')u' + cu)|u|^{p-2}\bar{u}\,dx$$
$$= -(p-2)\int a|u|^{p-4}(\operatorname{Re}(\bar{u}u'))^2\,dx$$
$$- \int a|u|^{p-2}|u'|^2\,dx$$
$$+ \frac{1}{p}\int (a'' - b' + pc)|u|^p\,dx \leqslant 0 \quad (3.2.11)$$

under hypotheses (3.2.8).

In the case $1 < p < 2$, a technical difficulty appears; even if $u \in \mathcal{D}$, it is not clear how $\theta(u)$ behaves at "non-analytic" zeros of u,—that is, at points x_0 where all the derivatives of u vanish but where u itself fails to vanish identically in any neighborhood of x_0.

To remedy this, we introduce the class $\mathcal{PC}^{(2)}$ of piecewise analytic, twice continuously differentiable complex valued functions with compact support in $-\infty < x < \infty$. If $u \in \mathcal{PC}^{(2)}$, it is clear that $|u|^{p-2}\bar{u}$ is infinitely differentiable between zeros of u, thus we only have to examine the behavior of its derivative

$$(|u|^{p-2}\bar{u})' = (p-2)|u|^{p-4}(u_1 u_1' + u_2 u_2')\bar{u} + |u|^{p-2}\bar{u}'$$

near an isolated zero x_0. Assuming that u is analytic at x_0, let $m \geqslant 0$ be an integer such that $u(x) = \cdots = u^{(m-1)}(0) = 0$, $u^{(m)}(0) \neq 0$. It follows from Taylor's formula that

$$|\theta(u)'(x)| = O(|x - x_0|^{m(p-1)-1})$$

near x_0, where $m(p-1) > 0$. If the point x_0 is an endpoint of some of the intervals where u is analytic, we simply use two one-sided estimations. It follows that $\theta(u)(x)$ is absolutely continuous and then that the basic inequality (3.2.11) holds for $1 < p < 2$ and $u \in \mathcal{PC}^{(2)}$. To see that the first two terms on the right-hand side add up to a non-positive number, we only have to notice that $|\mathrm{Re}(\bar{u}u')| \leqslant |u||u'|$. To check that $\mathrm{Re}\langle \theta(u), A_0 u \rangle \leqslant 0$ for $u \in \mathcal{D}$, we take a sequence $\{u_n\}$ in $\mathcal{PC}^{(2)}$ such that the supports of the u_n are contained in a bounded set and $u_n \to u$, $u_n' \to u'$, $u_n'' \to u''$ uniformly in $(-\infty, \infty)$, and notice that $A_0 u_n \to A_0 u$ and (passing if necessary to a subsequence) $\theta(u_n) \to \theta(u)$ weakly in $L^{p'}$. Of course we may avoid this last step by simply defining $D(A_0) = \mathcal{PC}^{(2)}$ when $p < 2$, which does not modify at all the ensuing theory.

We deal finally with the case $p = 1$. For $p > 1$ we will denote by θ_p in the next few lines the unique duality map $\theta_p : L^p \to L^{p'}$. It is plain that if $u \in \mathcal{D}$ and $\Theta(u) \in L^\infty$ is the duality set of u as an element of L^1, we have

$$\lim_{p \to 1+} \theta_p(u)(x) = u^*(x) \in \Theta(u) \tag{3.2.12}$$

almost everywhere in the support of u, whereas $\theta_p(u)(x)$ (say, for $1 < p \leqslant 2$) is uniformly bounded. It follows then from the dominated convergence theorem that

$$\int u^*(x) A_0 u(x)\, dx = \lim_{p \to 1+} \int \theta_p(u)(x) A_0 u(x)\, dx, \tag{3.2.13}$$

which, by virtue of inequality (3.2.11), shows $\mathrm{Re}\langle u^*, A_0 u \rangle \leqslant 0$ if conditions (3.2.8) hold with $p = 1$. That this inequality must hold for any $u^* \in \Theta(u)$ is obvious from the fact that two elements of $\Theta(u)$ coincide where $u \neq 0$.

The necessity of the first condition (3.2.8) for $p > 1$ follows from

3.2.3 Example. Let f, g be real-valued continuous functions, and let $p > 1$. Assume that

$$\int \left(fu'^2 |u|^{p-2} + g|u|^p \right) dx \geqslant 0 \tag{3.2.14}$$

for all $u \in \mathcal{D}_\mathbf{R}$ (or $u \in \mathcal{PC}_\mathbf{R}^{(2)}$) with support in some interval Ω. Then

$$f(x) \geqslant 0 \quad (x \in \Omega).$$

In fact, we only have to apply the result above to $f(x) = (p-1)a(x)$, $g(x) = -p^{-1}(a'' - b' + pc)$.

Finally, we show the necessity of conditions (3.2.8) for dissipativity of A_0 in L^1. Let u be a function in $\mathcal{D}_\mathbf{R}$ with support in an interval $[\alpha, \beta]$ and such that $u < 0$ in (α, x_0), $u > 0$ in (x_0, β) for some $x_0 \in (\alpha, \beta)$. Assuming for simplicity that $\|u\|_1 = 1$, if $u^* \in \Theta(u)$, then $u^*(x) = -1$ in (α, x_0), $u^*(x) = 1$ in (x_0, β). Since $u'(\alpha) = u'(\beta) = 0$, we have

$$\langle u^*, Au_0 \rangle = \int_\alpha^\beta \left((au')' + (b - a')u' + cu \right) u^*(x)\, dx$$

$$= -2a(x_0)u'(x_0) + \int_\alpha^\beta (a'' - b' + c)|u|\, dx, \tag{3.2.15}$$

and it is clear that we can manipulate α, β, u in such a way as to make (3.2.15) positive if either condition (3.2.8) is violated at some point.

3.2.4 Example. The second inequality 3.2.8 is not necessary for dissipativity of A_0 if $p > 1$.

An important role will be subsequently played by the *formal adjoint* of A_0, defined in the same domain as A_0 by the familiar expression

$$A_0'u = (au)'' - (bu)' + cu$$
$$= au'' + (2a' - b)u' + (a'' - b' + c)u$$
$$= \tilde{a}u'' + \tilde{b}u' + \tilde{c}u.$$

The operator A_0' satisfies the *Lagrange identity*

$$\int (A_0 u) v \, dx = \int u A_0' v \, dx \quad (u, v \in \mathcal{D}), \tag{3.2.16}$$

(and is uniquely determined by it). Clearly

$$(A_0')' = A_0.$$

It is easy to see that the two inequalities (3.2.8) are satisfied for an operator A_0 in L^p if and only if they are satisfied for the adjoint A_0' in $L^{p'}$ (in C_0 if $p = 1$).

We examine now the problem of extending A_0 in such a way that dissipativity is preserved and that (3.1.8) holds for some $\lambda > 0$ (in other words, the problem of finding m-dissipative extensions of A_0) in the spaces under consideration. We begin with the case $E = L^p$, $1 < p < \infty$. Let, as usual, $p'^{-1} + p^{-1} = 1$ and, given $\lambda > 0$, denote by $\mathcal{K}_{p', \lambda}$ the subspace of $L^{p'}(-\infty, \infty)$ consisting of all elements of the form $\lambda\varphi - A_0'\varphi$ ($\varphi \in \mathcal{D}$). If A_0' is dissipative in $L^{p'}$, it follows from the fundamental inequality (3.1.9) that

$$\|\varphi\|_{p'} \leq \frac{1}{\lambda} \|\lambda\varphi - A_0'\varphi\|_{p'} \quad (\varphi \in \mathcal{D}),$$

hence the functional

$$\Phi(\lambda\varphi - A_0'\varphi) = \langle v, \varphi \rangle \quad (\varphi \in \mathcal{D}),$$

where v is a fixed element of L^p, is well defined and $\|\Phi\| \leq \lambda^{-1}\|v\|_p$. By the Hahn-Banach theorem this functional can be extended to $L^{p'}$ without increasing its norm; accordingly, there exists a $u \in L^p$ with

$$\langle u, \lambda\varphi - A_0'\varphi \rangle = \langle v, \varphi \rangle \quad (\varphi \in \mathcal{D}) \tag{3.2.17}$$

and

$$\|u\|_p \leq \frac{1}{\lambda} \|v\|_p. \tag{3.2.18}$$

It is clear from the general definition of adjoint operators that (3.2.17) implies that $u \in D((A_0')^*)$ and

$$(\lambda I - (A_0')^*)u = v. \tag{3.2.19}$$

This proves that $(\lambda I - (A'_0)^*)D((A'_0)^*) = E$. In view of (3.2.18), if we could prove that $\lambda I - (A'_0)^*$ is one-to-one, it would follow that λ belongs to $\rho((A'_0)^*)$ and that

$$\|R(\lambda)\| < \frac{1}{\lambda} \quad (\lambda > 0), \qquad (3.2.20)$$

in other words, that $(A'_0)^*$ is m-dissipative. In order to do this we need more precise information about $(A'_0)^*$ and its domain. This information is contained in the following result, where no use is made of the dissipativity of A_0 or A'_0; however, we assume from now on that

$$a(x) > 0 \quad (-\infty < x < \infty). \qquad (3.2.21)$$

3.2.5 Lemma. (a) *Let f be a function that belongs to L^p_{loc} in an open interval Ω, $1 \leqslant p \leqslant \infty$. Let u be a locally summable function that satisfies*

$$au'' + bu' + cu = f \qquad (3.2.22)$$

in the sense of distributions, that is,

$$\int u A'_0 \varphi \, dx = \int f\varphi \, dx \quad (\varphi \in \mathcal{D}(\Omega)).$$

Then u coincides with a function in $W^{2,p}_{\text{loc}}$ and satisfies (3.2.22) almost everywhere. If f is continuous, then u coincides with a twice continuously differentiable function in Ω. (b) *Let v be a locally finite Borel measure in Ω, and assume that a locally finite Borel measure μ satisfies*

$$a\mu'' + b\mu' + c\mu = v \qquad (3.2.23)$$

in the sense of distributions, that is,

$$\int A'_0 \varphi \, d\mu = \int \varphi \, dv \quad (\varphi \in \mathcal{D}(\Omega)).$$

Then μ coincides with a continuous function in Ω.

The proof depends upon solving (3.2.22) by means of the usual "variation of constants" formula (Coddington [1961: 1, p. 68]). In fact, if u_1 and u_2 are two linearly independent solutions of the homogeneous equation $au'' + bu' + cu = 0$ and $W(\xi) = W(u_1, u_2; \xi)$ is their Wronskian, let $G(x, \xi) = (u_1(\xi)u_2(x) - u_1(x)u_2(\xi))/W(\xi)$. Then (3.2.22) has a solution v given by

$$v(x) = \int_{x_0}^x G(x, \xi)f(\xi) \, d\xi,$$

where x_0 is an arbitrary fixed point in Ω. Since G is twice continuously differentiable in x and $G(x, x) = 0$, then v belongs to $W^{2,p}_{\text{loc}}$ if f belongs to L^p_{loc}. If u is the solution of (3.2.22) postulated in (a), then $u - v$ satisfies the homogeneous equation in the sense of distributions ($\int (u - v)A_0\varphi \, dx = 0$ for all $\varphi \in \mathcal{D}$) and (a) will follow if we can prove that $u - v$ must coincide with an ordinary solution of the homogeneous equation. For a proof of this, see Dunford-Schwartz [1963: 1, p. 1291]; although infinitely differentiable coefficients are required there, the argument can be easily extended.

In case (b) we take

$$v(x) = \int_{x_0}^{x} G(x, \xi) \mu(d\xi),$$

which is immediately seen to be continuous, and argue as in (a).

Lemma 3.2.6 allows us to identify readily the adjoint of A_0'; $u \in D((A_0')^*)$ if and only if $u \in W_{loc}^{2,p}(-\infty, \infty) \cap L^p(-\infty, \infty)$ and $(A_0')^*u = au'' + bu' + cu \in L^p(-\infty, \infty)$. The converse part of this assertion follows from Lemma 3.2.6, while the direct half is obtained integrating by parts.

In order to study the uniqueness of the solution of (3.2.19) a further result is needed.

3.2.6 Lemma. *Let u be a solution in* $C^{(2)}(-\infty, \infty)$ *of*

$$au'' + bu' + cu = \lambda u \quad (x \geqslant \alpha), \tag{3.2.24}$$

where $a(x) \geqslant 0$, $c(x) \leqslant 0$ *in* $x \geqslant \alpha$ *and* $\lambda > 0$. *Then u is monotone* (*nonincreasing or nondecreasing*) *in* $x \geqslant \beta$ *for some* $\beta \geqslant \alpha$.

Proof. Assume not. Then both inequalities $u'(x) \leqslant 0$ and $u'(x) \geqslant 0$ are false in $x \geqslant \beta$ for any $\beta \geqslant \alpha$. This is easily seen to imply the existence of three numbers $\alpha < \beta_1 < \beta_2 < \beta_3$ with $u'(\beta_1) > 0$, $u'(\beta_2) < 0$, $u'(\beta_3) > 0$. Let ν_1 be a point in $[\beta_1, \beta_2]$ such that $u(\nu_1) = \max\{u(x); \beta_1 \leqslant x \leqslant \beta_2\}$. Then we have $u(\nu_1) > u(\beta_1), u(\beta_2)$ so that $\nu_1 \neq \beta_1, \beta_2$ and $u'(\nu_1) = 0$. Similarly, if $u(\nu_2) = \inf\{u(x); \beta_2 \leqslant x \leqslant \beta_3\}$, $u(\nu_2) < u(\beta_1), u(\beta_2)$ and $u'(\nu_2) = 0$. Since $u(\nu_1) > u(\nu_2)$, it is plain that we may assume that $u(\nu_1) > 0$ or that $u(\nu_2) < 0$. In the first case, as $u''(\nu_1) \leqslant 0$, we obtain from (3.2.24) that $0 = a(\nu_1)u''(\nu_1) + (c(\nu_1) - \lambda)u(\nu_1) < 0$, absurd. The second case is treated in the same way. This ends the proof.

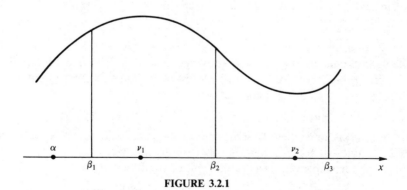

FIGURE 3.2.1

We return now to (3.2.19), assuming that

$$c(x) \leqslant 0. \tag{3.2.25}$$

If u_1, u_2 are two solutions of (3.2.19) in L^p, $u = u_1 - u_2$ satisfies $(A_0')^* u = \lambda u$. Making use of Lemma 3.2.5, we see that u is twice continuously differentiable and satisfies the equation

$$au'' + bu' + cu = \lambda u \qquad (3.2.26)$$

in $(-\infty, \infty)$. It follows from Lemma 3.2.6 applied to the real and imaginary parts of $u(x)$ and $u(-x)$ that they are monotonic for large $|x|$; since u belongs to L^p, this can only be possible if

$$\lim_{|x| \to \infty} |u(x)| = 0,$$

which shows that u belongs to the domain of the operator A_0 in C_0. We obtain then from (3.2.26), from the fact that A_0 is dissipative in C_0 and from inequality (3.1.9) that $u = 0$. This completes the proof that $(A_0')^*$ is m-dissipative in L^p.

We examine next the case $E = C_0(-\infty, \infty)$, under assumptions (3.2.21) and (3.2.25). By Lemma 3.2.1 these conditions imply dissipativity of A_0 in C_0. Our first concern is to show that A_0 (with $D(A_0)$ defined as in Lemma 3.2.1) is closed. In order to do this we consider the operator A_0' in L^1, with domain $\mathcal{D}$. Its adjoint $(A_0')^*$ is a closed operator in L^∞ and it is not difficult to see with the help of Lemma 3.2.5(a) that A_0 is just the restriction of $(A_0')^*$ to the space of all $u \in D((A_0')^*) \cap C_0$ such that $(A_0')^* u \in C_0$ and is therefore closed.

We are forced to assume now that A_0' is dissipative in $L^{p'}$ for some $p' > 1$. In view of the comments following (3.2.16), it suffices to assume that the second inequality (3.2.8) holds for some $p > 1$.

Let v be an element of C_0 with compact support. We can employ the argument previously used in the L^p case to produce a solution of (3.2.19) with $\lambda = 1$. By Lemma 3.2.5 u must then belong to $C^{(2)}$ and satisfy the homogeneous equation (3.2.26) for $|x|$ large enough, and it follows once again that $|u(x)| \to 0$ as $|x| \to \infty$. Hence $u \in D(A_0)$ and satisfies

$$(I - A_0) u = v. \qquad (3.2.27)$$

If v is an arbitrary element of C_0, we take a sequence $\{v_n\}$ in C_0 such that the v_n have compact support and converge to v in C_0. If u_n is the solution of $(I - A_0) u_n = v_n$ constructed in the previous comments, we obtain from dissipativity of A_0 in C_0 that

$$\|u_n - u_m\| \leq \|(I - A_0)(u_n - u_m)\| = \|v_n - v_m\|$$

so that $\{u_n\}$ is convergent to some $u \in C_0$; since A_0 is closed, $u \in D(A_0)$ and (3.2.27) holds. It follows then that A_0 is m-dissipative.

We have proved parts (a) and (b) of the following result.

3.2.7 Theorem. (a) *Let the coefficients of the operator A_0 satisfy*

$$a(x) > 0, \quad c(x) \leq 0 \quad (-\infty < x < \infty) \qquad (3.2.28)$$

and assume A_0' is dissipative in $L^{p'}$ for some $p' > 1$ (for which a sufficient *condition is*

$$c(x) + \frac{1}{p}(a''(x) - b'(x)) \leqslant 0 \quad (-\infty < x < \infty) \qquad (3.2.29)$$

for some $p > 1$). Let A be the operator defined by

$$Au(x) = a(x)u''(x) + b(x)u'(x) + c(x)u(x), \qquad (3.2.30)$$

where $D(A)$ is the set of all $u \in C^{(2)} \cap C_0$ such that (3.2.30) belongs to C_0. Then A is m-dissipative in C_0.

(b) *Let $1 < p < \infty$. Assume inequalities (3.2.28) hold, and suppose A'_0 is dissipative in $L^{p'}$ (for which (3.2.29) is a sufficient condition). Let A be the operator defined by (3.2.30), where $D(A)$ is the set of all $u \in W^{2,p}_{loc} \cap L^p$ such that the right-hand side belongs to L^p. Then A is m-dissipative in L^p.*

(c) *Assume inequalities (3.2.28) hold and*

$$a''(x) - b'(x) + c(x) \leqslant 0 \quad (-\infty < x < \infty). \qquad (3.2.31)$$

Then the operator A defined as in (b) with $p = 1$ is m-dissipative in L^1.

Part (c) follows from (a) through an application of the duality theory in Section 2.2. In fact, the assumptions on the coefficients of A_0 are equivalent to

$$\tilde{a}(x) > 0, \quad \tilde{c}(x) \leqslant 0, \quad \tilde{a}''(x) - \tilde{b}'(x) + \tilde{c}(x) \leqslant 0 \qquad (3.2.32)$$

in $(-\infty, \infty)$. Adding the second inequality to the third, we obtain $\tilde{a}''(x) - \tilde{b}'(x) + 2\tilde{c}(x) \leqslant 0$; hence we are within the assumptions in (a) and the operator A_0 (with the domain corresponding to the space C_0) is m-dissipative in C_0. We apply now Theorem 2.2.11; in order to do this we have to identify $C^\#_0 = \overline{D((A'_0)^*)} \subseteq C^*_0 = \Sigma(-\infty, \infty)$ (see Section 7) and the operator $(A'_0)^\#$. Assume $\mu \in \Sigma = C^*_0$ belongs to the domain of $(A'_0)^*$. Then there exists $\nu \in \Sigma$ such that

$$\int A_0 \varphi(x) \mu(dx) = \int \varphi(x) \nu(dx)$$

for all test functions $\varphi \in \mathcal{D}$. It follows from Lemma 3.2.5(b) that $\mu(dx) = u(x)\,dx$, where u is a continuous function in $\mathbb{R}$. Hence any element of $D((A'_0)^*)$ must be a continuous function. Conversely, it is obvious that if u belongs to $\mathcal{D}$, then $u \in D((A'_0)^*)$ (and $(A'_0)^* u = A_0 u$). Accordingly, the closure of $D((A'_0)^*)$ in the topology of Σ must be L^1 (recall that if $\mu(dx) = u(x)\,dx$, then $\|\mu\|_\Sigma = \|u\|_1$). We have then shown that

$$E^\# = L^1.$$

The operator $(A'_0)^\#$ (which is obviously an extension of A_0) must be m-dissipative in L^1. A precise characterization of $(A'_0)^\#$ can now be obtained. By definition of the #-adjoint and by Lemma 3.2.5, every $u \in D((A'_0)^\#)$ must belong to the domain of the operator A defined in (c) and satisfy $(A_0')^\# u = Au$; in other words,

$$(A_0')^\# \subseteq A.$$

Take now $u_0 \in D(A)$; since $(A_0')^\#$ is m-dissipative there exists $u_1 \in D((A_0')^\#) \subseteq D(A)$ with $(I - A)u_1 = (I - (A_0')^\#)u_1 = (I - A)u_0$. Then $u = u_1 - u_0$ is a solution of (3.2.26) with $\lambda = 1$. It follows from Lemma 3.2.5 that $u \in C^{(2)}$ and from Lemma 3.2.6 that u, being in L^1, must belong to C_0, a fortiori to $D(A_0)$ in C_0. But the hypotheses in (c) imply that A_0 is dissipative in C_0, hence $u = 0$. We have then shown that $(A_0')^\# = A$ and in this way completed the proof of Theorem 3.2.7.

We note that in (a) it is sufficient to assume that $A_0' - cI$ is dissipative in $L^{p'}$ for some c: we only have to take λ sufficiently large in (3.2.19) (see also Remark 3.4.4)

Some additional information on the operators A in Theorem 3.2.8 can be gleaned under supplementary hypotheses by making further use of the duality theory in Section 4. In fact, let $p > 1$ and let A_0 satisfy the assumptions in (c). It is plain that A_0 satisfies the hypotheses in (b) as an operator in L^p and A_0' satisfies the same hypotheses in $L^{p'}$. Applying Theorem 3.2.7, we see that $(A_0')^*$ is a m-dissipative extension of A_0 in L^p and A_0^* is a m-dissipative extension of A_0' in $L^{p'}$. However, it follows from Theorem 3.1.14 that A_0^{**} is also m-dissipative, and, since $A_0^* \supseteq A_0'$,

$$\bar{A}_0 = A_0^{**} \subseteq (A_0')^*.$$

Since $\rho(A_0^{**}) \cap \rho((A_0')^*)$ is not empty (both resolvent sets must contain the half plane $\operatorname{Re} \lambda > 0$), an argument similar to the one closing the proof of Lemma 2.3.9 shows that

$$A = (A_0')^* = \bar{A}_0. \tag{3.2.33}$$

We note finally the particularly interesting case where A_0 is formally self-adjoint,

$$A_0 u = (au')' + cu, \tag{3.2.34}$$

where the hypotheses in any of the three cases in Theorem 3.2.7 (hence also the ones leading to (3.2.33)) reduce to

$$a(x) > 0, \quad c(x) \leqslant 0.$$

It is easy to check that throughout all the theory in this section we have not actually used the fact that $a \in C^{(2)}$ and $b \in C^{(1)}$, but only that $a \in C^{(1)}$ and $a' - b \in C^{(1)}$. Because of this, it is perhaps more convenient to write A_0 in *variational form*,

$$A_0 = (au')' + \hat{b}u' + cu \tag{3.2.35}$$

with the requirement that $a, \hat{b} \in C^{(1)}$, $c \in C^{(0)}$. The whole theory unfolds in the same way with some simplifications in the writing of adjoints and dissipativity conditions.

3.2.8 Example. The dissipativity condition (3.2.2) in C_0 has the same form for (3.2.35) as for (3.2.1). The dissipativity conditions (3.2.8) in L^p, $1 \leqslant p < \infty$, become

$$a(x) \geqslant 0, \quad c(x) - \frac{1}{p}\hat{b}(x) \leqslant 0 \quad (-\infty < x < \infty). \tag{3.2.36}$$

3.2.9 Example. Let $A = A_p$ be the operator in L^p defined in Theorem 3.2.7(b). Denote by A'_p the operators corresponding to the formal adjoint A'_0. Then

$$A'_{p'} = A_p^* \quad (1 < p < \infty) \tag{3.2.37}$$

with $p'^{-1} + p^{-1} = 1$. (We adopt here the hypotheses in (c)).

3.3. ORDINARY DIFFERENTIAL OPERATORS IN A CLOSED INTERVAL. SEMI-INFINITE INTERVALS

We examine here the operator A_0 in a semi-infinite closed interval, which we may assume without loss of generality to be $[0, \infty)$. The treatment makes use essentially of the same technical tools employed in the previous section. The main novelty is that, in order to obtain a domain where the operator is m-dissipative, we will be forced to impose a boundary condition at the left endpoint.

We assume for the coefficients a, b, c the same degree of smoothness as in the previous section; a is twice continuously differentiable, b is continuously differentiable, and c is continuous in $x \geqslant 0$.

The boundary condition at the origin will be of one of the following two forms:
(I) $u'(0) = \gamma u(0)$ \hfill (3.3.1)

(where γ is an arbitrary real number) or

(II) $u(0) = 0.$ \hfill (3.3.2)

We begin by considering spaces of continuous functions. If the boundary condition is of type (I), the space is $E = C_0[0, \infty)$ (which consists of all continuous functions in $x \geqslant 0$ with $\lim_{x \to \infty} |u(x)| = 0$). When the boundary condition is (3.3.2), however, the use of the same space will prevent the domain of our differential operator from being dense in E. We use instead $E = C_{0,0}[0, \infty)$, that is, we require all the functions in the space to satisfy the boundary condition (II). (See Section 7).[2]

Let β denote one of the boundary conditions (3.3.1) or (3.3.2). The operator $A_0(\beta)$ is defined by

$$A_0(\beta)u(x) = a(x)u''(x) + b(x)u'(x) + c(x)u(x), \tag{3.3.3}$$

---

[2] Following strictly the notational conventions of Section 7 the correct notation would be $C_{0,(0)}$. We shall ignore, here and in other places, the distinction.

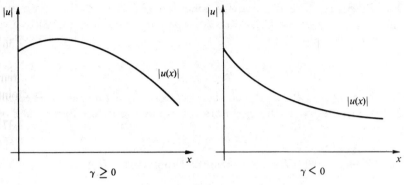

FIGURE 3.3.1

the domain of $A_0(\beta)$ consisting of all $u \in C^{(2)} \cap E$ satisfying the corresponding boundary condition and such that $au'' + bu' + cu \in E$. (The requirement that u satisfy the boundary condition is of course superfluous in case (II) since all functions in E do.)

3.3.1 Lemma. *The operator $A_0(\beta)$ is dissipative (dissipative with respect to a duality map θ) in E if and only if*

$$a(x) \geqslant 0, \quad c(x) \leqslant 0 \quad (x \geqslant 0) \tag{3.3.4}$$

with boundary condition (II). In case (I), $A_0(\beta)$ is dissipative if (3.3.4) holds and

$$\gamma \geqslant 0. \tag{3.3.5}$$

Conditions (3.3.4) are also necessary in this case, whereas (3.3.5) is necessary if $a(0) > 0$.

Proof. Case (II) yields to the argument used in Lemma 3.2.1 if we observe that the boundary condition (3.3.2) implies that $|u|$ cannot have a maximum at $x = 0$ unless it vanishes identically. The same reasoning would of course apply to boundary conditions (I) if we could show that (3.3.5) implies that, if $u \not\equiv 0$, $|u|$ may not have a maximum at $x = 0$. Assume this is not the case. Then

$$0 \geqslant \tfrac{1}{2}(|u|^2)'(0) = u_1(0)u_1'(0) + u_2(0)u_2'(0) = \gamma(u_1^2(0) + u_2^2(0)),$$

which implies $u(0) = 0$ if $\gamma > 0$, whence $u \equiv 0$, absurd. In case $\gamma = 0$, $|u|$ may have a maximum at $x = 0$ but we prove using Taylor's formula that $(|u|^2)''(0) \leqslant 0$. The necessity of (3.3.4) follows from the corresponding result for the whole line in the previous section (the argument yields the inequalities for $x > 0$ and they follow for $x = 0$ by continuity). To see the necessity of (3.3.5) when $a(0) > 0$, assume that $\gamma < 0$. Given an arbitrary number α, it is easy to see that we can construct a nonnegative function u in $D(A_0(\beta))$

having a single maximum at $x = 0$ and such that $u(0) = 1$, $u''(0) = \alpha$. Then $\Theta(u) = \{\delta\}$, δ the Dirac measure, and $\langle \delta, A_0 u \rangle = a(0)\alpha + \gamma b(0) + c(0)$, which can be made positive by adequate choice of α.

We consider next the case $E = L^p(0, \infty)$. Here $D(A_0(\beta))$ consists of all infinitely differentiable functions with compact support in $x \geq 0$ satisfying the corresponding boundary condition.

3.3.2   Lemma.   *The operator $A_0(\beta)$ is dissipative in L^p ($1 \leq p < \infty$) if*

$$a(x) \geq 0, \quad c(x) + \frac{1}{p}(a''(x) - b'(x)) \leq 0 \quad (x \geq 0) \qquad (3.3.6)$$

with boundary condition (II). *In case a boundary condition of type* (I) *is used, $A_0(\beta)$ is dissipative if both inequalities* (3.3.6) *and*

$$\gamma a(0) - \frac{1}{p}(a'(0) - b(0)) \geq 0 \qquad (3.3.7)$$

hold. The first inequality (3.3.6) *is necessary in all cases; the second, as well as* (3.3.7), *are necessary when $p = 1$.*

Proof.   We consider first the case $p > 1$, using integration by parts as in the proof of Lemma 3.2.2 but now taking boundary terms into account. Keeping in mind the observations about the case $1 < p < 2$ made in the proof of Lemma 3.2.2, we have

$$\|u\|^{p-2} \mathrm{Re} \langle \theta(u), A_0(\beta)u \rangle = \mathrm{Re} \int_0^\infty ((au')' + (b - a')u' + cu)|u|^{p-2} \bar{u} \, dx$$

$$= -\left\{ \gamma a(0) - \frac{1}{p}(a'(0) - b(0)) \right\} |u(0)|^p$$

$$- (p-2) \int_0^\infty a|u|^{p-4} (\mathrm{Re}(\bar{u}u'))^2 \, dx$$

$$- \int_0^\infty a|u|^{p-2}|u'|^2 \, dx$$

$$+ \frac{1}{p} \int_0^\infty (a'' - b' + pc)|u|^p \, dx \leq 0 \qquad (3.3.8)$$

if a boundary condition of type (I) is used; if the boundary condition is of type (II), the boundary term vanishes. The case $p = 1$ is settled through a limiting argument exactly as the one in the proof of Lemma 3.2.2.

The necessity of the first condition (3.3.6) for $p \geq 1$, as well as the necessity of the second for $p = 1$, follow as in the previous section; we only have to repeat the arguments there using functions that vanish near zero and thus satisfy any conceivable boundary condition. We show now that (3.3.7) is necessary when $p = 1$. Assume $\gamma a(0) - a'(0) + b(0) < 0$ and let u be an element of $D(A_0(\beta))$, which is positive in an interval $[0, \alpha)$, $\alpha > 0$, zero for $x \geqslant \alpha$, and has norm 1. Then any $u^* \in \Theta(u)$ equals 1 in $0 \leqslant x < \alpha$ and

$$\langle u^*, A_0(\beta)u \rangle = \int_0^\alpha ((au')' + (b - a')u' + cu)\, dx$$

$$= -(\gamma a(0) - a'(0) + b(0))u(0) + \int_0^\alpha (a'' - b' + c)u\, dx,$$

which clearly can be made positive by taking, say, $u(0)$ large and u sufficiently small off the origin.

3.3.3 Example. The second inequality (3.3.6) is not necessary for dissipativity of $A_0(\beta)$ for $p > 1$ regardless of the type of boundary condition used.

3.3.4. Example. Inequality (3.3.7) is not necessary for dissipativity of $A_0(\beta)$ for $p > 1$ when a boundary condition of type (I) is used.

We bring into play the adjoint A_0' defined as in the previous section. The Lagrange identity takes here the following form: if $u, v \in \mathcal{D}$,

$$\int_0^\infty (A_0 u)v\, dx = \int_0^\infty uA_0'v\, dx - a(0)(u'(0)v(0) - u(0)v'(0))$$

$$+ (a'(0) - b(0))u(0)v(0), \qquad (3.3.9)$$

where the functions u, v do not necessarily satisfy boundary conditions at the origin. Formula (3.3.9) allows us to compute the adjoints of differential operators when boundary conditions are included in the definition of the domains. (The notion of dissipativity does not play a role here.) To this end, we introduce the notion of *adjoint boundary condition* (with respect to A_0), assuming from now on that $a(0) > 0$. If the boundary condition β is of type (II), the adjoint boundary condition β' is β itself. If β is of type (I), then β' is

$$u'(0) = \{\gamma - a(0)^{-1}(a'(0) - b(0))\}u(0) = \gamma'u(0). \qquad (3.3.10)$$

The motivation of this definition is, clearly, that the boundary terms in the Lagrange formula (3.3.9) vanish when u satisfies the boundary condition β and v satisfies the boundary condition β'; in other words,

$$\int_0^\infty (A_0(\beta)u)v\, dx = \int_0^\infty uA_0'(\beta')v\, dx \quad (u \in D(A_0(\beta)), v \in D(A_0'(\beta'))).$$

$$(3.3.11)$$

It is clear from the definition of β' that if β is any boundary condition, $(\beta')' = \beta$ (note that in applying the definition of adjoint to β' the coefficients $\tilde{a}, \tilde{b}, \tilde{c}$ of the adjoint operator must be used: in other words, the second prime indicates adjoint with respect to A_0').

A simple computation shows that an operator $A_0(\beta)$ satisfies the assumptions in Lemma 3.3.2 in L^p if and only if $A_0'(\beta')$ satisfies the same assumptions in $L^{p'}$ (C_0 if $p = 1$).

In the following result no use is made of the dissipativity of A_0; however, we assume from now on that

$$a(x) > 0 \quad (x \geqslant 0). \tag{3.3.12}$$

3.3.5 Lemma.   (a) *Let $1 \leqslant p \leqslant \infty$ and let $f \in L^p_{loc}$ in $x \geqslant 0$. Assume that u is a locally summable function defined in $x \geqslant 0$ and such that*

$$\int_0^\infty u A_0'(\beta') \varphi \, dx = \int_0^\infty f\varphi \, dx \quad (\varphi \in D(A_0'(\beta'))). \tag{3.3.13}$$

Then u coincides with a function in $W^{2,p}_{loc}(0, \infty)$ that satisfies the boundary condition β and

$$au'' + bu' + cu = f \quad (x \geqslant 0). \tag{3.3.14}$$

If f is continuous, u is twice continuously differentiable in $x \geqslant 0$.

(b) *Let v be a locally finite Borel measure, and μ a locally finite Borel measure defined in $x \geqslant 0$ that satisfies*

$$\int_0^\infty A_0'(\beta') \varphi \, d\mu = \int_0^\infty \varphi \, dv \quad (\varphi \in D(A_0'(\beta'))). \tag{3.3.15}$$

Then u coincides with a continuous function in $t \geqslant 0$.

Proof.   We can continue the coefficients a, b, c to $x \leqslant 0$ preserving their smoothness and in such a way that the extended $a(\cdot)$ is positive; we denote the extensions by the same symbols. To prove (a), assume that the boundary condition is of type (I) and let φ_0 be a fixed function in $\mathfrak{D}$ with $\varphi_0(0) = 0$, $\varphi_0'(0) = 1$, and u_- a smooth (say, $C^{(2)}$) function in $x \leqslant 0$ with compact support and such that $u_-(0) = 1$ and

$$u_-'(0) = \gamma u_-(0) = \gamma. \tag{3.3.16}$$

We may and will assume in addition that

$$\int_{-\infty}^0 A_0 u_- \varphi_0 \, dx = \int_0^\infty f\varphi_0 \, dx = 0. \tag{3.3.17}$$

It follows from the first equality and the Lagrange identity (3.3.9) in $x \leqslant 0$ that

$$\int_{-\infty}^0 u_- A_0'\varphi_0 \, dx \neq 0.$$

Define now $u_1 = u$ for $x \geqslant 0$, $u_1 = \alpha u_-$ in $x < 0$, where α is such that

$$\int_{-\infty}^\infty u_1 A_0'\varphi_0 \, dx = 0. \tag{3.3.18}$$

If ψ is an arbitrary function in $\mathfrak{D}$, we can clearly write

$$\psi = \lambda\varphi_0 + \varphi \tag{3.3.19}$$

$(\lambda = \psi'(0) - \gamma'\psi(0))$, where φ satisfies the boundary condition β'. It is a

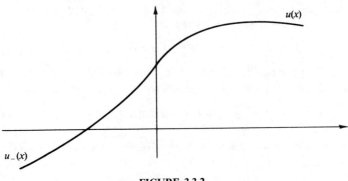

FIGURE 3.3.2

consequence of (3.3.16) and of the analogue of (3.3.11) in $x \leqslant 0$ that

$$\int_{-\infty}^{0} u_1 A_0' \varphi \, dx = \int_{-\infty}^{0} (A_0 u_1) \varphi \, dx. \tag{3.3.20}$$

We obtain, combining this equality with (3.3.19), (3.3.18) and (3.3.17), that we have

$$\int_{-\infty}^{\infty} u_1 A_0' \psi \, dx = \int_{-\infty}^{\infty} f_1 \varphi \, dx = \int_{-\infty}^{\infty} f_1 \psi \, dx \tag{3.3.21}$$

for all $\psi \in \mathcal{D}$, where $f_1 = f$ for $x \geqslant 0$, $f_1 = A_0 u_1$ for $x < 0$. Making use of Lemma 3.2.5 we see that $u_1 \in W^{2,p}_{\text{loc}}(-\infty, \infty)$ and that $au_1'' + bu_1' + cu_1 = f_1$ in $-\infty < x < \infty$, from which (3.3.14) follows. That u satisfies the boundary condition β results from the fact that u_- does. To prove the statement regarding f continuous, we need f_1 to be continuous. This can be achieved by imposing on u_- the additional boundary condition $A_0 u_-(0) = f_1(0)$. The case of boundary conditions of type (II) is handled in an entirely similar way and we omit the details.

The proof of (b) runs much along the same lines; the measure μ is continued to $x < 0$ in such a way that the analogue of (3.3.21) holds for an arbitrary test function ψ, and then we apply the second part of Lemma 3.2.5.

We note that an immediate consequence of part (a) is that the adjoint of the operator $A_0'(\beta')$ in $L^{p'}(0, \infty)$ $(1 \leqslant p < \infty)$ is the operator $A(\beta)$ in L^p, defined by $A(\beta)u = au'' + bu' + cu$, where $D(A(\beta))$ is the set of all $u \in W^{2,p}_{\text{loc}}(0, \infty) \cap L^p(0, \infty)$ such that $A(\beta)u \in L^p(0, \infty)$ and u satisfies the boundary condition β.

We examine now the problem of finding m-dissipative extensions of $A_0(\beta)$. The treatment is exceedingly similar to that in the previous section for $(-\infty, \infty)$ so that we limit ourselves to stating the main result and sketching the proof.

3.3.6 Theorem. (a) *Let the coefficients of the operator A_0 and the boundary condition β satisfy*

$$a(x) > 0, \quad c(x) \leqslant 0 \quad (x \geqslant 0) \tag{3.3.22}$$

and

$$\gamma \geqslant 0 \tag{3.3.23}$$

(the latter inequality if β is of type (I)), and assume $A_0'(\beta)$ is dissipative in $L^{p'}$ for some $p > 1$, which is the case if

$$c(x) + \frac{1}{p}(a''(x) - b'(x)) \leqslant 0 \quad (x \geqslant 0) \tag{3.3.24}$$

and

$$\gamma a(0) - \frac{1}{p}(a'(0) - b(0)) \geqslant 0 \tag{3.3.25}$$

if β is of type (I). Let $A(\beta)$ be the operator defined by

$$A(\beta)u(x) = a(x)u''(x) + b(x)u'(x) + c(x)u(x), \tag{3.3.26}$$

where $D(A(\beta))$ is the set of all $u \in C^{(2)}[0,\infty) \cap C_0[0,\infty)$ such that (3.3.26) belongs to $C_0[0,\infty)$ (if the boundary condition is of type (II), C_0 is replaced by $C_{0,0}$). Then $A(\beta)$ is m-dissipative in C_0 ($C_{0,0}$).

(b) *Let $1 < p < \infty$. Assume inequalities (3.3.22) and (3.3.23) hold and suppose A_0' is dissipative in $L^{p'}$ (for which (3.3.24) and (3.3.25) are sufficient conditions). Let $A(\beta)$ be the operator defined by (3.3.26), where $D(A(\beta))$ is the set of all $u \in W_{\mathrm{loc}}^{2,p}(0,\infty) \cap L^p(0,\infty)$ such that the right-hand side belongs to L^p. Then $A(\beta)$ is m-dissipative in L^p.*

(c) *Assume inequalities (3.3.22) and (3.3.23) hold, and that*

$$a''(x) - b'(x) + c(x) \leqslant 0, \tag{3.3.27}$$
$$\gamma a(0) - a'(0) + b(0) \geqslant 0. \tag{3.3.28}$$

Then the operator $A(\beta)$ defined as in (b) with $p = 1$ is m-dissipative in L^1.

To prove (b) let $v \in L^p$; we obtain, availing ourselves of the inequality $\|\varphi\|_{p'} \leqslant \lambda^{-1}\|\lambda\varphi - A_0'(\beta')\varphi\|_{p'}$ ($\varphi \in D(A_0'(\beta'))$) a $u \in L^p$ such that $\|u\|_p \leqslant \lambda^{-1}\|v\|_p$ and

$$(\lambda I - A_0'(\beta')^*)u = (\lambda I - A(\beta))u = v,$$

and the uniqueness of u follows from the fact that $A_0(\beta)$ is dissipative in $C_{0,0}$ or C_0, which is in turn guaranteed by conditions (3.3.22) and (3.3.23). Part (a) follows exactly in the same way as in Theorem 3.2.7. Finally, part (c) results from observing that inequalities (3.3.27) and (3.3.28) transform into $\tilde{c}(x) \leqslant 0$, $\gamma' \leqslant 0$ for the formal adjoint A_0' and the adjoint boundary condition β'. On the other hand, inequalities (3.3.22) are $\tilde{a}(x) > 0$, $\tilde{a}''(x) - \tilde{b}'(x) + \tilde{c}(x) \leqslant 0$, and (3.3.23) becomes $\gamma'\tilde{a}(0) - \tilde{a}'(0) + \tilde{b}(0) \geqslant 0$. We obtain the inequalities $\tilde{a}''(x) - \tilde{b}(x) + 2\tilde{c}(x) \leqslant 0$ and $2\gamma'\tilde{a}(0) - \tilde{a}'(0) + \tilde{b}(0) \geqslant 0$, which imply the assumptions in (a) for $A_0'(\beta')$. We then extend $A_0'(\beta')$ to a m-dissipative operator and apply the #-adjoint theory; in the final identification of the domain of $A_0'(\beta')^{\#}$, a decisive role is played by the dissipativity of $A_0(\beta)$.

The comments after the proof of Theorem 3.2.8 have a counterpart here; in particular, under the hypotheses in (c),

$$A(\beta) = A_0'(\beta')^* = \overline{A_0(\beta)}$$

for $1 < p < \infty$. We note finally that if we write A_0 in variational form (3.2.35), we only need to assume that $\hat{a}, \hat{b} \in C^{(1)}, \hat{c} \in C^{(0)}$, as in the case considered in Section 3.2.

3.3.7 Example. If A_0 is written in variational form, the dissipativity condition (3.3.25) becomes

$$\gamma\hat{a}(0) - \frac{1}{p}\hat{b}(0) \geqslant 0. \tag{3.3.29}$$

3.3.8 Example. Let β be an arbitrary boundary condition at $x = 0$, $A_p(\beta)$ the operator defined in Theorem 3.3.6(b). Then

$$A_{p'}'(\beta') = A_p(\beta)^* \quad (1 < p < \infty) \tag{3.3.30}$$

for $p'^{-1} + p^{-1} = 1$, where $A_{p'}'(\beta')$ is the operator in $L^{p'}$ corresponding to the formal adjoint A_0' and the adjoint boundary condition β'. (We adopt here the hypotheses in (c)).

3.4. ORDINARY DIFFERENTIAL OPERATORS IN A COMPACT INTERVAL

The problem of finding m-dissipative extensions of A_0 is considerably simpler in the case of a compact interval $[0, l]$ (at least when $a(\cdot)$ does not vanish in $0 \leqslant x \leqslant l$) since we have at our disposition the theory of Green functions of ordinary differential operators. We assume once again that a is twice continuously differentiable, b continuously differentiable, c continuous in $0 \leqslant x \leqslant l$. For the sake of simplicity we consider only separated boundary conditions at the endpoints, that is, a boundary condition β_0 at 0 and a boundary condition β_l at l, where β_x indicates one of the following two conditions:

(I) $u'(x) = \gamma_x u(x),$ (3.4.1)

(II) $u(x) = 0.$ (3.4.2)

In the treatment of the regular case ($a > 0$) we can dispense entirely with the operators A_0 of the past two sections. Given two boundary conditions β_0, β_l, we define an operator $A(\beta_0, \beta_l)$ as follows. In the case $E = L^p(0, l)$ ($1 \leqslant p < \infty$),

$$A(\beta_0, \beta_l)u(x) = a(x)u''(x) + b(x)u'(x) + c(x)u(x), \quad (3.4.3)$$

where the domain of $D(A(\beta_0, \beta_l))$ consists of all $u \in W^{2, p}(0, l)$ that satisfy the corresponding boundary condition at each endpoint. When E is a space of continuous functions, the definition of the space itself depends on the type of boundary conditions used. If both boundary conditions are of type (I), then $E = C[0, l]$ and $A(\beta_0, \beta_l)$ is defined by (3.4.3) with $D(A(\beta_0, \beta_l))$ consisting of all $u \in C^{(2)}[0, l]$ that satisfy the boundary conditions at each endpoint. If one of the boundary conditions (say, β_0) is of type (II), $E = C_0[0, l]$ (we use here the notation in Section 7)[3] consisting of all $u \in C[0, l]$ with $u(0) = 0$; $D(A(\beta_0, \beta_l))$ is defined in a way similar to the previous case, where now $A(\beta_0, \beta_l)u$ must belong to C_0. Similar restrictions of the space C are taken if β_l, or if both boundary conditions are of type (II); the corresponding spaces will be denoted $C_l[0, l], C_{0, l}[0, l]$.

The dissipativity results for the operators $A(\beta_0, \beta_l)$ in the different spaces considered can be proved essentially as in the case of one boundary condition (which was considered in the previous section) and will be for the most part left to the reader.

3.4.1 Lemma. *The operator $A(\beta_0, \beta_l)$ is dissipative in $C[0, l]$ (or in $C_0, C_l, C_{0, l}$ depending on the boundary conditions used) if and only if*

$$a(x) \geqslant 0, \quad c(x) \leqslant 0 \quad (0 \leqslant x \leqslant l) \quad (3.4.4)$$

and

$$\gamma_0 \geqslant 0, \quad \gamma_l \leqslant 0 \quad (3.4.5)$$

when boundary conditions of type (I) are used at either endpoint.

The case $E = L^p$ is dealt with by integration by parts for $1 < p < \infty$ and by means of the usual limiting argument when $p = 1$.

As pointed out in the two previous sections, there are difficulties with the differentiability of $\theta(u)$ when $1 < p < 2$, which can be avoided by performing the integration by parts for functions u_n in $\mathscr{P}\mathcal{C}^{(2)}$ (or simply polynomials) that satisfy the boundary conditions and then approximating any $u \in D(A(\beta_0, \beta_l))$ by a sequence $\{u_n\}$ in $\mathscr{P}\mathcal{C}^{(2)}$ in such a way that $u_n \to u, u_n' \to u'$ uniformly in $a \leqslant x \leqslant b$, $u_n'' \to u''$ in $L^p(0, l)$.

---

[3]See footnote 2, p. 135

The analogue of (3.3.8) is

$$\|u\|^{p-2}\mathrm{Re}\langle\theta(u), A(\beta_0,\beta_l)u\rangle$$

$$= \mathrm{Re}\int_0^l ((au')' + (b-a')u' + cu)|u|^{p-2}\bar{u}\,dx$$

$$= \left\{ \gamma_l a(l) - \frac{1}{p}(a'(l) - b(l)) \right\}|u(l)|^p$$

$$- \left\{ \gamma_0 a(0) - \frac{1}{p}(a'(0) - b(0)) \right\}|u(0)|^p$$

$$- (p-2)\int_0^l a|u|^{p-4}(\mathrm{Re}(\bar{u}u'))^2\,dx$$

$$- \int_0^l a|u|^{p-2}|u'|^2\,dx + \frac{1}{p}\int_0^l (a'' - b' + pc)|u|^p\,dx. \tag{3.4.6}$$

The boundary terms vanish when boundary conditions of type (II) are used. The preceding inequality plainly implies the following result.

3.4.2 Lemma. *The operator $A(\beta_0, \beta_l)$ is dissipative in the space L^p $(1 \leqslant p < \infty)$ if*

$$a(x) \geqslant 0, \quad c(x) + \frac{1}{p}(a''(x) - b'(x)) \leqslant 0 \quad (0 \leqslant x \leqslant l) \tag{3.4.7}$$

with boundary conditions (II). *If boundary conditions of type* (I) *are used instead at either endpoint, $A(\beta_0, \beta_l)$ is dissipative if* (3.4.7) *holds and*

$$\gamma_0 a(0) - \frac{1}{p}(a'(0) - b(0)) \geqslant 0, \quad \gamma_l a(l) - \frac{1}{p}(a'(l) - b(l)) \leqslant 0 \tag{3.4.8}$$

at the corresponding endpoint. The first inequality (3.4.7) *is necessary in all cases; the second, as well as both inequalities* (3.4.8), *is necessary when $p = 1$.*

The necessity proof for $p = 1$ is entirely similar to that in the previous section for the case of the half-line.

We turn now to the problem of solving the equation $(I - A(\beta_0, \beta_l))u = v$ in the different spaces under consideration. To this end we use the theory of Green functions for ordinary differential operators (an account of this theory, together with the results to be used below, can be found for instance in Coddington-Levinson [1955: 1].) Let $a(x) > 0$ in $0 \leqslant x \leqslant l$, u_0 a nontrivial solution of the equation

$$au'' + bu' + (c-1)u = 0 \tag{3.4.9}$$

in $0 \leqslant x \leqslant l$ that satisfies the boundary condition β_0, and let u_l be a nontrivial solution of (3.4.9) in $0 \leqslant x \leqslant l$ that satisfies the boundary condition β_l. The functions u_0 and u_l are linearly independent if $A(\beta_0, \beta_l)$ is dissipative in any of the spaces L^p, $1 \leqslant p < \infty$ or C. In fact, if they were not,

u_0 would be a nontrivial solution of (3.4.9) in $0 \leqslant x \leqslant l$ satisfying the boundary conditions at both endpoints. Then u_0 would belong to $D(A(\beta_0, \beta_l))$ in the space under consideration and (3.4.9) would be $A(\beta_0, \beta_l)u = u$, an equation that can only have the solution $u = 0$ in view of the dissipativity of $A(\beta_0, \beta_l)$.

The linear independence of u_0 and u_l implies of course that their Wronskian $W(x) = W(u_0, u_l, x)$ does not vanish at any point of the interval $0 \leqslant x \leqslant l$. We define then

$$G(x, \xi) = \begin{cases} u_0(x)u_l(\xi)/a(\xi)W(\xi) & (0 \leqslant x \leqslant \xi \leqslant l) \\ u_l(x)u_0(\xi)/a(\xi)W(\xi) & (0 \leqslant \xi \leqslant x \leqslant l). \end{cases} \quad (3.4.10)$$

It is an elementary matter to verify that if $f \in L^p(0, l)$, $1 \leqslant p < \infty$, then

$$u(x) = \int_0^l G(x, \xi)f(\xi)\,d\xi \quad (0 \leqslant x \leqslant l) \quad (3.4.11)$$

belongs to $D(A(\beta_0, \beta_l))$ and satisfies $A(\beta_0, \beta_l)u - u = f$. The same observation applies to the space $C[0, l]$ (or $C_0, C_l, C_{0,l}$ according to which boundary conditions are used). We have then proved the following

3.4.3   Theorem.   *Assume that the operator $A(\beta_0, \beta_l)$ is dissipative in $L^p(0, l)$ $(1 \leqslant p < \infty)$ or in $C[0, l]$ $(C_0, C_l, C_{0,l}$ depending on the boundary conditions) and that*

$$a(x) > 0 \quad (0 \leqslant x \leqslant l). \quad (3.4.12)$$

Then $A(\beta_0, \beta_l)$ is m-dissipative.

Observations similar to those in the last two sections apply to operators written in variational form (3.2.35).

3.4.4   Remark.   The results in the last three sections can be extended in an obvious way by replacing if necessary the operators A_0, $A_0(\beta)$ or $A_0(\beta_0, \beta_l)$ by $A_0 - \omega I$, $A_0(\beta) - \omega I$ or $A_0(\beta_0, \beta_l) - \omega I$ for ω sufficiently large; in this way, the second condition (3.2.28) can be weakened to

$$c(x) \leqslant \omega \quad (-\infty < x < \infty) \quad (3.4.13)$$

and (3.2.29) becomes

$$c(x) + \frac{1}{p}(a''(x) - b'(x)) \leqslant \omega \quad (-\infty < x < \infty). \quad (3.4.14)$$

The conclusion of Theorem 3.2.8 is in this case that $A - \omega I$ is dissipative in the various spaces considered there, so that $A \in \mathcal{C}_+(1, \omega)$. The same observations apply to the operators $A(\beta)$ in Theorem 3.3.9 and to the operators $A(\beta_0, \beta_l)$ in a compact interval $[0, l]$ in Theorem 3.4.3, and are especially interesting here: in fact, (3.4.13) and (3.4.14) will always be satisfied by ω large enough and it turns out that the only essential restriction on the coefficients (besides smoothness) is (3.4.12); if it holds it will always be true that $A \in \mathcal{C}_+$.

3.4.5   Example.   Let β_0 (resp. β_l) be a boundary condition at $x = 0$ (resp. at $x = l$), and let $A_p(\beta_0, \beta_l)$ $(1 \leqslant p < \infty)$ be the operator in L^p defined by (3.4.3): the corresponding operator in $C(C_0, C_l, C_{0,l})$ is denoted by $A(\beta_0, \beta_l)$. The operators defined with respect to the formal adjoint and the adjoint boundary conditions are $A_p'(\beta_0', \beta_l')$: we only assume that $a(x) > 0$ in $0 \leqslant x \leqslant l$ (see Remark 3.4.4) and conditions (3.4.8). Then

$$A_{p'}'(\beta_0', \beta_l') = A_p(\beta_0, \beta_l)^* \quad (1 < p < \infty), \tag{3.4.15}$$

where $p'^{-1} + p^{-1} = 1$. For $p = 1$ we have

$$A'(\beta_0', \beta_l') = A_1(\beta_0, \beta_l)^{\#} \tag{3.4.16}$$

and

$$A_1'(\beta_0', \beta_l') = A(\beta_0, \beta_l)^{\#}. \tag{3.4.17}$$

3.4.6   Example.   Denote by A_0 the operators defined in Section 3.2 in $L^p(-\infty, \infty)$ with domain $\mathcal{D}$, $A_0(\beta)$ the corresponding operators in $L^p(0, \infty)$. Using (3.4.15) and assuming that $a(x) > 0$, we can prove that if $1 < p < \infty$, then (a) A_0' is dissipative in $L^{p'}(-\infty, \infty)$ if and only if A_0 is dissipative in $L^p(-\infty, \infty)$ and (b) $A_0'(\beta')$ is dissipative in $L^{p'}(0, \infty)$ if and only if $A_0(\beta)$ is dissipative in $L^p(0, \infty)$. The general case $a(x) \geqslant 0$ can be handled using the previous result for $a(x) + \varepsilon$, $\varepsilon > 0$ and letting $\varepsilon \to 0$.

3.5.   SYMMETRIC HYPERBOLIC SYSTEMS IN THE WHOLE SPACE

Consider the differential operator

$$Lu = \sum_{k=1}^{m} A_k(x) D^k u + B(x) u.$$

Here $x = (x_1, \ldots, x_m)$ is a point in m-dimensional Euclidean space $\mathbb{R}^m$, $u(x) = (u_1(x), \ldots, u_\nu(x))$ a ν-dimensional vector function of x, $D^k = \partial / \partial x_k$, and the coefficients $A_k(x) = \{a_k^{ij}(x)\}$, $B(x) = \{b^{ij}(x)\}$ are $\nu \times \nu$ matrix functions defined in $\mathbb{R}^m$. Since no additional complications are caused by allowing the coefficients A_k, B to be complex-valued, we shall do so, assuming $A_1(\cdot), \ldots, A_m(\cdot)$ to be self-adjoint and continuously differentiable in $\mathbb{R}^m$ and $B(\cdot)$ to be continuous there. We shall work only in the complex space $L^2(\mathbb{R}^m)^\nu$ (see Section 7); in view of the result in Example 1.6.1, we scarcely need to apologize for lack of generality on this score. (We will consider in Chapter 5 the operator L in certain Sobolev spaces; see Section 1.6 for the constant coefficient case.)

Our goal in this section is to find a domain for L that makes it m-dissipative under adequate restrictions on its coefficients. This will be accomplished by defining $L = \mathcal{C}_0$ in a domain consisting of very smooth functions (where verification of dissipativity merely involves integration by parts) and then extending it to what essentially is its greatest possible domain. The fact that the extension $\mathcal{C}$ is also dissipative will follow from an approximation argument showing that $\mathcal{C}$ is the closure of $\mathcal{C}_0$ and from Lemma 3.1.11. Finally, m-dissipativity of $\mathcal{C}$ (that is, verification of (3.1.8)) will be a consequence of the maximality of the domain of $\mathcal{C}$ and of a duality argument.

We begin to carry out this program defining $\mathcal{C}_0$ by

$$\mathcal{C}_0 u = Lu. \tag{3.5.1}$$

The domain of $\mathcal{C}_0$ is by no means critical; we define it as the set $\mathcal{D}^\nu$ of all vector functions $u = (u_1, \ldots, u_\nu)$ whose components belong to $\mathcal{D}$. It follows from the symmetry of the A_k that

$$D^k\big(A_k(x)u(x), u(x)\big) = \big(D^k A_k(x)u(x), u(x)\big)$$
$$+ 2\operatorname{Re}\big(A_k(x)D^k u(x), u(x)\big).$$

Hence

$$\operatorname{Re}\big(\mathcal{C}_0 u(x), u(x)\big) = \tfrac{1}{2} \sum_{k=1}^{m} D^k\big(A_k(x)u(x), u(x)\big)$$

$$- \tfrac{1}{2} \sum_{k=1}^{m} \big(D^k A_k(x)u(x), u(x)\big)$$

$$+ \operatorname{Re}\big(B(x)u(x), u(x)\big) \quad (x \in \mathbb{R}^m),$$

thus

$$\operatorname{Re}(\mathcal{C}_0 u, u) = \operatorname{Re} \int \big(\mathcal{C}_0 u(x), u(x)\big)\, dx = \operatorname{Re} \int \big(K(x)u(x), u(x)\big)\, dx,$$

where

$$K(x) = B(x) - \tfrac{1}{2} \sum_{k=1}^{m} D^k A_k(x). \tag{3.5.2}$$

The following result is then immediate.

3.5.1 Lemma. *The operator $\mathcal{C}_0$ is dissipative in $E = L^2(\mathbb{R}^m)^\nu$ if and only if $K(x)$ is dissipative in $\mathbf{C}^\nu$ for all x, that is*

$$\operatorname{Re}(K(x)u, u) \leqslant 0 \quad (x \in \mathbb{R}^m, u \in \mathbf{C}^\nu). \tag{3.5.3}$$

The *formal adjoint* of $\mathcal{C}_0$ is defined by

$$\mathcal{C}_0' u = -\sum_{k=1}^{m} D^k\big(A_k(x)u\big) + B(x)' u$$

$$= -\sum_{k=1}^{m} A_k(x) D^k u + \left(B(x)' - \sum_{k=1}^{m} D^k A_k(x)\right) u$$

$$= \sum_{k=1}^{m} \tilde{A}_k(x) u + \tilde{B}(x) u, \tag{3.5.4}$$

where B' indicates the adjoint of B. The domain of $\mathcal{C}_0'$ is the same as that of

$\mathcal{C}_0$, and $\mathcal{C}_0'$ is the only operator satisfying the Lagrange identity $(\mathcal{C}_0 u, v) = (u, \mathcal{C}_0' v)$ or

$$\int \left(\mathcal{C}_0 u(x), v(x)\right) dx = \int \left(u(x), \mathcal{C}_0' v(x)\right) dx \qquad (3.5.5)$$

for all $u, v \in D(\mathcal{C}_0)$. If we denote by $\tilde{K}(x)$ the matrix associated with $\mathcal{C}_0'$ in the same way $K(x)$ is with $\mathcal{C}_0$, that is, if

$$\tilde{K}(x) = \tilde{B}(x) - \frac{1}{2} \sum_{k=1}^{m} D^k \tilde{A}_k(x),$$

a simple computation yields

$$\tilde{K}(x) = K(x)' \quad (x \in \mathbb{R}^m). \qquad (3.5.6)$$

The operator $\mathcal{C}$ is now simply defined as the adjoint of $\mathcal{C}_0'$,

$$\mathcal{C} = (\mathcal{C}_0')^*; \qquad (3.5.7)$$

in other words, $D(\mathcal{C})$ consists of all $u \in E$ such that there exists a $v \in E$ with

$$(u, \mathcal{C}_0' w) = (v, w) \quad (w \in D(\mathcal{C}_0')),$$

and for those u,

$$\mathcal{C} u = v.$$

Moreover, $\mathcal{C}$ is closed (Section 4). In view of the Lagrange identity it is plain that

$$\mathcal{C}_0 \subseteq (\mathcal{C}_0')^* = \mathcal{C}. \qquad (3.5.8)$$

We show next that

$$(I - \mathcal{C}) D(\mathcal{C}) = E. \qquad (3.5.9)$$

To this end, denote by $\mathcal{H}$ the subspace of E consisting of all elements of the form

$$(I - \mathcal{C}_0') u \quad (u \in D(\mathcal{C}_0')).$$

Let v be an arbitrary element of E; define a functional Φ in the subspace $\mathcal{H}$ by means of the formula

$$\Phi\left((I - \mathcal{C}_0') w\right) = (v, w) \quad (w \in D(\mathcal{C}_0')).$$

Since $\mathcal{C}_0'$ is dissipative, the fundamental inequality (3.1.9) implies that Φ is well defined and bounded (its norm does not surpass $\|v\|$). Extending this functional to the whole space E with the same norm and applying the Riesz representation theorem, we see that there exists a $u \in E$ such that

$$(u, w - \mathcal{C}_0' w) = (v, w) \quad (w \in D(\mathcal{C}_0')),$$

which clearly implies that $u \in D((\mathcal{C}_0')^*) = D(\mathcal{C})$ and that

$$(I - \mathcal{C}) u = v, \qquad (3.5.10)$$

proving (3.5.9). Moreover, it follows from our observation about the norm of Φ that $\|u\| \leqslant \|v\|$. But we stumble upon a serious difficulty here: in order to conclude that $\mathcal{Q}$ is m-dissipative from the information available, we need to show that u is the *only* solution of (3.5.10). We obviate this difficulty, as indicated earlier in this section, by showing that $\bar{\mathcal{Q}}_0 = \mathcal{Q}$; however, a growth condition on the matrices $A_1, \ldots, A_m$ will have to be postulated.

3.5.2 Theorem. *Assume that*

$$|A_k(x)| \leqslant C\rho(|x|) \quad (x \in \mathbb{R}^m, 1 \leqslant k \leqslant m), \qquad (3.5.11)$$

where ρ is a continuous positive function increasing so slowly that

$$\int^{\infty} \frac{dr}{\rho(r)} = \infty. \qquad (3.5.12)$$

Let $u \in D(\mathcal{Q})$. Then there exists a sequence $\{u_n\}$ in $D(\mathcal{Q}_0)$ such that

$$u_n \to u, \quad \mathcal{Q}_0 u_n \to \mathcal{Q} u$$

in E; in other words, $\mathcal{Q} = \bar{\mathcal{Q}}_0$.

We note that the dissipativity assumption (3.5.3) is irrelevant here. The proof will be carried out by means of an argument involving mollifiers (see Section 8). Recall that if β is a function in $\mathcal{D}$, which is nonnegative, vanishes in $|x| \geqslant 1$, and has integral 1, we define a sequence $\{J_n\}$ of operators in E by

$$J_n u(x) = (\beta_n * u)(x) = \int \beta_n(x - y) u(y) \, dy,$$

where $\beta_n(x) = n^m \beta(nx)$. Then each J_n is a bounded operator in E (precisely, $\|J_n\| \leqslant 1$ for all n). Moreover, J_n is self-adjoint and

$$J_n u \to u \quad (n \to \infty)$$

in the norm of E for any $u \in E$. Finally, $J_n u$ is a $C^{(\infty)}$ function for all $u \in E$. We shall need some additional information on the interaction of the J_n and $\mathcal{Q}$, which is contained in the following result. It will be necessary to assume temporarily that the A_k, their partial derivatives, and B are bounded:

$$|A_k(x)|, |D^j A_k(x)|, |B(x)| \leqslant C \quad (x \in \mathbb{R}^m, 1 \leqslant j, k \leqslant m). \quad (3.5.13)$$

3.5.3 Lemma. *Assume $A_1, \ldots, A_m, B$ satisfy (3.5.13). Then* (a)

$$\|J_n \mathcal{Q} u - \mathcal{Q} J_n u\| \leqslant C\|u\| \quad (u \in D(\mathcal{Q})) \qquad (3.5.14)$$

(in particular, $J_n u \in D(\mathcal{Q})$).
  (b) *If $u \in D(\mathcal{Q})$,*

$$\|J_n \mathcal{Q} u - \mathcal{Q} J_n u\| \to 0 \quad (n \to \infty). \qquad (3.5.15)$$

The proof does not require the dissipativity condition (3.5.3). To prove (3.5.14) (which only plays an auxiliary role with respect to the more fundamental relation (3.5.15)), we begin with the case $u \in D(\mathcal{C}_0)$. We have

$$J_n \mathcal{C}_0 u(x) = \sum_{k=1}^m \int \beta_n(x-y) A_k(y) D^k u(y)\, dy + \int \beta_n(x-y) B(y) u(y)\, dy$$

$$= - \sum_{k=1}^m \int D_y^k (\beta_n(x-y) A_k(y)) u(y)\, dy$$

$$+ \int \beta_n(x-y) B(y) u(y)\, dy$$

$$= \sum_{k=1}^m \int (D_x^k \beta_n(x-y) A_k(y) - \beta_n(x-y) D^k A_k(y)) u(y)\, dy$$

$$+ \int \beta_n(x-y) B(y) u(y). \qquad (3.5.16)$$

On the other hand, it is clear that $J_n u \in D(\mathcal{C}_0)$ and

$$\mathcal{C}_0 J_n u(x) = \sum_{k=1}^m \int D_x^k \beta_n(x-y) A_k(x) u(y)\, dy$$

$$+ \int \beta_n(x-y) B(x) u(y)\, dy. \qquad (3.5.17)$$

Hence

$$(J_n \mathcal{C}_0 u - \mathcal{C}_0 J_n u)(x) = \sum_{k=1}^m \int D_x^k \beta_n(x-y)(A_k(y) - A_k(x)) u(y)\, dy$$

$$- \sum_{k=1}^m \int \beta_n(x-y) D^k A_k(y) u(y)\, dy$$

$$+ \int \beta_n(x-y)(B(y) - B(x)) u(y)\, dy. \qquad (3.5.18)$$

We now estimate these integrals. The hypotheses on $A_1, \ldots, A_m$ clearly imply that each A_k is uniformly Lipschitz continuous in $\mathbb{R}^m$. Hence

$$|D_x^k \beta_n(x-y)(A_k(y) - A_k(x))| \le C|x-y|\,|D^k \beta_n(x-y)|$$

$$= \xi_{n,k}(x-y), \qquad (3.5.19)$$

and we also have[4]

$$|\beta_n(x-y) D^k A_k(y)| \le C \beta_n(x-y), \qquad (3.5.20)$$

$$|\beta_n(x-y)(B(y) - B(x))| \le C \beta_n(x-y) \qquad (3.5.21)$$

---

[4]Here and in other inequalities C denotes a constant, generally not the same in different inequalities.

for $x, y \in \mathbb{R}^m$. Then we can write

$$|(J_n \mathcal{Q}_0 u - \mathcal{Q}_0 J_n u)(x)| \leqslant \int \left(\sum_{k=1}^{m} \xi_{n,k}(x - y) + C\beta_n(x - y) \right) |u(y)| \, dy$$

$$= \sum_{k=1}^{m} (\xi_{n,k} * |u|)(x) + C(\beta_n * |u|)(x) \quad (x \in \mathbb{R}^m).$$

$$\tag{3.5.22}$$

We estimate the right-hand side of (3.5.22) with the help of Young's theorem (see Section 8). We have $\int \beta_n \, dx = 1$, whereas

$$\int \xi_{n,k}(x) \, dx = C \int_{|x| \leqslant 1/n} |x| \, |D^k \beta_n(x)| \, dx$$

$$= C \int_{|x| \leqslant 1} |x| \, |D^k \beta(x)| \, dx.$$

Accordingly we obtain

$$\|J_n \mathcal{Q}_0 u - \mathcal{Q}_0 J_n u\| \leqslant C \|u\| \tag{3.5.23}$$

for all $u \in D(\mathcal{Q}_0)$.

We prove next the corresponding particular case of (3.5.15). Let $u \in D(\mathcal{Q}_0)$, $K = \text{supp}(u)$. Integrate by parts the first m integrals on the right-hand side of (3.5.18) after making the substitution $D_x^k \beta_n(x - y) = -D_y^k \beta_n(x - y)$ in the integrand. The result is

$$(J_n \mathcal{Q}_0 u - \mathcal{Q}_0 J_n u)(x) = \sum_{k=1}^{m} \int \beta_n(x - y)(A_k(x) - A_k(y)) D^k u(y) \, dy$$

$$- \int \beta_n(x - y)(B(x) - B(y)) u(y) \, dy. \quad (3.5.24)$$

Now, if $x, y \in \mathbb{R}^m$,

$$\beta_n(x - y)|A_k(x) - A_k(y)| \leqslant C|x - y|\beta_n(x - y)$$

$$\leqslant 2Cn^{-1}\beta_n(x - y), \tag{3.5.25}$$

the last inequality resulting from the fact that the support of β_n is contained in $|x| \leqslant 1/n$. Observe next that B must be uniformly continuous in K; therefore, if $\varepsilon > 0$, there exists n_0 such that if x, y are arbitrary elements of K,

$$|\beta_n(x - y)(B(x) - B(y))| \leqslant \varepsilon \beta_n(x - y) \tag{3.5.26}$$

for $n \geqslant n_0$. Making use of (3.5.25) and (3.5.26) combined with Young's inequality in the estimate obtained taking absolute values in (3.5.24), we obtain

$$\|J_n \mathcal{Q}_0 u - \mathcal{Q}_0 J_n u\| \to 0 \quad (n \to \infty) \tag{3.5.27}$$

for any $u \in D(\mathcal{Q}_0)$.

To prove (3.5.14), we only have to observe that

$$J_n \mathcal{Q} - \mathcal{Q} J_n = J_n(\mathcal{Q}'_0)^* - (\mathcal{Q}'_0)^* J_n \subseteq (\mathcal{Q}'_0 J_n)^* - (J_n \mathcal{Q}'_0)^*$$
$$\subseteq (\mathcal{Q}'_0 J_n - J_n \mathcal{Q}'_0)^* \qquad (3.5.28)$$

(which is a consequence of (4.7) and (4.8)) and use (3.5.23) for $\mathcal{Q}'_0$. Finally, we show (3.5.15). To this end, we take $u \in D(\mathcal{Q})$ and a sequence $\{u_m\} \subset D(\mathcal{Q}_0)$ such that $u_m \to u$ (in the norm of E). Then

$$\|J_n \mathcal{Q} u - \mathcal{Q} J_n u\| \leqslant \|J_n \mathcal{Q}_0 u_m - \mathcal{Q}_0 J_n u_m\| + C\|u - u_m\|.$$

Using (3.5.27) the result follows easily.

In the next result the boundedness assumptions (3.5.13) are entirely given up in exchange for restrictions on u.

3.5.4   Lemma.   *Let $u \in D(\mathcal{Q})$ have compact support. Then* (3.5.15) *holds.*

Proof.   Let K be the support of u; for $a > 0$ denote by K_a the set of all $x \in \mathbb{R}^m$ with $\mathrm{dist}(x, K) \leqslant a$. Let $\hat{A}_1, \dots, \hat{A}_m$ be continuously differentiable $\nu \times \nu$ matrix functions in $\mathbb{R}^m$, zero outside of K_1, and coinciding with $A_1, \dots, A_m$ in K; likewise, extend the restriction of B to K to a $\nu \times \nu$ continuous matrix function $\hat{B}$ in $\mathbb{R}^m$ vanishing outside K_1. Let

$$\hat{L}u = \sum_{k=1}^m \hat{A}_k(x) D^k u + \hat{B}(x) u \qquad (3.5.29)$$

and $\hat{\mathcal{Q}}_0, \hat{\mathcal{Q}}$ the operators defined from $\hat{L}$ in the same way $\mathcal{Q}_0, \mathcal{Q}$ were defined from L. Since $\mathrm{supp}(J_n u) \subseteq K_1$, it can be easily verified that

$$\mathcal{Q} J_n u = \hat{\mathcal{Q}} J_n u, \quad J_n \mathcal{Q} u = J_n \hat{\mathcal{Q}} u,$$

hence the result follows readily from Lemma 3.5.3.

Proof of Theorem 3.5.2.   We may assume that ρ is infinitely differentiable. Let φ be another infinitely differentiable function of t with $\varphi(t) = 0$ if $t \leqslant 0$, $\varphi(t) = 1$ for $r \geqslant 1$. Given $0 < r < \infty$, define

$$f_r(s) = \begin{cases} 1 & (0 \leqslant s \leqslant r) \\ 1 - \displaystyle\int_r^s \frac{d\sigma}{\rho(\sigma)} & (r \leqslant s \leqslant s_r) \\ 0 & (s_r \leqslant s), \end{cases}$$

where s_r is such that $\rho(s)^{-1}$ has integral 1 in $r \leqslant s \leqslant s_r$. Although $f_r(s)$ itself is merely continuous, it is not difficult to see that

$$\chi_r(x) = \varphi(f_r(|x|))$$

belongs to $\mathcal{D}$, its support being contained in $|x| \leqslant s_r$; its first partials have support in $r \leqslant |x| \leqslant s_r$ and

$$|D^k \chi_r(x)| \leqslant C/\rho(|x|) \quad (x \in \mathbb{R}^m, 1 \leqslant k \leqslant m). \qquad (3.5.30)$$

If $u \in L^p(\mathbb{R}^m)^\nu$, it is clear that

$$\chi_r u \to u$$

as $r \to \infty$. On the other hand, if $u \in D(\mathcal{Q})$, a simple computation with adjoints shows that

$$\mathcal{Q}(\chi_r u) = \left(\sum A_k D^k \chi_r \right) u + \chi_r \mathcal{Q} u.$$

In view of (3.5.11) and (3.5.30), we obtain, taking into account that the $D^k \chi_r$ vanish for $|x| \leqslant r$,

$$\left\| \left(A_k D^k \chi_r \right) u \right\| \leqslant C \int_{|x| \geqslant r} |u|^2 \, dx,$$

which tends to zero as $r \to \infty$; accordingly,

$$\mathcal{Q}(\chi_r u) \to \mathcal{Q} u. \tag{3.5.31}$$

We combine this result with Lemma 3.5.4: the result is Theorem 3.5.2.

Observing that $\mathcal{Q}_0$ is dissipative if 3.5.3 holds, and applying Lemma 3.1.11 we obtain

3.5.5 Theorem. *Let $\mathcal{Q}_0, \mathcal{Q}_0'$ be the operators defined by (3.5.1) and (3.5.4), respectively, and let the matrix function*

$$K(x) = B(x) - \tfrac{1}{2} \sum_{k=1}^{m} D^k A_k(x) \tag{3.5.32}$$

be dissipative for all $x \in \mathbb{R}^m$. Assume in addition that (3.5.11) holds with a function ρ satisfying (3.5.12). Then the operator

$$\mathcal{Q} = \overline{\mathcal{Q}}_0 = \left(\mathcal{Q}_0' \right)^* \tag{3.5.33}$$

is m-dissipative

We note that there are no conditions on the growth of $D^k A_k, B$, other than the unilateral ones implicit in the requirement that the matrix K be dissipative (and which are of course necessary for dissipativity of $\mathcal{Q}_0$ or of any extension thereof).

On the other hand, although condition (3.5.11)–(3.5.12) is not necessary for the operator $\mathcal{Q} = (\mathcal{Q}_0')^*$ to be m-dissipative, it is best possible in a certain sense for the equality $\overline{\mathcal{Q}}_0 = \mathcal{Q}$ to hold.

*3.5.6 Example. Let ρ be positive, nondecreasing, and differentiable in $r \geqslant 0$. Assume that

$$\int_0^\infty \frac{dr}{\rho(r)} < \infty. \tag{3.5.34}$$

Define $\rho(r) = \rho(0)$ for $r < 0$. Then the operator

$$L = \rho(x_2) D^1 + \rho(x_1) D^2 \tag{3.5.35}$$

in $\mathbb{R}^2$ does not satisfy the conclusion of Theorem 3.5.2, that is, $\overline{\mathcal{Q}}_0 \neq \mathcal{Q}$. (However, $\mathcal{Q}$ is m-dissipative.) See the author, [1980: 1].

The following consequence of Theorem 3.5.5 is immediate considering the operators $\mathcal{Q}_0 - \omega I$ and $-\mathcal{Q}_0 - \omega I$.

3.5.7 Corollary. (a) *Let K satisfy*

$$\mathrm{Re}(K(x)u, u) \leqslant \omega |u|^2 \quad (x \in \mathbb{R}^m, u \in \mathbb{C}^\nu) \tag{3.5.36}$$

for some $\omega \geqslant 0$ *and assume that* (3.5.11)–(3.5.12) *holds. Then* $\mathcal{Q} = \bar{\mathcal{Q}}_0 = (\mathcal{Q}_0')^* \in \mathcal{C}_+(1, \omega)$.
  (b) *If* (3.5.36) *is strengthened to*

$$|\mathrm{Re}(K(x)u, u)| \leqslant \omega |u|^2 \quad (x \in \mathbb{R}^m, u \in \mathbb{C}^\nu), \tag{3.5.37}$$

then $\mathcal{Q} = \bar{\mathcal{Q}}_0 = (\mathcal{Q}_0')^* \in \mathcal{C}(1, \omega)$.

3.5.8 Remark. In all conditions involving the matrix $K(x)$, we can replace it by the "symmetrized" matrix $Q(x)$ defined by

$$2Q(x) = B(x) + B(x)' - \sum_{k=1}^{m} D^k A_k(x) \tag{3.5.38}$$

since $\mathrm{Re}(K(x)u, u) = (Q(x)u, u)$. An obvious advantage is the following: if
$\tilde{Q}(x)$ denotes the matrix corresponding to L', then

$$\tilde{Q} = Q. \tag{3.5.39}$$

3.6. ISOMETRIC PROPAGATORS AND CONSERVATIVE OPERATORS. DISSIPATIVE OPERATORS IN HILBERT SPACE

When the equation

$$u'(t) = Au(t) \tag{3.6.1}$$

models a physical system and the norm $\|\cdot\|$ is the energy of the system (as it
is the case, for instance, with Maxwell's equations in Section 1.6), one may
be led, on physical grounds, to conclude that the energy is constant; if the
model (3.6.1) is to be a reasonable one, we must have

$$\|S(t)u\| = \|u\| \quad (u \in E) \tag{3.6.2}$$

in $t \geqslant 0$. We say then that the propagator $S(\cdot)$ is *isometric* in $t \geqslant 0$. The
characterization of the operators $A \in \mathcal{C}_+$ for which the propagator of (3.6.1)
is isometric in $t \geqslant 0$ is rather simple.
  The operator A is said to be *conservative* if

$$\mathrm{Re}\langle u^*, Au \rangle = 0 \quad (u \in D(A), u^* \in \Theta(u)) \tag{3.6.3}$$

and *conservative with respect to* θ ($\theta: D(A) \to E^*$ a duality map) if

$$\mathrm{Re}\langle \theta(u), Au \rangle = 0 \quad (u \in D(A)). \tag{3.6.4}$$

3.6.1 Theorem. Let $A \in \mathcal{C}_+(1,0)$ and assume the propagator of (3.6.1) is isometric. Then A is conservative with respect to some duality map. Conversely, assume that A is densely defined, conservative with respect to some duality map $\theta : D(A) \to E^*$ and

$$(\lambda I - A)D(A) = E \qquad (3.6.5)$$

for some $\lambda > 0$. Then $A \in \mathcal{C}_+(1,0)$ and the propagator of (3.6.1) is isometric.

Proof. If A is conservative with respect to θ, it is also dissipative with respect to θ; in view of (3.6.5) we deduce from Theorem 3.1.8 that $A \in \mathcal{C}_+(1,0)$. Let now $S(\cdot)$ be the propagator of (3.6.1) and $u \in D(A^2)$. If $t, h \geq 0$, we have

$$S(t+h)u = S(t)u + hAS(t)u + \int_t^{t+h}(t+h-s)S(s)A^2 u\,ds$$

$$= S(t)u + hAS(t)u + h\rho(t,h), \qquad (3.6.6)$$

where $\|\rho(t,h)\| \to 0$ as $h \to 0$, uniformly in $t \geq 0$. Then

$$\|S(t)u\|\,\|S(t+h)u\| \geq |\langle \theta(S(t)u), S(t+h)u \rangle|$$

$$\geq \mathrm{Re}\langle \theta(S(t)u), S(t+h)u \rangle$$

$$= \|S(t)u\|^2 + h\,\mathrm{Re}\langle \theta(S(t)u), \rho(t,h)\rangle \quad (3.6.7)$$

in view of (3.6.6) and of the fact that A is conservative. Assuming that $u \neq 0$, consider the set

$$e = e(u) = \{t \geq 0;\ \|S(t)u\| = \|u\|\}.$$

Since $0 \in e$, e is nonempty; moreover, e is obviously a closed interval. If $e \neq [0, \infty)$, let t_0 be the right endpoint of e and $\alpha > 0$ so small that $\|S(t)u\|$ is bounded away from zero in $t_0 \leq t \leq t_0 + \alpha$. For any such t we write (3.6.7) and divide by $\|S(t)u\|$; the result is

$$\|S(t+h)u\| \geq \|S(t)u\| - h\eta(t,h), \qquad (3.6.8)$$

where η is nonnegative and $\eta(t,h) \to 0$ as $h \to 0$ uniformly in $t_0 \leq t \leq t_0 + \alpha$. Let now ε be a small positive number and $\delta > 0$ such that $|\eta(t,h)| \leq \varepsilon$ for $0 \leq h \leq \delta$ and $t_0 \leq t \leq t_0 + \alpha$. Let $t_0 < t_1 < \cdots < t_m = t_0 + \alpha$ be a partition of the interval $t_0 \leq t \leq t_0 + \alpha$ such that $t_j - t_{j-1} \leq \delta$ $(1 \leq j \leq m)$. Then

$$0 \leq \|S(t_0)u\| - \|S(t_0 + \alpha)u\|$$

$$\leq \sum_{j=1}^m \left(\|S(t_{j-1})u\| - \|S(t_j)u\| \right)$$

$$\leq \sum_{j=1}^m (t_j - t_{j-1})\eta(t_j, t_j - t_{j-1}) \leq \alpha\varepsilon.$$

Since ε is arbitrary, it follows that $\|S(t_0 + \alpha)u\| = \|S(t_0)u\|$, which contradicts the fact that t_0 is the right endpoint of e. Hence $e = [0, \infty)$ and

$t \to \|S(t)u\|$ is constant in $t \geqslant 0$. Using now the fact that $D(A^2)$ is dense in E, a standard approximation argument takes care of the case $u \in E$.

We prove now the opposite implication. Assume that $A \in \mathcal{C}_+(1,0)$ and let the solution operator of (3.6.1) be isometric. Take $t > 0$, $u \in D(A)$, and consider the scalar function

$$\varphi(s) = \mathrm{Re}\langle u_t^*, S(s)u \rangle \quad (s \geqslant 0),$$

where $u_t^* \in \Theta(S(t)u)$. Clearly φ is continuously differentiable and has a relative maximum at $s = t$. Hence

$$\mathrm{Re}\langle u_t^*, AS(t)u \rangle = 0.$$

This proves that (3.6.3) is satisfied for those $u \in D(A)$ of the form $S(t)u$, $u \in D(A)$. This would end the proof if $A \in \mathcal{C}$, but some work remains to be done in the general case. Let $u \in D(A)$, $\{t_n\}$ a sequence of positive numbers tending to zero, $u_n^* \in \Theta(S(t_n)u)$. By passing if necessary to a subsequence, we can assume that $u_n^* \to u^*$ weakly, where $u^* \in \Theta(u)$ (see the proof of Lemma 3.1.11) and

$$\langle u^*, Au \rangle = \lim\langle u_n^*, AS(t_n)u \rangle = 0.$$

This completes the proof.

The corresponding result for $A \in \mathcal{C}$ is Theorem 3.1.9 (see (3.1.11) and surrounding comments).

We have already noted in Section 3.1 that when E is a Hilbert space an operator A is dissipative if and only if

$$\mathrm{Re}(Au, u) = \mathrm{Re}(u, Au) \leqslant 0 \quad (u \in D(A)) \tag{3.6.9}$$

and the notion of dissipativity with respect to a duality map coincides also with (3.6.9). The Hilbert space theory, as it can be expected, is somewhat richer than the general theory and connects with some celebrated results of classical functional analysis. We begin with a characterization of m-dissipative operators; from now on $E = H$ is assumed to be a (complex) Hilbert space. A dissipative operator A is *maximal dissipative* if and only if A has no proper dissipative extension (if $A \subseteq B$ and B is dissipative, then $A = B$).

3.6.2 Theorem. *Let A be m-dissipative. Then A is maximal dissipative. Conversely, let A be maximal dissipative and closed (or densely defined). Then A is m-dissipative.*

Proof. Let A be m-dissipative, B a proper dissipative extension of A, and u an element of $D(B)\backslash D(A)$. Write $v = (I - A)^{-1}(I - B)u$. Then $v \in D(A)$ and $(I - A)v = (I - B)v = (I - B)u$; since $u \neq v$, this shows that $I - B$ is not one-to-one, contradicting inequality (3.1.9) for B. Then we have $A = B$, hence A must be maximal.

Conversely, assume A is maximal; note that if A is densely defined we may construct using Lemma 3.1.11 a closed dissipative extension of A, which must coincide with A by maximality. We can then assume in any case that A is closed. Let $K = (I - A)D(A) \neq E$. We have already observed (Remark 3.1.12) that K is closed, so that its orthogonal complement $K^{\perp}$ does not reduce to $\{0\}$. It is immediate that $K^{\perp}$ consists of all $v \in D(A^*)$ with

$$A^*v = v. \tag{3.6.10}$$

We note next that $D(A) \cap K^{\perp} = \{0\}$. In fact, let $u \in D(A) \cap K^{\perp}$. Then $((I - A)u, u) = 0$, or $(Au, u) = \|u\|^2$; since $\text{Re}(Au, u) \leqslant 0$, $u = 0$ as claimed. We define an operator $\tilde{A}$ in $D(\tilde{A}) = D(A) + K^{\perp}$ as follows:

$$\tilde{A}(u + v) = Au - v. \tag{3.6.11}$$

Clearly $\tilde{A}$ is an extension of A. Taking (3.6.10) into account, we have

$$(\tilde{A}(u + v), u + v) = (Au, u) - (v, v) + 2i\text{Im}(u, v) \tag{3.6.12}$$

for $u + v \in D(\tilde{A})$, which proves $\tilde{A}$ dissipative. This contradicts the assumption that A is maximal dissipative. Accordingly $(I - A)D(A) = E$, thus ending the proof.

Some additional information can be coaxed out of these arguments. Let A be as before a closed dissipative operator (if A is not closed but densely defined, we take its closure using Lemma 3.1.11). The operator $\tilde{A}$ can be constructed as above, and it is not difficult to show that $\tilde{A}$ is m-dissipative. In fact, take $w \in H$ and write $w = (I - A)u + 2v$, with $u \in D(A)$, $v \in K^{\perp}$. Then $(I - \tilde{A})(u + v) = (I - A)u + 2v = w$, which shows that $\tilde{A}$ is m-dissipative. (It follows from Theorem 3.6.2 that $\tilde{A}$ is as well maximal dissipative.) We have then proved

3.6.3 Lemma. *Let A_0 be dissipative in H and assume that A_0 is either closed or densely defined. Then there exists a m-dissipative extension of A.*

Lemma 3.6.3 has the following application to the theory of the Cauchy problem in Hilbert space: if A_0 is a closed or densely defined (but otherwise arbitrary) dissipative operator, there exists an extension $A \in \mathcal{C}_+(0, 1)$ of A_0, that is, the Cauchy problem for

$$u'(t) = Au(t)$$

is well posed in $t \geqslant 0$ and the propagator $S(\cdot)$ satisfies $\|S(t)\| \leqslant 1$. The requirements on A_0 are in general not difficult to meet: for instance, if A_0 is a differential operator and we take as $D(A_0)$ a space of sufficiently smooth functions in L^2 (satisfying boundary conditions if we work in a portion of Euclidean space), it is easy, using integration by parts, to establish conditions on the coefficients of A_0 and on the boundary conditions (if any) that guarantee dissipativity of A_0. On the other hand, since the elements of $D(A_0)$ are only restricted by smoothness assumptions and (possibly) by boundary conditions, it is plain that $D(A_0)$ must be dense in L^2. However, the result refers not to A_0 but to A, and the construction of A and of

$D(A)$ provided by Lemma 3.6.5 is far from explicit, since it presupposes the identification of the subspace K or, equivalently, of the nullspace of the operator $I - A_0^*$. Because of this, the usefulness of Lemma 3.6.5 is rather limited in this context.

3.6.4   Remark. The argument in Theorem 3.6.2 proving that a m-dissipative operator must be maximal dissipative extends without changes to the Banach space case. This is not the case with the converse, as the following counterexample shows.

3.6.5   Example (Lumer-Phillips [1961: 1]). Let $E = C_{0,1}[0, 1]_{\mathbb{R}}$ and $Au = u'$, where $D(A)$ is the set of all $u \in E$ such that $Au' \in E$. Then (a) A is maximal dissipative, and (b) $(I - A)D(A) \neq E$.

We note also that Lemma 3.6.3 fails in the Banach space case, since an operator as the one in Example 3.6.5 is not m-dissipative and, being maximal dissipative, does not admit dissipative extensions at all. We note in passing that every dissipative operator (not necessarily closed or densely defined) in an arbitrary Banach space E can be extended to a maximal dissipative operator by a standard application of Zorn's lemma. This is slightly more general than Lemma 3.6.3 even in the Hilbert space case, although the extension is even more esoteric.

Theorems 3.6.1 and 3.1.9 have interesting interpretations in Hilbert spaces. Observe that both relations (3.6.3) and (3.6.4) coalesce into

$$\text{Re}(Au, u) = 0 \quad (u \in D(A)). \tag{3.6.13}$$

It follows that the operator A is conservative if and only if $B = -iA$ is symmetric. In fact, assume B is symmetric. Then $(iBu, u) = -i(Bu, u)$, where (Bu, u) is real, thus iB is conservative. On the other hand, let A be conservative. Taking real parts in the identity $(A(u + v), u + v) = (Au, u) + (Au, v) + (Av, u) + (Av, v)$, we obtain

$$\text{Re}(Au, v) + \text{Re}(u, Av) = 0 \quad (u, v \in D(A)). \tag{3.6.14}$$

Applying (3.6.14) to iu, v and adding the two equalities, we obtain $(Au, v) + (v, Au) = 0$ or

$$(iAu, v) = (u, iAv), \tag{3.6.15}$$

which shows that $B = -iA$ is symmetric (of course, so is iA).

Condition (3.6.5) (say, for $\lambda = 1$) can be written in terms of B as follows:

$$(B + iI)D(B) = H, \tag{3.6.16}$$

while the same condition for $-A$ is

$$(B - iI)D(B) = H. \tag{3.6.17}$$

Clearly (3.6.17) is equivalent to the vanishing of the negative deficiency index d_- of B; likewise, (3.6.16) holds if and only if $d_+ = 0$ and any of the two conditions is equivalent to B being maximal symmetric (see Section 6). We obtain in this way the following result.

3.6.6   Theorem.   *Let $A \in \mathcal{C}_+(1,0)$ and let the propagator of* (3.6.1) *be isometric. Then $A = iB$, where B is maximal symmetric. Conversely, let B be maximal symmetric. Then either $A = iB$ (if $d_+ = 0$) or $A = -iB$ (if $d_- = 0$) belongs to $\mathcal{C}_+(1,0)$ and the propagator of* (3.6.1) *is isometric.*

Theorem 3.6.6 is the translation of Theorem 3.6.1 to the Hilbert space case. The corresponding version of Theorem 3.1.9 is obtained noting that both defficiency indices of a symmetric operator are zero if and only if the operator is self-adjoint.

3.6.7   Theorem.   *The operator A belongs to $\mathcal{C}(0,1)$ and the propagator of* (3.6.1) *is isometric in $-\infty < t < \infty$ if and only if $A = iB$, where B is a self-adjoint operator.*

We note that in this case each of the isometric operators $S(t)$ is invertible $(S(t)^{-1} = S(-t))$ thus $S(t)$ is *unitary*, that is, $S(t)^* = S(t)^{-1}$.

These results can be applied to the operators $\mathcal{C}$ of the previous section in an obvious fashion.

3.6.8   Example.   Let $A_1, \ldots, A_m, B$ satisfy the smoothness assumptions in Section 3.5, plus condition (3.5.11)–(3.5.12) and $\mathrm{Re}(K(x)u, u) = 0$ for all $x \in \mathbb{R}^m$, $u \in \mathbb{C}^\nu$, or

$$Q(x) = 0 \quad (x \in \mathbb{R}^m). \tag{3.6.18}$$

Then $i\mathcal{C} = i\overline{\mathcal{C}}_0 = i(\mathcal{C}'_0)^*$ is self-adjoint, so that $\mathcal{C} \in \mathcal{C}(1,0)$. Condition (3.6.18) is necessary in order that $\mathcal{C}$ be conservative.

On the other hand, if (3.5.11)–(3.5.12) is violated, condition (3.6.18) does no longer guarantee that $\mathcal{C}$ is conservative (the operator in Example 3.5.6 is not).

3.6.9   Example.   None of the operators A, $A(\beta)$ or $A(\beta_0, \beta_1)$ of Sections 3.2, 3.3, or 3.4 is conservative in the spaces L^p $(1 \leqslant p \leqslant \infty)$ or C (see the next section, however, for conservation of norm of nonnegative solutions in L^1).

3.6.10   Example.   Check Theorem 3.6.1 for the operators in Example 2.2.6.

3.7.   DIFFERENTIAL EQUATIONS IN BANACH LATTICES. POSITIVE SOLUTIONS. DISPERSIVE OPERATORS.

We have already observed in Section 1.3 that when the equation

$$\frac{\partial u}{\partial t} = \Delta u \tag{3.7.1}$$

is used as a model for heat or diffusion processes, the only solutions that can have any physical meaning must be positive. Also we have seen in a particular case that solutions of (3.7.1) that are initially positive stay positive forever. We introduce in this section an abstract framework to deal with this kind of situation.

Let E be a *real* Banach space with a partial order relation $\leqslant$ between its elements. Assume E is a *lattice* with respect to $\leqslant$, that is, that any two elements of E have a least upper bound $u \vee v$ and a greater lower bound $u \wedge v$. Following Birkhoff [1967: 1] (see also Schaefer [1974: 1]), we say that E is a *Banach lattice* if the following relations among the linear structure of E, its norm $\|\cdot\|$, and the order relation $\leqslant$, are satisfied:

(I)   If $u \leqslant v$, then $u + w \leqslant v + w$ $(w \in E)$.   (3.7.2)

(II)   If $u \leqslant v$, then $\lambda u \leqslant \lambda v$ $(\lambda \geqslant 0)$.   (3.7.3)

and

(III)   If $|u| \leqslant |v|$, then $\|u\| \leqslant \|v\|$,   (3.7.4)

where $|u|$, the *modulus* (or *absolute value*) of u, is the element of E defined by

$$|u| = u \vee (-u).$$

Given $u \in E$, let

$$u_+ = u \vee 0, \quad u_- = (-u) \vee 0.$$

We call u_+ (u_-) the *positive* (*negative*) part of u. It follows from (I) (taking $w = -u - v$) that $u \leqslant v$ implies $-v \leqslant -u$; we obtain easily from this that

$$u \vee v = -((-u) \wedge (-v)), \quad u \wedge v = -((-u) \vee (-v)) \quad (3.7.5)$$

and it is a direct consequence of (I) that

$$\begin{array}{cc} (u \vee v) + w = (u + w) \vee (v + w), \\ (u \wedge v) + w = (u + w) \wedge (v + w) \end{array} \quad (w \in E) \quad (3.7.6)$$

and of (II) that

$$\begin{array}{cc} \lambda(u \vee v) = (\lambda u) \vee (\lambda v), \\ \lambda(u \wedge v) = (\lambda u) \wedge (\lambda v) \end{array} \quad (\lambda \geqslant 0). \quad (3.7.7)$$

It follows from (3.7.5) and (3.7.6) that $u - u_+ = u - u \vee 0 = u + (-u) \wedge 0 = 0 \wedge u = -(0 \vee (-u)) = -u_-$, thus

$$u = u_+ - u_-. \quad (3.7.8)$$

On the other hand, $u + |u| = u + u \vee (-u) = 2u \vee 0 = 2u_+$, so that

$$|u| = u_+ + u_-. \quad (3.7.9)$$

Other useful relations, immediate consequences of the three postulates and of the previously shown equalities, are

$$|u| = 0 \text{ if and only if } u = 0, \tag{3.7.10}$$

$$|\lambda u| = |\lambda| \, |u|, \tag{3.7.11}$$

and the "triangle inequality"

$$|u + v| \leqslant |u| + |v| \tag{3.7.12}$$

valid for arbitrary $u, v \in E$ and $\lambda \in \mathbb{R}$. Note also that (3.7.4) implies

$$\| \, |u| \, \| = \|u\|. \tag{3.7.13}$$

We have $u = v + (u - v) \leqslant v_+ + |u - v|$, so that $u_+ \leqslant v_+ + |u - v|$. Interchanging u and v and combining the two inequalities, we obtain

$$|u_+ - v_+| \leqslant |u - v|,$$

which implies, in view of (III),

$$\|u_+ - v_+\| \leqslant \|u - v\|, \tag{3.7.14}$$

so that the map $u \to u_+$ is uniformly continuous. Naturally, the same is true of $u \to u_-$; it may be shown by the same elementary means that $(u, v) \to u \wedge v$ and $(u, v) \to u \vee v$ are uniformly continuous from $E \times E$ into E. As a consequence of these remarks, we see that the set

$$E_+ = \{u; \, u \geqslant 0\} = \{u; \, u_- = 0\}$$

of *nonnegative*[5] elements of E is closed.

All the real spaces hitherto used in our treatment of differential operators are Banach lattices under the natural order relations: in the space $C(K) = C(K)_{\mathbb{R}}$ we say that $u \leqslant v$ if $u(x) \leqslant v(x)$ for $x \in K$. The same definition applies to the spaces $C_0(K)$ and the various subspaces used. In the spaces $L^p(\Omega, \mu) = L^p(\Omega; \mu)_{\mathbb{R}}$ $(1 \leqslant p \leqslant \infty)$, $u \leqslant v$ if $u(x) \leqslant v(x)$ μ-almost everywhere in Ω.

A duality map $\theta: E \to E^*$ is called *proper* if it satisfies the following two conditions:

(IV)   If $u, v \geqslant 0$, then $\langle \theta(u), v \rangle \geqslant 0$. $\qquad\qquad$ (3.7.15)

(V) $\langle \theta(u_+), u \rangle = \|u_+\|^2$ $\quad (u \in E)$. $\qquad\qquad$ (3.7.16)

A cursory examination of the duality maps in question establishes the following result.

3.7.1 Lemma. (a) *Let K be a closed subset of $\mathbb{R}^m$ and $E = C_0(K)$, $C(K)$ (if K is compact), or a closed subspace thereof. Then any duality map $\theta: E \to E^*$ is proper.* (b) *Let Ω be a measurable subset of $\mathbb{R}^m$, $E = L^p(\Omega)$. Then if $1 < p < \infty$, the only duality map $\theta: E \to E^*$ is proper; if $p = 1$, a*

---

[5] In this context, "nonnegative" does not mean "not negative."

duality map θ is proper if and only if

$$\theta(u)(x) = 0 \text{ a.e. where } u(x) = 0.$$

An operator $B: E \to E$ is said to be *positive* if and only if

$$Bu \geqslant 0 \text{ whenever } u \geqslant 0, \tag{3.7.17}$$

and an operator $A: D(A) \to E$ will be called *dispersive* (with respect to a proper duality map θ) if and only if

$$\langle \theta(u_+), Au \rangle \leqslant 0 \quad (u \in D(A)). \tag{3.7.18}$$

Our goal is to characterize those operators $A \in \mathcal{C}_+(1,0)$ for which the propagator of

$$u'(t) = Au(t) \tag{3.7.19}$$

is positive for all t. This will be done in the following Theorem 3.7.3. We begin by establishing an auxiliary result of considerable interest in itself.

3.7.2 Lemma. *Let* $A \in \mathcal{C}_+(\omega)$. *Then the propagator of* (3.7.19) *is positive for all* $t \geqslant 0$ *if and only if* $R(\lambda)$ *is positive for all* $\lambda > \omega$.

Proof. We use (2.1.10) and (2.1.27), valid for real as well as for complex spaces by virtue of Remark 3.1.13:

$$R(\lambda)u = \int_0^\infty e^{-\lambda t} S(t)u\, dt \quad (\lambda > \omega), \tag{3.7.20}$$

$$S(t)u = \lim_{n \to \infty} \left(\frac{n}{t}\right)^n R\left(\frac{n}{t}\right)^n u \quad (t > 0). \tag{3.7.21}$$

The result follows then from the fact that the set $\{u; u \geqslant 0\}$ is closed.

3.7.3 Theorem. *Let* $A \in \mathcal{C}_+(0,1)$ *and assume the propagator* $S(\cdot)$ *of* (3.7.19) *is positive for all* t *(i.e., solutions with positive initial value remain positive forever). Then* A *is dispersive with respect to any proper duality map θ. Conversely, assume that* A *is dispersive with respect to some proper duality map θ and*

$$(\lambda I - A)D(A) = E \tag{3.7.22}$$

for some $\lambda > 0$. *Then* $A \in \mathcal{C}_+(0,1)$ *and the propagator of* (3.7.19) *is positive for all* $t \geqslant 0$.

Proof. Assume that $S(t)$ is positive for $t \geqslant 0$, and let θ be a proper (but otherwise arbitrary) duality map. If $u \in E$, $u_+, u_- \geqslant 0$; then $S(t)u_- \geqslant 0$, and, since θ is proper,

$$\langle \theta(u_+), S(t)u_- \rangle \geqslant 0.$$

Accordingly,

$$
\begin{aligned}
\langle \theta(u_+), u \rangle = \|u_+\|^2 &\geqslant \|S(t)u_+\| \|u_+\| \geqslant \langle \theta(u_+), S(t)u_+ \rangle \\
&\geqslant \langle \theta(u_+), S(t)u_+ \rangle - \langle \theta(u_+), S(t)u_- \rangle \\
&= \langle \theta(u_+), S(t)u \rangle.
\end{aligned}
$$

Taking $u \in D(A)$, we obtain

$$
\langle \theta(u_+), Au \rangle = \lim_{h \to 0+} h^{-1}(\langle \theta(u_+), S(h)u \rangle - \langle \theta(u_+), u \rangle) \leqslant 0,
$$

which proves that A is dispersive. In order to show the converse, let $\lambda > 0$ be such that (3.7.22) is satisfied and let $v \geqslant 0$, u the solution of

$$
\lambda u - Au = v. \tag{3.7.23}
$$

Then, if A is dispersive with respect to the proper duality map θ, we have

$$
\begin{aligned}
\lambda \|u_-\|^2 = \lambda \|(-u)_+\|^2 &= \lambda \langle \theta((-u)_+), -u \rangle \\
&\leqslant \lambda \langle \theta((-u)_+), -u \rangle - \langle \theta((-u)_+), A(-u) \rangle \\
&= -\langle \theta((-u)_+), v \rangle \leqslant 0.
\end{aligned}
$$

Clearly this implies that $u_- = 0$, so that $u \geqslant 0$ and therefore

$$
\begin{aligned}
\lambda \|u\|^2 = \lambda \|u_+\|^2 &= \lambda \langle \theta(u_+), u \rangle \leqslant \lambda \langle \theta(u_+), u \rangle - \langle \theta(u_+), Au \rangle \\
&= \langle \theta(u_+), v \rangle \leqslant \|u_+\| \|v\| = \|u\| \|v\|,
\end{aligned}
$$

whence we obtain

$$
\|u\| \leqslant \frac{1}{\lambda} \|v\|
$$

for the solution u of (3.7.22) under the assumption that $v \geqslant 0$. Applying the result to $v = 0$, we see that $\lambda I - A$ is one-to-one. Accordingly, $R(\lambda) = (\lambda I - A)^{-1}$ exists and the preceding arguments show that $R(\lambda)$ is a positive operator and

$$
\|R(\lambda)u\| \leqslant \frac{1}{\lambda} \|u\| \tag{3.7.24}
$$

if $u \geqslant 0$. To prove that this inequality actually holds for any $u \in E$, we only have to observe that

$$
\begin{aligned}
|R(\lambda)u| &= (R(\lambda)u)_+ + (R(\lambda)u)_- \\
&= R(\lambda)u_+ + R(\lambda)u_- \\
&= R(\lambda)(u_+ + u_-) \\
&= R(\lambda)|u|
\end{aligned}
$$

so that, in view of (3.7.24) and of the fact that $R(\lambda)$ is positive,

$$\|R(\lambda)u\| = \|R(\lambda)|u|\| \leqslant \lambda^{-1}\||u|\| = \lambda^{-1}\|u\|.$$

We have thus shown that

$$\|R(\lambda)\| \leqslant \frac{1}{\lambda} \quad (\lambda > 0). \tag{3.7.25}$$

We apply then Theorem 3.1.1 (as modified by Remark 3.1.13) and Lemma 3.7.2 to deduce that $A \in \mathcal{C}_+(1,0)$ and that $S(t)$ is positive for all t. This ends the proof.

We obtain as a byproduct of Theorem 3.7.3 that every dispersive operator that satisfies (3.7.22) must be dissipative. It is apparently unknown whether the implication holds in absence of (3.7.22). However, we have:

3.7.4 Example (Phillips [1962: 1]). (a) Let θ be a proper duality map such that: for every u there exist scalars $\alpha, \beta \geqslant 0$ (depending on u) with

$$\theta(u) = \alpha\theta(u_+) - \beta\theta(u_-). \tag{3.7.26}$$

Then if A is dispersive with respect to θ, it is also dissipative with respect to θ. (b) All the proper duality maps appearing in Lemma 3.7.1 (b) satisfy the requirements in (a).

We shall use most of the time Theorem 3.7.3 "at half power": we start with an operator A already known to belong to the class $\mathcal{C}_+$ and prove that A is in addition dispersive, so that $S(t)$ is positive for all t.

As a first application of the results in this section we reexamine the ordinary differential operators considered in Sections 2, 3, and 4 of this chapter, or rather their restrictions to the corresponding spaces of real-valued functions. To simplify the notation, these restrictions will be denoted by the same symbols. Consider the case $E = C_0(-\infty, \infty)$. If $u \in D(A)$ and $\mu \in \Theta(u_+)$, then μ is a positive measure and every point x in the support of μ is a maximum of u_+, hence of u; we have then $u'(x) = 0$, $u''(x) \leqslant 0$. Thus

$$\langle \mu, Au \rangle = \int (au'' + bu' + cu) \, d\mu \leqslant 0 \tag{3.7.27}$$

if the assumptions of Lemma 3.2.1 are satisfied, so that the corresponding A is dispersive. We deal next with the case $E = L^p(-\infty, \infty)$; for purposes of identification, we call A_p the operator defined in part (b) of Theorem 3.2.7. Direct verification of dispersivity is difficult in this case, since integration by parts is prevented by lack of sufficient information on the behavior of elements of $D(A)$ at infinity. We follow then a more roundabout way, examining directly the solution operators $S_p(\cdot)$ of $u'(t) = A_p u(t)$ and $S(\cdot)$ of $u'(t) = Au(t)$ (A the operator in C_0 in Theorem 3.2.7(a)). It was observed in the course of the proof of Theorem 3.2.7 that if $\lambda > 0$, v is a function, say, in $\mathcal{D}$, and if u is the solution of $\lambda u - A_p u = v$ in $D(A_p) \subseteq L^p$, then u

actually belongs to $D(A) \subseteq C_0$; it follows that

$$R(\lambda; A_p)v = R(\lambda; A)v. \qquad (3.7.28)$$

We note next that the preceding inequality extends to the powers of $R(\lambda)$, since (3.7.28) holds for all $\lambda > 0$ and powers of the resolvent can be expressed by means of derivatives with respect to λ (see Section 3). It follows then from (3.7.21) that

$$S_p(t)v = S(t)v$$

for all $t > 0$ and $v \in \mathcal{D}$.

Let now $u \in L^p, u \geqslant 0$; choose a sequence of nonnegative functions $\{v_n\}$ in $\mathcal{D}$ with $\|v_n - u\|_p \to 0$. Then $S(t)v_n = S_p(t)v_n \to S_p(t)u$ in L^p, all the functions $S(t)v_n$ being positive in $(-\infty, \infty)$. Since some subsequence must converge almost everywhere, it follows that $S_p(t)u(x) \geqslant 0$ almost everywhere, so that $S_p(t)$ is a positive operator for all $t \geqslant 0$ and all $p \geqslant 1$.

An entirely similar analysis can be carried out in the cases covered by Theorems 3.3.9 and 3.4.3. Details are left to the reader.

3.7.5 Theorem. *Let A be any of the operators considered in Theorems 3.2.7, 3.3.6, or 3.4.3. Then A is dispersive with respect to any proper duality map; equivalently, the solution operator of $u' = Au$ is positive for all $t \geqslant 0$.*

We close this section with some remarks on conservation of norms of nonnegative solutions. It was observed in Section 1.3 that, while the diffusion equation does not possess an isometric propagator in any L^p space (or in C), the L^1 norm of a solution with nonnegative initial value remains constant in time. This property admits an abstract formulation.

3.7.6 Theorem. *Let $A \in \mathcal{C}_+(0,1)$, and let $S(\cdot)$ be the solution operator of (3.7.19). Assume $N \subseteq E$, $N \cap D(A^2)$ is dense in N,*

$$S(t)N \subseteq N \quad (t \geqslant 0) \qquad (3.7.29)$$

and

$$\langle \theta(u), Au \rangle = 0, \quad u \in N \cap D(A^2), \qquad (3.7.30)$$

where $\theta: N \cap D(A^2) \to E^*$ is a duality map. Then,

$$\|S(t)u\| = \|u\| \quad (t \geqslant 0) \qquad (3.7.31)$$

for $u \in N$. Conversely, assume that N is a set satisfying (3.7.29), and (3.7.31) for $u \in N$. Then there exists a duality map $\theta: N \cap D(A) \to E^*$ such that (3.7.30) holds for $u \in N \cap D(A)$.

The proof of Theorem 3.7.6 can be read off that of Theorem 3.6.1 (where $N = E$) and will be left to the reader.

Theorem 3.7.6 can of course be applied to the differential operators considered in this chapter. We limit ourselves to the finite interval case.

3.7.7   Corollary.   *Let $A(\beta_0, \beta_l)$ be one of the operators in Lemma 3.4.2. Assume that*

$$a(x) > 0, \quad a''(x) - b'(x) + c(x) = 0 \quad (0 \leqslant x \leqslant l) \qquad (3.7.32)$$

and that the boundary conditions at 0 and at l are of type (I) *with*

$$\gamma_0 a(0) - a'(0) + b(0) = \gamma_l a(l) - a'(l) + b(l) = 0 \qquad (3.7.33)$$

Then $A(\beta_0, \beta_l)$ is m-dissipative in $L^1(0, l)$, dispersive with respect to any proper duality map, and (3.7.31) *holds for any $u \geqslant 0$.*

The proof applies Theorem 3.7.6 with $N = \{u; u \geqslant 0\}$. It is just as simple, however, to give a proof based on an argument similar to (1.3.17): it suffices to observe that

$$\frac{d}{dt} \int_0^l u(t, x) \, dx = 0$$

for every solution $u(t, x) = u(t)(x)$ of $u'(t) = A(\beta_0, \beta_l) u(t)$, which follows from (3.7.32) and (3.7.33), integrating by parts.

We note that when A_0 is written in variational form (3.2.35) the two conditions in Corollary 3.7.7 are:

$$a(x) > 0, \quad \hat{b}(x) = c(x) \quad (0 \leqslant x \leqslant l) \qquad (3.7.34)$$

$$\gamma_0 a(0) = \hat{b}(0), \quad \gamma_l a(l) = \hat{b}(l) \qquad (3.7.35)$$

3.7.8   Example.   Check Theorem 3.7.3 for the operators in Examples 2.2.4, 2.2.5, 2.2.6, and 2.2.7.

3.8.   MISCELLANEOUS COMMENTS

A result of Lyapunov [1892: 1] can be stated in a slightly generalized version as follows: the eigenvalues of the matrix A have negative real parts if and only if a positive definite matrix Y can be found satisfying

$$\mathrm{Re}(YAu, Yu) \leqslant -\kappa(u, u) \quad (u \in \mathbb{C}^m) \qquad (3.8.1)$$

for some $\kappa > 0$ (see Daleckiĭ-Kreĭn [1970: 1]). This property of the eigenvalues of A was recognized much before to be equivalent to

$$\|\exp(tA)\| \to 0 \text{ as } t \to +\infty; \qquad (3.8.2)$$

algebraic criteria were known in particular cases to physicists and engineers like Maxwell working on regulator theory and were given generality by Routh and Hurwitz during the last years of the nineteenth century. Thus, Lyapunov's result can be considered as a forerunner of the theory of dissipative operators presented in this chapter.

The Hilbert space theory was initiated and developed by Phillips [1959: 1] although dissipativity and closely associated ideas were used before for partial differential operators by Friedrichs [1954: 1], [1958: 1], Phillips himself [1957: 1], and others; we note that Friedrichs coined the word *accretive* to indicate the negative of a dissipative operator. The bulk of the theory of dissipative operators in Hilbert space as presented in this chapter can be found in Phillips [1959: 1]. Dissipative operators in Banach spaces were introduced by Lumer and Phillips [1961: 1] in a slightly different form using the "semi-inner products" introduced by Lumer [1961: 1] instead of duality maps; the definition was cast in the present form by Nelson [1964: 1]. The term "duality map" is taken from Goldstein [1970: 5]; the present treatment of dissipativity leans heavily on these lecture notes. We note a definition of dissipative operators in Banach space due to Maz'ja and Sobolevskiĭ [1962: 1] under certain smoothness conditions on the norm. The results in Section 3.1 are due to Lumer-Phillips [1961: 1] or are adaptations of earlier results of the Hilbert space theory. Theorem 3.6.7 is the theorem of Stone mentioned in Section 2.5, whereas Theorem 3.6.6 was discovered by J. L. B. Cooper [1947: 1]; both results were originally formulated in the semigroup-generator language. The version of Cooper's theorem for Banach spaces presented here (Theorem 3.6.1) seems to be a folk result. A related result can be found in Papini [1971: 1].

The present theory of semigroups of positive operators in Banach lattices is due to Phillips [1962: 1]; dispersive operators were introduced by W. J. Firey in a particular instance (see Phillips [1962: 1, p. 295]). Some of the ideas in Section 3.7 have interesting finite-dimensional ancestors; we refer the reader to Bellman [1970: 1, Ch. 14].

We comment finally on the results on ordinary and partial differential operators in this chapter. The theory of symmetric hyperbolic operators in Section 3.5 is due to Friedrichs [1944: 1]. His result on identity of $\overline{\mathcal{C}}_0$ (the *strong extension* of $\mathcal{C}_0$) and $(\mathcal{C}_0')^*$ (the *weak extension* of $\mathcal{C}_0$) has been extensively generalized (for instance, to situations involving boundary conditions as in Lax-Phillips [1960: 1]). We note that assumption (3.5.12) on the function ρ is nothing but Wintner's condition in [1945: 1] for global existence of solutions of the ordinary differential system $x'(t) = a(x(t))$, where $x(\cdot)$ is a m-dimensional vector function, $a(\cdot)$ maps $\mathbb{R}^m$ into itself, and $|a(x)| \leqslant \rho(|x|)$. This is not surprising since it has been shown by Povzner [1964: 1] that, under the assumption that $\sum D^k a_k = 0$, global existence is equivalent to the identity of the weak and strong extensions of $\mathcal{C}_0 = \sum a_k(x) D^k$; many other related results are proved there. For connected material see Lelong-Ferrand [1958: 1].

One of the results in the investigations of Weyl [1909: 1], [1910: 1], [1910: 2] on the spectral theory of ordinary differential operators in infinite intervals can be stated in modern terminology as follows: the operator

$$A_0 u = (p(x)u')' + q(x)u \qquad (3.8.3)$$

in $x \geqslant 0$ (with domain and smoothness assumptions as in Section 3.3) has a self-adjoint extension in $L^2(0, \infty)$ obtained by imposition of a boundary condition at zero if

$$p(x) > 0, q(x) \leqslant \omega \quad (-\infty < x < \infty). \tag{3.8.4}$$

(This corresponds to what Weyl calls the *limit point case*; in the *limit circle case*, to obtain a self-adjoint extension we must impose in addition a *boundary condition at infinity* of the form

$$\lim_{x \to +\infty} \left(\omega_0(x)u(x) + \omega_1(x)u'(x) \right) = 0$$

for suitable functions ω_0, ω_1.) Under conditions (3.8.4), the extension is semibounded above; weaker conditions on q, even if pertaining to the limit point case, produce self-adjoint extensions not necessarily semibounded above (for a complete account of the theory see Dunford-Schwartz [1963: 1, Ch. 13]). Roughly along the same lines, Yosida [1949: 2] and Hille [1954: 2] considered a general non-self-adjoint A_0 in relation to applications to unidimensional Markov processes. The problem is that of obtaining extensions of the operator A_0 in L^1 and in spaces of continuous functions by means of boundary conditions at finite points and at infinity; these extensions are required to be infinitesimal generators of positive semigroups. The works of Feller [1954: 1], [1954: 2], [1955: 1], [1955: 2], [1956: 1], and others proceed in the same direction but use extremely weak assumptions on the operators (see Mandl [1968: 1] or Itô-McKean [1974: 1] for modern treatments). The results in Sections 3.2 and 3.3 concentrate only on the simpler problem of obtaining m-dissipative extensions of A_0 imposing no boundary conditions at infinity in the case $-\infty < x < \infty$ or only one boundary condition at 0 in the case $x \geqslant 0$. The generation properties of ordinary differential operators in a finite interval were investigated by Mullikin [1959: 1] in L^p and C spaces; his results are more general than those in Section 3.4 since dissipativity conditions like (3.4.8) on the boundary conditions are absent, and will be met again (this time in almost full generality) in Section 4.3.

Many papers on the theory and applications of dissipative operators will be found later under other headings. Here we refer the reader to Hasegawa [1966: 1], [1969: 1], [1971: 1], Ljubič [1963: 1], [1965: 1], Mlak [1960: 1], [1960/61: 1], [1966: 1], Moore [1971: 1], [1971: 2], Nagumo [1956: 1], Olobummo [1963: 1], [1967: 1], Olobummo-Phillips [1965: 1], Papini [1969: 1], [1969: 2], [1971: 1], Phillips [1957: 1], [1959: 1], [1959: 3], [1961: 1], [1961: 2], and Crandall-Phillips [1968: 1].

(a) *Special Semigroups*. Partly along the lines of Stone's theorem, a large number of results dealing with semigroups $S(\cdot)$ where each $S(t)$ belongs to certain classes of operators in Hilbert or Banach spaces have been investigated. For $S(t)$ self-adjoint we have already pointed out in Section 2.5 the works of Hille [1938: 1] and Sz.-Nagy [1938: 1]; an extension to unbounded self-adjoint operators has been given by Devinatz [1954: 1]. Normal operators are treated in Sova [1968: 4], and spectral operators in McCarthy-Stampfli [1964: 1], Berkson [1966: 1], Panchapagesan [1969: 1], and Sourur [1974: 1], [1974: 2]. For generalized spectral operators in the sense of Colojoara-Foias [1968: 1], see Lutz [1977: 1].

Other works deal with, for instance, semigroups or groups of isometries in special spaces like L^p or H^p. See Berkson-Porta [1974: 1], [1976: 1], [1977: 1], Berkson-Kaufman-Porta [1974: 1], Berkson-Fleming-Jamison [1976: 1], and Goldstein [1973: 1]. Stone's theorem has also been generalized in the direction of allowing the parameter t to roam over groups more general than the real line. See Salehi [1972: 1].

(b) *Contraction Semigroups and Dissipative Operators in Hilbert Spaces.* Let H_+ be a Hilbert space, H a closed subspace of H_+, P the orthogonal projection of H_+ into H. If B_+ is an operator in H_+ we define pr B_+, the *projection of B_+ into H* by

$$Bu = (\operatorname{pr} B_+)u = PB_+ u \quad (u \in H).$$

It was proved by Sz.-Nagy [1953: 1], [1954: 1] that if $S(\cdot)$ is a strongly continuous semigroup in a Hilbert space such that

$$\|S(t)\| \leqslant 1 \tag{3.8.5}$$

for $t \geqslant 0$, then there exists a second Hilbert space H_+ having H as a closed subspace and a strongly continuous group $U(\cdot)$ of *unitary* operators such that

$$S(t) = \operatorname{pr} U(t) \quad (t \geqslant 0). \tag{3.8.6}$$

Moreover, the space H_+ is *minimal* (in the sense of being generated by all elements of the form $U(t)u, u \in H$) and uniquely determined (up to isomorphisms). For additional details, see Sz.-Nagy-Foias [1966: 1]. We note that the result above was proved by Cooper [1947: 1] in the particular case where each $S(t)$ is an isometry; in this setting, the theorem is a consequence of the well known fact that every symmetric operator A in H can always be extended to a self-adjoint operator in a larger space H_+.

Sz.-Nagy's result combined with Stone's theorem shows that (modulo the projection P) the study of strongly continuous contraction semigroups in a Hilbert space can be reduced to that of operator-valued functions of the form $\exp(itA)$, A self-adjoint, with a corresponding relation between their generators (maximal dissipative operators) and self-adjoint operators. We refer the reader to Sz.-Nagy and Foias [1969: 1] for a thorough exploitation of this line of approach and note the following discrete version of (3.8.6) also due to Sz.-Nagy [1953: 1]: if S is an operator in H with $\|S\| \leqslant 1$, then there exists H_+ containing H as a closed subspace and a unitary operator U in H_+ such that

$$S^n = \operatorname{pr} U^n \quad (n \geqslant 0). \tag{3.8.7}$$

The space H_+ and the operator U enjoy the same properties of minimality and uniqueness up to isomorphism as in the continuous case.

(c) *Uniformly Bounded Groups, Semigroups, and Cosine Functions in Hilbert Space.* In view of the results in the previous subsection, the following question is of considerable interest. Let H be a Hilbert space, $S(\cdot)$ a semigroup (group) of linear bounded operators in H such that

$$\|S(t)\| \leqslant C \tag{3.8.8}$$

in $t \geqslant 0$ (in $-\infty < t < \infty$). Can H be given an equivalent Hilbert norm in which (3.8.5) holds in $t \geqslant 0$ (in $-\infty < t < \infty$)?

For a group, the answer is in the affirmative, as shown by Sz.-Nagy [1947: 1]. We note that, since $S(\cdot)$ satisfies (3.8.5), each $S(t)$ is invertible and isometric, hence *unitary* in the new norm. Sz.-Nagy also proves the discrete version of the result above: if S is an operator such that

$$\|S^n\| \leqslant C \quad (n = \ldots, -1, 0, 1, \ldots), \tag{3.8.9}$$

then a similar renorming of the space makes S unitary. The corresponding theorems for uniformly bounded semigroups in $t \geqslant 0$ are false: in the discrete case, a counterexample was given by Foguel [1964: 1] and in the continuous case by Packel [1969: 1]. See Halmos [1964: 1] for additional information. An analogue of Sz.-Nagy's result for cosine functions was proved by the author [1970: 4]: under the hypothesis that $C(\cdot)$ is uniformly bounded in $-\infty < t < \infty$ (or, equivalently, in $t \geqslant 0$), an equivalent Hilbert norm can be found in which each $C(t)$ is *self-adjoint.* A discrete analogue holds for uniformly bounded families of operators $\{C_n\}$ satisfying $C_0 = I$, $C_{n+m} + C_{n-m} = 2C_n C_m$ $(m, n = \ldots, -1, 0, 1, \ldots)$. Sz.-Nagy's theorem has been generalized by Dixmier [1950: 1] to uniformly bounded representation of certain groups into the algebra of linear bounded operators in a Hilbert space. A similar generalization of the author's result was given by Kurepa [1972: 1].

We note that the problem of renorming a *Banach* space with an equivalent norm, which improves (3.8.8) into (3.8.5), is rather trivial; see Example 3.1.10.

(d) *Truly Nonlinear Equations and Semigroups.* The theory of contraction semigroups, and of dissipative operators as their infinitesimal generators, has been generalized in the last two decades to nonlinear operators. In the nonlinear realm, a *contraction semigroup* is a family $\{S(t); t \geqslant 0\}$ of nonlinear maps defined in a Banach space E or a subset X thereof and satisfying the semigroup equations (2.1.2). The nonlinear counterpart of (3.8.5) is

$$\|S(t)u - S(t)v\| \leqslant \|u - v\| \quad (u, v \in X),$$

while the dissipativity condition (3.1.5) reads

$$\text{Re}\langle w^*, Au - Av \rangle \leqslant 0 \quad (w^* \in \Theta(u - v)).$$

The relation between semigroup and generator roughly parallels the linear case, but many interesting new phenomena appear (for instance, the infinitesimal generator may be multiply valued). For an account of the theory and the applications we refer the reader to Barbu [1974: 1], Brézis [1973: 1], Ciorănescu [1973: 1], and Pazy [1970: 1]. A short exposition of some of the central points of the theory can be found on Yosida [1978: 1].

Chapter 4

Abstract Parabolic Equations: Applications to Second Order Parabolic Equations

An operator $A \in \mathcal{C}_+$ is *abstract elliptic* if the solutions of the differential equation $u'(t) = Au(t)$ are infinitely differentiable in $t > 0$. Particularly important examples of abstract elliptic operators are those in the class $\mathcal{C}$, where solutions can be analytically extended to a sector about the positive real axis. The class $\mathcal{C}$ is characterized in the first two sections of this chapter; the rest of the material is on applications to differential operators. Section 4.3 deals with the ordinary differential operators in compact intervals last seen in Section 3.4; in Sections 4.4 to 4.10 we study second order partial differential operators in smooth domains of Euclidean space, beginning with dissipativity and assignation of boundary conditions. Extensions in $\mathcal{C}_+$ of differential operators in various spaces are then constructed. We treat L^2 spaces in Sections 4.6 and 4.7, and L^p spaces and spaces of continuous functions in Section 4.8. Section 4.9 is devoted to the proof that the L^p extensions belong in fact to the class $\mathcal{C}$ (with attendant smoothness results for solutions of the equation $u'(t) = Au(t)$) and to the removal of certain dissipativity assumptions on the boundary conditions; these assumptions are only significant in L^1 and in spaces of continuous functions and play a strictly auxiliary role in other spaces. Finally, we study in Section 4.10 compactness of families of solutions and preservation of positivity of initial data.

For the ordinary differential operators in Section 4.3, the treatment is essentially the same in all spaces; in contrast, the L^2 case for partial differential operators is more elementary and is thus presented first. The reader interested only in the L^2 theory may proceed directly to Section 4.6 from Section 4.2, skipping most of the intermediate material.

4.1. ABSTRACT PARABOLIC EQUATIONS

Let A be an operator in $\mathcal{C}$ and let $u(t) = S(t)u$ be a generalized solution of

$$u'(t) = Au(t) \qquad (4.1.1)$$

in $(-\infty, \infty)$. We know (Section 2.2) that if the initial value $u(0)$ belongs to $D(A)$, then $u(\cdot)$ will be a genuine solution of (4.1.1); in particular, $u(\cdot)$ is continuously differentiable and belongs to $D(A)$ for all t. On the other hand, let $u(\cdot)$ be differentiable at t_0. Then, as $h^{-1}(S(t_0 + h) - S(t_0))u = h^{-1}(S(h) - I)S(t_0)u$, it is immediate that $S(t_0)u \in D(A)$; thus $u(0) = S(-t_0)S(t_0)u \in D(A)$ as well. In other words, only solutions that originate in $D(A)$ may be smooth. This property is based, roughly speaking, on the fact that the solution operator of (4.1.1) is defined for all t. We study in this section operators A having properties in a sense opposite to those of elements of the class $\mathcal{C}$.

The operator $A \in \mathcal{C}_+$ is said to be *abstract elliptic* (in symbols, $A \in \mathcal{C}^\infty$) and the equation (4.1.1) *abstract parabolic* if and only if every generalized solution of (4.1.1) is continuously differentiable in $t > 0$ (even if $u(0)$ does not belong to $D(A)$). The following result shows that the smoothness properties implied by this definition are stronger than apparent.

4.1.1 Lemma. *The following four statements are equivalent:*

(a) $S(\cdot)u$ *is continuously differentiable in* $t > 0$ *for all* $u \in E$.
(b) $S(\cdot)u$ *is infinitely differentiable in* $t > 0$ *for all* $u \in E$.
(c) $S(t)E \subseteq D(A)$ $(t > 0)$.
(d) $S(t)E \subseteq D(A^\infty) = \cap_{n=1}^\infty D(A^n)$ $(t > 0)$.

Proof. It is plain that (b) implies (a) and (d) implies (c). We show that (c) implies (d). If $t > 0$, we have $S(t)u = S(t/2)S(t/2)u$; since A and S commute, $AS(t)u = S(t/2)AS(t/2)u \in D(A)$ and thus $S(t)u \in D(A^2)$. The general statement is obtained on the basis of the identity $S(t) = S(t/n)^n$.

We show next the implication (a) $\Rightarrow$ (c). Let $0 < \alpha < \beta < \infty$ and consider the operator

$$u \to (S(t)u)' = \mathcal{S}u$$

from E into the space $C = C(\alpha, \beta; E)$ of all continuous E-valued functions $f(\cdot)$ defined in $\alpha \leqslant t \leqslant \beta$, endowed with the supremum norm. Assume

$u_n \to u$ in E and $\mathbb{S}u = (S(\cdot)u_n)' \to f(\cdot)$ in C. Then

$$S(t)u_n - S(\alpha)u_n \to \int_\alpha^t f(s)\, ds \quad (\alpha \leqslant t \leqslant \beta),$$

so that letting $n \to \infty$, we obtain $\mathbb{S}u = (S(t)u)' = f(t)$. This shows that $\mathbb{S}$ is closed, hence bounded because of the closed graph theorem. But if $u \in D(A)$, $(S(t)u)' = AS(t)u$; thus $AS(t)$ is bounded and it results from the closedness of A that $S(t) \subseteq D(A)$ for $\alpha \leqslant t \leqslant \beta$. Since α and β are arbitrary, (c) follows.

We prove finally that (b) is a consequence of (d). To this end we observe that if $u \in D(A^\infty)$, then $(S(t)u)' = AS(t)u = S(t)Au$, which is again continuously differentiable; then $(S(t)u)'' = (S(t)Au)' = S(t)A^2u, \ldots,$ and so on. It follows that $S(\cdot)u$ is infinitely differentiable; making use of (d) and writing $S(t) = S(t - t_0)S(t_0)u$ for $t > t_0$, we see that (b) holds. This ends the proof.

We have proved as well the fact that $AS(t) = S'(t)$ is a bounded operator and that $t \to AS(t)$ is strongly continuous for $t > 0$. Clearly the same is true of $A^m S(t) = S(t - t_0)(AS(t_0/m))^m$. We define

$$\gamma(t) = \|AS(t)\| = \|S'(t)\| \quad (t > 0). \tag{4.1.2}$$

Because of the strong continuity of $AS(\cdot)$ and of the uniform boundedness theorem, γ is bounded on compacts of $t > 0$. The same is true of

$$\gamma_m(t) = \|A^m S(t)\| = \|S^{(m)}(t)\| \quad (t > 0)$$

for any $m \geqslant 1$. If $0 < \alpha \leqslant t < t' \leqslant \beta < \infty$,

$$\|S^{(m)}(t')u - S^{(m)}(t)u\| \leqslant \int_t^{t'} \|S^{(m+1)}(s)u\|\, ds \leqslant C(t' - t)\|u\|,$$

where C only depends on α, β so that $t \to S^{(m)}(t)$ is continuous in $t > 0$ as a (E)-valued function. Since, on the other hand, we have

$$S^{(m)}(t') - S^{(m)}(t) = \int_t^{t'} S^{(m+1)}(s)\, ds$$

(as can be seen applying the operators on both sides to an arbitrary element $u \in E$), the following result is immediate.

4.1.2   Corollary.   Let $A \in \mathcal{C}^\infty$. Then $t \to S(t) \in (E)$ is infinitely differentiable in $t > 0$ and $S^{(m)}(t) = A^m S(t)$ $(t > 0)$.

We examine more closely the growth properties of the function γ. In view of the inequality $\|AS(t)\| \leqslant \|AS(t_0)\| \|S(t - t_0)\|$, we see that

$$\gamma(t) \leqslant C_{t_0} e^{\omega t} \quad (t \geqslant t_0) \tag{4.1.3}$$

if $A \in \mathcal{C}_+(\omega)$. In contrast, γ may grow arbitrarily fast when $t \to 0$.

The following result provides a complete characterization of the class $\mathcal{C}^\infty$.

4.1.3 Theorem.   *Let $A \in \mathcal{C}_+$. Assume that for every $\alpha > 0$ there exists*
$\beta = \beta(\alpha) > 0$ *such that $\rho(A)$ contains the region*

$$\Omega(\alpha, \beta) = \{\lambda; \operatorname{Re} \lambda \geqslant \beta - \alpha \log|\lambda|\} \qquad (4.1.4)$$

and

$$\|R(\lambda)\| \leqslant C|\lambda|^m \quad (\lambda \in \Omega(\alpha, \beta)) \qquad (4.1.5)$$

where m is an integer independent of α. Then $A \in \mathcal{C}^\infty$. Conversely, let $A \in \mathcal{C}^\infty$. Then for every α there exists β such that $\Omega(\alpha, \beta) \subseteq \rho(A)$ and

$$\|R(\lambda)\| \leqslant C|\lambda| \quad (\lambda \in \Omega(\alpha, \beta)). \qquad (4.1.6)$$

Theorem 4.1.3 is a particular case of the much more general Theorem 8.5.1, thus we omit the proof (see also Figure 8.5.1 for a region $\Omega(\alpha, \beta)$).

Several subclasses of $\mathcal{C}^\infty$ characterized by the growth of the function γ at the origin have been identified.

4.1.4 Example.   Assume γ is integrable near zero. Then A is bounded. In fact, we have

$$\|I - \lambda R(\lambda)\| \leqslant \int_0^\infty e^{-\lambda t} \gamma(t) \, dt$$

for λ large enough; for $\lambda \to \infty$ the right-hand side tends to zero and the result follows in the same way as Lemma 2.3.5.

An extremely important subclass of $\mathcal{C}^\infty$ (whose characterization will be completed in the next section) is the class $\mathcal{Q}$ of all $A \in \mathcal{C}^\infty$ such that

$$\gamma(t) = 0(t^{-1}) \text{ as } t \to 0. \qquad (4.1.7)$$

It is a remarkable consequence of (4.1.7) that S, which is only assumed to be differentiable, admits in fact an extension to part of the complex plane as an analytic function. To simplify the statement of this result (which will be proved later in this section), we introduce some notations.

Let $0 < \varphi \leqslant \pi/2$. An operator $A \in \mathcal{C}_+$ is said to *belong to* $\mathcal{Q}(\varphi)$ if S can be extended to a (E)-valued function $S(\zeta)$ defined and analytic in

$$\{\zeta; |\arg \zeta| < \varphi, \zeta \neq 0\} = \Sigma_+(\varphi)$$

(here we take arg in $(-\pi, \pi]$) and strongly continuous in

$$\{\zeta; |\arg \zeta| \leqslant \varphi\} = \Sigma(\varphi).$$

We note that the strong continuity assumption simply reduces to

$$\lim_{|\arg \zeta| \leqslant \varphi, \zeta \to 0} S(\zeta) u = u.$$

It is easy to see by means of standard analytic continuation arguments that the semigroup equation (2.1.2) holds in $\Sigma(\varphi)$, that is,

$$S(z + \zeta) = S(z)S(\zeta) \quad (z, \zeta \in \Sigma(\varphi)). \qquad (4.1.8)$$

In fact, if (2.1.2) holds for $s, t \geqslant 0$, it will hold for $s \geqslant 0$, $\zeta \in \Sigma_+(\varphi)$ (both sides of (4.1.8) are an analytic continuation to $\Sigma_+(\varphi)$ of $t \to S(s)S(t)$) and a similar argument takes care of the variable s.

It is a consequence of the uniform boundedness theorem that $\|S(\zeta)\|$ is bounded in, say, $|\arg \zeta| \leqslant \varphi$, $|\zeta| \leqslant a$. A reasoning based in this and on (4.1.8) and entirely similar to the one leading to (2.1.3) shows that

$$\|S(\zeta)\| \leqslant Ce^{\omega|\zeta|} \quad (\zeta \in \Sigma(\varphi)) \tag{4.1.9}$$

for some constants C, ω (not necessarily the same in (2.1.3)).

For $0 < \varphi \leqslant \pi/2$ we define

$$\mathcal{C}(\varphi-) = \bigcup_{0 \leqslant \varphi' < \varphi} \mathcal{C}(\varphi').$$

A real variable characterization of the classes $\mathcal{C}(\varphi-)$ is made explicit in the following result.

4.1.5   Theorem.   Let $C \geqslant 1/e$. Assume $A \in \mathcal{C}^\infty$ and

$$\|AS(t)\| \leqslant Ct^{-1} \tag{4.1.10}$$

for t near zero. Then

$$A \in \mathcal{C}(\varphi-),$$

where $\varphi = \arcsin(Ce)^{-1}$. *Conversely, let* $A \in \mathcal{C}(\varphi-)$. *Then* $A \in \mathcal{C}^\infty$ *and* (4.1.10) *holds near zero.*

Let $\mathcal{C}$ be the class of all $A \in \mathcal{C}^\infty$ satisfying (4.1.10) near zero. Then Theorem 4.1.5 can be restated in the single equality

$$\mathcal{C} = \bigcup_{0 \leqslant \varphi \leqslant \pi/2} \mathcal{C}(\varphi-) = \bigcup_{0 \leqslant \varphi < \pi/2} \mathcal{C}(\varphi).$$

Proof.   Assume $A \in \mathcal{C}$ so that (4.1.10) holds for $0 < t \leqslant \delta > 0$. We have

$$\|S^{(n)}(t)\| = \|A^n S(t)\|$$

$$= \|(AS(t/n))^n\| \leqslant (Cn/t)^n \quad (n \geqslant t/\delta, 0 < t < \infty). \tag{4.1.11}$$

Accordingly, given $t > 0$, the coefficients of the power series

$$\sum_{n=0}^{\infty} \frac{(\zeta - t)^n}{n!} S^{(n)}(t) \tag{4.1.12}$$

(except perhaps for a finite number), are bounded in the norm of (E) by

$$\frac{1}{n!} \left(\frac{Cn}{t} \right)^n = \frac{1}{\sqrt{2\pi n}} \left(\frac{Ce}{t} \right)^n (1 + o(1)) \quad (n \to \infty) \tag{4.1.13}$$

by virtue of Stirling's formula. Then the series (4.1.12) is actually convergent in the disk $|\zeta - t| < t/Ce$ and defines there a (E)-valued holomorphic extension of S. Carrying out this construction for all $t > 0$ and showing in

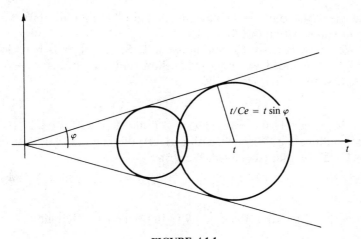

FIGURE 4.1.1

the usual way that extensions to overlapping disks coincide in their intersection, we see that S can be analytically extended to the open sector

$$\Sigma_+(\varphi -) = \left\{\zeta;\ |\arg\zeta| < \varphi = \arcsin(Ce)^{-1};\ \zeta \neq 0\right\}.$$

Let $0 < \varphi' < \varphi$, $\zeta \in \Sigma_+(\varphi')$, $|\zeta| \leq \delta\cos\varphi'$, $\varphi'' = \arg\zeta$ so that $t = |\zeta|/\cos\varphi'' \leq |\zeta|/\cos\varphi' \leq \delta$ and (4.1.10) holds. We have

$$\|S(\zeta)\| \leq C_0 + \sum_{n=1}^{\infty} \frac{|\zeta - t|^n}{n!}\left(\frac{Cn}{t}\right)^n = C_0 + \sum_{n=1}^{\infty} \frac{(Cn\sin\varphi'')^n}{n!}$$

$$\leq C_0 + C' \sum_{n=1}^{\infty} \frac{(Ce\sin\varphi')^n}{\sqrt{2\pi n}} \leq C_0 + C'' \frac{\sin\varphi'}{1 - Ce\sin\varphi'}. \quad (4.1.14)$$

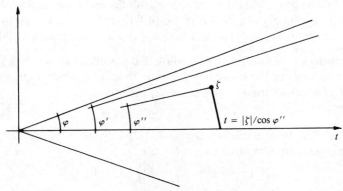

FIGURE 4.1.2

An estimate of the same sort can be obtained in $|\zeta| \leqslant a$ for arbitrary a using the semigroup equation (4.1.8).

We prove now strong continuity of S. Since $\|S(\zeta)\|$ is bounded in $\Sigma_+(\varphi')$ for $|\zeta|$ bounded, we only need to show continuity for, say, $u \in D(A)$. We note first that if $u \in D(A)$, then

$$S'(\zeta) u = S(\zeta) A u \quad (\zeta \in \Sigma_+(\varphi)). \tag{4.1.15}$$

In fact, this is clearly the case for $\zeta = t > 0$ and follows for $\zeta \in \Sigma_+(\varphi)$ by analytic continuation. Together with (4.1.14), this implies that $\|S'(\zeta) u\|$ is bounded in $\Sigma_+(\varphi')$ for ζ bounded. We write

$$S(\zeta) u - u = (S(\zeta) - S(|\zeta|)) u + S(|\zeta|) u - u \quad (u \in D(A)). \tag{4.1.16}$$

We have

$$\|(S(\zeta) - S(|\zeta|)) u\| \leqslant \int_{|\zeta|}^{\zeta} \|S'(z) u\| \, |dz| \leqslant C |\zeta - |\zeta|| \, \|u\|$$

in view of the boundedness of $\|S'(\zeta) u\|$. We obtain

$$\|S(\zeta) u - u\| \leqslant C |\zeta - |\zeta|| \, \|u\| + \|S(|\zeta|) u - u\|,$$

which, in view of the strong continuity of $S(t)$ for t real implies the desired result.

We prove now the converse. Assume that $A \in \mathcal{C}(\varphi -)$ for some $\varphi > 0$. Then, if $\varphi' < \varphi$ we have

$$AS(t) = S'(t)$$

$$= \frac{1}{2\pi i} \int_{|z - t| = t \sin \varphi'} \frac{S(z)}{(z - t)^2} \, dz \tag{4.1.17}$$

from which (4.1.10) readily follows.

4.1.6 Remark. Making use of the argument at the end of the proof of Theorem 4.1.5 we can prove that

$$\|AS(\zeta)\| \leqslant C / |\zeta| \quad (\zeta \in \Sigma_+(\varphi'), \ |\zeta| \leqslant a) \tag{4.1.18}$$

if $a > 0$, $\varphi' < \varphi$ (the constant C, of course, will depend on φ' and a). In fact, we only have to apply Cauchy's formula (4.1.17) for $t = \zeta \in \Sigma_+(\varphi')$ integrating over circles $|z - \zeta| = |\zeta| \sin \eta$, where $\eta < \varphi - \varphi'$. (See Figure 4.1.3).

4.1.7 Remark. Although we may allow the parameter φ in the definition of $\mathcal{C}(\varphi)$ to roam in the range $0 < \varphi \leqslant \pi$, it should be pointed out that the class $\mathcal{C}(\varphi)$ becomes trivial when $\varphi \geqslant \pi/2$; precisely,

$$\mathcal{C}(\varphi) = (E) \quad (\varphi \geqslant \pi/2). \tag{4.1.19}$$

In fact, let $S(\zeta)$ be analytic in a region $\Sigma_+(\varphi)$ with $\varphi \geqslant \pi/2$. Making use of the semigroup equation (4.1.8) we can write

$$S(\varepsilon) = S(\varepsilon + i) S(\varepsilon - i) \quad (\varepsilon > 0).$$

If we let $\varepsilon \to 0$, $S(\varepsilon + i) \to S(i)$, $S(\varepsilon - i) \to S(-i)$ in (E) so that $S(\varepsilon) \to I$ in (E). This shows, using Lemma 2.3.5, that A is bounded, thus proving

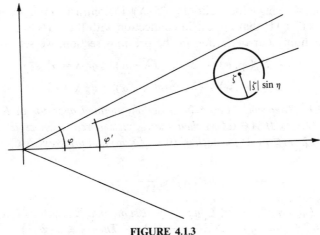

FIGURE 4.1.3

that equality (4.1.19) holds.

Using the argument in Theorem 4.1.5 we easily obtain:

4.1.8   Theorem.   *Assume* $A \in \mathcal{C}^{\infty}$ *and*

$$\limsup_{t \to 0+} t \|AS(t)\| < \frac{1}{e} = 0.367\ldots \tag{4.1.20}$$

Then A is bounded. The constant $1/e$ *in* (4.1.20) *is best possible.*

That $1/e$ is the best possible constant is seen as follows: let l^2 be the Hilbert space of all sequences $\xi = \{\xi_1, \xi_2, \ldots\}$, $\eta = \{\eta_1, \eta_2, \ldots\}, \ldots$ ($\xi_n, \eta_n \in \mathbb{C}$) such that $\Sigma |\xi_n|^2 < \infty$ endowed with the scalar product $(\xi, \eta) = \Sigma \bar{\xi}_n \eta_n$, and let A be the self-adjoint operator defined by $A\xi = \{-n\xi_n\}$. Then $A \in \mathcal{C}^{\infty}$ with $S(t)\xi = \{e^{-nt}\xi_n\}$ and $\limsup t \|AS(t)\| = 1/e$.

4.1.9   Example.   Let $A \in \mathcal{C} \cap \mathcal{C}^{\infty}$. Then A is bounded.

4.1.10   Example.   Let $A \in \mathcal{C}^{\infty}$. The assumption that $\|AS(\cdot)u\|$ is integrable near zero for every $u \in E$ does *not* imply that A is bounded (compare with Example 4.1.4). This can be verified with the same operator in Example 2.4.7, this time in the Banach space of all sequences $\xi = \{\xi_1, \xi_2, \ldots\}$ such that $\|\xi\| = \Sigma |\xi_n| < \infty$, endowed with the norm $\|\cdot\|$.

4.2.   ABSTRACT PARABOLIC EQUATIONS; ANALYTIC PROPAGATORS

We give in this section a complete characterization of operators in the class $\mathcal{C}$ in terms of the location of $\rho(A)$ and the growth of $\|R(\lambda)\|$ there. This characterization turns out to be fairly simple to apply, involving only the verification of a single inequality for $\|R(\lambda)\|$ in contrast to the infinite set of inequalities (2.1.11) that appear in connection with the class $\mathcal{C}_+$.

Extending our definition in the previous section, we set, for α real,

$$\Sigma_+(\varphi, \alpha) = \{\lambda; |\arg(\lambda - \alpha)| \leqslant \varphi, \lambda \neq \alpha\}$$

$$\Sigma_+(\varphi-, \alpha) = \{\lambda; |\arg(\lambda - \alpha)| < \varphi, \lambda \neq \alpha\}.$$

4.2.1   Theorem.   *Let A be a densely defined operator in E, and let $0 < \varphi < \pi/2$. (a) If $A \in \mathcal{C}(\varphi)$ then there exists a real number α such that $\Sigma_+(\varphi' + \pi/2, \alpha) \subseteq \rho(A)$ for each φ' $(0 < \varphi' < \varphi)$ and there exists $C = C_{\varphi'}$ such that*

$$\|R(\lambda)\| \leqslant \frac{C}{|\lambda - \alpha|} \tag{4.2.1}$$

for $\lambda \in \Sigma_+(\varphi' + \pi/2, \alpha)$. (b) Conversely, assume that $\Sigma_+(\varphi + \pi/2, \alpha) \subseteq \rho(A)$ and that (4.2.1) holds for $\lambda \in \Sigma_+(\varphi + \pi/2, \alpha)$. Then $A \in \mathcal{C}(\varphi-)$.

Proof.   Let $A \in \mathcal{C}(\varphi)$. If ω is the constant in (4.1.9), let $\alpha = \omega/\cos\varphi$ and $A_\alpha = A - \alpha I$, so that $S_\alpha(\zeta) = S(\zeta; A_\alpha) = e^{-\alpha\zeta}S(\zeta)$. Since $\mathrm{Re}\,\zeta \geqslant \cos\varphi|\zeta|$ in $\Sigma(\varphi)$, we have

$$\|S_\alpha(\zeta)\| \leqslant Ce^{-\alpha\mathrm{Re}\,\zeta + \omega|\zeta|} \leqslant C \quad (\zeta \in \Sigma(\varphi)). \tag{4.2.2}$$

Hence

$$R(\lambda)u = R(\lambda - \alpha; A_\alpha)u$$

$$= \int_0^\infty e^{-(\lambda-\alpha)t}S_\alpha(t)u\,dt \quad (u \in E) \tag{4.2.3}$$

for $\mathrm{Re}(\lambda - \alpha) > 0$. If $\mathrm{Re}(\lambda - \alpha)e^{i\varphi} > 0$ as well, we can deform the path of integration in (4.2.3) into the ray $\Gamma_- = \{\zeta; \arg\zeta = -\varphi, \mathrm{Re}\,\zeta \geqslant 0\}$ and thus extend analytically $R(\lambda)$ to the half plane $\mathrm{Re}(\lambda - \alpha)e^{i\varphi} > 0$. Likewise, if $\mathrm{Re}(\lambda - \alpha) > 0$ and $\mathrm{Re}(\lambda - \alpha)e^{-i\varphi} > 0$, we can integrate in the ray $\Gamma_+ = \{\zeta; \arg\zeta = \varphi, \mathrm{Re}\,\zeta \geqslant 0\}$ and extend $R(\lambda)$ to the half plane $\mathrm{Re}(\lambda - \alpha)e^{-i\varphi} > 0$. That these extensions are in fact $R(\lambda)$ in the new regions follows from Lemma 3.6. We have then proved that $R(\lambda)$ exists in $\Sigma_+(\varphi + \pi/2, \alpha)$. We estimate now $\|R(\lambda)\|$. If $\mathrm{Re}(\lambda - \alpha)e^{i\varphi} > 0$, we obtain, bounding the integral in Γ_-, that

$$\|R(\lambda)\| \leqslant C\int_0^\infty \exp\{-(\mathrm{Re}(\lambda - \alpha)e^{i\varphi})t\}\,dt$$

$$= C/\mathrm{Re}(\lambda - \alpha)e^{i\varphi} \quad (\mathrm{Re}(\lambda - \alpha)e^{i\varphi} > 0). \tag{4.2.4}$$

Similarly, if $\mathrm{Re}(\lambda - \alpha)e^{-i\varphi} > 0$, we integrate in Γ_+, obtaining

$$\|R(\lambda)\| \leqslant C/\mathrm{Re}(\lambda - \alpha)e^{-i\varphi} \quad (\mathrm{Re}(\lambda - \alpha)e^{-i\varphi} > 0). \tag{4.2.5}$$

Inequality (4.2.1) then results from the easily verified estimates

$$|\lambda - \alpha| \leqslant \mathrm{Re}(\lambda - \alpha)e^{i\varphi}/\sin(\varphi - \varphi')$$

$$(-(\varphi' + \pi/2) \leqslant \arg(\lambda - \alpha) \leqslant 0) \tag{4.2.6}$$

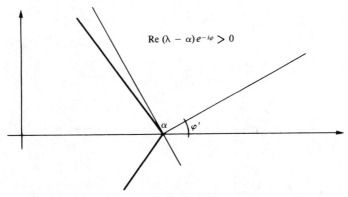

$$\text{Re}\,(\lambda - \alpha)e^{-i\varphi} > 0$$

FIGURE 4.2.1

and

$$|\lambda - \alpha| \leqslant \text{Re}(\lambda - \alpha)e^{-i\varphi}/\sin(\varphi - \varphi')$$

$$(0 \leqslant \arg(\lambda - \alpha) \leqslant \varphi' + \pi/2) \quad (4.2.7)$$

valid if $\varphi - \varphi' \leqslant \pi/2 - \varphi$. (The constants in (4.2.1) and (4.2.2) are of course different; the same observation applies to subsequent estimates.)

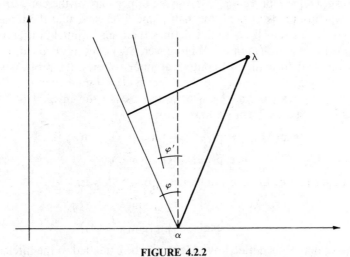

FIGURE 4.2.2

We prove now the converse. Suppose that $R(\lambda)$ exists in $\Sigma_+(\varphi + \pi/2, \alpha)$ and let (4.2.1) hold. By means of a translation of A similar to the one used in the first part of the proof, we may and will assume that $\alpha < 0$. Let $0 < \varphi_1, \varphi_2 \leqslant \varphi$ and let $\Gamma(\varphi_1, \varphi_2)$ be the contour consisting of the two rays $\arg \lambda = \varphi_1 + \pi/2$, $\text{Im}\,\lambda \geqslant 0$, and $\arg \lambda = -(\varphi_2 + \pi/2)$, $\text{Im}\,\lambda \leqslant 0$,

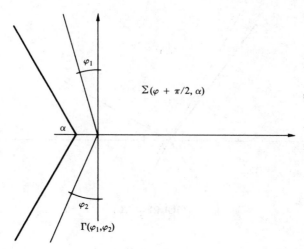

FIGURE 4.2.3

the entire contour oriented clockwise with respect to the right half plane.
    Define

$$S(t) = \frac{1}{2\pi i} \int_{\Gamma(\varphi_1, \varphi_2)} e^{\lambda t} R(\lambda) \, d\lambda \quad (t > 0). \qquad (4.2.8)$$

Since $|\exp(\lambda t)| = \exp(-\sin\varphi_1|\lambda|t)$ in the upper part of the contour and a similar equality holds in the lower half plane, it is clear that the integral in (4.2.8) converges for all $t > 0$ and defines there an infinitely differentiable (E)-valued function $S(t)$ in $t > 0$; moreover, $S(t)$ does not depend on φ_1 or φ_2 as a simple deformation-of-contour argument shows. We prove next that S can be analytically extended to the sector $\Sigma_+(\varphi')$ for every $\varphi' < \varphi$. To this end, take $t = \zeta$ with $0 \leqslant \arg\zeta \leqslant \varphi' < \varphi$ and use in the integral (4.2.8) the contour $\Gamma(\varphi - \varphi', \varphi)$. Then we have

$$|\exp(\lambda\zeta)| = \exp(-|\lambda| \, |\zeta| \sin(\varphi - \varphi' + \arg\zeta))$$
$$\leqslant \exp(-|\lambda| \, |\zeta| \sin(\varphi - \varphi'))$$

on the upper half of the contour, whereas on the lower,

$$|\exp(\lambda\zeta)| = \exp(-|\lambda| \, |\zeta| \sin(\varphi - \arg\zeta))$$
$$\leqslant \exp(-|\lambda| \, |\zeta| \sin(\varphi - \varphi')).$$

This shows that S, as defined by (4.2.8), can be extended to the intersection of $\Sigma_+(\varphi')$ with the upper half plane. A similar reasoning takes care of the lower half plane and a simple estimation of the integral (4.2.8) with judicious choice of the contour shows that

$$\|S(\zeta)\| \leqslant C \quad (\zeta \in \Sigma_+(\varphi'); \; |\zeta| \geqslant a) \qquad (4.2.9)$$

for any $a > 0$, where C may depend on a. However, when $\zeta \to 0$, the factor

$\exp(\lambda\zeta)$ in the integral (4.2.8) dissolves away and we need to bound in a different way. To do this, we take first ζ with $0 \leqslant \arg\zeta \leqslant \varphi'$, $\zeta \neq 0$, and use the contour $\Gamma(\varphi - \varphi', \varphi)$ in the integral in (4.2.8). After this we make the change of variables $z = \lambda\zeta$, obtaining

$$S(\zeta) = \frac{1}{2\pi i} \int_{\zeta\Gamma(\varphi - \varphi', \varphi)} \frac{e^z}{\zeta} R\left(\frac{z}{\zeta}\right) dz, \qquad (4.2.10)$$

where of course $\zeta\Gamma(\varphi - \varphi', \varphi) = \Gamma(\varphi - \varphi' + \arg\zeta, \varphi - \arg\zeta)$. We can now deform the contour to, say, $\Gamma(\varphi, \varphi - \varphi')$ and modify this contour in such a way that z passes to the right of the origin as it travels upwards. Call Γ' the contour thus obtained.

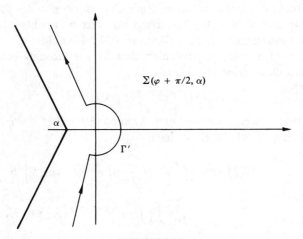

FIGURE 4.2.4

We make use of (4.2.1) to estimate the integrand in (4.2.10); there exist $a, C' > 0$ such that if $|\zeta| \leqslant a$,

$$\left\| R\left(\frac{z}{\zeta}\right) \right\| \leqslant C \left| \frac{z}{\zeta} - \alpha \right|^{-1} \leqslant C' \left| \frac{\zeta}{z} \right| \quad (z \in \Gamma') \qquad (4.2.11)$$

(note that if $z \in \Gamma'$ and $0 \leqslant \arg\zeta \leqslant \varphi'$, then $z\zeta^{-1} \in \Sigma_+(\varphi + \pi/2, \alpha)$). We obtain

$$\|S(\zeta)\| \leqslant C' \int_{\Gamma'} \left| \frac{e^z}{z} \right| |dz|. \qquad (4.2.12)$$

A symmetric argument takes care of the range $-\varphi' \leqslant \arg\zeta \leqslant 0$; we finally deduce that

$$\|S(\zeta)\| \leqslant C_{\varphi'} \quad (\zeta \in \Sigma_+(\varphi'), |\zeta| \leqslant a) \qquad (4.2.13)$$

We show next that S is strongly continuous at the origin, that is,

$$\lim_{|\arg\zeta| \leqslant \varphi', |\zeta| \to 0} S(\zeta)u = u \quad (u \in E) \qquad (4.2.14)$$

for every $\varphi' < \varphi$. In view of (4.2.13) we only have to prove this for u in a dense subset of E, say, in $D(A)$. As customary we treat separately the cases $0 \leqslant \arg\zeta \leqslant \varphi'$ and $-\varphi' \leqslant \arg\zeta \leqslant 0$. In the first case we use the contour $\Gamma(\varphi - \varphi', \varphi)$ modified near the origin in the same way as Γ' (we call Γ' again the resulting contour). Making use of the equality $R(\lambda)u = \lambda^{-1}R(\lambda)Au + \lambda^{-1}u$ in the integral (4.2.8), we obtain

$$S(\zeta)u = \frac{1}{2\pi i}\int_{\Gamma'} e^{\lambda\zeta}R(\lambda)u\, d\lambda$$

$$= \frac{1}{2\pi i}\int_{\Gamma'} \frac{e^{\lambda\zeta}}{\lambda}R(\lambda)Au\, d\lambda + \left(\frac{1}{2\pi i}\int_{\Gamma'} \frac{e^{\lambda\zeta}}{\lambda}d\lambda\right)u.$$

The first integral is continuous in $\Sigma(\varphi')$ and vanishes at $\zeta = 0$, as can be seen making use of (4.2.1). The second one can be readily computed by residues and evaluates to 1. This shows that (4.2.14) holds.

We have proved, in particular, that $S(\cdot)$ is strongly continuous in $t \geqslant 0$ and it follows from (4.2.13) that

$$\|S(t)\| \leqslant C \quad (t \geqslant 0).$$

To show that $A \in \mathcal{C}_+$, we use now Lemma 2.2.3. Let $0 \leqslant \varphi' < \varphi$, $\Gamma = \Gamma(\varphi', \varphi')$. If $\mu > 0$ and $u \in E$, we have

$$\int_0^\infty e^{-\mu t}S(t)u\, dt = \int_0^\infty e^{-\mu t}\left(\frac{1}{2\pi i}\int_\Gamma e^{\lambda t}R(\lambda)u\, d\lambda\right)dt$$

$$= \frac{1}{2\pi i}\int_\Gamma\left(\int_0^\infty e^{-(\mu-\lambda)t}\, dt\right)R(\lambda)u\, d\lambda$$

$$= -\frac{1}{2\pi i}\int_\Gamma \frac{1}{\lambda - \mu}R(\lambda)u\, d\lambda$$

$$= R(\mu)u, \qquad\qquad (4.2.15)$$

where the interchange of the order of integration can be easily justified estimating the integrand, and the last step is an application of Cauchy's formula in the region to the right of Γ, which is clearly valid in view of (4.2.1). This ends the proof of Theorem 4.2.1.

Theorem 4.2.1 can of course be cast in the form of an equivalence.

4.2.2 Theorem. *Let A be a densely defined operator in E, $0 \leqslant \varphi \leqslant \pi/2$. Then $A \in \mathcal{Q}(\varphi-)$ if and only if for every $\varphi' < \varphi$ there exists a real number $\alpha = \alpha(\varphi')$ such that $\Sigma(\varphi' + \pi/2, \alpha) \subseteq \rho(A)$ and (4.2.1) holds for $\lambda \in \Sigma_+(\varphi' + \pi/2, \alpha)$, with $C = C_{\varphi'}$.*

The next result shows that an inequality of the type of (4.2.1) need only be assumed in a half plane. A similar argument shows that $\mathcal{Q}(\varphi-)$ may be replaced by $\mathcal{Q}(\varphi)$ in part (b) of Theorem 2.4.1.

4.2.3   Lemma.   *Let A be a densely defined operator in E. Assume that $R(\lambda)$ exists in* $\operatorname{Re}\lambda > \alpha$ *and*

$$\|R(\lambda)\| \leqslant \frac{C}{|\lambda - \alpha|} \tag{4.2.16}$$

there. Then A satisfies the hypotheses of Theorem 4.2.1 for $\varphi < \varphi_0 = \arcsin(1/C)$.

Proof.   It follows from (4.2.16) that $\|R(\lambda)\|$ remains bounded as $\operatorname{Re}\lambda \to \alpha$ as long as $|\lambda - \alpha|$ does not approach zero. Then $R(\lambda)$ exists if $\operatorname{Re}\lambda = \alpha$, $\lambda \neq \alpha$. We develop $R(\lambda)$ in Taylor series about $\alpha + i\eta$, $|\eta| > 0$:

$$R(\lambda) = \sum_{n=0}^{\infty} (\alpha + i\eta - \lambda)^n R(\alpha + i\eta)^{n+1}. \tag{4.2.17}$$

In view of (4.2.16), the series is convergent in the region

$$|\lambda - (\alpha + i\eta)| < \eta/C. \tag{4.2.18}$$

The union of all circles defined by (4.2.18) with the half plane $\operatorname{Re}\lambda > \alpha$ evidently coincides with $\Sigma_+((\varphi_0 + \pi/2)-, \alpha)$. Take now $\varphi < \varphi_0$, $C' = 1/\sin\varphi$. If $\lambda \in \Sigma_+(\varphi + \pi/2, \alpha)$ and $\operatorname{Re}\lambda \leqslant \alpha$, $\operatorname{Im}\lambda \geqslant 0$, let $\eta = |\lambda - \alpha|/\cos\varphi$. Then $|\lambda - (\alpha + i\eta)| \leqslant (\sin\varphi)\eta = C'^{-1}\eta$ and we can estimate the series (4.2.17) as

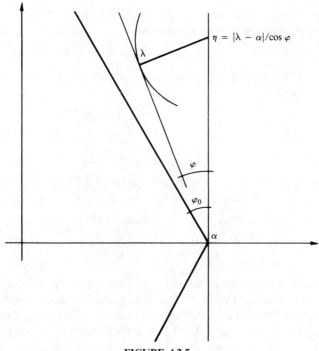

FIGURE 4.2.5

follows:

$$\|R(\lambda)\| \leqslant \frac{C}{\eta} \sum_{n=0}^{\infty} C^n \left| \frac{\lambda - (\alpha + i\eta)}{\eta} \right|^n$$

$$\leqslant \frac{C}{\eta} \sum_{n=0}^{\infty} \left(\frac{C}{C'} \right)^n = \frac{CC' \cos \varphi}{C' - C} \cdot \frac{1}{|\lambda - \alpha|} \quad (4.2.19)$$

A symmetric argument takes care of the case Im $\lambda \leqslant 0$.

The following result, which is a generalization of (half of) Theorem 3.1.8, will be useful in the characterization of operators in the class $\mathcal{C} \cap \mathcal{C}_+(0, 1)$.

4.2.4   Theorem. *Let A be a densely defined operator in E, $\theta: D(A) \to E^*$ a duality map. Define*

$$\mathfrak{s} = \mathfrak{s}(\theta; A)$$

$$= \{\lambda \in \mathbb{C}; \lambda = \langle \theta(u), Au \rangle; u \in D(A), u \neq 0\} \quad (4.2.20)$$

and let $\mathfrak{r} = \mathfrak{r}(\theta; A)$ be the complement of $\mathfrak{s}$. Then:

(a) *If $\lambda \in \mathfrak{r}$,*

$$\|u\| \leqslant \text{dist}(\lambda, \mathfrak{s})^{-1} \|\lambda u - Au\| \quad (u \in D(A)). \quad (4.2.21)$$

(b) *If A is closed, $(\lambda I - A)D(A)$ is closed for $\lambda \in \mathfrak{r}$.*

(c) *Assume that $\lambda_0 \in \mathfrak{r}$ is such that*

$$(\lambda_0 I - A)D(A) = E. \quad (4.2.22)$$

Then $\mathfrak{r}(\lambda_0)$, the connected component of $\mathfrak{r}$ that contains λ_0 satisfies

$$\mathfrak{r}(\lambda_0) \subseteq \rho(A) \quad (4.2.23)$$

and

$$\|R(\lambda)\| \leqslant \text{dist}(\lambda, \mathfrak{s})^{-1} \quad (\lambda \in \mathfrak{r}(\lambda_0)). \quad (4.2.24)$$

Proof.   Clearly we only have to prove (4.2.21) for $\|u\| = 1$. If $\lambda \in \mathfrak{r}$, we have

$$0 < \text{dist}(\lambda, \mathfrak{s}) \leqslant |\lambda - \langle \theta(u), Au \rangle|$$

$$= |\langle \theta(u), \lambda u - Au \rangle| \leqslant \|\theta(u)\| \|\lambda u - Au\|$$

$$= \|\lambda u - Au\|,$$

which proves (a). It is plain that (b) is an immediate consequence of (a). To prove (c), we only have to show that $\mathfrak{r}(\lambda_0) \cap \rho(A)$ is both open and closed in $\mathfrak{r}$. That $\mathfrak{r}(\lambda_0) \cap \rho(A)$ is open in $\mathfrak{r}$ is evident. Let $\{\lambda_n\}$ be a sequence in $\mathfrak{r}(\lambda_0) \cap \rho(A)$ such that $\lambda_n \to \lambda \in \mathfrak{r}$. Then there exists $\delta > 0$ with

$$\text{dist}(\lambda_n, \mathfrak{s}) \geqslant \delta$$

for n large enough. This shows, in view of (4.2.21) that $\|R(\lambda_n)\| \leqslant \delta^{-1}$. By virtue of Lemma 3.5, $\lambda \in \rho(A)$. This ends the proof, since (4.2.24) is an immediate consequence of (4.2.21).

4.2.5 Corollary. *Let A be a densely defined operator in E such that*
$$\mathrm{Re}\langle \theta(u), Au \rangle \leqslant -\delta |\mathrm{Im}\langle \theta(u), Au \rangle| \quad (u \in D(A)), \quad (4.2.25)$$
where $\theta: D(A) \to E^*$ is a duality map and $\delta \geqslant 0$. Assume, moreover, that
$$(\lambda I - A)D(A) = E \quad (4.2.26)$$
for some $\lambda > 0$. Then $A \in \mathcal{C}(\varphi -) \cap \mathcal{C}_+(1,0)$, where $\varphi = \mathrm{arc\,tg}\,\delta$. In particular, if (4.2.26) holds and $\mathrm{Im}\langle \theta(u), Au \rangle = 0$,
$$\langle \theta(u), Au \rangle \leqslant 0 \quad (u \in D(A)), \quad (4.2.27)$$
then $A \in \mathcal{C}((\pi/2)-) \cap \mathcal{C}_+(1,0)$.

Proof. Inequality (4.2.25) clearly implies that $\mathrm{r}(\theta, A)$ contains the sector $\Sigma_+((\varphi + \pi/2)-)$ with $\varphi = \mathrm{arc\,tg}\,\delta$. We obtain immediately from (4.2.24) that
$$\|R(\lambda)\| \leqslant 1/\mathrm{Re}\,\lambda e^{i\varphi}$$
in $\mathrm{Re}\,\lambda e^{i\varphi} > 0$; likewise, if $\mathrm{Re}\,\lambda e^{-i\varphi} > 0$,
$$\|R(\lambda)\| \leqslant 1/\mathrm{Re}\,\lambda e^{-i\varphi}$$
and the proof ends exactly like the first part of the proof of Theorem 4.2.1 by using inequalities (4.2.6) and (4.2.7).

We note that, under the hypotheses of Corollary 4.2.5, we can obtain very precise estimates for $\|S(\zeta)\|$ in the entire sector $\Sigma_+(\varphi)$. In fact, using (4.2.25) we deduce that
$$\mathrm{Re}\langle \theta(u), e^{i\psi}Au \rangle = (\cos\psi)\mathrm{Re}\langle \theta(u), Au \rangle - (\sin\psi)\mathrm{Im}\langle \theta(u), Au \rangle \leqslant 0$$
$$(4.2.28)$$
if $|\mathrm{tg}\,\psi| \leqslant \delta$. Accordingly, the operator $e^{i\psi}A$ is dissipative if $|\psi| \leqslant \varphi = \mathrm{arc\,tg}\,\delta$. Since $e^{i\psi}A$ generates the strongly continuous semigroup $t \to S(e^{i\psi}t)$ $(t \geqslant 0)$, we obtain from Theorem 3.1.8 that
$$\|S(\zeta)\| \leqslant 1 \quad (|\arg\zeta| \leqslant \varphi). \quad (4.2.29)$$
We also note that the argument leading to (4.2.29) implies in particular that $S(\cdot)$ has a strongly continuous extension to the *closed* sector $\Sigma(\varphi)$.

4.3. APPLICATIONS TO ORDINARY DIFFERENTIAL OPERATORS

We examine the differential operators
$$A(\beta_0, \beta_l)u(x) = a(x)u''(x) + b(x)u'(x) + c(x)u(x) \quad (4.3.1)$$
in a finite interval $0 \leqslant x \leqslant l$ in the light of the previous results on the class $\mathcal{C}$. As in Section 3.4, our standing assumptions are that a is twice continu-

ously differentiable, b is continuously differentiable, and c is continuous in $0 \leqslant x \leqslant l$. The boundary conditions at each endpoint are of type (I) or (II) (see (3.4.1) and (3.4.2)). Also, we assume that

$$a(x) > 0 \quad (0 \leqslant x \leqslant l). \tag{4.3.2}$$

The domain of $A(\beta_0, \beta_l)$ in $L^p(0, l)$ $(1 \leqslant p < \infty)$ consists of all $u \in W^{2, p}(0, l)$ that satisfy the boundary condition at each endpoint: in the space $C[0, l]$ $(C_0, C_l, C_{0, l}$ depending on the boundary conditions used), $D(A(\beta_0, \beta_l))$ is the set of all $u \in C^{(2)}[0, l]$ satisfying the given boundary condition at each endpoint with $A(\beta_0, \beta_l)u$ in the space. When several spaces are at play, we indicate by $A_p(\beta_0, \beta_l)$ the operator in L^p and by $A(\beta_0, \beta_l)$ the operator in C or subspaces thereof.

4.3.1   Theorem.   *The operator* $A_p(\beta_0, \beta_l)$ $(1 < p < \infty)$ *belongs to* $\mathcal{C}(\varphi_p -)$, *where*

$$\varphi_p = \operatorname{arc tg}\left\{ \left(\frac{p}{p-2} \right)^2 - 1 \right\}^{1/2}. \tag{4.3.3}$$

For every φ *with* $0 < \varphi < \varphi_p$, *there exists* $\omega = \omega(p, \varphi)$ *such that*

$$\|S_p(\zeta)\| \leqslant e^{\omega|\zeta|} \quad (|\arg \zeta| \leqslant \varphi), \tag{4.3.4}$$

where $S_p(\zeta) = \exp(\zeta A_p(\beta_0, \beta_l))$.

Proof.   The result is a consequence of Corollary 4.2.5. To check this we denote by θ the only duality map in L^p and perform an integration by parts similar to that in (3.4.6). Assume for the moment that $p \geqslant 2$, that both boundary conditions are of type (I), and take u twice continuously differentiable and satisfying the boundary conditions. For any δ we obtain

$$\|u\|^{p-2}\left(\operatorname{Re}\langle \theta(u), A_p(\beta_0, \beta_l)u \rangle \pm \delta \operatorname{Im}\langle \theta(u), A_p(\beta_0, \beta_l)u \rangle \right)$$

$$= \operatorname{Re} \int_0^l ((au')' + (b - a')u' + cu)|u|^{p-2}\bar{u}\, dx$$

$$\pm \delta \operatorname{Im} \int_0^l ((au')' + (b - a')u' + cu)|u|^{p-2}\bar{u}\, dx$$

$$= \left\{ \gamma_l a(l) - \frac{1}{p}(a'(l) - b(l)) \right\}|u(l)|^p$$

$$- \left\{ \gamma_0 a(0) - \frac{1}{p}(a'(0) - b(0)) \right\}|u(0)|^p$$

$$- (p - 2) \int_0^l a|u|^{p-4}(\operatorname{Re}(\bar{u}u'))^2\, dx$$

$$\mp \delta(p - 2) \int_0^l a|u|^{p-4}\operatorname{Re}(\bar{u}u')\operatorname{Im}(\bar{u}u')\, dx$$

$$- \int_0^l a|u|^{p-2}|u'|^2\, dx \pm \delta \int_0^l (b - a')|u|^{p-2}\operatorname{Im}(\bar{u}u')\, dx$$

$$+ \frac{1}{p} \int_0^l (a'' - b' + pc)|u|^p\, dx. \tag{4.3.5}$$

We transform the sum of the first three integrals on the right-hand side using the following result: given a real constant $\alpha > -1$,

$$|z|^2 + \alpha\big((\operatorname{Re} z)^2 \pm \delta(\operatorname{Re} z)(\operatorname{Im} z)\big) \geqslant 0 \qquad (4.3.6)$$

for every $z \in \mathbb{C}$ if and only if

$$1 + \delta^2 \leqslant \left(\frac{\alpha + 2}{\alpha}\right)^2. \qquad (4.3.7)$$

To prove this we begin by observing that (4.3.6) is homogeneous in z, thus we may assume that $z = e^{i\varphi}$, reducing the inequality to the trigonometric relation $1 + \alpha(\cos^2 \varphi \pm \delta \cos\varphi \sin\varphi) \geqslant 0$, or, equivalently (setting $\psi = 2\varphi$), $2 + \alpha(1 + \cos\psi \pm \delta\sin\psi) \geqslant 0$. Let $\alpha \geqslant 0$. Since the minimum of the function $g(\psi) = \cos\psi \pm \delta\sin\psi$ equals $-(1 + \delta^2)^{1/2}$, (4.3.6) will hold if and only if $2 - \alpha((1 + \delta^2)^{1/2} - 1) \geqslant 0$, which is (4.3.7).

On the other hand, the maximum of g is $(1 + \delta^2)^{1/2}$; if $-1 < \alpha < 0$ (4.3.6) will hold if and only if $2 - |\alpha|((1 + \delta^2)^{1/2} + 1) \geqslant 0$, which is again (4.3.7).

In view of the homogeneity of (4.3.6), it is obvious that if (4.3.7) is strict, there exists $\tau > 0$ (depending on δ) such that, for every $z \in \mathbb{C}$,

$$|z|^2 + \alpha\big((\operatorname{Re} z)^2 \pm \delta(\operatorname{Re} z)(\operatorname{Im} z)\big) \geqslant \tau|z|^2. \qquad (4.3.8)$$

We use (4.3.8) for $z = \bar{u}u'$:

$$-(p-2)\int_0^l a|u|^{p-4}(\operatorname{Re}(\bar{u}u'))^2\, dx \mp \delta(p-2)\int_0^l a|u|^{p-4}\operatorname{Re}(\bar{u}u')\operatorname{Im}(\bar{u}u')\, dx$$

$$- \int_0^l a|u|^{p-2}|u'|^2\, dx = -\int a|u|^{p-4}f\, dx, \qquad (4.3.9)$$

where

$$f = |\bar{u}u'|^2 + (p-2)\big\{(\operatorname{Re}(\bar{u}u'))^2 \pm \delta\operatorname{Re}(\bar{u}u')\operatorname{Im}(\bar{u}u')\big\}$$

$$\geqslant \tau|\bar{u}u'|^2 = \tau|u|^2|u'|^2 \qquad (4.3.10)$$

for some $\tau > 0$ if

$$0 \leqslant \delta < \left\{\left(\frac{p}{p-2}\right)^2 - 1\right\}^{1/2}. \qquad (4.3.11)$$

We must now estimate the other terms on the right-hand side of (4.3.5). To this end, consider a real-valued continuously differentiable function ρ in $0 \leqslant x \leqslant l$. Since $(\rho|u|^p)' = \rho'|u|^p + p\rho|u|^{p-2}\operatorname{Re}(\bar{u}u')$, we obtain, for any $\varepsilon > 0$,

$$|\rho(l)|u(l)|^p - \rho(0)|u(0)|^p|$$

$$\leqslant \frac{p\varepsilon^2}{2}\int_0^l |\rho|\,|u|^{p-2}|u'|^2\, dx + \int_0^l \left(|\rho'| + \frac{p\varepsilon^{-2}}{2}|\rho|\right)|u|^p\, dx, \qquad (4.3.12)$$

where we have applied the inequality

$$2|u|\,|u'| = 2\left(\varepsilon^{-1}|u|\right)\left(\varepsilon|u'|\right) \leqslant \varepsilon^2|u'|^2 + \varepsilon^{-2}|u|^2. \qquad (4.3.13)$$

We use (4.3.12) for any ρ such that

$$\rho(l) = \gamma_l a(l) - \frac{1}{p}\left(a'(l) - b(l)\right),$$

$$\rho(0) = \gamma_0 a(0) - \frac{1}{p}\left(a'(0) - b(0)\right) \qquad (4.3.14)$$

(we may take ρ linear) to estimate the first two terms on the right-hand side of (4.3.5); for the fourth integral we use the inequality (4.3.13), in both cases with ε sufficiently small (in function of the constant τ in (4.3.10)). We can then bound the right-hand side of (4.3.5) by an expression of the type

$$-c\int_0^l |u|^{p-2}|u'|^2\,dx + \omega\int_0^l |u|^p\,dx$$

where $\omega = \omega(\delta)$ and $c = c(\delta) > 0$. Upon dividing by $\|u\|^{p-2}$ we obtain

$$\mathrm{Re}\langle\theta(u), A_p(\beta_0, \beta_l)u\rangle \leqslant \pm\,\delta\,\mathrm{Im}\langle\theta(u), A_p(\beta_0, \beta_l)u\rangle + \omega\|u\|^2.$$
$$(4.3.15)$$

To extend (4.3.15) to any $u \in D(A_p(\beta_0, \beta_l))$, we use an obvious approximation argument. Inequality (4.3.15) is obtained in the same way when β_0, β_l (or both) are of type (II). In the last case the use of the function ρ is of course unnecessary.

When $1 \leqslant p \leqslant 2$ we prove (4.3.15), say, for a polynomial satisfying the boundary conditions and take limits as before. See the comments in the proof of Lemma 3.2.2.

Inequality (4.3.15) is none other than (4.2.25) for the operator $A_p(\beta_0, \beta_l) - \omega I$, thus Corollary 4.2.5 applies. The estimate (4.3.4) is a direct consequence of (4.3.15); in fact, if $\varphi' < \varphi_p$ and $|\varphi| \leqslant \varphi'$, we have

$$\mathrm{Re}\langle\theta(u), e^{i\varphi}A_p(\beta_0, \beta_l)u\rangle = (\cos\varphi)\,\mathrm{Re}\langle\theta(u), A_p(\beta_0, \beta_l)u\rangle$$

$$- (\sin\varphi)\,\mathrm{Im}\langle\theta(u), A_p(\beta_0, \beta_l)u\rangle$$

$$\leqslant \omega(\cos\varphi)\|u\|^2 \quad (u \in D(A)) \qquad (4.3.16)$$

so that $e^{i\varphi}A_p(\beta_0, \beta_l) - \omega\cos\varphi I$ is dissipative.

A careful examination of the result just obtained reveals that the restrictions on the domain of analyticity of $S_p(\zeta)$ (which become progressively worse when $p \to 1$ or $p \to \infty$ since $\varphi_p \to 0$ in those two cases) are actually restrictions on the validity of (4.3.4).

This makes clear the limitations of any analysis of operators in $\mathcal{C}$ based on Corollary 4.2.5; an operator $A \in \mathcal{C}(\varphi)$ such that $S(\zeta) = \exp(\zeta A)$ satisfies

$$\|S(\zeta)\| \leqslant Ce^{\omega|\zeta|} \quad (\zeta \in \Sigma(\varphi)) \qquad (4.3.17)$$

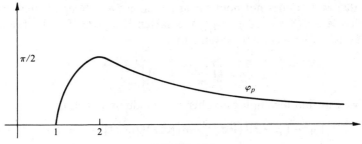

FIGURE 4.3.1

but not (4.3.4) in any sector may slip past Corollary 4.2.5 in the same way an operator in $\mathcal{C}_+(C, \omega)$ may fail to be detected by Theorem 3.1.8. These considerations are thrown into focus by the following

4.3.2   Example.   Let $A = A_p(\beta_0, \beta_l)$, where the basic interval is $0 \leqslant x \leqslant \pi$, $Au = u''$, and both boundary conditions are of type (II). The operator $S(\zeta) = \exp(\zeta A)$ can be explicitly computed by the separation-of-variables techniques in Sections 1.1 and 1.3: if $u \sim \Sigma a_n \sin nx$, then

$$S(\zeta)u = \sum_{n=1}^{\infty} e^{-n^2\zeta} a_n \sin nx$$

$$= (4\pi\zeta)^{-1/2} \int_{-\infty}^{\infty} e^{-(x-\xi)^2/4\zeta} u(\xi)\, d\xi, \qquad (4.3.18)$$

where in the last formula u has been continued to $-\infty < x < \infty$ in such a way that the extended function is odd about $\xi = 0$ and $\xi = \pi$. It is plain that if $\operatorname{Re}\zeta > 0$, the operator $S(\zeta)$ is bounded in $L^p(0, \pi)$, $1 \leqslant p < \infty$ and in $C_{0,\pi}[0, \pi]$. We estimate its norm in the latter space:

$$\|S(\zeta)u\|_C \leqslant (4\pi|\zeta|)^{-1/2} \left(\int_{-\infty}^{\infty} e^{-\xi^2 \operatorname{Re}\zeta/4|\zeta|^2}\, d\xi \right) \|u\|_C$$

$$= (|\zeta|/\operatorname{Re}\zeta)^{1/2} \|u\|_C \qquad (4.3.19)$$

so that

$$\|S(\zeta)\| \leqslant (|\zeta|/\operatorname{Re}\zeta)^{1/2}. \qquad (4.3.20)$$

A corresponding bound in L^p can be obtained by interpolation. But an estimate of the type of (4.3.4) will not hold in any sector $|\varphi| < \varphi_0$ if $E = L^1$ or if $E = C$. This follows from the next example and a simple interpolation argument.

It is of some interest to ask whether the angle φ_p in Theorem 4.3.1 is the best (i.e., the largest) such that (4.3.4) holds for every $\varphi < \varphi_p$. This is in fact the case.

4.3.3   Example.   Let $1 < p < \infty$, $A_p(\beta_0, \beta_l)$ a general differential operator. Then, if $|\varphi| > \varphi_p$, the operator $e^{i\varphi} A_p(\beta_0, \beta_l) - \omega I$ is not dissipative for any

ω, so that (4.3.4) does not hold for any such φ even though it may be the case that $A \in \mathcal{C}(\varphi')$ with $\varphi' > \varphi_p$. We sketch the proof for the boundary conditions $u'(0) = u'(l) = 0$. Assume that

$$\delta > \left\{ \left(\frac{p}{p-2} \right)^2 - 1 \right\}^{1/2}.$$

Then we can find a complex number z of modulus 1 such that

$$|z|^2 + (p-2)\big((\mathrm{Re}\, z)^2 - \delta(\mathrm{Re}\, z)(\mathrm{Im}\, z)\big) = -\mu < 0. \qquad (4.3.21)$$

Let η be a smooth real-valued function such that $\eta'(0) = \eta'(l) = 0$. Then $u(x) = \exp(z\eta(x)) \in D(A_p(\beta_0, \beta_l))$ and we have $\overline{u(x)}u'(x) = z\eta'(x)\exp(2(\mathrm{Re}\, z)\eta(x)) = z\psi(x)$ with ψ real. Accordingly, if f is the function in (4.3.10),

$$f = -\mu|u|^2|u'|^2 = -\mu\psi^2$$

Estimating the other terms in the same way as in Theorem 4.3.1, we obtain the inequality

$$\mathrm{Re}\langle \theta(u), A_p(\beta_0, \beta_l)u \rangle - \delta \mathrm{Im}\langle \theta(u), A_p(\beta_0, \beta_l)u \rangle$$

$$\geq c\|u\|^{2-p} \int |u|^{p-2}|u'|^2 \, dx - C\|u\|^2, \qquad (4.3.22)$$

where $c > 0$. If (4.3.15) holds for some ω, we deduce that there exists a constant C' such that

$$\int_0^l |u|^{p-2}|u'|^2 \, dx \leq C' \int_0^l |u|^p \, dx,$$

an inequality that can be easily disproved.

Theorem 4.3.1 can be used to obtain sharp bounds on the norm of certain multiplier operators. For instance, let $l = \pi$, $Au = u''$, β_0, β_l boundary conditions of type (I) with $\gamma_0 = \gamma_l = 0$. Then $S(\zeta) = \exp(\zeta A_p(\beta_0, \beta_l))$ is given by

$$S(\zeta)u = \sum_{n=0}^{\infty} e^{-n^2\zeta} a_n \cos nx$$

for $u \sim \Sigma a_n \cos nx$. Consider $S(\zeta)$ in the space L^p, $1 < p < \infty$. Since $S(\zeta)\cos x = \cos x$, $\|S(\zeta)\| \geq 1$. Examining (4.3.5), taking into account that $a = 1$, $b = c = 0$, and making use of (4.3.6) and (4.3.9), we see that $e^{i\varphi}A_p(\beta_0, \beta_l)$ is dissipative if $|\varphi| \leq \varphi_p$ so that

$$\|S(\zeta)\| = 1 \quad \text{if} \quad |\arg \zeta| \leq \varphi_p. \qquad (4.3.23)$$

On the other hand, it follows from Example 4.3.3 that the inequality does not extend to $|\arg \zeta| > \varphi_p$ even in the weakened form (4.3.4).

Since Theorem 4.3.1 does not provide any information on A_1 or A, we give later a sort of substitute that falls short of proving that these operators belong to $\mathcal{C}$ (Theorem 4.3.5); this result hinges upon the case $p = 2$, where the sector of analyticity in Theorem 4.3.1 is optimal (see Remark 4.1.7). As a preliminary step we prove that A_1 and $A \in \mathcal{C}_+$, which was only shown in Section 3.4 under dissipativity assumptions on the boundary conditions.

4.3.4   Theorem.   (a) *The operator* $A_1(\beta_0, \beta_l)$ *belongs to* $\mathcal{C}_+$ *in* $L^1(0, l)$. (b) *The operator* $A(\beta_0, \beta_l)$ *belongs to* $\mathcal{C}_+$ *in* $C[0, l]$ ($C_0, C_l, C_{0,l}$ *depending on the boundary conditions used*).

Proof.   Assume for the moment that both boundary conditions are of type (I). If the dissipativity conditions (3.4.8) are not satisfied for $p = 1$ there is obviously no hope to prove that $A_1(\beta_0, \beta_l) - \omega I$ is dissipative for any ω since they are necessary for dissipativity of *any* operator, $\tilde{A}(\beta_0, \beta_l)$ using these boundary conditions in L^1. The same comment applies to conditions (3.4.5) in C. However, a renorming of the space will remove this obstacle.

Let $1 \leqslant p < \infty$, ρ a continuous positive function in $0 \leqslant x \leqslant l$. Consider the norm

$$\|u\|_\rho = \left(\int_0^l |u(x)|^p \rho(x)^p \, dx \right)^{1/p} \tag{4.3.24}$$

in $L^p(0, l)$. Clearly $\|\cdot\|_\rho$ is equivalent to the original norm of L^p; we write $L^p(0, l)_\rho$ to indicate that L^p is equipped with $\|\cdot\|_\rho$ rather than with its original norm. The dual space $L^p(0, l)^*_\rho$ can be identified with $L^{p'}(0, l)$, $p'^{-1} + p^{-1} = 1$, *endowed with its usual norm*, an element $u^* \in L^{p'}(0, l)$ acting on $L^p(0, l)_\rho$ through the formula

$$\langle u^*, u \rangle_\rho = \int_0^l u^* u \rho \, dx. \tag{4.3.25}$$

If $p > 1$, there exists only one duality map $\theta_\rho : L^p_\rho \to L^{p'}$ given by $\theta_\rho(u) = \theta(\rho u)$, θ the duality map corresponding to the case $\rho \equiv 1$ (see Example 3.1.6). For $p = 1$ the duality set of an element $u \in L^1(0, l)_\rho$ coincides with the duality set of $u\rho$ as an element of $L^1(0, l)$ (see again Example 3.1.6). We take now u twice continuously differentiable and perform the customary integrations by parts, assuming that ρ is twice continuously differentiable as well:

$$\|u\rho\|^{p-2} \mathrm{Re}\langle \theta_\rho(u), A_p(\beta_0, \beta_l)u \rangle_\rho$$

$$= \|u\rho\|^{p-2} \mathrm{Re}\langle \rho\theta(\rho u), A_p(\beta_0, \beta_l)u \rangle$$

$$= \mathrm{Re} \int_0^l ((au')' + (b - a')u' + cu)|u|^{p-2} \bar{u} \rho^p \, dx$$

$$= \left\{ \gamma_l a(l) - \frac{1}{p}(a'(l) - b(l)) \right\} |u(l)|^p \rho(l)^p$$

$$-\left\{\gamma_0 a(0) - \frac{1}{p}(a'(0) - b(0))\right\}|u(0)|^p \rho(0)^p$$

$$-(p-2)\int_0^l a|u|^{p-4}(\text{Re}(\bar{u}u'))^2 \rho^p\,dx$$

$$-\int_0^l a|u|^{p-2}|u'|^2\rho^p\,dx - p\int_0^l a|u|^{p-2}\text{Re}(\bar{u}u')\rho^{p-1}\rho'\,dx$$

$$+\frac{1}{p}\int_0^l(a''-b'+pc)|u|^p\rho^p\,dx + \int_0^l(a'-b)|u|^p\rho^{p-1}\rho'\,dx$$

$$=\left\{\left(\gamma_l - \frac{\rho'(l)}{\rho(l)}\right)a(l) - \frac{1}{p}(a'(l)-b(l))\right\}|u(l)|^p\rho(l)^p$$

$$-\left\{\left(\gamma_0 - \frac{\rho'(0)}{\rho(0)}\right)a(0) - \frac{1}{p}(a'(0)-b(0))\right\}|u(0)|^p\rho(0)^p$$

$$-(p-2)\int_0^l a|u|^{p-4}(\text{Re}(\bar{u}u'))^2\rho^p\,dx$$

$$-\int_0^l a|u|^{p-2}|u'|^2\rho^p\,dx$$

$$+\int_0^l\left\{(a\rho^{p-1}\rho')' + \frac{1}{p}(a''-b'+pc)\rho^p + (a'-b)\rho^{p-1}\rho'\right\}|u|^p\,dx.$$

$$(4.3.26)$$

Since $\rho'(0)$ and $\rho'(l)$ can be chosen at will, we do so in such a way that the quantities between curly brackets in the first two terms on the right-hand side of (4.3.26) are nonpositive, say, for $1 < p \leqslant 2$. As the first two integrals together contribute a nonpositive amount, we can bound (4.3.26) by an expression of the form $\omega'\|u\|^p \leqslant \omega\|u\|_\rho^p$, *where ω does not depend on p.* Consider now the space $L^1(0, l)_\rho$. Again under the identification (4.3.25), the duality set $\Theta_\rho(u)$ of an element u consists of all $u^* \in L^\infty(0, l)$ with $u^*(x) = \|u\|_\rho|u(x)|^{-1}\bar{u}(x) = \|u\rho\||u(x)|^{-1}\bar{u}(x)$ where $u(x) \neq 0$ and $|u^*(x)| \leqslant \|u\|_\rho$ elsewhere. We can then take limits in (4.3.26) as $p \to 1$ in the same way as in Lemma 3.2.2 and obtain an inequality of the form

$$\text{Re}\langle u^*, A_1(\beta_0, \beta_l)u\rangle_\rho \leqslant \omega\|u\|_\rho^2 \qquad (4.3.27)$$

in L^1. The inequality is extended to arbitrary $u \in D(A_1(\beta_0, \beta_l))$ by means of the usual approximation argument. Now that $A_1(\beta_0, \beta_l) - \omega I$ has been shown to be dissipative, m-dissipativity is established using Green functions as in the end of Section 3.4. The case where one (or both) of the boundary conditions are of type (II) is treated in an entirely similar way; naturally, the use of the weight function is unnecessary in the latter case.

To prove a similar result for the operator A we renorm the space C or the corresponding subspace by means of

$$\|u\|_\rho = \max_{0 \leqslant x \leqslant l}|u(x)|\rho(x), \qquad (4.3.28)$$

where ρ is a positive, twice continuously differentiable function in $0 \leqslant x \leqslant l$. The use of the weight function ρ is again unnecessary when both boundary conditions are of type (II): we treat below in detail the case where β_0 and β_l are of type (I), the "mixed" case being essentially similar. Choose ρ in such a way that

$$\rho'(0) + \gamma_0 \rho(0) \geqslant 0 \quad (\text{resp. } \rho'(l) + \gamma_l \rho(l) \leqslant 0) \qquad (4.3.29)$$

if $\gamma_0 < 0$ (resp. $\gamma_l > 0$). The dual of $C[0, l]$ equipped with $\|\cdot\|_\rho$ can again be identified with $\Sigma[0, l]$, an element $\mu \in \Sigma[0, l]$ acting on functions $u \in C[0, l]$ through the formula

$$\langle \mu, u \rangle = \int_0^l u(x) \rho(x) \mu(dx). \qquad (4.3.30)$$

Accordingly, the norm of a measure $\mu \in \Sigma$ as an element of C^* is still $\|\mu\| = \int_0^l |\mu|(dx)$ and the identification of the duality sets $\Theta_\rho(u)$ can be easily adapted from Section 3.1; $\Theta_\rho(u)$ consists of all $\mu \in \Sigma$ with support in $m_\rho(u) = \{x; |u(x)|\rho(x)| = \|u\|_\rho\}$ and such that $u\rho\mu$ (or $u\mu$) is a positive measure in $m_\rho(u)$ with $\|\mu\| = \|u\|_\rho$. The same comments apply of course to the spaces $C_0, C_l, C_{0,l}$, where the corresponding measures are required to vanish at $0, l, 0$ and l.

We now show that $A(\beta_0, \beta_l) - \omega I$ is m-dissipative for ω large enough. Observe first that if $u'(0) = \gamma_0 u(0)$, $u'(l) = \gamma_l u(l)$, then $u\rho$ satisfies the boundary conditions

$$(u\rho)'(0) = \gamma_{0,\rho}(u\rho)(0), \quad (u\rho)'(l) = \gamma_{l,\rho}(u\rho)(l), \qquad (4.3.31)$$

where

$$\gamma_{0,\rho} = \gamma_0 + \rho'(0)\rho(0)^{-1} \geqslant 0, \quad \gamma_{l,\rho} = \gamma_l + \rho'(l)\rho(l)^{-1} \leqslant 0. \quad (4.3.32)$$

An argument already employed in Section 3.3 shows that for any $u \in D(A(\beta_0, \beta_l))$, $u \neq 0$ the set $m_\rho(u)$ does not contain either endpoint of the interval $[0, l]$ if both $\gamma_{0,\rho}, \gamma_{l,\rho} > 0$, so that

$$(|u\rho|^2)'(x) = 0, (|u\rho|^2)''(x) \leqslant 0 \quad (x \in m_\rho(u)). \qquad (4.3.33)$$

On the other hand, if either $\gamma_{0,\rho}$ or $\gamma_{l,\rho}$ vanish, $m_\rho(u)$ may contain the corresponding endpoint but arguing again as in Section 3.3, we see that (4.3.33) holds. Writing $\eta = \rho^2$, we have $(|u\rho|^2)' = 2(u_1 u_1' + u_2 u_2')\eta + (u_1^2 + u_2^2)\eta'$, $(|u\rho|^2)'' = 2(u_1 u_1'' + u_2 u_2'')\eta + 2(u_1'^2 + u_2'^2)\eta + 4(u_1 u_1' + u_2 u_2')\eta' + (u_1^2 + u_2^2)\eta''$. Hence, it follows from (4.3.33) that

$$\text{Re}(u^{-1}u') = \|u\|^{-2}(u_1 u_1' + u_2 u_2') = -\tfrac{1}{2}\|u\|^{-2}(u_1^2 + u_2^2)\eta^{-1}\eta'$$
$$= -\tfrac{1}{2}\eta^{-1}\eta', \qquad (4.3.34)$$

$$\text{Re}(u^{-1}u'') = \|u\|^{-2}(u_1 u_1'' + u_2 u_2'') \leqslant -\|u\|^{-2}(u_1'^2 + u_2'^2)$$
$$+ \|u\|^{-2}(u_1^2 + u_2^2)\eta^{-2}\eta'^2 - \tfrac{1}{2}\|u\|^{-2}(u_1^2 + u_2^2)\eta^{-1}\eta''$$
$$= -\|u\|^{-2}|u'|^2 + \eta^{-2}\eta'^2 - \tfrac{1}{2}\eta^{-1}\eta'' \qquad (4.3.35)$$

in $m_\rho(u)$. Accordingly, if $\mu \in \Theta_\rho(u)$, we have

$$\mathrm{Re}\langle \mu, A(\beta_0, \beta_l)u \rangle = \int_{m_\rho(u)} \mathrm{Re}(u^{-1}A(\beta_0, \beta_l)u)\rho u\, d\mu$$

$$\leqslant \omega\|u\|_\rho^2 \qquad (4.3.36)$$

for some constant ω, which shows that $A(\beta_0, \beta_l) - \omega I$ is dissipative. That $(\lambda I - A(\beta_0, \beta_l))u = v$ has a solution u for all v is once again shown by means of Green functions. This ends the proof of Theorem 4.3.4.

Theorem 4.3.1 only states that $u(t) = S_p(t)u$, the solution (or generalized solution) of $u'(t) = A_p(\beta_0, \beta_l)u(t)$ can be extended to a function $u(\zeta) = S_p(\zeta)$ "analytic in the L^p mean." However, more than this is true, since $u(t, x) = u(t)(x)$ can be shown to be pointwise analytic. This result extends to $p = 1$.

4.3.5   Theorem.   *Let $u \in L^p$ $(1 \leqslant p < \infty)$. Then $u(t, x) = (S_p(t)u)(x)$, after eventual modification in an x-null set for each t, can be extended to a function $u(\zeta, x)$ analytic in $\Sigma_+(\pi/2-)$. Moreover, $u(\zeta, \cdot) \in C$ $(C_0, C_l, C_{0,l}$ depending on the boundary conditions) for all $\zeta \in \Sigma_+(\pi/2-)$ and, for every φ with $0 < \varphi < \pi/2$ and every $\varepsilon > 0$ there exist constants $C = C(\varphi, \varepsilon)$, $\omega = \omega(\varphi, \varepsilon)$ such that*

$$\|S_p(\zeta)u\|_C \leqslant Ce^{\omega|\zeta|}\|u\| \qquad (\zeta \in \Sigma(\varphi), |\zeta| \geqslant \varepsilon), \qquad (4.3.37)$$

where the norm on the right-hand side is the L^p norm.

Proof.   Let p, q be such that $1 \leqslant p < q \leqslant \infty$. Since $A_p(\beta_0, \beta_l) \supseteq A_q(\beta_0, \beta_l)$ (both operators thought of as acting in L^q) we also have $R(\lambda; A_p(\beta_0, \beta_l)) \supseteq R(\lambda; A_q(\beta_0, \beta_l))$ and it follows from formula (2.1.27) that the same inclusion relation holds for the solution operators S_p; precisely, if $u \in L^q(0, l)$, we have

$$S_p(t)u = S_q(t)u \qquad (4.3.38)$$

for $t \geqslant 0$. Consider now a function u in $W^{1,p}(0, l)$ for some $p \geqslant 1$. We have

$$u(x) - u(x') = \int_{x'}^{x} u'(\xi)\, d\xi \qquad (4.3.39)$$

for almost all x, x', so that it follows from Hölder's inequality that, after eventual modification in a null set, u is continuous in $0 \leqslant x \leqslant l$. Moreover, setting (say) $x' = 0$ in (4.3.39) we obtain an inequality of the form

$$|u(x)| \leqslant |u(0)| + l^{1-1/p}\|u'\|_{L^p},$$

where the expression on the right-hand side is easily seen to be a norm equivalent to the original norm of $W^{1,p}$. It follows that

$$W^{1,p}(0, l) \subseteq C[0, l] \qquad (4.3.40)$$

with bounded inclusion.

We take now $u \in L^p(0, l)$ and λ sufficiently large. We have

$$S_p(t)u = R(\lambda; A_p(\beta_0, \beta_l))(\lambda I - A_p(\beta_0, \beta_l))^2 S_p(t) R(\lambda; A_p(\beta_0, \beta_l)) u$$
$$= R(\lambda; A_2(\beta_0, \beta_l))(\lambda I - A_2(\beta_0, \beta_l))^2 S_2(t) R(\lambda; A_p(\beta_0, \beta_l)) u,$$

$$(4.3.41)$$

the last equality justified by (4.3.40) and preceding considerations. In view of Theorem 4.3.1 for $p = 2$, $S_2(t)$ can be analytically extended to $\Sigma_+(\pi/2-)$ and satisfies (4.3.4) there for any φ with $0 < \varphi < \pi/2$. Noting that $A_2(\beta_0, \beta_l)^2 S_2(\zeta) = (A_2(\beta_0, \beta_l)S(\zeta/2))^2$ and using (4.1.18), we can show that for every φ with $0 < \varphi < \pi/2$ and for each $\varepsilon > 0$ there exist C, ω depending on φ and ε such that

$$\|(\lambda I - A_2(\beta_0, \beta_l))^2 S_2(\zeta)\| \leqslant Ce^{\omega|\zeta|} \quad (\zeta \in \Sigma(\varphi), |\zeta| \geqslant \varepsilon) \quad (4.3.42)$$

with C and ω depending on φ. Consider now $D(A_p(\beta_0, \beta_l))$, which consists of all $u \in W^{1,p}(0, l)$ that satisfy the boundary conditions at 0 and l. Since $D(A_p(\beta_0, \beta_l))$ is a Banach space both under the norm of $W^{1,p}$ and under (the norm equivalent to) the graph norm $\|u\|' = \|(\lambda I - A_p(\beta_0, \beta_l))u\|$ and the first norm dominates the second (times a constant), it follows from the closed graph theorem that both are equivalent, so that, taking norms in (4.3.41) (or, rather, in its analytic extension to $\mathrm{Re}\,\zeta > 0$), we obtain from (4.3.40) that $S_p(\zeta)$ can be extended to a $C[0, l]$-valued analytic function in $\Sigma_+(\pi/2-)$ and

$$\|S_p(\zeta)u\|_C \leqslant C'\|(\lambda I - A_2(\beta_0, \beta_l))^2 S_2(\zeta) R(\lambda; A_p(\beta_0, \beta_l))u\|_2$$
$$\leqslant C''e^{\omega|\zeta|}\|R(\lambda; A_p(\beta_0, \beta_l))u\|_2$$
$$\leqslant Ce^{\omega|\zeta|}\|u\|_p$$

under the conditions claimed in Theorem 4.3.5. This ends the proof.

As it may be expected, greater x-smoothness may be obtained if the coefficients of A_0 are assumed smoother. An auxiliary estimate will be needed.

4.3.6 Lemma. *Let a (resp. b, c) be $k + 2$ times (resp. $k + 1, k$ times) continuously differentiable in $0 \leqslant x \leqslant l$; let $1 \leqslant p \leqslant \infty$ and λ large enough. Then the only solution of*

$$(\lambda I - A_p(\beta_0, \beta_l))u = f \quad (4.3.43)$$

in $D(A(\beta_0, \beta_l))$ belongs to $W^{k+2,p}(0, l)$ if $f \in W^{k,p}(0, l)$. Moreover, there exists a constant $C = C_k$ independent of f such that

$$\|u\|_{W^{k+2,p}} \leqslant C\|f\|_{W^{k,p}}. \quad (4.3.44)$$

Proof. For $k = 0$ the result is of course a consequence of the identification of the domain of $A_p(\beta_0, \beta_l)$ in Section 3.4. Consider now the case $k = 1$. If u is an arbitrary solution (no boundary conditions) of the homogeneous equation $au'' + bu' + (c - \lambda)u = 0$, then $v = u'$ satisfies $av'' + bv' + (c - \lambda)v = g = -a'u'' - b'u' - c'u$ (where g is continuous in $0 \leqslant x \leqslant l$) in the sense of distributions. It follows then essentially as in Lemma 3.2.5 that v is twice continuously differentiable in $0 \leqslant x \leqslant l$, so that $u \in C^{(3)}[0, l]$. Let now $G(x, \xi)$ be the Green function of $A(\beta_0, \beta_l) - \lambda I$, where λ is large enough. Then we can write the solution of (4.3.43) in the form

$$u(x) = -\int_0^l G(x, \xi) f(\xi) d\xi$$

$$= -u_l(x) \int_0^x \frac{u_0(\xi)}{a(\xi)W(\xi)} f(\xi) d\xi - u_0(x) \int_x^l \frac{u_l(\xi)}{a(\xi)W(\xi)} f(\xi) d\xi$$

(see Section 3.4 for the description of all the functions involved). We then have

$$u'(x) = -u_l'(x) \int_0^x \frac{u_0(\xi)}{a(\xi)W(\xi)} f(\xi) d\xi - u_0'(x) \int_x^l \frac{u_l(\xi)}{a(\xi)W(\xi)} f(\xi) d\xi.$$

$$(4.3.45)$$

Since both a, W are nonzero and continuously differentiable and $u_0, u_1 \in C^{(3)}[0, l]$, it is clear that $u \in W^{3, p}(0, l)$ and that an inequality of the type of (4.3.44) holds.[1] The proof for $k = 2, 3, \ldots$ is similar.

To simplify the statement of the next result we assume infinite differentiability of the coefficients of A_0.

4.3.7 Theorem. *Let a, b, c be infinitely differentiable in $0 \leqslant x \leqslant l$. Then each generalized solution $u(t, x) = u(t)(x) = (S_p(t)u)(x)$ of $u'(t) = A_p(\beta_0, \beta_l)u(t)$, $1 \leqslant p < \infty$ (after eventual modification in an x-null set for each t) can be extended to a function $u(\zeta, x)$ infinitely differentiable in x and analytic in ζ for $\mathrm{Re}\,\zeta > 0$; moreover, for every $\varepsilon > 0$, every positive φ with $\varphi < \pi/2$ and every integer $k = 0, 1, \ldots$ there exist C, ω depending on ε, φ and k such that*

$$\|u(\zeta, \cdot)\|_{C^{(k)}} \leqslant Ce^{\omega|\zeta|}\|u\| \quad (\zeta \in \Sigma(\varphi), |\zeta| \geqslant \varepsilon), \quad (4.3.46)$$

where the norm on the right-hand side is the L^p norm. A similar statement holds for solutions in C.

Proof. We have already observed in the proof of Theorem 4.3.5 that the $W^{2, p}$ norm and the norm $\|u\|' = \|(\lambda I - A_p(\beta_0, \beta_l))u\|$ are equivalent in $D(A_p(\beta_0, \beta_l))$ (λ large enough). If $u \in D(A_p(\beta_0, \beta_l)^2)$, then $(\lambda I - A_p(\beta_0, \beta_l))u$ belongs to $W^{2, p}$, and it follows from Lemma 4.3.6 that

---

[1]Of course, that $u \in W^{3, p}(0, l)$ follows directly from Lemma 3.2.6. The present argument is used to establish (4.3.44).

$u \in W^{4,p}$ with

$$\|u\|_{W^{4,p}} \leqslant C\|(\lambda I - A_p(\beta_0, \beta_l))u\|_{W^{2,p}}$$

$$\leqslant C'\|(\lambda I - A_p(\beta_0, \beta_l))^2 u\|_{L^p}.$$

An inductive argument then shows in the same fashion that

$$\|u\|_{W^{2j,p}} \leqslant C_k\|(\lambda I - A_p(\beta_0, \beta_l))^j u\|_{L^p}. \tag{4.3.47}$$

Write $A_p(\beta_0, \beta_l)^j S(\zeta) = (A_p(\beta_0, \beta_l) S(\zeta/j))^j$. Using (4.1.18) for $|\zeta| \geqslant \varepsilon/j$ and (4.3.47), we deduce that $\|u(\zeta, \cdot)\|_{W^{2j,p}}$ satisfies an estimate of the required form. The result then follows from the fact (consequence of (4.3.40) applied to higher derivatives) that $W^{k+1,p}(0, l) \subseteq C^{(k)}[0, l]$ with bounded inclusion.

4.3.8 Remark. If the operator A is written in variational form (2.2.35) all the results in this section (except for Lemma 4.3.6 and Theorem 4.3.7) can be proved under the only assumption that $a, \hat{b}$ are continuously differentiable; the assumption on c is the same. In Lemma 4.3.6 it is enough to assume that $a, \hat{b}$ are $k + 1$ times continuously differentiable, c is k times continuously differentiable.

4.3.9 Example. The results in Example 3.4.5 can be extended to the operators in the present section (i.e., the dissipativity assumptions on the boundary conditions can be lifted).

4.4. SECOND ORDER PARTIAL DIFFERENTIAL OPERATORS. DISSIPATIVITY

Throughout the rest of this chapter we shall study operators of the form

$$A_0 u = \sum_{j=1}^{m} \sum_{k=1}^{m} a_{jk}(x) D^j D^k u + \sum_{j=1}^{m} b_j(x) D^j u + c(x) u \tag{4.4.1}$$

in a domain Ω in Euclidean space $\mathbb{R}^m$ with boundary Γ. The coefficients will be assumed to be real and, for the time being, as differentiable as required in the various computations. Note that if the a_{jk} are continuously differentiable in Ω, we may write A_0 in *divergence* or *variational* form,

$$A_0 u = \sum_{j=1}^{m} \sum_{k=1}^{m} D^j(a_{jk}(x) D^k u) + \sum_{j=1}^{m} \hat{b}_j(x) D^j u + c(x) u, \tag{4.4.2}$$

where

$$\hat{b}_j = b_j - \sum_{k=1}^{m} D^k a_{jk}. \tag{4.4.3}$$

In this section and the next we shall use either of the two forms as convenience dictates. We may and will assume that the matrix $\mathcal{C}(x) =$

$\{a_{jk}(x)\}$ is symmetric for all x, since we can always replace $a_{jk}(x)$ by $\frac{1}{2}\{a_{jk}(x)+a_{kj}(x)\}$, which does not modify the action of A_0 on functions.

Before discussing the assignation of boundary conditions, we examine in this section the behavior of A_0 applied to functions that vanish for large $|x|$ and near the boundary of Ω. To this end, we define $D(A_0)$ to be the space $\mathfrak{D}^{(2)}(\Omega)$ of all twice continuously differentiable functions on Ω with compact support contained in Ω (recall that supp u, the support of u, is the closure of the set $\{x \in \Omega;\ u(x) \neq 0\}$). We shall consider A_0 in the complex spaces $L^p(\Omega)$, $1 \leqslant p < \infty$ and $C_0(\overline{\Omega})$, the space of all functions continuous in $\overline{\Omega}$ that tend to zero as $|x| \to \infty$ if Ω is unbounded[2]; if Ω is bounded, $E = C(\overline{\Omega})$.

4.4.1   Lemma. *The operator A_0 is dissipative in $C_0(\overline{\Omega})$ (dissipative with respect to a duality map) if and only if*

$$\mathcal{Q}(x) \geqslant 0, \quad c(x) \leqslant 0 \quad (x \in \Omega), \tag{4.4.4}$$

where $\mathcal{Q}(x) \geqslant 0$ means the matrix $\mathcal{Q}(x)$ is positive definite,

$$\sum_{j=1}^{m} \sum_{k=1}^{m} a_{jk}(x)\xi_j\xi_k \geqslant 0 \quad (\xi \in \mathbb{R}^m). \tag{4.4.5}$$

The proof runs very much as in the one-dimensional case (Lemma 3.2.1) thus we omit some details. Let $u \in D(A_0)$, $u \neq 0$, $m(u) = \{x \in \Omega;\ |u(x)| = \|u\|\}$. Since $u(x) = 0$ for $x \in \Gamma$ and at infinity, $m(u)$ is bounded and $m(u) \cap \Gamma = \varnothing$. Any element μ in $\Theta(u) \subset \Sigma(\overline{\Omega})$ is supported by $m(u)$ and $u(x)\mu(dx)$ is a positive measure in $m(u)$. Since $|u|^2$ attains its maximum in $m(u)$, we must have

$$\tfrac{1}{2}D^j|u|^2 = u_1 D^j u_1 + u_2 D^j u_2 = 0 \quad (x \in m(u)) \tag{4.4.6}$$

for $j = 1, 2, \ldots, m$, (where $u = u_1 + iu_2$), and the Hessian matrix $\mathcal{H}(x; |u|^2) = \{D^j D^k |u(x)|^2\}$ is negative definite ($-\mathcal{H}$ is positive definite) for $x \in m(u)$. We note now that

$$\tfrac{1}{2}D^j D^k |u|^2 = u_1 D^j D^k u_1 + u_2 D^j D^k u_2 + D^j u_1 D^k u_1 + D^j u_2 D^k u_2$$

while, for $x \in m(u)$ we have

$$u^{-1}D^j D^k u = \|u\|^{-2}\left(u_1 D^j D^k u_1 + u_2 D^j D^k u_2\right)$$
$$+ i\|u\|^{-2}\left(u_1 D^j D^k u_2 - u_2 D^j D^k u_1\right)$$

and

$$u^{-1}D^j u = i\|u\|^{-2}\left(u_1 D^j u_2 - u_2 D^j u_1\right).$$

It follows immediately that $\mathrm{Re}(u^{-1}D^j u) = 0$ and $\mathrm{Re}(u^{-1}D^j D^k u) =$

---

[2] Naturally $D(A_0) = \mathfrak{D}^{(2)}(\Omega)$ is not dense in $E = C_0(\overline{\Omega})$. This has no bearing on dissipativity.

$\|u\|^{-2}(\frac{1}{2}D^jD^k|u|^2 - D^ju_1D^ku_1 - D^ju_2D^ku_2)$; accordingly,

$\operatorname{Re}\langle\mu, A_0u\rangle$

$$= \int_{m(u)} \{\|u\|^{-2}\sum\sum a_{jk}(\tfrac{1}{2}D^jD^k|u|^2 - D^ju_1D^ku_1 - D^ju_2D^ku_2)+c\}u\,d\mu \leqslant 0$$

(4.4.7)

in view of the hypotheses and of the following elementary result on linear algebra.

4.4.2 Lemma. *Let* $\mathcal{Q} = \{a_{jk}\}, \mathcal{K} = \{h_{jk}\}$ *be symmetric positive definite* $m \times m$ *matrices. Then*

$$\sum_{k=1}^{m}\sum_{j=1}^{m} a_{jk}h_{jk} \geqslant 0$$

(4.4.8)

Proof. Let $\mathcal{B} = \{b_{jk}\}$ be an orthogonal matrix such that $\mathcal{K} = \mathcal{B}'\mathcal{D}\mathcal{B}$, where $'$ indicates transpose and $\mathcal{D}$ is a diagonal matrix; since the elements $\nu_1,\ldots,\nu_m$ in the diagonal are eigenvalues of $\mathcal{K}$ they must be nonnegative. Clearly we have

$$\sum\sum a_{jk}h_{jk} = \sum\sum\sum a_{jk}b_{ij}\sqrt{\nu_i}\,b_{ik}\sqrt{\nu_i}\,,$$

which is obviously nonnegative. This ends the proof.

To prove that both conditions (4.4.4) are necessary, let ξ be an arbitrary vector in $\mathbb{R}^m$, α a positive number, x_0 a point in Ω. Since $\xi\otimes\xi = \{\xi_j\xi_k\}$ is a positive definite matrix we can find a real valued twice differentiable function with compact support contained in Ω (that is, a function in $D(A_0)$) having a single maximum at x_0 and such that

$$u(x_0) = \alpha, \qquad \mathcal{K}(x_0; u) = -\xi\otimes\xi.$$

The duality set $\Theta(u)$ than consists of the single element $\alpha\delta_0$ (δ_0 the Dirac measure at x_0) and

$$\langle\delta_0, Au\rangle = -(\mathcal{Q}\xi, \xi) + \alpha c(x_0),$$

(4.4.9)

which can be made positive by adequate choice of ξ, α if any of the two conditions (4.4.4) are violated.

Second order operators A_0 satisfying (4.4.5) are called *elliptic* in Ω. As we shall see next, the ellipticity condition is also necessary and sufficient (together with additional assumptions on the b_j and c) for dissipativity of A_0 in $L^p(\Omega)$.

4.4.3 Lemma. *The operator* A_0 *is dissipative in* L^p $(1 \leqslant p \leqslant \infty)$ *if*

$$\mathcal{Q}(x) \geqslant 0 \quad (x \in \Omega)$$

(4.4.10)

and

$$c(x) + \frac{1}{p}\left(\sum\sum D^jD^k a_{jk}(x) - \sum D^jb_j(x)\right) \leqslant 0 \quad (x \in \Omega).$$

(4.4.11)

The first inequality is necessary in all cases; the second is necessary when $p = 1$.

Proof. If $p > 1$ the only duality map $\theta: L^p \to L^{p'} (p'^{-1} + p^{-1} = 1)$ is $\theta(u) = \|u\|^{2-p}|u(x)|^{p-2}\bar{u}(x) (u \neq 0)$. For $p \geq 2$ we prove just as in the case $m = 1$ that if $u \in D(A_0)$, then $\theta(u)$ has continuous first partials and

$$\|u\|^{p-2}D^j\theta(u)(x) = (p-2)|u|^{p-4}\mathrm{Re}(\bar{u}D^ju)\bar{u} + |u|^{p-2}D^j\bar{u}$$

while, on the other hand,

$$D^j|u|^p = p|u|^{p-2}\mathrm{Re}(\bar{u}D^ju).$$

To compute $\mathrm{Re}\langle\theta(u), A_0u\rangle$ we use the divergence form (4.4.2) of the operator, obtaining

$$\|u\|^{p-2}\mathrm{Re}\langle\theta(u), A_0u\rangle$$

$$= \mathrm{Re}\int_\Omega\left\{\sum\sum D^j\left(a_{jk}D^ku\right) + \sum\left(b_j - \sum D^ka_{jk}\right)D^ju + cu\right\}|u|^{p-2}\bar{u}\,dx$$

$$= -(p-2)\int_\Omega|u|^{p-4}\left\{\sum\sum a_{jk}\mathrm{Re}(\bar{u}D^ju)\mathrm{Re}(\bar{u}D^ku)\right\}dx$$

$$\quad -\int_\Omega|u|^{p-2}\left\{\sum\sum a_{jk}D^juD^k\bar{u}\right\}dx$$

$$\quad +\frac{1}{p}\int_\Omega\left\{\sum\sum D^jD^ka_{jk} - \sum D^jb_j + pc\right\}|u|^p\,dx \leq 0 \qquad (4.4.12)$$

by means of integration by parts in each variable (since u vanishes near the boundary Γ no assumptions on applicability of the divergence theorem are necessary). Note that in establishing (4.4.12) we have made use of the inequality

$$\sum\sum a_{jk}\zeta_j\bar{\zeta}_k = \sum\sum a_{jk}\mathrm{Re}\,\zeta_j\mathrm{Re}\,\zeta_k + \sum\sum a_{jk}\mathrm{Im}\,\zeta_j\mathrm{Im}\,\zeta_k \geq 0 \quad (4.4.13)$$

valid for arbitrary complex $\zeta_1, \ldots, \zeta_m$.

In the case $1 < p < 2$ the computations are formally valid, but (as already pointed out when $m = 1$ in Section 3.2) it is not clear whether $\theta(u)$ is absolutely continuous in any one variable. We bring again into play the class $\mathcal{PC}^{(2)}$ of piecewise analytic, twice continuously differentiable functions of one variable with compact support. If $u \in D(A_0)$, then u can be uniformly approximated together with its derivatives of order ≤ 2 by a linear combination of functions of the form

$$u_1 \otimes \cdots \otimes u_m,$$

where the u_1 are twice continuously differentiable functions of one variable and the support of each term in the linear combination is contained in Ω (a theorem of this type is proved in Schwartz [1966: 1, p. 108]. A similar statement can be easily seen to hold with $u_j \in \mathcal{PC}^{(2)}$. A linear combination v of the type obtained has the property that $x_j \to v(x) (x_1, \ldots, x_{j-1}, x_{j+1}, \ldots, x_m$ fixed) belongs to $\mathcal{PC}^{(2)}$ so that $\theta(u)$ is absolutely continuous as a function of any one of its variables and therefore the manipulations leading to (4.4.12) are fully justified also for $1 < p < 2$.

The case $p = 1$ is again handled by a limiting argument as we did in Lemma 3.2.2; also, the treatment for $m = 1$ can be extended to show that (4.4.11) is necessary when $p = 1$.

Finally, the necessity of (4.4.10) for $1 < p < \infty$ follows from

4.4.4   Example.   Let $p > 1$, and let $\{a_{jk}\}$, g, be continuous in Ω. Assume that

$$\int_\Omega \left\{ |u|^{p-2} \sum \sum a_{jk} D^j u D^k u + g|u|^p \right\} dx \geq 0$$

for all $u \in \mathcal{D}(\Omega)_{\mathbb{R}}$. Then

$$\mathcal{C}(x) \geq 0 \quad (x \in \Omega).$$

An important role in the rest of the chapter will be played by the *formal adjoint* A_0' of A_0 defined by

$$A_0' u = \sum \sum D^j D^k \big(a_{jk}(x) u \big) - \sum D^j \big(b_j(x) u \big) + c(x) u. \quad (4.4.14).$$

The domain of A_0' is defined in the same way as $D(A_0)$. Integrating by parts, we obtain

$$\int_\Omega (A_0 u) v \, dx = \int_\Omega u A_0' v \, dx \qquad (4.4.15)$$

for $u, v \in D(A_0) = D(A_0')$, which equality can be taken as a definition of A_0'. We relate now the dissipativity conditions for A_0 and A_0'. Clearly (4.4.4) is the same either for A_0 or A_0' and it is easy to see that the inequality (4.4.11) holds for an operator A_0 in L^p if and only if it holds for A_0' in $L^{p'}$, $p'^{-1} + p^{-1} = 1$ (in C_0 if $p = 1$). The necessity of these conditions in L^1 and C_0 then shows that A_0 is dissipative in $L^1(\Omega)$ if and only if A_0' is dissipative in $C_0(\bar{\Omega})$.

From Section 4.6 onwards the operator A_0 will generally be written in divergence form (4.4.2). In this notation, the formal adjoint is given by

$$A_0' u = \sum \sum D^j \big(a_{jk}(x) D^k u \big) - \sum D^j \big(\hat{b}_j(x) u \big) + c(x) u \quad (4.4.16)$$

and the dissipativity condition (4.4.11) for $E = L^p(\Omega)$, $1 \leq p < \infty$, takes the simple form

$$c(x) - \frac{1}{p} \sum D^j \hat{b}_j(x) \leq 0. \qquad (4.4.17)$$

Obviously, the analogous dissipativity conditions for $C(\bar{\Omega})$ are the same irrespective of the form we use in writing A_0.

We note finally that the fact that (4.4.11) (or (4.4.17)) is not necessary for dissipativity of A_0 when $1 < p < \infty$ has already been pointed out in Section 2.3 for the case $m = 1$ (Example 3.2.4). A counterexample for the case $m > 1$ can be constructed in an obvious way using that result.

4.4.5   Example.   Show that inequality (4.4.11) is necessary for dissipativity of A_0 in $L^1(\Omega)$.

4.5.   SECOND ORDER PARTIAL DIFFERENTIAL OPERATORS. ASSIGNATION OF BOUNDARY CONDITIONS

We need now conditions on Ω, Γ, in order to be able to apply the divergence theorem in Ω,

$$\int_{\Gamma} (U, \nu)\, d\sigma = \int_{\Omega} \operatorname{div} U dx, \qquad (4.5.1)$$

to smooth vector fields U with compact support, where ν is the outer normal vector at Γ and $d\sigma$ denotes the hyperarea differential on Γ. We introduce a sequence of classes of sets where the conditions for validity of (4.5.1) are amply satisfied.

A domain $\Omega \subseteq \mathbb{R}^m$ is said to *of class* $C^{(k)}$ (k an integer $\geqslant 0$) if and only if given any point $\bar{x} \in \Gamma$ there exists an open neighborhood V of $\bar{x}$ in $\mathbb{R}^m$ and a map $\eta : \bar{V} \to \bar{\mathbb{S}}^m$ ($\mathbb{S}^m = \mathbb{S}^m(0,1) = \{\eta \in \mathbb{R}^m; |\eta| < 1\}$ the open unit sphere in $\mathbb{R}^m$) such that:

(a) *η is one-to-one and onto $\bar{\mathbb{S}}^m$ with $\eta(\bar{x}) = 0$*
(b) *The map η (resp. η^{-1}) possesses partial derivatives of order $\leqslant k$ in V (resp. in $\mathbb{S}^m$) which are continuous in $\bar{V}$ (resp. in $\bar{\mathbb{S}}^m$)*
(c)

$$\eta(V \cap \Omega) = \mathbb{S}_+^m = \{\eta \in \mathbb{S}^m; \eta_m > 0\} \qquad (4.5.2)$$

$$\eta(V \cap \Gamma) = \mathbb{S}_0^m = \{\eta \in \mathbb{S}^m; \eta_m = 0\}^3 \qquad (4.5.3)$$

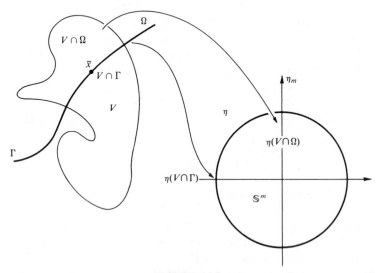

FIGURE 4.5.1

---

[3] For $k \geqslant 1$ it follows that Γ is a differential manifold of class $C^{(k)}$ and dimension $m - 1$. For $k = 0$, Γ is locally homeomorphic to $\mathbb{S}_0^m$; this kind of subset of $\mathbb{R}^m$ is usually called a *topological manifold* (of dimension $m - 1$).

We shall call any of the neighborhoods V postulated above a *boundary patch* with *associated map* η. An argument involving local compactness of $\mathbb{R}^m$ shows that we can find a sequence $V_1, V_2, \ldots$ of boundary patches and a sequence $\mathbb{S}_1, \mathbb{S}_2, \ldots$ of spheres $\mathbb{S}_n = \mathbb{S}^m(x_n, \rho_n) = \{x \in \mathbb{R}^m; |x - x_n| < \rho_n\}$ (called *interior patches*) with the following properties:

(d) *Let K be any compact subset of $\mathbb{R}^m$. Then*

$$K \cap V_n = \varnothing, \quad K \cap \mathbb{S}_n = \varnothing \tag{4.5.4}$$

except for a finite number of indices.

(e) *There exists $\rho', 0 < \rho' < 1$, and a sequence $\{\rho_n'\}$, $0 < \rho_n' < \rho_n$ such that $\bar{\Omega}$ is contained in the union of the $\mathbb{S}^m(x_n, \rho_n')$ and the $V_n' = \eta_n^{-1}(\mathbb{S}^m(0, \rho'))$ (η_n is the associated map of V_n).*

In intuitive terms, the $\{V_n\}$ and the $\{\mathbb{S}_n\}$ cover Ω with "something to spare near the boundaries."

We point out that in the case where Ω is bounded, the Heine-Borel theorem shows that the cover $\{V_n\} \cup \{\mathbb{S}_n\}$ can be chosen *finite*.

It is easy to see that, given $\rho'', \rho_1'', \rho_2'', \ldots$ with $\rho' < \rho'' < 1$, $\rho_n' < \rho_n'' < \rho_n$ we can find nonnegative functions $\zeta, \zeta_1, \zeta_2, \ldots$ in $\mathfrak{D}$ with $\zeta(x) = 1$ in $\mathbb{S}^m(0, \rho'), \zeta_n(x) = 1$ in $\mathbb{S}^m(x_n, \rho_n')$, $\zeta(x) = 0$ outside of $\mathbb{S}^m(0, \rho'')$, $\zeta_n(x) = 0$ outside of $\mathbb{S}^m(x_n, \rho_n'')$. Since, in view of (d), all the terms of the sum

$$\sum \zeta_n(x) + \sum \zeta(\eta_n(x)) = \tau(x), \tag{4.5.5}$$

except for a finite number, vanish on any given compact set it is clear that $\tau(\cdot)$ is well defined and k times continuously differentiable if Ω is of class $C^{(k)}$; moreover (e) implies that

$$\tau(x) > 0 \quad (x \in \bar{\Omega})$$

so that if we define $\chi_n^i(x) = \zeta_n(x)/\tau(x)$, $\chi_n^b(x) = \zeta(\eta_n(x))/\tau(x)$, we obtain a *partition of unity* $\{\chi_n^i\} \cup \{\chi_n^b\}$,

$$\sum \chi_n^i(x) + \sum \chi_n^b(x) = 1 \quad (x \in \bar{\Omega}) \tag{4.5.6}$$

subordinated to the cover $\{\mathbb{S}_n\} \cup \{V_n\}$ in the sense that the support of each χ is contained (with something to spare) in either a boundary or an interior patch. This partition of unity (which has some deluxe features not always needed) consists of k times continuously differentiable functions if Ω is of class $C^{(k)}$ and will allow us to piece together local results. In the sequel, interior patches $\mathbb{S}_n$ will be denoted by V_n^i and boundary patches by V_n^b.

We note, finally, that various additional conditions can be imposed on the V_n^i and the V_n^b without modifying the previous statements; for example, we may require these sets to be of diameter $\leqslant \delta$ (this will be essential in the proof of Theorem 4.8.4). With reference to Green's theorem (4.5.1), it is not difficult to see that it holds in a domain of class $C^{(1)}$ for vector fields U continuously differentiable on $\overline{\Omega}$ and having compact support; in particular, ν and $d\sigma$ are well defined everywhere. It is clear that the requirement that Ω be of class $C^{(1)}$ leaves out such everyday domains as cubes, where the divergence theorem holds; however, this is not very important since we shall be forced to suppose in most of our treatment that Ω is at least of class $C^{(2)}$. On the other hand, although tougher domains will be admitted in Section 4.6, no use of the results in this section will be made there.

We study now dissipativity of certain extensions of A_0, beginning with the L^p case. We denote by A_1 the operator (4.4.1) with domain $D(A_1) = \mathfrak{D}^{(2)}(\overline{\Omega})$ consisting of all functions u twice continuously differentiable in $\overline{\Omega}$ with compact support. We assume that Ω is of class $C^{(1)}$ and that the coefficients a_{jk}, b_j, c admit extensions to $\overline{\Omega}$ having as many continuous derivatives as needed in each statement.

We look for dissipative restrictions of A_1. Our first computation is an offspring of (4.4.12), taking into account the boundary terms. As before, we transform first A_0 to divergence form and then apply the divergence theorem, obtaining

$$\|u\|^{p-2}\,\mathrm{Re}\langle\theta(u), A_1 u\rangle = \mathrm{Re}\int_\Omega \left\{\sum\sum D^j\left(a_{jk}D^k u\right)\right\}|u|^{p-2}\bar{u}\,dx$$

$$+\,\mathrm{Re}\int_\Omega\left\{\sum\left(b_j - \sum D^k a_{jk}\right)D^j u\right\}|u|^{p-2}\bar{u}\,dx + \int_\Omega c|u|^p\,dx$$

$$= -(p-2)\int_\Omega |u|^{p-4}\left\{\sum\sum a_{jk}\mathrm{Re}(\bar{u}D^j u)\mathrm{Re}(\bar{u}D^k u)\right\}dx$$

$$-\int_\Omega |u|^{p-2}\left\{\sum\sum a_{jk}D^j u D^k\bar{u}\right\}dx$$

$$+\frac{1}{p}\int_\Omega\left\{\sum\sum D^j D^k a_{jk} - \sum D^j b_j + pc\right\}|u|^p\,dx$$

$$+\,\mathrm{Re}\int_\Gamma\left\{\sum\sum a_{jk}\nu_j D^k u\right\}|u|^{p-2}\bar{u}\,d\sigma$$

$$+\frac{1}{p}\int_\Gamma\left\{\sum\left(b_j - \sum D^k a_{jk}\right)\nu_j\right\}|u|^p\,d\sigma, \tag{4.5.7}$$

where $\nu_1,\ldots,\nu_m$ are the components of the outer normal vector ν. If hypotheses (4.4.10) and (4.4.11) in the previous section are satisfied it is clear that the integrals over Ω contribute a nonpositive amount to (4.5.7). To streamline the writing of the two boundary integrals we introduce the following notations:

$$D^{\tilde{\nu}} = \sum_{j=1}^m \sum_{k=1}^m a_{jk}\nu_j D^k u, \tag{4.5.8}$$

which, modulo a factor, is the derivative of u in the direction of the *conormal vector* $\tilde{\nu} = (\Sigma a_{j1}\nu_j, \dots, \Sigma a_{jm}\nu_j)$; naturally, $D^{\tilde{\nu}}u$ is called the *conormal derivative* of u at the boundary (with respect to A_0). We also write

$$b = \sum_{j=1}^{m} \left(b_j - \sum_{k=1}^{m} D^k a_{jk} \right) \nu_j. \tag{4.5.9}$$

The boundary terms in (4.5.7) can then be expressed thus:

$$\operatorname{Re} \int_{\Gamma} \left\{ (D^{\tilde{\nu}}u)|u|^{p-2}\bar{u} + \frac{1}{p} b|u|^p \right\} d\sigma. \tag{4.5.10}$$

We shall use the following linear algebra result:

4.5.1 Lemma. *Let $\mathcal{C}$ be a symmetric positive definite matrix. Assume that $(\mathcal{C}\xi, \xi) = 0$ for some $\xi \in \mathbb{R}^m$. Then*

$$\mathcal{C}\xi = 0.$$

The proof results immediately from writing ξ in terms of a basis of eigenvectors of $\mathcal{C}$.

To study (4.5.7) we divide now the boundary Γ into three disjoint pieces:

$$\Gamma^1 = \{x \in \Gamma; (\mathcal{C}(x)\nu, \nu) > 0\} \tag{4.5.11}$$

$$\Gamma^2 = \{x \in \Gamma; (\mathcal{C}(x)\nu, \nu) = 0, b(x) > 0\} \tag{4.5.12}$$

$$\Gamma^3 = \{x \in \Gamma; (\mathcal{C}(x)\nu, \nu) = 0, b(x) \leqslant 0\}. \tag{4.5.13}$$

At Γ^2 and Γ^3 we have $D^{\tilde{\nu}}u = 0$ for any u in view of Lemma 4.5.1 and the first term in the integrand of (4.5.10) vanishes. The second term contributes a nonnegative amount in Γ^3 irrespective of any boundary condition satisfied by u; to achieve the same in Γ^2 we require that

$$u(x) = 0 \quad (x \in \Gamma^2). \tag{4.5.14}$$

Finally, the contribution over Γ^1 will be nonpositive if

$$u(x) = 0 \quad (x \in \Gamma^1) \tag{4.5.15}$$

or

$$D^{\tilde{\nu}}u(x) = \gamma(x)u(x) \quad (x \in \Gamma^1), \tag{4.5.16}$$

where γ is a function with

$$\gamma(x) + \frac{1}{p}b(x) \leqslant 0 \quad (x \in \Gamma^1). \tag{4.5.17}$$

We have proved part of the following result:

4.5.2 Lemma. *The restriction A of A_1 to the space of all functions $u \in \mathcal{D}^{(2)}(\bar{\Omega})$ satisfying the Dirichlet boundary condition (4.5.14) on Γ^2 and the*

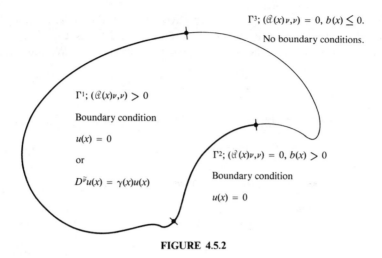

Γ^3; $(\mathcal{C}(x)\nu,\nu) = 0$, $b(x) \leq 0$.

No boundary conditions.

Γ^1; $(\mathcal{C}(x)\nu,\nu) > 0$

Boundary condition

$u(x) = 0$

or

$D^{\tilde{\nu}}u(x) = \gamma(x)u(x)$

Γ^2; $(\mathcal{C}(x)\nu,\nu) = 0$, $b(x) > 0$

Boundary condition

$u(x) = 0$

FIGURE 4.5.2

boundary condition (4.5.15) or (4.5.16) in Γ^3 is dissipative in $L^p(\Omega)$, $1 \leq p < \infty$
if

$$\mathcal{C}(x) \geq 0 \quad (x \in \overline{\Omega}), \tag{4.5.18}$$

$$c(x) + \frac{1}{p}\left(\sum\sum D^j D^k a_{jk}(x) - \sum D^j b_j(x)\right) \leq 0 \quad (x \in \overline{\Omega}), \tag{4.5.19}$$

and

$$\gamma(x) + \frac{1}{p}\sum\left(b_j(x) - \sum D^k a_{jk}(x)\right)\nu_j(x) \leq 0 \quad (x \in \Gamma^1), \tag{4.5.20}$$

(where $\nu(x) = (\nu_1(x),\ldots,\nu_m(x))$ is the outer normal vector at x) if (4.5.16) is
used on Γ^1. Condition (4.5.18) is necessary for dissipativity of any extension of
A_0 in $L^p(\Omega)$, $1 \leq p < \infty$; conditions (4.5.19) and (4.5.20) are necessary when
$p = 1$. Moreover, in this last case, the boundary condition (4.5.14) on Γ^2 is as
well necessary for dissipativity.

The sufficiency proof has already been completed, at least when
$p > 1$, and is extended to the case $p = 1$ by means of the usual limiting
argument. The necessity statements regarding (4.5.18) and (4.5.19) are
handled, just as in the previous section, by taking functions that vanish near
the boundary. Finally, the necessity of (4.5.20) for $p = 1$ follows very much
as in the one dimensional case (note that the proof can be "localized" by
taking u with small and conveniently located support).

The case $1 < p < 2$ presents the customary difficulties in that the calculation
(4.5.7) may not be directly justifiable for $u \in D(A_0)$; we must then approximate an
arbitrary $u \in D(A_0)$ uniformly together with its first two derivatives by a sequence
$\{u_n\}$ for which (4.5.7) is licit. We can do this by first extending u to a $C^{(2)}$ function $\tilde{u}$
in $\mathbb{R}^m$ with compact support (this can be easily handled by using a partition of unity
like (4.5.5)) and then approximating $\tilde{u}$ as in the previous section.

Before studying the case $E = C(\overline{\Omega})$ we examine the way A_1 transforms under a smooth, invertible change of variables $x \to \eta = \eta(x)$. Denote

by $\tilde{A}_1$ the operator in the new variables,

$$\tilde{A}_1 u = \sum_{j=1}^{m} \sum_{k=1}^{m} \tilde{a}_{jk}(\eta) D^j D^k u(\eta) + \sum_{j=1}^{m} \tilde{b}_j(\eta) D^j u(\eta) + \tilde{c}(\eta) u(\eta),$$

(4.5.21)

where, using the traditional notation for partial derivatives,

$$\tilde{a}_{jk} = \sum_{\iota=1}^{m} \sum_{\kappa=1}^{m} a_{\iota\kappa} \frac{\partial \eta_j}{\partial x_\iota} \frac{\partial \eta_k}{\partial x_\kappa},$$

(4.5.22)

$$\tilde{b}_j = \sum_{\iota=1}^{m} b_\iota \frac{\partial \eta_j}{\partial x_\iota} + \sum_{\iota=1}^{m} \sum_{\kappa=1}^{m} a_{\iota\kappa} \frac{\partial^2 \eta_j}{\partial x_\iota \partial x_\kappa}, \quad \tilde{c} = c.$$

(4.5.23)

Note that, if $\xi = (\xi_1, \ldots, \xi_m) \in \mathbb{R}^m$,

$$\sum\sum \tilde{a}_{jk} \tilde{\xi}_j \tilde{\xi}_k = \sum\sum a_{jk} \tilde{\xi}_j \tilde{\xi}_k,$$

where

$$\tilde{\xi}_j = \sum_{\iota=1}^{m} \frac{\partial \eta_\iota}{\partial x_j} \xi_\iota$$

(4.5.24)

so that the condition that $\mathcal{C}(x) \geq 0$ is invariant. If the transformation involves a piece of the boundary Γ, then the unit outer normal vector $\nu = (\nu_1, \ldots, \nu_m)$ at some point $\bar{x} \in \Gamma$ and the unit outer normal vector $n = (n_1, \ldots, n_m)$ at $\bar{\eta} = \eta(\bar{x})$ are related to each other by the equations

$$\rho^{-1} n_j = \sum_{\iota=1}^{m} \frac{\partial x_\iota}{\partial \eta_j} \nu_\iota, \quad \rho \nu_j = \sum_{\iota=1}^{m} \frac{\partial \eta_\iota}{\partial x_j} n_\iota,$$

(4.5.25)

where $\rho = \rho(\bar{x}) > 0$ is a normalization factor; then

$$\rho^{-2} \sum\sum \tilde{a}_{jk} n_j n_k = \sum\sum a_{jk} \nu_j \nu_k,$$

(4.5.26)

which shows that the definition of Γ^1 is invariant as well. A simple computation shows that, if

$$\tilde{b} = \sum_{j=1}^{m} \left(\tilde{b}_j - \sum_{k=1}^{m} \frac{\partial \tilde{a}_{jk}}{\partial \eta_k} \right) n_j,$$

(4.5.27)

then

$$\tilde{b} = \rho b$$

at all points $x \in \Gamma$ where $(\mathcal{C}(x)\nu, \nu) = 0$. Hence Γ^2 and Γ^3 are also invariant. Note that it follows from (4.5.22) and (4.5.25) that the conormal derivative is invariant in the following sense:

$$D^{\tilde{n}} u = \sum\sum \tilde{a}_{jk} n_j \frac{\partial u}{\partial \eta_k} = \rho \sum\sum a_{jk} \nu_j \frac{\partial u}{\partial x_k} = \rho D^{\tilde{\nu}} u.$$

(4.5.28)

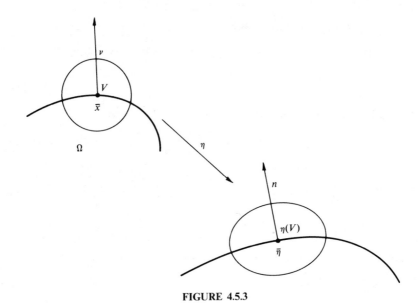

FIGURE 4.5.3

4.5.3 Lemma. *Let A be the restriction of A_1 to the space of all functions $u \in \mathcal{D}^{(2)}(\overline{\Omega})$ satisfying the Dirichlet boundary condition (4.5.14) on Γ^2 and the boundary condition (4.5.15) or (4.5.16) on Γ^1. Then A is dissipative in $C(\overline{\Omega})$ if and only if*

$$\mathcal{Q}(x) \geqslant 0, \quad c(x) \leqslant 0 \quad (x \in \overline{\Omega}), \tag{4.5.29}$$

and

$$\gamma(x) \leqslant 0 \quad (x \in \Gamma^1) \tag{4.5.30}$$

if the boundary condition (4.5.16) is used on Γ^1. The boundary condition (4.5.14) on Γ^2 is necessary for dissipativity.

We limit ourselves to proving the sufficiency part. Let $u \in D(A)$, $u \neq 0$ and $m(u) = \{x \in \overline{\Omega}: |u(x)| = \|u\|\}$. We have to prove that

$$\mathrm{Re} \int_{m(u)} Au(x)\mu(dx) \leqslant 0$$

for any $\mu \in \Theta(u)$; since $u(x)\mu(dx)$ is a positive measure, it is enough to prove that

$$\mathrm{Re}\, u(x)^{-1} Au(x) \leqslant 0 \quad (x \in m(u)). \tag{4.5.31}$$

To prove (4.5.31) at points $\bar{x} \in m(u) \cap \Omega$, we use the argument in Lemma 4.4.1, thus only elements of $m(u) \cap \Gamma$ (which were absent there) may be cause of trouble. Clearly $m(u) \cap \Gamma^2 = \varnothing$, but $m(u)$ may contain points of Γ^3 (and also of Γ^1 if the boundary condition is (4.5.16)).

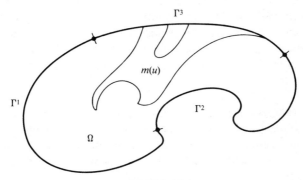

FIGURE 4.5.4

Let $\bar{x} \in m(u) \cap \Gamma^3$ and let V^b be a boundary patch with associated map η with $\eta(\bar{x}) = 0$ (see the beginning of this section). Through the change of coordinates $x \to \eta$, we may assume that, locally, $\bar{x} = 0$, Γ is the hyperplane $x_m = 0$ and Ω is the half space $x_m > 0$. Since u reaches its maximum at $\bar{x}$, we must have there

$$D^1|u|^2 = \cdots = D^{m-1}|u|^2 = 0, \quad D^m|u|^2 \leqslant 0, \qquad (4.5.32)$$

and the "partial Hessian" $\mathcal{H}_{m-1}(0; |u|^2)$ of $|u|^2$ with respect to $x_1, \ldots, x_{m-1}$ is negative definite at $\bar{x}$. Noting that the outer normal vector at 0 is $\nu = (\nu_1, \ldots, \nu_m) = (0, \ldots, 0, -1)$ and that we must have $(\mathcal{C}\nu, \nu) = \Sigma\Sigma a_{jk}(0)\nu_j\nu_k = 0$, we see that $a_{mm}(0) = 0$. It follows from positive definiteness of $\mathcal{C}$ that $a_{jm}(0) = a_{mj}(0) = 0$ $(1 \leqslant j \leqslant m - 1)$ as well; in fact, if this were not true, the vector $\xi = (0, \ldots, \tau, \ldots, 1)$ would make $(\mathcal{C}(0)\xi, \xi)$ negative for τ suitable placed and sized. We have then, at $x = 0$,

$$Au = \sum_{j=1}^{m-1} \sum_{k=1}^{m-1} a_{jk}D^jD^ku + \sum_{j=1}^{m-1} b_jD^ju + cu + b_mD^mu = A^mu + b_mD^mu,$$

$$(4.5.33)$$

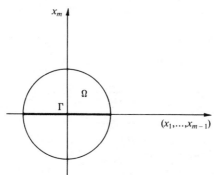

FIGURE 4.5.5

where A^m acts only on $x_1, \ldots, x_{m-1}$. Moreover, since $0 \in \Gamma^3$, we must have $b(0) \leqslant 0$, where, in the present coordinates, $b = \Sigma D^k a_{mk} - b_m$. Since $\mathscr{C}(x)$ is positive definite, $a_{mm}(x) \geqslant 0$ everywhere and thus a_{mm} has a minimum at 0; accordingly, $D^1 a_{mm} = \cdots = D^{m-1} a_{mm} = 0$, $D^m a_{mm} \geqslant 0$ there and

$$b_m(0) = D^m a_{mm}(0) - b(0) \geqslant 0. \tag{4.5.34}$$

We use now the decomposition (4.5.33) as follows. The first term involves the $(m-1)$-dimensional operator A^m and is treated as in Lemma 4.4.2 using the first $m-1$ equalities in (4.5.32) and the negative definiteness of the partial Hessian $\mathscr{H}_{m-1}(0; |u|^2)$. In the second term we use (4.5.34) together with the relation

$$\text{Re } u^{-1} D^m u = \tfrac{1}{2} \|u\|^{-2} D^m |u|^2$$

and the last inequality (4.5.32); adding the two terms we obtain (4.5.31) at $\bar{x}$.

The preceding argument settles completely the problem if the Dirichlet boundary condition is used in Γ^1, since in that case

$$m(u) \cap (\Gamma^1 \cup \Gamma^2) = \varnothing. \tag{4.5.35}$$

Clearly, we can also cover the boundary condition (4.5.16) if we show that (4.5.30) implies (4.5.35), which we do now. Since we already know that $m(u) \cap \Gamma^2 = \varnothing$, we only have to examine the case where $\bar{x} \in m(u) \cap \Gamma^1$. If $\gamma(\bar{x}) < 0$, we have

$$D^{\tilde{\nu}} |u|^2 = 2 \text{Re } \bar{u} D^{\tilde{\nu}} u = 2\gamma |u|^2 < 0 \tag{4.5.36}$$

at $\bar{x}$. Since $(\mathscr{C}\nu, \nu) = (\tilde{\nu}, \nu) > 0$ there, the conormal vector points away from Ω (at least locally); hence (4.5.36) implies that u must be increasing along $-\tilde{\nu}$ near $\bar{x}$, a contradiction. If $\gamma(\bar{x}) = 0$, however, this simple argument breaks down (and (4.5.35) may in fact be false). To handle this case, using again a boundary patch V^b and the associated map η we can assume locally that $\bar{x} = 0$, Γ is the hyperplane $x_m = 0$, Ω the half space $x_m > 0$. Since $\bar{x}$ is a maximum of $|u|^2$, $D^j |u|^2(\bar{x}) = 0$ ($1 \leqslant j \leqslant m-1$), but in this case it is also true that

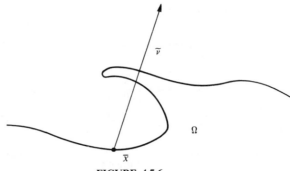

FIGURE 4.5.6

$$D^{\tilde{\nu}}|u|^2 = 2\operatorname{Re}\bar{u}D^{\tilde{\nu}}u = 0$$

at 0 and, since $(\mathcal{Q}\nu, \nu) > 0$, $\tilde{\nu}$ does not belong to the hyperplane $x_m = 0$; hence all the first partials of $|u|^2$ vanish at $\bar{x} = 0$:

$$D^1|u|^2 = \cdots = D^m|u|^2 = 0. \qquad (4.5.37)$$

We make use of Maclaurin's formula up to terms of order two,

$$|u(x)|^2 = |u(0)|^2 + \tfrac{1}{2}\big(\mathcal{K}(0; |u|^2)x, x\big) + o\big(|x|^2\big)$$

in the half space $x_m > 0$. Since 0 is a maximum of $|u|^2$, we must have $(\mathcal{K}(0, |u|^2)\xi, \xi) \leqslant 0$ in the half space $\xi_m \geqslant 0$, thus in $\mathbb{R}^m$, and we can handle $\bar{x}$ exactly as a point of $m(u) \cap \Omega$. This ends the proof.

The necessity arguments are conducted much as in the one-dimensional case (Lemma 4.4.3) and are left to the reader (Example 4.5.5).

As we did in last section, we translate the conditions obtained to operators in divergence form. Condition (4.5.19) is (4.4.17); the definition of b is now

$$b = \sum_{j=1}^{m} \hat{b}_j \nu_j \qquad (4.5.38)$$

and (4.5.20) accordingly becomes

$$\gamma(x) + \frac{1}{p}\sum \hat{b}_j(x)\nu_j \leqslant 0. \qquad (4.5.39)$$

We bring into play the formal adjoint A_0' defined in the previous section (see (4.4.14)). The Lagrange identity (4.4.15) takes the form

$$\int_{\Omega}(A_0 u)v\,dx = \int_{\Omega} u(A_0' v)\,dx + \int_{\Gamma}\big((D^{\tilde{\nu}}u)v - uD^{\tilde{\nu}}v + buv\big)\,dx,$$

$$(4.5.40)$$

where b is given by (4.5.9) and $u, v \in \mathcal{D}^{(2)}(\bar{\Omega})$. Assume, for the sake of simplicity, that $\Gamma^1 = \Gamma$ (i.e., that $(\mathcal{Q}(x)\nu, \nu) > 0$ for $x \in \Gamma$). If u satisfies the boundary condition (4.5.16), then

$$\int_{\Omega}(A_0 u)v\,dx = \int_{\Omega} u(A_0 v)\,dx \qquad (4.5.41)$$

if v satisfies the *adjoint boundary condition*

$$D^{\tilde{\nu}}v(x) = (\gamma(x) + b(x))v(x) \quad (x \in \Gamma^1). \qquad (4.5.42)$$

On the other hand, (4.5.41) holds if both u and v satisfy the Dirichlet boundary condition, which is thus its own adjoint.

4.5.4   Example.   Show that (4.5.20) is necessary for dissipativity in $L^1(\Omega)$ of the restriction of A_1 satisfying (4.5.16) on Γ^1. Show that the Dirichlet boundary condition (4.5.14) on Γ^2 is as well necessary.

4.5.5   Example.   Show that (4.5.30) is necessary for dissipativity in $C(\overline{\Omega})$ of the restriction of A_1 satisfying (4.5.16) on Γ^1. Show that the Dirichlet boundary condition (4.5.14) on Γ^2 is as well necessary.

4.6.   SECOND ORDER PARTIAL DIFFERENTIAL OPERATORS. CONSTRUCTION OF m-DISSIPATIVE EXTENSIONS IN L^2

We consider operators in divergence form

$$A_0 u = \sum_{j=1}^{m} \sum_{k=1}^{m} D^j\big(a_{jk}(x)D^k u\big) + \sum_{j=1}^{m} b_j(x)D^j u + c(x)u \quad (4.6.1)$$

in an arbitrary domain Ω in Euclidean space $\mathbb{R}^m$ (to simplify notations the coefficients $\hat{b}_j$ in (4.4.2) are called b_j here). A great part of the theory can be constructed without smoothness assumptions on the coefficients, thus we shall only suppose that the a_{jk}, the b_j and c are *measurable* and *bounded* in Ω. In this level of generality, of course, the first difficulty is how to *define* $D(A_0)$ (even if $u \in \mathcal{D}(\Omega), \Sigma a_{jk} D^k u$ may fail to have weak derivatives). We shall avoid this problem simply by not using A_0 in the general case (if the coefficients fulfill the differentiability assumptions in the previous two sections we set $D(A_0) = \mathcal{D}^{(2)}(\Omega)$.) Besides the measurability assumption, we require that A_0 be *uniformly elliptic*:

(A) There exists $\kappa > 0$ such that

$$\sum\sum a_{jk}(x)\xi_j\xi_k \geqslant \kappa|\xi|^2 \quad (\xi \in \mathbb{R}^m) \qquad (4.6.2)$$

a.e. in Ω.

Since conditions (4.5.19) and (4.5.20) for boundary conditions of type (I) are far from necessary for $p = 2$, we shall discard them and content ourselves with constructing from A_0 operators A in $\mathcal{C}_+(1, \omega)$ (i.e., such that $A - \omega I$ is m-dissipative) for some ω.

All the subsequent treatment is based on:

4.6.1   Lemma.   *Let H be a complex Hilbert space and let the functional $B: H \times H \to \mathbb{C}$ be sesquilinear (linear in the second variable, conjugate linear in the first). Assume that*

$$|B(u,v)| \leqslant C\|u\|\,\|v\|, \quad |B(u,u)| \geqslant c\|u\|^2 \quad (u, v \in H) \qquad (4.6.3)$$

for some $c, C > 0$. Then if Φ is a bounded linear functional in H, there exists a *unique* $u_0 \in H$ *with* $C^{-1}\|\Phi\| \leqslant \|u_0\| \leqslant c^{-1}\|\Phi\|$ *and such that*

$$\Phi(u) = B(u_0, u). \qquad (4.6.4)$$

Proof. It follows from the first condition (4.6.3) that $v \to B(u, v)$ (u fixed) is a bounded linear functional in H. Therefore, there exists $w = Qu$ such that

$$B(u, v) = (w, v), \qquad (4.6.5)$$

$(\cdot, \cdot)$ the scalar product in H. It is clear that the operator Q so defined is linear in H. Moreover, $\|w\|^2 = \|Qu\|^2 = (Qu, Qu) = B(u, Qu) = |B(u, Qu)| \leqslant C\|u\|\|Qu\|$ so that Q is bounded (with norm $\leqslant C$). On the other hand, $\|u\|\|Qu\| \geqslant |(Qu, u)| = |B(u, u)| \geqslant c\|u\|^2$ so that $\|Qu\| \geqslant c\|u\|$; hence Q maps H in a one-to-one fashion onto a closed subspace K of H. If $K \neq H$, there exists a nonzero w_0 such that $B(v, w_0) = (Qv, w_0) = 0$ for all $v \in H$. But then $B(w_0, w_0) = 0$, which contradicts the second inequality (4.6.3). It follows that $K = H$ and that Q is invertible. If $v_0 \in H$ is such that $\Phi(u) = (v_0, u)$, we have

$$\Phi(u) = B(Q^{-1}v_0, u),$$

which shows (4.6.4). The statements on the norm of $u_0 = Q^{-1}v_0$ are obvious from the previous argument.

We look now at the problem of obtaining extensions (properly speaking, "definitions") of A_0 that belong to $\mathcal{C}_+(1, \omega)$. These extensions will be obtained on the basis of boundary conditions β of one of the following types:

(I) $\qquad\qquad D^{\bar{\nu}}u(x) = \gamma(x)u(x) \quad (x \in \Gamma) \qquad\qquad (4.6.6)$

(II) $\qquad\qquad u(x) = 0 \quad (x \in \Gamma) \qquad\qquad\qquad (4.6.7)$

already considered in Section 4.5. As we may expect, these boundary conditions will only be satisfied in a suitably weak sense.

We begin with the Dirichlet boundary condition (4.6.7). The basic space in our treatment will be the closure $H_0^1(\Omega)$ of $\mathcal{D}(\Omega)$ in $H^1(\Omega)$ (see Section 8 for definitions); as usual, $H^1(\Omega)$ and $H_0^1(\Omega)$ are equipped with the scalar product

$$(u, v)_{H^1(\Omega)} = (u, v)_{L^2(\Omega)} + \sum_{j=1}^{m} (D^j u, D^j v)_{L^2(\Omega)}. \qquad (4.6.8)$$

For $\lambda \geqslant \text{ess sup } c + 1$ we define

$$(u, v)_\lambda = \int_\Omega \left\{ \sum\sum a_{jk}(x) D^j \bar{u} D^k v + (\lambda - c(x)) \bar{u} v \right\} dx. \qquad (4.6.9)$$

It is easy to see that each $(\cdot, \cdot)_\lambda$ possesses all the properties of a scalar product; moreover, the norm $\|\cdot\|_\lambda$ derived from $(\cdot, \cdot)_\lambda$ is equivalent to the norm provided by (4.6.8). In fact, assuming (as we may) that $\kappa \leqslant 1$ in (4.6.2), we obtain

$$\kappa\|u\|^2_{H^1_0(\Omega)} \leqslant \|u\|^2_\lambda \leqslant M\|u\|^2_{H^1_0(\Omega)} \quad \left(u \in H^1_0(\Omega)\right), \qquad (4.6.10)$$

where M is a constant depending on λ, on the a_{jk} and on c (recall these functions are bounded in Ω). We define a sesquilinear form in $H^1_0(\Omega)$ by the formula

$$B_\lambda(u, v) = \int_\Omega \left\{ \sum\sum a_{jk} D^j \bar{u} D^k v - \left(\sum b_j D^j \bar{u}\right) v + (\lambda - c)\bar{u}v \right\} dx$$

$$= (u, v)_\lambda - \int_\Omega \left(\sum b_j D^j \bar{u}\right) v\, dx. \qquad (4.6.11)$$

Assume that

$$\lambda - c(x) \geqslant K > 0 \quad \text{a.e. in } \Omega, \qquad (4.6.12)$$

where K is to be fixed later. Then

$$\left| \int_\Omega \left(\sum b_j D^j \bar{u}\right) v\, dx \right|$$

$$\leqslant \int_\Omega \left| \left(\sum b_j D^j \bar{u}\right) v \right| dx$$

$$\leqslant \left(\int_\Omega (\lambda - c(x))^{-1} \left|\sum b_j D^j u\right|^2 dx \right)^{1/2} \left(\int_\Omega (\lambda - c(x))|v|^2 dx \right)^{1/2}$$

$$\leqslant CK^{-1/2}\left(\sum\|D^j u\|\right)\|v\|_\lambda \leqslant Cm^{1/2}K^{-1/2}\left(\sum\|D^j u\|^2\right)^{1/2}\|v\|_\lambda$$

$$\leqslant Cm^{1/2}K^{-1/2}\|u\|_{H^1_0(\Omega)}\|v\|_\lambda \leqslant C'K^{-1/2}\|u\|_\lambda\|v\|_\lambda \quad \left(u, v \in H^1_0(\Omega)\right)$$

$$(4.6.13)$$

by the Schwarz inequality, where K is the constant in (4.6.12) and C' does not depend on λ. Taking K sufficiently large, we may assume that $C'K^{-1/2} = \alpha < 1$ so that

$$\text{Re } B_\lambda(u, u) = \|u\|^2_\lambda - \text{Re} \int_\Omega \left(\sum b_j D^j \bar{u}\right) u\, dx$$

$$\geqslant (1 - \alpha)\|u\|^2_\lambda \quad \left(u \in H^1_0(\Omega)\right) \qquad (4.6.14)$$

and the second inequality (4.6.3) holds for B_λ; that the first is verified as well follows from the Schwarz inequality for $(\cdot, \cdot)_\lambda$ and from (4.6.13) for any $K > 0$.

We define now an operator $A(\beta)$ in $L^2(\Omega)$ as follows. An element $u \in L^2(\Omega)$ belongs to the domain of A if and only if $u \in H^1_0(\Omega)$ and $w \to B_\lambda(u, w)$ is continuous in $H^1_0(\Omega)$ *in the norm of* $L^2(\Omega)$. Since $H^1_0(\Omega)$ is dense in $L^2(\Omega)$, we can extend the functional $w \to \lambda(u, w) - B_\lambda(u, w)$ continuously to all of $L^2(\Omega)$, and there exists a unique element $v \in L^2(\Omega)$ such that

$$\lambda(u,w) - B_\lambda(u,w) = (v,w) \quad (w \in H_0^1(\Omega)). \tag{4.6.15}$$

By definition,

$$v = A(\beta)u \tag{4.6.16}$$

and it is clear that $A(\beta)u$ is linear. We show that

$$(\lambda I - A(\beta))D(A(\beta)) = L^2(\Omega). \tag{4.6.17}$$

To do this, let $v \in L^2(\Omega)$. Since $\Phi(w) = (v,w)$ is continuous in $L^2(\Omega)$—thus in $H_0^1(\Omega)$—we use Lemma 5.6.1 to deduce the existence of a $u \in H_0^1(\Omega)$ such that $B_\lambda(u,w) = (v,w)$; in view of the definition of $A(\beta)$, $u \in D(A(\beta))$, and $(\lambda I - A(\beta))u = v$, thus proving (4.6.17). We have not shown yet that $D(A(\beta))$ is dense in $L^2(\Omega)$. Since $H_0^1(\Omega)$ is dense in $L^2(\Omega)$ it is enough to prove that $D(A(\beta))$ is dense in $H_0^1(\Omega)$. If this were not the case, applying Lemma 4.6.1 to $\bar{B}_\lambda$, we could find a nonzero $w \in H_0^1(\Omega)$ such that $B_\lambda(u,w) = 0$ for all $u \in D(A(\beta))$. In view of (4.6.15), this would imply that $((\lambda I - A(\beta))u, w) = 0$ for all such u, hence $w = 0$ by virtue of (4.6.17). It only remains to be shown that $A(\beta)$ belongs to $\mathcal{C}_+(1, \omega)$ for ω large enough. This is obvious from (4.6.14) since

$$\text{Re}(A(\beta)u, u) = \lambda\|u\|^2 - \text{Re}\,B_\lambda(u,u) \leqslant \lambda\|u\|^2 \quad (u \in D(A));$$
$$\tag{4.6.18}$$

it suffices to set $\omega = \lambda$.

We have completed the proof of:

4.6.2   Theorem.   *Let Ω be an arbitrary domain in $\mathbb{R}^m$ and let the coefficients of the differential operator A_0 be in $L^\infty(\Omega)$ and satisfy (4.6.2); assume that β is the Dirichlet boundary condition. Then the operator $A(\beta)$ defined by (4.6.15) and (4.6.16) belongs to $\mathcal{C}_+(1, \omega)$ for ω sufficiently large.*

The treatment of boundary conditions β of type (I) is slightly different. Here the basic space is $H^1(\Omega)$. To surmise the form of the functional B to be used now, we perform the following computation, where Γ is the boundary of Ω:

$$\int_\Omega \left\{ \sum\sum a_{jk} D^j \bar{u} D^k v - \left(\sum b_j D^j \bar{u}\right)v + (\lambda - c)\bar{u}v \right\} dx$$

$$= \int_\Gamma (D^{\tilde{\nu}}\bar{u})v\, d\sigma + \int_\Omega ((\lambda I - A_1)\bar{u})v\, dx,$$

thus if β is to be satisfied, the functional must be

$$B_{\lambda,\beta}(u,v) = \int_\Omega \left\{ \sum\sum a_{jk} D^j \bar{u} D^k v - \left(\sum b_j D^j \bar{u}\right)v + (\lambda - c)\bar{u}v \right\} dx$$

$$- \int_\Gamma \gamma \bar{u}v\, d\sigma = B_\lambda(u,v) - \int_\Gamma \gamma \bar{u}v\, dx, \tag{4.6.19}$$

and if the assumptions of Lemma 4.6.1 are to hold we need to postulate that:

  (B) Γ is smooth enough to allow definition of the hyperarea differential $d\sigma$.
  (C) The function γ is measurable and locally summable on Γ; there exists a positive constant $\alpha' < 1$ such that

$$\left| \int_\Gamma \gamma \bar{\varphi} \psi \, d\sigma \right| \leqslant \alpha' \|\varphi\|_\lambda \|\psi\|_\lambda \quad (\varphi, \psi \in \mathfrak{D}) \qquad (4.6.20)$$

  for λ large enough.

The functional $B_{\lambda,\beta}$ is defined by (4.6.19) (strictly speaking, first for $u, v \in \mathfrak{D}$ due to the surface integral term and then extended to $u, v \in H^1(\Omega)$ by continuity of $B_{\lambda,\beta}$ and denseness of $\mathfrak{D}$: see Theorem 4.6.5). To obtain an inequality of the type of (4.6.14) for $B_{\lambda,\beta}$ in $H^1(\Omega)$, it suffices to use (4.6.14) in $H^1(\Omega)$ for B_λ with $\alpha < 1 - \alpha'$.

The construction of $A(\beta)$ runs very much along the same lines as in the case of the Dirichlet boundary condition. Since $B_{\lambda,\beta}$ satisfies the assumptions of Lemma 4.6.1, we can define $A(\beta)$ by

$$\lambda(u, w) - B_{\lambda,\beta}(u, w) = (A(\beta)u, w), \qquad (4.6.21)$$

the domain of $A(\beta)$ defined as the set of all $u \in H^1(\Omega)$ such that $w \to B_{\lambda,\beta}(u, w)$ is continuous in the norm of $L^2(\Omega)$. Denseness of $D(A(\beta))$ and equality (4.6.17) are proved exactly in the same way. Finally, to show that $A(\beta) \in \mathcal{C}(1, \omega)$, we only have to note that an obvious analogue of (4.6.18) holds for $A(\beta)$ and $B_{\lambda,\beta}$. We have thus completed the proof of

4.6.3   Theorem.   *Let Ω be a domain in $\mathbb{R}^m$ with boundary Γ satisfying* (B). *Assume the coefficients of A_0 are in $L^\infty(\Omega)$ and satisfy the uniform ellipticity condition* (A). *Finally, let γ be a locally integrable function such that* (C) *holds. Then the operator $A(\beta)$ defined by (4.6.21) belongs to $\mathcal{C}_+(1, \omega)$ for ω large enough.*

It is natural to ask in which sense, if any, functions in the domains of the operators $A(\beta)$ in Theorems 4.6.2 and 4.6.3 satisfy the boundary condition β. The fact that every element of $D(A(\beta))$ (in fact, of $H_0^1(\Omega)$) will (almost) satisfy the Dirichlet boundary condition β if adequate conditions are imposed on Γ follows from the next result, which is stated in much more generality than needed now, with a view to future use.

4.6.4   Theorem.   *Let Ω be a bounded domain of class $C^{(k)}$, $k \geqslant 1$,[4] and let $1 \leqslant p < \infty$. Then* (a) *if*

$$1 \leqslant p < m/k, \quad 1 \leqslant q \leqslant (m-1)p/(m-kp) \qquad (4.6.22)$$

---

[4] Or, more generally, having the *uniform $C^{(k)}$-regularity property* and a *simple k-extension operator*: see Adams [1975: 1], especially pp. 67 and 83. It follows from the definition in p. 67 that a bounded $C^{(k)}$ domain has the uniform $C^{(k)}$-regularity property; on the other hand, since Γ is bounded, the existence of the simple k-extension operator is automatic (loc. cit. p. 84)

there exists a constant C (depending only on Ω, p, q) such that

$$\|u\|_{L^q(\Gamma)} = \left(\int_\Gamma |u|^q \, d\sigma\right)^{1/q} \leqslant C\|u\|_{W^{k,p}(\Omega)} \qquad (4.6.23)$$

for every $u \in C^{(k)}(\overline{\Omega})$. (b) If

$$p \geqslant m/k, \qquad (4.6.24)$$

then (4.6.23) holds for every $q \geqslant 1$.

The proof (of a more general result) can be seen in Adams [1975: 1, p. 114].

The next result we need is on approximation of elements of $W^{k,p}(\Omega)$ by functions of $\mathcal{D}$ (which was already used in the construction of $B_{\lambda,\beta}$ in (4.6.21)).

4.6.5   Theorem.   *Let Ω be a domain of class $C^{(0)}$, $1 \leqslant p < \infty$. Then $\mathcal{D}$ (or, rather, the set of restrictions of functions of $\mathcal{D}$ to Ω) is dense in $W^{k,p}(\Omega)$.*

For a proof see Adams [1975: 1, p. 54], where the assumptions on the domain Ω are less stringent.[5]

By means of Theorems 4.6.4 and 4.6.5 we can prove that elements of $H^1(\Omega) = W^{1,2}(\Omega)$ "have boundary values" in a precise sense. Let $u \in H^1(\Omega)$ and let $\{u_n\}$ be a sequence in $\mathcal{D}$ converging to u in $H^1(\Omega)$. Then it follows from (4.6.23) that the restrictions of $\{u_n\}$ to Γ (which we denote by the same symbol) converge in $L^2(\Gamma)$ (actually in $L^{2(m-1)/(m-2)}(\Omega)$) to an element of $L^2(\Gamma)$. We can thus define the *restriction* or *trace* of u to Γ by

$$u = \lim u_n$$

the limit understood in $L^2(\Gamma)$ (it is easy to see that it does not depend on the particular approximating sequence $\{u_n\} \subset \mathcal{D}$ used). It follows then that elements of $A(\beta) \subseteq H_0^1(\Omega)$ do vanish on Γ in the sense outlined above since the approximating sequence $\{u_n\}$ may (and must) be chosen in $\mathcal{D}(\Omega)$.

Treatment of a boundary condition of type (4.6.6) is not as simple since functions in $H^1(\Omega)$ do not "have derivatives at the boundary," and a more careful characterization of $D(A(\beta))$ is imperative. This will be done in the next section, under additional smoothness assumptions.

From now until the end of the chapter we place ourselves under more comfortable hypotheses than those hitherto used. The domain Ω will be bounded and at least of class $C^{(1)}$ (although a piecewise $C^{(1)}$ boundary, like that of a parallelepipedon, is acceptable). The coefficients a_{jk}, b_j will be supposed to be continuously differentiable in $\overline{\Omega}$, while c is continuous there; also, γ is continuous on Γ. We also assume that the uniform ellipticity

---

[5] The *segment property* suffices (see Adams [1975: 1, p. 54]).

assumption (4.6.2) holds everywhere in Ω for some $\kappa > 0$. In this framework, hypotheses (B) and (C) on the boundary Γ are automatically satisfied. This is obvious for (B); to prove (C) it is enough to apply Theorem 4.6.4 with $k = 1$, $p = q = 1$, obtaining

$$\left| \int_\Gamma \gamma \bar\varphi \psi \, d\sigma \right| \leq C' \|\bar\varphi \psi\|_{L^1(\Gamma)} \leq C \|\bar\varphi \psi\|_{W^{1,1}(\Omega)}.$$

But

$$\|\bar\varphi \psi\|_{W^{1,1}(\Omega)} = \|\bar\varphi \psi\|_{L^1(\Omega)} + \sum \|D^j(\bar\varphi \psi)\|_{L^1(\Omega)}$$

$$\leq \|\bar\varphi \psi\|_{L^1(\Omega)} + \sum \|(D^j \bar\varphi) \psi\|_{L^1(\Omega)} + \sum \|\bar\varphi D^j \psi\|_{L^1(\Omega)}$$

$$\leq \|\varphi\|_{L^2(\Omega)} \|\psi\|_{L^2(\Omega)} + \left(\sum \|D^j \varphi\|_{L^2(\Omega)} \right) \|\psi\|_{L^2(\Omega)}$$

$$\quad + \|\varphi\|_{L^2(\Omega)} \left(\sum \|D^j \psi\|_{L^2(\Omega)} \right)$$

$$\leq \|\varphi\|_{L^2(\Omega)} \|\psi\|_{L^2(\Omega)} + m^{1/2} \left(\sum \|D_j \varphi\|_{L^2(\Omega)}^2 \right)^{1/2} \|\psi\|_{L^2(\Omega)}$$

$$\quad + m^{1/2} \|\varphi\|_{L^2(\Omega)} \left(\sum \|D^j \psi\|_{L^2(\Omega)}^2 \right)^{1/2}$$

$$\leq \left(K^{-1} + 2m^{1/2} \kappa^{-1/2} K^{-1/2} \right) \|\varphi\|_\lambda \|\psi\|_\lambda, \tag{4.6.25}$$

where K is the constant in (4.6.12) and φ, ψ are arbitrary functions in $\mathcal{D}$.

Under the present assumptions, the operator A_0 can be applied to twice differentiable functions. Given a boundary condition β of any of the forms (4.6.6) or (4.6.7), we denote by $A_0(\beta)$ the operator A_0 acting on the space $C^{(2)}(\bar\Omega)_\beta$ of all functions $u \in C^{(2)}(\bar\Omega)$ which satisfy the boundary condition β. Since $\mathcal{D}(\Omega) \subseteq C^{(2)}(\bar\Omega)_\beta$, it is clear that $D(A_0(\beta))$ is dense in $L^2(\Omega)$ for any boundary condition β.

Assume for the moment that β is a boundary condition of type (I) and let A_0' be the formal adjoint of A_0,

$$A_0' u = \sum \sum D^j \left(a_{jk} D^k u \right) - \sum D^j \left(b_j(x) u \right) + c(x) u, \tag{4.6.26}$$

and β' the adjoint boundary condition

$$D^\nu u(x) = (\gamma(x) + b(x)) u(x). \tag{4.6.27}$$

We can apply the theory hitherto developed to A_0' and β'; the bilinear functional is now

$$B_{\lambda,\beta'}'(u,v) = \int_\Omega \left\{ \sum \sum a_{jk} D^j \bar u D^k v + \left(\sum b_j D^j \bar u \right) v + \left(\lambda - c + \sum D^j b_j \right) \bar u v \right\} dx$$

$$\quad - \int_\Gamma (\gamma + b) \bar u v \, d\sigma.$$

Assuming that u and $v \in \mathcal{D}$, we obtain applying the divergence theorem (see (4.5.40)) that

$$B_{\lambda,\beta'}'(u,v) = \overline{B_{\lambda,\beta}(v,u)} \tag{4.6.28}$$

for $u, v \in \mathfrak{D}$ and thus (by the usual approximation argument) for $u, v \in H^1(\Omega)$. The operator $A'(\beta')$ is defined by

$$\lambda(u, w) - B'_{\lambda, \beta'}(u, w) = (A'(\beta')u, w), \qquad (4.6.29)$$

the domain of $A'(\beta')$ consisting of all those $u \in H^1(\Omega)$ that make $B'_{\lambda, \beta'}$ $L^2(\Omega)$-continuous in $H^1_0(\Omega)$. It follows easily from this and from (4.6.28) that $(A'(\beta')u, w) = (u, A(\beta)w)$ if $u \in D(A'(\beta'))$ and $w \in D(A(\beta))$, so that $A'(\beta') \subseteq A(\beta)^*$. But $A'(\beta')$ belongs to $\mathcal{C}_+$, while $A(\beta)^*$, being the adjoint of an operator in $\mathcal{C}_+$, belongs itself to $\mathcal{C}_+$ (Theorem 3.1.14). Accordingly both $R(\lambda; A(\beta)^*)$ and $R(\lambda; A'(\beta'))$ exist for λ large enough and we prove as in the end of Theorem 2.2.1 that

$$A'(\beta') = A(\beta)^*. \qquad (4.6.30)$$

Equality (4.5.40) shows that $A_0(\beta) \subseteq A(\beta)$, $A'_0(\beta') \subseteq A'(\beta')$, thus we obtain upon taking adjoints,

$$A_0(\beta) \subseteq A(\beta) \subseteq A'_0(\beta')^*, \quad A'_0(\beta') \subseteq A'(\beta') \subseteq A_0(\beta)^*. \ (4.6.31)$$

As we shall see in the next section (under additional assumptions), the two double inclusion relations (4.6.31) are actually equalities (with $A_0(\beta)$, $A'_0(\beta')$ replaced by their closures).

It remains to deal with Dirichlet boundary conditions. Here $H^1(\Omega)$ must be replaced by $H^1_0(\Omega)$ and the adjoint boundary condition β' is of course β itself (see Section 4.5); we leave the details of the argument to the reader.

We collect some of the preceding results.

4.6.6 Theorem. *Let Ω be a bounded domain of class $C^{(1)}$ and let a_{jk}, b_j be continuously differentiable in $\bar{\Omega}$, c continuous in $\bar{\Omega}$; assume the uniform ellipticity condition (4.6.2) is satisfied everywhere in Ω for some $\kappa > 0$. Then (i) If β is the Dirichlet boundary condition, $A_0(\beta)$ is densely defined and possesses an extension $A(\beta) \in \mathcal{C}_+(1, \omega)$ defined from the functional B_λ for λ large enough by (4.6.15), (4.6.16). (ii) If β is an arbitrary boundary condition of type (I) with γ continuous on Γ then $A_0(\beta)$, again densely defined, possesses an extension $A(\beta) \in \mathcal{C}_+(1, \omega)$ defined as in (i) but using the functional $B_{\lambda, \beta}$. The duality relation (4.6.30) holds in all cases.*

4.6.7 Remark. All the developments in this section can be applied to the case $\Omega = \mathbb{R}^m$, where Γ is empty. The treatments for all boundary conditions coalesce.

4.6.8 Remark. The method used to construct the operator $A(\beta)$ can be made to function under much weaker assumptions. For instance, consider the operator

$$A_0 u = \sum \sum D^j \big(a_{jk}(x) D^k u \big) + c(x) u, \qquad (4.6.32)$$

where the a_{jk} and c are merely measurable and locally integrable (but not necessarily bounded) in an arbitrary domain $\Omega \subseteq \mathbb{R}^m$; we require in addition that c be essentially bounded above and that the symmetric matrix $\mathcal{C}(x) = \{a_{jk}(x)\}$ be positive definite, that is,

$$(\mathcal{C}\xi, \xi) = \sum \sum a_{jk}(x)\xi_j \xi_k \geq 0 \quad (\xi \in \mathbb{R}^m), \qquad (4.6.33)$$

a.e. in Ω (note that (4.6.33) does not prevent the quadratic form $\sum \sum a_{jk}(x)\xi_j \xi_k$ from degenerating anywhere in Ω, or even the a_{jk} from vanishing at the same time in parts of Ω). Consider for simplicity the case where β is the Dirichlet boundary condition. Instead of $H_0^1(\Omega)$, the basic space is now $\mathcal{H}_0(\Omega)$, the completion of $\mathcal{D}(\Omega)$ in the norm derived from the scalar product

$$(\varphi, \psi)_\lambda = \int_\Omega \left\{ \sum \sum a_{jk}(x) D^j \bar{\varphi} D^k \psi + (\lambda - c(x))\bar{\varphi}\psi \right\} dx \quad (4.6.34)$$

with $\lambda \geq \operatorname{ess\,sup} c + 1$. The functional B_λ is nothing but $B_\lambda(u, v) = (u, v)_\lambda$ (in which case Lemma 4.6.1 actually reduces to the representation theorem for bounded linear functionals in a Hilbert space) and the construction of the operator $A(\beta)$ proceeds without changes.

Addition of first-order terms $b_j D^j u$ to the operator A_0 is possible, but we must assume that, roughly speaking, "the b_j vanish where $\sum \sum a_{jk}\xi_j \xi_k$ degenerates." Precisely, we postulate the existence of constants $\lambda > 0$, $0 < \alpha < 1$ with

$$\left(\sum b_j(x)\xi_j\right)^2 \leq \alpha^2(\lambda - c(x))\sum \sum a_{jk}(x)\xi_j \xi_k \quad (\xi \in \mathbb{R}^m) \quad (4.6.35)$$

a.e. in Ω. This guarantees an inequality of the form (4.6.14), thus the construction of $A(\beta)$ is carried out in the same way.

Boundary conditions of type (I) are handled through assumptions (B) and (C); here we require the constant α' in (4.6.20) to satisfy $\alpha' < 1 - \alpha$, α the constant in (4.6.35). Of course, the question of in what sense the boundary conditions are satisfied admits an even less clean cut answer in this level of generality, even for the Dirichlet boundary condition: roughly speaking, the boundary condition $u = 0$ will be satisfied in the part of the boundary Γ where $\sum \sum a_{ij}(x)\xi_i \xi_j$ is nondegenerate (see the discussion in Section 4.5 on assignation of boundary conditions).

4.7. REGULARITY THEOREMS

We place ourselves here under (some of) the assumptions in the second half of Section 4.6. To begin with, we shall assume throughout the rest of the chapter, unless otherwise stated, that Ω is of class $C^{(2)}$. For the sake of simplicity, we also assume that Ω is *bounded*, although this is not essential in some results. The operator A_0 will be written in divergence form,

$$A_0 u = \sum_{j=1}^{m} \sum_{k=1}^{m} D^j \left(a_{jk} D^k u \right) + \sum_{j=1}^{m} b_j D^j u + cu, \qquad (4.7.1)$$

and to simplify future statements we introduce the following definition: A_0 is of class $C^{(k)}$, $k \geq 1$ in Ω (in symbols, $A_0 \in C^{(k)}(\overline{\Omega})$) if and only if $a_{jk} \in C^{(k)}(\overline{\Omega})$, $b_j, c \in C^{(k-1)}(\overline{\Omega})$. We shall assume throughout that A_0 is uniformly elliptic:

$$\sum\sum a_{jk}(x)\xi_j\xi_k \geq \kappa|\xi|^2 \quad (x \in \Omega, \xi \in \mathbb{R}^m) \qquad (4.7.2)$$

for some $\kappa > 0$ and write

$$K_n = \max_{x \in \overline{\Omega}} \left\{ |D^\alpha a_{jk}(x)|, |D^{\tilde{\alpha}} b_j(x)|, |D^{\tilde{\alpha}} c(x)| \right\}$$

for $1 \leq j, k \leq m$ and $|\alpha| \leq n$, $|\tilde{\alpha}| \leq n - 1$; clearly the definition makes sense if $A_0 \in C^{(k)}(\overline{\Omega})$ and $1 \leq n \leq k$.

Roughly speaking, the results in this section state that the solution of $(\lambda I - A(\beta))u = f$ is "two degrees smoother" than f in the L^2 sense, at least if the coefficients of A_0 are smooth enough. In all that follows, $\lambda > 0$ will be a sufficiently large parameter, this expression meaning that all the arguments in Section 4.6 function, in particular, the second inequality (4.6.3) is satisfied by B_λ in the case of Dirichlet boundary conditions or by $B_{\lambda,\beta}$ for boundary conditions of type (I).

Our first result holds with no assumptions on Ω.

4.7.1 Interior Regularity Theorem. *Let Ω be an arbitrary bounded domain in $\mathbb{R}^m$ and let A_0 be of class $C^{(1)}(\overline{\Omega})$ and uniformly elliptic. Assume $u \in H^1(\Omega)$ satisfies*

$$B_\lambda(u, \varphi) = (f, \varphi) = \int_\Omega f\varphi \, dx \quad (\varphi \in \mathcal{D}(\Omega)) \qquad (4.7.3)$$

for some $f \in L^2(\Omega)$, and let Ω' be a domain with $\overline{\Omega}' \subseteq \Omega$. Then $u \in H^2(\Omega')$ and there exists a constant C (depending only on Ω, Ω', K_1, κ, λ) such that

$$\|u\|_{H^2(\Omega')} \leq C \left(\|f\|_{L^2(\Omega)} + \|u\|_{H^1(\Omega)} \right). \qquad (4.7.4)$$

Proof. Let χ be a function in $\mathcal{D}(\Omega)$ such that $\chi(x) = 1$ in Ω' and let $v = \chi u$. Plainly $v \in H_0^1(\Omega)$ and a simple computation shows that v satisfies

$$B_\lambda(v, \varphi) = (g, \varphi) \quad (\varphi \in \mathcal{D}(\Omega)), \qquad (4.7.5)$$

where

$$g = \chi f - u \left(\sum\sum D^j \left(a_{jk} D^k \chi \right) + \sum b_j D^j \chi \right) - 2\sum\sum a_{jk} D^j u D^k \chi.$$

$$(4.7.6)$$

Let now h be a nonzero real number, i an integer $(1 \leq i \leq m)$, $e_i =$

$(0,\dots,0,1,0,\dots,0) \in \mathbb{R}^m$ (1 in the i-th place). Given an arbitrary function $w = w(x)$, we define

$$\tau_h^i w(x) = w(x + he_i), \quad \Delta_h^i w(x) = \big(\tau_h^i w(x) - w(x)\big)/h.$$

Obviously $\Delta_h^i w$ will approximate $D^i w$ for h small. The precise statement we need is:

4.7.2 Lemma. *Let $1 < p < \infty$, Ω an arbitrary domain in $\mathbb{R}^m$, u a function in $L^p(\Omega)$. Assume there exists a constant C such that*

$$\big\| \Delta_h^i w \big\|_{L^p(\Omega')} \leqslant C$$

for h small enough, where Ω' is any domain with $\overline{\Omega}' \subseteq \Omega$. Then the derivative $D^i u$ exists in the sense of distributions in Ω, belongs to $L^p(\Omega)$ and

$$\| D^i w \|_{L^p(\Omega)} \leqslant C. \tag{4.7.7}$$

For a proof, see Gilbarg-Trudinger [1977: 1, p. 162].

We note the following formal rules of computation with the operators τ_h^i and Δ_h^i:

$$D^j \Delta_h^i = \Delta_h^i D^j,$$

$$\int_\Omega (\Delta_h^i w) z \, dx = -\int_\Omega w \big(\Delta_{-h}^i z \big) \, dx,$$

$$\Delta_h^i (wz) = \big(\Delta_h^i w \big) \tau_h^i z + w \Delta_h^i z \tag{4.7.8}$$

(the second holds if w or z vanish near the boundary of Ω and h is sufficiently small). Using these rules, we take $\varphi \in \mathcal{D}(\Omega)$ and compute:

$$\hat{B}_\lambda \big(\Delta_h^i v, \varphi \big) = \int_\Omega \Big\{ \sum\sum a_{jk} \big(\Delta_h^i D^j \bar{v} \big) D^k \varphi + \lambda \big(\Delta_h^i \bar{v} \big) \varphi \Big\} \, dx$$

$$= -\int_\Omega \Big\{ \sum\sum \Delta_{-h}^i \big(a_{jk} D^k \varphi \big) D^j \bar{v} + \lambda \bar{v} \Delta_{-h}^i \varphi \Big\} \, dx$$

$$= -\hat{B}_\lambda \big(v, \Delta_{-h}^i \varphi \big)$$

$$\qquad - \int_\Omega \Big\{ \sum\sum \Delta_{-h}^i a_{jk} \tau_{-h}^i D^k \varphi D^j \bar{v} \Big\} \, dx, \tag{4.7.9}$$

where we denote by $\hat{B}_\lambda$ the functional corresponding to the homogeneous operator

$$A_0 u = \sum_{j=1}^m \sum_{k=1}^m D^j \big(a_{jk} D^k u \big). \tag{4.7.10}$$

Let g be the function in (4.7.5). Define

$$\hat{g} = g + \sum b_j D^j v + cv$$

We use now (4.7.9) keeping in mind that $\hat{B}_\lambda(v, \Delta^i_{-h}\varphi) = (\hat{g}, \Delta^i_{-h}\varphi)$ and that both inequalities (4.6.3) hold for $\hat{B}_\lambda$ (where c, C only depend on κ, λ, K_1) in the norm $\|\cdot\|_\lambda$ or (with different constants) in the norm of $H^1_0(\Omega)$. Note first that, since $\Delta^i_h v \in H^1_0(\Omega)$ for h sufficiently small, there exists $\varphi \in \mathcal{D}(\Omega)$ (depending on h) such that

$$\|\Delta^i_h v - \varphi\|_{H^1_0(\Omega)} \leqslant \frac{c}{2C}\|\Delta^i_h v\|_{H^1_0(\Omega)}$$

where c, C are, for the moment, the constants in (4.6.3). Hence

$$\left|\hat{B}_\lambda(\Delta^i_h v, \varphi)\right| \geqslant \left|\hat{B}_\lambda(\Delta^i_h v, \Delta^i_h v)\right| - \left|\hat{B}_\lambda(\Delta^i_h v, \varphi - \Delta^i_h v)\right|$$

$$\geqslant c\|\Delta^i_h v\|^2_{H^1_0(\Omega)} - C\|\Delta^i_h v\|_{H^1_0(\Omega)}\|\varphi - \Delta^i_h v\|_{H^1_0(\Omega)} \geqslant \frac{c}{2}\|\Delta^i_h v\|^2_{H^1_0(\Omega)}.$$

$$(4.7.11)$$

By virtue of (4.7.9) and (4.7.10),

$$|\hat{B}_\lambda(\Delta^i_h v, \varphi)| \leqslant C'\left(\|\hat{g}\|_{L^2(\Omega)}\|\Delta^i_{-h}\varphi\|_{L^2(\Omega)} + \|v\|_{H^1_0(\Omega)}\|\varphi\|_{H^1_0(\Omega)}\right), \quad (4.7.12)$$

where C' does not depend on h. Noticing that

$$\|\Delta^i_{-h}\varphi\|_{L^2(\Omega)} \leqslant \|\varphi\|_{H^1_0(\Omega)} \leqslant C''\|\Delta^i_h v\|_{H^1_0(\Omega)},$$

where C'' is likewise independent of h, we obtain combining (4.7.11) and (4.7.12) that

$$\|\Delta^i_h v\|_{H^1_0(\Omega)} \leqslant C\left(\|g\|_{L^2(\Omega)} + \|v\|_{H^1_0(\Omega)}\right)$$

or

$$\|\Delta^i_h D^j v\|_{L^2(\Omega)} \leqslant C\left(\|g\|_{L^2(\Omega)} + \|v\|_{H^1(\Omega)}\right)$$

for arbitrary i, j, thus the theorem results from Lemma 4.7.2 and the fact that $u = v$ in Ω'.

Under the assumptions of theorem 4.6.6, we obtain the important result that $A(\beta)$ (β any boundary condition) acts on elements of its domain "in the classical sense"; in other words, if $u \in D(A(\beta))$,

$$A(\beta)u = \sum\sum a_{jk}D^j D^k u + \sum\left(b_j + \sum D^k a_{jk}\right)D^j u + cu, \quad (4.7.13)$$

where all the derivatives of u involved exist[6] and belong to $L^2_{loc}(\Omega)$. To see this, let f be the right-hand side of (4.7.13); an integration by parts shows that $B_\lambda(u, \varphi) = (\lambda u - f, \varphi)$ for $\varphi \in \mathcal{D}(\Omega)$, thus by uniqueness, $A(\beta)u$ is given by (4.7.13).

Theorem 4.7.1 contains no information on the *global* behavior of u, precisely, on whether u belongs to $H^2(\Omega)$. To obtain results of this type, we begin with a local result for Dirichlet boundary conditions.

---

[6] Of course, in the sense of distributions.

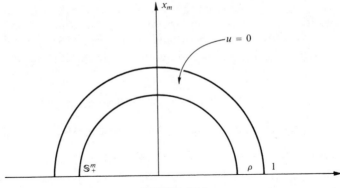

FIGURE 4.7.1

4.7.3 Local Boundary Regularity Lemma. *Let A_0 be of class $C^{(1)}$ in $\mathbb{S}_+^m$ and uniformly elliptic there. Assume $u \in H_0^1(\mathbb{S}_+^m)$ is zero for $|x| \leqslant \rho < 1$ and satisfies*

$$B_\lambda(u, \varphi) = (f, \varphi) \quad \left(\varphi \in \mathcal{D}(\mathbb{S}_+^m)\right), \tag{4.7.14}$$

where $f \in L^2(\mathbb{S}_+^m)$. Then $u \in H^2(\mathbb{S}_+^m)$ and there exists a constant C depending only on K_1, κ, λ such that

$$\|u\|_{H^2(\mathbb{S}_+^m)} \leqslant C\left(\|f\|_{L^2(\mathbb{S}_+^m)} + \|u\|_{H_0^1(\mathbb{S}_+^m)}\right). \tag{4.7.15}$$

The proof is a rather immediate adaptation of that of Theorem 4.7.1. Since u is already zero for $|x| \leqslant \rho < 1$, multiplication by χ is unnecessary. A moment's reflection shows that all the arguments of Theorem 4.7.1 apply as long as the operators τ_h^i, Δ_h^i act on directions parallel to the hyperplane $x_m = 0$, that is, if $1 \leqslant i \leqslant m - 1$. Accordingly, we can prove existence of

$$D^i D^j u \quad (1 \leqslant i \leqslant m - 1, \quad 1 \leqslant j \leqslant m) \tag{4.7.16}$$

in $L^2(\mathbb{S}_+^m)$ and establish for each an estimate of the required type. The only second derivative not included in (4.7.16) is $(D^m)^2 u$; however, by virtue of Theorem 4.7.1, $(D^m)^2 u$ exists in the sense of distributions in $\mathbb{S}_+^m$ and is locally square integrable there: in view of (4.7.13),

$$(D^m)^2 u = -a_{mm}^{-1}\left(\sum\sum a_{jk} D^j D^k u + \sum\left(b_j + \sum D^k a_{jk}\right) D^j u + cu\right)$$
$$+ a_{mm}^{-1}(\lambda u - f), \tag{4.7.17}$$

where the double sum is extended to indices $(j, k) \neq (m, m)$. Since $a_{mm}^{-1}(x) \leqslant \kappa^{-1}$ by virtue of the uniform ellipticity assumption (4.7.2), the corresponding estimate for $(D^m)^2$, hence (4.7.15), follows.

We piece together these results.

4.7.4 Global Regularity Theorem. *Let Ω be a bounded domain of class $C^{(2)}$, and let A_0 be of class $C^{(1)}$ in Ω and uniformly elliptic. Let*

$u \in D(A(\beta))$ (β the Dirichlet boundary condition), that is, let $u \in H_0^1(\Omega)$ satisfy

$$B_\lambda(u, w) = (f, w) \quad (w \in H_0^1(\Omega)) \tag{4.7.18}$$

for some $f \in L^2(\Omega)$ ($f = (\lambda I - A(\beta))u$). Then $u \in H^2(\Omega)$ and there exists a constant C (depending only on Ω, K_1, κ, λ) such that

$$\|u\|_{H^2(\Omega)} \leqslant C\big(\|f\|_{L^2(\Omega)} + \|u\|_{H_0^1(\Omega)}\big). \tag{4.7.19}$$

The central idea of the proof consists in dividing Ω into interior and boundary patches (in the sense of Section 4.5) and then using Theorem 4.7.1 for the interior patches and Corollary 4.7.3 (via a change of variables) for the others. The effect of these changes of variables is examined in the following result.

4.7.5 Lemma.   *Let V, V' be bounded domains in $\mathbb{R}^m$, $\eta : V \to V'$ an invertible map onto V' such that the derivatives $D^\alpha\eta$ (resp. $D^\alpha\eta^{-1}$) exist and are continuous in $\overline{V}$ (resp. $\overline{V}'$) for $|\alpha| \leqslant k$. Given $u \in W^{k,p}(V')$, define*

$$Nu(x) = u(\eta(x)).$$

Then N is a bounded, invertible linear operator from $W^{k,p}(V')$ onto $W^{k,p}(V)$, $1 \leqslant p < \infty$.

For a proof see Adams [1975: 1, p. 63].

Proof of Theorem 4.7.4.   Let $V_1^i, V_2^i, \ldots, V_1^b, V_2^b, \ldots$ be a finite cover of Ω by interior and boundary patches and let $\chi_1^i, \chi_2^i, \ldots, \chi_1^b, \chi_2^b, \ldots$ be a subordinated partition of unity in the sense of Section 4.5. We have

$$u = \sum \chi_n^i u + \sum \chi_n^b u = \sum u_n^i + \sum u_n^b. \tag{4.7.20}$$

Each of the functions u_n^i satisfies

$$B_\lambda(u_n^i, \varphi) = (g_n^i, \varphi) \quad (\varphi \in \mathcal{D}(V_n^i)), \tag{4.7.21}$$

where

$$g_n^i = \chi_n^i f - u\Big(\sum\sum D^j\big(a_{jk}D^k\chi_n^i\big) + \sum b_j D^j\chi_n^i\Big)$$
$$- 2\sum\sum a_{jk}D^j u D^k\chi_n^i. \tag{4.7.22}$$

Thus it follows from Theorem 4.7.1 that $u_n^i \in H^2(V_n^i)$. Accordingly, $u_n^i \in H^2(\Omega)$ and[7]

$$\|u_n^i\|_{H^2(\Omega)} \leqslant C\big(\|g_n^i\|_{L^2(\Omega)} + \|u_n^i\|_{H^1(\Omega)}\big)$$
$$\leqslant C'\big(\|f\|_{L^2(\Omega)} + \|u\|_{H^1(\Omega)}\big) \quad (n = 1, 2, \ldots) \tag{4.7.23}$$

---

[7]Naturally, all the interior patches can be lumped together. Also, multiplication by χ^i (and the subsequent calculation) can be easily avoided.

The treatment of boundary patches is not as immediate. It is still true that u_n^b satisfies (4.7.21) (where φ belongs now to $H_0^1(V_n^b \cap \Omega)$) with g_n^b given by an expression similar to (4.7.21). If η is the map associated with V_n^b, then we can show that $\tilde{u}_n^b$ (which is u_n^b as a function of $\eta_1, \ldots, \eta_m$) satisfies

$$\tilde{B}_\lambda(\tilde{u}_n^b, w) = (\tilde{g}_n^b, w) \quad (w \in H_0^1(\mathbb{S}_+^m)), \qquad (4.7.24)$$

where the tilde in g_n^b is interpreted in a similar way and $\tilde{B}_\lambda$ is the functional defined in terms of the transformed operator $\tilde{A}_0$ in the same way B_λ was defined using A_0. Evidently $\tilde{A}_0$ is of class $C^{(1)}(\overline{\mathbb{S}}_+^m)$ and it was already shown in Section 4.5 that

$$\sum\sum \tilde{a}_{jk}\xi_j\xi_k = \sum a_{jk}\tilde{\xi}_j\tilde{\xi}_k, \qquad (4.7.25)$$

where $\tilde{\xi} = J'\xi$, J the Jacobian matrix of η with respect to x; thus the uniform ellipticity condition (4.7.2) is satisfied (with a different constant $\tilde{\kappa}$ bounded below by a multiple of κ in each patch). Since $\tilde{u}_n^b \in H_0^1(\mathbb{S}_+^m)$ and it vanishes for $|x| \leqslant \rho < 1$, Lemma 4.7.3 applies to show that $\tilde{u}_n^b \in H^2(\mathbb{S}_+^m)$ and

$$\|\tilde{u}_n^b\|_{H^2(\mathbb{S}_+^m)} \leqslant C\left(\|\tilde{g}_n^b\|_{L^2(\mathbb{S}_+^m)} + \|\tilde{u}_n^b\|_{H^1(\mathbb{S}_+^m)}\right),$$

which implies, making use of Lemma 4.7.5, that $u_n^b \in H^2(\Omega)$ and that

$$\|u_n^b\|_{H^2(\Omega)} \leqslant C'\left(\|g_n^b\|_{L^2(\Omega)} + \|u_n^b\|_{H^1(\Omega)}\right)$$

$$\leqslant C''\left(\|f\|_{L^2(\Omega)} + \|u\|_{H^1(\Omega)}\right) \quad (n = 1, 2, \ldots).$$

Combining these inequalities with (4.7.20) and (4.7.23), we deduce that $u \in H^2(\Omega)$ and (4.7.19) holds, thus ending the proof of Theorem 4.7.4.

It is easy to prove that (4.7.19) implies

$$\|u\|_{H^2(\Omega)} \leqslant C\|(\lambda I - A(\beta))u\|_{L^2(\Omega)} \qquad (4.7.26)$$

with C depending only on Ω, K_1, κ, λ. To see this, we only have to observe that

$$\|u\|_{H_0^1(\Omega)} \leqslant C'\|f\|_{L^2(\Omega)} = C'\|(\lambda I - A(\beta)u)\|_{L^2(\Omega)},$$

which is an immediate consequence of the first inequality (4.6.5) for B_λ, since

$$c\|u\|_{H_0^1(\Omega)}\|u\|_{L^2(\Omega)} \leqslant c\|u\|_{H_0^1(\Omega)}\|u\|_{H_0^1(\Omega)} \leqslant |B_\lambda(u, u)| = |(f, u)|$$

$$\leqslant \|f\|_{L^2(\Omega)}\|u\|_{L^2(\Omega)} \quad (u \in H_0^1(\Omega)).$$

We consider now boundary conditions of type (I). The arguments are very similar, and the only piece of fresh information needed is a result analogous to Lemma 4.7.3 for these boundary conditions.

4.7.6   Local Boundary Regularity Lemma.   *Let A_0 be of class $C^{(1)}$ in $\mathbb{S}_+^m$ and uniformly elliptic there. Assume $u \in H^1(\mathbb{S}_+^m)$ is zero for $|x| \leqslant \rho < 1$*

and satisfies

$$B_{\lambda,\beta}(u,\varphi) = B_\lambda(u,\varphi) - \int_{\mathbb{S}_0^m} \gamma\bar{u}\varphi\, d\sigma = (f,\varphi) \quad (\varphi \in \mathcal{D}) \quad (4.7.27)$$

where γ is continuously differentiable in $\overline{\mathbb{S}}_0^m$ and $f \in L^2(\mathbb{S}_+^m)$. Then $u \in H^2(\mathbb{S}_+^m)$ and there exists a constant C (depending only on K_1, κ, λ, M_1, where

$$M_k = \max_{x \in \overline{\mathbb{S}}_0^m} \{|D^{\hat{\alpha}}\gamma(x)|\}$$

for $\hat{\alpha} = (\alpha_1,\ldots,\alpha_{m-1})$, $|\hat{\alpha}| = \alpha_1 + \cdots + \alpha_{m-1} \leqslant k$) such that

$$\|u\|_{H^2(\mathbb{S}_+^m)} \leqslant C\big(\|f\|_{L^2(\mathbb{S}_+^m)} + \|u\|_{H^1(\mathbb{S}_+^m)}\big). \quad (4.7.28)$$

Proof. Let $i = 1, 2, \ldots, m-1$. We compute the left-hand side of (4.7.27) with $\Delta_h^i u$ in place of u. The expression for $B_\lambda(\Delta_h^i u, \varphi)$ is exactly the same as that in (4.7.9). On the other hand,

$$\int_{\mathbb{S}_0^m} \gamma\Delta_h^i\bar{u}\varphi\, d\sigma = -\int_{\mathbb{S}_0^m} \bar{u}\Delta_{-h}^i(\gamma\varphi)\, d\sigma = -\int_{\mathbb{S}_0^m} \bar{u}\big(\Delta_{-h}^i\gamma\tau_{-h}^i\varphi + \gamma\Delta_{-h}^i\varphi\big)\, d\sigma,$$

thus, if $\hat{B}_\lambda$ is the functional in (4.7.9) and $\hat{B}_{\lambda,\beta}$ is similarly defined with respect to the homogeneous operator (4.7.10),

$$\hat{B}_{\lambda,\beta}\big(\Delta_h^i u, \varphi\big) = -\hat{B}_{\lambda,\beta}\big(u, \Delta_{-h}^i\varphi\big)$$

$$- \int_{\mathbb{S}_+^m} \big\{\sum\sum \Delta_{-h}^i a_{jk}\tau_{-h}^i D^k\varphi D^j\bar{u}\big\}\, dx$$

$$+ \int_{\mathbb{S}_0^m} \hat{u}\Delta_{-h}^i\gamma\tau_{-h}^i\varphi\, d\sigma, \quad (4.7.29)$$

where the hypersurface integral on the right-hand side of (4.7.29) is estimated with the help of the Schwarz inequality and Theorem 4.6.4:

$$\left(\int_{\mathbb{S}_0^m} \bar{u}\Delta_{-h}^i\gamma\tau_{-h}^i\varphi\, dx\right)^2 \leqslant M_1^2 \int_{\mathbb{S}_0^m} |u|^2\, d\sigma \int_{\mathbb{S}_0^m} |\tau_{-h}^i\varphi|^2\, d\sigma$$

$$\leqslant C\|u\|_{H^1(\mathbb{S}_+^m)}^2 \|\varphi\|_{H^1(\mathbb{S}_+^m)}^2.$$

We proceed now exactly as in the end of the proof of Theorem 4.7.1 to show existence of, and to estimate the second derivatives in (4.7.16); the missing derivative $(D^m)^2 u$ is accounted for by means of (4.7.17). Details are omitted.

4.7.7 Global Regularity Theorem. *Let Ω be a bounded domain of class $C^{(2)}$ and let A_0 be of class $C^{(1)}(\overline{\Omega})$ and uniformly elliptic in Ω. Let β be a boundary condition of type (I) with γ continuously differentiable on the* *boundary* Γ. *Assume that* $u \in H^1(\Omega)$ *belongs to* $D(A(\beta))$, *that is,*

$$B_{\lambda, \beta}(u, w) = (f, w) \quad (w \in H^1(\Omega)) \tag{4.7.30}$$

for some $f \in L^2(\Omega)$ ($f = (\lambda I - A(\beta))u$). *Then* $u \in H^2(\Omega)$ *and there exists a constant* C (*depending only on* Ω, K_1, κ, λ, M_1) *such that*

$$\|u\|_{H^2(\Omega)} \le C(\|f\|_{L^2(\Omega)} + \|u\|_{H^2(\Omega)}). \tag{4.7.31}$$

The proof is essentially similar to that of Theorem 4.7.4, the only slight difference appearing in the treatment of the boundary patches $V_1^b, V_2^b, \ldots$ and the functions $u_n^b = \chi_n^b u$. It is easy to see that each u_n^b satisfies

$$B_{\lambda, \beta}(u_n^b, \varphi) = B_\lambda(u_n^b, \varphi) - \int_{\Gamma \cap V_n^b} \gamma_n^b \bar{u}_n^b \varphi \, d\sigma = (g_n^b, \varphi) \quad (\varphi \in \mathcal{D}), \tag{4.7.32}$$

where g_n^b is given by an expression of the type of (4.7.22) and

$$\gamma_n^b = D^{\tilde{\nu}} \chi_n^b + \chi_n^b \gamma. \tag{4.7.33}$$

If η is the map associated with the boundary patch V_n^b and $\sim$ indicates, as before, that a function is considered to be a function of $\eta_1, \ldots, \eta_m$, a straightforward computation[8] shows that

$$\tilde{B}_{\lambda, \beta}(\tilde{u}_n^b, \varphi) = \tilde{B}_\lambda(\tilde{u}_n^b, \varphi) - \int_{S_0^m} \rho \tilde{\gamma}_n^b \bar{\tilde{u}}_n^b \varphi \, d\sigma = (\tilde{g}_n^b, \varphi) \quad (\varphi \in \mathcal{D})$$

where again $B_{\lambda, \beta}$, B_λ are the functionals corresponding to $\tilde{A}_0$ and ρ is the normalization factor in (4.5.28). We can then apply Lemma 4.7.6 and the proof runs much like that of Theorem 4.7.5.

In some instances it is advantageous to replace inequality (4.7.31) by one of the type of (4.7.26), that is, by

$$\|u\|_{H^2(\Omega)} \le C\|(\lambda I - A(\beta))u\|_{L^2(\Omega)} \tag{4.7.34}$$

with C depending only on Ω, K_1, κ, λ, M_1. The argument leading to (4.7.34) is very much the same as that justifying (4.7.26) in the case of Dirichlet boundary conditions.

An important consequence of the regularity results just proved is that the domains of the operators $A(\beta)$ can be now identified in a much more satisfactory way than hitherto possible.

4.7.8 Corollary. *Let* Ω *be a bounded domain of class* $C^{(2)}$, *and let* A_0 *be of class* $C^{(1)}$ *and uniformly elliptic in* Ω. *Finally, let* β *be a boundary condition of type* (I) *or* (II) (*in the first case, with* γ *continuously differentiable*

---

[8]In this, an important role is played by the invariance of the conormal vector (or derivative) shown in (4.5.28).

on Γ). *Then* $D(A(\beta))$ *consists of all* $u \in H^2(\Omega)$ *that satisfy the boundary condition* β *almost everywhere on* Γ. *When* β *is the Dirichlet boundary condition, we can also write*

$$D(A(\beta)) = H^2(\Omega) \cap H_0^1(\Omega). \tag{4.7.35}$$

Proof. We begin with the case of Dirichlet boundary conditions. The inclusion $D(A(\beta)) \subseteq H^2(\Omega)$ results from Theorem 4.7.4 and the fact that any $u \in D(A(\beta)) \subseteq H_0^1(\Omega)$ vanishes a.e. on Γ has already been discussed in the previous section (see the comments after Theorem 4.6.5). The converse is a consequence of the following result and of a simple application of the divergence theorem:

4.7.9 Lemma. *Let* Ω *be a bounded domain of class* $C^{(1)}$, $1 \leqslant p < \infty$, u *an element of* $W^{1,p}(\Omega)$ *such that* $u = 0$ *almost everywhere on the boundary* Γ. *Then* $u \in W_0^{1,p}(\Omega) =$ *closure of* $\mathcal{D}(\Omega)$ *in* $W^{1,p}(\Omega)$.

The proof is based on Theorem 4.6.5 and on the method used in the proof of Theorem 3.18 in Adams [1975: 1, p. 54]. We omit the details.

The case of boundary conditions of type (I) is handled as follows. Let u be an arbitrary function in $\mathcal{D}$. Making use of Theorem 4.6.4 for the first derivatives of u, we obtain

$$\|D^{\tilde{\nu}} u\|_{L^2(\Gamma)} \leqslant C \|u\|_{H^2(\Omega)} \tag{4.7.36}$$

(the index 2 on the left-hand side can be improved to $2(m-1)/(m-4)$). We take $u \in H^2(\Omega)$ and approximate it in the H^2 norm by a sequence $\{u_n\}$ in $\mathcal{D}$. Applying (4.7.36) to $u_n - u_m$, we can then define

$$D^{\tilde{\nu}} u = \lim D^{\tilde{\nu}} u_n, \tag{4.7.37}$$

the limit understood in the norm of $L^2(\Gamma)$ and (as easily seen) defined independently of the sequence $\{u_n\}$. Let β be a boundary condition of type (I) and $u \in D(A(\beta))$. We have already shown that $u \in H^2(\Omega)$ (Theorem 4.7.7), thus $D^{\tilde{\nu}}$ is defined in the sense indicated above. Applying the divergence theorem to each member u_n of an approximating sequence and to a $\varphi \in \mathcal{D}$ we obtain

$$B_{\lambda,\beta}(u_n, \varphi) = B_\lambda(u_n, \varphi) - \int_\Gamma D^{\tilde{\nu}} \bar{u}_n \varphi \, d\sigma = \int_\Omega \{(\lambda I - A_1) \bar{u}_n\} \varphi \, dx.$$

$$\tag{4.7.38}$$

Letting $n \to \infty$, we use the fact[9] that $A_1 \bar{u}_n \to A_1 \bar{u}$ in $L^2(\Omega)$ while $D^{\tilde{\nu}} u_n$ converges in $L^2(\Gamma)$. We take limits in (4.7.38) and conclude from the

---

[9]An abuse of notation is implicit here since A_1 was not originally defined in $H^2(\Omega)$ (see Section 4.5).

definition of $A(\beta)$ through $B_{\lambda,\beta}$ that

$$\int_\Gamma D^{\tilde{\nu}}\bar{u}\varphi\,d\sigma = \int_\Gamma \gamma\bar{u}\varphi\,d\sigma$$

for arbitrary $\varphi \in \mathcal{D}$; hence

$$D^{\tilde{\nu}}u = \gamma u \quad \text{a.e. on } \Gamma \tag{4.7.39}$$

as claimed. Conversely, let $u \in H^2(\Omega)$ such that (4.7.39) is satisfied a.e. on Γ and let $\{u_n\}$ be again an approximating sequence in $\mathcal{D}$. Using the divergence theorem and taking limits, we easily obtain that $B_{\lambda,\beta}(u,\varphi) = (f,\varphi)$ for every $\varphi \in \mathcal{D}$, where $f = (\lambda I - A_0)u$, thus $u \in D(A(\beta))$.

Stronger regularity results can be obtained under stronger assumptions on the coefficients of A_0 and the domain Ω. These results, which we prove below, will be the basis of the treatment in L^p spaces for $p > 2$ in the following sections.

4.7.10   Interior Regularity Theorem.   *Let the hypotheses of Theorem 4.7.1 be satisfied with A_0 of class $C^{(k+1)}$ in Ω and f in $H^k(\Omega)$. Then $u \in H^{(k+2)}(\Omega')$ and there exists a constant C (depending only on Ω, Ω', K_k, κ, λ) such that*

$$\|u\|_{H^{k+2}(\Omega')} \leqslant C\big(\|f\|_{H^k(\Omega)} + \|u\|_{H^1(\Omega)}\big). \tag{4.7.40}$$

Proof. We only consider the case $k = 1$ since extension to $k > 1$ involves no new ideas. Let Ω'' be a domain with $\bar{\Omega}' \subset \Omega''$, $\bar{\Omega}'' \subset \Omega$. By Theorem 4.7.1, $u \in H^2(\Omega'')$; it is then easy to see that $u_i = D^i u \in H^1(\Omega'')$ satisfies

$$B_\lambda(u_i,\varphi) = (f_i,\varphi) \tag{4.7.41}$$

for $\varphi \in \mathcal{D}(\Omega'')$, where

$$f_i = D^i f + \sum\sum D^i a_{jk} D^j D^k u + \sum D^i\Big(b_j + \sum D^k a_{jk}\Big) D^j u + (D^i c)u. \tag{4.7.42}$$

Clearly we have

$$\|f_i\|_{L^2(\Omega'')} \leqslant \|f\|_{H^1(\Omega'')} + C\|u\|_{H^2(\Omega'')},$$

where C depends only on Ω, Ω'', K_2, κ, λ. Inequality (4.7.40) then follows immediately from Theorem 4.7.1.

4.7.11   Local Boundary Regularity Lemma.   *Let the hypotheses of Lemma 4.7.3 be satisfied with A_0 of class $C^{(k+1)}$ in $\mathbb{S}_+^m$, and $f \in H^k(\mathbb{S}_+^m)$. Then $u \in H^{k+2}(\mathbb{S}_+^m)$ and there exists a constant C (depending only on K_k, κ, λ) such that*

$$\|u\|_{H^{k+2}(\mathbb{S}_+^m)} \leqslant C\big(\|f\|_{H^k(\mathbb{S}_+^m)} + \|u\|_{H^1(\mathbb{S}_+^m)}\big). \tag{4.7.43}$$

Proof. Again we only consider the case $k = 1$. Since $u \in H^2(\mathbb{S}_+^m)$, $u_i = D^i u \in H^1(\mathbb{S}_+^m)$ for any i. Moreover, if $i = 1, 2, \ldots, m-1$, we check easily that $u_i = 0$ a.e. on $\mathbb{S}_0^m$; thus, using Lemma 4.7.9, we see that $u_i = D^i u \in H_0^1(\mathbb{S}_+^m)$ for $i = 1, 2, \ldots, m-1$ and each u_i satisfies (4.7.41) with f_i given by (4.7.42) and vanishes in $|x| \leqslant \rho$. Hence Lemma 4.7.3 shows that all third generalized derivatives $D^i D^j D^k u$ exist for $1 \leqslant i \leqslant m-1$, $1 \leqslant j, k \leqslant m$ and their L^2 norm is bounded by the right-hand side of (4.7.43). The only derivative absent from the argument is $(D^m)^3 u$. However, we can express it by means of the others in the style of (4.7.4) keeping in mind that $\sum\sum a_{jk} D^j D^k (D^m u) + \sum (b_j + \sum D^k a_{jk}) D^j (D^m u) + c(D^m u) = \lambda D^m u - f_m$. This ends the proof.

4.7.12 Global Regularity Theorem. *Let the assumptions of Theorem 4.7.7 be satisfied with Ω of class $C^{(k+2)}$, A_0 of class $C^{(k+1)}$ in Ω and $f = (\lambda I - A(\beta))u$ in $H^k(\Omega)$. Then $u \in H^{k+2}(\Omega)$ and there exists a constant C (depending only on Ω, K_k, κ, λ) such that*

$$\|u\|_{H^{k+2}(\Omega)} \leqslant C\left(\|f\|_{H^k(\Omega)} + \|u\|_{H^1(\Omega)}\right). \tag{4.7.44}$$

The proof follows closely that of Theorem 4.7.4 and is thus omitted; in the boundary patches we use of course Lemma 4.7.11 in lieu of Lemma 4.7.3.

If λ is large enough, inequality (4.7.44) implies

$$\|u\|_{H^{k+2}(\Omega)} \leqslant C\|(\lambda I - A(\beta))u\|_{H^k(\Omega)} \tag{4.7.45}$$

with C depending only on Ω, K_k, κ, λ; to see this, we argue as in the proof of (4.7.26).

4.7.13 Local Boundary Regularity Lemma. *Let the hypotheses of Lemma 4.7.6 be satisfied with A_0 of class $C^{(k+1)}$ in $\mathbb{S}_+^m$, $\gamma \in C^{(k+1)}(\overline{\mathbb{S}}_0^m)$, $f \in H^k(\mathbb{S}_+^m)$. Then $u \in H^{k+2}(\mathbb{S}_+^m)$ and there exists a constant C (depending only on K_k, κ, λ, M_k) such that*

$$\|u\|_{H^{k+2}(\mathbb{S}_+^m)} \leqslant C\left(\|f\|_{H^k(\mathbb{S}_+^m)} + \|u\|_{H^1(\mathbb{S}_+^m)}\right). \tag{4.7.46}$$

Proof. Once again we only consider the case $k = 1$, and set $u_i = D^i u$, $1 \leqslant i \leqslant m-1$. The function u_i belongs to $H^1(\mathbb{S}_+^m)$ and satisfies

$$B_{\lambda,\beta}(u_i, \varphi) = B_\lambda(u_i, \varphi) - \int_{\mathbb{S}_0^m} \gamma \bar{u}_i \varphi \, d\sigma = (f_i, \varphi) + \int_{\mathbb{S}_0^m} \bar{h}_i \varphi \, d\sigma \tag{4.7.47}$$

for $\varphi \in \mathcal{D}(\Omega)$, where f_i is given by (4.7.42) and

$$h_i = (D^i \gamma)u + \left(\sum D^i a_{jm}\right) D^j u.$$

Strictly speaking, Lemma 4.7.6 is not directly applicable here due to the presence of the hypersurface integral on the right-hand side of (4.7.47). However, this additional term can be estimated with the help of Theorem

4.6.4:

$$\left| \int_{\mathbb{S}_0^m} h_i \varphi \, d\sigma \right| \leqslant \|h_i\|_{L^2(\mathbb{S}_0^m)} \|\varphi\|_{L^2(\mathbb{S}_0^m)} \leqslant C \|h_i\|_{H^1(\mathbb{S}_+^m)} \|\varphi\|_{H^1(\mathbb{S}_+^m)}$$

$$\leqslant C \Big(M_2 \|u\|_{H^1(\mathbb{S}_+^m)} + m K_2 \|u\|_{H^2(\mathbb{S}_+^m)} \Big) \|\varphi\|_{H^1(\mathbb{S}^m)}.$$

Using now (4.7.26) for the second term between parenthesis, we apply the argument in Lemma 4.7.6 to the difference quotients $\Delta_h^j u_i$, $1 \leqslant j \leqslant m-1$ and prove that each u_i belongs to $H^2(\Omega)$ and that the derivatives $D^i D^j D^k u$, $1 \leqslant i \leqslant m-1$, $1 \leqslant j, k \leqslant m$ satisfy an estimate of the type of (4.7.73). The derivative $(D^m)^3 u$ is captured as in the end of Lemma 4.7.9.

4.7.14 Global Regularity Theorem. *Let the assumptions of Theorem 4.7.7 be satisfied with Ω of class $C^{(k+2)}$, A_0 of class $C^{(k+1)}$ in Ω, $\gamma \in C^{(k+1)}(\Gamma)$, and $f = (\lambda I - A(\beta)) u$ in $H^k(\Omega)$. Then $u \in H^{k+2}(\Omega)$ and there exists a constant C (depending only on Ω, K_k, κ, λ, M_k) such that*

$$\|u\|_{H^{k+2}(\Omega)} \leqslant C \big(\|f\|_{H^k(\Omega)} + \|u\|_{H^1(\Omega)} \big). \tag{4.7.48}$$

The proof follows that of Theorem 4.7.12 and is thus left to the reader.

As in the case of Dirichlet boundary conditions we can show that (4.7.48) implies

$$\|u\|_{H^{k+2}(\Omega)} \leqslant C \|(\lambda I - A(\beta)) u\|_{H^k(\Omega)} \tag{4.7.49}$$

for λ large enough with C depending only on Ω, K_k, κ, λ, M_k.

With the help of the following result, a particular case of the Sobolev imbedding theorem, we can show that u must in fact have a certain number of derivatives (in the classical sense) if k is large enough.

4.7.15 Theorem. *Let Ω be a domain of class $C^{(1)}$ and let $u \in W^{k,p}(\Omega)$ with $1 \leqslant p < \infty$ and $(k-j)p > m$ for some $j \geqslant 0$. Then $u \in C^{(j)}(\bar{\Omega})$ (after eventual modification in a null set) and*

$$\|u\|_{C^{(j)}(\bar{\Omega})} \leqslant C \|u\|_{W^{k,p}(\Omega)}, \tag{4.7.50}$$

where the constant C does not depend on u.[10]

The result is a particular case of Theorem 5.4 in Adams [1975: 1, pp. 97–98] and yields the following

4.7.16 Corollary. *Let the assumptions of Theorems 4.7.12 or 4.7.14 be satisfied with $k + 2 - j > m/2$ for some $j > 0$. Then $u \in C^{(j)}(\bar{\Omega})$ and*

$$\|u\|_{C^{(j)}(\bar{\Omega})} \leqslant C \big(\|f\|_{H^k(\Omega)} + \|u\|_{H^1(\Omega)} \big), \tag{4.7.51}$$

---

[10] This result holds under the sole assumption that Ω possesses the *strong local Lipschitz property* (Adams [1975: 1, p. 66]), which holds for any domain of class $C^{(1)}$.

where C does not depend on u. If the assumptions are satisfied for all k, then $u \in C^{(\infty)}(\bar{\Omega})$.

4.8. CONSTRUCTION OF m-DISSIPATIVE EXTENSIONS IN $L^p(\Omega)$ AND $C(\bar{\Omega})$

The results in the previous section are sufficient to assemble suitable extensions in $L^p(\Omega)$; however, a precise characterization of the domains can only be obtained using a deep a priori estimate (Theorem 4.8.4). We assume again throughout this section (unless otherwise stated) that Ω is a bounded domain of class $C^{(2)}$ and that A_0 is an operator of class $C^{(1)}$ in Ω, although we shall occasionally be forced to work with smoother domains and operators in several results. The uniform ellipticity assumption (4.7.2) is assumed to hold. Since A_0 will be written both in the form (4.4.1) and in variational form, we shall denote by $\hat{b}_j$ in this section the first order coefficients in (4.4.2).

4.8.1 Lemma. *Let β be a boundary condition of type* (II) *or of type* (I) *with γ continuous on Γ, $1 \leqslant p < \infty$, $W^{2,p}(\Omega)_\beta$ the space of all $u \in W^{2,p}(\Omega)$ that satisfy β a.e on Γ, $C^{(2)}(\bar{\Omega})_\beta$ the space of all $u \in C^{(2)}(\bar{\Omega})$ that satisfy β (in the classical sense) everywhere on Γ. Then $C^{(2)}(\bar{\Omega})_\beta$ is dense in $W^{2,p}(\Omega)_\beta$ in the norm of $W^{2,p}(\Omega)$.*

We note that the boundary values and the boundary derivatives of a function $u \in W^{2,p}(\Omega)$ are defined in the same way as for $p = 2$ in Corollary 4.7.8:

$$u = \lim u_n \quad \text{in } L^p(\Gamma), \tag{4.8.1}$$

$$D^{\tilde{\nu}}u = \lim D^{\tilde{\nu}}u_n \quad \text{in } L^p(\Gamma), \tag{4.8.2}$$

where $\{u_n\}$ is an arbitrary sequence in $\mathcal{D}$ converging to u in $W^{2,p}(\Omega)$; existence of the limit follows from Theorem 4.6.4, and existence of the sequence $\{u_n\}$ from Theorem 4.6.5. The exponent p in the two limit relations (4.8.1) and (4.8.2) can be improved except when $p = 1$ (see again Theorem 4.6.4). The proof of Lemma 4.8.1 is fairly elementary and we omit it.

4.8.2 Corollary. *Under the assumptions of Theorem 4.6.6, we have*

$$\overline{A_0(\beta)} = A(\beta) = (A_0'(\beta'))^*, \tag{4.8.3}$$

the closure and adjoint understood in $L^2(\Omega)$.

Proof. Combining (4.7.13) with (4.7.26) for the Dirichlet boundary condition (or (4.7.34) for other boundary conditions), we obtain that the H^2 norm and the graph norm $\|(\lambda I - A(\beta))u\|$ (λ large enough) are comparable in $D(A(\beta))$. The first equality (4.8.3) then results immediately from Lemma

4.8.1 for $p = 2$, the second from the first applied to the formal adjoint $A_0'(\beta')$ and from (4.6.30).

We can now construct the desired extensions in $L^p(\Omega)$.

4.8.3   Theorem.   *Let* $1 \leqslant p \leqslant \infty$, Ω *a bounded domain of class* $C^{(2)}$ *if* $1 \leqslant p \leqslant 2$ *(of class* $C^{(k+2)}$ *with* $k > m/2$ *if* $2 < p \leqslant \infty$*),* A_0 *an operator of class* $C^{(1)}$ *if* $1 \leqslant p \leqslant 2$ *(of class* $C^{(k+1)}$ *if* $2 < p \leqslant \infty$*) satisfying the uniform ellipticity assumption* (4.7.2), β *a boundary condition of type* (II) *or of type* (I) *with* $\gamma \in C^{(1)}(\Gamma)$ *(*$\gamma \in C^{(k+1)}(\Gamma)$ *if* $2 < p \leqslant \infty$*) satisfying*

$$\gamma(x) + \frac{1}{p} \sum_{j=1}^{m} \hat{b}_j(x) \nu_j \leqslant 0 \quad (x \in \Gamma), \tag{4.8.4}$$

where $\nu = (\nu_1, \ldots, \nu_m)$ *is the outer normal vector at* x *(if* $p = \infty$*, this inequality is* $\gamma(x) \leqslant 0$*). Finally, let* $b_j \in C^{(1)}(\overline{\Omega})$ *and*

$$\omega_p = \max_{x \in \Omega} \left(c(x) - \frac{1}{p} \sum_{j=1}^{m} D^j \hat{b}_j(x) \right) \tag{4.8.5}$$

($\omega_p = \max c(x)$ *if* $p = \infty$*). If* $1 \leqslant p < \infty$*, the closure* $A_p(\beta) = \overline{A_0(\beta)}$ *in* $L^p(\Omega)$ *belongs to* $\mathcal{C}_+(1, \omega_p)$*. Moreover, if* $1 < p < \infty$*, we have*

$$A_p(\beta) = \overline{A_0(\beta)} = A_0'(\beta')^* = \left(\overline{A_0'(\beta')} \right)^*, \tag{4.8.6}$$

where $A_0'(\beta')$ *is thought of as an operator in* $L^{p'}(\Omega)$ *(*$p'^{-1} + p^{-1} = 1$*). Finally, if* $p = \infty$*, the closure* $A_\infty(\beta) = \overline{A_0(\beta)}$ *in* $C(\overline{\Omega})$ *(*$C_\Gamma(\overline{\Omega})$ *if* β *is of type* (II)*) belongs to* $\mathcal{C}_+(1, \omega_\infty)$*.*

Proof. We note that if ω_p', ω_∞' are defined by (4.8.5) in relation to the formal adjoint A_0', then we check easily that $\omega_p' = \omega_{p'}$ for $1 < p < \infty$, and that $\omega_\infty' = \omega_1$. Obviously we may assume that $\omega_p = 0$; otherwise we simply replace $A_0(\beta)$ by $A_0(\beta) - \omega_p I$. The same observation holds for ω_∞. We begin with the case $E = L^p(\Omega)$, $1 \leqslant p \leqslant 2$. It follows from Corollary 4.8.2 that if $\lambda > 0$ is large enough, $(\lambda I - A_0(\beta)) D(A_0(\beta))$ is dense in $(\lambda I - A(\beta)) D(A(\beta)) = L^2(\Omega)$. We see that $(\lambda I - A_0(\beta)) D(A_0(\beta))$ is dense in $L^p(\Omega)$, $1 \leqslant p \leqslant 2$, and by virtue of Lemma 3.1.11, $A_0(\beta)$ is m-dissipative in $L^p(\Omega)$ (recall that $A_0(\beta)$ is dissipative in view of Lemma 4.5.2). The case $p > 2$ is slightly more complex to handle. Let $f \in H^k(\Omega) \cap L^p(\Omega) = H^k(\Omega)$ and let $u \in D(A(\beta))$ be the solution of $(\lambda I - A(\beta)) u = f$ for sufficiently large $\lambda > 0$. In view of Theorem 4.7.14 (or Theorem 4.7.12 in the case of Dirichlet boundary conditions), $u \in H^{k+2}(\Omega)$, hence $u \in C^{(2)}(\overline{\Omega})$ by Theorem 4.7.15. It follows from Corollary 4.7.8 that u satisfies β almost everywhere (thus everywhere) on Γ, hence $u \in C^{(2)}(\overline{\Omega})_\beta = D(A_0(\beta))$. Since f is arbitrary we deduce that $(\lambda I - A_0(\beta)) D(A_0(\beta)) \supseteq H^k(\Omega)$, which is dense in L^p. As we already know that $A_0(\beta)$ is dissipative in L^p (Lemma 4.5.2), we obtain again from Lemma 3.1.11 and Remark 3.1.12 that $\overline{A_0(\beta)}$ is

m-dissipative in $L^p(\Omega)$. The case $E = C(\overline{\Omega})$ ($C_\Gamma(\overline{\Omega})$ if Dirichlet boundary conditions are used) is handled in the same way. It remains to prove the duality relations (4.8.6). From the Lagrange identity (4.5.40) we deduce that $A_0(\beta) \subseteq A_0'(\beta')^*$, thus $\overline{A_0(\beta)} \subseteq A_0'(\beta')^* = \overline{A_0'(\beta')}^*$. However, we know that $\overline{A_0(\beta)}$ is m-dissipative, and so is $A_0'(\beta')^*$ after Theorem 2.3.10, since $L^p(\Omega)$ is reflexive for $1 < p < \infty$. Hence the resolvent of both operators exists for $\lambda > 0$ and they must then coincide (see the end of the proof of Lemma 2.3.9). This completes the proof.

The precedent L^p theory is not entirely satisfactory. In fact, the characterization of the domain of

$$A_p(\beta) = \overline{A_0(\beta)} \quad (= A_0'(\beta')^* \text{ if } E = L^p(\Omega), 1 < p < \infty) \quad (4.8.7)$$

is somewhat imprecise (compare with the L^2 case in Corollary 4.7.8). This objection will be surmounted with the help of the following result.

4.8.4   Theorem. *Let Ω be a bounded domain of class $C^{(2)}$, and let*

$$A_0 u = \sum\sum a_{jk}(x) D^j D^k u + \sum b_j(x) D^j u + c(x) u, \quad (4.8.8)$$

where the coefficients a_{jk}, b_j, c, are continuous in $\overline{\Omega}$; assume A_0 is uniformly elliptic, that is, there exists $\kappa > 0$ such that

$$\sum\sum a_{jk}(x) \xi_j \xi_k \geqslant \kappa |\xi|^2 \quad (x \in \Omega, \xi \in \mathbb{R}^m) \quad (4.8.9)$$

and let β be the Dirichlet boundary condition. Finally, let $1 < p < \infty$. Then there exists a constant C (depending only on p, κ, Ω, the modulus of continuity of a_{jk} on Ω and the maximum of $|a_{jk}|$, $|b_j|$, $|c|$ on $\overline{\Omega}$) such that

$$\|u\|_{W^{2,p}(\Omega)} \leqslant C\big(\|A_0 u\|_{L^p(\Omega)} + \|u\|_{W^{1,p}(\Omega)}\big) \quad (4.8.10)$$

for all u in $C^{(2)}(\overline{\Omega})_\beta$, the set of all $u \in C^{(2)}(\overline{\Omega})$ that satisfy β everywhere on Γ. The same conclusion holds for boundary conditions β of type (I) with the added assumption that the a_{jk} are continuously differentiable in $\overline{\Omega}$ and that γ is continuously differentiable on Γ. In this case the constant C depends on p, κ, Ω, the modulus of continuity of a_{jk} and of $D^i a_{jk}$, the maximum of $|\gamma|$ and $|D^\mu \gamma|$ on Γ (D^μ any first-order derivative on Γ) and the maximum of $|a_{jk}|$, $|D^i a_{jk}|$, $|b_j|$, $|c|$ on $\overline{\Omega}$.

The basic ingredient in the proof will be the following result on Fourier multipliers. We denote by $\mathscr{F}u$ the Fourier transform of u (see Section 8 for definition and properties).

4.8.5   Theorem. *Let $\mathrm{m} = \mathrm{m}(\sigma)$ be a function k times continuously differentiable for $\sigma \neq 0$ in $\mathbb{R}^m_\sigma$, with $k > m/2$. Assume that for every differential monomial D^α, $\alpha = (\alpha_1, \ldots, \alpha_m)$ with $|\alpha| = \alpha_1 + \cdots + \alpha_m \leqslant k$, we have*

$$|D^\alpha \mathrm{m}(\sigma)| \leqslant B|\sigma|^{-|\alpha|}. \quad (4.8.11)$$

Finally, let $1 < p < \infty$. Then there exists a constant C (depending only on B

and p) such that

$$\|\mathscr{F}^{-1}(\mathfrak{m}\,\mathscr{F}u)\|_{L^p(\mathbb{R}^m)} \leqslant C\|u\|_{L^p(\mathbb{R}^m)} \qquad (4.8.12)$$

for every $u \in L^p(\mathbb{R}^m) \cap L^2(\mathbb{R}^m)$.

The requirement that $u \in L^2(\mathbb{R}^m)$ is made to assure existence of $\mathscr{F}u$: since (4.8.11) for $\alpha = 0$ implies that $\mathfrak{m}$ is bounded, $\mathfrak{m}\,\mathscr{F}u \in L^2(\mathbb{R}^m_\sigma)$ and $\mathscr{F}^{-1}(\mathfrak{m}\,\mathscr{F}u)$ exists as well. For a proof of Theorem 4.8.5 see Stein [1970: 1, p. 96] or Hörmander [1960: 1].

With the help of this result we establish several estimates for first and second derivatives of the inhomogeneous Laplace equation.[11]

4.8.6 Theorem. *Let u be twice continuously differentiable with support contained in* $|x| \leqslant M$ *and let* $1 < p < \infty$. *Then there exists a constant* C_0 *(depending only on p, M) such that*

$$\|u\|_{W^{2,p}(\mathbb{R}^m)} \leqslant C_0\|\Delta u\|_{L^p(\mathbb{R}^m)} \qquad (4.8.13)$$

where $\Delta = \Sigma(D^j)^2$ *is the Laplacian.*

Proof. Let $f = \Delta u$. We have

$$\mathscr{F}(D^j D^k u)(\sigma) = \frac{\sigma_j \sigma_k}{|\sigma|^2}\mathscr{F}f(\sigma)$$

for $1 \leqslant j, k \leqslant m$, where $\mathfrak{m}_{jk}(\sigma) = \sigma_j \sigma_k |\sigma|^{-2}$ is easily seen to satisfy (4.8.11), hence the estimates for the second derivatives follow from Theorem 4.8.5 (note, incidentally, that these estimates do not depend on M). To bound the first derivatives and u itself, it is obviously sufficient to prove that an estimate of the form

$$\|\varphi\|_{L^p(\mathbb{R}^m)} \leqslant C \sum_{j=1}^{m} \|D^j\varphi\|_{L^p(\mathbb{R}^m)} \qquad (4.8.14)$$

holds for φ continuously differentiable with support in $|x| < M$, with C depending only on p and M. To establish (4.8.14) we use the formula

$$\varphi(x) = \sum_{j=1}^{m} \frac{1}{\omega_m} \int_{\mathbb{R}^m} \frac{x_j - y_j}{|x - y|^m} D^j\varphi(y)\, dy \qquad (4.8.15)$$

(where $\omega_m = 2\pi^{m/2}/\Gamma(m/2)$ is the hyperarea of the unit sphere in $\mathbb{R}^m$), which can be easily established as follows (see Stein [1970: 1, p. 125]). Let $\xi \in \mathbb{R}^m$, $|\xi| = 1$. Then

$$\varphi(x) = \int_0^\infty (\operatorname{grad}\varphi(x - \xi t), \xi)\, dt.$$

---

[11]Theorem 4.8.6 can also be proved with the help of the Riesz transforms and the Calderón-Zygmund theorem (Stein [1970: 1, p. 59]).

Integrating this equality on $|\xi| = 1$, (4.8.15) results. We can rewrite this formula as follows:

$$\varphi(x) = \sum (P_j * \varphi)(x),$$

where $P_j(x) = \omega_m^{-1} \chi(x) x_j |x|^{-m}$, χ the characteristic function of $|x| \leqslant 2M$. Since each P_j belongs to $L^1(\mathbb{R}^m)$, the estimate follows from Young's Theorem 8.1. This ends the proof.

4.8.7 Corollary. Let $\mathbb{R}_+^m = \{x \in \mathbb{R}^m; x_m > 0\}$ and let u be a twice continuously differentiable function in $\bar{\mathbb{R}}_+^m$ with support contained in $|x| \leqslant M$ and such that $u(x) = 0$ for $x_m = 0$. Let $1 < p < \infty$. Then there exists a constant C_0 (depending only on p, M) such that

$$\|u\|_{W^{2,p}(\mathbb{R}_+^m)} \leqslant C_0 \|\Delta u\|_{L^p(\mathbb{R}_+^m)}. \qquad (4.8.16)$$

Proof. Extend the function u to $\mathbb{R}^m$ by setting $u(\hat{x}, x_m) = -u(\hat{x}, -x_m)$ for $x_m < 0$, where $\hat{x} = (x_1, \ldots, x_{m-1}) \in \mathbb{R}^{m-1}$. It is clear that u is once (but in general not twice) continuously differentiable; in fact, although $D^j D^k u$ exists and is continuous if $j \neq m$ or $k \neq m$, $(D^m)^2 u$ may be discontinuous as x passes across $x_m = 0$. However, if $\varphi \in \mathfrak{D}$,

$$\int_{\mathbb{R}^m} u(\hat{x}, x_m)(D^m)^2 \varphi(\hat{x}, x_m) \, d\hat{x} \, dx_m$$

$$= \int_{x_m \geqslant 0} u(\hat{x}, x_m)\left((D^m)^2 \varphi(\hat{x}, x_m) - (D^m)^2 \varphi(\hat{x}, -x_m)\right) d\hat{x} \, dx_m$$

$$= \int_{x_m \geqslant 0} (D^m)^2 u(\hat{x}, x_m)\left(\varphi(\hat{x}, x_m) - \varphi(\hat{x}, -x_m)\right) d\hat{x} \, dx_m$$

$$= \int_{\mathbb{R}^m} (D^m)^2 u(\hat{x}, x_m) \varphi(\hat{x}, x_m) \, d\hat{x} \, dx_m, \qquad (4.8.17)$$

where $(D^m)^2 u(x)$ has been extended to all of $\mathbb{R}^m$ in the same fashion as u; if $f = \Delta u$ is extended in a similar manner, we have

$$\Delta u = f \qquad (4.8.18)$$

in $\mathbb{R}^m$, in the sense of distributions. Since the extended u belongs to $W^{2,p}$, we can approximate it in the $W^{2,p}$-norm by a sequence $\{\varphi_n\}$ in $\mathfrak{D}$ such that $\varphi_n = 0$ if $|x| > M + 1$. Making use of Theorem 4.8.6 (note that $\Delta \varphi_n \to f = \Delta u$ in $L^p(\mathbb{R}^m)$) and taking limits, the result follows.

We treat next the case of boundary conditions of type (I).

4.8.8 Theorem. Let u (resp. ψ) be twice (resp. once) continuously differentiable in $\bar{\mathbb{R}}_+^m$ with supported contained in $|x| \leqslant M$. Assume $D^m u(\hat{x}, 0) = \psi(\hat{x}, 0)$, and let $1 < p < \infty$. Then there exists C_0 (depending only on p, M) such that

$$\|u\|_{W^{2,p}(\mathbb{R}_+^m)} \leqslant C_0\left(\|\Delta u\|_{L^p(\mathbb{R}_+^m)} + \|\psi\|_{W^{1,p}(\mathbb{R}_+^m)}\right). \qquad (4.8.19)$$

The proof is sketched at the end of this section (Example 4.8.21).

Proof of Theorem 4.8.4. We begin with the case where β is the Dirichlet boundary condition. Availing ourselves of the observations at the beginning of Section 4.5 we select a cover of Ω consisting of a finite number of interior patches $V_1^i, V_2^i, \ldots$ and boundary patches $V_1^b, V_2^b, \ldots$ and denote by $\chi_1^i, \chi_2^i, \ldots, \chi_1^b, \chi_2^b, \ldots$ the partition of unity subordinated to the cover $\{V_j^i, V_k^b\}$. Clearly we can choose both the interior and boundary patches so small that if x and x' belong to one of them,

$$|a_{jk}(x) - a_{jk}(x')| \leqslant \varepsilon, \tag{4.8.20}$$

where $\varepsilon > 0$ is a constant to be specified later. We examine the situation in each patch.

(a) *Interior patches.* At one of these, say $V^i = \mathbb{S}^m(\bar{x}, \rho)$, the function $u^i = \chi^i u$ satisfies

$$A_0 u^i = g^i, \tag{4.8.21}$$

where

$$g^i = \chi^i A_0 u + \left(\sum \sum a_{jk} D^j D^k \chi^i + \sum b_j D^j \chi^i \right) u + 2 \sum \sum a_{jk} D^j \chi^i D^k u. \tag{4.8.22}$$

Denote by $A_{0,\bar{x}}$ the principal part of the operator A_0 frozen at $\bar{x}$,

$$A_{0,\bar{x}} u = \sum_{j=1}^{m} \sum_{k=1}^{m} a_{jk}(\bar{x}) D^j D^k u. \tag{4.8.23}$$

We perform an affine change of variables $x \to \eta = L(x - \bar{x})$ in V^i in such a way that the quadratic form $\sum\sum a_{jk}(\bar{x})\xi_j \xi_k$ becomes $\sum \xi_j^2$ in the new variables; clearly this change can be made in such a way that

$$|D^j \eta_k|, |D^j \eta_k^{-1}| \leqslant N, \tag{4.8.24}$$

where N depends only on the ellipticity constant κ and the maximum of the $|a_{jk}|$. We can then write (4.8.21) in the new coordinates as follows:

$$\tilde{A}_0 \tilde{u}^i = \Delta \tilde{u}^i + \sum \sum \left(\tilde{a}_{jk}(\eta) - \tilde{a}_{jk}(0) \right) D^j D^k \tilde{u}^i + \sum \tilde{b}_j D^j \tilde{u}^i + \tilde{c} \tilde{u}^i = \tilde{g}^i, \tag{4.8.25}$$

where $\tilde{A}_0$ is the operator A_0 in the variables $\eta_1, \ldots, \eta_m$ (see (4.5.21)) and $\tilde{u}^i, \tilde{g}^i$ are u^i, g^i as functions of the η_j. We choose now the constant ε in (4.8.20) in such a way that $|\tilde{a}_{jk}(\eta) - \tilde{a}_{jk}(0)| \leqslant 1/2m^2 C_0$ when $\eta \in \tilde{V}^i = L(V^i - \bar{x})$, C_0 the constant in Theorem 4.8.6 (this is clearly possible since (4.8.24) holds independently of $\bar{x}$ and the a_{jk} are uniformly continuous in $\bar{\Omega}$). We make then use of (4.8.25) and of Theorem 4.8.6 to conclude that

$$\|\tilde{u}^i\|_{W^{2,p}(\tilde{V}^i)} \leqslant \tfrac{1}{2} \|\tilde{u}^i\|_{W^{2,p}(\tilde{V}^i)} + C' \left(\|\tilde{g}^i\|_{L^p(\tilde{V}^i)} + \|\tilde{u}^i\|_{W^{1,p}(\tilde{V}^i)} \right), \tag{4.8.26}$$

where C' is a positive constant. We obtain from this inequality that

$$\|\tilde{u}^i\|_{W^{2,p}(\tilde{V}^i)} \leqslant C''\left(\|\tilde{g}^i\|_{L^p(\tilde{V}^i)} + \|\tilde{u}^i\|_{W^{1,p}(\tilde{V}^i)}\right).$$

Hence, after a change of variables, an examination of (4.8.22) and an application of Lemma 4.7.5,

$$\|\chi^i u\|_{W^{2,p}(\Omega)} = \|\chi^i u\|_{W^{2,p}(V^i)} \leqslant C\left(\|A_0 u\|_{L^p(V^i)} + \|u\|_{W^{1,p}(V^i)}\right)$$
$$\leqslant C\left(\|A_0 u\|_{L^p(\Omega)} + \|u\|_{W^{1,p}(\Omega)}\right), \qquad (4.8.27)$$

where it is easy to see that C depends only on the constant N in (4.8.24), on the constant C_0, and on the maxima of the $|a_{jk}|$, the $|b_j|$ and $|c|$ on $\bar{\Omega}$ (see equality (4.8.25) and its consequence (4.8.26)).

   (b) *Boundary patches.*   At one of these (strictly speaking, in $V^b \cap \Omega$) the function $u^b = \chi^b u$ satisfies

$$A_0 u^b = g^b, \qquad (4.8.28)$$

where g^b can be expressed from χ^b and u in the same way as g^i from χ^i and u in (4.8.22), and

$$u^b(x) = 0 \quad (x \in \Gamma \cap V^b). \qquad (4.8.29)$$

Let now η be the map associated to the patch V^b and let $\tilde{A}_0$ be the operator A_0 in the variables $\eta_1, \dots, \eta_m$. Then $\tilde{A}_0$ is uniformly elliptic in $\mathbb{S}^m_+ = \mathbb{S}^m_+(0, 1)$ (with ellipticity constant $\tilde{\kappa}$ depending on κ and η as seen in (4.7.25) and following remarks). As in the case of A_0, we define $\tilde{A}_{0,0}$ to be the principal part of the operator $\tilde{A}_0$ frozen at $0 = \eta(\bar{x})$:

$$\tilde{A}_{0,0} = \sum_{j=1}^{m} \sum_{k=1}^{m} \tilde{a}_{jk}(0) D^j D^k. \qquad (4.8.30)$$

We perform now two additional changes of variables. The first is a linear change $\eta \to \xi' = L\xi$ reducing the quadratic form $\sum\sum \tilde{a}_{jk}(0)\xi_j\xi_k$ to canonical form $\sum \xi_j^2$; clearly (4.8.24) and following comments hold for L, the operator $\tilde{A}_{0,0}$ transforms into Δ and the hyperplane $\eta_m = 0$ into some hyperplane H. Finally, we select an orthogonal transformation $\eta' \to \hat{\eta} = U\eta'$, which transforms H back into $\hat{\eta}_m = 0$ (since the transformation is orthogonal, the principal part of the transformed operator will still be Δ). We consider then the total change of variables $\hat{\eta} = UL\eta$ under which the equation (4.8.30) transforms (in obvious notation) into

$$\hat{A}_0 \hat{u}^b = \hat{g}^b \quad (\hat{\eta}_m \geqslant 0), \qquad (4.8.31)$$
$$\hat{u}^b = 0 \quad (\hat{\eta}_m = 0), \qquad (4.8.32)$$

with the support of $\hat{u}$ contained in $|\hat{\eta}| \leqslant M$ (M only depending on χ^b, L). We can argue then essentially as after (4.8.25) (but using now Corollary 4.8.7) to obtain

$$\|\chi^b u\|_{W^{2,p}(\Omega)} \leqslant C\left(\|A_0 u\|_{L^p(\Omega)} + \|u\|_{W^{1,p}(\Omega)}\right). \qquad (4.8.33)$$

We note that at the heart of this argument lies the fact that there exists an absolute constant N such that Γ can be covered with boundary patches of *arbitrarily small diameter* (chosen independently of N) and such that $|D^j\eta_k|$, $|D^j\eta_k^{-1}| \leqslant N$ for the associated maps η. This follows easily from compactness of Γ and will also be used in the treatment of other boundary conditions.

We deal next with boundary conditions of type (I). The treatment of interior patches is obviously the same, thus we only have to figure out what happens at the boundary patches. We take them so small that if x and x' belong to one of them we have

$$|a_{jk}(x) - a_{jk}(x')|, |D^i a_{jk}(x) - D^i a_{jk}(x')| \leqslant \varepsilon. \qquad (4.8.34)$$

The constant $\varepsilon > 0$ will be specified later.

(c) *Boundary patches* (*boundary conditions of type* (I)). We use again here the change of variables $\hat{\eta}$ (but drop the hat from η for simplicity of notation); as before, $\hat{}$ over a function denotes the action of the change of variables. We argue as in (b) but we must examine carefully the (local) transformation of the boundary condition β. Because of the invariance of the conormal derivative (see (4.5.28)), the function $\hat{u}$ satisfies

$$D^{\tilde{n}}\hat{u}^b(\eta) = \rho(\eta)\hat{\gamma}(\eta)\hat{u}^b(\eta) \quad (\eta_m = 0), \qquad (4.8.35)$$

where $\rho(\eta)$ is the normalization factor in (4.5.28) and $\tilde{n}$ is the conormal vector (with respect to $\hat{A}_0$) at the corresponding point of the hyperplane $\eta_m = 0$. Now, since the principal part of $\hat{A}_0$ at $\eta = 0$ is the Laplacian, $\tilde{n} = (0,\ldots,0,-1)$ and $D^{\tilde{n}} = -D^m$ at $\eta = 0$. We have

$$-D^m\hat{u}^b(\eta) = -(D^{\tilde{n}} + D^m)\hat{u}^b(\eta) + \rho(\eta)\hat{\gamma}(\eta)\hat{u}^b(\eta)$$
$$= h_1(\eta) + h_2(\eta) \quad (\eta_m = 0), \qquad (4.8.36)$$

where $D^{\tilde{n}}\hat{u}^b(\eta) = -\Sigma\hat{a}_{jm}(\eta)D^j\hat{u}^b(\eta)$. We extend h_1 and h_2 to $\eta_m > 0$ setting $\hat{\gamma}(\eta_1,\ldots,\eta_{m-1},\eta_m) = \hat{\gamma}(\eta_1,\ldots,\eta_{m-1},0)$ and continue ρ in the same fashion. Then, taking ε sufficiently small in (4.8.34) we can assure that

$$\|h_1\|_{W^{1,p}(\mathbf{R}_+^m)} = \|(D^{\tilde{n}} + D^m)\hat{u}^b\|_{W^{1,p}(\mathbf{R}_+^m)} \leqslant (1/4C_0)\|\hat{u}^b\|_{W^{2,p}(\mathbf{R}_+^m)},$$

$$(4.8.37)$$

where C_0 is the constant in Theorem 4.8.8. On the other hand, there obviously exists a constant C' such that

$$\|h_2\|_{W^{1,p}(\mathbf{R}_+^m)} \leqslant \|\rho\hat{\gamma}\hat{u}^b\|_{W^{1,p}(\mathbf{R}_+^m)} \leqslant C'\|\hat{u}^b\|_{W^{1,p}(\mathbf{R}_+^m)}. \qquad (4.8.38)$$

We combine the last two inequalities with (4.8.25) and Theorem 4.8.8 and obtain an inequality of the type of (4.8.26) for $\tilde{u}^b$. After the inverse change of variables, (4.8.33) results, the constant C satisfying the stipulations in the statement of Theorem 4.8.4. Putting together inequalities (4.8.27) for the interior patches and (4.8.33) for the boundary patches, we obtain (4.8.10), thus completing the proof of Theorem 4.8.4.

Inequality (4.8.10) can be cast into a more useful form if the boundary condition β is of type (II) or if β is of type (I) and the dissipativity condition (4.8.4) is satisfied for the p in question. We shall use the following result:

4.8.9 Lemma. *Let Ω be a bounded domain of class $C = C^{(1)}$, $1 < p < \infty$, $\varepsilon > 0$. Then there exists a constant C depending only on Ω, p, ε such that*

$$\|u\|_{W^{1,p}(\Omega)} \leqslant \varepsilon \|u\|_{W^{2,p}(\Omega)} + C\|u\|_{L^p(\Omega)}.$$

For a proof (under considerably weaker hypotheses) see Adams [1975: 1, p. 75], where a more general result involving higher order Sobolev norms can also be found.

We make use of Lemma 4.8.9 with $\varepsilon < 1/C$, C the constant in (4.8.10), and obtain (with a different constant)

$$\|u\|_{W^{2,p}(\Omega)} \leqslant C\big(\|A_0 u\|_{L^p(\Omega)} + \|u\|_{L^p(\Omega)}\big). \tag{4.8.39}$$

We note next that $A_0 - \omega_p I$ is dissipative if ω_p is given by (4.8.5); therefore, if $\lambda > \omega_p$,

$$\|u\|_{L^p(\Omega)} \leqslant \big(\lambda - \omega_p\big)^{-1}\|(\lambda I - A_0)u\|_{L^p(\Omega)}$$

(see (3.1.9)). Combining these two last inequalities, we obtain, again with a different constant,

$$\|u\|_{W^{2,p}(\Omega)} \leqslant C\|(\lambda I - A_0)u\|_{L^p(\Omega)} \tag{4.8.40}$$

(λ large enough) for $u \in C^{(2)}(\overline{\Omega})_\beta$. In view of Lemma 4.8.1, inequality (4.8.40) (as well as (4.8.10), of course) can be extended to $u \in W^{2,p}(\Omega)_\beta$ using an obvious approximation argument.

The main application of Theorem 4.8.4 will be a precise characterization of the domain of $A_p(\beta)$, the closure of $A_0(\beta)$ in L^p, $1 < p < \infty$. As a bonus, we shall also obtain a similar characterization in the $C(\overline{\Omega})$ case.

4.8.10 Corollary. *Let Ω, A_0 satisfy the assumptions in Theorem 4.8.3. Then*

$$D\big(A_p(\beta)\big) = D\big(\overline{A_0(\beta)}\big) = W^{2,p}(\Omega)_\beta,$$

where $W^{2,p}(\Omega)_\beta$ is the space of all $u \in W^{2,p}(\Omega)$ that satisfy the boundary condition β at Γ in the sense of Lemma 4.8.1.

Proof. The fact that β makes sense for any $u \in W^{2,p}(\Omega)$ has been already established (see Lemma 4.8.1); it is easy to see that $W^{2,p}(\Omega)_\beta$ is a closed subspace of $W^{2,p}(\Omega)$. We make now use of the fact that $C^{(2)}(\overline{\Omega})_\beta = D(A_0(\beta))$ is dense in $W^{2,p}(\Omega)_\beta$ (Lemma 4.8.1) and of the equivalence of the $W^{2,p}(\Omega)$-norm and the graph norm $\|(\lambda I - A_0)u\|_{L^p(\Omega)}$ in $D(A_0(\beta))$, which is an easy consequence of (4.8.40). The remaining details are very similar to those for the case $p = 2$ (Corollary 4.7.8) and are left to the reader.

We note that $A_p(\beta)$ acts on elements of its domain in the classical sense, that is, (4.7.13) holds.

We can also identify in a fairly satisfactory way the domain of the closure of $A_0(\beta)$ in $C(\overline{\Omega})$ ($C_\Gamma(\overline{\Omega})$ if β is the Dirichlet boundary condition).

4.8.11   Theorem.   *Let the assumptions of Theorem 4.8.3 be satisfied for $p = \infty$. Assume that (4.8.4) holds for $p = 1, \infty$, and let $A_\infty(\beta)$ be the closure of $A_0(\beta)$ in $C(\overline{\Omega})$ (in $C_\Gamma(\overline{\Omega})$ if β is of type (II)). Then $D(A_\infty(\beta))$ consists of all $u \in \cap_{p \geqslant 1} W^{2,p}(\Omega)_\beta$ such that*[12]

$$A_0(\beta)u \in C(\overline{\Omega}) \quad (C_\Gamma(\overline{\Omega}) \text{ if } \beta \text{ is of type (II)}). \tag{4.8.41}$$

Every $u \in D(A_\infty(\beta))$ belongs to $C^{(1)}(\overline{\Omega})$ and satisfies the boundary condition β in the classical sense everywhere on Γ. The following alternate characterization of $A_\infty(\beta)$ also holds: $D(A_\infty(\beta))$ consists of all $u \in C(\overline{\Omega})$ $(C_\Gamma(\overline{\Omega}))$ such that $A_\infty(\beta)$, understood in the sense of distributions, belongs to $C(\overline{\Omega})$ $(C_\Gamma(\overline{\Omega}))$, i.e., such that there exists $v = A_\infty(\beta)u$ in $C(\overline{\Omega})$ $(C_\Gamma(\overline{\Omega}))$ with

$$\int_\Omega uA_0'(\beta')w\,dx = \int_\Omega vw\,dx \quad \left(w \in D(A_0'(\beta')) = C^{(2)}(\overline{\Omega})_{\beta'}\right).$$

The proof is based on the #-adjoint theory of Section 2.3. This theory will be applied to the space $E = L^1(\Omega)$ and the operator $A = A_1'(\beta') = A_0'(\beta')$. The first task is to identify the space $E^\# = D(A_1'(\beta')^*) = D(A_0'(\beta')^*)$ in $E^* = L^\infty(\Omega)$, where A_0' is the formal adjoint of A_0 and β' is the adjoint boundary condition. We shall presently show that

$$E^\# = \begin{cases} C(\overline{\Omega}) \text{ if } \beta \text{ is of type (I),} \\ C_\Gamma(\overline{\Omega}) \text{ if } \beta \text{ is of type (II).} \end{cases} \tag{4.8.42}$$

Consider first the case of Dirichlet boundary conditions. Application of the Lagrange identity (4.5.40) shows that $A_0(\beta) \subseteq A_0'(\beta')^*$ so that $D(A_0(\beta)) = C^{(2)}(\overline{\Omega}) \cap C_\Gamma(\overline{\Omega}) \subseteq E^\#$ and, taking closures, $C_\Gamma(\overline{\Omega}) \subseteq E^\#$. The opposite inclusion is not nearly as trivial. Let $u \in D(A_0'(\beta')^*)$ (as before, $A_0'(\beta')$ considered as an operator in $L^1(\Omega)$). Then there exists $f \in L^\infty(\Omega)$ such that

$$\int_\Omega uA_0'(\beta')w\,dx = \int_\Omega fw\,dx \quad (w \in D(A_0'(\beta'))). \tag{4.8.43}$$

But this implies that $u \in D(A_p'(\beta')^*) = D(A_0'(\beta')^*)$, with $A_0'(\beta')$ thought of as an operator in $L^p(\Omega)$, $p > 1$ and $A_0'(\beta')^*u = f \in L^p(\Omega)$. Since $A_0'(\beta')^* = A_0(\beta) = A_p(\beta)$, it follows from the previous Corollary 4.8.10 that $u \in W^{2,p}(\Omega)_\beta$ for all $p \geqslant 1$. Taking $p > m$, we deduce that $u \in C^{(1)}(\overline{\Omega})$ from Theorem 4.7.15; since u satisfies the boundary condition β on Γ in the generalized sense of Lemma 4.8.1, it is not difficult to show that u must

---

[12] The definition of A_0 is slightly stretched here.

satisfy β in the classical sense. As each $u \in D(A_0'(\beta')^*)$ belongs then to $C_\Gamma(\bar{\Omega})$ the same must be true of each $u \in E^\#$ (the L^∞ and the C norm coincide in $C(\bar{\Omega})$) and the proof of the second equality in (4.8.42) is complete. The first can be handled in an entirely similar way and details are omitted.

The identification of $A_1'(\beta')^\# = \overline{A_0'(\beta')}^\#$ is essentially the same as that of $A_1'(\beta')^*$ examined above (with the only difference that $f \in C_\Gamma(\bar{\Omega})$ or $C(\bar{\Omega})$ instead of L^∞). In particular, every $u \in D(A_0(\beta)^\#)$ satisfies the stipulations set forth in Theorem 4.8.11. Conversely, let u be a function enjoying these privileges. Then it is obvious that u must satisfy (4.8.43) with $f = A_0(\beta)u$ and thus belongs to $D(A_0'(\beta')^\#)$ with $A_0'(\beta')^\# u = f$.

We note finally that since $A_0(\beta) \subseteq A_0'(\beta')^\#$ in $C_\Gamma(\bar{\Omega})$ or $C(\bar{\Omega})$, we must have $A_\infty(\beta) = \overline{A_0(\beta)} \subseteq A_0'(\beta')^\#$, hence using once again the argument at the end of Lemma 2.1.3 we deduce that

$$A_\infty(\beta) = A_1'(\beta')^\# = A_0'(\beta')^\# \tag{4.8.44}$$

(Note that, strictly speaking, the expression $A_0'(\beta')^\#$ does not make sense since $A_0'(\beta')$ itself is not a semigroup generator; however, the meaning is clear in the present context.)

The computation of the $\#$-adjoint of $A_\infty(\beta)$ is considerably more complex and will require additional L^p estimates. We shall only consider the Dirichlet boundary condition.

Let $f_0, f_1, \ldots, f_m \in L^p(\Omega)$ for some $p \geq 1$. We shall examine in what follows solutions of the equation

$$A_0 u = f_0 + \sum D^j f_j \tag{4.8.45}$$

satisfying a boundary condition of type (II). The main result on this score is the following Theorem 4.8.12, where both u and f_j are smooth and the solution of (4.8.45) is therefore understood in the classical sense.

4.8.12 Theorem. *Let Ω be a bounded domain of class $C^{(2)}$ and let A_0 be given by (4.8.8) with a_{jk}, b_j continuously differentiable, c continuous in $\bar{\Omega}$. Let β be the Dirichlet boundary condition. Finally, let $1 < p < \infty$. Then there exists a constant C (depending only on p, κ, Ω, the modulus of continuity of the a_{jk}, $D^i a_{jk}$ and the maximum of $|a_{jk}|$, $|D^i a_{jk}|$, $|b_j|$, $|D^i b_j|$, $|c|$ on $\bar{\Omega}$) such that if $u \in C^{(2)}(\bar{\Omega})_\beta$ and (4.8.45) holds with $f_0 \in C(\bar{\Omega})$, $f_1, \ldots, f_m \in C^{(1)}(\bar{\Omega})$, we have*

$$\|u\|_{W^{1,p}(\Omega)} \leq C \left(\sum_{j=0}^m \|f_j\|_{L^p(\Omega)} + \|u\|_{L^p(\Omega)} \right). \tag{4.8.46}$$

The proof runs along the "local reduction to the Laplacian" lines of Theorem 4.8.4, thus we only sketch the main points leaving some details to the reader. For the local treatment of the Dirichlet boundary condition, we need analogues of Theorem 4.8.6 and Corollary 4.8.7. These are:

4.8.13   Theorem.   *Let u be twice continuously differentiable, f_0 continuous, $f_1, \ldots, f_m$ continuously differentiable, all functions will support in $|x| \leqslant M$, and let $1 < p < \infty$. Assume that*

$$\Delta u = f_0 + \sum D^j f_j. \tag{4.8.47}$$

Then there exists a constant C_0 depending only on p and M such that

$$\|u\|_{W^{1,p}(\mathbb{R}^m)} \leqslant C_0 \sum_{j=0}^{m} \|f_j\|_{L^p(\mathbb{R}^m)}. \tag{4.8.48}$$

Proof.   It follows from (4.8.47) that

$$\mathcal{F}(D^k u) = \frac{i\sigma_k}{|\sigma|^2} \mathcal{F}f_0(\sigma) + \sum_{j=1}^{m} \frac{\sigma_j \sigma_k}{|\sigma|^2} \mathcal{F}f_j(\sigma) = I_k f_0 + \sum_{j=1}^{m} I_{jk} f_j. \tag{4.8.49}$$

Making use of Theorem 4.8.5, we obtain

$$\|I_{jk} f_j\|_{L^p(\mathbb{R}^m)} \leqslant C_{jk} \|f_j\|_{L^p(\mathbb{R}^m)} \tag{4.8.50}$$

with C_{jk} only depending on p. To estimate I_k for $m \geqslant 3$, we note that if ω_m is the hyperarea of the unit sphere in R^m,

$$I_k f_0(x) = \frac{1}{(2-m)\omega_m} \int_{\mathbb{R}^m} \left\{ D_x^k \frac{1}{|x-y|^{m-2}} \right\} f_0(y) \, dy$$

(this follows essentially from Stein [1957: 1, p. 126]; for the case $m = 2$ see Example 4.8.23), and bound this integral using Young's inequality as in the end of the proof of Theorem 4.8.6, obtaining

$$\|I_k f_0\|_{L^p(\Omega)} \leqslant C_k \|f_0\|_{L^p(\Omega)}, \tag{4.8.51}$$

where C_k only depends on p and M. The desired inequality is now obtained from this one and (4.8.50), the L^p norm of u estimated in terms of the L^p norm of its derivatives again as in Theorem 4.8.6.

4.8.14   Corollary.   *Let $u \in C^{(2)}(\overline{\mathbb{R}}_+^m)$, $f_0 \in C(\overline{\mathbb{R}}_+^m)$, $f_1, \ldots, f_m \in C^{(1)}(\overline{\mathbb{R}}_+^m)$ be functions with support in $|x| \leqslant M$. Assume that u satisfies (4.8.47) in $x_m > 0$ and that $u(x) = 0$ for $x_m = 0$. Then there exists $C_0 > 0$ depending only on p, M such that*

$$\|u\|_{W^{1,p}(\mathbb{R}_+^m)} \leqslant C_0 \sum_{j=0}^{m} \|f_j\|_{L^p(\mathbb{R}_+^m)}. \tag{4.8.52}$$

Proof.   We extend u to the entire space setting $u(\hat{x}, x_m) = -u(\hat{x}, -x_m)$ $(x_m < 0)$, where $\hat{x} = (x_1, \ldots, x_{m-1})$ as in the proof of Corollary 4.8.7. The same extension is used for $f_0, \ldots, f_{m-1}$; for f_m we use instead the even extension $f_m(\hat{x}, x_m) = f(\hat{x}, -x_m)$. Although the extended functions satisfy (4.8.47) off the hyperplane $x_m = 0$, they do not fit into the hypotheses of Theorem 4.8.13 since $(D^m)^2 u$ and $f_0, D^1 f_1, \ldots, D^m f_m$ are discontinuous across $x_m = 0$. However, the Fourier transform argument is easily seen to extend to the present situation.

We prove Theorem 4.8.12 by means of a cover of Ω by interior and boundary patches. The notations in the proof of Theorem 4.8.4 will be freely used here.

(a) *Interior patches.* At one of these the function $u^i = \chi^i u$ satisfies

$$A_0 u^i = g^i, \tag{4.8.53}$$

where

$$
\begin{aligned}
g^i &= \chi^i\Big(f_0 + \sum D^j f_j\Big) + \Big(\sum\sum a_{jk} D^j D^k \chi^i + \sum b_j D^j \chi^i\Big) u \\
&\quad + 2\sum\sum a_{jk} D^j \chi^i D^k u \\
&= \chi^i f_0 - \Big(\sum D^j \chi^i\Big) f_j - \Big(\sum\sum a_{jk} D^j D^k \chi^i - \sum b_j D^j \chi^i\Big) u \\
&\quad - \Big(2\sum\sum D^k a_{jk} D^j \chi^i\Big) u + \sum D^j\big(\chi^i f_j\big) + 2\sum D^k\Big(\big(\sum a_{jk} D^j \chi^i\big) u\Big) \\
&= g_0 + \sum D^j g_j.
\end{aligned}
$$

After the affine change of variables $x \to \eta = L(x - \bar{x})$ necessary to transform the principal part of A_0 at $\bar{x}$ into the Laplacian, we obtain

$$
\begin{aligned}
\tilde{A}_0 \tilde{u}^i &= \Delta \tilde{u}^i + \sum\sum \big(\tilde{a}_{jk}(\eta) - \tilde{a}_{jk}(0)\big) D^j D^k \tilde{u}^i + \sum \tilde{b}_j D^j \tilde{u}^i + c\tilde{u}^i \\
&= \tilde{g}_0 + \sum\sum l_{kj} D^k \tilde{g}_j,
\end{aligned}
$$

where $\langle l_{jk}\rangle = L$. Observing that

$$
\begin{aligned}
\sum\sum \big(\tilde{a}_{jk}(\eta) - \tilde{a}_{jk}(0)\big) D^j D^k \tilde{u}^i &= \sum\sum D^j\big\{\big(\tilde{a}_{jk}(\eta) - \tilde{a}_{jk}(0)\big) D^k \tilde{u}^i\big\} \\
&\quad - \sum\sum \big(D^j \tilde{a}_{jk}(\eta) - D^j \tilde{a}_{jk}(0)\big) D^k \tilde{u}^i \\
&\quad - \sum D^j \tilde{a}_{jk}(0) D^k \tilde{u}^i
\end{aligned}
$$

and that

$$\sum \tilde{b}_j D^j \tilde{u}^i = \sum D^j\big(\tilde{b}_j \tilde{u}^i\big) - \Big(\sum D^j \tilde{b}_j\Big) \tilde{u}^i,$$

we see that, if V^i is small enough,

$$\|\tilde{u}^i\|_{W^{1,p}(\tilde{V}^i)} \leqslant \tfrac{1}{2}\|\tilde{u}^i\|_{W^{1,p}(\tilde{V}^i)} + C'\left(\sum_{j=0}^{m} \|\tilde{g}^i\|_{L^p(\tilde{V}^i)} + \|\tilde{u}^i\|_{L^p(\tilde{V}^i)}\right) \tag{4.8.54}$$

from which we obtain

$$\|\chi^i u\|_{W^{1,p}(\Omega)} \leqslant C\left(\sum_{j=0}^{m} \|f_j\|_{L^p(\Omega)} + \|u\|_{L^p(\Omega)}\right). \tag{4.8.55}$$

(b) *Boundary patches.* The treatment is much the same, the only difference being that the change of coordinates $x \to \eta$ is now in general nonlinear; the function $g_0 + \sum D^j g_j$ transforms then into $\tilde{g}_0 + \sum\sum l_{kj} D^k \tilde{g}_k$, $\{l_{jk}\}$ the Jacobian matrix of $\eta(x)$. We write this last expression in the form $\tilde{g}_0 + \sum D^k (\sum l_{jk} \tilde{g}_k) - (\sum\sum D^k l_{jk}) g_k$ and proceed as above, applying this time Corollary 4.8.14.

We shall need to define below what is meant by a solution of (4.8.45) when $u, f_0, f_1, \ldots, f_m$ are no longer smooth. This will be done in the style of Section 4.6: given $f_0, f_1, \ldots, f_m \in L^1(\Omega)$ (Ω an arbitrary domain) a function $u \in W^{1,1}(\Omega)$ is declared a *solution* of (4.8.45) satisfying the Dirichlet boundary condition if and only if $u \in W_0^{1,1}(\Omega)$ and

$$\int_\Omega \left\{ \sum\sum a_{jk} D^j u D^k \varphi - \left(\sum \hat{b}_j D^j u \right) \varphi - c u \varphi \right\} dx$$

$$= - \int_\Omega \left(f_0 \varphi - \sum f_j D^j \varphi \right) dx \quad (\varphi \in \mathcal{D}(\Omega))^{13} \qquad (4.8.56)$$

4.8.15   Theorem.   *Let Ω be a bounded domain of class $C^{(k+2)}$, A an operator of class $C^{(k+1)}$ with $k > m/2$ satisfying the uniform ellipticity assumption (4.8.9),[14] β a boundary condition of type (II). Let $1 < p < \infty$, $f_0, f_1, \ldots, f_m \in L^p(\Omega)$, $\lambda > 0$ large enough. Then there exists a (unique) solution[15] $u \in W_0^{1,p}(\Omega)$ of*

$$(\lambda I - A_0) u = f_0 + \sum D^j f_j \qquad (4.8.57)$$

Moreover, there exists a constant $C > 0$ independent of $f_0, f_1, \ldots, f_m$ such that

$$\|u\|_{W^{1,p}(\Omega)} \leqslant C \sum_{j=0}^m \|f_j\|_{L^p(\Omega)}. \qquad (4.8.58)$$

The proof hinges upon the case $p = 2$. Here the solution is constructed by means of the sesquilinear functional B_λ in Section 4.6 in the space $H_0^1(\Omega) = W_0^{1,2}(\Omega)$. It is sufficient to observe that the linear functional

$$\Phi(v) = \int_\Omega \left(f_0 v - \sum f_j D^j v \right) dx \qquad (4.8.59)$$

is continuous in $H^1(\Omega)$ and apply Lemma 4.6.1 to B_λ in $H_0^1(\Omega)$. We deduce that the solution satisfies the inequality

$$\|u\|_{W_0^{1,2}(\Omega)} \leqslant C \sum_{j=0}^m \|f_j\|_{L^2(\Omega)}. \qquad (4.8.60)$$

The next step is to extend this result to $p \geqslant 2$. To do this, another particular case of the Sobolev embedding theorem will be needed.

---

[13] Here $\hat{b}_1, \ldots, \hat{b}_m$ are the first order coefficients of the operator A_0 in divergence form (see (4.4.3)) and $W_0^{j,p}(\Omega) = \overline{\mathcal{D}(\Omega)} \subseteq W^{j,p}(\Omega)$.

[14] The concept of operator of class $C^{(k)}$ was only defined (in Section 4.7) for operators written in divergence form (4.4.2). Of course, the definition is the same for operators written in the form (4.8.8).

[15] See footnote 12, p. 244.

4.8.16 Theorem. *Let* $\Omega \subseteq \mathbb{R}^m$ *be a bounded domain of class* $C^{(1)}$, $p \geqslant 1$, k *an integer* $\geqslant 1$. (a) *If* $kp < m$, *then every* $u \in W^{k,p}(\Omega)$ *belongs to* $L^q(\Omega)$ *if*

$$1 \leqslant q \leqslant mp/(m - kp).$$

(b) *If* $kp = m$, *then every* $u \in W^{k,p}(\Omega)$ *belongs to* $L^q(\Omega)$ *for*

$$1 \leqslant q < \infty.$$

In both cases there exists a constant $C = C(p, q, \Omega)$ *such that*

$$\|u\|_{L^q(\Omega)} \leqslant C\|u\|_{W^{k,p}(\Omega)}. \tag{4.8.61}$$

For a proof see Adams [1975: 1, p. 97].[16]

We continue with the proof of Theorem 4.8.15. Let $p > 2$, $f_0, f_1, \ldots, f_m$ elements of $L^p(\Omega)$. Since $L^p(\Omega) \subset L^2(\Omega)$, there exists $u \in W_0^{1,2}(\Omega)$ satisfying (4.8.57) and (4.8.60). Consider m sequences $\{f_0^{(n)}\}, \ldots, \{f_m^{(n)}\}$ of smooth functions (say, in $\mathcal{D}(\Omega)$) such that

$$\|f_j^{(n)} - f_j\|_{L^p(\Omega)} \to 0 \quad (0 \leqslant j \leqslant m) \tag{4.8.62}$$

as $n \to \infty$. Since $f_0 + \Sigma D^j f_j \in \mathcal{D}(\Omega)$, it follows from Corollary 4.7.16 that the solution of

$$(\lambda I - A_0) u^{(n)} = f_0^{(n)} + \sum D^j f_j^{(n)}$$

in $W_0^{1,2}(\Omega)$ actually belongs to $C^{(2)}(\bar{\Omega}) \cap C_\Gamma(\bar{\Omega})$ so that Theorem 4.8.12 applies. We use now inequality (4.8.60) for $u^{(n)} - u^{(k)}$ deducing that $\{u^{(n)}\}$ is a Cauchy sequence in $W_0^{1,2}(\Omega)$, hence, by Theorem 4.8.17, in $L^p(\Omega)$ for all $p \geqslant 2$ if $m = 2$ or for

$$1 \leqslant p \leqslant 2m/(m - 2) \tag{4.8.63}$$

if $m \geqslant 3$. Combining this with inequality (4.8.46) (for the operator $A_0 - \lambda I$), we deduce that $\{u^{(n)}\}$ is a Cauchy sequence in the space $W_0^{1,p}(\Omega)$. Taking limits in (4.8.56), we obtain a solution of (4.8.57) in $W_0^{1,p}(\Omega)$, but we still must show that u satisfies (4.8.58) instead of the weaker inequality (4.8.46). To this end we consider the "solution operator" of the equation (4.8.57),

$$\mathfrak{S}(f_0, f_1, \ldots, f_m) = u$$

mapping $L^p(\Omega)^{m+1}$ into $W^{1,p}(\Omega)$ in the range of p given by (4.8.63). Obviously $\mathfrak{S}$ is well defined (uniqueness of u for $p > 2$ follows from uniqueness for $p = 2$) and it follows immediately from (4.8.46) and (4.8.56) that $\mathfrak{S}$ is closed; hence we obtain from the closed graph theorem (Section 1) that $\mathfrak{S}$ is bounded, so that (4.8.58) holds. A completely similar argument pivoting on $p = 2m/(m - 2)$ extends the result to the range $p \geqslant 2$ if $m \leqslant 4$

---

[16] This result is proved there for arbitrary domains satisfying the *cone property* (Adams [1975: 1, pp. 76 and 66]).

or

$$2 \leqslant p \leqslant 2m/(m-4)$$

if $m \geqslant 5$; arguing repeatedly in the same way, we prove Theorem 4.8.16 in the range $p \geqslant 2$ for arbitrary values of m.

The case $1 < p \leqslant 2$ (which, incidentally, is the one of interest for us) is handled by means of a duality argument. Let $f_0, f_1, \ldots, f_m$ be again smooth functions so that the solution $u \in W_0^{1,2}(\Omega)$ of (4.8.57) belongs to $C^{(2)}(\overline{\Omega}) \cap C_\Gamma(\overline{\Omega})$ and let $g_0, g_1, \ldots, g_m$ be arbitrary functions in $L^{p'}(\Omega)$ ($p'^{-1} + p^{-1} = 1$). Since $p' \geqslant 2$, if λ is large enough there exists a (unique) $v \in W_0^{1,p'}(\Omega)$ such that

$$(\lambda I - A_0')v = g_0 + \sum D^j g_j \qquad (4.8.64)$$

(where A_0' is the formal adjoint of A_0) satisfying the estimate

$$\|v\|_{W_0^{1,p'}(\Omega)} \leqslant C \sum_{j=0}^{m} \|g_j\|_{L^{p'}(\Omega)}. \qquad (4.8.65)$$

An approximation argument shows that (4.8.56) can be used for functions $\varphi \in C^{(1)}(\overline{\Omega}) \cap C_\Gamma(\overline{\Omega})$. Taking $\varphi = u$, we obtain

$$\int_\Omega \left(g_0 u - \sum g_j D^j u \right) dx$$

$$= \int_\Omega \left\{ \sum\sum a_{jk} D^j v D^k u + \left(\sum D^j (b_j v) \right) u \right) + (\lambda - c) vu \right\} dx$$

$$= \int_\Omega \left\{ \sum\sum a_{jk} D^j v D^k u - v \left(\sum b_j D^j u \right) + (\lambda - c) vu \right\} dx$$

$$= \int_\Omega \left(f_0 v - \sum f_j D^j v \right) dx, \qquad (4.8.66)$$

hence, using Hölder's inequality,

$$\left| \int_\Omega \left(g_0 u + \sum g_j D^j u \right) dx \right| \leqslant \left(\sum_{j=0}^{m} \|f_j\|_{L^p(\Omega)} \right) \|v\|_{W_0^{1,p'}(\Omega)}$$

$$\leqslant C \left(\sum_{j=0}^{m} \|f_j\|_{L^p(\Omega)} \right) \left(\sum_{j=0}^{m} \|g_j\|_{L^{p'}(\Omega)} \right) \qquad (4.8.67)$$

by (4.8.65). Since the $g_0, \ldots, g_m$ are arbitrary, inequality (4.8.58) is obtained for $u, f_0, \ldots, f_m$.

Let now $f_0, f_1, \ldots, f_m$ be arbitrary elements of $L^p(\Omega)$; for each j select a sequence $\{f_j^{(n)}\}$ in $\mathcal{D}(\Omega)$ such that $f_j^{(n)} \to f$ in $L^p(\Omega)$ and let $u^{(n)} \in C^{(2)}(\overline{\Omega}) \cap C_\Gamma(\overline{\Omega})$ be the solution of $(\lambda I - A_0)u^{(n)} = f_0^{(n)} + \sum D^j f_j^{(n)}$. Then it follows from the previous considerations that $u^{(n)}$ is Cauchy, hence convergent in $W_0^{1,p}(\Omega)$ so that $u = \lim u^{(n)}$ is a solution of (4.8.57) satisfying (4.8.58).

It remains to be shown that the solution of (4.8.57) is unique also for $1 < p \leqslant 2$. This is easiest seen as follows. If $u \in W_0^{1,p}(\Omega)$ satisfies (4.8.57) with $f_0 = f_1 = \cdots = f_m = 0$, then a simple integration by parts shows that $u \in D(A_0'(\beta')^*)$ and $(\lambda I - A_0'(\beta')^*)u = 0$, where $A_0'(\beta')$ is thought of a an operator in $L^{p'}(\Omega)$ and $\beta' = \beta$ is the Dirichlet boundary condition. Since $A_p(\beta) - \omega I = A_0'(\beta')^* - \omega I$ is m-dissipative in $L^p(\Omega)$ for ω sufficiently large, $u = 0$ and uniqueness follows.

We use the results obtained for an alternate characterization of the domain of $A_1(\beta)$, the closure of $A_0(\beta)$ in $L^1(\Omega)$.

4.8.17 Theorem. *Let C be a bounded domain of class $C^{(k+2)}$, A_0 an operator of class $C^{(k+1)}$ in Ω with $k > m/2$ satisfying the uniform ellipticity assumption (4.8.9)[17] β the Dirichlet boundary condition. Then $D(A_1(\beta))$ consists of all $u \in L^1(\Omega)$ such that $A_1(\beta)u$ (understood in the sense of distributions) belongs to $L^1(\Omega)$, i.e., such that there exists $v \, (= A_1(\beta)u)$ in $L^1(\Omega)$ satisfying*

$$\int_\Omega u A_0'(\beta')w \, dx = \int_\Omega vw \, dx \quad \left(w \in D(A_0'(\beta')) = C^{(2)}(\bar{\Omega}) \cap C_\Gamma(\bar{\Omega}) \right) \quad (4.8.68)$$

Every $u \in D(A(\beta))$ belongs to $W_0^{1,q}(\Omega)$ if $q < m/(m-1)$. For any such q there exists a constant $C = C(q)$ such that

$$\|u\|_{W^{1,q}(\Omega)} \leqslant C\|(\lambda I - A_1(\beta))u\|_{L^1(\Omega)} \quad \left(u \in D(A_1(\beta))\right), \quad (4.8.69)$$

where $\lambda > 0$ is independent of q.

The proof is based on the theory of the #-adjoint in Section 2.2 applied to the operator $A_\infty'(\beta')$ in $C_\Gamma(\bar{\Omega})$. We begin with the identification of $C_\Gamma(\bar{\Omega})^\# = \overline{D\left(A_\infty'(\beta')^*\right)} = \overline{D\left(A_0'(\beta')^*\right)} \subseteq C_\Gamma(\bar{\Omega})^* = \Sigma_\Gamma(\bar{\Omega})$. Let ν be an arbitrary measure in $\Sigma(\bar{\Omega}) = C(\bar{\Omega})^*$ (see Section 7). According to Theorem 4.7.16, if $p > m$ every function $u \in W^{1,p}(\Omega)$ is (equivalent to) a continuous function in $\bar{\Omega}$ and the identity map from $W^{1,p}(\Omega)$ into $C(\bar{\Omega})$ is bounded. We can thus define a continuous linear functional Φ in $W^{1,p}(\Omega)$ by means of the formula

$$\Phi(u) = \int_{\bar{\Omega}} u(x)\nu(dx). \quad (4.8.70)$$

It is plain that $W^{1,p}(\Omega)$ can be linearly and isometrically embedded in the Cartesian product $L^p(\Omega)^{m+1}$ of $L^p(\Omega)$ with itself $m+1$ times by means of the assignation

$$u \to (u, -D^1 u, \ldots, -D^m u)$$

---

[17]See footnote 14, p. 248.

(see Section 7) and we can thus extend Φ by the Hahn-Banach Theorem 2.1 to $L^p(\Omega)^{m+1}$ with the same norm; hence there exists an element $f = (f_0, f_1, \ldots, f_m)$ of $L^{p'}(\Omega)^{m+1}$, $p'^{-1} + p^{-1} = 1$, such that

$$\|f\|_{L^{p'}(\Omega)^{m+1}} = \left(\sum_{j=0}^{m} \|f_j\|_{L^{p'}(\Omega)}^{p'} \right)^{1/p'} \leqslant C\|\nu\|_{\Sigma(\bar{\Omega})} \tag{4.8.71}$$

(where C does not depend on ν and $\|\nu\|_{\Sigma(\bar{\Omega})} = $ total variation of ν) and

$$\Phi(u) = \int_{\Omega} \left(f_0 u - \sum_{j=1}^{m} f_j D^j u \right) dx \quad (u \in W^{1,p}(\Omega)). \tag{4.8.72}$$

Let $\mu \in \Sigma_\Gamma(\bar{\Omega})$ be an element of $D(A_0'(\beta')^*) = D((\lambda I - A_0'(\beta'))^*)$, so that there exists $\nu = (\lambda I - A_0'(\beta'))\mu \in \Sigma_\Gamma(\bar{\Omega})$ with

$$\int_{\bar{\Omega}} (\lambda I - A_0'(\beta')) u(x) \mu(dx) = \int_{\bar{\Omega}} u(x) \nu(dx) \quad \left(u \in C^{(2)}(\bar{\Omega}) \cap C_\Gamma(\bar{\Omega}) \right) \tag{4.8.73}$$

We apply the argument above to the measure ν. Once the functions $f_0, f_1, \ldots, f_m \in L^{p'}(\Omega)$ in (4.8.72) are manufactured, we use Theorem 4.8.15 to construct a solution $u' \in W_0^{1,p'}(\Omega)$ of (4.8.57). It follows from this equation and (4.8.73) after an integration by parts that the measure $\sigma(dx) = \mu(dx) - u'(x)\, dx$ satisfies

$$\int_{\bar{\Omega}} (\lambda I - A_0'(\beta')) u(x) \sigma(dx) = 0 \quad \left(u \in C^{(2)}(\bar{\Omega}) \cap C_\Gamma(\bar{\Omega}) \right) \tag{4.8.74}$$

We choose now $\lambda > \omega_\infty$ (see (4.8.5)); since $(\lambda I - A_0'(\beta'))(C^{(2)}(\bar{\Omega}) \cap C_\Gamma(\bar{\Omega}))$ is dense in $C_\Gamma(\bar{\Omega})$, it follows instantly that σ vanishes identically, so that $\mu(dx) = u'(x)\, dx$ and $D(A_0'(\beta')^*) \subseteq W_0^{1,p'}(\Omega)$, in particular, $D(A_0'(\beta')^*) \subseteq L^1(\Omega)$. Since any $u \in C^{(2)}(\bar{\Omega})_\beta$ belongs to $D(A_0'(\beta')^*)$ and this space is dense in $L^1(\Omega)$, it follows that

$$C_\Gamma(\bar{\Omega})^\# = L^1(\Omega),$$

and $A_\infty'(\beta')^\#$ is defined as the operator with domain consisting of all $u \in L^1(\Omega)$ such that there exists $v = A_\infty'(\beta')^\#$ satisfying (4.8.68). Obviously, $A_0(\beta) \subseteq A_\infty'(\beta')^\#$ so that $A_1(\beta) = \overline{A_0(\beta)} \subseteq A_\infty'(\beta')^\#$; since both operators belong to $\mathcal{C}_+$,

$$A_1(\beta) = A_\infty'(\beta')^\# = A_0'(\beta')^\#, \tag{4.8.75}$$

and the proof of Theorem 4.8.17 for boundary conditions of type (II) is complete (see the comments after (4.8.44) in relation to the second equality (4.8.75)). Inequality (4.8.69) is an obvious consequence of (4.8.71) and (4.8.58).

4.8.18 Remark. We show in the next section that the last vestige of dissipativity requirements (namely, (4.8.4) on boundary conditions of type

(I)) can be removed. In the case $E = L^p(\Omega)$ the argument involves no more than a sharpening of inequality (4.5.7) and shows that $A(\beta) - \omega I$ is m-dissipative for ω large enough (possibly $\omega > \omega_p$) when the boundary condition β is of type (I) and (4.8.4) does not hold. As a bonus, we shall deduce that $A_p(\beta) \in \mathcal{C}$. (Theorem 4.9.1)

The cases $E = L^1(\Omega)$ and $E = C(\overline{\Omega})$, $C_\Gamma(\overline{\Omega})$ are far more delicate since (4.8.4) for $p = 1$ is necessary for dissipativity of *any* operator $A(\beta)$ (in particular, of $A(\beta) - \omega I$ with ω arbitrary). The same comment applies to the cases $E = C(\overline{\Omega})$, $C_\Gamma(\overline{\Omega})$ (see Lemma 4.5.3). However, a renorming of the space will do the trick. (Theorem 4.9.3)

In Example 4.8.19 through Example 4.8.21 a proof of Theorem 4.8.8 is given. We assume that $m \geqslant 3$ in all examples except in Examples 4.8.22, 4.8.23, and 4.8.24.

4.8.19 Example. Let ψ be a function in $W^{1,p}(\mathbb{R}^m_+)$ $(1 < p < \infty)$ with support in $|x| \leqslant M$. Then there exists a function $\varphi \in W^{2,p}(\mathbb{R}^m_+)$ with support in $|x| \leqslant 2M$ such that $D^m\varphi(\hat{x},0) = \psi(\hat{x},0)$ (boundary values and derivatives understood as in Lemma 4.8.1) and

$$\|\varphi\|_{W^{2,p}(\mathbb{R}^m_+)} \leqslant C_0 \|\psi\|_{W^{1,p}(\mathbb{R}^m_+)}, \qquad (4.8.76)$$

where C_0 depends only on p and M.

We prove this result with the help of the Neumann kernel $K(\hat{x}, x_m)$ $= (2/(2-m)\omega_m)(|\hat{x}|^2 + x_m^2)^{-(m-2)/2}$ of $\mathbb{R}^m_+$, ω_m the hyperarea of the unit sphere in $\mathbb{R}^m$. Using an obvious approximation argument, we may and will assume that ψ is infinitely differentiable. Define

$$\theta(x) = \theta(\hat{x}, x_m) = \int_{\xi_m = 0} \psi(\hat{\xi},0) K(\hat{x} - \hat{\xi}, x_m)\, d\hat{\xi} \qquad (4.8.77)$$

with $\hat{\xi} = (\xi_1, \ldots, \xi_{m-1})$. Obviously θ is infinitely differentiable in $x_m > 0$. Observe that

$$\theta(x) = -\int_{\xi_m = 0} \left\{ \int_0^\infty D_\xi^m \big(\psi(\hat{\xi}, \xi_m) K(\hat{x} - \hat{\xi}, x_m - \xi_m)\big) d\xi_m \right\} d\hat{\xi}$$

$$= -\int_{\xi_m \geqslant 0} D^m \psi(\hat{\xi}, \xi_m) K(\hat{x} - \hat{\xi}, x_m - \xi_m)\, d\hat{\xi}\, d\xi_m$$

$$+ \int_{\xi_m \geqslant 0} \psi(\hat{\xi}, \xi_m) D_x^m K(\hat{x} - \hat{\xi}, x_m - \xi_m)\, d\hat{\xi}\, d\xi_m, \qquad (4.8.78)$$

hence, if $1 \leqslant k \leqslant m - 1$,

$$D^k\theta(x) = \frac{2}{(m-2)\omega_m} \int_{\xi_m \geqslant 0} D^m\psi(\xi) D_x^k |x - \xi|^{-(m-2)}\, d\xi$$

$$- \frac{2}{(m-2)\omega_m} \int_{\xi_m \leqslant x_m} \psi(x - \xi) D^k D^m |\xi|^{-(m-2)}\, d\xi, \qquad (4.8.79)$$

and another differentiation produces the formula

$$D^j D^k \theta(x) = \frac{2}{(m-2)\omega_m} \int_{\xi_m \geq 0} D^m \psi(\xi) D_x^j D_x^k |x - \xi|^{-(m-2)} \, d\xi$$

$$- \frac{2}{(m-2)\omega_m} \int_{\xi_m \geq 0} D^j \psi(\xi) D_x^k D_x^m |x - \xi|^{-(m-2)} \, d\xi$$

$$(4.8.80)$$

if $j \neq m$; for $j = m$ we have instead

$$D^m D^k \theta(x) = \frac{2}{(m-2)\omega_m} \int_{\xi_m \geq 0} D^m \psi(\xi) D_x^m D_x^k |x - \xi|^{-(m-2)} \, d\xi$$

$$- \frac{2}{(m-2)\omega_m} \int_{\xi_m \geq 0} D^m \psi(\xi) D_x^k D_x^m |x - \xi|^{-(m-2)} \, d\xi$$

$$- \frac{2}{(m-2)\omega_m} \int_{\xi_m = 0} \psi(\hat{\xi}, x_m) D_x^k D_x^m (|\hat{x} - \hat{\xi}|^2 + x_m^2)^{-(m-2)/2} \, d\hat{\xi}$$

$$(4.8.81)$$

To deal with the two integrals in (4.8.80) or the first two integrals in (4.8.81) we note that the Fourier transform of (the distribution) $|x|^{-(m-2)}$ equals $-C_m |\sigma|^{-2}$ (C_m a constant), thus the convolutions amount to multiplication of Fourier transforms by $C_m \sigma_j \sigma_k |\sigma|^{-2}$ and $C_m \sigma_k \sigma_m |\sigma|^{-2}$, respectively, and can be handled using Theorem 4.8.5.

To deal with the third integral, we integrate by parts, obtaining

$$\int_{\mathbb{R}^{m-1}} D^k \psi(\hat{\xi}, x_m) H(\hat{\xi} - \hat{x}, x_m) \, d\hat{\xi} \tag{4.8.82}$$

where the expression $H(\hat{x}, x_m) = (2/(2-m)\omega_m) D^m (|\hat{x}|^2 + x_m)^{-(m-2)/2} = (2/\omega_m) x_m (|\hat{x}|^2 + x_m^2)^{-m/2}$ is the Dirichlet kernel of $\mathbb{R}_+^m$. But it is a classical result (see Courant-Hilbert [1962: 1, p. 268]) that

$$\int_{\mathbb{R}^{m-1}} H(\hat{x}, x_m) \, d\hat{x} = 1 \quad (x_m > 0). \tag{4.8.83}$$

Since $H \geq 0$, the integral (4.8.82) can be estimated with the help of Young's Theorem 8.1. A bound of the form $\|D^k \psi(\cdot, x_m)\|_{L^p(\mathbb{R}^{m-1})}$ is obtained, thus we only have to take the p-th power and integrate with respect to x_m. The missing derivative $(D^m)^2$ is included observing that the function θ, being defined by (4.8.77), is harmonic in $\mathbb{R}_+^m$ so that $-(D^m)^2 = \Sigma(D^j)^2$. Derivatives of order < 2 are accounted for using (4.8.15) as in Theorem 4.8.6. Finally, the fact that $D^m \theta(\hat{x}, 0) = \psi(\hat{x}, 0)$ follows from an argument familiar in the theory of mollifiers using (4.8.83) and the fact that the integral of $H(\hat{x}, x_m)$ on $|\hat{x}| \geq \delta > 0$ tends to zero as $x_m \to 0$. We have then shown that θ satisfies all the properties required of φ except for the location of the support. To achieve this we simply define $\varphi = \chi \theta$, χ a function in $\mathcal{D}$ with support in $|x| \leq 2M$ such that $\chi(x) = 1$ for $|x| \leq M$.

4.8.20   Example.   Let $u \in W^{2,p}(\mathbb{R}^m_+)$ $(1 < p < \infty)$ with support in $|x| \leqslant M$ and such that $D^m u(\hat{x}, 0) = 0$. Then there exists a constant C_0 (depending only on p, M) such that (4.8.16) holds in $\mathbb{R}^m_+$.

The proof is essentially similar to that of Corollary 4.8.7: we use now the *even* extensions $u(\hat{x}, x_m) = u(\hat{x}, -x_m)$, $f(\hat{x}, x_m) = f(\hat{x}, -x_m)$ $(x_m < 0)$, where $f = \Delta u$. We apply then Theorem 4.8.6, generalized to functions in $W^{2,p}(\mathbb{R}^m)$ through an obvious approximation argument.

4.8.21   Example.   Prove Theorem 4.8.8. (*Hint:* Let φ be the function constructed in Example 4.8.20. Apply Example 4.8.20 to $u - \varphi$.)

4.8.22   Example.   Prove the results in Examples 4.8.19 to 4.8.21 for $m = 2$. (The only notable difference is that the Neumann kernel of $\overline{\mathbb{R}}^2_+$ is $K(x_1, x_2)$ $= (1/2\pi)\log(x_1^2 + x_2^2)$. The Dirichlet kernel of $\overline{\mathbb{R}}^2_+$ is the same: $H(x_1, x_2) = D^2 K(x_1, x_2) = (1/\pi)x_2(x_1^2 + x_2^2)^{-1} = (2/\omega_2)x_2(x_1^2 + x_2^2)^{-1}$.)

4.8.23   Example.   Show how the proof of Theorem 4.8.13 must be modified for $m = 2$.

4.8.24   Example.   Show that the characterizations of $D(A_\infty(\beta))$ in Theorem 4.8.11 and of $D(A_1(\beta))$ in Theorem 4.8.17 cannot be substantially improved in the sense that, in general, $D(A_\infty(\beta)) \neq C^{(2)}(\overline{\Omega})_\beta$ and $D(A_1(\beta))$ $\neq W^{2,1}(\Omega)_\beta$ when $m \geqslant 2$ (take, for instance, $\Omega = \mathbb{S}^m = \{x; |x| \leqslant 1\}$ and solve $\Delta u = f$, $u = 0$ on $\Gamma = \{x; |x| = 1\}$ by means of the Green function $G(x, y)$ of the unit sphere. The second derivatives are expressed by singular integral operators

$$D^j D^k u = \int_{\mathbb{S}^m} D^j D^k G(x, y) f(y) \, dy,$$

which do *not* map $C_\Gamma(\overline{\Omega})$ into $C_\Gamma(\overline{\Omega})$ or $L^1(\Omega)$ into $L^1(\Omega)$).

4.9.   ANALYTICITY OF SOLUTION OPERATORS

We apply the theory in Sections 4.1 and 4.2 to the differential operators in the last five sections. Although treatment of operators with measurable coefficients in $L^2(\Omega)$ is possible, only "analyticity in the mean" is obtained and we limit ourselves to the operators in Sections 4.7 and 4.8. Most of the results are obvious generalizations of those for the one-dimensional case in Section 4.3, thus some of the details will be omitted. The operator A_0 will be written in divergence form,[18]

$$A_0 = \sum\sum D^j\big(a_{jk}(x)D^k u\big) + \sum b_j(x) D^j u + c(x)u. \qquad (4.9.1)$$

---

[18]As in other places, we write b_j instead of $\hat{b}_j$ for simplicity of notation.

We shall show below that the operators $A_p(\beta) = \overline{A_0(\beta)}$, $1 < p < \infty$, defined in the previous section belong to $\mathcal{C}$; as an added bonus the dissipativity assumptions (4.8.4) on the boundary condition can be abandoned.

4.9.1   Theorem. *Let $1 < p < \infty$, Ω a bounded domain of class $C^{(2)}$ for $1 < p \leqslant 2$ (of class $C^{(k+2)}$ with $k > m/2$ if $p > 2$), A_0 an operator of class $C^{(1)}$ in Ω for $1 < p \leqslant 2$) (of class $C^{(k+1)}$ with $k > m/2$ if $p > 2$) satisfying the uniform ellipticity assumption (4.8.9), β a boundary condition of type (II) or of type (I) with $\gamma \in C^{(1)}(\Gamma)$ ($\gamma \in C^{(k+1)}(\Gamma)$ if $p > 2$). Then the operator $A_p(\beta) = \overline{A_0(\beta)}$ belongs to $\mathcal{C}(\varphi_p -)$ in $L^p(\Omega)$, where*

$$\varphi_p = \operatorname{arc tg}\left\{\left(\frac{p}{p-2}\right)^2 - 1\right\}^{1/2}. \tag{4.9.2}$$

For every φ with $0 < \varphi < \varphi_p$ there exists $\omega = \omega(p, \varphi)$ such that

$$\|S_p(\zeta)\| \leqslant e^{\omega|\zeta|} \quad (|\arg \zeta| \leqslant \varphi), \tag{4.9.3}$$

where $S_p(\zeta) = \exp(\zeta A_p(\beta))$.

Proof. Let θ be the only duality map in $L^p(\Omega)$. We perform an integration by parts similar to that in (4.5.7), but keeping track of imaginary parts: for $p \geqslant 2$ the calculation makes sense for arbitrary u in $C^{(2)}(\overline{\Omega})_\beta$, whereas for $1 < p \leqslant 2$ we use the customary approximation argument.

$$\|u\|^{p-2}\left(\operatorname{Re}\langle\theta(u), A_p(\beta)u\rangle \pm \delta\operatorname{Im}\langle\theta(u), A_p(\beta)u\rangle\right)$$

$$= \operatorname{Re}\int_\Omega \left\{\sum\sum D^j\left(a_{jk}D^k u\right) + \sum b_j D^j u + cu\right\}|u|^{p-2}\bar{u}\,dx$$

$$\pm \delta\operatorname{Im}\int_\Omega \left\{\sum\sum D^j\left(a_{jk}D^k u\right) + \sum b_j D^j u + cu\right\}|u|^{p-2}\bar{u}\,dx$$

$$= \int_\Gamma\left(\gamma + \frac{1}{p}b\right)|u|^p\,d\sigma$$

$$- (p-2)\int_\Omega |u|^{p-4}\left\{\sum\sum a_{jk}\operatorname{Re}(\bar{u}D^j u)\operatorname{Re}(\bar{u}D^k u)\right\}dx$$

$$\mp \delta(p-2)\int |u|^{p-4}\left\{\sum\sum a_{jk}\operatorname{Re}(\bar{u}D^j u)\operatorname{Im}(\bar{u}D^k u)\right\}dx$$

$$- \int_\Omega |u|^{p-2}\left\{\sum\sum a_{jk}D^j u D^k\bar{u}\right\}dx \pm \delta\int_\Omega |u|^{p-2}\left\{\sum b_j\operatorname{Im}(\bar{u}D^j u)\right\}dx$$

$$+ \frac{1}{p}\int_\Omega \left\{pc - \sum D^j b_j\right\}|u|^p\,dx. \tag{4.9.4}$$

We have

$$-(p-2)\int_\Omega |u|^{p-4}\Big\{\sum\sum a_{jk}\,\mathrm{Re}(\bar u D^j u)\,\mathrm{Re}(\bar u D^k u)\Big\}\,dx$$

$$\mp\,\delta(p-2)\int_\Omega |u|^{p-4}\Big\{\sum\sum a_{jk}\,\mathrm{Re}(\bar u D^j u)\,\mathrm{Im}(\bar u D^k u)\Big\}\,dx$$

$$-\int_\Omega |u|^{p-2}\Big\{\sum\sum a_{jk}\,D^j u D^k \bar u\Big\}\,dx$$

$$=-\int_\Omega |u|^{p-4}\Big\{\sum\sum a_{jk} f_{jk}\Big\}\,dx, \tag{4.9.5}$$

where

$$f_{jk}=(\bar u D^j u)(u D^k \bar u)$$

$$+(p-2)\{\mathrm{Re}(\bar u D^j u)\,\mathrm{Re}(\bar u D^k u)\pm\delta\,\mathrm{Re}(\bar u D^j u)\,\mathrm{Im}(\bar u D^k u)\}. \tag{4.9.6}$$

Let $z_1,\dots,z_m$ be arbitrary complex numbers. Consider the matrix $Z=\{z_{jk}\}$ with elements

$$z_{jk}=z_j\bar z_k+\alpha\{(\mathrm{Re}\,z_j)(\mathrm{Re}\,z_k)\pm\delta(\mathrm{Re}\,z_j)(\mathrm{Im}\,z_k)\},$$

where α,δ are real constants with $\alpha>-1$. If $\xi_1,\dots,\xi_m$ are real, we have

$$\sum\sum z_{jk}\xi_j\xi_k=\Big(\sum z_j\xi_j\Big)\overline{\Big(\sum z_j\xi_j\Big)}$$

$$+\alpha\Big\{\mathrm{Re}\Big(\sum z_j\xi_j\Big)\,\mathrm{Re}\Big(\sum z_k\xi_k\Big)\pm\delta\,\mathrm{Re}\Big(\sum z_j\xi_j\Big)\,\mathrm{Im}\Big(\sum z_k\xi_k\Big)\Big\},$$

which is nonnegative if

$$1+\delta^2\leqslant\left(\frac{\alpha+2}{\alpha}\right)^2 \tag{4.9.7}$$

in view of (4.3.6) and following comments; it follows then that Z is positive definite. As in the one-dimensional case, it is easy to see that if the inequality (4.9.7) is strict, then Z will satisfy $\sum\sum z_{jk}\xi_j\xi_k\geqslant\tau|z|^2|\xi|^2$ for all $z_1,\dots,z_m,\xi_1,\dots,\xi_m$, where $\tau>0$. We apply this to the matrix $F=\{f_{jk}\}$ in (4.9.6): the uniform ellipticity assumption and a slight modification of Lemma 4.4.2 imply[19] that if δ satisfies (4.9.7) strictly with $\alpha=p-2$,

$$\sum\sum a_{jk} f_{jk}\geqslant\tau|u|^2\sum|D^j u|^2. \tag{4.9.8}$$

We estimate next the other terms in (4.9.4). Consider a real vector field $\rho=(\rho_1,\dots,\rho_m)$ continuously differentiable in $\bar\Omega$ and such that $(\rho,\nu)=$

---

[19]Lemma 4.4.1 is applied here to the "symmetrized" matrix $f_{jk}=\{\tfrac12(f_{jk}+\bar f_{kj})\}$ rather than to $\{f_{jk}\}$ itself.

$\gamma + p^{-1}b$ on Γ, where ν is the outer normal vector on Γ. Then $\operatorname{div}(|u|^p \rho) = |u|^p \operatorname{div}\rho + p|u|^{p-2}\Sigma\rho_j \operatorname{Re}(\bar{u}D^j u)$. By virtue of the divergence theorem,

$$\left| \int_\Gamma \left(\gamma + \frac{1}{p}b \right) |u|^p \, d\sigma \right|$$

$$= \left| \int_\Gamma (|u|^p \rho, \nu) \, d\sigma \right|$$

$$\leqslant \int_\Omega |\operatorname{div}\rho| \, |u|^p \, dx + p \int_\Omega |u|^{p-2} \left(\sum |\rho_j| \, |\bar{u}D^j u| \right) dx$$

$$\leqslant \frac{p\varepsilon^2}{2} \int_\Omega |u|^{p-2} \left(\sum |\rho_j| \, |D^j u|^2 \right) dx + \int_\Omega \left(|\operatorname{div}\rho| + \frac{p\varepsilon^{-2}}{2} \sum |\rho_j| \right) |u|^p \, dx,$$

$$(4.9.9)$$

where we have used the fact that

$$2|u| \, |D^j u| \leqslant \varepsilon^2 |D^j u|^2 + \varepsilon^{-2} |u|^2. \qquad (4.9.10)$$

This inequality is also used in an obvious way to estimate the fourth volume integral in (4.9.4). Taking $\varepsilon > 0$ sufficiently small in the resulting inequality and in (4.9.9) and dividing by $\|u\|^{p-2}$, we obtain an inequality of the form

$$\operatorname{Re}\langle \theta(u), A_0(\beta)u \rangle \leqslant \pm \delta \operatorname{Im}\langle \theta(u), A_0(\beta)u \rangle + \omega\|u\|^2, \quad (4.9.11)$$

which is then extended to $D(A_p(\beta))$ in the usual way. The fact that $(\lambda I - A_p(\beta))D(A_p(\beta)) = L^p(\Omega)$ is established in the same way as in Theorem 4.8.3 (note that the dissipativity conditions there play no role in this). Inequality (4.9.3) is proved like in the case $m = 1$.

It can be shown as in Section 4.3 that an estimate of the type of (4.9.3) will not hold if $|\varphi| > \varphi_p$.

4.9.2 Remark. The characterization of the domains of the operators $A_p(\beta)$ in Corollary 4.8.10 extends to the present case. The proof is the same.

We extend next the analyticity results in Section 4.3 for the cases $p = 1$ and $p = \infty$ to the present multidimensional situation, beginning with the following auxiliary result.

4.9.3 Theorem. *Under the assumptions in Theorem* 4.9.1 *for* $p > 2$ (a) *the operator* $A_1(\beta) = \overline{A_0(\beta)}$ *belongs to* $\mathcal{C}_+$ *in* $L^1(\Omega)$; (b) *the operator* $A_\infty(\beta) = \overline{A_0(\beta)}$ *belongs to* $\mathcal{C}_+$ *in* $C(\overline{\Omega})$ ($C_\Gamma(\overline{\Omega})$ *if the boundary condition is of type* (II)).

Proof. As in the case $m = 1$, (a) is proved by renorming the space $L^p(\Omega)$ by means of a continuous, positive weight function ρ defined in $\overline{\Omega}$:

$$\|u\|_\rho = \left(\int_\Omega |u(x)|^p \rho(x)^p \, dx \right)^{1/p}. \qquad (4.9.12)$$

The space $L^p(\Omega)$ equipped with this norm will be denoted $L^p(\Omega)_\rho$. The identification of the dual space is the same as that in the proof of Theorem 4.3.4; we use the same notation for duality maps and for application of functionals to elements of $L^p(\Omega)_\rho$. Assuming that ρ is twice continuously differentiable, we obtain

$$\|u\rho\|^{p-2}\mathrm{Re}\langle\theta_\rho(u), A_0(\beta)u\rangle_\rho = \|u\rho\|^{p-2}\mathrm{Re}\langle\rho\theta(\rho u), A_0(\beta)u\rangle$$

$$= \mathrm{Re}\int_\Omega\Big\{\sum\sum D^j\big(a_{jk}D^k u\big)+\sum b_j D^j u + cu\Big\}|u|^{p-2}\bar{u}\rho^p\,dx$$

$$= \int_\Gamma\Big(\gamma+\frac{1}{p}b\Big)|u|^p\rho^p\,d\sigma$$

$$-(p-2)\int_\Omega|u|^{p-4}\Big\{\sum\sum a_{jk}\mathrm{Re}(\bar{u}D^j u)\mathrm{Re}(\bar{u}D^k u)\Big\}\rho^p\,dx$$

$$-\int_\Omega|u|^{p-2}\Big\{\sum\sum a_{jk}D^j uD^k\bar{u}\Big\}\rho^p\,dx$$

$$-p\int_\Omega|u|^{p-2}\Big\{\sum\sum a_{jk}\mathrm{Re}(\bar{u}D^j u)D^k\rho\Big\}\rho^{p-1}\,dx$$

$$+\frac{1}{p}\int_\Omega\Big(pc-\sum D^j b_j\Big)|u|^p\rho^p\,dx - \int_\Omega\Big(\sum b_j D^j\rho\Big)|u|^p\rho^{p-1}\,dx. \quad (4.9.13)$$

We transform now the third volume integral keeping in mind that $p|u|^{p-2}\mathrm{Re}(\bar{u}D^j u)=D^j|u|^p$ and using the divergence theorem for the vector $U=(U_j)$ of components

$$U_j = |u|^p\rho^{p-1}\sum a_{jk}D^k\rho \quad (1\leqslant j\leqslant m).$$

Once this is done, the right-hand side of (4.9.13) can be written in the form

$$\int_\Gamma\Big(\gamma-\frac{1}{\rho}D^{\tilde{\nu}}\rho+\frac{1}{p}b\Big)|u|^p\rho^p\,d\sigma$$

$$-(p-2)\int_\Omega|u|^{p-4}\Big\{\sum\sum a_{jk}\mathrm{Re}(\bar{u}D^j u)\mathrm{Re}(\bar{u}D^k u)\Big\}\rho^p\,dx$$

$$-\int_\Omega|u|^{p-2}\Big\{\sum\sum a_{jk}D^j uD^k\bar{u}\Big\}\rho^p\,dx$$

$$+\int_\Omega\Big\{\sum\sum D^j\big(a_{jk}\rho^{p-1}D^k\rho\big)$$

$$+\Big(c-\frac{1}{p}\sum D^j b_j\Big)\rho^p - \Big(\sum b_j D^j\rho\Big)\rho^{p-1}\Big\}|u|^p\,dx. \quad (4.9.14)$$

We can now choose $D^{\tilde{\nu}}\rho$ at the boundary in such a way that the quantity between curly brackets in the surface integral in (4.9.14) is nonpositive. Noting that the first two volume integrals combine to yield a nonpositive amount we can bound the right-hand side of (4.9.14) by an expression of the form $\omega\|u\|_\rho$, where ω does not depend on p. Using a limiting argument as in the one-dimensional case, we obtain

$$\operatorname{Re}\langle u^*, A_0(\beta)u\rangle_\rho \leqslant \omega\|u\|_\rho^2, \qquad (4.9.15)$$

where the expression between brackets indicates application of the functional $u^* \in \Theta(u) \subset L^\infty(\Omega)$ to $A_0(\beta)u \in L^1(\Omega)_\rho$ and $\|\cdot\|_\rho$ is the norm of $L^1(\Omega)_\rho$. The customary approximation argument shows that (4.9.15) holds for $A_1(\beta) = \overline{A_0(\beta)}$. The fact that $(\lambda I - A_1(\beta))D(A_1(\beta)) = L^1(\Omega)$ for $\lambda > 0$ large enough is shown as in Theorem 4.8.3. Renorming of the space $L^1(\Omega)$ is of course unnecessary when β is the Dirichlet boundary condition.

We prove next (b) renorming the spaces $C(\overline{\Omega})$ or $C_\Gamma(\overline{\Omega})$ by means of

$$\|u\|_\rho = \max_{x \in \Omega} |u(x)|\rho(x), \qquad (4.9.16)$$

where ρ is a continuous positive function in $\overline{\Omega}$; the identification of the duals is achieved along the lines of Section 4.3. If β is of type (I) and condition (4.8.4) (for $p = \infty$) is not satisfied, we select ρ twice continuously differentiable and such that

$$D^{\tilde{\nu}}\rho(x) + \gamma(x)\rho(x) \leqslant 0 \quad (x \in \Gamma). \qquad (4.9.17)$$

Let $u \in C^{(2)}(\overline{\Omega})_\beta$, $u \neq 0$ and $m_\rho(u) = \{x;\ |u(x)|\rho(x) = \|u\|_\rho\}$ so that any $\mu \in \Theta_\rho(u)$ is supported by $m_\rho(u)$ and is such that $u\rho\mu$ (or $u\mu$) is a positive measure in $m_\rho(u)$. Since $D^{\tilde{\nu}}u(x) = \gamma(x)u(x)$ at the boundary, $u\rho$ satisfies the boundary condition

$$D^{\tilde{\nu}}(u(x)\rho(x)) = \gamma_\rho(x)(u(x)\rho(x)) \quad (x \in \Gamma), \qquad (4.9.18)$$

where

$$\gamma_\rho(x) = \gamma(x) + \frac{1}{\rho(x)}D^{\tilde{\nu}}\rho(x) \leqslant 0 \quad (x \in \Gamma). \qquad (4.9.19)$$

The argument employed in Section 4.5 shows that $m_\rho(u)$ does not meet the boundary Γ if $\gamma_\rho(x) < 0$ everywhere on Γ so that

$$D^j|u\rho|^2(x) = 0, \quad -\mathcal{H}(x;\ |u\rho|^2) \geqslant 0 \quad (x \in m_\rho(u)). \qquad (4.9.20)$$

On the other hand, if $\gamma_\rho(x) = 0$ for some $x \in \Gamma$, $m_\rho(u)$ may contain points of Γ, but we can prove in the same way as in Section 4.5 that (4.9.20) holds. Writing $\eta = \rho^2$ and $u = u_1 + iu_2$ with u_1, u_2 real, we obtain

$$D^j|u\rho|^2 = 2(u_1 D^j u_1 + u_2 D^j u_2)\eta + (u_1^2 + u_2^2)D^j\eta,$$

$$D^j D^k|u\rho|^2 = 2(u_1 D^j D^k u_1 + u_2 D^j D^k u_2)\eta + 2(D^j u_1 D^k u_1 + D^j u_2 D^k u_2)\eta$$
$$+ 2(u_1 D^j u_1 + u_2 D^j u_2)D^k\eta + 2(u_1 D^k u_1 + u_2 D^k u_2)D^j\eta$$
$$+ (u_1^2 + u_2^2)D^j D^k\eta.$$

Accordingly, it follows from (4.9.20) that if $x \in m_\rho(u)$,

$$\mathrm{Re}(u^{-1}D^j u) = \|u\|^{-2}(u_1 D^j u_1 + u_2 D^j u_2) = -\tfrac{1}{2}\|u\|^{-2}|u|^2\eta^{-1}D^j\eta$$
$$= -\tfrac{1}{2}\eta^{-1}D^j\eta. \qquad (4.9.21)$$

On the other hand, again for $x \in m_\rho(u)$, we have

$$\mathrm{Re}(u^{-1}D^j D^k u) = \|u\|^{-2}(u_1 D^j D^k u_1 + u_2 D^j D^k u_2)$$
$$= \tfrac{1}{2}\|u\|^{-2}\eta^{-1}D^j D^k|u\rho|^2 - \|u\|^{-2}(D^j u_1 D^k u_1 + D^j u_2 D^k u_2)$$
$$- \|u\|^{-2}(u_1 D^j u_1 + u_2 D^j u_2)\eta^{-1}D^k\eta$$
$$- \|u\|^{-2}(u_1 D^k u_1 + u_2 D^k u_2)\eta^{-1}D^j\eta$$
$$- \tfrac{1}{2}\|u\|^{-2}(u_1^2 + u_2^2)\eta^{-1}D^j D^k\eta$$
$$= \tfrac{1}{2}\|u\|^{-2}\eta^{-1}D^j D^k|u\rho|^2 - \|u\|^{-2}(D^j u_1 D^k u_1 + D^j u_2 D^k u_2)$$
$$+ \eta^{-2}D^j\eta D^k\eta - \tfrac{1}{2}\eta^{-1}D^j D^k\eta. \qquad (4.9.22)$$

Accordingly, if $\mu \in \Theta_\rho(u)$,

$$\mathrm{Re}\langle\mu, A_0(\beta)u\rangle = \int_{m_\rho(u)} \mathrm{Re}(u^{-1}A_0(\beta)u)\rho u \, d\mu \leq \omega\|u\|_\rho^2 \quad (4.9.23)$$

showing that $A_0 - \omega I$ is dissipative. The same conclusion then holds for $A_\infty(\beta) = \overline{A_0(\beta)}$, and we prove in the same fashion as in Theorem 4.8.3 that $(\lambda I - A_\infty(\beta))D(A_\infty(\beta)) = C(\overline{\Omega})$ (or $C_\Gamma(\overline{\Omega})$ for the Dirichlet boundary condition).

4.9.4 Remark. The characterization of the domain of $A_\infty(\beta)$ in Theorem 4.8.11 extends to the present case. So does the identification of $D(A_1(\beta))$ in Theorem 4.8.17.

The following result generalizes the pointwise analyticity conclusions of Theorem 4.3.5 to the present case.

4.9.5 Theorem. *Let Ω be a bounded domain of class $C^{(\infty)}$, A_0 an operator of class $C^{(\infty)}$ in Ω satisfying the uniform ellipticity assumption (4.8.9), β a boundary condition of type (II) or of type (I) with $\gamma \in C^\infty(\Gamma)$. Finally, let u be an element of $L^p(\Omega)$, $1 \leq p < \infty$ ($1 < p < \infty$ if β is of type (I)) or of $C(\overline{\Omega})$ ($C_\Gamma(\overline{\Omega})$ if β is of type (II)). Then $u(t, x) = (S_p(t)u)(x) = \{\exp(tA_p(\beta))\}u(x)$, after eventual modification in an x-null set for each t, can be extended to a function $u(\zeta, x)$ analytic in $\Sigma_+(\pi/2-)$. Moreover $u(\zeta, \cdot) \in C^{(\infty)}(\overline{\Omega})$ ($C^{(\infty)}(\overline{\Omega}) \cap C_\Gamma(\overline{\Omega})$ when β is the Dirichlet boundary condition) and for every $\varepsilon > 0$, every φ with $0 < \varphi < \pi/2$ and every integer $k = 0, 1, \ldots$, there exist constants C, ω depending on ε, φ, k, such that*

$$\|u(\zeta, \cdot)\|_{C^{(k)}(\overline{\Omega})} \leq Ce^{\omega|\zeta|}\|u\| \quad (\zeta \in \Sigma(\varphi), |\zeta| \geq \varepsilon), \qquad (4.9.24)$$

where the norm on the right-hand side is the L^p norm if $1 \leqslant p < \infty$ or the C norm if $p = \infty$.

It is obviously sufficient to prove the theorem in the most unfavorable situation, that where $u \in L^1(\Omega)$; as in the one-dimensional case the argument is based on repeated application of Sobolev's imbedding theorem, although the details are somewhat more complicated. We begin by selecting two integers j and l such that

$$j > m/4, \quad 2l > k + m/2. \tag{4.9.25}$$

Let $\lambda > 0$ be sufficiently large (so that $R(\lambda; A_1(\beta))$ exists). We have

$$S_1(t)u = R(\lambda; A_1(\beta))^l (\lambda I - A_1(\beta))^{j+l} S_1(t) R(\lambda; A_1(\beta))^j u. \tag{4.9.26}$$

It was shown in Theorem 4.8.18 that if $p < m/(m-1)$, then $D(A_1(\beta)) \subseteq W^{1,p}(\Omega)$ and inequality (4.8.69) holds. By Theorem 4.8.16, $W^{1,p}(\Omega) \subseteq L^{mp/(m-p)}(\Omega)$. Arguing exactly as in the proof of Theorem 4.3.5, we show that if $1 \leqslant r < s \leqslant \infty$, we have $A_r(\beta) \supseteq A_s(\beta)$ (both operators thought of as acting in $L^r(\Omega)$), hence $R(\lambda; A_s(\beta))$ is the restriction of $R(\lambda; A_r(\beta))$ to $L^s(\Omega)$ (to $C(\overline{\Omega})$ or $C_\Gamma(\overline{\Omega})$ if $s = \infty$). We can then write $R(\lambda; A_1(\beta))^2 u = R(\lambda; A_q(\beta)) R(\lambda; A_1(\beta)) u$ with $q = mp/(m-p)$, and it follows then from Corollary 4.8.10 (see inequality (4.8.40)) that $R(\lambda; A_1(\beta))^2 u \in W^{2,q}(\Omega)$ and

$$\|R(\lambda; A_1(\beta))^2 u\|_{W^{2,q}(\Omega)} \leqslant C\|u\|_{L^1(\Omega)}. \tag{4.9.27}$$

Applying again Theorem 4.8.16 we see that $R(\lambda; A_1(\beta))^2 u \in L^r(\Omega)$ with $r = mq/(m-2q) = mp/(m-3p)$ and

$$\|R(\lambda; A_1(\beta))^2 u\|_{L^r(\Omega)} \leqslant C\|u\|_{L^1(\Omega)}. \tag{4.9.28}$$

Arguing repeatedly in the same way, we deduce that $R(\lambda; A_1(\beta))^j u \in L^s(\Omega)$ with $s = mp/(m-(2j-1)p)$ and

$$\|R(\lambda; A_1(\beta))^j u\|_{L^s(\Omega)} \leqslant C\|u\|_{L^1(\Omega)}. \tag{4.9.29}$$

Since $p < m/(m-1)$ can be taken arbitrarily close to $m/(m-1)$, we may assume $mp/(m-(2j-1)p)$ is as close as we wish to $m/(m-2j)$ so that, in view of the first inequality (4.9.25), $s \geqslant 2$; hence $R(\lambda; A_1(\beta))^j u \in L^2(\Omega)$, and we may take $s = 2$ in (4.9.29). This means that $S_1(t)$ on the right-hand side of (4.9.26) can be replaced by $S_2(t)$ and we can then use Theorem 4.9.1 in the case $p = 2$. Writing $(\lambda I - A_1(\beta))^{j+l} S_1(t) = (\lambda I - A_2(\beta))^{j+l} S_2(t) = \{(\lambda I - A_2(\beta)) S_2(t/(j+l))\}^{j+l}$ and making use of (4.1.18), we obtain an inequality of the type of (4.9.24) for $(\lambda I - A_1(\beta))^{j+l} S_1(t)u$, but in the $L^2(\Omega)$ norm. Inequality (4.9.24) is now obtained as follows. It results from Theorem (4.7.12) that if $v \in L^2(\Omega)$, then $R(\lambda; A_2(\beta))^l v \in W^{2l,2}(\Omega) = H^{2l}(\Omega)$ and

$$\|R(\lambda; A_2(\beta))^l v\|_{H^{2l}(\Omega)} \leqslant C\|v\|_{L^2(\Omega)}.$$

We apply then Corollary 4.7.16. The treatment of boundary conditions of type (I) is entirely similar.

4.10. POSITIVITY AND COMPACTNESS OF SOLUTION OPERATORS

As pointed out in Section 3.7 the *real* spaces $L^p(\Omega)$ $(1 \leqslant p \leqslant \infty)$ and $C(\overline{\Omega}), C_\Gamma(\overline{\Omega})$ are Banach lattices with respect to their natural order relations $(u \leqslant v$ if $u(x) \leqslant v(x)$ everywhere in $\overline{\Omega}$ in the spaces $C(\overline{\Omega})$ and $C_\Gamma(\overline{\Omega})$ with "almost everywhere" instead of "everywhere" in the spaces $L^p(\Omega)$). We study in the sequel positivity of the solution operators S_p in $L^p(\Omega)$ $(1 \leqslant p < \infty)$ and of S in $C(\Omega)$ or $C_\Gamma(\Omega)$ using the theory of dispersive operators in Section 3.7. As in the case $m = 1$, we center the treatment in the case $E = C(\overline{\Omega})$ where verification of dispersivity is immediate. We have seen (Lemma 3.7.1) that all duality maps in $C(\overline{\Omega})$ (hence in $C_\Gamma(\overline{\Omega})$) are proper. Actually, we shall use this result not in $C(\overline{\Omega})$ with its original supremum norm but in relation to the weighted norm (4.9.16); the proof is the same.

4.10.1 Theorem. *Let* $k > m/2$, Ω *a bounded domain of class* $C^{(k+2)}$, A_0 *an operator of class* $C^{(k+1)}$ *in* Ω *satisfying the uniform ellipticity assumption* (4.8.9), β *a boundary condition of type* (II) *or of type* (I) *with* $\gamma \in C^{(k+1)}(\Gamma)$. *Then if* $A_p(\beta) = \overline{A_0(\beta)}$ $(A_0(\beta)$ *thought of as an operator in the real spaces* $L^p(\Omega)$ *and* $C(\overline{\Omega}))$ *the operator* $S_p(t) = \exp(tA_p(\beta))$ *is positive for all* $t \geqslant 0$, *i.e., for every* $u \in L^p(\Omega)$ $(C(\overline{\Omega}), C_\Gamma(\overline{\Omega}))$ *such that* $u(x) \geqslant 0$ *a.e. in* Ω $(u(x) \geqslant 0$ *in* $\overline{\Omega})$, *we have*

$$S_p(t)u(x) \geqslant 0 \text{ a.e. in } \Omega \quad (S(t)u(x) \geqslant 0 \text{ in } \overline{\Omega}).$$

The same positivity properly holds for the resolvents $R(\lambda; A_p(\beta))$ $(1 \leqslant p \leqslant \infty)$.

Proof. We begin by showing that (a translation of) the operator $A_0(\beta)$ is dispersive in $C(\overline{\Omega})$ (see Section 3.7, especially (3.7.27) and preceding comments), where we assume $C(\overline{\Omega})$ normed with (4.9.16), the weight function ρ satisfying condition (4.9.17). The subsequent computation is essentially the same as that leading to (4.9.23), but considerably simplified due to the use of real functions throughout, thus we include the details below. Let u be a function in $C^{(2)}(\overline{\Omega})_\beta$ such that $u_+ \neq 0$ (i.e., such that $u(x) > 0$ at some point of $\overline{\Omega}$)) and let $\mu \in \Theta_\rho(u_+)$ so that $\mu \in \Sigma(\overline{\Omega})$ is a positive measure with support in $m_\rho(u_+) = \{x \in \overline{\Omega}; u_+(x)\rho(x) = \|u_+\|_\rho\}$. If $x \in m_\rho(u_+)$, then $D^j(u\rho) = (D^ju)\rho + uD^j\rho = 0$ and the Hessian matrix $\mathcal{K}(x; u\rho)$ is negative definite. Since $D^jD^k(u\rho) = (D^jD^ku)\rho + D^juD^k\rho + D^kuD^j\rho + uD^jD^k\rho$, we easily show that

$$\langle \mu, A_0u \rangle = \int_{m_\rho(u_+)} \left\{ \sum\sum a_{jk}D^jD^ku + \sum b_jD^ju + cu \right\} \rho \, d\mu$$

$$\leqslant \omega \|u_+\|_\rho^2 \qquad (4.10.1)$$

showing that $A_0(\beta) - \omega I$ is dispersive with respect to any duality map. Since we have already shown in Theorem 4.9.3 that $A_\infty(\beta) - \omega I$ is m-dissipative for ω large enough (the passage from complex to real spaces effected along the lines of the remarks at the end of Section 3.1) it results from Theorem 3.7.3 that $S_\infty(t) = \exp(t A_\infty(\beta))$ is a positive operator for all $t \geq 0$. The case $p < \infty$ is handled through an approximation argument. Let v be a nonnegative element of $L^p(\Omega)$, $\{v_n\}$ a sequence of nonnegative functions in $\mathfrak{D}$ such that $v_n \to v$ in $L^p(\Omega)$, $\lambda > 0$ large enough, u (resp. u_n) the solution of $(\lambda I - A_p(\beta))u = v$ (resp. $(\lambda I - A_p(\beta))u_n = v_n$) in $L^p(\Omega)$. Since $A_\infty(\beta) \subset A_p(\beta)$ (see Section 4.9), it follows that $u_n \in D(A_\infty(\beta))$ and that $(\lambda I - A_\infty(\beta))u_n = v_n$. Now, since $R(\lambda; A_\infty(\beta))$ is a positive operator, $u_n \geq 0$ for all n, and it follows from continuity of $R(\lambda; A_p(\beta))$ that $u_n \to u$ in $L^p(\Omega)$. Hence $u \geq 0$ a.e. in Ω, showing that $R(\lambda; A_p(\beta))$ is a positive operator. The fact that $S_p(t)$ is a positive operator follows then from Theorem 3.7.2. This ends the proof of Theorem 4.10.1.

Going back to complex spaces, we study the compactness of the solution operators $S_p(\cdot)$. We limit ourselves to the following simple result.

4.10.2   Theorem. *Let the assumptions of Theorem 4.9.5 be satisfied, and let the complex number ζ ($Re\,\zeta > 0$) be fixed. If $\{u_k\}$ is a sequence in $L^p(\Omega)$ with $\|u_k\| \leq C$, then the sequence $\{S_p(\zeta, \cdot)u\}$ contains a subsequence convergent in the norm of $C(\overline{\Omega})$. The same property holds for $S_\infty(\zeta)$ in $C(\overline{\Omega})$ or $C_\Gamma(\overline{\Omega})$.*

The proof is an immediate consequence of the Arzelà-Ascoli theorem (see Lemma 1.3.3).

Theorem 4.10.2 implies that $S_p(t)$ is a compact operator in $L^p(\Omega)$ for every $t > 0$ and that $S_\infty(t)$ is compact in $C(\overline{\Omega})$ or $C_\Gamma(\overline{\Omega})$. This yields significant information on spectral properties of $A_p(\beta)$ and $A_\infty(\beta)$ through known properties of compact operators (Example 3.10) and the following result:

4.10.3   Example. Let A be the infinitesimal generator of a strongly continuous semigroup $S(\cdot)$ in an arbitrary Banach space E. Then: (a) For each $t > 0$ we have $\{\mu; \mu = e^{t\lambda}, \lambda \in \sigma(A)\} \subseteq \sigma(S(t))$ (see Hille-Phillips [1957: 1, p. 467]). (b) If $\mu \neq 0$ is an eigenvalue of $S(t)$, at least one of the (countably many) roots $\{\lambda_n\}$ of the equation

$$e^{t\lambda} = \mu \tag{4.10.2}$$

is an eigenvalue of A. (c) The generalized eigenspace of $S(t)$ corresponding to μ is the subspace generated by the generalized eigenspaces of A corresponding to all solutions of (4.10.2) which are eigenvalues of A.

4.10.4   Example. Let $A_p(\beta)$ be the operators in Theorems 4.9.5. Then (a) $\sigma(A)$ consists of a sequence $\{\lambda_1, \lambda_2, \ldots\}$ such that $|\lambda_n| \to \infty$ as $n \to \infty$. (b) For any φ, $0 < \varphi < \pi$, $\sigma(A)$ is contained in the sector $|\arg \lambda| \geq \varphi$ (except for a finite number of elements). (c) Each λ_n is an eigenvalue of A with finite generalized multiplicity (see Section 3).

4.10.5 Remark. Throughout the last three sections, we have been forced to impose additional smoothness conditions on the domain and on the operator A_0 when constructing extensions in $L^p(\Omega)$ ($p = 1$ or $p > 2$) or in $C(\overline{\Omega}), C_\Gamma(\overline{\Omega})$. These requirements can in fact be dispensed with, taking careful note of the "uniform" character of Theorems 4.8.4 and 4.8.12 and suitably approximating Ω by smoother domains and A_0 by smoother operators. In this way, even the assumptions for the range $1 < p \leqslant 2$ can be substantially weakened. We give some details below.

Let Ω be a bounded domain of class $C^{(2)}$, and let A_0 be the operator

$$A_0 u = \sum \sum a_{jk}(x) D^j D^k u + \sum b_j(x) u + c(x) u \qquad (4.10.3)$$

Then the conclusion of Theorem 4.9.1 (with identification of the domain of $A_p(\beta)$ as in Corollary 4.8.9) holds under the assumptions that $A_0 \in C^{(1)}(\overline{\Omega})$ and that $\gamma \in C^{(1)}(\Gamma)$ if β is a boundary condition of type (1). The same conditions insure the validity of Theorem 4.9.3 on the operator $A_\infty(\beta)$, with identification of domain as in Theorem 4.8.17. In the case of the space $L^1(\Omega)$, Theorem 4.9.3 holds if, in addition, $b_j \in C^{(1)}(\overline{\Omega})$ in (4.10.3). Even weaker assumptions can be handled through perturbation methods; see Section 5.5.

4.11. MISCELLANEOUS COMMENTS

The opening remarks of Section 4.1 (in particular, Lemma 4.1.1 and Corollary 4.1.2) are due to Hille [1950: 1]. In particular, Hille proved Theorem 4.1.8. Theorem 4.1.3 was discovered by Pazy [1968: 2]; earlier sufficient conditions for a semigroup to be in the class $\mathcal{C}^\infty$ were obtained by Hille [1950: 1] and Yosida [1958: 1]. The rest of the theory in Section 4.1 is due to Hille [1950: 1], but the relation between (4.1.10) and the possibility of extending S to a sector in the complex plane was discovered by Yosida [1958: 1]. Theorem 4.2.1 is due to Hille [1948: 1]. Theorem 4.2.4 was given by Pazy in his lecture notes [1974: 1]. The theory pertaining to the class $\mathcal{C}$ was extended to locally convex spaces by Yosida [1963: 2]; a similar extension of Theorem 4.1.3 is due to Watanabe [1972: 1]. We include below some bibliography on semigroups in the classes $\mathcal{C}$ and $\mathcal{C}^\infty$, abstract parabolic equations and related matters. See Neuberger [1964: 1], [1970: 1], [1973: 1], [1973: 3], Prozorovskaja [1967: 1], Certain [1974: 1], Solomjak [1958: 1], [1959: 1], [1960: 1], Beurling [1970: 1], Pazy [1971: 2], Hasegawa [1967: 1], Efgrafov [1961: 1], Valikov [1964: 1], and Kato [1970: 2].

     The treatment in Section 4.3 uses suggestions in Pazy [1974: 1], in particular that of showing analyticity by means of Theorem 4.2.4. The characterization of the "angle of dissipativity" φ_p in (4.3.3) is in the author

[1983: 1]. The idea of reducing the proof of Theorem 4.3.4 to the dissipative case by means of weight functions was inspired by the treatment of Mullikin [1959: 1].

Many (if not all) of the results in the treatment of partial differential operators in the rest of the chapter are well known to specialists, although it seems difficult to trace all to their sources. Others are folk theorems. The theory of assignation of boundary conditions in Section 4.5 is essentially due to Fichera in [1960: 1] and previous works; in particular, we owe to him the division of the boundary Γ according to the type of boundary condition (if any) imposed on each piece Γ^i. An elementary account of the Fichera theory can be found in Lieberstein [1972: 1]. The only novelty in our treatment is perhaps the emphasis on obtaining dissipative extensions of A_0.

Lemma 4.6.1 is the well-known Lax-Milgram lemma ([1954: 1]). The rest of Section 4.6.2 follows closely that paper, with the simplifications due to the fact that we only deal with a second-order elliptic operator. In the degenerate case, sketched in Remark 4.6.8, the use of "weighted Sobolev spaces" such as $\mathcal{H}_0(\Omega)$ is standard (see Oleinik-Radkevič [1971: 1]). In Section 4.7 we have followed closely Gilbarg-Trudinger [1977: 1], where historical references to the material in this and the previous section can be found. The treatment of the operator $A_p(\beta)$ in $L^p(\Omega)$, $p \neq 2$ as the closure of $A_0(\beta)$ is closely related with Phillips [1959: 1]. Theorem 4.8.4 is a very particular case of general estimates for higher order elliptic equations due to Agmon-Douglis-Nirenberg [1959: 1]; the proof here incorporates some obvious simplifications. Theorem 4.8.5 is due to Hörmander [1960: 1]. That the Agmon-Douglis-Nirenberg estimates imply the characterization of $D(A_p(\beta))$ in Corollary 4.8.10 for $1 < p < \infty$ is already pointed out in the work of these authors. The characterization of the operators $A_p(\beta)$ in the cases $p = 1, \infty$ has been apparently known for some time, but we have not attempted to trace each result back to its origin. The case $p = 1$ is treated in Brézis-Strauss [1973: 1] (see also Massey [1976: 1]). For the case $E = C$, see Stewart [1974: 1], and references therein; we note that in this last paper operators of arbitrary order are considered and the emphasis lies on showing that the operators in question belong to the class $\mathcal{C}$, a task that we have avoided. The results in Section 4.9 are multidimensional versions of those in Section 4.4 and can also be found in the author [1983: 1].

Chapter 5

Perturbation and Approximation of Abstract Differential Equations

Many differential equations of importance in applications (such as the Schrödinger equation) are of the form $u'(t) = (A + P)u(t)$, where the operator A is known to belong to $\mathcal{C}_+$ or to a subclass thereof and P is a perturbation operator. It is then of interest to determine conditions on P that guarantee that $A + P$ belongs to the same class.

Two results of this kind are proved in Sections 5.1 and 5.3, the first for the class $\mathcal{C}_+$ and the second for m-dissipative operators; there are also theorems for the class $\mathcal{C}$ and for the case where P is bounded. Applications to the neutron transport equation in Section 5.2 and to the Schrödinger and Dirac equations with potentials in Section 5.4 are examined in detail; also, it is shown in Section 5.5 how to relax some of the assumptions imposed on second order elliptic operators in Chapter 4. Symmetric hyperbolic operators, last seen in Chapter 3, reappear in Section 5.6; the objective is to study the Cauchy problem in Sobolev spaces generalizing the results pertaining to the constant coefficient case in Chapter 1.

A basic problem in the numerical treatment of the abstract differential equation $u'(t) = Au(t)$ is that of approximating its solutions by those of the equation $u'_n(t) = A_n u_n(t)$, where $A_n \to A$ in a suitable sense, or by those of a difference equation like $u(t + \tau) = \tau A_n u_n(t) + u_n(t)$. Sections 5.7 and 5.8 contain a study of this problem, with applications to a parabolic equation.

267

5.1. A PERTURBATION RESULT

It was already observed in Section 2.1 that direct application of Theorem 2.1.1 is difficult in general, except perhaps in the dissipative case treated in Chapter 3. It is many times useful to have at our disposition perturbation results, that is, theorems of the type "if A belongs to $\mathcal{C}_+$ (or to $\mathcal{C}, \mathcal{C}^\infty, \mathcal{Q}$) and P belongs to a certain class of operators $\mathcal{P}(A)$, then $A + P$ belongs to $\mathcal{C}_+$ (or to $\mathcal{C}, \mathcal{C}^\infty, \mathcal{Q}$). A straightforward but useful result in that direction is the following.

5.1.1   Theorem.   *Let $A \in \mathcal{C}_+$, $S(t) = S(t; A)$ the propagator of*

$$u'(t) = Au(t) \quad (t \geqslant 0). \tag{5.1.1}$$

Let P be a closed, densely defined operator such that $D(P) \supseteq D(A)$ and

$$\|PS(t)u\| \leqslant \alpha(t)\|u\| \quad (u \in D(A), t > 0), \tag{5.1.2}$$

where α is a finite, measurable function with

$$\int_0^1 \alpha(t)\, dt < \infty. \tag{5.1.3}$$

Then $A + P$ (with domain $D(A + P) = D(A)$) belongs to $\mathcal{C}_+$.

Proof.   Since we wish to show that $A + P \in \mathcal{C}_+$, we must deal with the equation

$$u'(t) = (A + P)u(t) = Au(t) + Pu(t). \tag{5.1.4}$$

Formally, we can solve (5.1.4) with initial condition

$$u(0) = u_0 \tag{5.1.5}$$

by means of the following successive approximation scheme: let $u_0(t) = S(t)u_0$ and $u_n(t)$ the solution of

$$u'_n(t) = Au_n(t) + Pu_{n-1}(t) \quad (t \geqslant 0) \tag{5.1.6}$$

$$u_n(0) = u_0. \tag{5.1.7}$$

Then $u(t) = \lim u_n(t)$ will be a solution of (5.1.4), (5.1.5). The initial value problem (5.1.6), (5.1.7) can of course be solved by means of the theory developed in Section 2.4; in view of (2.4.3) we must have

$$u_n(t) = S(t)u_0 + \int_0^t S(t-s)Pu_{n-1}(s)\, ds \quad (n \geqslant 1)$$

so that

$$u = (S + S * PS + S * (PS)^{*2} + \cdots)u_0, \tag{5.1.8}$$

where $*$ indicates convolution product and $*n$ the n-th convolution power. To justify formula (5.1.8) directly would be cumbersome, thus we will only take it as a heuristic guide.

In what follows $f, g, h, \ldots$ will be (E)-valued functions defined and strongly continuous in $t > 0$ such that $\|f\|, \|g\|, \ldots$ are integrable near zero. (Note that, by the uniform boundedness theorem, $\|f\|$ is bounded on compact subsets of $t > 0$; also, since $\|f(t)\| = \sup\{\|f(t)u\|; \|u\| \leqslant 1\}$, $\|f\|$ is lower semicontinuous, hence measurable). We denote by $\mathcal{F}$ the space of all such functions.

The convolution of f and $g \in \mathcal{F}$ is defined by the familiar formula

$$(f * g)(t)u = \int_0^t f(t - s)g(s)u\, ds. \tag{5.1.9}$$

The following properties of the convolution are either well known or easily deduced and left to the reader.

(a) *If $f, g \in \mathcal{F}$, then $f * g \in \mathcal{F}$ and*

$$\|(f * g)(t)\| \leqslant (\|f\| * \|g\|)(t) \quad (t > 0). \tag{5.1.10}$$

(b) $\qquad\qquad\qquad f *(g * h) = (f * g)* h.$

(c) *If f or g are strongly continuous in $t \geqslant 0$, so is $f * g$.*

Clearly we can use (5.1.10) combined with Young's Theorem 8.1 to estimate convolution powers: if $f \in \mathcal{F}$, $f^{*n} = f * f * \cdots * f$ and

$$\int_0^\infty \|f(t)\|\, dt = \gamma < \infty, \tag{5.1.11}$$

then

$$\int_0^\infty \|f^{*n}(t)\|\, dt \leqslant \gamma^n < \infty \quad (n \geqslant 1). \tag{5.1.12}$$

If g is bounded,

$$\|g(t)\| \leqslant C \quad (t > 0),$$

plainly we have

$$\|(g * f^{*n})(t)\| \leqslant C\gamma^n \quad (n \geqslant 1, t > 0). \tag{5.1.13}$$

We apply these observations to our perturbation problem. It follows from (5.1.2) that for each $t > 0$, $PS(t)$—which is in principle only defined in $D(A)$—admits an (E)-valued extension $\overline{PS(t)}$ that satisfies

$$\|\overline{PS(t)}\| \leqslant \alpha(t) \quad (t > 0). \tag{5.1.14}$$

Since

$$\overline{PS(t)} = \overline{PS(t_0)}S(t - t_0), \tag{5.1.15}$$

it is clear that $\overline{PS(\cdot)}$ is strongly continuous in $t > 0$; in view of (5.1.14), $\overline{PS}$ belongs to $\mathcal{F}$. On the other hand, it also follows from (5.1.15) that

$$\|\overline{PS(t)}\| \leqslant C\alpha(t_0)e^{-\omega t_0}e^{\omega t} \quad (t \geqslant t_0),$$

where C, ω are constants such that

$$\|S(t)\| \leqslant Ce^{\omega t} \quad (t \geqslant 0). \tag{5.1.16}$$

We choose now $\omega' > \omega$ so large that

$$\int_0^\infty e^{-\omega't}\|\overline{PS(t)}\|\,dt = \gamma < 1 \qquad (5.1.17)$$

and avail ourselves of the equality

$$\exp(-\omega't)(f * g) = \exp(-\omega't)f * \exp(-\omega't)g$$

to estimate

$$S_*(\overline{PS})^n = \exp(\omega't)\{\exp(-\omega't)S * \exp(-\omega't)(\overline{PS})^{*n}\},$$

making use of (5.1.16), (5.1.17), and (5.1.13). We obtain

$$\|S_*(\overline{PS})^{*n}(t)\| \leqslant C\gamma^n e^{\omega't} \quad (t \geqslant 0). \qquad (5.1.18)$$

This clearly means that the series

$$S_1(t) = \sum_{n=0}^\infty S_*(\overline{PS})^{*n}(t) \qquad (5.1.19)$$

converges in (E) uniformly on compact subsets of $t \geqslant 0$ and

$$\|S_1(t)\| \leqslant C(1-\gamma)^{-1}e^{\omega't} \quad (t \geqslant 0). \qquad (5.1.20)$$

Since reiterated applications of (c) show that every term of (5.1.19) is strongly continuous in $t \geqslant 0$, $S_1(t)$ is likewise strongly continuous in $t \geqslant 0$.

Let now f be a function in $\mathcal{F}$ growing at most exponentially at infinity and assume that $\omega' > \omega$ is so large that $\exp(-\omega't)\|f(t)\|$ is summable in $t \geqslant 0$. Let $\mathcal{L}f$ be the Laplace transform of f,

$$\mathcal{L}f(\lambda)u = \int_0^\infty e^{-\lambda t}f(t)u\,dt,$$

which exists in $\mathrm{Re}\,\lambda > \omega'$. It can be shown just as in the scalar case that if g is another function in $\mathcal{F}$ with $\exp(-\omega't)\|g\| \in L^1(0, \infty)$, then $\exp(-\omega't)\|f * g\| \in L^1(0, \infty)$ as well and

$$\mathcal{L}(f * g)(\lambda) = \mathcal{L}f(\lambda)\mathcal{L}g(\lambda) \quad (\mathrm{Re}\,\lambda > \omega'). \qquad (5.1.21)$$

On the other hand, it follows from Lemma 3.4 and an approximation argument that

$$\mathcal{L}(\overline{PSu}) = P\mathcal{L}S(\lambda)u = PR(\lambda)u \quad (\mathrm{Re}\,\lambda > \omega', u \in E). \qquad (5.1.22)$$

Let $\mathrm{Re}\,\lambda > \omega'$. Multiply the series (5.1.19) by $\exp(-\lambda t)$. In view of (5.1.18), the resulting series will converge uniformly in $t \geqslant 0$, the partial sums being bounded in norm by a constant times $\exp(-(\mathrm{Re}\,\lambda - \omega')t)$. We can then integrate term by term and make use of (5.1.21) and (5.1.22) to obtain

$$Q(\lambda) = \mathcal{L}S_1(\lambda) = R(\lambda)\sum_{n=0}^\infty (PR(\lambda))^n \quad (\mathrm{Re}\,\lambda > \omega'),$$

the series convergent in (E). Clearly $Q(\lambda)E \subseteq R(\lambda)E = D(A) = D(A + P)$

and

$$(\lambda I - A - P)Q(\lambda) = \sum_{n=0}^{\infty} (PR(\lambda))^n - \sum_{n=1}^{\infty} (PR(\lambda))^n = I;$$

in applying $\lambda I - A$ and P term by term to the series, we make use of the fact that both operators are closed. On the other hand, if $u \in D(A)$,

$$Q(\lambda)(\lambda I - A - P)u = R(\lambda) \sum_{n=0}^{\infty} (PR(\lambda))^n (\lambda I - A - P)u$$

$$= \sum_{n=0}^{\infty} (R(\lambda)P)^n u - \sum_{n=1}^{\infty} (R(\lambda)P)^n u = u$$

so that $Q(\lambda) = R(\lambda; A + P)$. The proof of Theorem 5.1.1 follows now from the fact that $Q(\lambda) = \mathcal{L}S_1(\lambda)$ and from Lemma 2.2.3.

A particularly interesting case of Theorem 5.1.1 is that where P is everywhere defined and bounded (all the hypotheses are then automatically satisfied). Since some additional precision can be gained, we prove this case separately.

5.1.2 Theorem. *Let $A \in \mathcal{C}_+(C, \omega)$ and let P be a bounded operator. Then $A + P \in \mathcal{C}_+(C, \omega + C\|P\|)$.*

Proof. By virtue of Theorem 2.1.1, $R(\lambda) = R(\lambda; A)$ exists for $\lambda > \omega$ and

$$\|R(\lambda; A)^n\| \leqslant C(\lambda - \omega)^{-n} \quad (\lambda > \omega, n \geqslant 1). \tag{5.1.23}$$

In particular, if

$$\lambda > \omega + C\|P\|,$$

then

$$\|PR(\lambda)\| \leqslant C\|P\|(\lambda - \omega)^{-1} < 1, \tag{5.1.24}$$

hence the series

$$Q(\lambda) = R(\lambda) \sum_{n=0}^{\infty} (PR(\lambda))^n$$

is convergent in (E). It can be proved exactly as in Theorem 5.1.1 that

$$Q(\lambda) = R(\lambda; A + P).$$

We estimate the powers of $R(\lambda; A + P)$. We have

$$R(\lambda; A + P)^m = \left\{ R(\lambda) \sum_{n=0}^{\infty} (PR(\lambda))^n \right\}^m$$

$$= \sum_{n=0}^{\infty} \sum_{j_1 + j_2 + \cdots + j_m = n} R(\lambda)(PR(\lambda))^{j_1}$$

$$\times R(\lambda)(PR(\lambda))^{j_2} \cdots R(\lambda)(PR(\lambda))^{j_m}.$$

Now, in view of (5.1.23) and (5.1.24),

$$\|R(\lambda; A + P)^m\| \leqslant \frac{C}{(\lambda - \omega)^m} \left\{ \frac{1}{1 - C\|P\|(\lambda - \omega)^{-1}} \right\}^m$$

$$= C(\lambda - (\omega + C\|P\|))^{-m} \quad (\lambda > \omega + C\|P\|),$$

which, in view of Theorem 2.1.1 as modified by Remark 2.1.4, ends the proof.

The following result relates Theorem 5.1.1 and Theorem 5.1.2 with the results in Section 3.7.

5.1.3 Corollary. *Let E be a Banach lattice, $A \in \mathcal{C}_+$, and assume that $S(t) = S(t; A)$ is positive for all $t \geqslant 0$. Let P be an operator satisfying the conditions in Theorem 5.1.1 and let P itself be positive (or, more generally, assume there exists a constant β such that*

$$Pu \geqslant -\beta u \quad (u \in D(A), u \geqslant 0)). \tag{5.1.25}$$

Finally, assume that $E_+ \cap D(A)$ is dense in E_+ (E_+ is the set of nonnegative elements of E). Then $S(t; A + P)$ is positive for all t.

Proof. Assume P is positive ($\beta = 0$ in (5.1.25)). Then $PS(t)u \geqslant 0$ if $u \in E_+ \cap D(A)$. The hypothesis on $E_+ \cap D(A)$ and the fact that E_+ is closed implies that $\overline{PS(t)}$ is positive for all t. Using again the closedness of E_+ we see that $S * \overline{PS}, S * (\overline{PS})^{*2}, \ldots$ are positive for all t and thus the same is true of $S(t; A + P)$, sum of the series (5.1.19). If P is not positive but only satisfies (5.1.25), we only have to apply the previous comments to $P + \beta I$ and use the formula $S(t; A + P) = e^{-\beta t}S(t; A + (P + \beta I))$.

We note that when P is bounded and everywhere defined the hypothesis that $E_+ \cap D(A)$ be dense in E_+ can be discarded.

To close this section we note the following result, that can be easily obtained applying Theorem 5.1.2 to A and $-A$:

5.1.4 Corollary. *Let $A \in \mathcal{C}(C, \omega)$ and let P be bounded. Then $A + P \in \mathcal{C}(C, \omega + C\|P\|)$.*

5.2. THE NEUTRON TRANSPORT EQUATION

As an application of the results in the previous section, we consider the *neutron transport equation*

$$\frac{\partial u}{\partial t} + (v, \text{grad}_x u) + \sigma(x, v)u = \int_V K(x, v, v')u(x, v', t) \, dv'. \tag{5.2.1}$$

Here $x = (x_1, x_2, x_3)$, $v = (v_1, v_2, v_3)$, and $u = u(x, v, t)$. This equation describes the migration and multiplication of neutrons in a body occupying a domain D of x-space: the nonnegative function u is the density at time t of neutrons of velocity v at x. The nonnegative function σ represents the absorption coefficient at x for neutrons of velocity v, whereas the integral operator on the right-hand side of (5.2.1) describes the production of neutrons at x by scattering and nuclear fission. We shall assume that D is open and connected in x-space $\mathbb{R}^3_x$ (although connectedness is not essential) and that the velocities v belong to some measurable set V in v-space $\mathbb{R}^3_v$.

Let $x \in \Gamma$ = boundary of D, and $v \in V$. We say that v is *incoming at x* (in symbols, $v \in \Im(x)$) if and only if there exists $\varepsilon > 0$ with

$$x + tv \in D \quad (0 < t < \varepsilon).$$

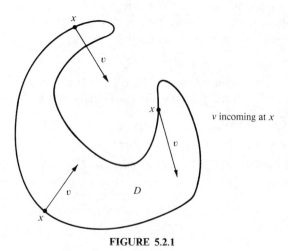

FIGURE 5.2.1

The solution of (5.2.1) is assumed to satisfy the following boundary condition:

$$u(x, v, t) = 0 \quad (x \in \Gamma, v \in \Im(x), t \geq 0). \tag{5.2.2}$$

Physically, this simply means that no neutrons enter the domain D from outside. This is natural if D is a convex region in empty space of if D is surrounded with absorbing material so that neutrons abandoning D are captured and unable to reenter. The initial neutron density is given:

$$u(x, v, 0) = u_0(x, v) \quad (x \in D, v \in V). \tag{5.2.3}$$

Let $\Omega = D \times V$. In view of the interpretation of u as a density, it is clear that the space $E = L^1(\Omega)_\mathbb{R}$ is the natural setting to study (5.2.1) as an abstract differential equation. However, since no additional difficulties appear, we shall take $E = L^p(\Omega)_\mathbb{R}$ with $1 \leq p < \infty$. With some additional restrictions on D we shall also consider the case where E is a suitable subspace of $C(\overline{\Omega})_{\mathbb{R}}$. Since only nonnegative solutions of (5.2.1) have physical meaning, it will be essential in each case to show that the solution remains nonnegative if the initial datum (5.2.3) is nonnegative.

We examine first (5.2.1) when $K = 0$ in the space $E = L^p(\Omega)$, $1 \le p < \infty$. We assume that σ is measurable and

$$0 \le \sigma(x, v) \le C \quad \text{a.e. in } \Omega. \tag{5.2.4}$$

Define

$$\mathfrak{A}_0 u(x, v) = -(v, \operatorname{grad}_x u(x, v)) - \sigma(x, v) u(x, v). \tag{5.2.5}$$

The domain of $\mathfrak{A}_0$ is described as follows. For each $x \in \mathbb{R}^3_x$, $v \in \mathbb{R}^3_v$, let

$$D(x, v) = D \cap \xi(x, v),$$

where $\xi(x, v)$ is the line $\{y; y = x + sv, -\infty < s < \infty\}$. Since D is open, $D(x, v)$ is open in $\xi(x, v)$, thus we can express $D(x, v)$ as the union of a finite or countable family of disjoint open intervals,

$$D(x, v) = J_1 \cup J_2 \cup \cdots,$$

which we call *component intervals* of $D(x, v)$. A function $u \in L^p(\Omega)$ belongs to $D(\mathfrak{A}_0)$ if and only if for almost all (x, v) in $\Omega = D \times V$

(a) *The function*

$$f(s; x, v) = u(x + sv, v)$$

can be extended in each component interval J_n of $D(x, v)$ to a function

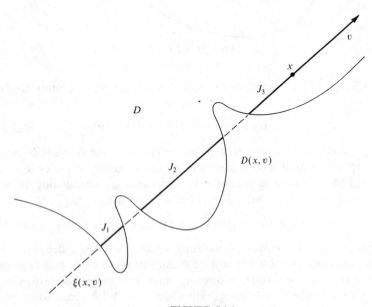

FIGURE 5.2.2

absolutely continuous in the closure $\bar{J}_n$ (locally absolutely continuous in $\bar{J}_n$ if J_n is infinite). (Note that if two intervals have a common endpoint, the two extensions may not coincide there.)

(b)

$$f = 0$$

at the left endpoint of each component interval of $D(x, v)$ (except if the left endpoint is $x - \infty v$).

(c) *The directional derivative*

$$f'(s) = (v, \text{grad}_x \mathfrak{u}) \tag{5.2.6}$$

belongs to $E = L^p(\Omega)$ (as a function of x and v).

Note that the expression on the right-hand side of (5.2.6) for the directional derivative is purely formal, since we do not assume the existence of the partial derivatives. Also, in view of (a), for almost all (x, v) the derivative (5.2.6) exists almost everywhere in $D(x, v)$, so that (5.2.6) is defined almost everywhere in Ω.

If φ is, say, an infinitely differentiable function of x with support contained in D and ψ is a function in $L^p(V)$, it is immediate that $\varphi(x)\psi(v) \in D(\mathfrak{A}_0)$. But the set $\mathfrak{D}$ of linear combinations of such functions is dense[1] in $L^p(\Omega)$ so that $D(\mathfrak{A}_0)$ is dense as well. We compute next the resolvent of $\mathfrak{A}_0$. To this end, we define a function $\eta = \eta(x, v)$ in Ω as follows:

$$\eta(x, v) = \sup\{t \geqslant 0; x - sv \in D, 0 < s < t\}[2]. \tag{5.2.7}$$

Since D is open, $\eta(x, v) > 0$ for all $(x, v) \in \Omega$ and, if $\eta(x, v) < \infty$, $x - tv$ belongs to D for $t < \eta(x, v)$ but not for $t = \eta(x, v)$. In more intuitive terms, $\eta(x, v)$ is the time taken by the point $x - tv$ to abandon D (for the first time) starting at $t = 0$.

5.2.1 Lemma. *η is lower semicontinuous in Ω.*

Proof. Let $(x_0, v_0) \in \Omega$ with $\eta(x_0, v_0) < \infty$. Since $x_0 - tv_0 \in D$ for $0 < t < \eta(x_0, v_0)$, a continuity argument can used to show that, given $\varepsilon > 0$ there exists $\delta > 0$ such that the circular cylinder consisting of all the points $x - tv_0$ with x roaming over the disk of radius δ centered at x_0 and perpendicular to the line $\xi(x_0, v_0)$, and $-\delta \leqslant t \leqslant \eta(x_0, v_0) - \varepsilon$ is entirely

---

[1]It is obviously sufficient to show that $\mathfrak{D}(D)$ is dense in $L^p(D)$ and that linear combinations of functions $f(x)g(x)$ ($f \in L^p(D)$, $g \in L^p(V)$) are dense in $L^p(D \times V)$. The final result can be proved with the help of mollifiers and the second by the Stone-Weierstrass theorem (Adams [1975: 1, p. 10]).

[2]We do not define η for $x \notin \Omega$, in particular for $x \in \Gamma$. We shall define it later, however (see p. 280) .

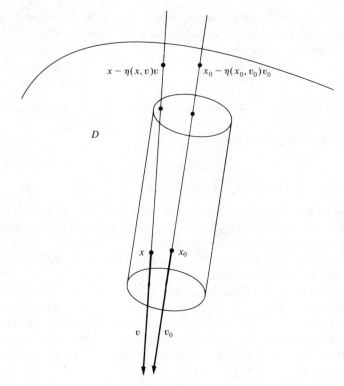

FIGURE 5.2.3

contained in D. It follows then that if (x, v) is sufficiently close to (x_0, v_0), we must have

$$\eta(x, v) \geqslant \eta(x_0, v_0) - 2\varepsilon.$$

A similar argument takes care of the case $\eta(x_0, v_0) = \infty$; here we show that $\eta(x, v)$ can be made arbitrarily large if (x, v) is sufficiently close to (x_0, v_0).

For $x \in D$, $v \in V$ define

$$S(x, v, t) = \begin{cases} \exp\left(-\int_0^t \sigma(x - sv, v)\, ds \right) & (0 \leqslant t < \eta(x, v)) \\ S(x, v, \eta(x, v)) & (t \geqslant \eta(x, v)). \end{cases} \tag{5.2.8}$$

Denote by D_n the intersection of D with the sphere of radius n centered at the origin and define V_n accordingly. If $v \in V_n$, we have

$$\int_{D_n} \sigma(x - sv, v)\, dx \leqslant \int_{D_{(a+1)n}} \sigma(x, v)\, dx$$

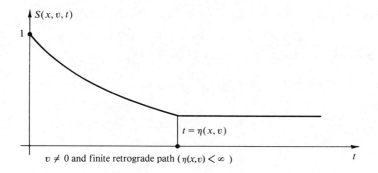

$v \neq 0$ and finite retrograde path ($\eta(x,v) < \infty$)

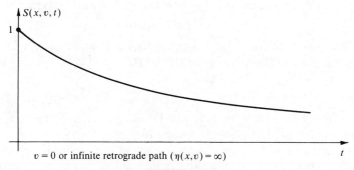

$v = 0$ or infinite retrograde path ($\eta(x,v) = \infty$)

FIGURE 5.2.4

for $v \in V_n$, $0 \leqslant s \leqslant a$; accordingly the integral of $\sigma(x - sv, v)$ with respect to (x, v, t) in $D_n \times V_n \times [0, n]$ is finite for every $n = 1, 2, \ldots$ and we deduce:

(d) $s \to \sigma(\tilde{x} - s\tilde{v}, \tilde{v})$ *is locally integrable in* $s \geqslant 0$ *for all* $(\tilde{x}, \tilde{v}) \in \tilde{\Omega}$, *where* $\tilde{\Omega}$ *is a subset of* Ω *whose complement has measure zero.*[3]

As an immediate consequence of (d) we obtain that the function defined by (5.2.8) exists for $(\tilde{x}, \tilde{v}, t)$ if $(\tilde{x}, \tilde{v}) \in \tilde{\Omega}$ for all $t > 0$; moreover, $S(\cdot, \cdot, t)$ is measurable in Ω for $t \geqslant 0$ fixed, since the set Ω_c of all $(x, v) \in \Omega$ with $S(x, v, t) \geqslant c > 0$ coincides with the set defined by the inequality

$$\int_0^t \sigma(x - sv, v)\, ds \leqslant -\log c,$$

where we set $\sigma = 0$ outside of Ω. Note also that for each $(\tilde{x}, \tilde{v}) \in \tilde{\Omega}$, the function $t \to S(\tilde{x}, \tilde{v}, t)$ is absolutely continuous in $t \geqslant 0$. Finally, we have

$$|S(x, v, t)| \leqslant 1 \quad \text{a.e. in } \Omega \times (0, \infty).$$

---

[3] The set $\tilde{\Omega}$ is obviously "translation invariant" in the following sense: if $(\tilde{x}, \tilde{v}) \in \tilde{\Omega}$ and $\tilde{x} + t\tilde{v} \in D$, then $(\tilde{x} + t\tilde{v}, \tilde{v}) \in \tilde{\Omega}$ (since $D(\tilde{x} + t\tilde{v}, \tilde{v}) = D(\tilde{x}, \tilde{v})$). Properties (d) and (e) below are also translation invariant in this sense.

Define an operator $\mathfrak{S}(t)$ $(t \geqslant 0)$ in $L^p(\Omega)$ by the formula

$$(\mathfrak{S}_0(t)\mathfrak{u})(x,v) = S(x,v,t)\mathfrak{u}(x - tv, v) \quad (\mathfrak{u} \in E), \qquad (5.2.9)$$

where we assume that $\mathfrak{u} = 0$ outside of Ω. We have

$$\|\mathfrak{S}_0(t)\mathfrak{u}\|^p \leqslant \int_\Omega S(x,v,t)^p |\mathfrak{u}(x - tv, v)|^p \, dx \, dv$$

$$\leqslant \int_V \left(\int_{D^*(v,t)} |\mathfrak{u}(x,v)|^p \, dx \right) dv$$

$$\leqslant \int_V \left(\int_D |\mathfrak{u}(x,v)|^p \, dx \right) dv = \|\mathfrak{u}\|^p,$$

where the set $D^*(v,t)$ consists of all $x \in D$ with $x + tv \in D$. Hence

$$\|\mathfrak{S}_0(t)\| \leqslant 1 \quad (t \geqslant 0). \qquad (5.2.10)$$

We show next that $\mathfrak{S}_0(\cdot)$ is strongly continuous. Let $\mathfrak{u}$ be a function in the set $\mathfrak{D}$ used to prove the denseness of $D(\mathfrak{A}_0)$. It is easy to see that if $t_0 \geqslant 0$, then

$$S(x,v,t)\mathfrak{u}(x - tv, v) \to S(x,v,t_0)\mathfrak{u}(x - t_0 v, v) \qquad (5.2.11)$$

as $t \to t_0$ for all $(\tilde{x}, \tilde{v}) \in \tilde{\Omega}$; in fact, (5.2.11) is immediate if $t_0 < \eta(x,v)$ or $t_0 > \eta(x,v)$ (see the comments following (d)) and results for $t_0 = \eta(x,v)$ observing that $\mathfrak{u}(x - \eta(x,v)v, v) = 0$ since $x - \eta(x,v)v$ must be a boundary point of D. We can then apply the dominated convergence theorem to deduce that

$$\|(\mathfrak{S}_0(t) - \mathfrak{S}_0(t_0))\mathfrak{u}\| \to 0 \quad (t \to t_0).$$

But in view of (5.2.10) and of the denseness of $\mathfrak{D}$, strong continuity follows (see Theorem 1.1).

We consider now the operator

$$\mathfrak{G}(\lambda)\mathfrak{v} = \int_0^\infty e^{-\lambda t} \mathfrak{S}_0(t)\mathfrak{v} \, dt \quad (\operatorname{Re}\lambda \leqslant 0, \mathfrak{v} \in E) \qquad (5.2.12)$$

(integration understood in the vector sense). An argument very similar to the one leading to (d) (but where the use of D_n, V_n is unnecessary due to the fact that $|\mathfrak{v}(x,v)|^p$ is integrable in $D \times V$) yields:

(e) *For any* $\mathfrak{v} \in E$, *the function* $s \to |\mathfrak{v}(x - sv, v)|^p$ *is integrable in* s *for all* $(\tilde{x}, \tilde{v}) \in \tilde{\Omega}$, *where* $\tilde{\Omega}$ *is a set of full measure in* Ω *and we have set* $\mathfrak{v} = 0$ *outside of* Ω. Plainly we may (and will) assume that the sets $\tilde{\Omega}$ in (d) and (e) are the same.

Let $\varphi = \varphi(x,v)$ be an arbitrary smooth function of compact support, thought of as an element of $L^{p'}(\Omega)$ $(p^{-1} + p'^{-1} = 1)$. Then

$$\langle \varphi, \mathfrak{G}(\lambda)\mathfrak{v} \rangle = \int_0^\infty e^{-\lambda t} \langle \varphi, \mathfrak{S}_0(t)\mathfrak{v} \rangle \, dt$$

$$= \int_\Omega \left(\int_0^{\eta(x,v)} e^{-\lambda t} S(x,v,t)\mathfrak{v}(x - tv, v) \, dt \right) \varphi(x,v) \, dx \, dv.$$

It follows from this equality that there exists a set $\tilde{\Omega}$ of full measure in Ω (which may be taken to coincide with the sets in (d) and (e)) such that, after eventual modification of $\mathfrak{u}$ in a null set,

$$\mathfrak{u}(x, \tilde{v}) = (\mathfrak{G}(\lambda)\mathfrak{v})(x, \tilde{v}) = \int_0^{\eta(x, \tilde{v})} e^{-\lambda s} S(x, \tilde{v}, s)\mathfrak{v}(x - s\tilde{v}, \tilde{v})\, ds$$

(5.2.13)

for $(\tilde{x}, \tilde{v}) \in \tilde{\Omega}$ and $x \in D(\tilde{x}, \tilde{v})$.[4] Let now x be a point in the closure of some component interval J_n of $D(\tilde{x}, \tilde{v})$ (we can take x to be the left endpoint if J_n does not contain $\tilde{x} - \infty\tilde{v}$, in which case $\eta(x, \tilde{v}) = 0$). Since $\eta(x + t\tilde{v}, v) = \eta(x, v) + t$, we obtain, after two elementary changes of variable, that if $0 \leqslant t \leqslant \eta(x, \tilde{v})$, then

$$\mathfrak{u}(x + t\tilde{v}, \tilde{v})$$

$$= \int_{-\eta(x, \tilde{v})}^t \left\{ \exp\left(-\lambda(t - s) - \int_s^t \sigma(x + r\tilde{v}, \tilde{v})\, dr \right) \right\} \mathfrak{v}(x + s\tilde{v}, \tilde{v})\, ds$$

so that (a) and (b) in the definition of $D(\mathfrak{A}_0)$ are satisfied. Moreover,

$$D_t\mathfrak{u}(x + t\tilde{v}, \tilde{v}) = \mathfrak{v}(x + t\tilde{v}, \tilde{v})$$

$$- (\lambda + \sigma(x + t\tilde{v}, \tilde{v}))\mathfrak{u}(x + t\tilde{v}, \tilde{v}) \quad (x + t\tilde{v} \in \bar{J}_n),$$

hence $\mathfrak{u} \in D(\mathfrak{A}_0)$ and

$$(\lambda I - \mathfrak{A}_0)\mathfrak{u} = \mathfrak{v}. \tag{5.2.14}$$

If $\mathfrak{v} = 0$, it is not difficult to see from the boundary condition (b) and from the fact that $\mathfrak{u}(x - t\mathfrak{v}, v)$ must be absolutely continuous in t for almost all (x, v) that any solution of (5.2.14) in $L^p(\Omega)$ must be zero almost everywhere. Since $\mathfrak{G}(\lambda)$ is bounded, we have shown that

$$\mathfrak{G}(\lambda) = R(\lambda; \mathfrak{A}_0) \quad (\operatorname{Re}\lambda > 0),$$

which equality, in view of (5.2.12) and of Lemma 2.2.3 implies that $B \in \mathcal{C}_+(1, 0)$ and that $\mathfrak{S}(\cdot)$, defined by (5.2.9), is the propagator of

$$\mathfrak{u}'(t) = \mathfrak{A}_0\mathfrak{u}(t). \tag{5.2.15}$$

We study now the full equation (5.2.1). To apply the results in Section 5.1 here, it is obviously sufficient to require that the integral operator

$$\mathfrak{R}u(x, v) = \int_V K(x, v, v')\mathfrak{u}(x, v')\, dv' \tag{5.2.16}$$

---

[4]A continuity argument extends (5.2.13) to x in $\bar{D}(\tilde{x}, \tilde{v})$, the closure of $D(\tilde{x}, \tilde{v})$. To see this we observe that $\eta(\cdot, \tilde{v})$ and $S(\cdot, \tilde{v}, t)$ can be extended to continuous functions in $\bar{D}(\tilde{x}, \tilde{v})$; on the other hand, $t \to \mathfrak{v}(x - t\tilde{v}, \tilde{v})$ belongs to L^p.

be bounded in $L^p(\Omega)$. A sufficient condition based on Hölder's inequality is

$$\sup_{x \in D} \int_V \left(\int_V |K(x,v,v')|^{p/(p-1)} \, dv' \right)^{p-1} dv < \infty \qquad (5.2.17)$$

for the case $p > 1$ and

$$\int_V \left(\sup_{x \in D} \sup_{v' \in V} |K(x,v,v')| \right) dv < \infty \qquad (5.2.18)$$

when $p = 1$, both of which are satisfied if D and V are bounded and K is measurable and bounded in Ω. It must be pointed out, however, that conditions like these may not be verified in some applications (see Larsen [1975: 1]).

5.2.2   Theorem.   *Let D be an open set in $\mathbb{R}_x^3$, V a measurable set in $\mathbb{R}_v^3$, σ a measurable, essentially bounded, nonnegative function in $D \times V$, K a nonnegative measurable function in $D \times V \times V$ such that the operator $\mathfrak{K}$ in* (5.2.16) *is bounded in $L^p(\Omega)$ for some p, $1 \leqslant p < \infty$. Then the operator $\mathfrak{A} = \mathfrak{A}_0 + \mathfrak{K}$, ($\mathfrak{A}_0$ defined by* (5.2.5) *and following comments) belongs to $\mathcal{C}_+(1, \|\mathfrak{K}\|)$ in $L^p(\Omega)$ and the propagator of*

$$\mathfrak{u}'(t) = \mathfrak{A}\mathfrak{u}(t) \qquad (5.2.19)$$

satisfies

$$\mathfrak{S}(t) \geqslant 0 \quad (t \geqslant 0). \qquad (5.2.20)$$

Theorem 5.2.2 immediately follows from Theorem 5.1.2; the result on positivity of $\mathfrak{S}(\cdot)$ is a consequence of Corollary 5.1.3, of the fact that $\mathfrak{S}_0(t)$ is positive for all $t \geqslant 0$ (which results directly from its definition) and from positivity of $\mathfrak{K}$, which is obvious since K is nonnegative.

The preceding treatment can be generalized in various ways; for instance, the space $L^p(\Omega)$ can be replaced by $L^p(\Omega, \mu)$, where $\mu(dx \, dv) = dx \, \nu(dv)$, the assumptions on σ can be relaxed to boundedness from below and integrability in Ω, K need only be bounded below, and so on.

In studying (5.2.1) in the space $C(\bar{\Omega})$ we stumble upon the following difficulty: if η is not continuous, the function $\mathfrak{S}_0(t)\mathfrak{u}(x,v)$ defined by (5.2.9) may not be continuous in (x,v) for some values of t even if $\mathfrak{u}$ satisfies the boundary condition (5.2.2). It is not difficult to surmise that lack of continuity of η is related to lack of convexity of D. (See Figure 5.2.5).

We assume in what follows that D is *open*, *convex*, and *bounded* and that V is *closed* and *bounded*. To begin with, we extend the function η to all $x \in \bar{D}$, adopting the convention that $\eta = 0$ if the set on the right-hand side of (5.2.7) is empty. It is easy to extend Lemma 5.2.1 to show that $\eta(x,v)$ (with (x,v) roaming in $\bar{\Omega} = \bar{D} \times V$) is lower semicontinuous at any $(x_0, v_0) \in \bar{D} \times V$, where $\eta(x_0, v_0) > 0$. (Of course, lower semicontinuity at (x_0, y_0) is obvious if $\eta(x_0, y_0) = 0$.) However, we can prove much more for

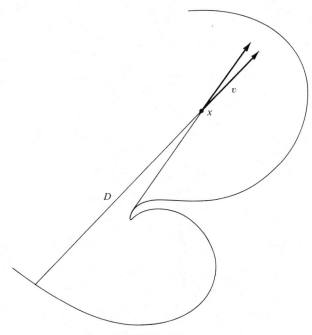

FIGURE 5.2.5

a convex set:

5.2.3   Lemma.   *Let D be convex and let $x_0 \in \overline{D}$, $v_0 \in V$ be such that $0 < \eta(x_0, v_0) < \infty$. Then η is continuous at (x_0, v_0).*

Note that $\eta(x_0, v_0) > 0$ if $x_0 \in D$ and v_0 is any element of V; hence as a particular case of Lemma 5.2.3, we obtain that $\eta(x, v)$ is continuous in Ω if D is convex.

The proof of Lemma 5.2.3 runs as follows. Since η is lower semicontinuous at the point in question, we only have to prove upper semicontinuity. If η is not upper semicontinuous at (x_0, v_0) there exist $\varepsilon > 0$ and two sequences $\{x_n\} \in \overline{D}$, $\{v_n\} \in V$ such that $x_n \to x_0$, $v_n \to v_0$ and

$$\eta(x_n, v_n) \geqslant \eta(x_0, v_0) + 2\varepsilon.$$

Let $0 < \delta < \eta(x_0, v_0)$. Since the point $x_\delta = x_0 - \delta v_0$ belongs to D, there exists a circular disk Δ contained in D with center at x_δ and perpendicular to the line joining x_0 and x_δ. On the other hand, $x_n - (\eta(x_n, v_n) - \varepsilon)v_n$ belongs to D and approaches the half-line

$$\{y; y = x_0 - tv_0, \eta(x_0, v_0) + \varepsilon \leqslant t < \infty\}$$

as $n \to \infty$. Using then the convexity of D as indicated in the next Figure 5.2.6, we deduce that $x_0 - tv_0 \in D$ for $\eta(x_0, v_0) \leqslant t < \eta(x_0, v_0) + \varepsilon$, which is absurd.

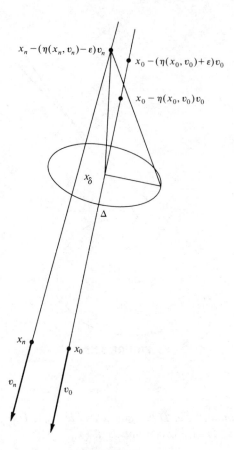

FIGURE 5.2.6

It is easy to see that the continuity property just proved extends to points (x_0, v_0) with $x_0 \in \Gamma$ and v_0 incoming at x_0, although we must have $\eta(x_0, v_0) = 0$ there. However, η may be discontinuous at points (x_0, v_0) where neither v_0 nor $-v_0$ belongs to $\Im(x_0)$; as the following example shows this cannot be remedied by modifying the definition of η at (x_0, v_0). Let D be the unit cube $|x_j| < 1, j = 1, 2, 3$ and assume that V contains a sufficiently large sphere. Let $0 < \alpha \leqslant 1$, $\varepsilon > 0$ small, $x_\varepsilon = (1 - \varepsilon, 0, 0)$, $v_\varepsilon = (-\varepsilon/\alpha, 0, -1)$. Then $x_\varepsilon \to x_0 = (1, 0, 0)$, $v_\varepsilon \to v_0 = (0, 0, -1)$ as $\varepsilon \to 0$, but $\eta(x_\varepsilon, v_\varepsilon) = \alpha$ for all ε; since α is arbitrary, there is no way to extend η continuously to (x_0, v_0). (See Figure 5.2.7).

We assume in the sequel that the function σ in equation (5.2.1) is continuous and nonnegative in $\overline{\Omega} = \overline{D} \times V$ so that inequalities (5.2.4) hold there. The space E is now the subspace of $C(\overline{\Omega})$ consisting of all u that satisfy the boundary condition (5.2.2). Clearly E is closed, thus a Banach

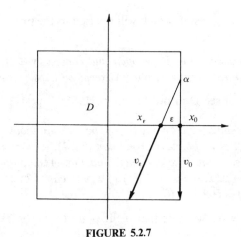

FIGURE 5.2.7

space in its own right. The operator $\mathfrak{A}_0$ is again given by (5.2.5), its domain described in the following way: $\mathfrak{u} \in E$ belongs to $D(\mathfrak{A}_0)$ if and only if for every $(x, v) \in \overline{D} \times V$ the function $s \to \mathfrak{u}(x + sv, v)$ is continuously differentiable in the closed interval $\overline{J} = \overline{D}(x, v)^5$, vanishes at the left endpoint and the directional derivative $(v, \mathrm{grad}_x \mathfrak{u})$ belongs to E as a function of x and v.

For each $t > 0$ and $\mathfrak{u} \in E$, extend $\mathfrak{u}$ setting $\mathfrak{u} = 0$ outside of $\overline{\Omega}$ and define

$$(\mathfrak{S}_0(t)\mathfrak{u})(x, v) = S(x, v, t)\mathfrak{u}(x - tv, v) \quad (\mathfrak{u} \in E) \quad (5.2.21)$$

(note that σ, thus S, is defined in $\overline{D} \times V$). However, it is not immediately obvious that $\mathfrak{S}_0(t)\mathfrak{u}$ will be continuous in $\overline{D} \times V$; in fact, if $\eta(x, v)$ is discontinuous at (x_0, v_0), then $S(x, v, t)$ may be discontinuous at $(x_0, v_0, \eta(x_0, v_0))$. However, one may suspect that at these points the possible discontinuity of S will be cancelled out by the boundary condition satisfied by $\mathfrak{u}$. This is in fact the case, at least if we rule out isolated velocity vectors be means of the following assumption:

(f) (i) *Let $v \in V$, $v \neq 0$, $\varepsilon > 0$. Then, on every half line sufficiently close to that determined by v there exist vectors $v' \in V$ with*

$$|v' - v| \leqslant \varepsilon.$$

(ii) *If $0 \in V$, 0 is an interior point of V.*

We note that in problems of physical interest V has spherical symmetry (if $v \in V$ and $|v'| = |v|$, then $v' \in V$) so that Assumption (f) is trivially satisfied if $0 \notin V$.

---

[5]Here $\overline{D}(x, v) = \overline{D} \cap \xi(x, v)$; since $\overline{D}$ is convex, $\overline{D}(x, v)$ consists of a single closed interval.

The following result, which will be used in the proof of Lemma 5.2.5, is well known.

5.2.4   Lemma.   *Let D be open and convex, $x \in D$, $y \in \bar{D}$. Then all points in the segment joining x and y (except perhaps y itself) belong to D.*

For a proof see Dunford-Schwartz [1958: 1, p. 413].

5.2.5   Lemma.   *Let $u = u(x, v)$ be a continuous function in $\bar{\Omega} = \bar{D} \times V$ that satisfies the boundary condition (5.2.2). Assume that D is convex, open and bounded, that V is compact, and that Assumption* (f) *is satisfied. Then the function $u(x, v, t) = (\mathfrak{S}_0(t)u)(x, v)$ defined by (5.2.21) is continuous in $\bar{\Omega} \times [0, \infty) = \bar{D} \times V \times [0, \infty)$.*

To begin with, we note that if the point $(x_0, v_0, t_0) \in \bar{D} \times V \times [0, \infty)$ is such that

$$u(x_0 - t_0 v_0, v_0) = 0 \tag{5.2.22}$$

(in particular, if $x_0 - t_0 v_0 \in \bar{D}$), then continuity of $u(x, v, t)$ at (x_0, v_0, t_0) is a consequence of the continuity of $u(x, v)$ regardless of behavior of S, since $u(x - tv, v)$ will be continuous *as a function defined in* $\mathbb{R}_x^3 \times V \times [0, \infty)$ at (x_0, v_0, t_0) (recall that u is extended outside of $\bar{\Omega}$ setting $u = 0$ there).

We consider three separate alternatives, assuming for the moment that $v_0 \neq 0$.

(I) $(x_0, v_0, t_0) \in \bar{D} \times V \times [0, \infty)$ *and* $\eta(x_0, v_0) > 0$ (this happens if and only if $x_0 \in D$ or $x_0 \in \Gamma$ with $-v_0$ incoming at x_0). Clearly u will be continuous at (x_0, v_0, t_0) if $t_0 < \eta(x_0, v_0)$ or if $t_0 > \eta(x_0, v_0)$; on the other hand, if $t_0 = \eta(x_0, v_0)$ the point $x_0 - t_0 v_0$ belongs to the boundary Γ and v_0 is incoming at $x_0 - t_0 v_0$ so that (5.2.22) holds.

(II) $(x_0, v_0, t_0) \in \Gamma \times V \times [0, \infty)$ *with* v_0 *incoming at* x_0. In this case,

$$u(x_0, v_0) = 0$$

so that continuity of $u(x, v, t)$ at $t_0 = 0$ results from continuity of $u(x, v)$. Assume then that $t_0 > 0$. Since v_0 is incoming at x_0, $x_0 + \varepsilon v_0 \in D$ if $\varepsilon > 0$ is sufficiently small. Hence, if $x_0 - t_0 v_0 \in \bar{D}$ we deduce from Lemma 5.2.4 that $x_0 \in D$, which is absurd. Accordingly, $x_0 - t_0 v_0 \notin \bar{D}$ and (5.2.22) holds.

(III) $(x_0, v_0, t) \in \Gamma \times V \times [0, \infty)$ *but neither* v_0 *nor* $-v_0$ *are incoming at* x_0. Our objective is again to show that (5.2.22) holds, and we may assume that

$$x_0 - t_0 v_0 \in \Gamma. \tag{5.2.23}$$

In fact, if $x_0 - t_0 v_0 \notin \bar{D}$, (5.2.22) is automatically satisfied; on the other hand, if $x_0 - t_0 v_0 \in D$, the vector $-v_0$ is incoming at x_0 against the hypotheses.

Let x be a point in D not in the line determined by x_0 and v_0. By Lemma 5.2.4 the whole (two-dimensional) interior of the triangle de-

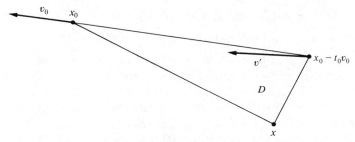

FIGURE 5.2.8

termined by the points x, x_0 and $x_0 - t_0 v_0$ lies in D. Making use of assumption (f) we deduce the existence of vectors v', arbitrarily close to v_0 and such that v' is incoming at $x_0 - t_0 v_0$; since $\mathfrak{u}(x_0 - t_0 v_0, v') = 0$ it follows from continuity of $\mathfrak{u}$ that (5.2.22) holds.

The case $v_0 = 0$ is simple. If $x_0 \in D$, $\eta(x_0, v_0) = \infty$ and we use alternative (I). On the other hand, if $x_0 \in \Gamma$, it follows from the second part of Assumption (f) that there exist arbitrarily small incoming vectors at x_0 so that $\mathfrak{u}(x_0, v_0) = \mathfrak{u}(x_0 - t_0 v_0, v_0) = 0$ for any $t_0 \geqslant 0$ and (5.2.22) holds. This ends the proof.

We can complete now the treatment of (5.2.1) in the space E by slight modifications of the L^p arguments. If $\mathfrak{u} \in E$, we have shown in Theorem 5.2.5 that $\mathfrak{S}_0(t)\mathfrak{u}$ is continuous in $\bar{\Omega} \times [0, \infty)$, hence uniformly continuous in each cylinder $\bar{\Omega} \times [0, a]$, $a > 0$. It is plain that $\mathfrak{S}_0(t)\mathfrak{u}$ satisfies (5.2.2) for all t if $\mathfrak{u}$ does; hence (5.2.21) defines an operator in (E) for all $t \geqslant 0$ and $\mathfrak{S}_0(\cdot)$ is strongly continuous in $t \geqslant 0$. Moreover,

$$\|\mathfrak{S}_0(t)\| \leqslant 1 \quad (t \geqslant 0)$$

since S is nonnegative and bounded by 1. Every λ with $\operatorname{Re} \lambda > 0$ belongs to $\rho(\mathfrak{A}_0)$ and

$$(R(\lambda; \mathfrak{A}_0)\mathfrak{v})(x, v) = \int_0^{\eta(x,v)} e^{-\lambda t} S(x, v, t) \mathfrak{v}(x - tv, v)\, dt$$

for all $(x, v) \in \bar{D} \times V$; since $\mathfrak{v} = 0$ outside of D,

$$R(\lambda; A_0)\mathfrak{v} = \int_0^\infty e^{-\lambda t} \mathfrak{S}_0(t) \mathfrak{v}\, dt \quad (\operatorname{Re} \lambda > 0, \mathfrak{v} \in E)$$

as an (E)-valued integral, so that $\mathfrak{A}_0 \in \mathcal{C}_+(1, 0)$ and $\mathfrak{S}_0(t) = \exp(t\mathfrak{A}_0)$, $t \geqslant 0$; again $\mathfrak{S}_0$ is positive for all t. Making use of Theorem 5.1.2 and Corollary 5.1.3, we obtain:

5.2.6 Theorem. *Let D be open and bounded in $\mathbb{R}^3_x$, V a compact set in $\mathbb{R}^3_v$ satisfying Assumption (f), σ a nonnegative continuous function in $D \times V$, K a nonnegative continuous function in $D \times V \times V$ making (5.2.16) a bounded operator in E. Then $\mathfrak{A} = \mathfrak{A}_0 + \mathfrak{K}$ belongs to $\mathcal{C}_+(1, \|\mathfrak{K}\|)$ and $\exp(t\mathfrak{A})$ is positive for all $t \geqslant 0$.*

We note that $\Re$ maps E into E if K is continuous in $\bar{D} \times V \times V$ and

$$K(x, v, v') = 0 \quad (x \in \Gamma, v \in \Im(x), v' \in V). \qquad (5.2.24)$$

In some treatments of the transport equation it is assumed that the neutrons belong to *velocity groups* characterized by $|v| = k_i > 0$, $i = 1, \ldots, m$. The $C(\bar{\Omega})$ treatment applies fully to this case; here V is the union of the spheres $V_i = \{v : |v| = k_i\}$. For the L^p treatment the appropriate space is $L^p(D \times V_1 \times \cdots \times V_m)$ with measure $\mu = dx \, d\sigma_1 \cdots d\sigma_m$, $d\sigma_i$ the area differential in V_i. Note that assumption (f) is satisfied.

5.2.7   Example.   (i) Let D be a bounded domain, V a measurable set such that

$$|v| \geqslant c > 0 \quad (v \in V). \qquad (5.2.25)$$

Then there exists $t_0 > 0$ such that

$$\mathfrak{S}_0(t) = 0 \quad (t \geqslant t_0). \qquad (5.2.26)$$

(ii) If the operator $\Re$ in (5.2.16) is compact, then $\mathfrak{S}(t)$ is compact and continuous in the norm of (E) in $L^p(\Omega)$ $(1 \leqslant p < \infty)$ for $t \geqslant t_0$ (this follows from (5.2.26) and the perturbation series (5.1.19)).

(iii) The same conclusion holds in $C(\bar{\Omega})$ (here we assume D open, convex, and bounded and V closed, bounded, and satisfying (5.2.25)).

Example 5.2.7 makes possible to obtain information on the spectral behavior of $\mathfrak{A}$ using Example 4.10.3:

5.2.8   Example.   Under the assumptions above, $\sigma(\mathfrak{A})$, the spectrum of $\mathfrak{A}$, is empty or consists of a (finite or infinite) sequence $\lambda_1, \lambda_2, \ldots$ of eigenvalues such that $\mathrm{Re}\,\lambda_k \to -\infty$ if the sequence is infinite. The generalized eigenspaces corresponding to each eigenvalue (Section 3) are finite dimensional.

5.3.   PERTURBATION RESULTS FOR OPERATORS IN $\mathcal{C}_+(1,0)$ AND IN $\mathcal{C}$

In the perturbation theorems considered in Section 5.1, no hypotheses were made on P except that P should be "dominated by A" in the sense specified there. In Theorem 5.1.2 the hypothesis on P is quite stringent; this is perhaps not the case with those in Theorem 5.1.1, but (5.1.2) is not easy to verify as it involves rather complete knowledge of $S(t)$. Note also that when $A \in \mathcal{C}$, the assumptions in Theorems 5.1.1 and 5.1.2 coincide; if (5.1.2) holds, then, for $t > 0$ and $u \in E$,

$$\|Pu\| = \|PS(t)S(-t)u\| \leqslant \alpha(t)\|S(-t)u\| \leqslant C\alpha(t)e^{\omega t}\|u\|$$

so that P must be bounded.

It seems natural to expect that if we assume that P itself or some extension thereof belongs to $\mathcal{C}_+$, then $A + P$ will belong to $\mathcal{C}_+$. However, this is false in general and to place additional assumptions on A, P that will make this true is far from trivial (see the comments in Section 5.9). We present a result in this direction.

5.3.1   Theorem.   *Let $A \in \mathcal{C}_+(1,0)$ (i.e., let A be m-dissipative) and let P be a dissipative operator with $D(P) \supseteq D(A)$ and*

$$\|Pu\| \leqslant a\|Au\| + b\|u\| \quad (u \in D(A)) \tag{5.3.1}$$

with $0 \leqslant a < 1$, $b \geqslant 0$. Then $A + P$ (with domain $D(A)$) is m-dissipative.

Proof.   It is plain that $A + P$ is dissipative, so that we only have to prove that

$$(\lambda I - (A + P))D(A) = E \tag{5.3.2}$$

for some $\lambda > 0$. We consider two cases.

(i) $a < \frac{1}{2}$. Since $(\lambda I - (A + P))D(A) = ((\lambda I - A) - P)R(\lambda)E = (I - PR(\lambda))E$, (5.3.2) will hold if

$$\|PR(\lambda)\| < 1. \tag{5.3.3}$$

But

$$\|PR(\lambda)u\| \leqslant a\|AR(\lambda)u\| + b\|R(\lambda)u\| = a\|(\lambda R(\lambda) - I)u\| + b\|R(\lambda)u\|$$

$$\leqslant \left(2a + \frac{b}{\lambda}\right)\|u\| \quad (\lambda > 0)$$

so that (5.3.3) (hence (5.3.2)) holds if $\lambda > b/(1 - 2a)$.

(ii) $a < 1$. Let $0 < \gamma \leqslant 1$. We have

$$\|(A + \gamma P)u\| \geqslant \|Au\| - \gamma\|Pu\| \geqslant \|Au\| - \|Pu\|$$

$$\geqslant (1 - a)\|Au\| - b\|u\|.$$

Let n be an integer so large that $a/n \leqslant (1 - a)/3$. Then

$$\|\frac{1}{n}Pu\| \leqslant \frac{a}{n}\|Au\| + \frac{b}{n}\|u\|$$

$$\leqslant \frac{1 - a}{3}\|Au\| + \frac{b}{n}\|u\|$$

$$\leqslant \frac{1}{3}\|(A + \gamma P)\| + b\left(\frac{1}{3} + \frac{1}{n}\right)\|u\|.$$

Applying the result in (i) for $\gamma = 0$, we see that $A + (1/n)P$ is m-dissipative; taking then $\gamma = 1/n$, $\gamma = 2/n, \ldots, \gamma = (n-1)/n$, and applying (i) in each case, we prove at the last step that $A + P$ is m-dissipative.[6]

---

[6]Of course, the argument proves (taking $\gamma = 1$) that $A + (1 + n^{-1})P$ is m-dissipative. This is not unexpected since $(1 + n^{-1})P$ satisfies (5.3.1) with different constants $a', b', a' < 1$.

It is plain that Theorem 5.3.1 must fail if $a = 1$. To see this, take $A \in \mathcal{C}_+(1,0)$ with $D(A) \neq E$ and $P = -A$; here $A + P$ is the restriction of 0 to $D(A)$, which is not even closed. However, one may ask whether it is always true that some extension of $A + P$ is m-dissipative, as is the case in the preceding example. The answer is in the affirmative but an additional assumption is needed.

5.3.2   Corollary. Let A be m-dissipative and let P be a dissipative operator with $D(P) \supseteq D(A)$ and

$$\|Pu\| \leqslant \|Au\| + b\|u\| \tag{5.3.4}$$

for some $b \geqslant 0$. Assume in addition that P^*, the adjoint of P is densely defined in E^*. Then $\overline{A + P}$, the closure of $A + P$, is m-dissipative.

Proof. Since $A + P$ is densely defined and dissipative, it is closable and $\overline{A + P}$ is dissipative with respect to some duality map $\theta : D(\overline{A + P}) \to E^*$ (Lemma 3.1.11). It only remains to prove that

$$\left(I - \overline{A + P} \right) D\left(\overline{A + P}\right) = E. \tag{5.3.5}$$

Since the range of a closed dissipative operator must necessarily be closed, we only have to show that $X = (I - \overline{A + P}) D(\overline{A + P})$ is dense in E. Assume this is not the case. Then there exists a nonzero functional $u^* \in E^*$ with

$$\langle u^*, u \rangle = 0 \quad (u \in X).$$

Let $v \in E$ with

$$\|u^*\| \leqslant \langle u^*, v \rangle. \tag{5.3.6}$$

By virtue of Theorem 5.3.1 we know that $A + \gamma P$ is m-dissipative for $0 \leqslant \gamma < 1$. Accordingly, the equation

$$u - (A + \gamma P)u = v \tag{5.3.7}$$

has a solution $u = u_\gamma$ for every γ, $0 \leqslant \gamma < 1$ with

$$\|u_\gamma\| \leqslant \|v\|. \tag{5.3.8}$$

It follows from (5.3.4) that

$$\|Pu_\gamma\| \leqslant \|Au_\gamma\| + b\|u_\gamma\|$$

$$\leqslant \|(A + \gamma P)u_\gamma\| + \gamma\|Pu_\gamma\| + b\|u_\gamma\|$$

$$= \|v - u_\gamma\| + \gamma\|Pu_\gamma\| + b\|u_\gamma\|,$$

thus

$$(1 - \gamma)\|Pu_\gamma\| \leqslant \|v - u_\gamma\| + b\|u_\gamma\|$$

$$\leqslant (b + 2)\|v\|. \tag{5.3.9}$$

Let now $v^* \in D(P^*)$. We have

$$|\langle v^*, (1-\gamma)Pu_\gamma \rangle| = (1-\gamma)|\langle P^*v^*, u_\gamma \rangle|$$

$$\leqslant (1-\gamma)\|P^*u^*\|\,\|v\| \to 0 \quad \text{as } \gamma \to 0.$$

Since $D(P^*)$ is dense in E^* and $\{(1-\gamma)Pu_\gamma; \ 0 \leqslant \gamma < 1\}$ is bounded in E, it follows that

$$(1-\gamma)Pu_\gamma \to 0 \quad \text{as } \gamma \to 1-$$

E^*-weakly. Then we have

$$\|u^*\| \leqslant \langle u^*, v \rangle = \langle u^*, u_\gamma - Au_\gamma - \gamma Pu_\gamma \rangle$$

$$= \langle u^*, (1-\gamma)Pu_\gamma \rangle + \langle u^*, u_\gamma - Au_\gamma - Pu_\gamma \rangle$$

$$= \langle u^*, (1-\gamma)Pu_\gamma \rangle \to 0 \quad \text{as } \gamma \to 1-,$$

which is absurd since we have assumed $u^* \neq 0$. This ends the proof of Corollary 5.3.2.

5.3.3 Remark. Since P is assumed densely defined and dissipative it must be closable by Lemma 3.1.11; thus by virtue of Theorem 4.4, if E is reflexive (in particular if E is a Hilbert space), P^* will always be densely defined. Note also that we only need to assume P dissipative with respect to some duality map $\theta: D(P) \to E^*$; the same observation holds, of course, for Theorem 5.3.1.

The assumptions on A in Corollary 5.3.2 can be relaxed in the following way. Assume that A is only densely defined and dissipative, but that $\bar{A}$ is m-dissipative, and that (5.3.4) holds for P dissipative with $D(P) \supseteq D(A)$. Let $u \in D(\bar{A})$, $\{u_n\}$ a sequence in $D(A)$ with $u_n \to u$, $Au_n \to \bar{A}u$. Writing (5.3.4) for $u_n - u_m$, we see that $\{Pu_n\}$ is as well convergent, thus $u \in D(\bar{P})$ and

$$\|\bar{P}u\| \leqslant \|\bar{A}u\| + b\|u\|. \tag{5.3.10}$$

We can then apply Corollary 5.3.2 to $\bar{A}$ and $\bar{P}$ and deduce that $\overline{\bar{A} + \bar{P}}$ is m-dissipative. But if follows easily from the definition of closure and from (5.3.10) that

$$\overline{\bar{A} + \bar{P}} = \overline{A+P} \tag{5.3.11}$$

so that the conclusion of Corollary 5.3.2 remains unchanged. We note an important byproduct of the preceding arguments. Let $u \in D(\bar{A})$, $\{u_n\}$ a sequence in $D(A)$ with $u_n \to u$, $Au_n \to \bar{A}u$. We have shown that Pu_n is convergent, so that $u \in D(\overline{A+P})$; in other words,

$$D(\bar{A}) \subseteq D(\overline{A+P}).$$

As the example $P = -A$ shows, the opposite inclusion may not be true. However, if P satisfies (5.3.1) with $a < 1$, we have $\|(A+P)u\| \geqslant \|Au\| - \|Pu\|$

$\geqslant (1-a)\|Au\| - b\|u\|$ for $u \in D(A)$ so that

$$\|Pu\| \leqslant a(1-a)^{-1}\|(A+P)u\| + b'\|u\| \quad (u \in D(A)). \quad (5.3.12)$$

Hence $D(\overline{A+P}) \subseteq D(\overline{A})$ results as well and

$$D(\overline{A}) = D(\overline{A+P}). \quad (5.3.13)$$

We collect those observations.

5.3.4 Corollary. *Let A be densely defined and dissipative with $\overline{A}$ m-dissipative, and let P be a dissipative operator with $D(P) \supseteq D(A)$, $D(P^*)$ dense in E^* and*

$$\|Pu\| \leqslant a\|Au\| + b\|u\| \quad (u \in D(A)) \quad (5.3.14)$$

with $0 \leqslant a \leqslant 1, b \geqslant 0$. Then $\overline{A+P}$ is m-dissipative. If $a < 1$, then (i) the domain of $\overline{A+P}$ coincides with the domain of $\overline{A}$; (ii) if $A = \overline{A}$ is itself m-dissipative, then $A + P$ is m-dissipative and $D(A+P) = D(A)$. The assumption on $D(P^*)$ is unnecessary when $a < 1$.

Part (ii) of the case $a < 1$ is of course a restatement of Theorem 5.3.1.

Corollary 5.3.4 has important applications to self-adjoint operators in a Hilbert space H. Call an operator A in H *essentially self-adjoint* if (a) A is densely defined and symmetric. (b) $\overline{A}$ is self-adjoint.[7] Using the facts, proved in Section 3.6, that A is self-adjoint if and only if both iA and $-iA$ are m-dissipative (that is, if $iA \in \mathcal{C}(1,0)$) and that P is symmetric if and only if iP and $-iP$ are dissipative, we obtain the following result:

5.3.5 Corollary. *Let A be essentially self-adjoint, P a symmetric operator with $D(P) \supseteq D(A)$ satisfying (5.3.14) with $0 \leqslant a \leqslant 1$, $b \geqslant 0$. Then $\overline{A+P}$ is self-adjoint, that is, $A + P$ is essentially self-adjoint. If $a < 1$, then (i) $D(\overline{A}) = D(\overline{A+P})$, (ii) if A is self-adjoint, then $A + P$ is self-adjoint and $D(A+P) = D(A)$.*

The following result applies to operators in $\mathcal{C}$ and is roughly of the same type as Theorem 5.3.1; although P is not assumed here to belong to $\mathcal{C}$ or any subclass thereof, we must use a reinforced version of (5.3.1).

5.3.6 Theorem. *Let $A \in \mathcal{C}$, and let P be an operator with $D(P) \subseteq D(A)$ and such that, for every $a > 0$, no matter how small, there exists $b = b(a)$ with*

$$\|Pu\| \leqslant a\|Au\| + b\|u\| \quad (u \in D(A)). \quad (5.3.15)$$

Then $A + P$ (with domain $D(A)$) belongs to $\mathcal{C}$.

---

[7]It was observed in Section 6 that if A is symmetric, iA is dissipative, hence $\overline{A}$ exists by Lemma 3.1.11.

Proof. If $A \in \mathcal{C}$, then (Theorem 4.2.1) there exists α such that $R(\lambda)$ exists for $\operatorname{Re} \lambda > \alpha$ and

$$\|R(\lambda)\| \leqslant \frac{C}{|\lambda - \alpha|} \tag{5.3.16}$$

there. Clearly we have

$$\lambda I - A - P = (I - PR(\lambda))(\lambda I - A)$$

in $D(A)$; accordingly, if $\|PR(\lambda)\| < 1$, $\lambda \in \rho(A + P)$, and

$$R(\lambda; A + P) = R(\lambda)(I - PR(\lambda))^{-1}. \tag{5.3.17}$$

We estimate $\|PR(\lambda)\|$ using (5.3.15):

$$\|PR(\lambda)\| \leqslant a\|AR(\lambda)\| + b\|R(\lambda)\|$$

$$\leqslant a\left(1 + \frac{C|\lambda|}{|\lambda - \alpha|}\right) + \frac{bC}{|\lambda - \alpha|}. \tag{5.3.18}$$

Take now $\alpha' > \alpha$. Then $|\lambda/(\lambda - \alpha)|$ is bounded in $\operatorname{Re} \lambda \geqslant \alpha'$; therefore we may choose a so small that

$$a\left(1 + \frac{C|\lambda|}{|\lambda - \alpha|}\right) \leqslant \gamma < \frac{1}{2} \quad (\operatorname{Re} \lambda \geqslant \alpha').$$

Taking α' larger if necessary, we may also assume that

$$\frac{bC}{|\lambda - \alpha|} \leqslant \gamma \quad (\operatorname{Re} \lambda \geqslant \alpha').$$

Combining these two estimates with (5.3.18), we see that $\|PR(\lambda)\| \leqslant 2\gamma < 1$ in $\operatorname{Re} \lambda \geqslant \alpha'$; therefore, $(I - PR(\lambda))^{-1}$ exists and $\|(I - PR(\lambda))^{-1}\| \leqslant (1 - 2\gamma)^{-1}$ there. In view of (5.3.16) and (5.3.17), we obtain

$$\|R(\lambda; A + P)\| \leqslant \frac{C'}{|\lambda - \alpha|} \leqslant \frac{C'}{|\lambda - \alpha'|} \quad (\operatorname{Re} \lambda \geqslant \alpha'),$$

which, using Lemma 4.2.3, shows that $A + P \in \mathcal{C}$ as claimed.

Note that we only need to assume the existence of an estimate of the type of (5.3.15) for some $a < 1/(2 + 2C)$.

5.3.7 Corollary. *Let $A \in \mathcal{C}$ and let P be bounded. Then $A + P \in \mathcal{C}$.*

Proof. (5.3.15) holds for any a and $b = \|P\|$.

5.3.8 Remark. The comments on Remark 5.3.3 apply here with some modifications. Assume that A is closable and that $\bar{A} \in \mathcal{C}$, all the other hypotheses in Theorem 5.3.5 remaining unchanged except that we *assume* now that P is closable. Hence (5.3.15) holds for $\bar{A}$ and $\bar{P}$. Theorem 5.3.5 then implies that $\bar{A} + \bar{P} \in \mathcal{C}$. If $A + P$ were closable, we would have

$$\bar{A} + \bar{P} = \overline{\bar{A} + \bar{P}} = \overline{A + P} \tag{5.3.19}$$

(the first equality follows from the fact that $\overline{A} + \overline{P}$, which belongs to $\mathcal{C}$, must be closed). It is remarkable that (5.3.19) will always be true, as (5.3.15) and the fact that A and P are closable imply that $A + P$ is closable. To see this, we take $a < 1$ in (5.3.15) and use its consequence (5.3.12). Let $\{u_n\}$ be a sequence in $D(A)$ with $u_n \to 0$, $(A + P)u_n \to v$. Then it follows from (5.3.12) applied to $u_n - u_m$ that $Pu_n \to w \in E$. Since P is closable, $w = 0$; then $Au_n \to v$, and we deduce that $v = 0$.

5.3.9   Corollary.   *Let A be a closable, densely defined operator with $\overline{A} \in \mathcal{C}$, P a closable operator with $D(P) \supseteq D(A)$ satisfying (5.3.15) for a arbitrarily small. Then $A + P$ (with domain $D(A)$) is closable and $\overline{A + P} \in \mathcal{C}$.*

5.4.   PERTURBATION RESULTS FOR THE SCHRÖDINGER AND DIRAC EQUATIONS

We retake here the study of the Schrödinger equation (1.4.1), this time for a potential $U \neq 0$. Setting $q = -h^{-1}U$, we may write (1.4.1) as an equation in the Hilbert space $E = L^2(\mathbb{R}^3)_{\mathbb{C}}$ in the form

$$u'(t) = (\mathcal{C} + \mathcal{Q})u(t), \tag{5.4.1}$$

where $\mathcal{C} = i\kappa\Delta$ is the operator described in Section 1.4 with $\kappa = h/2\mathfrak{m}$ and $\mathcal{Q}$ is the multiplication operator

$$(\mathcal{Q}u)(x) = iq(x)u(x) \tag{5.4.2}$$

(the domain of $\mathcal{Q}$ will be precised later). It was proved in Section 1.4 that the operator $\mathcal{C}$ belongs to $\mathcal{C}(1,0)$, thus by the results in Section 3.6, $i\mathcal{C}$ is self-adjoint in E. (This was also shown directly in Example 1.4.6.) In view of the interpretation of the solution ψ of (1.4.1) sketched in Section 1.4, it is of great importance to find conditions on $\mathcal{Q}$ (that is, on q) that make $i(\mathcal{C} + \mathcal{Q})$ a self-adjoint, or at least an essentially self-adjoint operator. For one result in this direction, we use (part of) Corollary 5.3.5 with $P = \mathcal{Q}$.

5.4.1   Theorem.   *Let*

$$q(x) = q_0(x) + q_1(x), \tag{5.4.3}$$

where $q_0 \in L^\infty(\mathbb{R}^3)_{\mathbb{R}}$ and $q_1 \in L^2(\mathbb{R}^3)_{\mathbb{R}}$, and let the domain of the operator $\mathcal{Q}$ in (5.4.2) consist of all $u \in L^2(\mathbb{R}^3)$ such that $qu \in L^2(\mathbb{R}^3)$. Then $D(\mathcal{Q}) \supseteq D(\mathcal{C})$ and the operator $i(\mathcal{C} + \mathcal{Q})$ is self-adjoint.

Proof.   To show that $D(\mathcal{Q}) \supseteq D(\mathcal{C})$, it is obviously sufficient to prove that every $u \in D(\mathcal{C})$ is bounded in $\mathbb{R}^3$. This follows from the Fourier analysis manipulations at the end of Section 1.6, but a slightly improved version is essential here. Using the notations in Section 8 for Fourier transforms, let $u \in D(\mathcal{Q})$ and $\rho > 0$. Then

$$\mathcal{F}u(\sigma) = \frac{1}{|\sigma|^2 + \rho^2}\left(|\sigma|^2 + \rho^2\right)\mathcal{F}u(\sigma)$$

$$= \frac{1}{|\sigma|^2 + \rho^2}\left(-\mathcal{F}(\Delta u)(\sigma) + \rho^2\mathcal{F}u(\sigma)\right). \tag{5.4.4}$$

Making use of the Schwarz inequality, we see that $\mathcal{F}u \in L_\sigma^1(\mathbb{R}^3)$ and

$$\|\mathcal{F}u\|_{L_\sigma^1} \leqslant \pi\rho^{-1/2}\|\left(-\Delta + \rho^2\right)u\|_{L_x^2}. \tag{5.4.5}$$

Accordingly, after eventual modification in a null set, $u \in L^2(\mathbb{R}^3) \cap C_0(\mathbb{R}^3) \subseteq D(\mathcal{Q})$ and[8]

$$\|u\|_{C_0} \leqslant C\left(\rho^{-1/2}\|\Delta u\| + \rho^{3/2}\|u\|\right),$$

where C does not depend on ρ. Moreover, $\|\Delta u\| = \kappa^{-1}\|\mathcal{Q}u\|$, hence

$$\|\mathcal{Q}u\| \leqslant \|q_0\|_{L^\infty}\|u\| + \|q_1\|_{L^2}\|u\|_{C_0}$$

$$\leqslant C'\left(\rho^{-1/2}\|\mathcal{Q}u\| + \left(\rho^{3/2} + 1\right)\|u\|\right) \tag{5.4.6}$$

so that (5.3.14) holds for any $a > 0$ if ρ is sufficiently large and Corollary 5.3.5 applies (note that $i\mathcal{Q}$ is symmetric since q is real-valued). This completes the proof.

Observe that Theorem 5.4.1 applies to the *Coulomb potential* $U(x) = c|x|^{-1}$ for any value of c. Potentials of the form $U(x) = V(|x - x_0|)$ correspond to motion of particles in *central attractive* (or *repulsive*) fields (see Schiff [1955:1]). Generalizing the previous observation, Theorem 5.4.1 applies if V is bounded at infinity and $r^2V(r)^2$ is integrable near zero (for instance, if $U(r) = cr^{-\beta}$ with $\beta < \frac{3}{2}$). Also, it should be pointed out that even if q is complex-valued, inequality (5.4.6) holds although $\mathcal{Q}$ fails of course to be symmetric. Nevertheless, if we assume that

$$|\text{Im}\, q(x)| \leqslant \omega \quad \text{a.e. in } \mathbb{R}^3,$$

it is easy to see that Corollary 5.3.4 applies both to $\mathcal{Q}, \mathcal{Q} - \omega I$ and to $-\mathcal{Q}, -\mathcal{Q} - \omega I$ showing that $\mathcal{Q} + \mathcal{Q} \in \mathcal{C}(1, \omega)$.

5.4.2 Example. *The Schrödinger equation of a system of n particles.* The quantum mechanical description of the motion of a particle sketched in Section 1.4 extends to the case of n particles. The Schrödinger equation

---

[8]Here $\|\cdot\|$, without subindex, indicates the L^2 norm. The same observation holds in forthcoming inequalities.

takes the form

$$-\frac{h}{i}\frac{\partial \psi}{\partial t} = -\sum_{j=1}^{n}\frac{h^2}{2m_j}\Delta_j \psi + U\psi \qquad (5.4.7)$$

where the function ψ depends on the $3n$-dimensional vector $x = (x^1,\ldots,x^n)$ $(x^j = (x_1^j, x_2^j, x_3^j),\ 1 \le j \le n), \Delta_j$ denotes the Laplacian in the variables x^j and the potential U depends on x; m_j is the mass of the jth particle. Again $|\psi|^2$ is interpreted as a probability density in $\mathbb{R}^{3n}$, hence the $L^2(\mathbb{R}^{3n})$ norm of ψ must remain constant. The Fourier analysis of the unperturbed case $U = 0$ is essentially covered by the results in Section 1.4; we define, for $\kappa_1 = h/2m_1,\ldots,\kappa_n = h/2m_n$,

$$\mathcal{C}u = i(\kappa_1\Delta_1 + \cdots + \kappa_n\Delta_n)$$

with domain u consisting of all $u \in L^2(\mathbb{R}^{3n})$ such that $\Sigma\kappa_j\Delta_j$ (understood in the sense of distributions) belongs to L^2. An element $u \in L^2(\mathbb{R}^{3n})$ belongs to $D(\mathcal{C})$ if and only if its Fourier transform $\mathcal{F}u$ satisfies $|\sigma|^2\mathcal{F}u \in L^2(\mathbb{R}_\sigma^{3n})$ (here $\sigma = (\sigma^1,\ldots,\sigma^n)$, $\sigma^j = (\sigma_1^j, \sigma_2^j, \sigma_3^j)$, $1 \le j \le n$) and $\mathcal{F}\mathcal{C}u = (-i\Sigma\kappa_j|\sigma^j|^2)\mathcal{F}u$. The operator $i\mathcal{C}$ is self-adjoint (this follows from Example 1.4.6). It can be shown arguing as in Remark 1.4.3 that $D(\mathcal{C}) = H^2(\mathbb{R}^{3n})$ and that the norm of $H^2(\mathbb{R}^{3n})$ is equivalent to the graph norm $\|u\| + \|\mathcal{C}u\|$ or to its Hilbert version $(\|u\|^2 + \|\mathcal{C}u\|^2)^{1/2}$ (which are instantly seen to be equivalent themselves).

We examine the perturbation of $\mathcal{C}$ by an operator $\mathfrak{Q}$ given by (5.4.2), with $q = -h^{-1}U$. An important instance is that where the particles move under the influence of Coulomb forces, both due to a central field and to the other particles; here

$$q(x) = \sum_j \frac{c_j}{|x^j|} + \sum\sum_{j<k} \frac{c_{jk}}{|x^j - x^k|}, \qquad (5.4.8)$$

where c_j, c_{jk} are real constants. Theorem 5.4.1 does not apply now: for instance, if $n \ge 2$, the function $1/|x^1|$ does not admit a decomposition of the form (5.4.3); moreover the elements of $D(\mathcal{C})$ are no longer continuous or even bounded in $\mathbb{R}^{3n}$ since $2 \le 3n/2$ (see Section 1.6). We define a set $I_1,\ldots,I_n$ of (nonlinear) operators from $L^2(\mathbb{R}^{3n})$ into $L^2(\mathbb{R}^3)$ as follows:

$$(I_ju)(x^j) = \left\{\int_{\mathbb{R}^{3(n-1)}} |u(x^1,\ldots,x^n)|^2\,dx^1 \cdots dx^{j-1}\,dx^{j+1} \cdots dx^n\right\}^{1/2}.$$

Clearly $\|I_ju\| = \|u\|$ for all $u \in L^2$. We shall show that if $u \in D(\mathcal{C})$ and $a > 0$ is arbitrary, there exists $b \ge 0$ such that

$$|(I_ju)(x^j)|^2 \le a\|\mathcal{C}u\|^2 + b\|u\|^2 \quad (x^j \in \mathbb{R}^3, u \in D(\mathcal{C})). \quad (5.4.9)$$

To prove (5.4.9) we may obviously assume that $j = 1$. Denote (as before) by

$\mathcal{F}$ the Fourier transform in $\mathbb{R}^{3n}$ and by $\mathcal{F}_1$ (resp. $\mathcal{F}_{1-}$) the Fourier transform in x^1 (resp. in $x^2, \ldots, x^n$). If $u \in \mathcal{D}(\mathbb{R}^{3n})$, we have

$$|I_1 u(x^1)|^2 = \int |u(x)|^2 \, dx^2 \cdots dx^n$$

$$= \int |\mathcal{F}_{1-} u(x^1, \sigma^2, \ldots, \sigma^n)|^2 \, d\sigma^2 \cdots d\sigma^n$$

$$\leqslant \frac{1}{(2\pi)^3} \int \left\{ \int |\mathcal{F}u(\sigma^1, \ldots, \sigma^n)| \, d\sigma^1 \right\}^2 d\sigma^2 \cdots d\sigma^n.$$

$$(5.4.10)$$

We apply now (5.4.5) to $\mathcal{F}_{1-} u(x^1, \sigma^2, \ldots, \sigma^n)$ for $\rho > 0$, obtaining

$$\|\mathcal{F}u(\sigma)\|_{L^1_{\sigma^1}} \leqslant \pi \rho^{-1/2} \|(|\sigma^1|^2 + \rho^2) \mathcal{F}u(\sigma)\|_{L^2_{\sigma^1}}.$$

Replacing in (5.4.10), we obtain

$$|I_1 u(x^1)|^2 \leqslant C(\rho^{-1} \|\mathcal{C}u\|^2 + \rho^3 \|u\|^2), \tag{5.4.11}$$

therefore (5.4.9) results for arbitrarily small a. Using the fact that $\mathcal{D}$ is dense in $D(\mathcal{C}) = H^2$ in the norm of H^2, hence in the graph norm of $\mathcal{C}$ (see Remark 1.4.3), the inequality is seen to hold for any $u \in D(\mathcal{C})$ (note that $|I_j u(x^j) - I_j v(x^j)| \leqslant |I_j(u - v)(x^j)|$, hence passage to the limit is licit). Inequality (5.4.9) is used for the potential (5.4.8) as follows. The domain of $\mathcal{Q}$ is defined again as the set of all $u \in L^2$ such that $qu \in L^2$; $\mathcal{Q}$ is symmetric since the c_j and the c_{jk} are real. For terms of the form $c_j |x^j|^{-1}$, we note that

$$\int |x^j|^{-2} |u(x)|^2 \, dx = \int |x^j|^{-2} |I_j u(x^j)|^2 \, dx^j. \tag{5.4.12}$$

We divide the domain of integration in $|x^1| \geqslant 1$ and $|x^1| \leqslant 1$; each of the integrals is estimated as follows:

$$\int_{|x^j| \geqslant 1} |x^j|^{-2} |I_j u(x^j)|^2 \, dx^j \leqslant \int_{|x^j| \geqslant 1} |I_j u(x^j)|^2 \, dx^j$$

$$\leqslant \int_{\mathbb{R}^{3n}} |u(x)|^2 \, dx = \|u\|^2,$$

$$\int_{|x^j| \leqslant 1} |x^j|^{-2} |I_j u(x^j)|^2 \, dx^j \leqslant (a\|\mathcal{C}u\|^2 + b\|u\|^2) \int_{|x^j| \leqslant 1} |x^j|^{-2} \, dx^j$$

$$\leqslant (a'\|\mathcal{C}u\| + b'\|u\|)^2,$$

where a' can be taken as small as desired (we have used here the inequality $\alpha^2 + \beta^2 \leqslant (\alpha + \beta)^2$).

Consider next any of the other terms, say $c_{12} |x^1 - x^2|^{-1}$. The change of variables $y^1 = x^1 - x^2$, $y^2 = x^2, \ldots, y^n = x^n$ yields

$$\int |x^1 - x^2|^{-2} |u(x)|^2 \, dx = \int |x^1|^{-2} |v(x)|^2 \, dx, \tag{5.4.13}$$

where $v(x^1, \ldots, x^n) = u(x^1 + x^2, x^2, \ldots, x^n)$ is easily seen to belong to $D(\mathcal{Q})$ if u does, with $\|\mathcal{Q}v\| \leqslant C\|\mathcal{Q}u\|$. Collecting our observations, we see that $\mathcal{Q}, \mathcal{Q}$ satisfy (5.3.1) with a arbitrarily small, so that $i(\mathcal{Q} + \mathcal{Q})$ with domain $D(\mathcal{Q})$ is self-adjoint.

5.4.3 Example. *Perturbation theory for the Dirac equation.* As a consequence of the results in Example 1.6.6, we know that the operator

$$\mathcal{Q}u = c\sum \alpha_j D^j u + i\lambda\beta u \qquad (5.4.14)$$

in $L^2(\mathbb{R}^3)^4$ ($\lambda = mc^2/h$) whose domain consists of all $u = (u_1, u_2, u_3, u_4)$ in $H^1(\mathbb{R}^3)^4$, belongs to $\mathcal{C}(1,0)$; equivalently, $i\mathcal{Q}$ is self-adjoint. We consider now the case $U \neq 0$ in (1.6.34); as for the Schrödinger equation, we write $q = -h^{-1}U$, but we limit ourselves to examine here the Coulomb potential

$$q(x) = \frac{d}{|x - x^0|}, \qquad (5.4.15)$$

where x^0 is a point in three-dimensional Euclidean space. A key role in the following argument is played by the inequality

$$\int |x - x^0|^{-2}|u(x)|^2 \, dx \leqslant \frac{4}{(m-2)^2} \int |\mathrm{grad}\, u(x)|^2 \, dx \qquad (5.4.16)$$

valid for any function $u \in H^1(\mathbb{R}^m)$, $m \geqslant 3$ and any $x^0 \in \mathbb{R}^m$. We prove it first for a smooth function assuming (as we may) that $x^0 = 0$. Let $\varphi = \varphi(\rho)$ be a function in $\mathcal{D}(\mathbb{R}^1)$. Then

$$\left(\rho^{m-2}|\varphi|^2\right)' = (m-2)\rho^{m-3}|\varphi|^2 + 2\rho^{m-2}\mathrm{Re}(\varphi'\bar{\varphi}).$$

Integrating in $0 \leqslant \rho < \infty$,

$$\int \rho^{m-3}|\varphi|^2 \, d\rho = -\frac{2}{(m-2)} \int \rho^{m-2}\mathrm{Re}(\varphi'\bar{\varphi}) \, d\rho$$

$$\leqslant \frac{2}{m-2}\left(\int \rho^{m-3}|\varphi|^2 \, d\rho\right)^{1/2}\left(\int \rho^{m-1}|\varphi'|^2 \, d\rho\right)^{1/2}$$

by the Schwarz inequality, so that

$$\int \rho^{m-3}|\varphi|^2 \, d\rho \leqslant \left(\frac{2}{m-2}\right)^2 \int \rho^{m-1}|\varphi'|^2 \, d\rho.$$

We take now $u \in \mathcal{D}(\mathbb{R}^m)$ and apply this inequality to $\varphi(\rho) = u(\rho x)$, $|x| = 1$; observing that $|\varphi'(\rho)| = |\Sigma x_k D^k u(\rho x)| \leqslant |x|\,|\mathrm{grad}\, u(\rho x)|$ and integrating the resulting inequality in $|x| = 1$, (5.4.16) results. To establish the inequality for an arbitrary $u \in H^1(\mathbb{R}^m)$, we approximate u in the H^1 norm by a sequence $\{\varphi_n\}$ in $\mathcal{D}$ and apply (5.4.16) to each φ_n. Obviously we can take limits on the right-hand side; on the left-hand side, we select if necessary a subsequence so that $\{\varphi_n\}$ is pointwise convergent to u and apply Fatou's theorem.

We apply inequality (5.4.16) for $m = 3$ to the function $q(x)$ in (5.4.15) and to each coordinate of $u \in D(\mathcal{Q}) = H^1(\mathbb{R}^3)^4$. Adding up we obtain

$$\int q(x)^2 |u(x)|^2 \, dx = d^2 \int |x - x^0|^{-2} |u(x)|^2 \, dx$$

$$\leq 4d^2 \int |\operatorname{grad} u(x)|^2 \, dx$$

$$\leq 4c^{-2}d^2 \|\mathcal{Q}u\|^2 \tag{5.4.17}$$

The last inequality requires some explanation. For the operator (5.4.14), we have $\|\mathcal{Q}u\|^2 = \|c\Sigma\alpha_j D^j u + i\lambda\beta u\|^2 = \|(c\Sigma\sigma_j\alpha_j - \lambda\beta)\mathcal{F}u\|^2 = \|(c\Sigma\sigma_j\alpha_j)\mathcal{F}u\|^2 + |\lambda|^2\|u\|^2$. We make use of (1.6.38) plus the explicit computation of the constant θ in (1.6.39): in this way, (5.4.17) follows.

We apply Corollary 5.3.5 to the self-adjoint operator $i\mathcal{Q}$ and the symmetric operator $i\mathcal{Q}$: if

$$|d| \leq c/2, \tag{5.4.18}$$

the operator $\overline{i(\mathcal{Q} + \mathcal{Q})}$ is self-adjoint, while if the inequality (5.4.18) is strict, the operator $i(\mathcal{Q} + \mathcal{Q})$ (with domain $D(\mathcal{Q})$) is self-adjoint. Since (5.4.18) imposes a definite limitation on the size of $|d|$, it is interesting to translate it directly for the potential U in the equation (1.6.27). Writing $U = e|x - x^0|^{-1}$, the inequality turns out to be $|e| \leq ch/2$.

Obviously, this perturbation analysis can be applied to the general symmetric hyperbolic operator $\mathcal{Q}$ in (1.6.22) if condition (1.6.36) is satisfied.

5.4.4 Remark. When the potential function U in the Schrödinger equation (1.4.1) is "too singular" it may be the case that $i(\mathcal{Q} + \mathcal{Q})$ fails to be essentially self-adjoint. It is then important to construct self adjoint extensions $i(\mathcal{Q} + \mathcal{Q})_s$ of $i(\mathcal{Q} + \mathcal{Q})$ (so that the Schrödinger equation can be "embedded" in the equation

$$u'(t) = (\mathcal{Q} + \mathcal{Q})_s u(t), \tag{5.4.19}$$

which possesses a unitary propagator). As pointed out in Section 6, these extensions will exist if and only if both defficiency indices of $\mathcal{Q} + \mathcal{Q}$ coincide. However, mere existence of $(\mathcal{Q} + \mathcal{Q})_s$ is of little physical interest; at the very least, the domain of $(\mathcal{Q} + \mathcal{Q})_s$ should be clearly identified. We present here results generalizing Theorem 5.4.1 that combine perturbation theory with the methods used in Section 4.6 to deal with parabolic equations with rough coefficients. In the nomenclature of Section 4.6 we consider the differential operator

$$A_0 = \Delta + q(x) \tag{5.4.20}$$

in $\Omega = \mathbb{R}^m$, Δ the m-dimensional Laplacian and q real, locally integrable, and bounded above in $\mathbb{R}^m$. The operator A_0 is not obviously defined for any nonzero function (qu may fail to be square integrable even for $u \in \mathfrak{D}$). However, an "extension" can be defined as follows. Let $\lambda \geq \text{ess.\,sup}\, q + 1$ and $\mathcal{H} \subseteq L^2(\mathbb{R}^m)$ the completion of $\mathfrak{D}(\mathbb{R}^m)$ with respect to the scalar product

$$(\varphi, \psi)_\lambda = \int_{\mathbb{R}^m} \{(\text{grad}\,\varphi, \text{grad}\,\psi) + (\lambda - q(x))\bar{\varphi}\psi\}\, dx. \quad (5.4.21)$$

Then an element $u \in \mathcal{H}$ belongs to $D(A)$ if and only if $w \to (u,w)_\lambda$ is continuous in $\mathcal{H}$ *in the norm of* L^2, and Au is the unique element of L^2 that satisfies

$$\lambda(u,w) - (u,w)_\lambda = (Au,w) \quad (w \in \mathcal{H}) \quad (5.4.22)$$

(note that in the present situation we have $B_\lambda(u,w) = (u,w)_\lambda$). The fact that A is self-adjoint follows from (4.6.30) and Remark 4.6.8 (note that the formal adjoint A_0' coincides with A_0). We can then set $(\mathfrak{A} + \mathfrak{D})_s = iA$ and obtain a solution of our problem. The characterization of $D(A) = D(\mathfrak{A} + \mathfrak{D})$, however, is too vague for comfort and we try to obtain a better one. Note first that the uniform ellipticity assumption (4.6.2) is satisfied so that $D(A) \subseteq \mathcal{H} \subseteq H^1(\mathbb{R}^m)$. On the other hand, if $u \in D(A)$, we obtain integrating by parts that

$$\int u\Delta\varphi\, dx = \int (Au - qu)\varphi\, dx$$

for $\varphi \in \mathfrak{D}$, so that $\Delta u + qu = Au \in L^2(\mathbb{R}^m)$. (Note that $|q|u = \sqrt{|q|}\, u\sqrt{|q|}$ is locally integrable by the Schwartz inequality so that $\Delta u \in L^1_{\text{loc}}(\mathbb{R}^m)$.) Conversely, if $u \in \mathcal{H}$ is such that $\Delta u + qu \in L^2(\mathbb{R}^m)$ and $\varphi \in \mathfrak{D}$, we have, after another integration by parts,

$$(u,\varphi)_\lambda = -\int u\Delta\varphi\, dx + \int (\lambda - q)u\varphi\, dx$$

$$= -\int (\Delta u + qu - \lambda u)\varphi\, dx.$$

Hence $(u,w)_\lambda$ is continuous in the norm of L^2 for $w \in \mathfrak{D}$ (a fortiori, for $w \in \mathcal{H}$) and $u \in D(A)$ with $Au = \Delta u + qu$. We have then shown that

$$D((\mathfrak{A} + \mathfrak{D})_s) = D(A) = \{u \in \mathcal{H}; \Delta u + qu \in L^2\}. \quad (5.4.23)$$

This result is not entirely satisfactory, however, since functions in $\mathcal{H}$ are not easy to characterize. A better identification can be achieved at the cost of some restrictions on q. Surprisingly enough, these bear on the presence of singularities of q at finite points rather than on its behavior at infinity.

5.4.5   Theorem.   *Let q be real, locally integrable, and bounded above in* $\mathbb{R}^m$. *Let*

$$\mathfrak{m}_{\rho, R}(q; x) = \int_{|y - x| \leqslant R} |y - x|^{2 - m - \rho} |q(y)| \, dy \qquad (5.4.24)$$

and assume there exist $\rho, R > 0$ *such that* $\mathfrak{m}_{\rho, R}$ *is locally bounded.*[9] *Then* $D(A)$ *consists of all* $u \in H^1(\mathbb{R}^m)$ *such that* $\sqrt{|q|}\, u$ *and* $\Delta u + qu$ *belong to* $L^2(\mathbb{R}^m)$.

For the proof we need a simple inequality.

5.4.6   Lemma.   *Let p be locally integrable and a.e. nonnegative,* ψ *continuously differentiable in* $\mathbb{R}^m$ *with compact support. Then, if* $R > 0$, $0 < \rho < 2$,

$$\int_{\mathbb{R}^m} p(x) |\psi(x)|^2 \, dx$$

$$\leqslant \frac{2}{\omega_m \rho} \int_{\mathbb{R}^m} \left(R^\rho |\mathrm{grad}\, \psi(x)|^2 + R^{\rho - 2} |\psi(x)|^2 \right) \mathfrak{m}_{\rho, R}(p; x) \, dx,$$

$$(5.4.25)$$

where $\mathfrak{m}_{\rho, R}$ *is defined as in* (5.4.24) *and* ω_m *denotes the hyperarea of the unit sphere in* $\mathbb{R}^m$.

Proof.   Let $\eta(r)$ be continuously differentiable in $r \geqslant 0$. Obviously there exists s, $0 < s \leqslant R$, such that the value of $|\eta|$ at s does not surpass its average over $0 \leqslant r \leqslant R$. Expressing $\eta(s) - \eta(0)$ as the integral of $\eta'(r)$ in $0 \leqslant r \leqslant s$, we obtain

$$|\eta(0)| \leqslant \int_0^R \left(|\eta'(r)| + R^{-1} |\eta(r)| \right) dr.$$

Apply this inequality to $\eta(r) = \psi(x - r(y - x))$, afterwards integrating in $|y - x| = 1$. An interchange of the order of integration combined with the inequality $|\Sigma(y_k - x_k) D^k \psi| \leqslant |y - x| |\mathrm{grad}\, \psi|$ easily yields

$$|\psi(x)| \leqslant \frac{1}{\omega_m} \int_{|y - x| \leqslant R} |y - x|^{1 - m} \left(|\mathrm{grad}\, \psi(y)| + R^{-1} |\psi(y)| \right) dy.$$

We estimate now by means of the Schwarz inequality writing $|y - x|^{1 - m} = |y - x|^{(\rho - m)/2} |y - x|^{(2 - m - \rho)/2}$ and noting that $(a + b)^2 \leqslant 2(a^2 + b^2)$. We obtain

$$|\psi(x)|^2 \leqslant \frac{2 R^\rho}{\omega_m \rho} \int_{|y - x| \leqslant R} |y - x|^{2 - m - \rho} \left(|\mathrm{grad}\, \psi(y)|^2 + R^{-2} |\psi(y)|^2 \right) dy.$$

---

[9]An application of the Fubini theorem shows that $\mathfrak{m}_{\rho, R}$ exists a.e. and is locally integrable if $\rho < 2$. We only use the fact that q is locally integrable.

Finally, we multiply by $p(x)$ and integrate in $\mathbb{R}^m$; after an obvious interchange of the order of integration, (5.4.25) results.

Proof of Theorem 5.4.5. In view of (5.4.23), to show that any u satisfying the assumptions belongs to $D(A)$ it is enough to prove that if λ is as in (5.4.21), then any u satisfying

$$u \in H^1(\mathbb{R}^m), \quad (\lambda - q)^{1/2} u \in L^2(\mathbb{R}^m), \qquad (5.4.26)$$

belongs to $\mathcal{H}$, that is, there exists a sequence $\{\varphi_n\}$ in $\mathcal{D}$ such that

$$\|\varphi_n - u\|_{H^1} \to 0, \quad \|(\lambda - q)^{1/2}(\varphi_n - u)\|_{L^2} \to 0, \qquad (5.4.27)$$

as $n \to \infty$. To this end, let $\{\chi_n\}$ be a sequence in $\mathcal{D}$ such that $\chi_n(x) = 1$ for $|x| \leq n$, $\chi_n(x) = 0$ for $|x| \geq n + 1$ and having uniformly bounded first partials. It is a simple matter to show that

$$\|\chi_n u - u\|_{H^1} \to 0, \quad \|(\lambda - q)^{1/2}(\chi_n u - u)\|_{L^2} \to 0,$$

hence we may assume that the function u itself has compact support, say, contained in the ball $|x| \leq a$. Let $\{\beta_n\}$ be a mollifying sequence in $\mathcal{D}$ (see Section 8), J_n the operator of convolution by β_n. It is immediate that

$$D^k(J_n u) = J_n(D^k u),$$

hence it follows from Theorem 8.2 that $\varphi_n = J_n u = \beta_n * u$ converges to u in H^1, while each φ_n belongs to $\mathcal{D}$ with support contained in $|x| \leq a + 1$. We apply inequality (5.4.25) with $p = \lambda - q$ and $\psi = \varphi_n - \varphi_m$: since ψ vanishes for $|x| > a + 1$,

$$\int (\lambda - q(x))|\varphi_n(x) - \varphi_m(x)|^2 \, dx \leq C\|\varphi_n - \varphi_m\|_{H^1}, \qquad (5.4.28)$$

where we have used the fact that $\mathfrak{m}_{\rho, R}(q; x)$ is bounded in $|x| \leq a + 1$. Letting $m \to \infty$, we obtain the second relation (5.4.27), thus $u \in \mathcal{H}$ as claimed. The opposite implication follows from (5.4.23) and the fact that $u \in \mathcal{H} \subseteq H^1$; that $\sqrt{|q|}\, u \in L^2$ is a consequence of the definition of the norm of $\mathcal{H}$.

Note that the assumption on $\mathfrak{m}_{\rho, R}$ is independent of R. Observe also that an operator A with domain described as in Remark 4.6.8 could be defined without any special assumption on q by simply replacing $\mathcal{H}$ by the space of all $u \in H^1$ with $\sqrt{q}\, u \in L^2$ endowed with the scalar product (5.4.21). Thus, the importance of the result lies in the approximation property (5.4.27) and the subsequent identification of $D(A)$.

The result just proved, although partly more general than Theorem 5.4.1, does not include, say, potentials having quadratic singularities $c|x|^{-2}$ in $\mathbb{R}^3$. Attempts to apply Corollary 5.3.5 fail since the domain of A (even after Theorem 5.4.5) is not easy to work with; for instance, the Fourier transforms of the elements of $D(A)$ allow no easy characterization. On the other hand, inequalities of the type $(Pu, u) \leq a|(Au, u)| + b(u, u)$ are easier to establish (as we shall see in the next Theorem 5.4.9) and it is then of interest to have a perturbation theory based on these. We limit ourselves to a simple result involving perturbations of symmetric operators.

5.4.7 Theorem. *Let $\mathfrak{A}$ be a densely defined symmetric operator in a Hilbert space H such that*

$$(\mathfrak{A}u, u) \geqslant 0 \quad (u \in D(\mathfrak{A})),$$ (5.4.29)

and let $\mathfrak{B}$ be another symmetric operator such that $D(\mathfrak{B}) \supseteq D(\mathfrak{A})$ and

$$(\mathfrak{B}u, u) \leqslant a(\mathfrak{A}u, u) + b(u, u) \quad (u \in D(\mathfrak{A}))$$ (5.4.30)

with $0 \leqslant a \leqslant 1$, $b \geqslant 0$. Then there exists a self-adjoint extension $(\mathfrak{A} - \mathfrak{B})_s$ of $\mathfrak{A} - \mathfrak{B}$ with

$$((\mathfrak{A} - \mathfrak{B})_s u, u) \geqslant -b(u, u) \quad (u \in D(\mathfrak{A} - \mathfrak{B})_s).$$ (5.4.31)

Proof. Define

$$(u, v)_1 = ((\mathfrak{A} - \mathfrak{B})u, v) + (1 + b)(u, v)$$

for $u, v \in D(\mathfrak{A})$. Since

$$((\mathfrak{A} - \mathfrak{B})u, u) \geqslant (\mathfrak{A}u, u) - (\mathfrak{B}u, u) \geqslant (1 - a)(\mathfrak{A}u, u) - b(u, u),$$ (5.4.32)

it follows that $(\cdot, \cdot)_1$ has all the properties pertaining to a scalar product and we can complete $D(\mathfrak{A})$ with respect to the norm $\|\cdot\|_1 = (\cdot, \cdot)_1^{1/2}$. Since

$$\|u\|_1 \geqslant \|u\|,$$ (5.4.33)

the completion $H_1 \supseteq D(\mathfrak{A})$ is a subspace of H. We define an operator $\mathfrak{B}$ in H as follows: the domain of $\mathfrak{B}$ consists of all $u \in H_1$ that make the linear functional $w \to (u, w)_1$ continuous *in the norm* of H; since H_1 is dense in H, we may extend the functional to H and obtain a unique element $v \in H$ satisfying

$$(u, w)_1 = (v, w).$$ (5.4.34)

We define $\mathfrak{B}u = v$. Let $u, v \in D(\mathfrak{B})$; then

$$(\mathfrak{B}u, v) = (u, v)_1 = \overline{(v, u)_1} = \overline{(\mathfrak{B}v, u)} = (u, \mathfrak{B}v)$$

so that $\mathfrak{B}$ is symmetric. We prove next that $\mathfrak{B}^{-1} \in (H)$. Since $\|\mathfrak{B}u\|\|u\| \geqslant (\mathfrak{B}u, u) = (u, u)_1 \geqslant \|u\|^2$, we have $\|\mathfrak{B}u\| \geqslant \|u\|$ for $u \in D(\mathfrak{B})$, thus $\mathfrak{B}$ is one-to-one with bounded inverse $\mathfrak{B}^{-1}$; it only remains to be shown that $\mathfrak{B}D(\mathfrak{B}) = E$. To see this we take $v \in E$ and consider the functional $w \to (v, w)$. Obviously, this functional is continuous in H_1 so that there exists u satisfying (5.4.34); by definition of $\mathfrak{B}$, $u \in D(\mathfrak{B})$ and $\mathfrak{B}u = v$ showing that $\mathfrak{B}$ is onto as claimed. It is obvious that $\mathfrak{B}^{-1}$ is also symmetric, hence self-adjoint; that $\mathfrak{B}$ is self-adjoint follows easily from the formula

$$(\mathfrak{B}^*)^{-1} = (\mathfrak{B}^{-1})^*,$$ (5.4.35)

whose verification is elementary. It is also obvious that $(\mathfrak{A} - \mathfrak{B})_s = \mathfrak{B} - (1 + b)I$ is an extension of $\mathfrak{A} - \mathfrak{B}$ and that (5.4.31) holds. This concludes the proof.

Note that we have obtained a reasonably precise characterization of $D((\mathfrak{A} - \mathfrak{B}_s)$; another look at the proof of Theorem 5.4.7 shows that

$$D((\mathfrak{A} - \mathfrak{B})_s) = H_1 \cap D((\mathfrak{A} - \mathfrak{B})^*)$$ (5.4.36)

while $(\mathfrak{A} - \mathfrak{B})_s$ is the restriction of $(\mathfrak{A} - \mathfrak{B})^*$ to H_1. It is also interesting to note that if (5.4.30) holds with $a < 1$ and if

$$(\mathfrak{P}u, u) \geq -a'(\mathfrak{A}u, u) - b'(u, u) \quad (u \in D(\mathfrak{A})) \tag{5.4.37}$$

for some $a', b' \geq 0$ (without restrictions on their size), then H_1 is independent of $\mathfrak{P}$ in the following sense. Define

$$(u, v)_0 = (\mathfrak{A}u, v) + (u, v). \tag{5.4.38}$$

Then it follows from (5.4.32) that

$$(u, u)_1 \geq (1 - a)(u, u)_0,$$

while on the other hand,

$$(u, u)_1 \leq (1 + a')(\mathfrak{A}u, u) + (1 + b + b')(u, u) \leq C(u, u)_0,$$

hence we may use $(\cdot, \cdot)_0$ (which does not depend on $\mathfrak{P}$) in the definition of H_1. We note also that the argument in the proof of Theorem 5.4.7 can be adapted to show that any symmetric operator satisfying (5.4.29) has a self-adjoint extension with domain $H_1 \cap D(\mathfrak{A}^*)$. Finally, we point out that Theorem 5.4.7 is a sort of abstract version of the reasoning used in Section 4.6 to construct the operators $A(\beta)$ there; an even more similar method would be to define H_1 using $(\cdot, \cdot)_0$ and apply Lemma 4.6.1 to the functional $B(u, v) = (u, v)_1$.

In many applications (such as in Theorem 5.4.9 below) a generalization of Theorem 5.4.7 must be used. In it, the role of $(\mathfrak{P}u, v)$ is taken over by $\mathfrak{h}(u, v)$, where $\mathfrak{h}$ is a *symmetric sesquilinear form* with domain $D(\mathfrak{h})$, that is, a map from $D(\mathfrak{h}) \times D(\mathfrak{h})$ into $\mathbb{C}$ ($D(\mathfrak{h})$ a subspace of E) which is linear in v for u fixed and satisfies $\mathfrak{h}(v, u) = \overline{\mathfrak{h}(u, v)}$. The result is:

5.4.8   Theorem .   *Let $\mathfrak{A}$ be a densely defined symmetric operator such that (5.4.29) holds, and let $\mathfrak{h}$ be a symmetric sesquilinear form with domain $D(\mathfrak{h}) \supseteq D(\mathfrak{A})$ such that*

$$\mathfrak{h}(u, u) \leq a(\mathfrak{A}u, u) + b(u, u) \quad (u \in D(\mathfrak{A})) \tag{5.4.39}$$

with $0 \leq a \leq 1$, $b \geq 0$. Then there exists a self-adjoint operator $\mathfrak{C}$ with domain $D(\mathfrak{C}) \supseteq D(\mathfrak{A})$ such that

$$(\mathfrak{C}u, v) = (\mathfrak{A}u, v) - \mathfrak{h}(u, v) \quad (u, v \in D(\mathfrak{A})) \tag{5.4.40}$$

and

$$(\mathfrak{C}u, u) \geq -b(u, u). \tag{5.4.41}$$

If $a < 1$ in (5.4.39) and if there exists $a', b' \geq 0$ with

$$\mathfrak{h}(u, u) \geq -a'(\mathfrak{A}u, u) - b'(u, u), \tag{5.4.42}$$

then $\mathfrak{C}$ can be described as follows: the domain of $\mathfrak{C}$ consists of all $u \in H_1$ (H_1 the completion of $D(\mathfrak{A})$ with respect to (5.4.38)) that make the linear functional $w \to (u, w)_1$ continuous in the norm of H, where $(u, w)_1$ is (the extension to H_1) of

$$(u, w)_1' = (\mathfrak{A}u, w) - \mathfrak{h}(u, w)$$

and

$$(u, w)_1' = (\mathfrak{C}u, w) \quad (u \in D(\mathfrak{C}), w \in H_1).$$

The proof is essentially the same as that of Theorem 5.4.8 and we omit the details.

Theorem 5.4.8 will be applied to A, the operator defined in (5.4.22) and $\mathfrak{h}(u, v) = \mathfrak{h}_1(u, v) + \mathfrak{h}_2(u, v)$, where $\mathfrak{h}_j(u, v)$ "is" $(\mathfrak{P}_j u, v)$, $\mathfrak{P}_j$ the operator of multiplication by $q_j(x)$, $j = 1, 2$. However, the highly singular behavior of the q_j (see (i) and (ii) below) precludes the direct definition of the operators $\mathfrak{P}_j$. We make the following assumptions:

(i) q_1 is locally integrable with $\mathfrak{m}_{\rho, R}(q_1; x)$ bounded in $\mathbb{R}^m$ for some $\rho > 0$.
(ii) q_2 is a finite sum of terms of the form $c_j |x - x^j|^{-2}$.

The sesquilinear form $\mathfrak{h}_j$ has domain $\mathfrak{D}$ and is defined by

$$\mathfrak{h}_j(u, v) = \int q_j(x) \bar{u}(x) v(x)\, dx \quad (j = 1, 2). \tag{5.4.43}$$

Since $|\bar{u}(x) v(x)| \leq \frac{1}{2}(|u(x)|^2 + |v(x)|^2)$, we only have to prove that $\mathfrak{h}_j(u, v)$ exists for $u = v$. For $j = 1$, this is an obvious consequence of inequality (5.4.25): in fact, under hypothesis (i),

$$\mathfrak{h}_1(u, u) \leq C\rho^{-1} R^\rho \|\operatorname{grad} u\|^2 + C'(R) \|u\|^2, \tag{5.4.44}$$

where C does not depend on ρ or R. Note that by simply interchanging q_1 by $-q_1$ an estimate of the same type can be obtained for $-\mathfrak{h}_1(u, u)$, thus (5.4.42) holds for $\mathfrak{h}_1$.

The second sesquilinear form is estimated in the same way, but we use (5.4.16):

$$\mathfrak{h}_2(u, u) \leq \sum_{c_j > 0} c_j \int |x - x^j|^{-2} |u(x)|^2\, dx$$

$$\leq \left\{ \frac{4}{(m-2)^2} \sum_{c_j > 0} c_j \right\} \|\operatorname{grad} u\|^2. \tag{5.4.45}$$

It is again obvious that a similar inequality (with a different constant) can be established for $-\mathfrak{h}_2(u, u)$.

5.4.9   Theorem.   *Let* $m \geq 3$,

$$q = q_0 + q_1 + q_2,$$

where the q_j *are real. Let* q_0 *be locally integrable and bounded above, and let* q_1 *(resp.* q_2*) satisfy assumption (i) (resp. (ii) with*

$$\sum_{c_j > 0} c_j < (m-2)^2/4. \tag{5.4.46}$$

Then the operator A *described below is self-adjoint:* $D(A)$ *consists of all* $u \in \mathcal{K}_0$ *(the completion of* $\mathfrak{D}$ *in any of the norms*

$$\|u\|_\lambda = \int \left(|\operatorname{grad} u(x)|^2 + (\lambda - q_0(x))|u(x)|^2 \right) dx \tag{5.4.47}$$

with $\lambda \geq \operatorname{ess\, sup} q_0 + 1$*) such that* $\Delta u + qu \in L^2(\mathbb{R}^m)$ *and*

$$Au = \Delta u + qu. \tag{5.4.48}$$

If $\mathfrak{m}_{\rho, R}(q_0; x)$ *is locally bounded for some* $\rho > 0$*, then* $D(A)$ *consists of all* $u \in H^1(\mathbb{R}^m)$ *with* $\sqrt{|q_0|}\, u$ *and* $\Delta u + qu$ *in* $L^2(\mathbb{R}^m)$.

Proof. Let A^0 be the operator defined in (5.4.22) with $q = q_0$, $\mathfrak{A}$ the restriction of $\lambda I - A^0$ to $\mathfrak{D}$. We have

$$(\mathfrak{A}u, u) \geqslant (u, u)_\lambda \geqslant \|\mathrm{grad}\, u\|^2. \tag{5.4.49}$$

Combining this with (5.4.44) for R sufficiently small, we obtain (5.4.39) for $\mathfrak{h}_1$ with arbitrary $a > 0$ (note that if assumption (i) is satisfied for $R > 0$, it is satisfied as well with any $R' < R$). On the other hand, (5.4.39) for $\mathfrak{h}_2$ follows from (5.4.45) and (5.4.46). We have already noted that (5.4.42) holds for $\mathfrak{h}_1$ and $\mathfrak{h}_2$. Accordingly, we can apply Theorem 5.4.9 to the operator $\mathfrak{A}$ and to the sesquilinear form $\mathfrak{h} = \mathfrak{h}_1 + \mathfrak{h}_2$. Let $\mathfrak{C}$ be the self-adjoint operator in the conclusion. Since the norms derived from (5.4.38) and (5.4.47) are equivalent in $D(A)$, H_1 and $\mathcal{K}_0$ coincide. Let $u \in \mathfrak{D}$. Then we have

$$(\mathfrak{A}u, \varphi) - \mathfrak{h}(u, \varphi) = (\mathfrak{C}u, \varphi) \quad (\varphi \in \mathfrak{D}). \tag{5.4.50}$$

Since u belongs to $\mathfrak{D}$, $q_0 u$, $q_1 u$, and $q_2 u$ belong to $L^1(\mathbb{R}^m)$, thus it follows that

$$\mathfrak{C}u = \lambda u - \Delta u - qu \tag{5.4.51}$$

since φ is arbitrary. On the other hand, let u be an element of $D(\mathfrak{C})$, $\{u_n\}$ a sequence in $\mathfrak{D}$ with $u_n \to u$ in L^2. In view of (5.4.51), if $\varphi \in \mathfrak{D}$, we have $(\mathfrak{C}u, \varphi) = (u, \mathfrak{C}\varphi) = \lim(u_n, \mathfrak{C}\varphi) = \lim(\mathfrak{C}u_n, \varphi) = \lim(\lambda u_n - \Delta u_n - qu_n, \varphi) = (\lambda u - \Delta u - qu, \varphi)$, the last limit relation valid because $\lambda u_n - \Delta u_n - qu_n \to \lambda u - \Delta u - qu$ in the sense of distributions. Accordingly, (5.4.51) holds for every $u \in D(\mathfrak{C})$. Conversely, if $u \in \mathcal{K}_0$ is such that $\Delta u + qu \in L^2$, then we show using the definition that $u \in D(\mathfrak{C})$ with $\mathfrak{C}u$ given by (5.4.51). This ends the proof of Theorem 5.4.9; the additional statement concerning $\mathfrak{m}_{\rho, R}(q_0; x)$ locally bounded follows from Theorem 5.4.5.

The translation of (5.4.46) in physical units for $m = 3$ and the Schrödinger operator $\mathfrak{A} = i(h/2\mathfrak{m})\Delta u - i(1/h)U$ is this: if $U_2(x) = \Sigma e_j |x - x^j|^{-2}$ is the portion of the potential corresponding to q_2, we must have

$$\sum_{e_j < 0} |e_j| < \frac{h^2}{8\mathfrak{m}}.$$

5.4.10 Remark. The applications of Corollary 5.3.4 presented so far have been restricted to the self-adjoint case. For an application to the Schrödinger equation where the resulting semigroup is not unitary, see Nelson [1964: 1].

5.4.11 Example. Develop analogues of the perturbation results for the Schrödinger equation in dimension $m = 1, 2$ (see Kato [1976: 1, Ch. VI]).

5.5. SECOND ORDER DIFFERENTIAL OPERATORS

We apply the theory in Section 5.3 to the operators in Chapter 4; for the sake of simplicity, we limit ourselves to an example involving Dirichlet boundary conditions.

Let A_0 be the operator defined by (4.4.1), that is,

$$A_0 u = \sum \sum a_{jk}(x) D^j D^k u + \sum b_j(x) D^j u + c(x) u. \qquad (5.5.1)$$

5.5.1 Theorem. *Let $1 < p < \infty$, Ω a bounded domain of class $C^{(2)}$ for $1 \leqslant p \leqslant 2$ (of class $C^{(k+2)}$ with $k > m/2$ if $2 < p < \infty$). Assume the a_{jk} are continuous in $\bar{\Omega}$ and satisfy the uniform ellipticity condition*

$$\sum \sum a_{jk}(x) \xi_j \xi_k \geqslant \kappa |\xi|^2 \quad (x \in \Omega, \xi \in \mathbb{R}^m) \qquad (5.5.2)$$

and that the b_j, c are measurable and bounded in Ω. Let β be the Dirichlet boundary condition. Then the operator $A(\beta)$ defined by

$$A(\beta) u = A_0 u \qquad (5.5.3)$$

with domain $D(A(\beta)) = W^{2,p}(\Omega)_\beta$ (consisting of all $u \in W^{2,p}(\Omega)$ that satisfy the boundary condition β on Γ in the sense of Lemma 4.8.1) belongs to $\mathcal{C}$.

Proof. It is not difficult to see that given $\varepsilon > 0$ we can find functions a_{jk}^ε infinitely differentiable in $\bar{\Omega}$, having modulus of continuity uniformly bounded with respect to ε and such that

$$|a_{jk}(x) - a_{jk}^\varepsilon(x)| \leqslant \varepsilon \quad (x \in \bar{\Omega}, 1 \leqslant j, k \leqslant m). \qquad (5.5.4)$$

This condition clearly implies that the a_j^ε are uniformly bounded in $\bar{\Omega}$ independently of ε; moreover, if $0 < \kappa' < \kappa$, we shall have

$$\sum \sum a_{jk}^\varepsilon(x) \xi_j \xi_k \geqslant \kappa' |\xi|^2 \quad (x \in \Omega, \xi \in \mathbb{R}^m) \qquad (5.5.5)$$

for ε sufficiently small. Let

$$A_\varepsilon(\beta) u = \sum \sum a_j^\varepsilon(x) D^j D^k u \qquad (5.5.6)$$

with domain $W^{2,p}(\Omega)_\beta = W^{2,p}(\Omega) \cap W_0^{1,p}(\Omega)$. Then it follows from Theorem 4.8.4 and an approximation argument (see Lemma 4.8.1) that there exists $C > 0$ independent of ε (for ε sufficiently small) such that

$$\|u\|_{W^{2,p}(\Omega)} \leqslant C \left(\|A_\varepsilon(\beta) u\|_{L^p(\Omega)} + \|u\|_{L^p(\Omega)} \right) \quad (u \in D(A_\varepsilon(\beta))). \qquad (5.5.7)$$

On the other hand, if $P_\varepsilon(\beta) = A(\beta) - A_\varepsilon(\beta)$ and C' is a bound for $|b_1|, \ldots, |b_m|$ and $|c|$,

$$\|P_\varepsilon(\beta) u\|_{L^p(\Omega)} \leqslant m^2 \varepsilon \|u\|_{W^{2,p}(\Omega)} + (m+1) C' \|u\|_{W^{1,p}(\Omega)}. \qquad (5.5.8)$$

Combining this inequality with (5.5.7) and Lemma 4.8.9, we deduce that for any $a > 0$ there exists $b \geqslant 0$ such that

$$\|P_\varepsilon(\beta) u\|_{L^p(\Omega)} \leqslant a \|A_\varepsilon(\beta)\|_{L^p(\Omega)} + b \|u\|_{L^p(\Omega)}.$$

The result then follows from Corollary 4.8.10 and Lemma 5.3.6.

5.5.2 Example. State and prove a version of Theorem 5.5.1 for boundary conditions of type (I).

5.6. SYMMETRIC HYPERBOLIC SYSTEMS
IN SOBOLEV SPACES

We consider again the differential operator

$$Lu = \sum_{k=1}^{m} A_k(x)D^k u + B(x)u \qquad (5.6.1)$$

of Section 3.5, using the notations employed therein; in particular, $\mathcal{C}_0$ is defined by

$$\mathcal{C}_0 u = Lu, \quad D(\mathcal{C}_0) = \mathfrak{D}^\nu$$

and $\mathcal{C}_0'$ is defined in the same fashion using the formal adjoint. The results of Section 3.5 establish that, under adequate assumptions on the coefficients of (5.6.1), the Cauchy problem for the symmetric hyperbolic equation

$$u'(t) = \mathcal{C}u(t), \qquad (5.6.2)$$

where $\mathcal{C} = (\mathcal{C}_0')^*$, is properly posed in $L^2(\mathbb{R}^m)^\nu$. However, the results there leave us in the dark about regularity of the solutions of (5.6.2): if, say, the initial value $u_0 = u(0)$ of a solution $u(\cdot)$ has a certain number of partial derivatives with respect to $x_1, \ldots, x_m$, does $u(t)$ enjoy a similar property for all t? We know of course the answer to this question when L has constant coefficients (see Section 1.6); from this particular case, we can surmise that the "right" treatment for the problem is to examine (5.6.2) in the Sobolev spaces $H^s = H^s(\mathbb{R}^m)^\nu$. We do this beginning with the case $s = 1$, and we will find it convenient to introduce an abstract framework to handle this and other similar situations.

Let E, F be Banach spaces; when necessary, norms in E or F will be subindexed to avoid confusion. We write

$$F \to E$$

to indicate that: (a) F *is a dense subspace of* E, (b) *there exists a constant C such that*

$$\|u\|_E \leqslant C\|u\|_F \quad (u \in F). \qquad (5.6.3)$$

A simple example of this arrangement is that where $F = D(A)$, (A a closed, densely defined operator in E) and the norm of F is $\|u\|_F = \|u\|_E + \|Au\|_E$, the "graph norm" of $D(A)$.

Since at least two spaces will be at play simultaneously in the succeeding treatment, we shall write, for instance $A \in \mathcal{C}_+(C, \omega; E)$ to indicate that A belongs to $\mathcal{C}_+(C, \omega)$ *in the space* E (A, or restrictions thereof, may be defined in different spaces). Expressions like "$A \in \mathcal{C}_+(E)$", "$A \in \mathcal{C}(\omega; E)$",... and so on will be correspondingly understood.

Given an operator $A \in \mathcal{C}_+(E)$ and another Banach space F with $F \to E$, we say that F is *A-admissible* if and only if

$$(\exp(tA))F = S(t; A)F \subseteq F \quad (t \geqslant 0) \qquad (5.6.4)$$

and $t \to S(t; A)$ is strongly continuous *in F* for $t \geqslant 0$. (Motivation for this definition is obviously provided by the previous remarks on the equation (5.6.2), where $E = L^2$, $F = H^1$.)

If A is an *arbitrary* operator in E with domain $D(A)$, we denote by A_F the largest restriction of A to F with range in F, that is, the restriction of A with domain

$$D(A_F) = \{u \in D(A) \cap F; \, Au \in F\}. \qquad (5.6.5)$$

5.6.1 Example. (a) Let, as in Example 2.2.6, $E = L^2(-\infty, 0)$, and define $Au = u'$ (with $D(A)$ the set of all $u \in E$ such that u' exists in the sense of distributions, belongs to E and satisfies $u(0) = 0$), and let F be the Sobolev space $H^1(-\infty, 0)$ (see Section 8). Then F is not A-admissible. (b) Let E and A be as in (a), but take F as the subspace $H_0^1(-\infty, 0)$ of $H^1(-\infty, 0)$ defined by the condition $u(0) = 0$. Then F is A-admissible. (c) Let, as in Example 2.2.7, $E = L^2(0, \infty)$, $Au = u'$ defined as in (a) but without boundary conditions at 0. If $F = H^1(0, \infty)$, then F is A-admissible.

In Example (a), $A_F u = u'$, where $D(A_F)$ consists of all $u \in E$ with $u', u'' \in E$. The characterization of A_F in (b) and (c) is similar. We note that the two affirmative examples in (b) and (c) are consequences of the following result for $m = 1$: (d) Let $A \in \mathcal{C}_+(E)$, $m \geqslant 1$, $F = D(A^m)$, $\|u\|_F = \|(\lambda I - A)^m u\|_E$, where $\lambda \in \rho(A)$ is fixed. Then F is A-admissible.

5.6.2 Lemma. *Let $F \to E$, $A \in \mathcal{C}_+(E)$. Then F is A-admissible if and only if $R(\lambda; A)$ exists for sufficiently large λ and (a) $R(\lambda; A)F \subseteq F$ with*

$$\|R(\lambda; A)^n\|_{(F)} \leqslant C(\lambda - \omega)^{-n} \quad (n = 1, 2, \ldots). \qquad (5.6.6)$$

(b) $R(\lambda; A)F$ *is dense in F. Conversely, if (a) and (b) hold, the operator A_F belongs to $\mathcal{C}_+(F)$ and*

$$S(t; A_F) = S(t; A)_F \quad (t \geqslant 0). \qquad (5.6.7)$$

Condition (b) is unnecessary if F is reflexive.

Proof. We observe first that, if F is A-admissible, the definition (5.6.5) means that the operator on the right-hand side of (5.6.7) is none other than the restriction of $S(t; A)$ to F. Also, $S(t; A)_F$ is a strongly continuous semigroup in F; let $\hat{A}$ be its infinitesimal generator and C, ω two constants so large that both inequalities $\|S(t; \hat{A})\|_{(F)} \leqslant Ce^{\omega t}$ and $\|S(t; A)\|_{(E)} \leqslant Ce^{\omega t}$ hold. If $u \in F$, we have $S(t; \hat{A})u = S(t; A)u$ so that, in view of formula (2.1.10) with $n = 1$ and of the fact that $F \to E$, $R(\lambda; \hat{A})u = R(\lambda; A)u$ for $\lambda > \omega$. It results that $R(\lambda; \hat{A})$ is simply the restriction of

$R(\lambda; A)$ to F,

$$R(\lambda; \hat{A}) = R(\lambda; A)_F. \tag{5.6.8}$$

If $u \in D(\hat{A})$, we have

$$\hat{A}u = F - \lim_{h \to 0+} h^{-1}\big(S(t; \hat{A})u - u\big)$$

$$= E - \lim_{h \to 0+} h^{-1}\big(S(t; \hat{A})u - u\big)$$

$$= E - \lim_{h \to 0+} h^{-1}\big(S(t; A)u - u\big) = Au,$$

which shows that $\hat{A} \subseteq A_F$. On the other hand, let $u \in D(A_F)$; then $u = R(\lambda; A)v = R(\lambda; A)_F v$ for some $v \in F$, and in view of (5.6.8), u must belong to $D(\hat{A})$. Thus $\hat{A} = A_F$ and

$$R(\lambda; A_F) = R(\lambda; A)_F. \tag{5.6.9}$$

It is obvious then that (5.6.6) follows from (5.6.9) and the fact that $R(\lambda; A_F)$ satisfies inequalities (2.1.29); moreover, since $R(\lambda; A)F = R(\lambda; A_F)F$ by (5.6.9), the denseness condition (b) follows.

Conversely, assume that (a) and (b) hold. If $u \in F$ and λ is large enough, $v = R(\lambda; A)u \in F$; moreover, $Av = \lambda R(\lambda; A)u - u \in F$ so that $v \in D(A_F)$ and

$$(\lambda I - A_F)R(\lambda; A)u = u.$$

This obviously shows that $(\lambda I - A_F)D(A_F) = F$; since, on the other hand, $\lambda I - A_F$ is one-to-one because $\lambda I - A$ is, it results that $R(\lambda; A_F)$ exists and (5.6.9) holds. Condition (b) plainly insures denseness of $D(A_F)$, and inequalities (5.6.6) imply, via Theorem 2.1.1 and Remark 2.1.4, that $A_F \in \mathcal{C}_+(F)$. It only remains to show that

$$S(t; A_F) = S(t; A)_F$$

and this can be done making use of (5.6.9) and expressing each semigroup by means of the resolvent of its infinitesimal generator; in fact, using formula (2.1.27),

$$S(t; A_F)u = F - \lim_{n \to \infty} \left(\frac{n}{t}\right)^n R\left(\frac{n}{t}; A_F\right)^n u$$

$$= E - \lim_{n \to \infty} \left(\frac{n}{t}\right)^n R\left(\frac{n}{t}; A\right)^n u$$

$$= S(t; A)u$$

for $u \in F$ as claimed.

It remains to be shown that the denseness condition (b) can be dispensed with when F is reflexive. Observe first that (5.6.9) has been established without use of (b). If $R(\lambda; A)F = R(\lambda; A_F)F = D(A_F)$ is not dense in F, there exists some $u \in F$, $u \neq 0$ that does not belong to the closure

$\overline{D(A_F)}$. Consider

$$u_n = \lambda_n R(\lambda_n; A_F)u = \lambda_n R(\lambda_n; A)u,$$

where $\lambda_n \to \infty$. In view of the first inequality (5.6.6), $\{u_n\}$ is bounded in F and, since F is reflexive, we may assume that $\{u_n\}$ is F^*-weakly convergent in F to some $v \in F$.[10] But $\overline{D(A_F)}$ is F^*-weakly closed in F, so that $v \in \overline{D(A_F)}$. Now,

$$\lambda_n R(\lambda_n; A_F)R(\lambda_1; A_F)u = R(\lambda_1; A_F)\lambda_n R(\lambda_n; A_F)u \to R(\lambda_1; A_F)v$$

$$(5.6.10)$$

F^*-weakly in F whereas, by the second resolvent equation and the first inequality (5.6.6),

$$(\lambda_n R(\lambda_n; A_F) - I)R(\lambda_1; A_F)u$$

$$= (\lambda_1 - \lambda_n)^{-1}(\lambda_n R(\lambda_n; A_F) - \lambda_1 R(\lambda_1; A_F))u \to 0$$

as $n \to \infty$ showing, in combination with (5.6.10), that

$$R(\lambda_1; A_F)u = R(\lambda_1; A_F)v, \qquad (5.6.11)$$

which implies $u = v$, a contradiction since v belongs to $\overline{D(A_F)}$ and u does not. This completes the proof of Lemma 5.6.2.

Another useful criterion for A-admissibility of a subspace is

5.6.3 Lemma. *Let $F \to E$, $A \in \mathcal{C}_+(E)$, $K: F \to E$ an operator in $(F; E)$ such that the inverse K^{-1} exists and belongs to $(E; F)$. In order that F be A-admissible, it is necessary and sufficient that $A_1 = KAK^{-1}$ belong to $\mathcal{C}_+(E)$. In this case, we have*

$$S(t; A_1) = KS(t; A)K^{-1}. \qquad (5.6.12)$$

Proof. By definition, $D(A_1) = D(\lambda I - A_1)$ is the set of all $u \in E$ such that $K^{-1}u \in D(A)$ and $(\lambda I - A)K^{-1}u \in F$; accordingly,

$$D((\lambda I - A_1)) = KR(\lambda; A)F. \qquad (5.6.13)$$

Moreover, $(\lambda I - A_1)^{-1} = K(\lambda I - A)^{-1}K^{-1}$; thus if $\lambda \in \rho(A)$, we have

$$(\lambda I - A_1)^{-1} = KR(\lambda; A)K^{-1}. \qquad (5.6.14)$$

Note, however, that we have *not* written $R(\lambda; A_1)$ on the left-hand side of (5.6.14) as we cannot state without further analysis that $(\lambda I - A_1)^{-1} \in (E)$ (since K may not be continuous as an operator from E into E). Assume now that F is A-admissible. We have already seen in Lemma 5.6.2(b) that $R(\lambda; A)F$ is a dense subset of F and that $R(\lambda; A)_F \in (F)$; this plainly implies that $D(A_1)$ is dense in E (because of (5.6.13)) while it results from

---

[10]It should be remembered that the first inequality (5.6.6) does *not* guarantee that $\lambda R(\lambda; A_F)u \to u$ as $\lambda \to \infty$ except if u is known to be in $\overline{D(A_F)}$ (see Example 2.1.6).

(5.6.14) that $(\lambda I - A_1)^{-1} \in (E)$, thus $\lambda \in \rho(A_1)$. Taking powers in both sides of (5.6.14), we see that

$$R(\lambda; A_1)^n = KR(\lambda; A)^n K^{-1}, \tag{5.6.15}$$

thus $R(\lambda; A_1)$ satisfies as well inequalities (2.1.29) and consequently $A_1 \in \mathcal{C}_+(E)$. Equality (5.6.12) follows from (5.6.15) and the exponential formula (2.1.27) (see the end of the proof of Lemma 5.6.2(a)).

We prove now the converse. Assume that $A_1 = KAK^{-1} \in \mathcal{C}_+(E)$. Reading (5.6.15) backwards, we see that $R(\lambda; A)^n = K^{-1}R(\lambda; A_1)^n K$, hence $R(\lambda; A)F \subseteq F$ and (5.6.6) is satisfied for λ sufficiently large. On the other hand, $R(\lambda; A)F = R(\lambda; A)K^{-1}E = K^{-1}R(\lambda; A_1)E$, thus $K^{-1}D(A_1)$ must be dense in F since $D(A_1)$ is dense in E. Hence the hypotheses of Lemma 5.6.2 are satisfied.

In view of Theorem 5.1.2, the assumption that $KAK^{-1} \in \mathcal{C}_+(E)$ will hold if

$$KAK^{-1} = A + Q \tag{5.6.16}$$

with Q everywhere defined and bounded. This is the form in which Lemma 5.6.3 will be applied to the equation (5.6.2).

5.6.4 Theorem. *Let the matrices $A_1, \ldots, A_m, B$ be continuously differentiable in $\mathbb{R}^m$, and assume all the functions*

$$|A_k(x)|, |D^j A_k(x)|, |B(x)|, |D^j B(x)| \quad (1 \leqslant j, k \leqslant m) \tag{5.6.17}$$

are bounded in $\mathbb{R}^m$ by a constant M_1. Then (a) The operator $\mathcal{Q} = (\mathcal{Q}_0')^*$ defined in Section 3.5 belongs to $\mathcal{C}(1, \omega; E)$ with $E = L^2(\mathbb{R}^m)^\nu$ for some ω. (b) If $S(\cdot)$ is the propagator of

$$u'(t) = \mathcal{Q}u(t) \quad (-\infty < t < \infty), \tag{5.6.18}$$

we have

$$S(t)H^1 \subseteq H^1 \quad (-\infty < t < \infty), \tag{5.6.19}$$

where $H^1 = H^1(R^m)^\nu$. Finally, if $\mathcal{Q}_{H^1}$ is the largest restriction of $\mathcal{Q}$ to H^1 with range in H^1, the operator $\mathcal{Q}_{H^1}$ belongs to $\mathcal{C}_+(H^1)$; in particular, any solution $u(\cdot)$ of (5.6.18) with initial value $u(0)$ in H^1 remains in H^1 for all t, is continuous in the norm of H^1, and satisfies

$$\|u(t)\|_{H^1} \leqslant C_1 e^{\omega_1 |t|} \|u(0)\|_{H^1} \quad (-\infty < t < \infty), \tag{5.6.20}$$

for constants C_1, ω_1 independent of $u(\cdot)$.

Proof. The statements regarding the behavior of the operator $\mathcal{Q}$ in L^2 are immediate consequences of Corollary 3.5.5.[11] The rest of Theorem 5.6.4 will of course result if we show that $F = H^1$ is both $\mathcal{Q}$-admissible and

---

[11] Boundedness of $|D^k A(x)|$ and $|B(x)|$ insure inequality (3.5.37).

$(-\mathcal{Q})$-admissible. Since $-\mathcal{Q}$ satisfies assumptions of exactly the same type as $\mathcal{Q}$, we need only consider the latter operator. We use Lemma 5.6.3 as follows. Consider the operator $\mathfrak{R} = (I - \Delta)^{1/2}$, where Δ is the Laplacian; that is, if $\mathfrak{F}$ indicates, as in Section 8, the Fourier-Plancherel transform,

$$\mathfrak{R}u = \mathfrak{F}^{-1}\left(\left(1 + |\sigma|^2\right)^{1/2}\mathfrak{F}u(\sigma)\right).$$

In view of the Fourier transform characterization of Sobolev spaces over $\mathbb{R}^m$ (see again Section 8), it is obvious that $\mathfrak{R}$ is well defined and bounded as an operator from H^1 into L^2, and the same can be said of its inverse

$$\mathfrak{R}^{-1}u = \mathfrak{F}^{-1}\left(\left(1 + |\sigma|^2\right)^{-1/2}\mathfrak{F}u(\sigma)\right)$$

as an operator from L^2 into H^1. We note also that both $\mathfrak{R}$ and $\mathfrak{R}^{-1}$ map the space $\mathfrak{S}$ into itself. We introduce next the operator

$$\mathfrak{R}u = \mathfrak{F}^{-1}(|\sigma|\mathfrak{F}u(\sigma)) \tag{5.6.21}$$

from H^1 into L^2. It is obvious that if $u \in H^2$,

$$\|\mathfrak{R}u - \mathfrak{R}u\|_{L^2} \leqslant C\|u\|_{L^2}. \tag{5.6.22}$$

We note that $\mathfrak{R}$ (resp. $\mathfrak{R}$) is known as the *Bessel* (resp. *Riesz*) *potential operator of order* -1; see Stein [1970: 1, p. 117]. We shall also make use of the *Riesz transforms* $\mathfrak{R}_k$ defined by

$$\mathfrak{R}_k u = \mathfrak{F}^{-1}((i\sigma_k/|\sigma|)\mathfrak{F}u(\sigma)) \quad (1 \leqslant k \leqslant m) \,^{[12]} \tag{5.6.23}$$

It is plain that each $\mathfrak{R}_k$ is a bounded operator from L^2 into L^2 and that

$$\mathfrak{R} = \sum_{k=1}^{m} D^k \mathfrak{R}_k = \sum_{k=1}^{m} \mathfrak{R}_k D^k, \tag{5.6.24}$$

where, as usual, D^k is the operator of differentiation with respect to x_k. The Riesz transforms can be expressed by means of singular integral operators as follows:

$$\mathfrak{R}_k u(x) = \lim_{\varepsilon \to 0} C_m \int_{|x-y| \geqslant \varepsilon} \frac{x_k - y_k}{|x - y|^{m+1}} u(y)\, dy \tag{5.6.25}$$

$(C_m = 2^{m/2}\pi^{-1/2}\Gamma((m+1)/2)))$ for each $u \in L^2(\mathbb{R}^m)^\nu$, the limit understood in the norm of L^2; if u is smooth, the limit also exists pointwise (See Stein [1970: 1, p. 57].)

5.6.5 Lemma. *Let $A(\cdot)$ be a continuously differentiable $\nu \times \nu$ matrix-valued function defined in $\mathbb{R}^m$, and let*

$$\{\mathfrak{R}, A\}u(x) = \mathfrak{R}(A(x)u(x)) - A(x)(\mathfrak{R}u(x)). \,^{[13]} \tag{5.6.26}$$

---

[12] These definitions coincide with the ones in Stein modulo constants unimportant for our purposes.

[13] Expressions like $\{\mathfrak{R}, A\}$, $\{\mathfrak{R}, A_k\}$, and so on, are defined in a corresponding way.

Then, if $u \in \mathbb{S}^\nu$,

$$\{\mathfrak{R}, A\} u(x) = \lim_{\varepsilon \to 0} C_m \int_{|x-y| \geqslant \varepsilon} \frac{A(x) - A(y)}{|x-y|^{m+1}} u(y)\, dy, \quad (5.6.27)$$

where C_m is the constant in (5.6.25).

Proof. In view of (5.6.24), we have

$$\{\mathfrak{R}, A\} u = \sum_{k=1}^{m} \left(\mathfrak{R}_k D^k(Au) - A\mathfrak{R}_k D^k u \right)$$

$$= \sum_{k=1}^{m} (\mathfrak{R}_k A - A\mathfrak{R}_k) D^k u + \sum_{k=1}^{m} \mathfrak{R}_k D^k Au. \quad (5.6.28)$$

Now,

$$(\mathfrak{R}_k A - A\mathfrak{R}_k) D^k u(x)$$

$$= \lim_{\varepsilon \to 0} C_m \int_{|x-y| \geqslant \varepsilon} \frac{x_k - y_k}{|x-y|^{m+1}} (A(y) - A(x)) D^k u(y)\, dy$$

(the limit understood pointwise). Observe that using the divergence theorem in the region $|x-y| \geqslant \varepsilon$, we obtain

$$\int_{|x-y| \geqslant \varepsilon} \left\{ \sum_{k=1}^{m} \frac{x_k - y_k}{|x-y|^{m+1}} (A(y) - A(x)) D^k u(y) \right\} dy$$

$$+ \int_{|x-y| \geqslant \varepsilon} \left\{ \sum_{k=1}^{m} \frac{x_k - y_k}{|x-y|^{m+1}} D^k A(y) u(y) \right\} dy$$

$$+ \int_{|x-y| \geqslant \varepsilon} \frac{A(y) - A(x)}{|x-y|^{m+1}} u(y)\, dy$$

$$= \int_{|x-y| = \varepsilon} \frac{A(y) - A(x)}{|x-y|^{m}} u(y)\, d\sigma, \quad (5.6.29)$$

where $d\sigma$ is the hyperarea differential on $|x-y| = \varepsilon$. Since $A(y) = A(x) + \Sigma D^k A(x)(y_k - x_k) + o(|y - x|)$ as $y \to x$ and the integral of $(x_k - y_k)/|x-y|^m$ over $|x-y| = \varepsilon$ is zero, it is not difficult to show that the last integral in (5.6.29) tends to zero with ε. Taking advantage of (5.6.28), (5.6.27) follows.

We make now use of the following powerful result on singular integrals.

5.6.6 Theorem . *Let $A(\cdot)$ be as in Lemma 5.6.5; assume in addition that the derivatives $D^k A(x)$, $1 \leqslant k \leqslant m$ are bounded in $\mathbb{R}^m$ by a constant M. Let $\mathfrak{h}$ be an even, homogeneous function of degree $-(m+1)$ locally integrable* *in $|x| > 0$. Finally, let $1 < p < \infty$, and define*

$$\mathfrak{C}_\varepsilon u(x) = \int_{|x-y| \geqslant \varepsilon} \mathfrak{h}(x-y)(A(x) - A(y))u(y)\, dy. \quad (5.6.30)$$

Then $\mathfrak{C}_\varepsilon$ maps continuously $L^p(\mathbb{R}^m)^\nu$ into itself and satisfies

$$\|\mathfrak{C}_\varepsilon u\|_p \leqslant CM\|u\|_p, \quad (5.6.31)$$

where C depends on p and $\mathfrak{h}$, but not on u or ε. Furthermore, $\mathfrak{C}_\varepsilon u$ converges in L^p as $\varepsilon \to 0$; thus the operator

$$\mathfrak{C}u = \lim_{\varepsilon \to 0} \mathfrak{C}_\varepsilon u$$

satisfies

$$\|\mathfrak{C}u\|_p \leqslant CM\|u\|_p. \quad (5.6.32)$$

For the proof (of a more general result) see Calderón [1965: 1], where additional information is given.

We can combine Lemma 5.6.5 and Theorem 5.6.6 to deduce that if M_1 is a bound for $D^k A$ in $\mathbb{R}^m$, then

$$\|\{\mathfrak{R}, A\}u\|_{L^2} \leqslant CM_1\|u\|_{L^2} \quad (u \in \mathcal{S}^\nu).$$

Now, $\{\mathfrak{R}, A\} = \{\mathfrak{R}, A\} + \{\mathfrak{R} - \mathfrak{R}, A\}$, where in view of (5.6.22), $\mathfrak{R} - \mathfrak{R}$ is bounded in L^2, and the operator of multiplication by A is also bounded; it then follows that

$$\|\{\mathfrak{R}, A\}u\|_{L^2} \leqslant CM_1\|u\|_{L^2}, \quad (5.6.33)$$

where the constants C in both inequalities do not depend on $A(\cdot)$ (this is not particularly important now but will be vital in later applications in Section 7.8).

Let u, as before, be a function in $\mathcal{S}^\nu$. A simple calculation shows that

$$\mathfrak{R}\mathfrak{C}\mathfrak{R}^{-1}u = \mathfrak{C}u + \sum_{k=1}^m \{\mathfrak{R}, A_k\}D^k\mathfrak{R}^{-1}u + \{\mathfrak{R}, B\}\mathfrak{R}^{-1}u. \quad (5.6.34)$$

Making use of Theorem 5.6.6 and of the easily verified fact that

$$\|D^k\mathfrak{R}^{-1}u\|_{L^2} \leqslant \|u\|_{L^2}, \quad (5.6.35)$$

we see that there exists a bounded operator Q_1 in L^2 such that

$$\sum_{k=1}^m \{\mathfrak{R}, A_k\}D^k\mathfrak{R}^{-1}u = Q_1u \quad (u \in \mathcal{S}^\nu). \quad (5.6.36)$$

On the other hand, noting that $\mathfrak{R} = \mathfrak{R}^{-1} - \sum D^k\mathfrak{R}^{-1}D^k$, we obtain

$$\{\mathfrak{R}, B\}\mathfrak{R}^{-1}u = \left(\mathfrak{R}^{-1} - \sum_{k=1}^m D^k\mathfrak{R}^{-1}D^k\right)B\mathfrak{R}^{-1}u - Bu \quad (5.6.37)$$

On account of the fact that each $D^k \Re^{-1}$ is bounded (see (5.6.35)) and of the differentiability properties postulated for B, there exists a second bounded operator Q_2 in L^2 such that[14]

$$\{\Re, B\} \Re^{-1} u = Q_2 u \quad (u \in \mathbb{S}^{\nu}). \tag{5.6.38}$$

To prove (5.6.16) for $A = \mathcal{Q}$ and $K = \Re$ we still have to examine the domains of the unbounded operators involved. Let $u \in D(\mathcal{Q})$. Making use of Lemma 3.5.2, we can choose a sequence $\{u_n\}$ in $\mathbb{S}^{\nu}$ such that

$$u_n \to u, \quad \mathcal{Q} u_n \to \mathcal{Q} u$$

in L^2. Set

$$Q = Q_1 + Q_2,$$

write (5.6.34) for u_n, and take limits using (5.6.36) and (5.6.38); we obtain that $\Re \mathcal{Q} \Re^{-1} u_n \to \mathcal{Q} u + Q u$, thus $\mathcal{Q} \Re^{-1} u_n \to \Re^{-1}(\mathcal{Q} u + Q u)$, and it follows from the closedness of $\mathcal{Q}$ that $\Re^{-1} u \in D(\mathcal{Q})$ and $\mathcal{Q} \Re^{-1} u = \Re^{-1}(\mathcal{Q} u + Q u) \in H^1$; thus $\Re \mathcal{Q} \Re^{-1} u = \mathcal{Q} u + Q u$, showing that

$$\Re \mathcal{Q} \Re^{-1} \supseteq \mathcal{Q} + Q. \tag{5.6.39}$$

It follows that $\Re(\mathcal{Q} - \lambda I) \Re^{-1} \supseteq \mathcal{Q} + Q - \lambda I$; a fortiori, if inverses exist,

$$\Re(\mathcal{Q} - \lambda I)^{-1} \Re^{-1} \supseteq (\mathcal{Q} + Q - \lambda I)^{-1}. \tag{5.6.40}$$

But if λ is sufficiently large, it must belong to both $\rho(\mathcal{Q})$ and $\rho(\mathcal{Q} + Q)$ (Theorem 5.1.2), thus the operator on the right-hand side is everywhere defined and (5.6.40) is actually an equality. The same must be true of (5.6.39), that is,

$$\Re \mathcal{Q} \Re^{-1} = \mathcal{Q} + Q. \tag{5.6.41}$$

We obtain then from Lemma 5.6.3 that H^1 is $\mathcal{Q}$-admissible. This ends the proof of Theorem 5.6.4.

As it can be expected, results of the type of Theorem 5.6.4 can be obtained in Sobolev spaces H^r, $r \geq 1$, if additional smoothness assumptions are imposed on the coefficients of L. In fact, we have:

5.6.7 Theorem. *Let the matrices $A_1, \ldots, A_m$, B be r times continuously differentiable in $\mathbb{R}^m$ and assume all the functions*

$$|D^{\alpha} A_k(x)|, |D^{\alpha} B(x)| \quad (0 \leq |\alpha| \leq r, 1 \leq k \leq m) \tag{5.6.42}$$

are bounded in $\mathbb{R}^m$ by a constant M_r. Then conclusion (a) of Theorem 5.6.4 is valid and (b) holds replacing H^1 by H^r. In particular (5.6.19) becomes

$$S(t) H^r \subseteq H^r \quad (-\infty < t < \infty) \tag{5.6.43}$$

---

[14] Boundedness of $\{\Re, B\} \Re^{-1}$ (in fact, of $\{\Re, B\}$ itself) can be of course proved using Theorem 5.6.6. However, the present argument will be significant in Section 7.6.

and every solution $u(\cdot)$ of (5.6.18) with initial value in H^r remains in H^r for all t, is continuous in the norm of H^r and satisfies

$$\|u(t)\|_{H^r} \leqslant C_r e^{\omega_r |t|} \|u(0)\|_{H^r} \quad (-\infty < t < \infty)$$

for constants C_r, ω_r independent of $u(\cdot)$.

We limit ourselves to prove Theorem 5.6.7 in the case $r = 2$, since no new ideas are involved in the passage to the range $r > 2$. The role played when $r = 1$ by $\Re$ will now be taken over by $\Re^2$, which is a bounded operator from H^2 into L^2 with $(\Re^2)^{-1} = \Re^{-2} : L^2 \to H^2$ bounded as well. In view of (5.6.41), we have

$$\Re^2 \mathfrak{A} \Re^{-2} = \Re (\Re \mathfrak{A} \Re^{-1}) \Re^{-1} = \Re \mathfrak{A} \Re^{-1} + \Re Q \Re^{-1} = \mathfrak{A} + Q + \Re Q \Re^{-1},$$

thus our only task is to show that $\Re Q \Re^{-1}$ is everywhere defined and bounded; this will obviously hold if we prove that Q is a bounded operator in H^1. We do this next. Let A be a $\nu \times \nu$ matrix-valued function satisfying the hypotheses on A_k in Theorem 5.6.7 for $r = 2$, $\mathfrak{h}$ a function as in Theorem 5.6.6. Writing, as we may,

$$\mathfrak{C}_\varepsilon u(x) = \int_{|y| \geqslant \varepsilon} \mathfrak{h}(y)(A(x) - A(x-y)) u(x-y)\, dy,$$

we see that if $u \in \mathcal{S}^\nu$, $\mathfrak{C}_\varepsilon u$ is differentiable and

$$D^k \mathfrak{C}_\varepsilon u(x) = \int_{|x-y| \geqslant \varepsilon} \mathfrak{h}(x-y)(D^k A(x) - D^k A(y)) u(y)\, dy$$

$$+ \int_{|x-y| \geqslant \varepsilon} \mathfrak{h}(x-y)(A(x) - A(y)) D^k u(y)\, dy$$

$$(0 \leqslant k \leqslant m). \quad (5.6.44)$$

We can then use Theorem 5.6.6 to deduce that each $D^k \mathfrak{C}_\varepsilon u$ converges in L^2 as $\varepsilon \to 0$ to a function $v \in L^2$. Since $\mathfrak{C}_\varepsilon u \to \mathfrak{C} u$ in L^2, it follows that $D^k \mathfrak{C} u = v_k$ (in the sense of distributions) belongs to L^2, thus $\mathfrak{C} u \in H^1$. In view of the estimate (5.6.32), we have

$$\|\mathfrak{C} u\|_{H^1} \leqslant C M_2 \|u\|_{H^1} \quad (u \in \mathcal{S}^\nu). \quad (5.6.45)$$

We apply now these remarks to the operators $\{\Re, A_k\}$; in view of (5.6.27), it follows from (5.6.45) that

$$\|\{\Re, A_k\} u\|_{H^1} \leqslant C M_2 \|u\|_{H^1} \quad (u \in \mathcal{S}^\nu). \quad (5.6.46)$$

Observing that

$$\|\Re u - \Re u\|_{H^1} \leqslant C \|u\|_{H^1} \quad (u \in H^1)$$

and arguing as in the comments following Theorem 5.6.6, we see that (5.6.46) implies

$$\|\{\Re, A_k\} u\|_{H^1} \leqslant C M_2 \|u\|_{H^1} \quad (u \in \mathcal{S}^\nu), \quad (5.6.47)$$

hence, in view of (5.6.36) and of the fact that $\mathbb{S}^{\nu}(\mathbb{R}^m)$ is dense in $H^1(\mathbb{R}^m)^{\nu}$, Q_1 is a bounded operator in H^1. As for Q_2, defined by (5.6.38), it follows directly from (5.6.37), the hypotheses on B for $r = 2$ and again the denseness of $\mathbb{S}^{\nu}$, that it is a bounded operator in H^1. Hence $Q \in (H^1)$ and the proof of Theorem 5.6.7 for $r = 2$ is complete.

5.7. APPROXIMATION OF ABSTRACT DIFFERENTIAL EQUATIONS

It is often the case that solutions of an initial value problem

$$u'(t) = Au(t) \quad (t \geq 0), \quad u(0) = u_0, \tag{5.7.1}$$

where $A \in \mathcal{C}_+$, can be approximated by the solutions of

$$u_n'(t) = A_n u_n(t) \quad (t \geq 0), \quad u(0) = u_n, \tag{5.7.2}$$

with $u_n \to u$ and $A_n \to A$ in some way or other. Usually, A_n is an operator in a different Banach space E_n; one such instance is that where A is a differential operator in a space of functions defined in a domain $\Omega \subseteq \mathbb{R}^m$ and A_n is a finite difference approximation to A, in which case E_n may be a space of "discrete" functions defined only at certain grid points; moreover, instead of solving (5.7.2) (which would correspond to the "method of lines" in numerical analysis), we may also discretize the time variable, so that the differential equation in (5.7.2) may be discarded in favor of some finite difference approximation like

$$u_n(t + \tau) = \tau A_n u_n(t) + u_n(t). \tag{5.7.3}$$

We examine in this section and the next an abstract scheme including these instances.

In what follows, E and E_n $(n \geq 1)$ are (real or complex) Banach spaces. The first result is a sort of generalization of (a particular case of) Theorem 1.2. We shall henceforth denote by $\|\cdot\|$ the norms of E and of $(E; E)$ and by $\|\cdot\|_n$ the norms of E_n, of $(E; E_n)$ and of $(E_n; E_n)$, except when more precision is necessary.

5.7.1 Theorem. *Let the operators $Q_n \in (E; E_n)$ $(n \geq 1)$ be such that $\{\|Q_n u\|_n; \, n \geq 1\}$ is bounded for any $u \in E$. Then there exists $M \geq 0$ such that*

$$\|Q_n\|_n \leq M \quad (n \geq 1). \tag{5.7.4}$$

Proof. Denote by $\mathfrak{E}$ the space consisting of all sequences $u = (u_1, u_2, \ldots)$, $u_n \in E_n$ such that $\|u_n\|_n$ is bounded for $n \geq 1$. It is plain that the norm

$$\|u\| = \|u\|_{\mathfrak{E}} = \sup_{n \geq 1} \|u_n\|_n$$

makes $\mathfrak{E}$ a Banach space. The operator $Q: E \to \mathfrak{E}$ defined by

$$Qu = (Q_1 u, Q_2 u, \dots) \qquad (5.7.5)$$

in linear and everywhere defined, and it is also easily seen to be closed. It follows then from Theorem 3.1 that Q is bounded, that is, $\|Qu\|_{\mathfrak{E}} \leqslant M\|u\|_E$ for all $u \in E$. This is equivalent to (5.7.4).

We say that the sequence $\{E_n, P_n\}$, where $P_n \in (E; E_n)$, *approximates* E if and only if

$$\lim_{n \to \infty} \|P_n u\|_n = \|u\| \quad (u \in E). \qquad (5.7.6)$$

It follows from Theorem 5.7.1 that

$$\|P_n\|_n \leqslant M \quad (n \geqslant 1). \qquad (5.7.7)$$

A sequence $\{u_n\}$, $u_n \in E_n$ is said to *converge to* $u \in E$ (in symbols, $u_n \Rightarrow u$) if and only if

$$\|u_n - P_n u\|_n \to 0 \text{ as } n \to \infty. \qquad (5.7.8)$$

We note that the convergence relation (5.7.8) defines the element u uniquely: in fact, if $u_n \Rightarrow u$, $u_n \Rightarrow v$, we have, in view of (5.7.6),

$$\|u - v\| = \lim_{n \to \infty} \|P_n(u - v)\|_n$$
$$\leqslant \lim_{n \to \infty} \|u_n - P_n u\|_n + \lim_{n \to \infty} \|u_n - P_n v\|_n = 0.$$

A similar argument shows that

$$\|u\| = \lim_{n \to \infty} \|u_n\|_n. \qquad (5.7.9)$$

5.7.2 Example. The sequence $\{E_n, P_n\}$, where $E_n = E$ and $P_n = I$ obviously approximates E. Convergence according to (5.7.8) is nothing but convergence in E.

5.7.3 Example. Let Ω be a bounded domain in $\mathbb{R}^m$. For each $n = 1, 2, \dots$ divide Ω into a finite number of disjoint sets

$$\Omega = \Omega_{n1} \cup \Omega_{n2} \cup \cdots \cup \Omega_{n,r(n)} \qquad (5.7.10)$$

in such a way that

$$\lim_{n \to \infty} \left(\max_{1 \leqslant j \leqslant r(n)} \operatorname{diam} \Omega_{nj} \right) = 0 \qquad (5.7.11)$$

and select $x_{nj} \in \Omega_{nj}$ in an entirely arbitrary way. Let $E = C(\overline{\Omega})_{\mathbb{R}}$, $E_n = \mathbb{R}^{r(n)}$ $= r(n)$-dimensional Euclidean space endowed with the supremum norm $\|\{\xi_j\}\|_n = \sup|\xi_j|$ and let $P_n: C(\overline{\Omega}) \to \mathbb{R}^{r(n)}$ be defined by

$$P_n u = \big(u(x_{n1}), u(x_{n2}), \dots, u(x_{n,r(n)})\big).$$

Then $\{E_n, P_n\}$ approximates E. A similar result holds for $C(\overline{\Omega})_{\mathbb{C}}$.

5.7.4 Example. Let Ω, Ω_{nj} be as in Example 5.7.3; assume, moreover, that each Ω_{nj} is measurable. Let $E = L^p(\Omega)$ $(1 \leqslant p < \infty)$ and let E_n be the subspace of $L^p(\Omega)$ consisting of all functions that are constant in each Ω_{nj} with the norm inherited from $L^p(\Omega)$. Define an operator $P_n: L^p(\Omega) \to E_n$ by

$$(P_n u)(x) = m_{nj}(u) \quad (x \in \Omega_{nj}),$$

where m_{nj} is the average of u in Ω_{nj},

$$m_{nj}(u) = \frac{1}{\mu_{nj}} \int_{\Omega_{nj}} u(x) \, dx, \tag{5.7.12}$$

μ_{nj} the hypervolume of Ω_{nj}. By Hölder's inequality,

$$|m_{nj}(u)| \leq \frac{1}{\mu_{nj}^{1/p}} \left(\int_{\Omega_{nj}} |u(x)|^p \, dx \right)^{1/p},$$

so that each P_n is a bounded operator; in fact, $\|P_n u\|_n = \|u\|$, hence we have

$$\|P_n\| \leq 1. \tag{5.7.13}$$

and the fact that $\{E_n; P_n\}$ approximates E is obvious.

The main problem considered in this section is the following. Let $\{E_n, P_n\}$ approximate E, and let $A \in \mathcal{C}_+(E)$, $A_n \in \mathcal{C}_+(E_n)$. Under which conditions can we assure that

$$S_n(t)u \Rightarrow S(t)u? \tag{5.7.14}$$

where $S_n(t) = \exp(tA_n)$, $S(t) = \exp(tA)$, and convergence is understood in the sense of (5.7.8), uniformly with respect to t in one way or other. A solution is given in the following result; from now on we assume that $\{E_n, P_n\}$ approximates E.

5.7.5 Theorem. *Let* $A \in \mathcal{C}_+(C, \omega; E)$, $A_n \in \mathcal{C}_+(C, \omega; E_n)$ *(where* C, ω *do not depend on* n*). (a) Assume that for each* $u \in E$

$$\|S_n(t)P_n u - P_n S(t)u\|_n \to 0 \tag{5.7.15}$$

uniformly on compact subsets of $t \geq 0$, *where* $S_n(t) = \exp(tA_n)$, $S(t) = \exp(tA)$. *Then, for each* $u \in E$,

$$\|R(\lambda; A_n)P_n u - P_n R(\lambda; A)u\|_n \to 0 \tag{5.7.16}$$

uniformly in $\operatorname{Re} \lambda \geq \omega'$ *for each* $\omega' > \omega$. *(b) Conversely, assume that (5.7.16) holds for some* λ *with* $\operatorname{Re} \lambda > \omega$. *Then (5.7.15) holds uniformly on compact subsets of* $t \geq 0$.

Proof. Assume that (5.7.15) holds uniformly on compact subsets of $t \geq 0$. Let $\operatorname{Re} \lambda \geq \omega' > \omega$. Using formula (2.1.10), we obtain

$$\|R(\lambda; A_n)P_n u - P_n R(\lambda; A)u\|_n \leq \int_0^\infty e^{-\omega' t} \|S_n(t)P_n u - P_n S(t)u\|_n \, dt. \tag{5.7.17}$$

Divide the interval of integration at some sufficiently large $c > 0$; in $t > c$ we use the fact that $\|S_n(t)P_n u - P_n S(t)u\|_n \leqslant Ce^{\omega t}$ independently of n, while we take profit of the uniform convergence assumption in $t \leqslant c$. This completes the proof of (a).

To prove (b), let λ be such that (5.7.16) holds and consider the $(E; E_n)$-valued function

$$H(s) = S_n(t-s)R(\lambda; A_n)P_n S(s)R(\lambda; A)$$

in $0 \leqslant s \leqslant t$. It is easily seen that H is continuously differentiable with derivative

$$\begin{aligned}
H'(s) = &- S_n(t-s)A_n R(\lambda; A_n)P_n S(s)R(\lambda; A) \\
&+ S_n(t-s)R(\lambda; A_n)P_n S(s)AR(\lambda; A) \\
= &\, S_n(t-s)\big(P_n R(\lambda; A) - R(\lambda; A_n)P_n\big)S(s).
\end{aligned}$$

We apply H' to an element $u \in E$ and integrate in $0 \leqslant s \leqslant t$; after norms are taken, the inequality

$$\begin{aligned}
\|R(\lambda; A_n)&(P_n S(t) - S_n(t)P_n)R(\lambda; A)u\|_n \\
&\leqslant Ce^{\omega t}\int_0^t e^{-\omega s}\|(P_n R(\lambda; A) - R(\lambda; A_n)P_n)S(s)u\|_n \, ds
\end{aligned}$$

(5.7.18)

results. Noting that $\|S(s)\| \leqslant Ce^{\omega t}$, using (5.7.16), and applying the dominated convergence theorem to the integral, we see that if $v \in D(A)$ (so that $v = R(\lambda; A)u$),

$$\|R(\lambda; A_n)(P_n S(t) - S_n(t)P_n)v\|_n \to 0 \tag{5.7.19}$$

as $n \to \infty$, uniformly on compacts of $t \geqslant 0$. But the operators acting on v in (5.7.19) are uniformly bounded on compact subsets of $t \geqslant 0$ (by a constant times $(\lambda - \omega)^{-1}$), hence the limit relation must actually hold for all $u \in E$, uniformly on compact subsets of $t \geqslant 0$.

On the other hand, we have

$$R(\lambda; A_n)S_n(t)P_n - S_n(t)P_n R(\lambda; A) = S_n(t)\big(R(\lambda; A_n)P_n - P_n R(\lambda; A)\big) \tag{5.7.20}$$

and

$$R(\lambda; A_n)P_n S(t) - P_n S(t)R(\lambda; A) = \big(R(\lambda; A_n)P_n - P_n R(\lambda; A)\big)S(t). \tag{5.7.21}$$

We obtain then from (5.7.20) that, for $u \in E$,

$$\|(R(\lambda; A_n)S_n(t)P_n - S_n(t)P_n R(\lambda; A))u\|_n \to 0 \tag{5.7.22}$$

uniformly on compacts of $t \geqslant 0$. Likewise, it follows from (5.7.21) that, again for arbitrary u,

$$\|(R(\lambda; A_n)P_n S(t) - P_n S(t)R(\lambda; A))u\|_n \to 0 \tag{5.7.23}$$

in $t \geqslant 0$. To see that convergence is uniform on compact subsets of $t \geqslant 0$, we observe that $S(\cdot)u$, being continuous in $t \geqslant 0$, must be uniformly continuous on compact intervals $0 \leqslant t \leqslant c < \infty$; accordingly, given $\varepsilon > 0$, we can find a partition $0 = t_0 < t_1 < \cdots < t_n = c$ of $[0, c]$ such that $\|S(t)u - S(t')u\| \leqslant \varepsilon$ for any two points in the same interval; writing then $S(t) = S(t_j)u + (S(t_j)u - S(t)u)$ in (5.7.21) and using the uniform boundedness of $\|(R(\lambda; A_n)P_n - P_nR(\lambda; A))S(t)\|$, the required uniformity property follows. We combine now (5.7.19), (5.7.22), and (5.7.23) and obtain that

$$\|(P_nS(t) - S_n(t)P_n)R(\lambda; A)u\|_n \to 0 \qquad (5.7.24)$$

uniformly on compacts of $t \geqslant 0$ so that (5.7.15) follows for any $v \in R(\lambda; A)E = D(A)$ and thus, by denseness of $D(A)$ and uniform boundedness of $\|P_nS(t) - S_n(t)P_n\|$, it holds for any $u \in E$ uniformly on compact subsets of $t \geqslant 0$.

The following result shows that the operator A need not even be introduced in the statement of Theorem 5.7.5 if we impose additional assumptions on the A_n.

5.7.6 Theorem. *Let* $A_n \in \mathcal{C}_+(C, \omega; E_n)$, *where* C, ω *do not depend on* n. *Assume that there exists some* λ *with* $\mathrm{Re}\,\lambda > \omega$ *such that for each* $u \in E$ *there exists* $v \in E$ *with*

$$\|R(\lambda; A_n)P_nu - P_nv\|_n \to 0. \qquad (5.7.25)$$

Assume, moveover that

$$\lim_{\lambda \to \infty} \|\lambda R(\lambda; A_n)P_nu - P_nu\|_n = 0 \qquad (5.7.26)$$

for each $u \in E$, *uniformly with respect to* n. *Then there exists* $A \in \mathcal{C}_+(C, \omega; E)$ *such that* (5.7.16) *holds; a fortiori,* (5.7.15) *is satisfied uniformly on compacts of* $t \geqslant 0$.

Proof. Taking into account the observations after (5.7.8) and the equality (5.7.9), we deduce that the element v in (5.7.25) is unique and that

$$\|v\| \leqslant \liminf\|R(\lambda; A_n)\|_n\|u\| \leqslant C(\lambda - \omega)^{-1}\|u\| \qquad (5.7.27)$$

so that $v = Q(\lambda)u$, $Q(\lambda)$ a linear bounded operator in E; so far we are only certain of the existence of $Q(\lambda)$ for the value of λ postulated in Theorem 5.7.6. That it actually exists for other values of λ follows from

5.7.7 Lemma. *Let the assumptions of Theorem 5.7.6 be satisfied. Then* (5.7.25) *holds for every* λ *in* $\mathrm{Re}\,\lambda > \omega$ *(with* v *depending on* λ*).*

Proof. Let λ be such that (5.7.25) is verified. Then

$$\|R(\lambda; A_n)^2P_nu - R(\lambda; A_n)P_nv\|_n \leqslant C(\lambda - \omega)^{-1}\|R(\lambda; A_n)P_nu - P_nv\|_n \to 0.$$

Since $\|R(\lambda; A_n)P_nv - P_nw\|_n \to 0$ with $w = Q(\lambda)v = Q(\lambda)^2u$, we have

$\|R(\lambda; A_n)^2 P_n u - P_n Q(\lambda)^2 u\|_n \to 0$ as $n \to \infty$. Arguing further in the same way, we obtain

$$\|R(\lambda; A_n)^k P_n u - P_n Q(\lambda)^k u\|_n \to 0 \qquad (5.7.28)$$

as $n \to \infty$ for each fixed $u \in E$, $k = 1, 2, \ldots$. We use now formula (3.4): if $|\mu - \lambda| < \operatorname{Re}\lambda - \omega$,

$$R(\mu; A_n) = \sum_{k=0}^{\infty} (\lambda - \mu)^k R(\lambda; A_n)^{k+1}.$$

Making use of all the relations (5.7.28) and of the uniform bound on the $\|R(\lambda; A_n)^k\|$, we see that (5.7.25) actually holds in $|\lambda - \mu| < \operatorname{Re}\lambda - \omega$. Carrying out this "extension to circles" repeatedly in an obvious way, Lemma 5.7.7 follows. A consequence of the argument is that (5.7.25) holds uniformly on compact subsets of $\operatorname{Re}\lambda > \omega$.

Proof of Theorem 5.7.6. It follows from the second resolvent equation (3.6) for each $R(\lambda; A_n)$ that

$$Q(\lambda) - Q(\mu) = (\mu - \lambda) Q(\lambda) Q(\mu) \qquad (5.7.29)$$

for $\operatorname{Re}\lambda, \operatorname{Re}\mu > \omega$; incidentally, this equality shows that $Q(\lambda)$ and $Q(\mu)$ commute. Let $\mathfrak{R}(\lambda) = Q(\lambda)E$ and let $\mathfrak{N}(\lambda)$ be the nullspace of $Q(\lambda)$. It follows easily from (5.7.29) that both $\mathfrak{R}$ and $\mathfrak{N}$ are constant, that is,

$$\mathfrak{R}(\lambda) = \mathfrak{R}, \quad \mathfrak{N}(\lambda) = \mathfrak{N} \quad (\operatorname{Re}\lambda > \omega). \qquad (5.7.30)$$

We combine now (5.7.25) and (5.7.26), obtaining

$$\lim_{\lambda \to \infty} \|\lambda Q(\lambda) u - u\| = \lim_{\lambda \to \infty} \left(\lim_{n \to \infty} \|P_n(\lambda Q(\lambda) u - u)\|_n \right)$$

$$\leqslant \lim_{\lambda \to \infty} \left(\lim_{n \to \infty} \|(\lambda R(\lambda; A_n) P_n u - P_n u)\|_n \right)$$

$$+ \lim_{\lambda \to \infty} \left(\lim_{n \to \infty} \|\lambda R(\lambda; A_n) P_n u - \lambda P_n Q(\lambda) u\|_n \right)$$

$$= 0 \qquad (5.7.31)$$

for all $u \in E$. Accordingly, if $u \in \mathfrak{N}$, $u = \lim \lambda Q(\lambda) u = 0$; in other words,

$$\mathfrak{N} = \{0\} \qquad (5.7.32)$$

so that each $Q(\lambda)$ is one-to-one. The limit relation (5.7.31) also implies that $\mathfrak{R}$ is dense in E since u can be approximated by $\lambda Q(\lambda) u \in \mathfrak{R}$. The operators $A_\lambda = Q(\lambda)^{-1}$ are then well defined and have as a common domain the dense subspace $\mathfrak{R}$. We show finally that

$$\lambda I - A_\lambda = \mu I - A_\mu = A \quad (\operatorname{Re}\lambda, \operatorname{Re}\mu > \omega); \qquad (5.7.33)$$

to see this we apply $A_\lambda A_\mu = A_\mu A_\lambda$ to both sides of (5.6.29). It follows immediately from the definition of A that every λ with $\operatorname{Re}\lambda > \omega$ belongs to

$\rho(A)$ and

$$R(\lambda; A) = Q(\lambda).$$

Using (5.6.28) for any such λ,

$$\|R(\lambda; A)^k u\| = \lim_{n \to \infty} \|P_n Q(\lambda)^k u\|_n$$

$$= \lim_{n \to \infty} \|R(\lambda; A_n)^k P_n u\|_n$$

$$\leqslant C(\operatorname{Re} \lambda - \omega)^{-k} \|u\| \quad (\operatorname{Re} \lambda > \omega, k = 1, 2, \ldots),$$

$$(5.7.34)$$

where C does not depend on k. Hence $A \in \mathcal{C}_+(C, \omega; E)$ and the assumptions of Theorem 5.7.5(b) are satisfied. This ends the proof.

We note a curiosity pertaining to the case where $E_n = E$, $P_n = I$.

5.7.8 Example. Let $S_n(\cdot)$ be a sequence of strongly continuous semigroups such that for each $u \in E$ the sequence $S_n(t)u$ converges uniformly in $0 \leqslant t \leqslant \delta > 0$. Then there exists C, ω independent of n such that $\|S_n(t)\| \leqslant Ce^{\omega t}$.

It is of great interest to have conditions insuring the convergence of the propagators S_n that bear on the operators A_n themselves rather than on their resolvents $R(\lambda; A_n)$. Such conditions are given below (Theorem 5.7.11). Given a sequence $\{E_n; P_n\}$ approximating E and an operator A_n in each E_n with domain $D(A_n)$, we define $\mathcal{C}$, the *extended limit* of the A_n (in symbols, ex-$\lim_{n \to \infty} A_n$ or simply ex-$\lim A_n$) as follows: an element $u \in E$ belongs to $D(\mathcal{C})$ if and only if there exists a sequence $\{u_n\}$, $u_n \in D(A_n)$ and a $v = \mathcal{C}u \in E$ such that $u_n \Rightarrow u$ and $A_n u_n \Rightarrow v$, that is,

$$\|u_n - P_n u\|_n \to 0, \quad \|A_n u_n - P_n v\|_n \to 0 \qquad (5.7.35)$$

as $n \to \infty$. In general, we cannot expect $\mathcal{C}$ to be single valued, even in very special situations.

5.7.9 Example. If $E_n = E$, $P_n = I$, $A_n = A$, then ex-$\lim A_n$ is single valued if and only if A is closable; in that case, ex-$\lim A_n = \bar{A}$.

The following result establishes conditions under which $\mathcal{C}$ is single valued.

5.7.10 Lemma. *Assume that there exists* $\omega > 0$ *such that, for each* $u_n \in D(A_n)$,

$$\|(\lambda I - A_n)u_n\|_n \geqslant c(\lambda - \omega)\|u_n\|_n \qquad (5.7.36)$$

for $\lambda > \omega$ *and* $n = 1, 2, \ldots$, *the constant* $c > 0$ *independent of* n. *Assume, moreover, that* $D(\mathcal{C})$ *is dense in* E. *Then* $\mathcal{C}$ *is single valued and*

$$\|(\lambda I - \mathcal{C})u\| \geqslant c(\lambda - \omega)\|u\| \quad (\lambda > \omega). \qquad (5.7.37)$$

Proof. If $\mathcal{C}$ is not single valued, there exists $\bar{v} \in E$, $\bar{v} \neq 0$ and a sequence $\{\bar{u}_n\}$, $\bar{u}_n \in D(A_n)$ such that

$$\|\bar{u}_n\|_n \to 0, \quad \|A_n\bar{u}_n - P_n\bar{v}\|_n \to 0$$

as $n \to \infty$. On the other hand, if u is an arbitrary element of $D(\mathcal{C})$, there exists some sequence $\{u_n\}$ and some v (not necessarily unique) such that (5.7.35) holds. Since

$$\|(\lambda I - A_n)(u_n + \lambda \bar{u}_n)\|_n \geq c(\lambda - \omega)\|u_n + \lambda \bar{u}_n\|_n \quad \text{for} \quad \lambda > \omega$$

and

$$\|(\lambda I - A_n)(u_n + \lambda \bar{u}_n) - P_n(\lambda u - v - \lambda \bar{v})\|_n \to 0 \quad \text{as} \quad n \to \infty,$$

we have

$$\begin{aligned}
\|\lambda u - v - \lambda \bar{v}\| &= \lim \|P_n(\lambda u - v - \lambda \bar{v})\|_n \\
&= \lim \|(\lambda I - A_n)(u_n + \lambda \bar{u}_n)\|_n \geq \lim c(\lambda - \omega)\|u_n + \lambda \bar{u}_n\|_n \\
&= c(\lambda - \omega)\lim \|u_n\|_n = c(\lambda - \omega)\lim \|P_n u\|_n = c(\lambda - \omega)\|u\|.
\end{aligned}$$
(5.7.38)

Dividing by λ and letting $\lambda \to \infty$, we obtain

$$\|u - \bar{v}\| \geq c\|u\|.$$
(5.7.39)

Since $D(\mathcal{C})$ is dense in E, we can choose u with $\|u - \bar{v}\| < c\|u\|$, which contradicts (5.7.39) and shows that $\mathcal{C}$ is single valued. To obtain inequality (5.7.37), it suffices to read (5.7.38) for $\bar{v} = 0$.

5.7.11 Theorem. *Let $A_n \in \mathcal{C}_+(C, \omega; E_n)$, where C, ω do not depend on n. Assume, moreover, that $\mathcal{C} = \text{ex-lim } A_n$ is densely defined in E and that $(\lambda I - \mathcal{C})D(\mathcal{C})$ is dense in E for some $\lambda > \omega$. Then $\mathcal{C} \in \mathcal{C}_+(C, \omega; E)$ (in particular, it is single valued) and for each $u \in E$*

$$S_n(t)u \Rightarrow S(t)u$$
(5.7.40)

uniformly on compact subsets of $t \geq 0$, where $S(t) = \exp(t\mathcal{C})$. Conversely, if for each $u \in E$ the sequence $\{S_n(t)u\}$ converges uniformly on compact subsets of $t \geq 0$, $\mathcal{C} = \text{ex-lim } A_n$ is single valued, belongs to $\mathcal{C}_+(C, \omega; E)$ and (5.7.40) holds, with $S(t) = \exp(t\mathcal{C})$.

Proof. Since each A_n belongs to $\mathcal{C}_+(C, \omega; E_n)$ inequalities (5.7.36) hold with $c = 1/C$; taking into account the fact that $\mathcal{C}$ is densely defined, it follows from Lemma 5.7.10 that $\mathcal{C}$ is single valued and (5.7.37) holds. We show next that $\mathcal{C}$ is closed. Let $\{u_m\}$ be a sequence in $D(\mathcal{C})$ with $u_m \to u \in E$, $\mathcal{C}u_m \to v \in E$. Making use of the definition of $\mathcal{C}$, we can find for each m a sequence $\{u_{mn}; n \geq 1\}$, $u_{mn} \in D(A_n)$, such that

$$\lim_{n \to \infty} \|u_{mn} - P_n u_m\|_n = 0, \quad \lim_{n \to \infty} \|A_n u_{nm} - P_n \mathcal{C}u_m\|_n = 0.$$

It is easy to see that this implies the existence of a sequence $\{u_n\}$, $u_n \in E_n$ such that (5.7.35) holds, showing that $u \in D(\mathcal{C})$ and $\mathcal{C}u = v$. The closedness of $\mathcal{C}$ and inequality (5.6.37) obviously imply that

$$(\lambda I - \mathcal{C})D(\mathcal{C})$$

is a closed subspace of E for each $\lambda > \omega$. Since $(\lambda I - \mathcal{C})D(\mathcal{C})$ was assumed dense in E for some $\lambda > \omega$, $(\lambda I - \mathcal{C})D(\mathcal{C}) = E$ for that value of λ and $R(\lambda; \mathcal{C}) = (\lambda I - \mathcal{C})^{-1}$ exists. Let v be an arbitrary element of E; write $v = (\lambda I - \mathcal{C})u$, $u \in D(\mathcal{C})$, and select a sequence $\{u_n\}$, $u_n \in D(A_n)$ such that $\|u_n - P_n u\|_n \to 0$, $\|A_n u_n - P_n \mathcal{C} u\|_n = 0$. Then

$$\lim \| R(\lambda; A_n)P_n v - P_n R(\lambda; \mathcal{C})v \|_n = \lim \| R(\lambda; A_n)P_n(\lambda I - \mathcal{C})u - P_n u \|_n$$

$$= \lim \| R(\lambda; A_n)(\lambda I - A_n)u_n - u_n \|_n = 0,$$

$$(5.7.41)$$

hence the assumptions of Theorem 5.7.6 are satisfied with $v = Q(\lambda)u = R(\lambda; \mathcal{C})u$,[15] thus $\mathcal{C} \in \mathcal{C}_+(C, \omega; E)$ and (5.7.40) holds uniformly on compact subsets of $t \geq 0$. Conversely, assume that for each $u \in E$ and each $t \geq 0$ there exists $v = v(t)$ such

$$\| S_n(t)P_n u - P_n v(t) \|_n \to 0 \qquad (5.7.42)$$

uniformly on compact subsets of $t \geq 0$. Arguing much in the same fashion as in the construction of $Q(\lambda)$ in Theorem 5.7.6, we can show that $v(t) = S(t)u$, $S(t)$ a semigroup with $\|S(t)\| \leq Ce^{\omega t}$ ($t \geq 0$); also, since

$$\| S(t')u - S(t)u \| = \lim \| P_n(S(t')u - S(t)u) \|_n$$

with

$$\| P_n(S(t')u - S(t)u) \|_n \leq \| S_n(t')P_n u - P_n S(t')u \|$$

$$+ \| (S_n(t') - S_n(t))P_n u \| + \| S_n(t)P_n u - P_n S(t)u \|$$

and the convergence is uniform on compact subsets of $t \geq 0$, $S(\cdot)$ is strongly continuous. If A is its infinitesimal generator, it only remains to show that

$$A = \mathcal{C} = \text{ex-lim } A_n,$$

which we do now. Let $u \in D(A)$; write $v = (\lambda I - A)u$ for some $\lambda > \omega$ so that

$$u = R(\lambda; A)v = \int_0^\infty e^{-\lambda t}S(t)v\, dt$$

and define

$$u_n = R(\lambda; A_n)P_n v = \int_0^\infty e^{-\lambda t}S_n(t)P_n v\, dt.$$

---

[15] Assumption (5.7.26) is only used to show that $v = Q(\lambda)u = R(\lambda; A)u$, where A is a densely defined operator with $\lambda \in \rho(A)$; this is unnecessary here since $Q(\lambda)u = R(\lambda; \mathcal{C})$.

Then we have

$$\|u_n - P_n u\|_n \leqslant \int_0^\infty e^{-\lambda t} \|S_n(t)P_n v - P_n S(t)v\|_n \, dt \qquad (5.7.43)$$

and we show that $\|u_n - P_n u\|_n \to 0$ as in the proof of Theorem 5.7.5(a). On the other hand, $Au = \lambda u - v$, $A_n u_n = \lambda u_n - P_n v$ so that

$$\|A_n u_n - P_n Au\|_n \leqslant \lambda \|u_n - P_n u\|_n, \qquad (5.7.44)$$

and it follows that $u \in D(\mathcal{A})$ with $\mathcal{A}u = Au$, in other words, that $A \subseteq \mathcal{A}$ (note that $\mathcal{A}$, being densely defined must be single valued by Lemma 5.7.9). We can then apply (a) to show that $\mathcal{A} \in \mathcal{C}_+$ and an argument similar to that ending Lemma 2.3.9 shows that $A = \mathcal{A}$ as claimed.

5.7.12 Example. *The method of lines for a parabolic equation.* Let Ω be a bounded domain of class $C^{(\infty)}$ in $\mathbb{R}^m$,

$$A = \Delta + c(x), \qquad (5.7.45)$$

where $c \in C^{(\infty)}(\bar\Omega)$. As a very particular case of the results in Chapter 4 (Theorem 4.8.11), we know that the operator A with domain consisting of all $u \in C_\Gamma(\bar\Omega)$ such that $u \in W^{2,p}(\Omega)$ for all $p < \infty$ and $Au \in C_\Gamma(\bar\Omega)$ ($\Gamma =$ boundary of Ω) belongs to $\mathcal{C}_+(1, \omega)$ in $E = C_\Gamma(\bar\Omega)$ if

$$\omega = \max_{x \in \Omega} c(x).$$

(We consider only real spaces and the Dirichlet boundary condition.) The choice of approximating spaces is essentially that of Example 5.7.3, but the subdivision of Ω is precisely specified as follows. Assuming Ω is contained in the hypercube H defined by $|x_j| \leqslant a$ ($1 \leqslant j \leqslant m$), for each $n = 1, 2, \ldots$ we divide the hypercube into $(2n)^m$ hypercubes determined by division of each side of H into $2n$ equal segments. These hypercubes will be denoted as follows: if $\alpha = (\alpha_1, \ldots, \alpha_m)$ is a multi-index of integers with $-n \leqslant \alpha_j \leqslant n - 1$, $H_{n,\alpha}$ denotes the hypercube defined by the inequalities

$$\alpha_j a/n \leqslant x_j \leqslant (\alpha_j + 1)a/n. \qquad (5.7.46)$$

Also, $x_{n,\alpha}$ denotes the point $(\alpha_1 a/n, \ldots, \alpha_m a/n)$ and for each n, J_n is the set of all multi-indices α with $x_{n,\alpha} \in \Omega$, X_n the set of all $x_{n,\alpha}$ in Ω, $r(n)$ the number of elements in J_n. The approximating spaces will be $E_n = \mathbb{R}^{r(n)}$ whose elements are denoted by ξ, $\{\xi_{n,\alpha}; \alpha \in J_n\}$ or simply $\{\xi_{n,\alpha}\}$; each of these spaces is endowed with the supremum norm

$$\|\xi\|_n = \max_{\alpha \in J_n} |\xi_{n,\alpha}|. \qquad (5.7.47)$$

The operators P_n are defined by

$$P_n u = \{u(x_{n,\alpha}); \alpha \in J_n\}. \qquad (5.7.48)$$

The operators A_n will be finite difference approximations to the partial differential operator A. Obviously there is a great latitude in the choice of

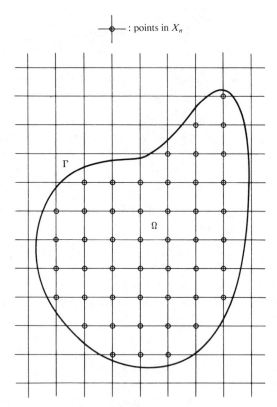

FIGURE 5.7.1

the A_n; we select the approximation

$$D_{h,k}f(x) = 2\frac{kf(x+h)-(k+h)f(x)+hf(x-k)}{hk(h+k)} \qquad (5.7.49)$$

with $h, k > 0$ for the second derivatives $(D^j)^2$, which has the important property of being nonpositive at maxima of f. Denote by $e(j)$ $(1 \leq j \leq m)$ the multi-index having zero components except for the j-th, which equals 1, and define A_n by

$$(A_n \xi)_{n,\alpha}$$

$$= \sum_{j=1}^{m} \frac{2}{h_{n,\alpha}^j k_{n,\alpha}^j \left(h_{n,\alpha}^j + k_{n,\alpha}^j\right)} \left(k_{n,\alpha}^j \sigma_{n,\alpha}^j - \left(h_{n,\alpha}^j + k_{n,\alpha}^j\right)\xi_{n,\alpha} + h_{n,\alpha}^j \zeta_{n,\alpha}^j\right)$$

$$+ c(x_{n,\alpha})\xi_{n,\alpha}, \qquad (5.7.50)$$

where the $h_{n,\alpha}^j, k_{n,\alpha}^j, \sigma_{n,\alpha}^j, \zeta_{n,\alpha}^j$ are defined as follows. If the (closed) segment I joining $x_{n,\alpha}$ and $x_{n,\alpha+e(j)}$ does not contain points of Γ, then $\sigma_{n,\alpha}^j = \xi_{n,\alpha+e(j)}$

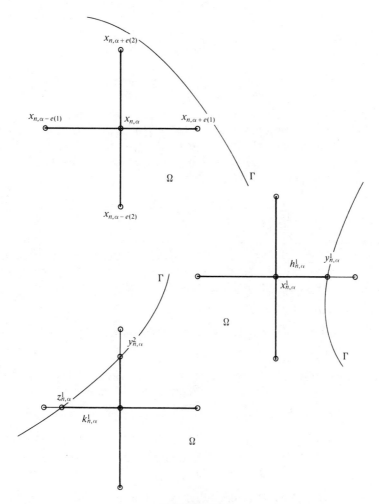

FIGURE 5.7.2

and $h_{n,\alpha}^j = a/n = \text{dist}(x_{n,\alpha}, x_{n,\alpha+e(j)})$. Otherwise, we call $y_{n,\alpha}^j$ the point in $I \cap \Gamma$ closest to $x_{n,\alpha}$ and define $h_{n,\alpha}^j = \text{dist}(x_{n,\alpha}, y_{n,\alpha})$ and $\sigma_{n,\alpha}^j = 0$. The point $z_{n,\alpha}^j$ and the numbers $k_{n,\alpha}^j, \zeta_{n,\alpha}^j$ are defined in a mirrorlike way with respect to the segment J joining $x_{n,\alpha}$ and $x_{n,\alpha-e(j)}$.

We show next that each A_n belongs to $\mathcal{C}_+(1, \omega)$ in $E_n = \mathbb{R}^{r(n)}$. It is plain that the dual space of $\mathbb{R}^{r(n)}$ is $\mathbb{R}^{r(n)}$ itself equipped with the norm

$$\|\eta\|_n = \sum_{\alpha \in J_n} |\eta_{n,\alpha}|,$$

where $\eta = \{\eta_{n,\alpha}\}$; application of the functional η to an element $\xi = \{\xi_{n,\alpha}\}$ is expressed by

$$\langle \eta, \xi \rangle = \sum_{\alpha \in J_n} \eta_{n,\alpha} \xi_{n,\alpha}.$$

The duality set $\Theta(\xi)$ of any $\xi \in E_n$ can be described as follows. Let $m(\xi) = \{\alpha \in J_n;\ |\xi_{n,\alpha}| = \|\xi\|_n\}$. Then $\Theta(\xi)$ consists of all $\eta = \{\eta_{n,\alpha}\}$ such that $\Sigma |\eta_{n,\alpha}| = \|\xi\|_n$, $\eta_{n,\alpha} = 0$ for $\alpha \notin m(\xi)$, $\eta_{n,\alpha}\xi_{n,\alpha} \geqslant 0$ for all α.

Consider an arbitrary nonzero $\xi \in \mathbb{R}^{r(n)}$ and let $\alpha \in m(\xi)$. If $\xi_{n,\alpha} > 0$, then $\xi_{n,\alpha}$ is a maximum of $\{\xi_{n,\alpha}\}$ and $(A_n\xi)_{n,\alpha} \leqslant \omega|\xi_{n,\alpha}| \leqslant \omega\|\xi\|$; on the other hand, if $\xi_{n,\alpha} < 0$, we obtain arguing in the same way that $(A_n\xi)_{n,\alpha} \geqslant -\omega|\xi_{n,\alpha}|$; in any case, if $\eta \in \Theta(\xi)$ we have

$$\langle \eta, A_n\xi \rangle \leqslant \omega\|\xi\| \quad (\xi \in \mathbb{R}^{r(n)}), \tag{5.7.51}$$

thus proving that $A_n \in \mathcal{C}_+(1, \omega; E_n)$.

To apply Theorem 5.7.11, we must identify $\mathcal{C} = $ ex-$\lim A_n$. To this end, we take $u \in C^{(2)}(\overline{\Omega}) \cap C_\Gamma(\overline{\Omega})$ and compute $A_n P_n u$; to avoid notational complications, we only do the calculation with (5.7.49). Using the Taylor formula for f up to second order terms, we see that

$$|D_{h,k}f(x) - f''(x)| \leqslant |f''(x+\sigma) - f''(x)| + |f''(x) - f''(x-\tau)|,$$

where $0 \leqslant \sigma \leqslant h$, $0 \leqslant \tau \leqslant k$, thus we easily verify that

$$\|A_n P_n u - P_n A u\|_n \to 0. \tag{5.7.52}$$

Since $C^{(2)}(\overline{\Omega}) \cap C_\Gamma(\overline{\Omega})$ is dense in $C_\Gamma(\overline{\Omega})$, we see that $\mathcal{C} = $ ex-$\lim A_n$ is densely defined (thus single valued by Lemma 5.7.10) and

$$\mathcal{C}u = Au \quad (u \in C^{(2)}(\overline{\Omega}) \cap C_\Gamma(\overline{\Omega})). \tag{5.7.53}$$

Let v be a smooth function (say, in $\mathcal{D}(\Omega)$). Then if λ is large enough, the solution $u \in D(A)$ of

$$(\lambda I - A)u = v \tag{5.7.54}$$

belongs to all $H^k(\Omega)$ by Theorem 4.7.12, a fortiori to $C^{(2)}(\overline{\Omega})$ by Corollary 4.7.15. This fact, combined with (5.7.52), shows that

$$(\lambda I - \mathcal{C})\big(C^{(2)}(\overline{\Omega}) \cap C_\Gamma(\overline{\Omega})\big) = (\lambda I - A)\big(C^{(2)}(\overline{\Omega}) \cap C_\Gamma(\overline{\Omega})\big)$$

is dense in $E = C_\Gamma(\overline{\Omega})$. It follows that the assumptions of Theorem 5.7.11 are fulfilled and thus that $\mathcal{C} \in \mathcal{C}_+(1, \omega; E)$. Moreover, it is a consequence of (5.7.53) and following comments that, if $v \in \mathcal{D}(\Omega)$, then $R(\lambda; \mathcal{C})v = R(\lambda; A)v$ so that $R(\lambda; \mathcal{C}) = R(\lambda; A)$ (by denseness of $\mathcal{D}(\Omega)$ in $C_\Gamma(\overline{\Omega})$), and we deduce that

$$\mathcal{C} = A. \tag{5.7.55}$$

We obtain from Theorem 5.7.11 that if $u(\cdot)$ is an arbitrary solution of the initial-value problem

$$u'(t) = Au(t), \quad u(0) = u, \tag{5.7.56}$$

then

$$\lim_{n \to \infty} \|\exp(tA_n)P_nu - P_nu(t)\|_n = 0 \qquad (5.7.57)$$

uniformly on compact subsets of $t \geqslant 0$ (since u may not belong to $D(A)$, $u(\cdot)$ may be a generalized solution; see Section 1.2).

5.8. APPROXIMATION OF ABSTRACT DIFFERENTIAL EQUATIONS BY FINITE DIFFERENCE EQUATIONS.

A *discrete semigroup* with *time scale* $\tau > 0$ in a Banach space E is a (E)-valued function $\mathfrak{S}$ defined in the set $\{l\tau;\ l = 0, 1, \ldots\}$ (in other words, a sequence of bounded operators $\{\mathfrak{S}(l\tau);\ l = 0, 1, \ldots\}$) with $\mathfrak{S}(0) = I$ and $\mathfrak{S}((l+m)\tau) = \mathfrak{S}(l\tau)\mathfrak{S}(m\tau)$ $(l, m = 0, 1, \ldots)$. Obviously, we have

$$\mathfrak{S}(l\tau) = \mathfrak{S}^l \quad (l \geqslant 0), \qquad (5.8.1)$$

where $\mathfrak{S} = \mathfrak{S}(\tau)$; hence

$$\|\mathfrak{S}(l\tau)\| \leqslant \|\mathfrak{S}\|^l = e^{\omega l\tau} \quad (l \geqslant 0) \qquad (5.8.2)$$

with $\omega = \tau^{-1}\log\|\mathfrak{S}\|$. If $\{\mathfrak{S}(t);\ t \geqslant 0\}$ is an arbitrary semigroup, then $\{\mathfrak{S}(l\tau);\ l \geqslant 0\}$ is a discrete semigroup with time scale τ. A theory of discrete semigroups roughly parallel to that of uniformly continuous semigroups (but much simpler in nature) can be easily developed. The *generator* of $\{\mathfrak{S}(l\tau);\ l \geqslant 0\}$ is by definition

$$A = \frac{1}{\tau}\left(\mathfrak{S}(\tau) - I\right) \qquad (5.8.3)$$

and we can express $\mathfrak{S}(\cdot)$ from A by means of the formula

$$\mathfrak{S}(l\tau) = (\tau A + I)^l \quad (l \geqslant 0). \qquad (5.8.4)$$

Conversely, each bounded operator A is the generator of a discrete semigroup $\mathfrak{S}(\cdot)$ with time scale τ; it suffices to take formula (5.8.4) as the definition of $\mathfrak{S}(l\tau)$.

There is a close relation between $\{\mathfrak{S}(l\tau)\}$ and the uniformly continuous semigroup $\{S(t)\}$ defined by $S(t) = e^{tA}$. To avoid inessential difficulties, we shall assume from now on that the operators $\mathfrak{S}(l\tau)$ are uniformly bounded.

5.8.1 Lemma. (a) *Assume*

$$\|\mathfrak{S}(l\tau)\| \leqslant C \quad (l \geqslant 0). \qquad (5.8.5)$$

Then

$$\|S(t)\| \leqslant C \quad (t \geqslant 0) \qquad (5.8.6)$$

and

$$\|\mathfrak{S}(l\tau) - S(l\tau)\| \leqslant \frac{C}{2} l\tau^2 \|A^2 u\|. \tag{5.8.7}$$

Proof. Noting that $tA = (t/\tau)\mathfrak{S}(\tau) - (t/\tau)I$, we have

$$\mathfrak{S}(l\tau)S(t) = e^{-t/\tau}\mathfrak{S}(l\tau)\sum_{m=0}^{\infty}\frac{(t/\tau)^m}{m!}\mathfrak{S}(m\tau)$$

$$= e^{-t/\tau}\sum_{m=0}^{\infty}\frac{(t/\tau)^m}{m!}\mathfrak{S}((l+m)\tau) \tag{5.8.8}$$

so that

$$\|\mathfrak{S}(l\tau)S(t)\| \leqslant Ce^{-t/\tau}\sum_{m=0}^{\infty}\frac{(t/\tau)^m}{m!} = C. \tag{5.8.9}$$

Taking $l = 0$, (5.8.6) follows.

Since $\mathfrak{S}$ and S obviously commute,

$$\mathfrak{S}(l\tau) - S(l\tau) = \left\{\sum_{j=0}^{l-1}\mathfrak{S}((l-j-1)\tau)S(j\tau)\right\}(\mathfrak{S}(\tau) - S(\tau)).$$

On the other hand,

$$S(\tau) - \mathfrak{S}(\tau) = S(\tau) - I - \tau A = \int_0^\tau (\tau - t)A^2 S(t)\, dt$$

so that, if $0 \leqslant j \leqslant l-1$,

$$\mathfrak{S}((l-j-1)\tau)S(j\tau)(\mathfrak{S}(\tau) - S(\tau))u$$
$$= \int_0^t (\tau - t)\mathfrak{S}((l-j-1)\tau)S(j\tau + t)A^2 u\, dt.$$

Making use of (5.8.9), we obtain

$$\|\mathfrak{S}((l-j-1)\tau)S(j\tau)(\mathfrak{S}(\tau) - S(\tau))u\| \leqslant \frac{C}{2}\tau^2\|A^2 u\|,$$

whence (5.8.7) follows immediately.

It is sometimes convenient to extend $\mathfrak{S}$ to all values of t by defining $\mathfrak{S}(t) = \mathfrak{S}(l\tau)$ in the interval $[l\tau, (l+1)\tau)$; in other words, we set

$$\mathfrak{S}(t) = \mathfrak{S}([t/\tau]\tau) \quad (t \geqslant 0), \tag{5.8.10}$$

where $[s]$ is the greatest integer $\leqslant s$. Then

$$\|\mathfrak{S}(t)u - S(t)u\| \leqslant \|\mathfrak{S}([t/\tau]\tau)u - S([t/\tau]\tau)u\| + \|S([t/\tau]\tau)u - S(t)\|$$
$$\leqslant \frac{C}{2}\left[\frac{t}{\tau}\right]\tau^2\|A^2 u\| + C\tau\|Au\| \leqslant C\tau\left(\frac{t}{2}\|A^2 u\| + \|Au\|\right),$$
$$\tag{5.8.11}$$

where we have estimated in an obvious way the equality

$$S(t')u - S(t)u = \int_t^{t'} S(s)Au\,ds.$$

We examine in this section the following situation. Let $\{E_n, P_n\}$ approximate E in the sense of Section 5.7 and let $A \in \mathcal{C}_+(E)$. A discrete semigroup $\{\mathfrak{S}_n(l\tau_n)\}$ with time scale $\tau_n \to 0$ is given in each E_n, and we seek to approximate S by the $\mathfrak{S}_n$ in the following sense:

$$\|\mathfrak{S}_n(l_n\tau_n)P_n u - P_n S(t)u\|_n \to 0, \tag{5.8.12}$$

where $\{l_n\}$ is a sequence of nonnegative integers such that

$$l_n\tau_n \to t.$$

Each of the results in Section 5.7 has an exact counterpart for discrete semigroups in relation with the present definition of approximation. We begin with that of Theorem 5.7.5; here and afterwards $\{\tau_n\}$ is a sequence of positive numbers tending to zero.

5.8.2 Theorem. Let $A \in \mathcal{C}_+(C, 0; E)$, $A_n \in (E_n)$, $\{\mathfrak{S}_n(l\tau_n); \ l \geqslant 0\}$ the discrete semigroup with time scale $\tau_n \to 0$ generated by A_n. Let, moreover,

$$\|\mathfrak{S}_n(l\tau_n)\| \leqslant C \quad (l \geqslant 0, n \geqslant 1) \tag{5.8.13}$$

for some C not depending on n. (a) Assume that for each $u \in E$

$$\|\mathfrak{S}_n([t/\tau_n]\tau_n)P_n u - P_n S(t)u\|_n \to 0 \tag{5.8.14}$$

uniformly on compacts of $t \geqslant 0$, where $[s]$ = greatest integer $\leqslant s$ and $S_n(t) = \exp(tA)$. Then

$$\|R(\lambda; A_n)P_n u - P_n R(\lambda; A)u\|_n \to 0 \tag{5.8.15}$$

uniformly on compact subsets of $\mathrm{Re}\,\lambda > 0$. (b) Conversely, assume that (5.8.15) holds for some λ with $\mathrm{Re}\,\lambda > 0$. Then (5.8.14) holds uniformly on compact subsets of $t \geqslant 0$.

Proof. If $\lambda > 0$,

$$
\begin{aligned}
R(\lambda; A_n) &= (\lambda I - A_n)^{-1} \\
&= \left(\lambda I - \tau_n^{-1}(\mathfrak{S}_n(\tau_n) - I)\right)^{-1} \\
&= \tau_n\left((\lambda\tau_n + 1)I - \mathfrak{S}_n(\tau_n)\right)^{-1} \\
&= \tau_n \sum_{l=0}^{\infty} (\lambda\tau_n + 1)^{-(l+1)} \mathfrak{S}_n(l\tau_n). \tag{5.8.16}
\end{aligned}
$$

Observe that the sum on the right-hand side of (5.8.16) is nothing but the integral in $t \geqslant 0$ of the function

$$F_n(t, \lambda) = (\lambda\tau_n + 1)^{-[t/\tau_n]-1} \mathfrak{S}_n([t/\tau_n]\tau_n). \tag{5.8.17}$$

Now, since $(1 + z/n)^{-n} \to e^{-z}$ as $n \to \infty$ uniformly on compacts, it follows that for each $\omega, \delta > 0$,

$$(1 + \lambda \tau_n)^{-[t/\tau_n]-1} \to e^{-\lambda t} \qquad (5.8.18)$$

uniformly on compact subsets of $\text{Re}\,\lambda \geq \omega$, $t \geq \delta$. Note next that if $x \geq 0$, then $(1 + x)^\alpha \geq 1 + \alpha x + \frac{1}{2}\alpha(\alpha - 1)x^2 \geq 1 + \alpha x + \frac{1}{4}\alpha^2 x^2$ for $\alpha \geq 2$. Accordingly, if $\text{Re}\,\lambda \geq \omega$,

$$
\begin{aligned}
\left| (1 + \lambda \tau_n)^{[t/\tau_n]+1} \right| &\geq (1 + \omega \tau_n)^{[t/\tau_n]+1} \\
&\geq (1 + \omega \tau_n)^{(t/\tau_n)} \\
&\geq 1 + \omega t + \tfrac{1}{4}\omega^2 t^2 \qquad (5.8.19)
\end{aligned}
$$

for $t \geq \delta$ and n sufficiently large. Collecting these observations, we deduce from (5.8.14) that

$$\| F_n(t, \lambda) P_n u - P_n e^{-\lambda t} S(t) \|_n \to 0 \qquad (5.8.20)$$

uniformly on compacts of $\text{Re}\,\lambda \geq \omega$ for $t \geq \delta$, while in view of (5.8.19),

$$\| F_n(t, \lambda) P_n u - P_n e^{-\lambda t} S(t) u \|_n \leq \beta(t) \quad (t \geq 0) \qquad (5.8.21)$$

with $\beta(\cdot)$ summable in $t \geq 0$. Integrating,

$$\| R(\lambda; A_n) P_n u - P_n R(\lambda; A) u \|_n \leq \int_0^\infty \| F_n(t; \lambda) P_n u - P_n e^{-\lambda t} S(t) u \|_n \, dt,$$

$$(5.8.22)$$

and the proof of (5.8.15) ends by division of the interval of integration and application of the dominated convergence theorem.

To prove (b) we begin by using Lemma 5.8.1 to deduce that the semigroups $\{ S_n(t) \} = \{ e^{tA_n} \}$ satisfy

$$\| S_n(t) \|_n \leq C \qquad (5.8.23)$$

and apply Theorem 5.7.5(b) to show that

$$\| S_n(t) P_n u - P_n S(t) u \|_n \to 0 \qquad (5.8.24)$$

uniformly on compacts of $t \geq 0$. We make then use of (5.8.11) for each $\mathfrak{S}_n$ and S_n; if $\text{Re}\,\lambda > 0$,

$$\| (\mathfrak{S}_n([t/\tau_n]\tau_n) - S_n(t)) R(\lambda; A_n)^2 \|_n$$

$$\leq C\tau_n \left(\frac{t}{2} \| A_n^2 R(\lambda; A_n)^2 \|_n + \| A_n R(\lambda; A_n)^2 \|_n \right). \qquad (5.8.25)$$

Since $A_n R(\lambda; A_n) = \lambda R(\lambda; A_n) - I$ is uniformly bounded in norm, the right-hand side of (5.8.25) tends to zero uniformly on compacts of $t \geq 0$. Select now $u = R(\lambda; A)^2 v$ in $D(A^2)$. Then

$$
\begin{aligned}
P_n u &= P_n R(\lambda; A)^2 v \\
&= R(\lambda; A_n)^2 P_n v + \left(P_n R(\lambda; A)^2 v - R(\lambda; A_n)^2 P_n v \right). \quad (5.8.26)
\end{aligned}
$$

Reasoning as in the proof of Lemma 5.7.7, we can show that $\|P_n R(\lambda; A)^2 v - R(\lambda; A_n)^2 P_n v\|_n \to 0$. Combining this with (5.8.23) and (5.8.24), we obtain (5.8.14) for $u \in D(A^2)$, thus for $u \in E$ by uniform boundedness of $\mathfrak{S}_n$ and S.

The following result is an analogue of Theorem 5.7.6.

5.8.3 Theorem. Let $\{\mathfrak{S}_n(l\tau_n); \ l \geq 0\}$ be a discrete semigroup in E_n with time scale $\tau_n \to 0$ and generator A_n. Assume that

$$\|\mathfrak{S}_n(l\tau_n)\| \leq C \quad (l \geq 0, n \geq 1) \tag{5.8.27}$$

with C independent of n. Suppose there exists some λ with $\mathrm{Re}\,\lambda > 0$ such that for each $u \in E$ there is a $v \in E$ with

$$\|R(\lambda; A_n) P_n u - P_n v\|_n \to 0. \tag{5.8.28}$$

Assume, moreover, that

$$\lim_{\lambda \to \infty} \|\lambda R(\lambda; A_n) P_n u - P_n u\|_n = 0 \tag{5.8.29}$$

for each $u \in E$ uniformly with respect to n. Then there exists $A \in \mathcal{C}_+(C, 0; E)$ such that (5.8.15) holds; a fortiori (5.8.14) is satisfied uniformly on compacts of $t \geq 0$.

Proof. Making use of Lemma 5.8.1 we deduce that

$$\|S_n(t)\|_n \leq C \quad (t \geq 0), \tag{5.8.30}$$

hence Theorem 5.7.6 can be applied to show the existence of A and the relation

$$\|S_n(t) P_n u - P_n S(t) u\|_n \to 0 \tag{5.8.31}$$

uniformly on compacts of $t \geq 0$. The corresponding relation for the $\mathfrak{S}_n$ is then derived as in the end of the proof of Theorem 5.8.2.

Finally, we prove a counterpart of Theorem 5.7.11.

5.8.4 Theorem. Let $\{\mathfrak{S}_n(l\tau_n); \ l \geq 0\}$ be a discrete semigroup in E_n with time scale $\tau_n \to 0$ and generator A_n such that (5.8.27) holds. (a) Assume that $\mathcal{C} = \text{ex-lim } A_n$ is densely defined in E and that $(\lambda I - \mathcal{C}) D(\mathcal{C})$ is dense in E for some $\lambda > 0$. Then $\mathcal{C} \in \mathcal{C}_+(C, 0; E)$ and for each $u \in E$,

$$\mathfrak{S}_n([t/\tau_n]\tau_n) u \Rightarrow S(t) u \tag{5.8.32}$$

uniformly on compact subsets of $t \geq 0$, where $S(t) = \exp(t\mathcal{C})$. (b) Conversely, if $\mathfrak{S}_n([t/\tau_n]\tau_n)$ converges strongly, uniformly on compact subsets of $t \geq 0$ to a strongly continuous semigroup $S(\cdot)$ with

$$\|S(t)\| \leq C \quad (t \geq 0), \tag{5.8.33}$$

then $\mathcal{C} = \text{ex-lim } A_n$ is single valued and coincides with the infinitesimal generator of $S(\cdot)$.

The proof of (a) follows again from application of Theorem 5.7.11 to the A_n (Lemma 5.8.1 implies that $\|S_n(t)\| \leq C$) whereby (5.8.24) results; the proof is then ended as that of Theorem 5.8.2. To show (b), let A be the infinitesimal generator of S, $u \in D(A)$, $v = (\lambda I - A)u$ for $\mathrm{Re}\,\lambda > 0$ and $u_n = R(\lambda; A_n)P_n v$. We make then use of (5.8.22) instead of (5.7.43) to show that $\|u_n - P_n u\|_n \to 0$; since $Au = \lambda u - v$ and $A_n u_n = \lambda u_n - P_n v$, $\|A_n u_n - P_n Au\|_n \leqslant \lambda \|u_n - P_n u\|_n \to 0$, hence $u \in D(\mathcal{Q})$ and $\mathcal{Q}u = Au$. The proof is completed like that of Theorem 5.7.11.

5.8.5 Example. *Finite difference methods for a parabolic equation.* We apply the results to the operator A in (5.7.45), using the notations and definitions in the previous section. The operators A_n are defined in (5.7.50); we replace the semigroups $\{S_n(t)\} = \{e^{tA_n}\}$ by the discrete semigroups with time scale τ_n generated by A_n, that is, by the $\mathfrak{S}_n$ defined by

$$\mathfrak{S}_n(l\tau_n) = (\tau_n A_n + I)^l. \tag{5.8.34}$$

This amounts to replacing the differential equation $u_n'(t) = A_n u(t)$ by the difference equation $\tau_n^{-1}(u_n(t + \tau_n) - u_n(t)) = A_n u(t)$ or, equivalently, $u_n(t + \tau_n) = (\tau_n A_n + I)u_n(t)$. It is also obvious that condition (5.8.27) for the discrete semigroups $\mathfrak{S}_n$ will be satisfied if and only if we can select the time scales $\tau_n \to 0$ in such a way that

$$\|(\tau_n A_n + I)^l\|_n \leqslant C \quad (l \geqslant 0, n \geqslant 1). \tag{5.8.35}$$

If $\xi = \{\xi_{n,\alpha}\} \in E_n = \mathbb{R}^{r(n)}$, we have

$$((\tau_n A_n + I)\xi)_{n,\alpha} = \sum_{j=1}^{m} \frac{2\tau_n}{h_{n,\alpha}^j k_{n,\alpha}^j (h_{n,\alpha}^j + k_{n,\alpha}^j)} (k_{n,\alpha}^j \sigma_{n,\alpha}^j + h_{n,\alpha}^j \zeta_{n,\alpha}^j)$$
$$+ \left(1 - 2\tau_n \sum_{j=1}^{m} \frac{1}{h_{n,\alpha}^j k_{n,\alpha}^j} + \tau_n c(x_{n,\alpha})\right) \xi_{n,\alpha}. \tag{5.8.36}$$

Hence

$$\|(\tau_n A_n + I)\xi\|_n \leqslant \left(2\tau_n \sum_{j=1}^{m} \frac{1}{h_{n,\alpha}^j k_{n,\alpha}^j}\right) \|\xi\|_n$$
$$+ \left(1 - 2\tau_n \sum_{j=1}^{m} \frac{1}{h_{n,\alpha}^j k_{n,\alpha}^j} + \tau_n c(x_{n,\alpha})\right) \|\xi\|_n$$
$$\leqslant \|\xi\|_n \tag{5.8.37}$$

for n large enough if

$$c(x) \leqslant 0 \tag{5.8.38}$$

and

$$\tau_n \leqslant \frac{1}{4} \left(\sum_{j=1}^{m} \frac{1}{h_{n,\alpha}^j k_{n,\alpha}^j}\right)^{-1} \quad (n \geqslant 1). \tag{5.8.39}$$

Obviously (5.8.37) implies (5.8.35). The fact that $\mathcal{C} = \text{ex-lim } A_n = A$ was already shown in Example 5.7.12, and it follows then from Theorem 5.8.4(a) that if $u(\cdot)$ is an arbitrary (genuine or generalized) solution of (5.7.56), then

$$\lim_{n \to \infty} \left\| \mathfrak{S}_n([t/\tau_n]\tau_n) P_n u - P_n u(t) \right\|_n$$

$$= \lim_{n \to \infty} \left\| (\tau_n A_n + I)^{[t/\tau_n]} P_n u - P_n u(t) \right\|_n = 0 \qquad (5.8.40)$$

uniformly on compact subsets of $t \geqslant 0$ if condition (5.8.39) on the τ_n holds for all n.

Inequality (5.8.39) is rather restrictive since very small $h_{n,\alpha}^j$, $k_{n,\alpha}^j$ may appear for any n. Even if this can be avoided due to the geometry of the boundary Γ, (5.8.39) implies

$$\tau_n \leqslant \frac{1}{4m} \left(\frac{a}{n} \right)^2 \quad (n \geqslant 1), \qquad (5.8.41)$$

which forces us to choose time scales τ_n much smaller than a/n, the "space scale," and makes then necessary the computation of very high powers of $\tau_n A_n + I$. One way out of this difficulty is to approximate (5.7.56) by the implicit finite difference equation $\tau_n^{-1}(u(t + \tau_n) - u(t)) = A_n u(t + \tau_n)$, or, in explicit form, $u(t + \tau_n) = (I - \tau_n A_n)^{-1} u(t)$, which corresponds to the discrete semigroup

$$\tilde{\mathfrak{S}}_n(l\tau_n) = (I - \tau_n A_n)^{-l} \quad (l \geqslant 0) \qquad (5.8.42)$$

with generator

$$\tilde{A}_n = \tau_n^{-1} \left((I - \tau_n A_n)^{-1} - I \right)$$

$$= (I - \tau_n A_n)^{-1} A_n. \qquad (5.8.43)$$

The existence of the inverse of $I - \tau_n A_n$ is assured since each A_n is dissipative; precisely, we have

$$\| (I - \tau_n A_n) \xi \|_n \geqslant \| \xi \|_n \quad (\xi \in \mathbb{R}^{r(n)}), \qquad (5.8.44)$$

hence $\| (I - \tau_n A_n)^{-1} \|_n \leqslant 1$ and $\tilde{\mathfrak{S}}_n$ satisfies (5.8.27) *with no conditions on the time scales* τ_n. However, we must also check that

$$A = \text{ex-lim } \tilde{A}_n. \qquad (5.8.45)$$

To do this we select $u \in C^{(4)}(\overline{\Omega}) \cap C_\Gamma(\overline{\Omega})$ with $Au \in C_\Gamma(\overline{\Omega})$ and write

$$\tilde{A}_n P_n u - P_n A u = (I - \tau_n A_n)^{-1} (A_n P_n u - (I - \tau_n A_n) P_n A u)$$

$$= (I - \tau_n A_n)^{-1} (A_n P_n u - P_n A u + \tau_n A_n P_n A u) \qquad (5.8.46)$$

It was already proved in Example 5.7.12 that $\| A_n P_n u - P_n A u \|_n \to 0$ (under the only assumption that $u \in C^{(2)}(\overline{\Omega}) \cap C_\Gamma(\overline{\Omega})$). The same argument applied to Au shows that $\|A_n P_n A u - P_n A^2 u\|_n \to 0$, hence $\|A_n P_n A u\|_n$ remains bounded as $n \to \infty$ so that $\|\tau_n A_n P_n A u\|_n \to 0$ as $\tau_n \to 0$, and it follows from (5.8.46) that

$$\|\tilde{A}_n P_n u - P_n A u\|_n \to 0 \qquad (5.8.47)$$

as $n \to \infty$. It results again from Theorem 4.7.12 and Corollary 4.7.15 that the solution u of $(\lambda I - A)u = v \in \mathcal{D}(\Omega)$ belongs to $C^{(4)}(\bar{\Omega}) \cap C_\Gamma(\bar{\Omega})$; moreover, $Au = \lambda u - v \in C_\Gamma(\bar{\Omega})$, thus we deduce from (5.8.46) that $\mathcal{C}u = Au$. Arguing as in Example 5.7.12, we conclude that $A = \text{ex-lim }\tilde{A}_n$ as claimed.

We obtain from Theorem 5.8.4 that

$$\lim_{n \to \infty} \|(I - \tau_n A_n)^{-[t/\tau_n]} P_n u - P_n u(t)\|_n \to 0 \qquad (5.8.48)$$

uniformly on compacts of $t \geqslant 0$ under the only condition that $\tau_n \to 0$; note that although (5.8.39) assumes more of the τ_n, (5.8.48) requires inversion of the matrix $(I - \tau_n A_n)$.

5.9. MISCELLANEOUS COMMENTS

Theorem 5.1.1 proved by Phillips [1955: 1] in a more general version. Theorem 5.1.2 was also proved by Phillips [1953: 1]. The neutron transport equation lends itself admirably to application of the theory of abstract differential equations. For space dimension one, this was done by Lehner and Wing [1956: 1]; the treatment in arbitrary dimension is due to Jörgens [1958: 1]. Many expositions in the same style and new results have been subsequently published (see for instance Vidav [1968: 1], [1970: 1], K. W. Reed [1965: 1], [1966: 1], Di Blasio [1973: 1] and Larsen [1975: 1]. Some details of the present treatment in spaces of continuous functions seem to be new.

Theorem 5.3.1 has a long history. It was proved by Trotter [1959: 1] for an unspecified sufficiently small a (but with additional information on $S(t; A + P)$; see (a) below), extended by Nelson [1964: 1] to the range $a < \frac{1}{2}$ and by Gustafson [1966: 1] to $a < 1$. We have followed Goldstein [1970: 5] in the proof. Corollary 5.3.2 is due to Chernoff [1972: 2] and independently to Okazawa [1971: 1] in the particular case where E is reflexive (where denseness of $D(P^*)$ is automatically verified; see Section 4). Perturbation theorems for self adjoint operators are of earlier date: Corollary 5.3.5 is due to Rellich [1939: 1] for the case $a < 1$ and A self-adjoint and was proved by Kato [1951: 1] under the assumption that

$$\|Pu\|^2 \leqslant \|Au\|^2 + b\|u\|^2 \quad (u \in D(A)) \qquad (5.9.1)$$

for some $b \geqslant 0$. We note that, since $(\alpha + \beta)^{1/2} \leqslant \alpha^{1/2} + \beta^{1/2}$, (5.9.1) implies (5.3.14) with $a = 1$. The opposite implication is false: however, if (5.3.14) is satisfied, we deduce that $\|Pu\|^2 \leqslant a^2\|Au\|^2 + 2ab\|Au\|\|u\| + b^2\|u\|^2 \leqslant$

$a^2(1 + \varepsilon^2)\|Au\|^2 + b^2(1 + \varepsilon^{-2})\|u\|^2$ for any $\varepsilon > 0$ so that (5.9.1) lies in between the versions of (5.3.14) for $a < 1$ and for $a = 1$. The proof of Corollary 5.3.5 in full generality is due independently to Wüst [1971: 1].

Theorem 5.3.6 is due to Hille [1948: 1] (see also Hille-Phillips [1957: 1, p. 417]), although instead of (5.3.15) it is postulated there that $\|PR(\lambda)\| < 1$ for some sufficiently large λ; in the present version, this is a consequence of the assumptions.

Theorem 5.4.1 and Examples 5.4.2 and 5.4.3 are due to Kato [1951: 1]; the characterization of the domain of A in Theorem 5.4.5 we have taken from Kato [1976: 1]. Theorem 5.4.7 is (essentially) the result of Stone [1932: 1] and Friedrichs [1934: 1] to the effect that any semi-bounded symmetric operator possesses a self-adjoint extension. The method used here is that due to Friedrichs, and can be considered as a forerunner of the Lax-Milgram lemma (Lemma 4.6.1) that covers a somewhat similar situation. We refer the reader to the treatise of Kato [1976: 1] for many additional results on the Schrödinger, Dirac, and related equations. The theory of "perturbation by sesquilinear forms," of which Theorem 5.4.8 offers a glimpse, has been greatly extended, in particular to nonsymmetric forms: a thorough exposition can be found in Kato [1976: 1]. Other works on perturbation of differential operators are Schechter [1971: 1] (where other references are given), Jörgens-Weidmann [1973: 1], Goodall-Michael [1973: 1], Gustafson-Rejto [1973: 1], Kato [1972: 1], [1974: 1], Agmon [1975: 1], Semenov [1977: 1], Bezverhniĭ [1975: 1], and Powers-Radin [1976: 1].

The material in Section 5.6 is due to Kato [1970: 1] and is basic in his treatment of time-dependent symmetric hyperbolic systems (to be found in Sections 7.7 and 7.8).

Approximation of the abstract differential equation (5.7.1) by difference equations like (5.7.3) was discussed for the first time by Lax (see Lax-Richtmyer [1956: 1] and references there). In this version, approximation proceeds in the space E and the operator A_n is required to "have the limit A" in a suitable subspace of E. Its most celebrated result is the *Lax equivalence theorem* (Lax-Richtmyer [1956: 1], Richtmyer-Morton [1967: 1]) of which Theorem 5.8.4 is a generalized version. Another treatment of the approximation problem is that of Trotter [1958: 1], which we have followed here. In this work the approximating sequences $\{E_n, P_n\}$ are introduced and convergence of the semigroups S_n (and of their discrete counterparts $\mathfrak{S}_n$) is related with convergence of the resolvents (Theorems 5.7.5, 5.7.6, 5.8.2, and 5.8.3). Theorem 5.7.6 is known as the Trotter-Kato theorem (see Trotter [1958: 1], Kato [1959: 1]). The notion of extended limit of an operator and the results on them proved here are due to Kurtz [1969: 1], [1970: 1], in particular those linking extended limits of the A_n and convergence of the semigroups S_n and the discrete semigroups $\mathfrak{S}_n$ (Theorems 5.7.11 and 5.8.4). For additional literature on discrete semigroups, see Gibson [1972: 1] and Packel [1972: 1].

The treatment of continuous and discrete approximations we have followed, although amply sufficient for our purposes, obscures the fact that what is at play here is the convergence of a sequence of vector-valued distributions given convergence of their inverses or their Laplace transforms. This point of view leads to an extension of the theory to equations much more general than (5.7.1) (such as integrodifferential or difference-differential equations); see Section 8.6 for additional information.

We note finally that the results on numerical treatment of the equation $u_t = \Delta u + cu$ (Examples 5.7.12 and 5.8.5) are not understood as "real" applications, especially in that the way we deal with the boundary condition $u = 0$ would make any numerical analyst shudder. Actual finite-difference schemes taking boundary conditions into account are usually far more sophisticated in that the grids at which the function is computed "follow the boundary" suitably. For examples, see Forsythe-Wasow [1960: 1] or Richtmyer-Morton [1967: 1].

We comment below on some developments related with the material in this chapter.

(a) *Addition and Perturbation of Infinitesimal Generators.*   Let A, B be two square matrices of the same dimension. A classical formula (that can be traced back to Lie in a more general form) states that

$$\lim_{n \to \infty} \left(\exp\left(\frac{t}{n}A\right)\exp\left(\frac{t}{n}B\right) \right)^n = \exp(t(A+B)) \qquad (5.9.2)$$

(see Cohn [1961: 1, p. 112]). It is natural to inquire whether an infinite-dimensional extension is possible. The first question is, of course, whether $A + B \in \mathcal{C}_+$ whenever A and B belong to $\mathcal{C}_+$. In this form the question is too restrictive to admit any reasonable answer; in fact, if A is an unbounded operator in $\mathcal{C}$ and $B = -A$, then $A + B$ is not closed, hence does not belong to $\mathcal{C}_+$. The following formulation avoids trivialities of this sort.

Addition Problem: Let A, B belong to $\mathcal{C}_+$ (or to a subclass thereof). Does $A + B$, with domain $D(A) \cap D(B)$, possess an extension $(A + B)_e$ in $\mathcal{C}_+$ (or in a subclass thereof)? If the answer is in the affirmative, does $(A + B)_e$ coincide with the closure $\overline{A + B}$?

Representation Problem: Assuming the answer to the addition problem is in the affirmative, can $\exp(t(A + B)_e)$ be obtained by means of formula (5.9.2)?

Perturbation Problem: Let A be an operator in $\mathcal{C}_+$ (or in a subclass thereof). Find conditions on the operator P in order that $A + P$ (with domain $D(A) \cap D(P)$) possess an extension $(A + P)_e$ in $\mathcal{C}_+$ (or a subclass thereof). Can $\exp(t(A + P)_e)$ be obtained directly from $\exp(tA)$ (say, as in (5.1.19))?

Obviously, the perturbation problem is more general than the addition problem in that P itself may not belong to $\mathcal{C}_+$.

The first systematic consideration of the addition problem is that of Trotter [1959: 1], whose work in relation to Theorem 5.3.1 was already noted above. Among other results in [1959: 1], the following negative answer to the addition problem is given.

5.9.1 Example. Let φ be a function defined in $-\infty < x < \infty$, absolutely continuous and with derivative φ' satisfying

$$0 < c \leqslant \varphi'(x) \leqslant C \quad \text{a.e.} \tag{5.9.3}$$

It follows from the first inequality that φ^{-1} exists and is likewise absolutely continuous. Define

$$S_\varphi(t)u(x) = u\big(\varphi(\varphi^{-1}(x)+t)\big)$$

for $u \in E = L^2(-\infty,\infty)$ and $-\infty < t < \infty$. Some simple arguments involving no more than change of variables show that $S_\varphi(\cdot)$ is a strongly continuous group with

$$\|S_\varphi(t)\| \leqslant C/c \quad (-\infty < t < \infty)$$

and infinitesimal generator

$$(A_\varphi u)(x) = \varphi'(\varphi^{-1}(x))u'(x),$$

where $D(A_\varphi)$ consists of all absolutely continuous $u \in E$ such that $u' \in E$ (hence $D(A_\varphi)$ does not depend on φ). Let $A = A_\varphi$, $B = -A_\psi$ with $\varphi(x) = x$ in $\infty < x < \infty$, $\psi(x) = x$ for $x \leqslant 0$, $\psi(x) = 2x$ for $x \geqslant 0$. Then

$$(A+B)u(x) = -h(x)u'(x), \tag{5.9.4}$$

where $h(x) = 1$ for $x \geqslant 0$, $h(x) = 0$ for $x \leqslant 0$. It is easy to see that $\overline{A+B}$ is again defined by (5.9.4), this time with domain consisting of all $u \in E$ such that u is absolutely continuous in $x \geqslant 0$ and $u' \in L^2(0,\infty)$. It results then that for any $\lambda > 0$ the function $u(x) = 0$ for $x < 0$, $u(x) = e^{-\lambda x}$ for $x \geqslant 0$ belongs to $D(\overline{A+B})$ and satisfies

$$\lambda u - (\overline{A+B})u = 0. \tag{5.9.5}$$

If $A + B$ possesses an extension $(A+B)_e \subseteq \mathcal{C}_+$, we must have $\overline{A+B} \subset (A+B)_e$, which shows, in view of (5.9.5), that every $\lambda > 0$ belongs to $\sigma((A+B)_e)$, a contradiction.

We note incidentally that, when E is a Hilbert space (as in the example just expounded) the addition problem always has a solution when both A and B belong to $\mathcal{C}_+(1,0)$ and $D(A) \cap D(B)$ is dense in E; in fact, since $A + B$ is dissipative, by Lemma 3.6.3 there always exists an m-dissipative extension $(A+B)_e$. In a general Banach space this extension may not exist (Example 3.6.5), but a partial result can be derived from Remark 3.1.12: $\overline{A+B} \in \mathcal{C}_+(1,0)$ if an only if $(\lambda I - (A+B))D(A+B)$ is dense in E. Under these hypotheses, it was also shown by Trotter [1959: 1] that the representation problem has an affirmative answer, that is, (5.9.2) holds. The limit is here understood in the strong sense, uniformly on compact subsets of $t \geqslant 0$. We refer the reader to Chernoff [1974: 1] for a deep study of the addition and representation problems and additional references. Other works on the subject and on the perturbation problem are Angelescu-Nenciu-Bundaru [1975: 1], Uhlenbrock [1971: 1], Belyĭ-Semenov [1975: 1], Chernoff [1976: 1], Dorroh [1966: 1], Da Prato [1968: 2], [1968: 3], [1968: 4], [1968: 5], [1968: 6], [1968: 7], [1969: 1], [1970: 1], [1974: 1], [1975: 1], Goldstein [1972: 1], Gustafson [1966: 1], [1968: 1], [1968: 2], [1968: 3], [1969: 1], Gustafson-Lumer [1972: 1], Gustafson-Sato [1969: 1], Kurtz [1972: 1], [1973: 1], [1975: 1], [1976: 1], [1977: 1], Lenard [1971: 1], Lovelady [1975: 1], Lumer [1974: 1], [1975: 1], [1975: 3], Miyadera [1966: 1], [1966: 2], Mlak [1961: 1], Okazawa [1973: 2], Rao [1970: 1], Showalter [1973: 1], Suzuki [1976: 1], Sunouchi [1970: 1], Vainikko-Šlapikiene [1971: 1], Webb [1972: 1], Semenov [1977: 2], and Yosida [1965: 2].

(b) *Approximation of Abstract Differential Equations.* In relation with the material in Sections 5.7 and 5.8, we mention the scheme developed by Jakut [1963: 1], [1963: 2], wherein the spaces E_n in Trotter's theory are quotient spaces of E by suitable subspaces and time-dependent operators are allowed. For an exposition of this theory see the treatise of Kreĭn [1967: 1]. See also Gudovič [1966: 1] for another abstract scheme. References to the earlier bibliography can be found in the book of Richtmyer-Morton [1967: 1]. We list below some of the more recent literature, comprising both continuous and discrete approximation; some of this material refers to specific types of partial differential equations (treated as abstract differential equations in suitable function spaces). See Belleni-Morante-Vitocolonna [1974: 1], Crouzeix-Raviart [1976: 1], Gašimov [1975: 1], Cannon [1975: 1], Gegečkori-Demidov [1973: 1], Geymonat [1972: 1], Grabmüller [1972: 1], [1975: 1], Grădinaru [1973: 1], Gröger [1976: 1], Gudovič-Gudovič [1970: 1], Ibragimov [1969: 1], [1972: 1], Ibragimov-Ismailov [1970: 1], Ismailov-Mamedov [1975: 1], Nečas [1974: 1], Nowak [1973: 1], Oja [1974: 1], Pavel [1974: 1], Ponomarev [1972: 1], Rastrenin [1976: 1], Šapatava [1972: 1], Seidman [1970: 1], Showalter [1976: 1], Topoljanskiĭ-Zaprudskiĭ [1974: 1], Vainikko-Oja [1975: 1], Veliev [1972: 1], [1973: 1], [1973: 2], [1975: 1], Veliev-Mamedov [1973: 1], [1974: 1], Zarubin [1970: 1], [1970: 2], Zarubin-Tiunčik [1973: 1], Takahashi-Oharu [1972: 1], Ujishima [1975/76: 1], Raviart [1967: 1], [1967: 2], Goldstein [1974: 2], Oharu-Sunouchi [1970: 1], Piskarev [1979: 1], [1979: 3], the author [1983: 1].

(c) *Singular Perturbations.* Roughly speaking, a *singular perturbation* of an abstract (not necessarily differential) equation is one that changes the character of the equation in a radical way. For instance, the algebraic time-dependent equation $Au(t) + f(t) = 0$ becomes the differential equation $\varepsilon u'_\varepsilon(t) = Au_\varepsilon(t) + f(t)$ by addition of the "small term" $\varepsilon u'_\varepsilon(t)$ to the right-hand side; similarly, the first order differential equation $u'(t) = Au(t) + f(t)$ may be singularly perturbed into the second order equation $\varepsilon u''_\varepsilon(t) + u'_\varepsilon(t) = Au_\varepsilon(t) + f(t)$. In these and other similar cases, the subject of study is the behavior of the solution $u_\varepsilon(\cdot)$ as $\varepsilon \to 0$ (hopefully, the limit will exist and equal the solution of the unperturbed equation). Another instance of singular perturbation is that considered in Example 5.9.4, where a Cauchy problem that is not properly posed is approximated by properly posed problems in a suitable sense. We limit ourselves to a few examples (which exhibit some of the typical features of the theory) and some references.

5.9.2 Example. Let A be an operator in $\mathcal{C}_+(C, -\omega)$ for some $\omega > 0$ (so that, in particular, $0 \in \rho(A)$), and let $f(\cdot)$ be a E-valued function continuous in $t \geqslant 0$. The solution u_ε of the initial-value problem

$$\varepsilon u'_\varepsilon(t) = Au_\varepsilon(t) + f(t) \quad (t \geqslant 0), \quad u(0) = u \in E \qquad (5.9.6)$$

with $\varepsilon > 0$ is given by

$$u_\varepsilon(t) = S(t/\varepsilon)u + \frac{1}{\varepsilon}\int_0^t S(s/\varepsilon)f(t-s)\,ds$$

$$= S(t/\varepsilon)u + \frac{1}{\varepsilon}\int_0^t S(s/\varepsilon)(f(t-s) - f(t))\,ds$$

$$+ S(t/\varepsilon)A^{-1}f(t) - A^{-1}f(t) \qquad (5.9.7)$$

(strictly speaking, u_ε will be a genuine solution of (5.9.6) only if $u \in D(A)$ and f satisfies, say, one of the two sets of assumptions in Lemma 2.4.2: under the present hypotheses u_ε is just a generalized solution in the sense of Section 2.4). To estimate the integral on the right-hand side of (5.9.7) in an interval $0 \leqslant t \leqslant T$, we take $r > 0$, ε sufficiently small and split the interval of integration $(0, t)$ in the two subintervals $(0, r\varepsilon)$ and $(\varepsilon r, t)$; after the change of variable $s/\varepsilon = \sigma$, we see that the norm of the integral does not surpass

$$C\int_0^r e^{-\omega\sigma}\|f(t-\varepsilon\sigma) - f(t)\|\,d\sigma + C' \max_{0 \leqslant s \leqslant T}\|f(s)\|e^{-\omega r} \qquad (5.9.8)$$

uniformly in $0 \leqslant t \leqslant T$. Clearly, a judicious choice of r and ε (in that order) will make (5.9.8) arbitrarily small. It follows from (5.9.7) that for each $\delta > 0$, $u_\varepsilon(\cdot)$ converges uniformly in $\delta \leqslant t \leqslant T$ to $u_0(t) = -A^{-1}f(t)$. (The lack of uniform convergence in $0 \leqslant t \leqslant T$ is of course unavoidable since $u_\varepsilon(0) = u$ while $u_0(0) = A^{-1}f(0)$.) More generally, it can be shown that convergence is uniform on compacts of $t \geqslant t(\varepsilon)$ if $t(\varepsilon)/\varepsilon \to \infty$ (see definition below). For additional information on this problem see Kreĭn [1967: 1, Ch. 4].

5.9.3 Example. Let A be the infinitesimal generator of a cosine function $C(\cdot)$ in a Banach space E (see 2.5(c)). The solution of the initial-value problem

$$\varepsilon^2 u''_\varepsilon(t) + u'_\varepsilon(t) = Au_\varepsilon(t) + f_\varepsilon(t), \quad u_\varepsilon(0) = u_\varepsilon, \quad u'_\varepsilon(0) = v_\varepsilon \quad (5.9.9)$$

can be written in the following form:

$$u_\varepsilon(t) = e^{-t/2\varepsilon^2}C(t/\varepsilon)u_\varepsilon + \Re(t;\varepsilon)\left(\tfrac{1}{2}u_\varepsilon\right) + \mathfrak{S}(t;\varepsilon)\left(\tfrac{1}{2}u_\varepsilon + \varepsilon^2 v_\varepsilon\right)$$

$$+ \int_0^t \mathfrak{S}(t-s;\varepsilon)f_\varepsilon(s)\,ds, \tag{5.9.10}$$

where the operator-valued functions $\Re(t;\varepsilon)$ and $\mathfrak{S}(t;\varepsilon)$ are defined as follows:

$$\Re(t;\varepsilon)u = \frac{te^{-t/2\varepsilon^2}}{\varepsilon^2} \int_0^{t/\varepsilon} \frac{I_1\left(\left((t/\varepsilon)^2 - s^2\right)^{1/2}/2\varepsilon\right)}{\left((t/\varepsilon)^2 - s^2\right)^{1/2}} C(s)u\,ds, \tag{5.9.11}$$

$$\mathfrak{S}(t;\varepsilon)u = \frac{e^{-t/2\varepsilon^2}}{\varepsilon} \int_0^{t/\varepsilon} I_0\left(\left((t/\varepsilon)^2 - s^2\right)^{1/2}/2\varepsilon\right)C(s)u\,ds \tag{5.9.12}$$

(see the author [1983: 4]); these formulas, which are examples of transmutation formulas (see Section 5(f)), are due to Kisyński [1970: 1]). On the other hand, the solution of the initial value problem

$$u_0'(t) = Au_0(t) + f(t), \quad u_0(0) = u, \tag{5.9.13}$$

is given by

$$u_0(t) = S(t)u + \int_0^t S(t-s)f(s)\,ds \tag{5.9.14}$$

with

$$S(t)u = \frac{1}{\sqrt{\pi t}} \int_0^\infty e^{-s^2/4t}C(s)u\,ds \tag{5.9.15}$$

(see 2.5(f)). As in the previous example, "solution" is understood here in the generalized sense. Under suitable assumptions on $u_\varepsilon, v_\varepsilon$, it can be proved that $u_\varepsilon(\cdot)$ converges to $u_0(\cdot)$ in various ways. We state a few results below.

Given a function $\varepsilon \to t(\varepsilon) > 0$ defined for $\varepsilon > 0$, we say that a family of functions $g_\varepsilon(\cdot)$ converges *uniformly on compacts of $t \geq t(\varepsilon)$* (as $\varepsilon \to 0$) to $g(\cdot)$ if and only if

$$\lim_{\varepsilon \to 0} \sup_{t(\varepsilon) \leq t \leq a} \|g_\varepsilon(t) - g(t)\| = 0$$

for every $a > 0$.

(i) (The author [1983: 4].) Consider the homogeneous equation ($f_\varepsilon = 0$), and assume that

$$u_\varepsilon \to w, \quad \varepsilon^2 v_\varepsilon \to u - w \quad (\varepsilon \to 0), \tag{5.9.16}$$

where w is arbitrary and u is the initial condition in (5.9.13). Then

$$u_\varepsilon(t) \to u_0(t) \quad (\varepsilon \to 0) \tag{5.9.17}$$

uniformly on compacts of $t \geqslant t(\varepsilon)$, uniformly with respect to u bounded, if

$$t(\varepsilon)/\varepsilon^2 \to \infty \quad (\varepsilon \to 0). \tag{5.9.18}$$

Condition (5.9.18) is necessary in order that the convergence result above be valid.

Obviously, uniform convergence on compact subsets of $t \geqslant 0$ will not hold in general since $u_\varepsilon \nrightarrow u$. If this condition is assumed and if $u \in D(A)$, Kisyński [1970: 1] has obtained a very precise estimate, reproduced in (ii) below in a slightly weaker form. It implies uniform convergence on compacts of $t \geqslant 0$ if $u_\varepsilon \to u$ and $\varepsilon^2 v_\varepsilon \to 0$. To expedite the formulation denote by C, ω two constants such that

$$\|C(t)\| \leqslant Ce^{\omega|t|} \quad (-\infty < t < \infty).$$

(ii) Let, as before, $u_\varepsilon(\cdot)$ (resp. $u_0(\cdot)$) be the solution of (5.9.9) (resp. (5.9.13)). Then, if $u \in D(A)$, we have

$$\|u_\varepsilon(t) - u_0(t)\| \leqslant C\varepsilon^2 (1 + \omega^2 t) e^{\omega^2 t} \|Au\|$$
$$+ Ce^{\omega^2 t}\|u_\varepsilon - u\| + C\varepsilon^2 e^{\omega^2 t}\|v_\varepsilon\| \quad (t \geqslant 0). \tag{5.9.19}$$

Uniform convergence on compacts of $t \geqslant 0$ can also be obtained under "crossover" of initial conditions (see (5.9.16)) if correction terms are added; these terms neutralize the contribution of v_ε to u, the initial datum in (5.9.13). A typical result is:

(iii) (The author, [1983: 4].) Let $u_\varepsilon(\cdot)$ be the solution of (5.9.9) with $u_\varepsilon = u + O(\varepsilon)$, $v_\varepsilon = \varepsilon^{-2}v + O(\varepsilon^{-1})$, and let $v_\varepsilon(\cdot)$ be the solution of the initial value problem

$$\varepsilon^2 v_\varepsilon''(t) + v_\varepsilon'(t) = 0, \quad v_\varepsilon'(0) = \varepsilon^{-2}v \tag{5.9.20}$$

that dies down as $t \to \infty$ (so that $v_\varepsilon(t) = -e^{-t/\varepsilon^2}v$). Then, if $u, v \in D(A)$, we have

$$u_\varepsilon(t) = u_0(t) + v_\varepsilon(t) + O(\varepsilon) \tag{5.9.21}$$

uniformly on compacts of $t \geqslant 0$.

This is the prototype of a family of results where asymptotic development in powers of ε are obtained for $u_\varepsilon(\cdot)$; in general, we assume that $u_\varepsilon = u_0 + \varepsilon u_1 + \cdots + \varepsilon^N u_N + O(\varepsilon^{N+1})$ and $v_\varepsilon = \varepsilon^{-2}v_0 + \varepsilon^{-1}v_1 + \cdots + \varepsilon^{N-2}v_{N-2} + O(\varepsilon^{N-1})$ and obtain an asymptotic development of the form

$$u_\varepsilon(t) = u_0(t) + \varepsilon u_1(t) + \cdots + \varepsilon^N u_N(t)$$
$$+ v_0(t/\varepsilon^2) + \varepsilon v_1(t/\varepsilon^2) + \cdots + \varepsilon^N v_N(t/\varepsilon^2) + O(\varepsilon^{N+1}),$$
$$\tag{5.9.22}$$

where $v_0 = v_\varepsilon$ is the correction term in (5.9.21) and $v_1, \ldots, v_N$ are higher order correctors.

Among the convergence results for the nonhomogeneous equation we have (iv) below, which involves the derivative $u'_\varepsilon(\cdot)$; here $u_\varepsilon(\cdot)$ is the solution of (5.9.9) with $u_\varepsilon(0) = u'_\varepsilon(0) = 0$.

(iv) Let E be a Hilbert space, $1 < p < \infty$, $f_\varepsilon, f \in L^p((0, T); E)$, $T > 0$. Assume that $f_\varepsilon(\cdot) \to f(\cdot)$ in L^p as $\varepsilon \to 0$. Then $u_\varepsilon(t) \to u_0(t)$ uniformly in $0 \leqslant t \leqslant T$, $u'_\varepsilon(\cdot) \to u'_0(\cdot)$ in L^p as $\varepsilon \to 0$, where $u_0(\cdot)$ is the solution of (5.9.13) with $u_0(0) = 0$. Here the derivative of u_0 is understood as in Example 2.5.8. (See the author [1983: 4]).

For a rather complete survey of the results available in this perturbation problem as well as for additional references, see the author [1983: 4], where some applications to partial differential equations are also covered. Other works on the subject are Dettman [1973: 1], Friedman [1969: 2], Griego-Hersh [1971: 1], Kisyński [1963: 1], Latil [1968: 1], Nur [1971: 1], Schoene [1970: 1], Smoller [1965: 1], [1965: 2], and Sova [1970: 2], [1972: 1].

5.9.4 **Example.** Consider the heat equation

$$u_t = \kappa \Delta u \quad (t \geqslant 0) \tag{5.9.23}$$

in $\mathbb{R}^m$, with $\kappa > 0$: we look at (5.9.23) in the space $E = L^2(\mathbb{R}^m)$, the domain $D(\Delta)$ consisting of all $u \in E$ such that $\Delta u \in E$. Working as in Section 1.4, we can show that Δ belongs to $\mathcal{C}_+$ (in fact, to $\mathcal{Q}$), the propagator of (5.9.23) given by

$$S(t)u = \mathcal{F}^{-1}\left(e^{-\kappa|\sigma|^2 t}\mathcal{F}u\right), \tag{5.9.24}$$

where $\mathcal{F}$ indicates Fourier transform. Let $T > 0$, $\eta > 0$, and $v \in E$ be given. Consider the (control) problem of finding $u \in E$ such that

$$\|S(T)u - v\| \leqslant \eta. \tag{5.9.25}$$

The "obvious" solution (taking $u = S(-T)v$) is meaningless since

$$S(-T)v = \mathcal{F}^{-1}\left(e^{\kappa|\sigma|^2 T}\mathcal{F}v\right) \tag{5.9.26}$$

may not belong to E (or even be defined in any reasonable sense) for an arbitrary $v \in E$. However, $S(-T)v \in E$ if $v \in \mathcal{F}^{-1}\mathcal{D}(\mathbb{R}^m)$ thus we may select $w \in \mathcal{F}^{-1}\mathcal{D}(\mathbb{R}^m)$ with $\|w - v\| \leqslant \eta$ and define $u = S(-T)w$. Unfortunately, this solution is far from satisfactory from the computational point of view, since very small changes in w (in the L^2 norm) may produce enormous variations in $S(-T)w$.

A more advantageous approach is the following. Consider the equation

$$u_t = \kappa \Delta u + \varepsilon \Delta^2 u, \tag{5.9.27}$$

where $\varepsilon > 0$. Although the operator $\kappa \Delta + \varepsilon \Delta^2$ does not belong to $\mathcal{C}_+$, $-(\kappa \Delta + \varepsilon \Delta^2)$ does. If we denote by $S_\varepsilon(\cdot)v$ the (in general weak) solution of

(5.9.27) in $t \leqslant 0$ satisfying $S_\varepsilon(0)v = v$, we have

$$S_\varepsilon(-t)v = \mathfrak{F}^{-1}\left(e^{-(\varepsilon|\sigma|^4 - \kappa|\sigma|^2)t}\mathfrak{F}v\right). \tag{5.9.28}$$

It follows from (5.9.24) and (5.9.28) that

$$S(T)S_\varepsilon(-T)v \to v \text{ as } \varepsilon \to 0$$

for any $v \in E$, hence an element $u \in E$ satisfying (5.9.25) will be obtained by setting $u = S_\varepsilon(-T)v$ for ε sufficiently small; since $S_\varepsilon(-T)$ is a bounded operator in E, small variations of v (with ε fixed) will only produce small variations of u and its numerical computation will pose no problem.

The argument above is a typical instance of a *quasi-reversibility* method. These methods were introduced and systematically examined by Lattès and Lions [1967: 1], although particular examples were used previously by Tikhonov and other workers in the field of ill-posed problems (see the forthcoming Section 6.6)). Other references on quasi-reversibility methods are V. K. Ivanov [1974: 1], Kononova [1974: 1], Lagnese [1976: 1], Melnikova [1975: 1], and Showalter [1974: 1]. See also Payne [1975: 1] for additional information and references.

The works below are on diverse singular perturbation problems for abstract differential equations. See Belov [1976: 1], [1976: 2], Bobisud [1975: 1], Bobisud-Boron-Calvert [1972: 1], Bobisud-Calvert [1970: 1], [1974: 1], Groza [1970: 1], [1973: 1], Janusz [1973: 1], Kato [1975: 2], Mika [1977: 1], Mustafaev [1975: 1], Osipov [1970: 1], Veliev [1975: 1], and Yoshikawa [1972: 1]. For an exposition of singular perturbation theory and associated asymptotic methods, see the treatise of Kreĭn [1967: 1], where additional references can also be found.

Chapter 6

Some Improperly Posed Cauchy Problems

Certain physical phenomena lend themselves to modeling by non-standard Cauchy problems for abstract differential equations, such as the reversed Cauchy problem for abstract parabolic equations (Section 6.2) or the second order Cauchy problem in Section 6.5, where one of the initial conditions is replaced by a growth restriction at infinity. If these problems are correctly formulated, it is shown that they behave not unlike properly posed Cauchy problems: solutions exist and depend continuously on their data.

6.1. IMPROPERLY POSED PROBLEMS

Many physical phenomena are described by models that are not properly posed in any reasonable sense; certain initial conditions may fail to produce a solution in the model (although the phenomenon itself certainly has an outcome) and/or solutions may not depend continuously on their initial data. One such phenomenon is, say, the tossing of a coin on a table. Assuming we could measure with great precision the initial position and velocity of the coin, we could integrate the differential equations describing the motion and predict the outcome (head or tails); however, it is obvious that for certain trajectories, arbitrarily small errors in the measurement of initial position and velocity will produce radical changes in the predicted outcome. In studying phenomena like these we are then forced to abandon

346

the deterministic point of view and to obtain information by other means (e.g., by probabilistic arguments). This was known to the creators of probability theory and was explicitly stated by Poincaré almost a century ago. In the words of Hadamard [1923: 1, p. 38]. "But in any concrete application, "known," of course, signifies "known with a certain approximation," all kinds of errors being possible, provided that their magnitude remains smaller than a certain quantity; and, on the other hand, we have seen that the mere replacing of the value zero for u_1 by the (however small) value (15) [here Hadamard refers to the "initial velocity" u_1 in the Cauchy problem for the Laplace equation] changes the solution not by very small but by very great quantities. Everything takes place, physically speaking, as if the knowledge of Cauchy's data would *not* determine the unknown function."

"This shows how very differently things behave in this case and in those that correspond to physical questions. If a physical phenomenon were to be dependent on such an analytical problem as Cauchy's for $\nabla^2 u = 0$, it would appear to us as being governed by pure chance (which, since Poincaré, has been known to consist precisely in such a discontinuity in determinism) and not obeying any law whatever."[1] Earlier in [1923: 1] (p. 32) Hadamard concludes: "But it is remarkable, on the other hand, that a sure guide is found in physical interpretation: an analytical problem always being correctly set, in our use of the phrase, when it is the translation of some mechanical or physical question; and we have seen this to be the case for Cauchy's problem in the instances quoted in the first place."

"On the contrary, none of the physical problems connected with $\nabla^2 u = 0$ is formulated analytically in Cauchy's way."[2]

A look at the vast amount of literature produced during the last two decades on *deterministic* treatment of improperly posed problems (including numerical schemes for the computation of solutions) would appear to prove Hadamard's dictum wrong. However, it may be said to remain true in the sense that, many times, a physical phenomenon appears to be improperly posed not due to its intrinsic character but to unjustified use of the model that describes it; for instance, the model may have "unphysical" solutions in addition to the ones representing actual trajectories of the system, bounds and constraints implicit in the phenomenon may be ignored in the model, the initial or boundary value problems imposed on the equation may be incorrect translations of physical requirements, and so on. We present in this section a number of examples where seemingly improperly posed problems result from neglecting physical considerations, but where the "discontinuity in determinism" can be removed by careful examination of

---

[1] From J. Hadamard, *Lectures on Cauchy's Problem in Linear Partial Differential Equations* (New Haven: Yale University Press, 1923).

[2] *Ibid.*

the relations between model and phenomenon. These examples motivate the theory in the rest of the chapter.

6.1.1 Example. *Solutions of the heat equation that cease to exist in finite time.* Consider the heat-diffusion equation

$$u_t = \kappa u_{xx} \qquad (6.1.1)$$

in the whole line $-\infty < x < \infty$. We can arrange an abstract Cauchy problem for it as follows. Let $a > 0$, $E = C_{a,2}$, where $C_{a,\alpha}$ is the space of all functions u defined in $-\infty < x < \infty$ and such that

$$\|u\| = \sup_{-\infty < x < \infty} |u(x)|(1 + |x|)\exp(-a|x|^{\alpha}) < \infty.$$

Obviously, each $C_{a,\alpha}$ is a Banach space. The operator A is defined by $Au = u''$ with domain $D(A)$ consisting of all twice differentiable u in E with $u'' \in E$. Let $0 < b < a$,

$$u(x, t) = \frac{1}{\sqrt{1 - 4b\kappa t}} e^{bx^2/(1 - 4b\kappa t)}.$$

Then the E-valued function $t \to u(\cdot, t)$ is a solution of

$$u'(t) = Au(t) \qquad (6.1.2)$$

in $-\infty < t < T = (a - b)/4ab\kappa$ with initial condition $u(x, 0) = e^{bx^2}$, but it ceases to exist[3] for $t = T$. This type of solutions of an abstract differential equation (6.1.2) have been called *explosive solutions* by Hille [1953: 3]. If (6.1.1) is to be taken as a model for, say, a heat propagation or diffusion process, the existence of explosive solutions is embarrassing since it would imply that certain heat or diffusion processes "blow themselves up" without any obvious cause (such as internal chemical reactions). However, an actual temperature distribution must be bounded, while the integral of a density (the total amount of matter) is finite. Both conditions are violated by the elements of the space E. We must then conclude that the misbehavior of the model is due to inadequate choice of the "state space" E; indeed, it is well known that explosive solutions of (6.1.1) do not exist if we take $E = BC(-\infty, \infty)$ (the space of bounded continuous functions endowed with the supremum norm) or $E = L^1(-\infty, \infty)$. In fact, in both these cases the solution corresponding to the initial condition $u(x)$ is given by the Weierstrass formula

$$u(x, t) = \frac{1}{2\sqrt{\pi \kappa t}} \int_{-\infty}^{\infty} e^{-(x - \xi)^2/4\kappa t} u(\xi) \, d\xi$$

for all $t > 0$ (for uniqueness in both cases see John [1978: 1, Ch. 7], where the m-dimensional case is also covered).

---

[3] That is, $\lim \|u(t)\| = \infty$ as $t \to T-$.

6.1.2 Example. *Nonuniqueness for the heat equation.* Let $g(t)$ be infinitely differentiable in $-\infty < t < \infty$. The function

$$u(x,t) = \sum_{n=0}^{\infty} \frac{g^{(n)}(t)}{(2n)!} x^{2n} \qquad (6.1.3)$$

is a formal solution of (6.1.1) (with $\kappa = 1$). If

$$g(t) = \begin{cases} \exp(-t^{-2}) & (t > 0) \\ 0 & (t \leqslant 0), \end{cases}$$

the series given by (6.1.3) is uniformly convergent (together with all its derivatives) on compact subsets of the (x, t)-plane; moreover, there exists a constant θ, $0 < \theta < 1$ such that

$$|u(x,t)| \leqslant \exp\left(\frac{x^2}{\theta t} - \frac{1}{2t^2}\right) \qquad (-\infty < x < \infty, t > 0). \qquad (6.1.4)$$

(For this result and related information see John [1978: 1, p. 172].) We note next that

$$\frac{x^2}{\theta t} - \frac{1}{2t^2} \leqslant \frac{x^4}{2\theta^2} \qquad (-\infty < x < \infty, t > 0).$$

Let $E = C_{a,4}$ with $a > 1/2\theta^2$, the operator A defined as in the previous example. It is not difficult to see that the function $t \to u(\cdot\,; t)$ is infinitely differentiable in the sense of the norm of E and solves (6.1.2) for all t. Since $u(0) = 0$, solutions of this equation are not uniquely determined by their initial data. This undesirable behavior of (6.1.1) as a model can again be traced to the unphysical nature of the space E in this Example.

It is remarkable that lack of uniqueness for solutions of (6.1.1) is also related to the neglect of the condition $u \geqslant 0$, which must hold for a density or an absolute temperature. In fact, it was shown by Widder [1944: 1] (see also [1975: 1, p. 157]) that an arbitrary solution of (6.1.1) in $-\infty < x < \infty$, $t \geqslant 0$ that vanishes for $t = 0$ and is nonnegative everywhere must be identically zero.

We note finally that all the anomalies related to nonuniqueness are consequence of the unavoidable fact that the line $t = 0$, carrier of the initial data, is a characteristic of the equation (6.1.1). For general results on nonuniqueness in characteristic Cauchy problems see Hörmander [1969: 1, Ch. 5].

For an abstract differential equation $u'(t) = Au(t)$, nonuniqueness, as well as existence of explosive solutions, are closely related to certain spectral properties of A. See Hille-Phillips [1957: 1, p. 620] and references therein for additional information.

6.1.3   Example. *The reversed Cauchy problem for the heat equation.* Consider the heat equation

$$u_t = \kappa(u_{xx} + u_{yy}) \tag{6.1.5}$$

in the square $\Omega = \{(x, y); \ 0 < x, y < \pi\}$ with Dirichlet boundary condition

$$u = 0 \tag{6.1.6}$$

on the boundary Γ. The initial-value problem for (6.1.5), (6.1.6) as an abstract Cauchy problem in the spaces $C_\Gamma(\bar{\Omega})$ and $L^p(\Omega)$, $1 \leqslant p < \infty$ was already examined in Sections 1.1 and 1.3. For the sake of definiteness, we choose $E = C_\Gamma(\bar{\Omega})$, which is natural if we wish to model a heat propagation process by means of (6.1.5) (6.1.6). Let $t' \leqslant T$ and assume we know the temperature distribution u at time T,

$$u_T(x, y) = u(x, y, T). \tag{6.1.7}$$

On the basis of this information, we wish to compute the state of the system at time t', or more generally, the past states

$$u(x, y, t) \quad (t' \leqslant t \leqslant T).$$

This *reversed Cauchy problem* is equivalent (changing t by $-t$) to the ordinary Cauchy problem for the equation

$$u_t = -\kappa(u_{xx} + u_{yy}) \tag{6.1.8}$$

with boundary condition (6.1.6). Although the reversed Cauchy problem makes perfect physical sense, it is far from being well posed in the sense of Section 1.2. It is still true that solutions will exist for a dense set of "final data" u_T; we only have to take u_T to be a trigonometric polynomial

$$u_T(x, y) = \sum_{m=1}^{M} \sum_{n=1}^{N} a_{mn} \sin mx \sin ny$$

in which case the solution (which is defined for all t) is

$$u(x, y, t) = \sum_{m=1}^{M} \sum_{n=1}^{N} e^{-\kappa(m^2 + n^2)(t - T)} a_{mn} \sin mx \sin ny$$

(see Section 1.1). However, if we take, say, $u_{T,n}(x, y) = \sin nx \sin ny$, we have $\|u_{T,n}\| = 1$, but, if u_n is the corresponding solution,

$$\|u_n(\cdot, \cdot, t')\| = e^{2\kappa n^2(T - t')} \quad (n \geqslant 1). \tag{6.1.9}$$

This invalidates in practice our model, since small errors in the measurement of u_T will foil any attempt at computation of $u(\cdot, \cdot, t')$. Unlike in the case of Maxwell's equations (Section 1.6) or the Schrödinger equation (Section 1.4) in L^p, a "reasonable" renorming of the space of final data will not help: if we measure u_T in, say, the maximum of $|u_T|$ and of the modulus of its derivatives of order $\leqslant k$, $\|n^{-(k+1)} u_{T,n}\| = n^{-1}$ while

$$\|n^{-(k+1)} u_n(\cdot, \cdot, t')\| = n^{-1} e^{2\kappa n^2(T - t')}.$$

This simply means that the Cauchy problem is not even properly posed in the sense of Hadamard (see Section 1.7). A way out is found, however, examining more carefully the physical process involved. We add the following hypotheses:

(a) *The system was in existence at some time $t'' < t'$* (we may obviously assume, by time independence of equation and boundary condition, that $t'' = 0$).

(b) *Ever since the time $t'' = 0$ an a priori bound on the temperature distribution has been in effect: there exists $C > 0$ such that*

$$\|u(\cdot, \cdot, t)\| \leqslant C \quad (0 \leqslant t \leqslant T), \tag{6.1.10}$$

where $\|\cdot\|$ again indicates the norm of $C_\Gamma(\bar{\Omega})$. We must then eliminate from consideration every "final state" giving rise to a solution that violates (6.1.10). It is plain that the class $\mathscr{F} = \mathscr{F}(T, C)$ of "admissible final states" consists of all $u_T \in C_\Gamma(\bar{\Omega})$ whose sine Fourier coefficients $\{a_{mn}\}$ satisfy

$$\left| \sum_{m=1}^{\infty} \sum_{n=1}^{\infty} e^{\kappa(m^2+n^2)t} a_{mn} \sin mx \sin ny \right| \leqslant C \quad ((x, y) \in \bar{\Omega}, 0 \leqslant t \leqslant T). \tag{6.1.11}$$

Although the class $\mathscr{F}$ is not simply characterized, it follows from the Parseval equality for Fourier series that for any $u_T \in \mathscr{F}$,

$$|a_{mn}| \leqslant 2Ce^{-\kappa(m^2+n^2)T}. \tag{6.1.12}$$

Let now $\varepsilon > 0$ and let u_T be an element of $\mathscr{F}$ satisfying $|u_T| \leqslant \delta$, where δ is to be determined later. Then the Fourier coefficients $\{a_{mn}\}$ of u_T satisfy

$$|a_{mn}| \leqslant 2\delta. \tag{6.1.13}$$

We have

$$|u(x, y, t)| = \left| \sum_{m=1}^{\infty} \sum_{n=1}^{\infty} e^{\kappa(m^2+n^2)(T-t)} a_{mn} \sin mx \sin ny \right|$$

$$\leqslant \sum_{m=1}^{\infty} \sum_{n=1}^{\infty} e^{\kappa(m^2+n^2)(T-t)} |a_{mn}|.$$

Dividing the sum in two parts, one corresponding to $m, n \leqslant N$, it is easy to see using (6.1.12) and (6.1.13) that if δ is sufficiently small we shall have

$$|u(x, y, t)| \leqslant \varepsilon \quad ((x, y) \in \Omega, t' \leqslant t \leqslant T)$$

so that a problem which is in some sense properly posed is obtained. It should be noted that the misbehavior of the model has not been eliminated, since the condition that $u_T \in \mathscr{F}$ remains highly unstable with respect to perturbations. However, numerical methods for treatment of the problem are available (see the references in Section 6.6).

We note, finally, that the two assumptions (a) and (b) are reasonable, even indispensable in our model: (a) is (almost) necessary for the reversed Cauchy problem to make sense, while (b) must hold in any heat propagation process with C (at the very least) the melting point of the material in question. Hence, the difficulties pointed out at the beginning of this example can be blamed not on improper modeling but on the neglect of available information.

6.1.4   Example.   *An incomplete Cauchy problem for the Laplace equation.* In the theory of probability, equations of the form $u_{tt} = -Au$ appear, with A an elliptic differential operator in a domain $\Omega \subseteq \mathbb{R}^m$ satisfying adequate boundary conditions (see Balakrishnan [1958: 1]); the spaces indicated for the treatment are $L^1(\Omega)$ and $C(\overline{\Omega})$ or subspaces thereof. To fix ideas, we examine the equation

$$u_{tt} = -(u_{xx} + u_{yy}) \qquad (6.1.14)$$

in the square $\Omega = \{(x, y);\ 0 < x,\ y < \pi\}$, with Dirichlet boundary condition

$$u = 0 \quad (x \in \Gamma). \qquad (6.1.15)$$

Consider two arbitrary trigonometric polynomials

$$u_0(x, y) = \sum \sum a_{mn}^0 \sin mx \sin ny, \quad u_1(x, y) = \sum \sum a_{mn}^1 \sin mx \sin ny.$$

The function

$$u(x, y, t) = \sum_{m=1}^{\infty} \sum_{n=1}^{\infty} \left(\cosh(m^2 + n^2)^{1/2} t\right) a_{mn}^0 \sin mx \sin ny$$

$$+ \sum_{m=1}^{\infty} \sum_{n=1}^{\infty} \frac{\left(\sinh(m^2 + n^2)^{1/2} t\right)}{(m^2 + n^2)^{1/2}} a_{mn}^1 \sin mx \sin ny$$

$$(6.1.16)$$

is a solution of (6.1.14), (6.1.15) satisfying the initial conditions

$$u(x, y, 0) = u_0(x, y), \quad u_t(x, y, 0) = u_1(x, y).$$

However, the solution does not depend continuously on its initial data. To see this we take $u_{0,n} = n^{-(k+1)} \sin nx \sin ny$ (which tends to zero uniformly in $\overline{\Omega}$ together with its partial derivatives of order $\leqslant k$), $u_{1,n} = 0$, and note that the corresponding solution is $u(x, y, t) = (n^{-(k+1)} \cosh\sqrt{2}\, nt) \sin nx \sin ny$. In fact, this is essentially the celebrated counterexample given by Hadamard [1923: 1] in relation to his formulation of properly posed problems (see the comments in Section 1.7). However, the subsidiary conditions imposed on the solutions of (6.1.14), (6.1.15) by their probabilistic interpretation are not both initial conditions u_0, u_1, but rather only one of them,

$$u(x, y, 0) = u_0(x, y), \qquad (6.1.17)$$

and the boundedness condition

$$\sup_{t \geqslant 0} \|u(\cdot, \cdot, t)\| < \infty \qquad (6.1.18)$$

(a third requirement is also present, see Balakrishnan [1958: 1], but we leave it aside for simplicity). This "incomplete Cauchy problem" is easily analyzed along the lines of Section 1.3; to simplify, we take $E = L^1(\Omega)$. If A denotes the operator defined in Section 1.3 (with $\kappa = 1$) and $t \to u(\cdot, \cdot, t)$ solves the equation

$$u''(t) + Au(t) = 0 \quad (t \geqslant 0), \qquad (6.1.19)$$

a calculation entirely similar to (1.3.4) shows that the Fourier series of $u(x, y, t)$ (for t fixed) must have the form (6.1.16). If we assume in addition that

$$\sup_{t \geqslant 0} \|u(t)\| < \infty, \qquad (6.1.20)$$

the same sort of argument reveals that $a_{mn}^1 = -(m^2 + n^2)^{1/2} a_{mn}^0$, so that

$$u(x, y, t) \sim \sum_{n=1}^{\infty} \sum_{m=1}^{\infty} e^{-(m^2 + n^2)^{1/2} t} a_{mn}^0 \sin mx \sin ny. \qquad (6.1.21)$$

Among other things, this shows that bounded solutions of (6.1.19) are uniquely determined by their initial value

$$u(0) = u_0. \qquad (6.1.22)$$

On the other hand, it is easily checked that the series on the right-hand side of (6.1.21) converges and defines a E-valued solution of (6.1.19) satisfying (6.1.20) and (6.1.22) for any u_0 in the subspace $D \subset L^1(\Omega)$ consisting of all functions whose Fourier coefficients $\{a_{mn}^0\}$ satisfy

$$\sum_{m=1}^{\infty} \sum_{n=1}^{\infty} (m^2 + n^2)|a_{mn}^0| < \infty. \qquad (6.1.23)$$

If u is an arbitrary solution, we can write

$$u(x, y, t) = \int_{\Omega} K(x, y, \xi, \eta, t) u_0(\xi, \eta) \, d\xi \, d\eta, \qquad (6.1.24)$$

where

$$K(x, y, \xi, \eta, t) = \frac{4}{\pi^2} \sum_{m=1}^{\infty} \sum_{n=1}^{\infty} e^{-(m^2 + n^2)^{1/2} t} \sin mx \sin m\xi \sin ny \sin n\eta$$

$$(6.1.25)$$

for $t > 0$. The function $u(x, y, t)$ is a classical solution of the Laplace equation $u_{xx} + u_{yy} + u_{tt} = 0$ in $\Omega \times (0, \infty)$ and tends to zero uniformly when $t \to +\infty$. Arguing as in Section 1.3, but this time on the basis of the maximum principle for harmonic functions, we can easily show that

$$K(x, y, \xi, \eta, t) \geqslant 0 \quad (t > 0, (x, y), (\xi, \eta) \in \overline{\Omega}) \qquad (6.1.26)$$

and

$$\int_{\Omega} K(x, y, \xi, \eta, t) \, d\xi \, d\eta \leqslant 1 \quad (t > 0, (x, y) \in \overline{\Omega}), \qquad (6.1.27)$$

hence (see (1.3.11)) if $u(\cdot)$ is an arbitrary solution of (6.1.19), (6.1.20),

$$\|u(t)\| \leqslant \|u(0)\| \quad (t \geqslant 0). \qquad (6.1.28)$$

Similar inequalities can be easily obtained in the norms of $L^p(\Omega)$ and $C_\Gamma(\overline{\Omega})$.

Taking these facts as motivation, we formulate the following *incomplete Cauchy problem*. Let A be a densely defined operator in a (real or complex) Banach space E. The incomplete Cauchy problem for (6.1.19) is *properly posed* in $t \geqslant 0$ if and only if:

(a) *There exists a dense subspace D of E such that, for any $u_0 \in D$ there exists a solution $u(\cdot)$ of (6.1.19) satisfying (6.1.20) and (6.1.22). (A solution is of course any twice continuously differentiable E-valued function $u(\cdot)$ satisfying the equation in $t \geqslant 0$.)*

(b) *There exists a nondecreasing, nonnegative function $C(t)$ defined in $t \geqslant 0$ such that*

$$\|u(t)\| \leqslant C(t)\|u(0)\| \quad (t \geqslant 0)$$

for any solution of (6.1.19), (6.1.20), (6.1.22).

Sufficient conditions on the operator A insuring validity of (a) and (b) will be given in Section 6.5. Many variants of this basic definition are possible: (6.1.20) may be replaced by other types of bound at infinity, $u'(0)$ may be preassigned instead of $u(0)$, and so on. Some of these will be mentioned in Section 6.6.

6.1.5 Example. Let A be a self-adjoint operator in a Hilbert space H. Assume $-A$ is positive definite ($((Au, u) \leqslant 0$ for $u \in D(A))$). Then the incomplete Cauchy problem for (6.1.19) is properly posed in $t \geqslant 0$ with $D = D(A)$. Using the functional calculus (see Section 6), we can easily deduce an explicit formula for the solutions: if $u \in D(A)$, the (unique) solution of (6.1.19), (6.1.20), (6.1.22) is

$$u(t) = \int_{-\infty}^{0^+} e^{-(-\lambda)^{1/2}t} P(d\lambda) u,$$

where $P(d\lambda)$ is the resolution of the identity associated with A.

6.2. THE REVERSED CAUCHY PROBLEM FOR ABSTRACT PARABOLIC EQUATIONS

It was pointed out in the previous section (Example 6.1.3) that if A is the Laplacian in the space $C_\Gamma(\overline{\Omega})$, Ω the square defined by $0 < x, y < \pi$, then the

Cauchy problem for the equation

$$u'(t) = Au(t) \quad (0 \leqslant t \leqslant T) \tag{6.2.1}$$

with the value of u prescribed at $t = T$,

$$u(T) = u_T \tag{6.2.2}$$

is in general not properly posed (the solution of (6.2.1), (6.2.2), if it exists at all, does not depend continuously on u_T); however, a well-posed problem results if we restrict our attention to solutions satisfying the a priori bound

$$\|u(t)\| \leqslant C \quad (0 \leqslant t \leqslant T). \tag{6.2.3}$$

We prove here a generalization of this result to operators in the class $\mathcal{C}$ in an arbitrary Banach space E. To simplify ensuing statements we make the following definition.

The reversed Cauchy problem for (6.2.1) *is properly posed for bounded solutions* in $0 \leqslant t \leqslant T$ if and only if: *for every* ε, C, t' *such that* $\varepsilon, C > 0$ *and* $0 < t' \leqslant T$ *there exists* $\delta = \delta(\varepsilon, C, t') > 0$ *such that if* $u(\cdot)$ *is any solution of* (6.2.1) *satisfying* (6.2.3) *and*

$$\|u(T)\| \leqslant \delta, \tag{6.2.4}$$

we have

$$\|u(t)\| \leqslant \varepsilon \quad (t' \leqslant t \leqslant T). \tag{6.2.5}$$

6.2.1   Theorem.   *Let* $A \in \mathcal{C}$. *Then the reversed Cauchy problem for* (6.2.1) *is properly posed for bounded solutions.*

For the proof we shall use some elementary results on E-valued analytic functions. Let Ω be a bounded domain in $\mathbb{C}$ whose boundary Γ consists of the union of a finite number of disjoint smooth arcs,[4] and let Γ_1 be a subset of Γ such that both Γ_1 and $\Gamma \setminus \Gamma_1$ are themselves finite unions of disjoint smooth arcs. The function $\omega(z, \Gamma_1; \Omega)$, which is harmonic in Ω and assumes the values 1 on Γ_1, 0 on $\Gamma \setminus \Gamma_1$, is called the *harmonic measure of* Γ_1 *with respect to* Ω. If Γ is the union of disjoint arcs,

$$\Gamma = \Gamma_1 \cup \Gamma_2 \cup \cdots \cup \Gamma_n, \tag{6.2.6}$$

where each Γ_j satisfies the conditions imposed earlier on Γ_1, it is obvious that

$$\sum_{j=1}^{n} \omega(z, \Gamma_j, \Omega) = 1 \quad (z \in \Omega) \tag{6.2.7}$$

since the left-hand side is a harmonic function that is identically 1 on Γ. A generalization of (6.2.7) is

---

[4] Here "disjoint" means "having no points other than endpoints in common" or, in other words, "having disjoint interiors." A smooth arc is by definition the continuously differentiable image of a closed interval.

6.2.2 The N Constants Theorem. *Let $\Gamma_1, \Gamma_2, \ldots, \Gamma_n$ be as in the comments preceding (6.2.6). Let f be a E-valued function, analytic in Ω and continuous in $\overline{\Omega}$.[5] Assume that*

$$\|f(z)\| \leqslant M_j \quad (z \in \Gamma_j, j = 1, 2, \ldots, n). \tag{6.2.8}$$

Then

$$\|f(z)\| \leqslant \prod_{j=1}^{n} M_j^{\omega(z, \Gamma_j; \Omega)} \quad (z \in \overline{\Omega}). \tag{6.2.9}$$

The proof can be achieved by applying the n constants theorem for complex-valued functions (see Nevanlinna [1936: 1, p. 41] or Hille [1962: 1, p. 409]) to the function $\langle f(z), u^* \rangle$ for all $u^* \in E^*$, $\|u^*\| \leqslant 1$. Details are left to the reader.

Proof of Theorem 6.2.1. Since $A \in \mathcal{C}$, there exist $\varphi, \alpha, C' > 0$ such that any solution $u(t) = S(t)u(0)$ of (6.2.1) can be extended to a function $u(\zeta)$ analytic in the sector $\Sigma_+(\varphi) = \{\zeta; |\arg \zeta| \leqslant \varphi, \zeta \neq 0\}$, continuous at $\zeta = 0$ and satisfying

$$\|u(\zeta)\| \leqslant C' e^{\alpha|\zeta|} \|u(0)\| \quad (\zeta \in \Sigma_+(\varphi)). \tag{6.2.10}$$

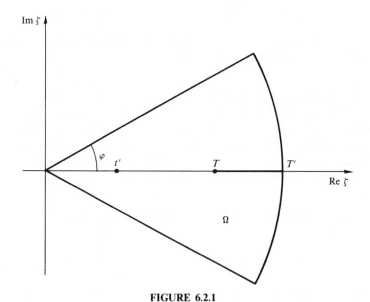

FIGURE 6.2.1

---

[5]Of course, discontinuities may be allowed at the junction of different arcs, although this is irrelevant for the present application.

Let now $T' > T$ and consider the domain Ω defined by the inequalities

$$|\arg \zeta| < \varphi, \quad \zeta \neq 0, \quad |\zeta| < T'$$

minus the segment $[T, T']$. In view of (6.2.10),

$$\|u(\zeta)\| \leqslant C' e^{\alpha T'} \|u(0)\|$$

when ζ moves over the outer boundary of Ω. On the other hand, if $\zeta \in [T, T']$,

$$\begin{aligned}
\|u(\zeta)\| &= \|S(\zeta - T) u(T)\| \\
&\leqslant C' e^{\alpha|\zeta - T|} \|u(T)\| \\
&\leqslant C' e^{\alpha(T' - T)} \|u(T)\|.
\end{aligned}$$

Applying the n constants theorem,

$$\begin{aligned}
\|u(\zeta)\| &\leqslant C' e^{\alpha T'(1 - \omega(\zeta))} e^{\alpha(T' - T)\omega(\zeta)} \|u(0)\|^{1 - \omega(\zeta)} \|u(T)\|^{\omega(\zeta)} \\
&\leqslant C' e^{\alpha T' - \alpha T\omega(\zeta)} \|u(0)\|^{1 - \omega(\zeta)} \|u(T)\|^{\omega(\zeta)} \quad (\zeta \in \Omega),
\end{aligned}$$

where ω is the harmonic measure of the segment $T \leqslant \zeta \leqslant T'$ with respect to Ω. Taking into account that $\omega(\zeta) \geqslant 0$ and assuming (as we may) that $C \geqslant 1$ in (6.2.3),

$$\|u(t)\| \leqslant CC' e^{\alpha T'} \|u(T)\|^{m(t')} \quad (0 < t' \leqslant t \leqslant T)$$

if $\|u(T)\| \leqslant 1$, where $m(t') = \inf\{\omega(t); \ t' \leqslant t \leqslant T\} > 0$ in view of the maximum principle for harmonic functions. Consequently, (6.2.5) will be satisfied if we take $\delta = (\varepsilon e^{-\alpha T'}/CC')^{1/m(t')}$. This ends the proof.

6.2.3 Remark. Theorem 6.2.1 can be proved under weaker hypotheses. In fact, let A be an operator in $\mathcal{C}_+$ such that $S(t) = \exp(tA)$ admits an analytic extension $S(\zeta)$ to a sector $\Sigma_+(\varphi)$ satisfying estimates of the type of (4.9.24). It suffices to excise from Ω all those ζ with $|\zeta| \leqslant t''$ for some $t'' < t'$. This implies that the results in this section can be applied to the operators $A_1(\beta)$, $A_\infty(\beta)$ in Chapter 4 as well as to their one-dimensional versions in Section 4.3.

6.3. FRACTIONAL POWERS OF CERTAIN UNBOUNDED OPERATORS

Throughout this and following sections, A will be a densely defined linear operator such that $\lambda \in \rho(A)$ if $\lambda > 0$ and

$$\|R(\lambda)\| \leqslant C/\lambda \quad (\lambda > 0). \tag{6.3.1}$$

If $A \in \mathcal{C}_+(C, 0)$, then (6.3.1) is satisfied. The converse is false (see Example 2.1.5).

Our objective is to define fractional powers A^α of the operator A (or, rather, of $-A$) with a view towards the study of bounded solutions of the second order equation (6.1.19). To this end we shall employ (variants of) the formula

$$(-A)^\alpha u = \frac{\sin \alpha \pi}{\pi} \int_0^\infty \lambda^{\alpha-1} R(\lambda)(-A)u \, d\lambda. \qquad (6.3.2)$$

The motivation of (6.3.2) is evident: if A "is a number" a not in $\lambda \geq 0$ (that is, if A is the operator of multiplication by a in one-dimensional Banach space) and $0 < \operatorname{Re}\alpha < 1$, then (6.3.2) becomes an identity, at least if the branches of $\lambda^{\alpha-1}$ and $(-a)^\alpha$ are adequately specified, to wit: $\lambda^{\alpha-1} > 0$ for $\lambda > 0$, with a similar choice of $(-a)^\alpha$. Two difficulties are obvious: on the one hand, the formula (6.3.2) will define $(-A)^\alpha u$ only for $u \in D(A)$ while we may expect $D((-A)^\alpha)$ to contain strictly $D(A)$ if, say, $0 < \alpha < 1$. On the other hand, (6.3.2) only makes sense for $0 < \operatorname{Re}\alpha < 1$. This last objection is not serious, since this range of α (precisely, the interval $0 < \alpha < 1$) is by far the most significant in applications; however, for the sake of completeness we shall construct $(-A)^\alpha$ for any α by appropriate modifications of (6.3.2). The formal definition follows.

Let $0 < \operatorname{Re}\alpha < 1$. The operator K_α (with domain $D(K_\alpha) = D(A)$) is defined by (6.3.2), that is,

$$K_\alpha u = \frac{\sin \alpha \pi}{\pi} \int_0^\infty \lambda^{\alpha-1} R(\lambda)(-A)u \, d\lambda \qquad (6.3.3)$$

with the choice of branch of $\lambda^{\alpha-1}$ precised above. Convergence of (6.3.3) at infinity is obvious from (6.3.1); near zero we use the identity

$$R(\lambda)(-A)u = u - \lambda R(\lambda)u \qquad (6.3.4)$$

and (6.3.1). The operators K_α are extended to other values of α as follows. If $u \in D(A^2)$, we can write, making again use of (6.3.4) (this time backwards):

$$\begin{aligned}
K_\alpha u &= \frac{\sin \alpha \pi}{\pi} \int_0^\infty \lambda^{\alpha-1} R(\lambda)(-A)u \, d\lambda \\
&= \frac{\sin \alpha \pi}{\pi} \int_0^1 \lambda^{\alpha-1} R(\lambda)(-A)u \, d\lambda \\
&\quad - \frac{\sin \alpha \pi}{\pi} \int_1^\infty \lambda^{\alpha-2} R(\lambda)A^2 u \, d\lambda - \frac{\sin \alpha \pi}{(1-\alpha)\pi} Au. \qquad (6.3.5)
\end{aligned}$$

But the integrals on the right-hand side actually converge in $0 < \operatorname{Re}\alpha < 2$, thus (6.3.5) is an analytic extension of $K_\alpha u$ to $0 < \operatorname{Re}\alpha < 2$. Moreover, if $1 < \operatorname{Re}\alpha < 2$ and $u \in D(A^2)$, another application of (6.3.4) yields

$$\begin{aligned}
K_{\alpha-1}(-A)u &= \frac{\sin(\alpha-1)\pi}{\pi} \int_0^\infty \lambda^{\alpha-2} R(\lambda)A^2 u \, d\lambda \\
&= \frac{\sin \alpha \pi}{\pi} \int_0^1 \left(\lambda^{\alpha-1} R(\lambda)(-A)u + \lambda^{\alpha-2} Au \right) d\lambda \\
&\quad - \frac{\sin \alpha \pi}{\pi} \int_1^\infty \lambda^{\alpha-2} R(\lambda)A^2 u \, d\lambda = K_\alpha u. \qquad (6.3.6)
\end{aligned}$$

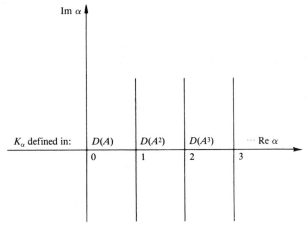

FIGURE 6.3.1

in view of (6.3.5). The operators K_α are then defined thus: if $n - 1 < \mathrm{Re}\,\alpha < n$, $D(K_\alpha) = D(A^n)$, and

$$K_\alpha u = K_{\alpha - n + 1}(-A)^{n-1}u \qquad (6.3.7)$$

for $u \in D(A^n)$, $n = 1, 2, \ldots$. On the other hand, if $\mathrm{Re}\,\alpha = n$, we use also (6.3.7) for $u \in D(A^{n+1})$, $K_{\alpha - n + 1}$ given by (6.3.5). It follows from the preceding considerations that these step-by-step definitions of K_α are consistent with each other: in other words, if $n - 1 < \mathrm{Re}\,\alpha \leqslant n$ and $u \in D(A^{n+1})$, $K_\alpha u$, defined by (6.3.7), can be analytically extended to $n - 1 < \mathrm{Re}\,\alpha < n + 1$ and the extension coincides with (6.3.7) in $n \leqslant \mathrm{Re}\,\alpha < n + 1$.

6.3.1 Lemma. *For each α ($\mathrm{Re}\,\alpha > 0$), K_α is closable.*

Proof. Let $0 < \mathrm{Re}\,\alpha < 1$, $\{u_k\}$ a sequence in $D(A)$ with $u_k \to 0$, $K_\alpha u_k \to u \in E$. Write

$$v_k = R(\mu) K_\alpha u_k \qquad (6.3.8)$$

for some fixed $\mu > 0$. It is an immediate consequence of its definition that K_α commutes with $R(\mu)$, that is, $R(\mu)E = D(A) = D(K_\alpha)$ and

$$R(\mu) K_\alpha u = K_\alpha R(\mu) u \quad (u \in D(A)). \qquad (6.3.9)$$

Moreover, it is not difficult to see that $K_\alpha R(\mu)$ is everywhere defined and bounded; in fact, we have

$$K_\alpha R(\mu) = \frac{\sin \alpha \pi}{\pi} R(\mu) \int_0^1 \lambda^{\alpha - 1}(I - \lambda R(\lambda)) \, d\lambda$$

$$- \frac{\sin \alpha \pi}{\pi} A R(\mu) \int_1^\infty \lambda^{\alpha - 1} R(\lambda) \, d\lambda. \qquad (6.3.10)$$

Accordingly, we obtain taking limits in (6.3.8) that $v \to 0$, hence $R(\mu)u = 0$. It follows that $u = 0$, which ends the proof in the range $0 < \operatorname{Re} \alpha < 1$. The result for $n - 1 \leqslant \operatorname{Re} \alpha < n$ follows recursively from (6.3.7) and from the facts that K_α and A commute and that $\rho(A) \neq \varnothing$. Details are left to the reader.

Making use of Lemma 6.3.1 we define

$$(-A)^\alpha = \overline{K}_\alpha \quad (\operatorname{Re} \alpha > 0). \tag{6.3.11}$$

(Note that condition (6.3.1) does not imply that A^{-1} exists, or even that A is one-to-one; thus we cannot reasonably expect to be able to define $(-A)^\alpha$ for $\operatorname{Re} \alpha < 0$.) We prove next some results that show that $(-A)^\alpha$ satisfies (at least to some extent) properties that should be expected of a fractional power.

6.3.2 Lemma.

$$(-A)^1 = -A.$$

Proof. Clearly we only have to prove that $K_1 u = -Au$ for $u \in D(A^2)$. We prove instead the stronger statement

$$\lim K_\alpha u = u \quad (u \in D(A)). \tag{6.3.12}$$

as $\alpha \to 1$ in any sector $|\arg(\alpha - 1)| \geqslant \varphi > \pi/2$. To see this we note that

$$K_\alpha u - (-A)u = \frac{\sin \alpha \pi}{\pi} \int_0^\infty \lambda^{\alpha - 1} \left(R(\lambda) - \frac{1}{\lambda + 1} I \right)(-A)u \, d\lambda$$

$$= I_1 + I_2 \tag{6.3.13}$$

dividing the interval of integration at $\lambda = a$. Plainly,

$$\|I_2\| \leqslant \frac{|\sin \alpha \pi|}{\sin(\operatorname{Re} \alpha)\pi} \sup_{\lambda \geqslant a} \|((\lambda + 1)R(\lambda) - I)Au\|.$$

Since $\lambda R(\lambda)v \to v$ as $\lambda \to \infty$ (see Example 2.1.6), $\|I_2\|$ can be made small independently of α if only we take a large enough. As for I_1, it is easy to see using (6.3.4) that the integrand is bounded by $C\lambda^{\operatorname{Re}\alpha - 1}$, thus in view of the factor $\sin \alpha \pi$, $\|I_1\| \to 0$ as $\alpha \to 1$. This ends the proof.

The corresponding equality for $\alpha = 0$ $((-A)^0 = I)$ cannot hold in general, since $K_\alpha u = 0$ whenever $Au = 0$. However, the equality becomes true when A^{-1} exists. Precisely, we have the following more general statement.

6.3.3 Lemma. *Let $u \in D(A)$ such that $\lambda R(\lambda)u \to 0$ as $\lambda \to 0+$. Then*

$$\lim K_\alpha u = u$$

as $\alpha \to 0$ in any sector $|\arg \alpha| \leqslant \varphi < \pi/2$.

Proof. We have

$$K_\alpha u - u = -\frac{\sin \alpha\pi}{\pi} \int_0^\infty \lambda^{\alpha-1} \left(R(\lambda) Au + \frac{1}{\lambda+1} u \right) d\lambda$$

$$= -\frac{\sin \alpha\pi}{\pi} \int_0^\infty \frac{\lambda^\alpha}{\lambda+1} ((\lambda+1)R(\lambda)u - u) \, d\lambda$$

$$= J_1 + J_2,$$

where J_1 and J_2 arise once again from division of the interval of integration at $\lambda = a$. To estimate the second integral we observe that $(\lambda+1)R(\lambda)u - u = R(\lambda)(Au + u)$ and use (6.3.1): it follows that $\|J_2\| \to 0$ as $\alpha \to 0$ due to the factor $|\sin \alpha\pi|$. In the matter of J_1, we observe that

$$\|(\lambda+1)R(\lambda)u - u\| \leqslant C + \|R(\lambda)u\|.$$

The portion of the estimate for J_1 corresponding to the constant C obviously tends to zero when $\alpha \to 0$ in the sector. On the other hand,

$$\frac{|\sin \alpha\pi|}{\pi} \int_0^a \frac{\lambda^{\text{Re}\,\alpha}}{\lambda+1} \|R(\lambda)u\| \, d\lambda = \frac{|\sin \alpha\pi|}{\pi} \int_0^a \frac{\lambda^{\text{Re}\,\alpha-1}}{\lambda+1} \|\lambda R(\lambda)u\| \, d\lambda$$

$$\leqslant \frac{|\sin \alpha\pi|}{\sin(\text{Re}\,\alpha)\pi} \max_{0 \leqslant \lambda \leqslant a} \|\lambda R(\lambda)u\|,$$

thus we can make $\|J_1\|$ arbitrarily small by taking a small enough. This completes the proof.

An essential property in any definition of fractional power is the additivity relation $(-A)^{\alpha+\beta} = (-A)^\alpha(-A)^\beta$. In the present level of generality this identity may not necessarily hold; however, we have the following result, which is nearly as strong.

6.3.4   Lemma.   *Let* $\text{Re}\,\alpha, \text{Re}\,\beta > 0$. *Then*

$$(-A)^{\alpha+\beta} = \overline{(-A)^\alpha(-A)^\beta}. \tag{6.3.14}$$

Proof. We start by proving that

$$K_\alpha K_\beta u = K_{\alpha+\beta} u \quad (u \in D(A^2), \text{Re}(\alpha+\beta) < 1). \tag{6.3.15}$$

Noting that $K_\beta u \in D(A) = D(K_\alpha)$, we can write

$$K_\alpha K_\beta u = \frac{\sin \alpha\pi}{\pi} \frac{\sin \beta\pi}{\pi} \int_0^\infty \int_0^\infty \lambda^{\alpha-1} \mu^{\beta-1} R(\lambda) R(\mu) A^2 u \, d\lambda \, d\mu,$$

$$\tag{6.3.16}$$

the integral being absolutely convergent in the sector $\lambda, \mu \geqslant 0$ (to estimate the integrand near the origin we use the the resolvent equation as we did in (6.3.3). We make now a few modifications in (6.3.16). The first one is described as follows. Divide the domain of integration into the two infinite triangles $\mu \leqslant \lambda$ and $\lambda \leqslant \mu$ and perform the change of variables $\mu = \lambda \sigma$ in the first integral, $\lambda = \sigma \mu$ in the second. Interchanging λ and μ in the second integral, we obtain

$$K_\alpha K_\beta u = \frac{\sin \alpha \pi}{\pi} \frac{\sin \beta \pi}{\pi} \int_0^\infty \int_0^1 (\sigma^{\alpha-1} + \sigma^{\beta-1}) \lambda^{\alpha+\beta-1} R(\lambda \sigma) R(\lambda) A^2 u \, d\sigma \, d\lambda.$$

$$(6.3.17)$$

We use next the second resolvent equation (3.6) and (6.3.4), obtaining

$$R(\lambda \sigma) R(\lambda) A^2 u = \frac{1}{1-\sigma} (R(\lambda) - \sigma R(\lambda \sigma))(-A) u \qquad (6.3.18)$$

and replace in the integrand. Due to the factor $(1-\sigma)^{-1}$, the two resulting integrals cannot be separated; however, we can write (6.3.17) as the limit when $\rho \to 1-$ of

$$\frac{\sin \alpha \pi}{\pi} \frac{\sin \beta \pi}{\pi} \int_0^\infty \int_0^\rho \frac{\sigma^{\alpha-1} + \sigma^{\beta-1}}{1-\sigma} \lambda^{\alpha+\beta-1} R(\lambda)(-A) u \, d\sigma \, d\lambda$$

$$- \frac{\sin \alpha \pi}{\pi} \frac{\sin \beta \pi}{\pi} \int_0^\infty \int_0^\rho \frac{\sigma^\alpha + \sigma^\beta}{1-\sigma} \lambda^{\alpha+\beta-1} R(\lambda \sigma)(-A) u \, d\sigma \, d\lambda.$$

$$(6.3.19)$$

Finally, we integrate first in λ in the second integral (6.3.19) making the change of variable $\lambda \sigma = \mu$ and again changing μ by λ. We obtain

$$K_\alpha K_\beta u = C(\alpha, \beta) \int_0^\infty \lambda^{\alpha+\beta-1} R(\lambda)(-A) u \, d\lambda, \qquad (6.3.20)$$

where

$$C(\alpha, \beta) = \frac{\sin \alpha \pi}{\pi} \frac{\sin \beta \pi}{\pi} \int_0^1 \frac{\sigma^{\alpha-1} + \sigma^{\beta-1} - \sigma^{-\alpha} - \sigma^{-\beta}}{1-\sigma} \, d\sigma.$$

It remains to evaluate $C(\alpha, \beta)$. This can be done directly (using the geometric series for $(1-\sigma)^{-1}$) or better yet by means of the following trick: if we take $E = \mathbb{C}$, $A = -1$, then $K_\alpha = 1$ for all α; thus $K_\alpha K_\beta = 1$ and it follows from (6.3.20) that

$$C(\alpha, \beta) = \frac{\sin(\alpha + \beta) \pi}{\pi},$$

whence (6.3.15) follows. Since $K_\alpha u$ is an analytic function of α in $0 < \operatorname{Re} \alpha < n$ for $u \in D(A^n)$, it is clear that (6.3.15) can be extended to $0 < \operatorname{Re} \alpha < 1$, $0 < \operatorname{Re} \beta < 1$ (with no conditions on $\alpha + \beta$) for $u \in D(A^2)$. More generally, if $0 < \operatorname{Re} \alpha, \operatorname{Re} \beta < n$ $(n \geqslant 1)$, then

$$K_\alpha K_\beta u = K_{\alpha+\beta} u \quad (u \in D(A^{2n})). \qquad (6.3.21)$$

Given α $(n-1 < \operatorname{Re} \alpha < n)$ and $m \geqslant n$ fixed, denote by K'_α the restriction of K_α from its rightful domain $D(A^n)$ to $D(A^m)$. Clearly K'_α is closable and $\overline{K}'_\alpha \subseteq \overline{K}_\alpha$. We actually have

$$\overline{K}'_\alpha = \overline{K}_\alpha = (-A)^\alpha. \tag{6.3.22}$$

To see this, let $u \in D(\overline{K}_\alpha)$, $\{u_k\}$ a sequence in $D(A^n)$ with $u_k \to u$, $K_\alpha u_k \to \overline{K}_\alpha u$, and $\{\mu_k\}$ a sequence of real numbers with $\mu_k \to \infty$. Making use of the fact that $\{\mu_k R(\mu_k)\}$ is uniformly bounded and converges strongly to the identity as $k \to \infty$, it is not difficult to see that

$$u'_k = (\mu_k R(\mu_k))^{m-n} u_k \to u_k,$$
$$K'_\alpha u'_k = K'_\alpha(\mu_k R(\mu_k))^{m-n} u_k$$
$$= (\mu_k R(\mu_k))^{m-n} K_\alpha u_k \to \overline{K}_\alpha u$$

as $k \to \infty$; since $u_k \in D(A^m) = D(K'_\alpha)$, $u \in D(\overline{K}'_\alpha)$, and $\overline{K}'_\alpha u = \overline{K}_\alpha u$, showing that (6.3.22) holds.

We have already noted in (6.3.10) that the operator $K_\alpha R(\mu)$ ($\mu > 0$) is everywhere defined and bounded when $\mathrm{Re}\,\alpha < 1$. A look at the definition of K_α for general α in (6.3.7) shows that $K_\alpha R(\mu)^n$ is everywhere defined and bounded when $\mathrm{Re}\,\alpha < n$. It follows easily from (6.3.9), (6.3.7) and the definition of $(-A)^\alpha$ that if $u \in D((-A)^\alpha) = D(\overline{K}_\alpha)$, then

$$K_\alpha R(\mu)^n = (-A)^\alpha R(\mu)^n u = R(\mu)^n (-A)^\alpha u, \tag{6.3.23}$$

again for $\mathrm{Re}\,\alpha < n$.

We can now complete the proof of (6.3.14). Let $\mathrm{Re}\,\alpha, \mathrm{Re}\,\beta < n$ (so that (6.3.21) holds) and let $u \in D((-A)^\alpha(-A)^\beta)$, $\{\mu_k\}$ a sequence of positive numbers with $\mu_k \to \infty$. Since $(\mu_k R(\mu_k))^{2n} \to I$ strongly as $k \to \infty$, it follows that $(\mu_k R(\mu_k))^{2n} u \to u$, whereas, in view of (6.3.21) and (6.3.23),

$$K_{\alpha+\beta}(\mu_k R(\mu_k))^{2n} u = K_\alpha K_\beta (\mu_k R(\mu_k))^{2n} u$$
$$= K_\alpha (\mu_k R(\mu_k))^n K_\beta (\mu_k R(\mu_k))^n u$$
$$= (\mu_k R(\mu_k))^n K_\alpha (\mu_k R(\mu_k))^n (-A)^\beta u$$
$$= (\mu_k R(\mu_k))^{2n} (-A)^\alpha (-A)^\beta u \to (-A)^\alpha (-A)^\beta u$$
$$\text{as } k \to \infty.$$

It follows that $u \in D((-A)^{\alpha+\beta})$ and $(-A)^{\alpha+\beta} u = (-A)^\alpha(-A)^\beta u$ so that $(-A)^\alpha(-A)^\beta \subseteq (-A)^{\alpha+\beta}$; taking closures, it results that $\overline{(-A)^\alpha(-A)^\beta} \subseteq (-A)^{\alpha+\beta}$. The reverse inclusion follows if we note that $\overline{(-A)^\alpha(-A)^\beta} \supseteq K_\alpha K'_\beta = K'_{\alpha+\beta}$ and take the closure of the right-hand side.

We note an obvious consequence of Lemma 6.3.4; if we set $\alpha = \beta = 1$ and make use of Lemma 6.3.2 and of the fact that (since $\rho(A) \neq \varnothing$) A^2 is closed, we obtain $\overline{K}_2 = A^2 = (-A)^2$. Similarly,

$$\overline{K}_n = (-A)^n \tag{6.3.24}$$

as we have the right to require from any definition of fractional power.

Some improvements on several of the preceding results are possible when $0 \in \rho(A)$ (which is not implied by (6.3.1)). Also (as we shall see in next section), fractional powers can be defined in this case in a less ad hoc way by means of the functional calculus for unbounded operators (see Example 3.12).

The first obvious consequence of the existence of A^{-1} is that we may construct fractional powers of $-A$ for exponents in the left half plane: in fact, if $n - 1 \leqslant \operatorname{Re} \alpha < n$, $n = 0, 1, \ldots$, we define

$$(-A)^{-\alpha} = (-A)^{n-\alpha}(-A)^{-n}. \tag{6.3.25}$$

It follows easily from (6.3.10) that $(-A)^{\beta}A^{-1}$ is bounded if $0 < \operatorname{Re} \beta < 1$, thus $(-A)^{-\alpha}$ is bounded if $\operatorname{Re} \alpha < 0$. As a consequence of (6.3.24), $(-A)^{-\alpha}$ assumes the correct values for $\alpha = 1, 2, \ldots$.

In view of (6.3.25) the domain of the operator $(-A)^{\alpha}(-A)^{-\alpha}$ contains $D(A)$ (note that $D((-A)^{\beta}) \supseteq D((-A)^k)$ for $\operatorname{Re}\beta \leqslant k$). Applying Lemma 6.3.4, we obtain $(-A)^{\alpha}(-A)^{-\alpha} = (-A)^{\alpha}(-A)^{n-\alpha}(-A)^n \subseteq (-A)^n(-A)^{-n} = I$. Since $(-A)^{-\alpha}$ is bounded and $(-A)^{\alpha}$ is closed, $(-A)^{-\alpha}E \subseteq D((-A)^{\alpha})$ and

$$(-A)^{\alpha}(-A)^{-\alpha} = I. \tag{6.3.26}$$

We use now the fact (already pointed out in the proof of Lemma 6.3.1) that K_{α} and A^{-1} (thus any power of A^{-1}) commute; then, if $u \in D(A^n) \subseteq D((-A)^{\alpha})$, $(-A)^{-\alpha}K_{\alpha}u = (-A)^{-\alpha}(-A)^{\alpha}u = u$; since $(-A)^{\alpha} = \overline{K}_{\alpha}$, this equality must actually hold for $u \in D((-A)^{\alpha})$, that is

$$(-A)^{-\alpha}(-A)^{\alpha}u = u \quad \left(u \in D\left((-A)^{\alpha}\right) \right).$$

Taking (6.3.26) into account,

$$(-A)^{-\alpha} = \left((-A)^{\alpha}\right)^{-1} \quad (\operatorname{Re} \alpha \geqslant 0). \tag{6.3.27}$$

We can improve (6.3.14) significantly in the present case as follows. Let B_1, B_2 be two operators possessing everywhere defined inverses B_1^{-1} and B_2^{-1}, and let $B_1 \subseteq B_2$. Then we show easily that $B_1 = B_2$. If $\operatorname{Re} \alpha > 0$, $\operatorname{Re} \beta > 0$, we can apply this argument to $B_1 = (-A)^{\alpha}(-A)^{\beta}$ $(B^{-1} = (-A)^{-\beta}(-A)^{-\alpha})$ and $B_2 = (-A)^{\alpha+\beta}$ $(B_2^{-1} = (-A)^{-(\alpha+\beta)})$ observing that (6.3.14) implies that $B_1 \subseteq B_2$. We obtain

$$(-A)^{\alpha+\beta} = (-A)^{\alpha}(-A)^{\beta} \tag{6.3.28}$$

for $\operatorname{Re} \alpha, \operatorname{Re} \beta > 0$. We note that (6.3.28) *cannot* be extended to all values of α and β: while $(-A)^{-\alpha+\alpha} = (-A)^0 = I$, $(-A)^{-\alpha}(-A)^{\alpha}$ is only defined in $D((-A)^{\alpha})$. The equality holds, of course, if $\operatorname{Re} \alpha, \operatorname{Re} \beta < 0$.

For the proof of the following results we only assume (6.3.1) (plus additional hypotheses as stated).

6.3.5 Example. Let A be self-adjoint in the Hilbert space H, $P(d\lambda)$ the resolution of the identity associated with A (note that (6.3.1) simply means in this case that $\sigma(A) \subseteq (-\infty, 0])$. Then

$$(-A)^{\alpha}u = \int_{-\infty}^{0+} (-\lambda)^{\alpha}P(d\lambda)u \quad (u \in E, \operatorname{Re}\alpha > 0), \quad (6.3.29)$$

where $(-\lambda)^{\alpha}$ is defined as in the beginning of this section and (6.3.29) is interpreted by means of the functional calculus for unbounded functions of λ (Dunford-Schwartz [1963: 1, Ch. X]). Formula (6.3.29) defines $(-A)^{\alpha}$ as well when $\operatorname{Re}\alpha \le 0$, although $(-A)^{\alpha}$ may be unbounded too. $((-A)^{\alpha}$ is bounded for $\operatorname{Re}\alpha < 0$ if and only if A is invertible with A^{-1} bounded.) Equality (6.3.28) holds if α, β belong to the right (left) half plane.

6.3.6 Example. *Moment inequalities.* Let $0 < \alpha < \beta < \gamma$. Then there exists a constant $C = C(\alpha, \beta, \gamma)$ such that

$$\|(-A)^{\beta}u\| \le C\|(-A)^{\gamma}u\|^{(\beta-\alpha)/(\gamma-\alpha)}\|(-A)^{\alpha}u\|^{(\gamma-\beta)/(\gamma-\alpha)} \quad (6.3.30)$$

for $u \in D((-A)^{\gamma})$. In particular, if we place ourselves under the hypotheses of Example 6.3.5, inequality (6.3.30) holds with $C = 1$.

6.3.7 Example. Let A be bounded (resp. compact). Then $(-A)^{\alpha}$ is bounded (resp. compact) for $\operatorname{Re}\alpha > 0$.

6.4. FRACTIONAL POWERS OF CERTAIN UNBOUNDED OPERATORS (CONTINUATION)

We place ourselves again under the sole hypothesis (6.3.1), that is, we do not assume existence of A^{-1}.

6.4.1 Lemma. *Assume that A satisfies (6.3.1) for some $C > 0$. Then $R(\lambda)$ exists in the sector*

$$\Sigma_{+}(\varphi-) = \{\lambda; |\arg\lambda| < \varphi, \lambda \ne 0\},$$

where $\varphi = \arcsin(1/C)$, and for every φ', $0 < \varphi' < \varphi$ there exists a constant C' such that

$$\|R(\lambda)\| \le C'/|\lambda| \quad (\lambda \in \Sigma_{+}(\varphi')). \quad (6.4.1)$$

The proof is rather similar to that of Lemma 4.2.3. The power series of $R(\lambda)$ about $\lambda_0 > 0$,

$$R(\lambda) = \sum_{j=0}^{\infty} (\lambda_0 - \lambda)^j R(\lambda_0)^{j+1},$$

converges in $|\lambda - \lambda_0| < \lambda_0/C$ and the union of all these circles is the sector $\Sigma_{+}(\varphi)$. Now, let $\varphi' < \varphi$, $\lambda \in \Sigma_{+}(\varphi')$, $\lambda_0 = |\lambda|/\cos\arg\lambda$. Then we have

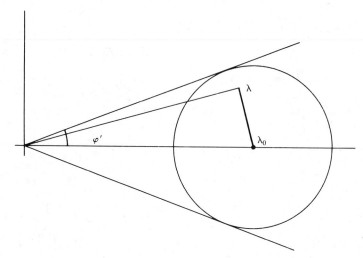

FIGURE 6.4.1

$|\lambda - \lambda_0| \leqslant \lambda_0 \sin \varphi'$, so that

$$\|R(\lambda)\| \leqslant \sum_{j=0}^{\infty} |\lambda - \lambda_0|^j \left(\frac{C}{\lambda_0}\right)^{j+1} = \frac{C}{\lambda_0(1 - C\sin\varphi')}$$

$$= \frac{C''}{\lambda_0} \leqslant \frac{C'}{|\lambda|}$$

as claimed.

We return briefly to the case $0 \in \rho(A)$ in quest of motivation for some of the forthcoming results. Let φ be the angle in Lemma 6.4.1, $\varphi' < \varphi$, and let Γ be a contour contained in $\rho(A)$, coming from infinity along the ray $\arg \lambda = -\varphi'$, $\text{Im } \lambda < 0$, leaving the origin to its right as it passes from this ray to $\arg \lambda = \varphi'$, $\text{Im } \lambda > 0$ and then going off to infinity along this last ray. Fractional powers of A can now be constructed using the functional calculus for unbounded operators (see Section 3); for $\text{Re }\alpha > 0$ we define

$$(-A)^{-\alpha} = e^{i\pi\alpha}A^{-\alpha} = \frac{e^{i\pi\alpha}}{2\pi i}\int_{\Gamma} \lambda^{-\alpha}R(\lambda)\, d\lambda. \qquad (6.4.2)$$

According to the rules pertaining to the functional calculus, the function $f(\lambda) = \lambda^{-\alpha}$ must be holomorphic in $\sigma(A)$; we take then in (6.4.2) the branch of $\lambda^{-\alpha}$, which is real for λ and α real and is discontinuous along the positive real axis. Taking into account the properties of the functional calculus discussed in Section 3, we obtain

$$(-A)^{-1} = -A^{-1}, \quad (-A)^{-(\alpha+\beta)} = (-A)^{-\alpha}(-A)^{-\beta}. \qquad (6.4.3)$$

We examine the behavior of $(-A)^{-\alpha}$ when $\alpha \to 0$. To this end, taking advantage of the independence with respect to Γ of the integral (6.4.2), we let both legs of the

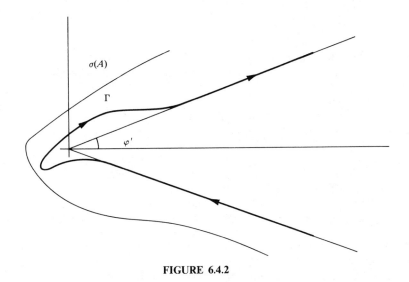

FIGURE 6.4.2

contour Γ collapse into the real axis; after a calculation of $\lambda^{-\alpha}$ above and below $\lambda \geq 0$, we obtain

$$(-A)^{-\alpha} = \frac{\sin \alpha\pi}{\pi} \int_0^\infty \lambda^{-\alpha} R(\lambda)\, d\lambda. \tag{6.4.4}$$

Since $0 \in \rho(A)$, there exists a constant C such that

$$\|R(\lambda)\| \leq C/(\lambda + 1) \quad (0 \leq \lambda < \infty).$$

Consequently, if $|\arg \alpha| \leq \varphi < \pi/2$, $0 < \operatorname{Re} \alpha < 1$, we have

$$\|A^{-\alpha}\| \leq C \frac{|\sin \alpha\pi|}{\pi} \int_0^\infty \frac{\lambda^{-\operatorname{Re}\alpha}}{\lambda + 1}\, d\lambda = C \frac{|\sin \alpha\pi|}{\sin(\operatorname{Re}\alpha)\pi}. \tag{6.4.5}$$

Take now $u \in D(A)$. A manipulation very similar to the one in Lemma 6.3.3 shows that $A^{-\alpha}u \to u$ as $\alpha \to 0$; combining this with (6.4.5), we obtain

$$(-A)^{-\alpha} \to I \tag{6.4.6}$$

strongly as $\alpha \to 0$, if $|\arg \alpha| \leq \varphi < \pi/2$.

Observe next that $(-A)^{-\alpha}$ is a (E)-valued holomorphic function of α in $\operatorname{Re} \alpha > 0$; this follows easily from (6.4.4) and an interchanging of limits and integrals. This implies that all the operators $(-A)^{-\alpha}$, $\operatorname{Re} \alpha > 0$, are one-to-one. In fact, let $u \in E$ be such that $(-A)^{-\alpha}u = 0$ for some α. In view of (6.4.3), $(-A)^{-\beta}u = 0$ for $\operatorname{Re}\beta \geq \operatorname{Re}\alpha$ and, by analyticity, this must actually hold for all β, $\operatorname{Re}\beta > 0$. Making then use of (6.4.6), we obtain $u = 0$.

We prove now that $(-A)^{-\alpha}E$ is dense in E for any α, $\operatorname{Re} \alpha > 0$. To see this, let $n - 1 \leq \operatorname{Re}\alpha < n$. If $u \in D(A^n)$, then $u = (-A)^{-n}v$ for some $v \in E$. Let $w =$

$(-A)^{\alpha-n}v$. Then

$$(-A)^{-\alpha}w = (-A)^{-\alpha}(-A)^{\alpha-n}v = (-A)^{-n}v = u.$$

Since $D(A^n)$ is dense in E our assertion follows.

We complete finally our definition of fractional powers by setting

$$(-A)^\alpha = ((-A)^{-\alpha})^{-1} \tag{6.4.7}$$

for $\operatorname{Re}\alpha > 0$; in view of the preceding observations, $(-A)^\alpha$ is closed and densely defined and all the properties proved in the previous section (e.g., (6.3.28)) follow immediately.

For α near $\operatorname{Re}\alpha = 0$, we define $(-A)^\alpha = (-A)(-A)^{\alpha-1}$; it is easy to see that this definition is consistent with the previous one wherever their ranges of applicability overlap.

We note, finally, that this construction of fractional powers coincides with the more general one developed in the previous section whenever both can be applied; this can be immediately seen comparing (6.4.4) with (6.3.3).

It turns out that the assumption that $0 \in \rho(A)$ leaves aside some important applications and is better omitted. However, the functional calculus just developed in this particular instance is an important heuristic guide even in the general case. For instance, assume that we want to show that $-(-A)^\alpha \in \mathcal{C}_+$ by defining $S(t; -(-A)^\alpha)$ directly in terms of $R(\lambda)$. Clearly, the right definition would be

$$S(t; -(-A)^\alpha) = S(t; -e^{-i\pi\alpha}A^\alpha)$$

$$= \frac{1}{2\pi i}\int_\Gamma \exp(-te^{-i\pi\alpha}\lambda^\alpha)R(\lambda)\,d\lambda \tag{6.4.8}$$

which makes sense, say, if $0 < \alpha < \frac{1}{2}$ (and the angle φ' is small enough). It is not difficult to see that we can collapse the path of integration into the real axis in the same way used to obtain (6.4.4); the result is formula (6.4.9) below, which has a meaning as well when $0 \in \sigma(A)$.

Let α be real. Define an operator-valued function by

$$S_\alpha(\zeta) = \frac{1}{\pi}\int_0^\infty \exp(-\zeta\lambda^\alpha\cos\pi\alpha)\sin(\zeta\lambda^\alpha\sin\pi\alpha)R(\lambda)\,d\lambda, \tag{6.4.9}$$

for ζ complex. In view of (6.4.1) the integrand is $O(|\lambda|^{\alpha-1})$ near the origin, and dies down exponentially at infinity if

$$(\cos\pi\alpha)\operatorname{Re}\zeta > (\sin\pi\alpha)|\operatorname{Im}\zeta|. \tag{6.4.10}$$

If $S_\alpha(\zeta)$ is to be defined at least for ζ real, we must have

$$0 \leqslant \alpha < \frac{1}{2}. \tag{6.4.11}$$

On the other hand, if (6.4.11) is verified, inequality (6.4.10) will hold in the sector $|\arg\zeta| < \psi$, where

$$\psi = \operatorname{arc\,tg}(\cos\pi\alpha/\sin\pi\alpha) = \pi(\tfrac{1}{2} - \alpha). \tag{6.4.12}$$

The most important case in applications, however, is $\alpha = \frac{1}{2}$. Here (6.4.9) becomes

$$S_{1/2}(\zeta) = \frac{1}{\pi}\int_0^\infty \sin\zeta\lambda^{1/2}R(\lambda)\,d\lambda, \tag{6.4.13}$$

but the integral is divergent at infinity even if $\zeta = t$ is real. To overcome this difficulty we modify (6.4.13) by an integration by parts as follows:

$$S_{1/2}(t) = \int_0^\infty h(t, \lambda) R(\lambda)^2 \, d\lambda \quad (t > 0), \tag{6.4.14}$$

where

$$h(t, \lambda) = \frac{1}{\pi} \int_0^\lambda \sin t\sigma^{1/2} \, d\sigma$$

$$= \frac{2}{\pi t^2} (\sin \lambda^{1/2} t - t\lambda^{1/2}\cos \lambda^{1/2} t),$$

and it is easy to see that, if $\delta > 0$ there exists a constant $C = C_\delta$ such that

$$|h(t, \lambda)| \leqslant C\lambda^{1/2} \quad (\lambda \geqslant 0), \quad |h(t, \lambda)| \leqslant Ct\lambda^{3/2} \quad (0 \leqslant \lambda \leqslant 1) \tag{6.4.15}$$

for $t \geqslant \delta$. This obviously implies that

$$\|S_{1/2}(t)\| \leqslant C + C't \quad (t \geqslant \delta \geqslant 0), \tag{6.4.16}$$

where C, C' may in principle depend on δ. In case $0 \in \rho(A)$, we may use the first estimate (6.4.15) only, thus $C' = 0$ in (6.4.6).

Taking advantage of the estimates (6.4.15), continuity of h with respect to t and the dominated convergence theorem in (6.4.14), we can prove continuity of $S_{1/2}(t)$ in $t > 0$ in the norm of (E). On the other hand, if $u \in D(A)$, we have

$$S_{1/2}(t)u - u = \lim_{N \to \infty} \frac{1}{\pi} \int_0^N \sin t\lambda^{1/2} \left(R(\lambda)u - \frac{1}{\lambda}u \right) d\lambda$$

$$= \frac{1}{\pi} \int_0^1 \sin t\lambda^{1/2} \left(R(\lambda)u - \frac{1}{\lambda}u \right) d\lambda$$

$$+ \frac{1}{\pi} \int_1^\infty \frac{1}{\lambda} \sin t\lambda^{1/2} R(\lambda) Au \, d\lambda \to 0 \quad \text{as } t \to 0. \tag{6.4.17}$$

Although $S_{1/2}(\cdot)$ is actually strongly continuous at the origin, we will not prove this directly, as a much stronger result follows from estimates on the resolvent of $-(-A)^{1/2}$ and the theory in Chapter 4.

We proceed to compute the resolvent. Let μ be a nonzero complex number in the sector $(\pi - \varphi)/2 < \arg \mu \leqslant \pi/2$ so that $-\mu^2$ belongs to the sector $\Sigma_+(\varphi -)$ in the statement of Lemma 6.4.1. Write

$$Q(\mu) = -\left(\mu I - (-A)^{1/2} \right) R(-\mu^2). \tag{6.4.18}$$

Since $D((-A)^{1/2}) \supseteq D(A)$, $Q(\mu)$ is everywhere defined and bounded. It was observed in the proof of Lemma 6.3.1 that $R(\lambda)$ and K_α commute for any α $(0 < \alpha < 1)$ and this implies that $R(\lambda)$ and $(-A)^\alpha = \overline{K}_\alpha$ commute as well; hence, if $u \in D((-A)^{1/2})$,

$$Q(\mu)u = -R(-\mu^2)\left(\mu I - (-A)^{1/2} \right)u. \tag{6.4.19}$$

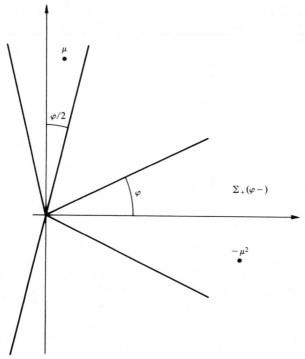

FIGURE 6.4.3

Let now $u \in D(((-A)^{1/2})^2)$. In view of the preceding observations, $R(-\mu^2)u \in D(((-A)^{1/2})^2)$ and

$$\left(\mu I + (-A)^{1/2}\right)Q(\mu)u = \left(-\mu^2 I - A\right)R\left(-\mu^2\right)u = u \quad (6.4.20)$$

since $((-A)^{1/2})^2 \subseteq -A$ (see (6.3.14)). On the other hand, $D(((-A)^{1/2})^2) \supseteq D(A^2)$ is dense in E; this and the facts that $Q(\mu)$ is bounded and $(-A)^{1/2}$ closed imply that $Q(\mu)E \subseteq D((-A)^{1/2})$ and

$$\left(\mu I + (-A)^{1/2}\right)Q(\mu) = I \quad (6.4.21)$$

so that $\mu I + (-A)^{1/2}$ is *onto*. Assume there exists $u \in D((-A)^{1/2})$ such that $(\mu I - (-A)^{1/2})u = 0$; we obtain from (6.4.19) that $Q(\mu)u = 0$, and then from (6.4.21) that $u = 0$, thus proving that $\mu I - (-A)^{1/2}$ is *one-to-one*.

We can repeat the entire argument replacing μ by $-\mu$, obtaining similar results for $\mu I - (-A)^{1/2}$ and $\mu I + (-A)^{1/2}$. It follows in particular that $\mu I + (-A)^{1/2}$ is invertible, and, in view of (6.4.21),

$$Q(\mu) = \left(\mu I + (-A)^{1/2}\right)^{-1} = R\left(\mu; -(-A)^{1/2}\right).$$

We try now to find a different formula for $R(\mu; -(-A)^{1/2})$. Making use of

(6.4.18) and (6.3.3), we obtain

$$R\left(\mu; -(-A)^{1/2}\right) = -\mu R(-\mu^2) - \frac{1}{\pi}\int_0^\infty \lambda^{-1/2}R(\lambda)AR(-\mu^2)\,d\lambda.$$

Using the second resolvent equation in the form

$$R(\lambda)AR(-\mu^2) = -(\lambda+\mu^2)^{-1}(\lambda R(\lambda)+\mu^2 R(-\mu^2))$$

and an elementary integration by residues, the formula

$$R\left(\mu; -(-A)^{1/2}\right) = \frac{1}{\pi}\int_0^\infty \frac{\lambda^{1/2}}{\lambda+\mu^2}R(\lambda)\,d\lambda \qquad (6.4.22)$$

results. Since the right-hand side of (6.4.22) can be extended analytically to $\operatorname{Re}\mu > 0$ from its original domain of definition, we obtain by an application of Lemma 3.6 that every μ in the right half plane belongs to $\rho(-(-A)^{1/2})$ and that (6.4.22) holds in this larger domain. The region of existence of $R(\mu; -(-A)^{1/2})$ can be further enlarged as follows. Let φ be the angle in Lemma 6.4.1, $0 < \varphi' < \varphi$; in view of (6.4.1) the domain of integration in (6.4.22) can be deformed into the ray $\arg\lambda = -\varphi'$, $\operatorname{Re}\lambda > 0$, thus extending analytically the right-hand side to the slanted half plane $\operatorname{Re}(\mu e^{i\varphi'/2}) > 0$; since $\varphi' < \varphi$ is arbitrary, an extension to $\operatorname{Re}(\mu e^{i\varphi/2}) > 0$ is obtained. A similar argument involving rays $\arg\lambda = \varphi'$, $\operatorname{Re}\lambda \geqslant 0$ takes care of the region $\operatorname{Re}(\mu e^{-i\varphi/2}) < 0$. Finally, an estimation of the integrals yields

$$\left\|R\left(\mu; -(-A)^{1/2}\right)\right\| \leqslant C/|\mu| \quad (\mu\in\Sigma_+((\pi+\varphi')/2)) \qquad (6.4.23)$$

for every φ', $0 < \varphi' < \varphi$, where C may depend on φ'. Details are very similar to those in the proof of Theorem 4.2.1 and are thus omitted.

We apply now Theorem 4.2.1 itself: according to it, $-(-A)^{1/2}\in \mathcal{C}((\varphi/2)-)$. Let $S(t; -(-A)^{1/2})$ be, in the usual notation, the semigroup generated by $-(-A)^{1/2}$. We wish to prove that

$$S\left(t; -(-A)^{1/2}\right) = S_{1/2}(t) \quad (t\geqslant 0). \qquad (6.4.24)$$

In order to do this, we take $u\in D(A)$ and make use of (6.4.16) and (6.4.17) (which implies that $S_{1/2}(t)u$ is continuous in $t\geqslant 0$.) An easily justifiable interchange of order of integration shows that for any $\mu > 0$,

$$\int_0^\infty e^{-\mu t}S_{1/2}(t)u\,dt = \frac{1}{\mu}u + \frac{1}{\pi}\int_0^\infty\left(\int_0^\infty e^{-\mu t}\sin t\lambda^{1/2}\,dt\right)\left(R(\lambda)u - \frac{1}{\lambda}u\right)d\lambda$$

$$= \frac{1}{\pi}\int_0^\infty \frac{\lambda^{1/2}}{\lambda+\mu^2}R(\lambda)u\,d\lambda$$

$$= R\left(\mu; -(-A)^{1/2}\right)u. \qquad (6.4.25)$$

Since the same equality must hold for $S(t; -(-A)^{1/2})u$, we obtain that

$S(t; -(-A)^{1/2})u$ and $S_{1/2}(t)u$ coincide for $u \in D(A)$ for $t > 0$. As both operators are continuous ($S_{1/2}(t)u$ because of the comments following (6.4.17)), we obtain (6.4.24) if we define $S_{1/2}(0) = I$, using uniqueness of Laplace transforms and denseness of $D(A)$.

We collect our results.

6.4.2   Theorem.   Let A satisfy (6.3.1). Then $-(-A)^{1/2} \in \mathcal{C}((\varphi/2)-)$, where $\varphi = \arcsin(1/C)$. If $0 \in \rho(A)$, we have

$$\|S(t; -(-A)^{1/2})\| \leqslant C' \quad (t \geqslant 0). \tag{6.4.26}$$

Equality (6.3.14) can be improved for $\alpha = \frac{1}{2}$ to:

$$((-A)^{1/2})^2 = (-A)^{1/2}(-A)^{1/2} = -A. \tag{6.4.27}$$

In fact, if μ is a complex number such that $-\mu^2 \in \rho(A)$, we have seen that both μ and $-\mu$ belong to $\rho(-(-A)^{1/2})$; thus

$$D(((-A)^{1/2})^2) = R(\mu; -(-A)^{1/2}) R(-\mu; -(-A)^{1/2}) E$$
$$= -R(-\mu^2; A) E = D(A).$$

This equality, combined with (6.3.14), implies (6.4.29).

6.4.3   Example.   Let A be dissipative (so that A is m-dissipative by virtue of Assumption 6.3.1). Then $-(-A)^{1/2}$ is m-dissipative.

6.4.4   Example.   Let E be a Banach lattice (see Section 3.7) and assume that A is dispersive with respect to a (any) proper duality map θ. Then the same is true of $-(-A)^{1/2}$.

6.5.   AN APPLICATION: THE INCOMPLETE CAUCHY PROBLEM

It was pointed out in Section 6.1 that if A is a self-adjoint operator in a Hilbert space such that $-A$ is positive definite (see Example 6.1.5), then for every $u \in D((-A)^{1/2})$ there exists a unique solution of

$$u''(t) + Au(t) = 0, \tag{6.5.1}$$

$$u(0) = u, \quad \sup_{t \geqslant 0} \|u(t)\| < \infty, \tag{6.5.2}$$

which is given by $u(t) = \exp(-t(-A)^{1/2})u$. We prove here a partial generalization of that result with the help of the theory of fractional powers developed in Sections 6.3 and 6.4.

Let A be an invertible operator satisfying the assumptions in Section 6.3; precisely, let $[0, \infty) \subseteq \rho(A)$ and

$$\|R(\lambda)\| \leqslant C/\lambda \quad (\lambda > 0). \tag{6.5.3}$$

Take $u \in D(((-A)^{1/2})^2) = D(A)$ (see (6.4.27)) and let

$$u(t) = S_{1/2}(t)u, \tag{6.5.4}$$

where $S_{1/2}(\cdot)$ is the semigroup defined in Section 6.4. In view of (6.4.16), $u(\cdot)$ is a solution of (6.5.1), (6.5.2). Moreover, the solution is unique. To prove this, let $v(\cdot)$ be an arbitrary bounded solution of (6.5.1) such that $v(0) = 0$, and define $w(t) = R(\mu)v(t)$ for some $\mu > 0$ fixed. Then $w(t)$ is bounded in $t \geqslant 0$ together with $w''(t) = -Aw(t) = -AR(\mu)v(t)$ and $(-A)^{1/2}w(t) = (-A)^{1/2}R(\mu)v(t)$; clearly, $\|w'(t)\| = O(t)$ as $t \to \infty$. Define $z(t) = w'(t) + (-A)^{1/2}w(t)$. Then it is easy to see that $z(t)$ is a genuine solution of

$$z'(t) = (-A)^{1/2}z(t) \quad (t \geqslant 0), \tag{6.5.5}$$

where $z(0) = w_1 = w'(0)$ and $\|z(t)\| = O(t)$, $\|z'(t)\| = O(t)$ as $t \to \infty$ in view of the preceding estimates on $w(\cdot)$. Define

$$\hat{z}(\lambda) = \int_0^\infty e^{\lambda t}z(t)\,dt \quad (\operatorname{Re}\lambda < 0). \tag{6.5.6}$$

The function $\hat{z}$ is obviously analytic in the left half plane. Integrating by parts and making use of (6.5.5) and of the bounds on z, z', it is easy to see that $\hat{z}(\lambda) \in D((-A)^{1/2})$ and

$$\left(\lambda I + (-A)^{1/2}\right)\hat{z}(\lambda) = -w_1 \quad (\operatorname{Re}\lambda < 0). \tag{6.5.7}$$

Hence the function $z(\lambda)$ is an analytic continuation of $-R(\lambda; -(-A)^{1/2})w_1$ (which is defined in $\Sigma_+(\frac{1}{2}(\pi + \varphi)-)$) to the whole complex plane minus the origin. In view of inequality (6.4.23), this continuation may have at most a pole of order 1 at $\lambda = 0$, so that we have

$$R\left(\lambda; -(-A)^{1/2}\right)w_1 = f(\lambda) + u/\lambda \quad (\lambda \neq 0),$$

where f is an entire E-valued function and $u \in E$. By virtue of (6.4.23) and of the inequality

$$\|\hat{z}(\lambda)\| \leqslant C/|\operatorname{Re}\lambda| \quad (\operatorname{Re}\lambda < 0),$$

which follows immediately from (6.5.6), we see that f dies down at infinity (hence vanishes identically), thus $R(\lambda; -(-A)^{1/2})w_1 = u/\lambda$. Making use of the definition of resolvent, we deduce that

$$\left(\lambda I + (-A)^{1/2}\right)u/\lambda = u + (-A)^{1/2}u/\lambda = w_1$$

obtaining that $(-A)^{1/2}u = 0$ and $u = w_1$, hence $(-A)^{1/2}w_1 = 0$. It follows that

$$S_{1/2}(t)w_1 = w_1 \quad (t \geqslant 0). \tag{6.5.8}$$

We consider now the function $y(t) = w'(t) - (-A)^{1/2}w(t)$. Since (as we easily verify) y is a genuine solution of $y'(t) = -(-A)^{1/2}y(t)$ in $t \geqslant 0$,

$y(0) = w_1$, we obtain

$$y(t) = w'(t) - (-A)^{1/2}w(t) = S_{1/2}(t)w_1 = w_1 \quad (t \geqslant 0). \quad (6.5.9)$$

Applying $S_{1/2}(t)$ to both sides,

$$\left(S_{1/2}(t)w(t)\right)' = w_1 \quad (t \geqslant 0),$$

hence

$$S_{1/2}(t)w(t) = tw_1, \quad (6.5.10)$$

keeping in mind that $w(0) = 0$. Since the left-hand side of (6.5.10) must be bounded, $w_1 = 0$, thus $S_{1/2}(t)w(t) = 0$ in $t \geqslant 0$. By virtue of Remark 4.2.2, each $S_{1/2}(t)$ is one-to-one and we finally obtain that $w(t) = 0$, hence $v(t) = 0$ in $t \geqslant 0$. This proves uniqueness of solutions of (6.5.1), (6.5.2).

We collect the results obtained.

6.5.1 Theorem. *Let A be a densely defined, invertible operator satisfying (6.5.3), and let $u \in D(A)$. Then there exists a unique solution $u(\cdot)$ of (6.5.1), (6.5.2). This solution is given by the formula*

$$u(t) = S_{1/2}(t)u, \quad (6.5.11)$$

where $S_{1/2}$ is a strongly continuous, uniformly bounded semigroup analytic in the sector $\Sigma_+((\varphi/2)-)$, $\varphi = \arcsin(1/C)$ (C the constant in (6.5.3). The generator of $S_{1/2}$ is the operator $-(-A)^{1/2}$.

We note that Theorem 6.5.1 obviously implies that the incomplete Cauchy problem (6.5.1), (6.5.2) is well posed as defined in Section 6.1: the solution of this problem satisfies

$$\|u(t)\| \leqslant C\|u\| \quad (t \geqslant 0), \quad (6.5.12)$$

where C does not depend on u.

6.6. MISCELLANEOUS COMMENTS

Properly and improperly posed problems have been commented on in Sections 1.7 and 6.1. Although the importance of problems that are improperly posed in one sense or other began to be realized in the forties (see, for instance, Tikhonov [1944: 1]), substantial progress in their treatment had to wait until the late fifties. Thus we read in Courant-Hilbert [1962: 1, p. 230]: "Nonlinear phenomena, quantum theory, and the advent of powerful numerical methods have shown that "properly posed" problems are by far not the only ones which appropriately reflect real phenomena. So far, unfortunately, little mathematical progress has been made in the important task of solving or even identifying and formulating such problems that are not "properly posed" but still are important and motivated by realistic situations." Since then, a true explosion of research activity has occurred, as attested by the size of the bibliography in Payne's monograph [1975: 1], which is almost exclusively devoted to several types of improperly posed Cauchy problems.

With the possible exception of the reversed Cauchy problem in Section 6.1, none of the problems and examples in this chapter would probably be recognized by specialists as improperly posed, but rather as properly posed problems whose correct formulation is somewhat nonstandard. We refer the reader to Payne [1975: 1] for examples, results and general information on ill-posed Cauchy problems, and we limit ourselves to a few comments on the material in this chapter.

(a) *Section 6.2.* Theorem 6.2.1, as well as the definition of problems that are properly posed for bounded solutions, are due to Kreĭn and Prozorovskaya [1960: 1]. The fact that an improperly posed problem may be "stabilized" by a priori restrictions on its solutions was already realized by Tikhonov [1944: 1]. The reversed Cauchy problem for the heat equation with a priori bounds on the solutions was considered for the first time by John [1955: 1] (where the bound takes the form of a nonnegativity condition). The Cauchy problem for elliptic equations was studied in the same vein by John [1955: 2], Pucci [1955: 1], [1958: 1], and Lavrentiev [1956: 1], [1957: 1].

The inequalities leading to Theorem 6.2.1 can be considerably sharpened under more precise assumptions on the operator A. For instance, we have

6.6.1 Example. Let A be a symmetric operator in a Hilbert space H, $u(\cdot)$ a solution of

$$u'(t) = Au(t) \quad (0 \leqslant t \leqslant T).$$

Then $\log \|u(t)\|$ is convex in $0 \leqslant t \leqslant T$; in particular,

$$\|u(t)\| \leqslant \|u(0)\|^{1-t/T}\|u(T)\|^{t/T} \quad (0 \leqslant t \leqslant T). \qquad (6.6.1)$$

To see this assume first that $u(t) \neq 0$ in $0 \leqslant t \leqslant T$. Then

$$\left(\log\|u(t)\|^2\right)'' = 2\{\|u(t)\|^{-2}(Au(t), u(t))\}'$$

$$= -4\|u(t)\|^{-4}(Au(t), u(t))^2 + 4\|u\|^{-2}\|Au(t)\|^2,$$

which is nonnegative in view of the Schwarz inequality. To show (6.6.1) in general it suffices to prove that $u(t)$ vanishes identically if $u(T) = 0$. If this is not true, we may obtain by translation an interval $0 \leqslant t \leqslant h$, where $u(h) = 0$, $u(t) \neq 0$ for $0 \leqslant t < h$, a contradiction in view of (6.6.1) applied in $[0, h']$, $h' < h$.

Similar *logarithmic convexity* results have been obtained for various functions $V(u(t))$ of solutions of various classes of differential equations.

For some references on the subject see Knops-Payne [1971: 1], Levine [1970: 1], [1970: 2], [1972: 1], [1973: 1], [1973: 2], Agmon [1966: 1], and Agmon-Nirenberg [1963: 1], [1967: 1]. An overview of the subject and numerous other references can be found in Payne [1975: 1].

As general references in the field of ill-posed problems (some of them not connected with the Cauchy problem), see Lavrentiev [1967: 1], Lavrentiev-Romanov-Vasiliev [1970: 1], Tikhonov-Arsenin [1977: 1], and Morozov [1973: 1].

(b) *Sections 6.3, 6.4, 6.5.* The definition of fractional powers presented here is due to Balakrishnan [1959: 1], [1960: 1]. It generalizes earlier definitions of Bochner [1949: 1] and Phillips [1952: 1], where $A \in \mathcal{C}_+(C,0)$ for some C (so that $\rho(A) \supset (0, \infty)$ and $\|R(\lambda)^n\| \leqslant C\lambda^{-n}$ for $\lambda > 0$). Here the semigroups generated by $-(-A)^\alpha$ can be explicitly written in terms of the semigroup generated by A by means of the formula

$$\exp\left(-t(-A)^\alpha\right)u = \int_0^\infty f_\alpha(t, s)\exp(sA)u\,ds, \qquad (6.6.2)$$

where $f_\alpha(t, s)$ is the function that makes (6.6.2) an identity in the scalar case:

$$f_\alpha(t, s) = \frac{1}{2\pi i}\int_{\omega - i\infty}^{\omega + i\infty} e^{\lambda s}e^{-t\lambda^\alpha}\,d\lambda. \qquad (6.6.3)$$

Here $\omega > 0$ and the branch of λ^α is chosen in such a way that λ^α is analytic in $\mathrm{Re}\,\lambda > 0$ and $\mathrm{Re}\,\lambda^\alpha > 0$ there. See Yosida [1978: 1] for justification of (6.6.2) and additional details.

The particular case of Balakrishnan's definition where $A^{-1} \in (E)$ was discovered independently by Krasnosel'skiĭ and Sobolevskiĭ [1959: 1]. A vast literature on diverse properties of fractional powers and on their applications to the abstract Cauchy problem exists. See Komatsu [1966: 1], [1967: 1], [1969: 1] [1969: 2], [1970: 1], [1972: 1], Kozlov [1958: 1], Guzmán [1976: 1], Hovel-Westphal [1972: 1], Kato [1960: 1], [1961: 2], [1961: 3], [1962: 1], Sobolevskiĭ [1960: 1], [1961: 2], [1964: 5], [1966: 2], [1967: 1], [1967: 3], [1972: 1], [1977: 1], Kalugina [1977: 1], Staffney [1976: 1], Volkov [1973: 1], Watanabe [1977: 1], Westphal [1968: 1], [1970: 1], [1970: 2], [1974: 1], [1974: 2], Yoshikawa [1971: 1], Yoshinaga [1971: 1], Emami-Rad [1975: 1], Balabane [1976: 1], Sloan [1975: 1], Krasnosel'skiĭ-Sobolevskiĭ [1962: 1], [1964: 1], Krasnosel'skiĭ-Kreĭn [1957: 1], [1964: 1], Langer [1962: 1], Nollau [1975: 1], and Lions [1962: 1]. For general information on fractional powers and additional references see the treatises of Kreĭn [1967: 1], Tanabe [1979: 1], and Yosida [1978: 1]. We note that the theory of fractional powers is a particular instance of an *operational calculus*, where one tries to define $f(A)$ for a class of functions wide enough while conserving (inasmuch as possible) desirable properties like $(fg)(A) =$

$f(A)g(A)$. An instance of an operational calculus for a fairly arbitrary unbounded operator A was sketched in Example 3.12: as seen in Section 6.4, this calculus is essentially sufficient to define fractional powers of certain operators (but not of those satisfying only (6.3.1)). For information on operational calculi, see Hille-Phillips [1957: 1], Dunford-Schwartz [1958: 1], Balakrishnan [1959: 1], Kantorovitz [1970: 1], Faraut [1970: 1], and Hirsch [1972: 2]. In the Hilbert space case, advantage can be taken of the "reduction to self-adjoint operators" outlined in Section 3.8(b)): see Sz.-Nagy-Foias [1969: 1].

In view of some of the applications, it is important to compute the fractional powers of ordinary and partial differential operators, or at least to identify their domains. This is no easy task, and the results are sometimes less than intuitively obvious.

6.6.2 Example (Yosida [1978: 1]). Let $E = L^2(-\infty, \infty)$, $Au = u''$ with domain consisting of all $u \in E$ such that u'' (taken in the sense of distributions) belongs to E. Then A satisfies (6.3.1) (in fact, it belongs to $\mathcal{C}_+(1, 0)$). The operator $-(-A)^{1/2}$ is the singular integral operator

$$-(-A)^{1/2}u(x) = \lim_{h \to 0+} \frac{1}{\pi} \int_{-\infty}^{\infty} \frac{u(x - \xi) - u(x)}{\xi^2 + h^2} d\xi \qquad (6.6.4)$$

with domain $D((-A)^{1/2})$ consisting of all $u \in E$ such that the limit exists in the L^2 sense (and not the first derivative operator $Bu = iu'$ as one may expect). However, we do have

$$D\big((-A)^{1/2}\big) = D(B).$$

To prove these statements we do not need the general theory of fractional powers since A is a self-adjoint operator. Through the Fourier-Plancherel transform, A is equivalent to the multiplication operator $\tilde{A}u(\sigma) = -\sigma^2 u(\sigma)$ in L_σ^2, thus $(-\tilde{A})^{1/2}u(\sigma) = |\sigma|u(\sigma)$ and formula (6.6.4) is an immediate consequence of the fact that

$$\int_{-\infty}^{\infty} \frac{e^{i\sigma x}}{x^2 + h^2} dx = \frac{\pi}{h} e^{-h|\sigma|}.$$

We note that

$$B^2 = -A$$

and the "discrepancy" between $(-A)^{1/2}$ and B is in fact a consequence of the fact that if a is a real number, $\sqrt{a^2}$ equals $|a|$ and not a. In fact, we have (in the sense of the functional calculus for self-adjoint operators)

$$(-A)^{1/2} = |B|.$$

For results on the identification of fractional powers of differential operators see the treatise of Triebel [1978: 1] and bibliography therein, Evzerov

[1977: 1], Evzerov-Sobolevskiĭ [1973: 1], [1976: 1], Krasnosel'skiĭ-Pustyl'nik-Zabreĭko [1965: 1], [1965: 2], Segal-Goodman [1965: 1], Masuda [1972: 2], Murata [1973: 1], Sobolevskiĭ [1972: 1], [1977: 1], and Koledov [1976: 1].

(c) *Section 6.5.* The results here are due to Balakrishnan [1960: 1]. Generalizations to the higher-order equation

$$u^{(n)}(t) = Au(t) \tag{6.6.5}$$

were carried out by Radnitz [1970: 1] and the author and Radnitz [1971: 2]. The problem here is that of existence and uniqueness of solutions of (6.6.5) that satisfy the estimate

$$\|u(t)\| = O(e^{\omega t}) \quad \text{as } t \to \infty \tag{6.6.6}$$

and the incomplete initial conditions

$$u^{(k)}(0) = u_k \in E \quad (k \in \alpha), \tag{6.6.7}$$

where α is a predetermined subset of the set $\langle 0, 1, \ldots, n-1 \rangle$; also, the dependence of the solution on the initial data (6.6.7) is examined. The author's paper [1973: 1] proceeds along similar lines, but the selection of the solution of (6.6.5) satisfying the incomplete initial conditions (6.6.7) is made not necessarily on the basis of the growth condition (6.6.6) but in an unspecified way producing existence, uniqueness, and linear and continuous dependence on the initial data. In a sense, these investigations are related to Hille's results on the higher-order Cauchy problem (see 2.5(c)) and with results for particular higher-order equations due to Lions [1961: 1].

A problem of a somewhat similar sort is the *boundary value problem* for a higher order equation

$$u^{(n)}(t) + A_1(t)u^{(n-1)}(t) + \cdots + A_{n-1}(t)u'(t) + A_n(t)u(t) = f(t). \tag{6.6.8}$$

Here, solutions are sought in a finite interval (say, $0 \leqslant t \leqslant T$) and boundary conditions are given at each endpoint:

$$u^{(k)}(0) = u_{k,0} \quad (k \in \alpha_0), \quad u^{(k)}(T) = u_{k,T} \quad (k \in \alpha_T), \tag{6.6.9}$$

where α_0, α_T are fixed subsets of $\langle 0, 1, \ldots, n-1 \rangle$. (We may also have "mixed" boundary conditions involving both extremes.) We seek conditions on the A_j, α_0, α_T, which make the problem *properly posed* in the sense that existence, uniqueness, and continuous dependence on the boundary data and on f hold. To give an idea of the difficulties involved, consider the following

6.6.3 Example. Let $E = L^2(0, \pi)$, $Au = u''$ with domain $D(A)$ consisting of all $u \in E$ with $u'' \in E$ and $u(0) = u(\pi) = 0$. Any solution $u(t)(x) = u(x, t)$ of the abstract differential equation

$$u''(t) = Au(t) \tag{6.6.10}$$

can be written in the form

$$u(x,t) = \sum_{n=1}^{\infty} (a_n\cos nt + b_n\sin nt)\sin nx.$$

Assume we impose the initial and final conditions

$$u(x,0) = 0, \quad u(x,T) = u(x) = \sum_{n=1}^{\infty} c_n\sin nx \in E. \qquad (6.6.11)$$

Then $a_n = 0$ and

$$b_n = c_n/\sin nT \quad (n \geqslant 1). \qquad (6.6.12)$$

If $\sin nT = 0$ for some n (that is, if T/π is rational), condition (6.6.12) forces c_n to be zero, thus the subspace D of final data for which the problem has solutions is not dense in E. On the other hand, if T/π is irrational, D includes all u with finite Fourier series and is thus dense in E. However, in order to obtain a "continuous dependence" estimate of the type

$$\|u(t)\| \leqslant C\|u(T)\| \quad (0 \leqslant t \leqslant T), \qquad (6.6.13)$$

we must have $|\sin nT| \geqslant \varepsilon > 0$ for all $n \geqslant 1$, in other words

$$(nT) \geqslant \delta > 0 \quad (n \geqslant 1), \qquad (6.6.14)$$

where (nT) indicates the distance from nT to the nearest point of the form $k\pi$. But it is an immediate consequence of Kronecker's theorem in number theory (Hardy-Wright [1960: 1, p. 376]) that (6.6.14) never happens when T/π is irrational, thus a bound of the type of (6.6.13) has no chance to exist. Since the bound does hold for $u \in D$ when T/π is rational, existence of solutions to the boundary value problem for "sufficiently many" boundary data and continuous dependence clash in an obvious sense. A sort of way out can be attempted by replacing the inequality (6.6.13) by the weaker version

$$\|u(t)\| \leqslant C\|(D_x)^k u(T)\| \quad (0 \leqslant t \leqslant T). \qquad (6.6.15)$$

To obtain this, we only need to show that $|\sin nT| \geqslant \varepsilon n^{-k}$ for all $n \geqslant 1$ for some $\varepsilon > 0$, or, equivalently,

$$(nT) \geqslant \delta n^{-k} \quad (n \geqslant 1) \qquad (6.6.16)$$

for some $\delta > 0$. According to a theorem of Liouville (Hardy-Wright [1960: 1, p. 161]), this is the case if T/π is an algebraic number of degree $k + 1$. The preceding considerations, which are particular cases of those in Bourgin-Duffin [1939: 1] (and which can be easily adapted to the case $u(x,0) \neq 0$) provide an answer to the problem of choosing T in order to make the boundary value problem (6.6.10), (6.6.11) properly posed (although the answer would probably not be appreciated very much by a physicist or engineer, dealing with the very practical problem of "interpolating" the movement of a string, its position known at two times $t = 0$ and $t = T$).

Although no necessary and sufficient conditions on the $A_j(t)$, T and the sets α_0, α_T in order that the boundary value problem be well posed are apparently known, even in the particular case where the A_j do not depend on time, numerous results in both directions exist. Necessary conditions regarding the number of boundary conditions and their distribution between the two endpoints were found by the author [1974: 1] for the equation

$$u^{(n)}(t) = Au(t) \qquad (6.6.17)$$

using a modification of an idea of Radnitz [1970: 1] to obtain the resolvent of A in terms of the solutions of (6.6.17), (6.6.9). The following references deal mostly with *sufficient* conditions in order that the boundary value problem (6.6.8), (6.6.9) be properly posed in various senses; most of the material is on second-order equations with one boundary condition at each endpoint, or with mixed boundary conditions involving both endpoints. See the treatise of Kreĭn [1967: 1] and Kreĭn-Laptev [1962: 1], [1966: 1], [1966: 2], Favini [1975: 2], Gorbačuk-Kočubeĭ [1971: 1], Gorbačuk-Gorbačuk [1976: 1], Hönig [1973: 1], Jakubov [1973: 1], Jurčuc [1974: 1], [1976: 1], [1976: 2], Kislov [1972: 1], [1975: 1], Kutovoĭ [1976: 1], McNabb [1972: 1], McNabb-Schumitzky [1973: 1], [1974: 1], Mihaĭlec [1974: 1], [1975: 1], Misnaevskiĭ [1976: 1], Orudzev [1976: 1], Pavec [1971: 1], Romanko [1976: 1], [1977: 1], Venni [1975: 1], Laptev [1966: 1], [1968: 1], J. M. Cooper [1971: 1], and Karasik [1976: 1].

Chapter 7

The Abstract Cauchy Problem for Time-Dependent Equations

Many deep and useful results for the time-dependent equation $u'(t) = A(t)u(t)$ can be obtained by extensions and modifications of the theory pertaining to the equation $u'(t) = Au(t)$. There are two main avenues of approach. In the first (Sections 7.1 to 7.6) one assumes that each $A(t)$ belongs to the class $\mathcal{C}$ introduced in Chapter 4, with a uniform bound on the norm of $R(\lambda; A(t))$. The Cauchy problem is then shown to be properly posed, the definition of solution depending on the smoothness of the function $t \to A(t)^{-1}$. There are applications to the time-dependent version of the parabolic equations considered in Chapter 4.

In the second approach we only assume that each $A(t)$ belongs to $\mathcal{C}_+$; the hypotheses on the function $t \to A(t)$ make possible the constructions of the solutions in "product integral" form, corresponding to approximation of $A(\cdot)$ by a piecewise constant operator-valued function. The natural application here is to time-dependent symmetric hyperbolic equations; many of the auxiliary tools needed were already developed in Section 5.6.

7.1. THE ABSTRACT CAUCHY PROBLEM FOR TIME-DEPENDENT EQUATIONS

Let $0 < T \leqslant \infty$ and $\{A(t); 0 \leqslant t < T\}$ a family of operators in an arbitrary Banach space E. *Strong* or *genuine* solutions of the equation

$$u'(t) = A(t)u(t) \tag{7.1.1}$$

in any interval I contained in $0 \leqslant t < T$ are defined, as in the time-independent case, to be continuously differentiable functions $u(\cdot)$ such that $u(t) \in D(A)$ and (7.1.1) is satisfied everywhere in I.

We shall say that the Cauchy problem for (7.1.1) is *well posed* (or *properly posed*) in $0 \leqslant t < T$ if the following two assumptions are satisfied:

(a) (Existence of solutions for sufficiently many initial data) *There exists a dense subspace D of E such that for every s, $0 \leqslant s < T$ and every $u_0 \in D$ there exists a solution $t \to u(t, s)$ of (7.1.1) in $s \leqslant t < T$ with*

$$u(s, s) = u_0. \qquad (7.1.2)$$

(b) (Continuous dependence of solutions on their initial data) *There exists a strongly continuous (E)-valued function $S(t, s)$ defined in the triangle $0 \leqslant s \leqslant t < T$ such that if $u(t, \cdot)$ is a solution of (7.1.1) in $s \leqslant t < T$, we have*

$$u(t, s) = S(t, s) u(s, s). \qquad (7.1.3)$$

The operator S will be called the *propagator* (or *evolution operator* or *solution operator*) of (7.1.1). Because of the uniform boundedness theorem, (b) implies

$$\|S(t, s)\| \leqslant C \quad (0 \leqslant s \leqslant t \leqslant T') \qquad (7.1.4)$$

for any $T' < T$, so that the label given to (b) is justified. Note that solutions depend continuously not only on their initial data, but also on the time s at which these data are imposed. It is clear that when $A(t) \equiv A$ and $T = \infty$, our definition is equivalent to that in Section 1.2 for the time-independent Cauchy problem. Just as in there, if $0 \leqslant s < T$ and $u_0 \in E$, we call the function

$$u(t) = S(t, s) u_0 \qquad (7.1.5)$$

a *generalized solution* of (7.1.1). We shall discuss later the relation among generalized solutions of (7.1.1) and weak solutions (that is, solutions in the sense of distributions).

Note that (b) implies that D is contained in $D(A(s))$ for any s in $[0, T)$, thus the set

$$D(A(\cdot)) = \bigcap_{0 \leqslant t < T} D(A(t)) \qquad (7.1.6)$$

must be dense in E. This imposition can be avoided if need be by requiring the solution $u(\cdot, s)$ in (a) to satisfy (7.1.1) only in $0 < s < T$ (but, naturally, retaining continuity in $0 \leqslant s < T$ to give sense to the initial condition) or by allowing D to depend on s, in which case $D(s) \subseteq D(A(s))$ and denseness of each $D(s)$ only implies denseness of the domain of each $A(s)$, hardly an extravagant assumption. Unlike in the time-independent case, the definition of solution of (7.1.1) is critical, and several different notions will have to be used below.

The following two equalities are obtained in the same way as (2.1.2):

$$S(s,s) = I \quad (0 \leqslant s < T), \tag{7.1.7}$$

$$S(t,r)S(r,s) = S(t,s) \quad (0 \leqslant s \leqslant r \leqslant t < T). \tag{7.1.8}$$

We note finally that the notion of properly posed Cauchy problem can be formulated equally well in intervals other than $[0, T)$.

The case where the operators $A(t)$ are everywhere defined and depend continuously of t in one way or another is of scarce interest to us for the reasons already pointed out in the time-independent case in Section 1.2. However, the results are considerably less trivial when $A(t)$ is not a constant function and we include one of them, which is an infinite-dimensional extension of a classical theorem on existence and uniqueness of solutions of systems of ordinary differential equations.

7.1.1 Theorem. *Let $A(t)$ be bounded and everywhere defined in $0 \leqslant t < T$; moreover, let $t \to A(t) \in (E)$ be strongly continuous in $0 \leqslant t < T$. Then the Cauchy problem for (7.1.1) is properly posed in $0 \leqslant t < T$ and the function $(s, t) \to S(t, s) \in (E)$ is continuous in $0 \leqslant s \leqslant t < T$.*

Proof. The proof is a standard application of the successive approximation method. We begin by observing that if $u_0 \in E$, the differential equation (7.1.1) and the initial condition (7.1.2) are equivalent to the single integral equation[1]

$$u(t,s) = u_0 + \int_s^t A(\tau) u(\tau, s) \, d\tau. \tag{7.1.9}$$

We take $u_0(t) \equiv u_0$ as initial approximation and define the following ones recursively by

$$u_n(t,s) = u_0 + \int_s^t A(\tau) u_{n-1}(\tau, s) \, d\tau, \quad (n = 1, 2, \ldots). \tag{7.1.10}$$

By the uniform boundedness theorem, if $T' < T$

$$\|A(t)\| \leqslant C \quad (0 \leqslant t \leqslant T'), \tag{7.1.11}$$

which yields, via a simple inductive argument applied to (7.1.10), the

---

[1]If $v(\cdot)$ is a continuous E-valued function, $A(\cdot)v(\cdot)$ is continuous. This is a simple consequence of the strong continuity of $A(\cdot)$ and of the bound (7.1.11). This will be used many times in the sequel.

sequence of inequalities

$$\|u_n(t,s)-u_{n-1}(t,s)\| \leqslant \frac{C^n(t-s)^n}{n!}\|u_0\| \quad (n=1,2,\dots) \quad (7.1.12)$$

in $s \leqslant t \leqslant T'$. Accordingly, the series $\Sigma\|u_n(t,s)-u_{n-1}(t,s)\|$ converges (and thus $u(t,s)=\lim u_n(t,s)$ exists) uniformly on $0 \leqslant s \leqslant t \leqslant T'$. Taking limits in (7.1.10), we see that $u(\cdot,s)$ is a solution of (7.1.9) and then a solution of the initial-value problem (7.1.1), (7.1.2).

To prove uniqueness of $u(t,s)$ and continuity of the solution operator, we make use of the following result (Gronwall's lemma), which we state in a version more general than necessary now, with a view to future use.

7.1.2 Lemma. *Let β be locally integrable and nonnegative in $t \geqslant a$. Assume the nonnegative continuous function $\eta(t)$ satisfies*

$$\eta(t) \leqslant \alpha + \int_a^t \beta(s)\eta(s)\,ds \tag{7.1.13}$$

in $t \geqslant a$, where $\alpha \geqslant 0$. Then

$$\eta(t) \leqslant \alpha\exp\left(\int_a^t \beta(s)\,ds\right) \tag{7.1.14}$$

in $t \geqslant a$.

Proof. We assume first that $\alpha > 0$. It follows from (7.1.13) that

$$\beta(t)\eta(t)\left(\alpha+\int_a^t \beta(s)\eta(s)\,ds\right)^{-1} \leqslant \beta(t) \quad (t \geqslant a).$$

Integrating from a to t gives

$$\log\left(\alpha+\int_a^t \beta(s)\eta(s)\,ds\right)-\log\alpha \leqslant \int_a^t \beta(s)\,ds,$$

whence (7.1.14) follows using again (7.1.13). The case $\alpha = 0$ is dealt with taking $\alpha \to 0+$.

We go back to the uniqueness problem. Let $u(t,s)$ be a solution of (7.1.9) in $s \leqslant t \leqslant T'$. If we set $\eta(t)=\|u(t,s)\|$, we have

$$\eta(t) \leqslant \|u_0\|+C\int_s^t \eta(s)\,ds, \tag{7.1.15}$$

thus it follows from Lemma 7.1.2 that

$$\|u(t,s)\| \leqslant e^{C(t-s)}\|u_0\| \quad (s \leqslant t \leqslant T'), \tag{7.1.16}$$

which shows that $u(t,s) \equiv 0$ if $u_0 = 0$. Since $T' < T$ is arbitrary, existence and uniqueness of the solution is established in $s \leqslant t < T$ for any $s < T$. We note, incidentally, that the arguments leading to existence and uniqueness of the solutions of (7.1.1) work just as well for $t \leqslant s$ as for $t \geqslant s$ so that solutions actually exist in $0 \leqslant t \leqslant T'$ (i.e., the Cauchy problem can be solved not only forward but also backwards). Hence the propagator $S(t, s)$ can be defined in the square $0 \leqslant s, t < T$ rather than only in the triangle $0 \leqslant s \leqslant t < T$ and satisfies in each square $0 \leqslant s \leqslant t \leqslant T'$ an estimate of the form

$$\|S(t, s)\| \leqslant e^{C|t-s|}. \tag{7.1.17}$$

We prove finally that S is continuous in the sense of the norm of (E). To this end, we note that

$$\|S(t', s') - S(t, s)\| \leqslant \|S(t', s') - S(t, s')\| + \|S(t, s') - S(t, s)\| \tag{7.1.18}$$

and proceed to estimate the two terms on the right for $0 \leqslant s \leqslant s' \leqslant T'$ and $0 \leqslant t \leqslant t' \leqslant T'$, where $T' < T$. In view of (7.1.11) and (7.1.17),

$$\|(S(t', s') - S(t, s'))u\| \leqslant \int_t^{t'} \|A(\tau)\| \|S(\tau, s')u\| \, d\tau$$

$$\leqslant C(t' - t)e^{CT'}\|u\| \tag{7.1.19}$$

for any $u \in E$. The second term on the right-hand side of (7.1.18) is also estimated by means of (7.1.11) and (7.1.17). If $u \in E$, we have

$$\|(S(t, s') - S(t, s))u\| \leqslant \int_s^{s'} \|A(\tau)\| \|S(\tau, s)u\| \, ds$$

$$+ \int_s^t \|A(\tau)\| \|(S(\tau, s') - S(\tau, s))u\| \, d\tau$$

$$\leqslant C(s' - s)e^{CT'} + C \int_s^t \|(S(\tau, s') - S(\tau, s))u\| \, ds.$$

Applying Lemma 7.1.2 to $\eta(t) = \|(S(t, s') - S(t, s))u\|$, we obtain

$$\|(S(t, s') - S(t, s))u\| \leqslant C(s' - s)e^{2CT'}\|u\| \tag{7.1.20}$$

for any $u \in E$, which ends the proof of Theorem 7.1.1.

Motivated by some of the results above (see the observations following (7.1.16)), we introduce a definition that will also find application in Section 7.7 and following sections. We say that the Cauchy problem for (7.1.1) is *well posed* (or *properly posed*) *forward and backwards*, or *in both senses of time* if the solutions required in (a) actually exist in $0 \leqslant t < T$, rather than only in $s \leqslant t < T$ and if the operator $S(t, s)$ exists and is strongly continuous not only in the triangle $0 \leqslant s \leqslant t < T$ but in the square $0 \leqslant s, t < T$. This notion can be formulated equally well for an operator-valued function $A(\cdot)$ defined in an arbitrary interval I, not necessarily of the form $[0, T)$ (say $[0, T]$, $(-T, T)$, etc.). It is clear that this definition generalizes to the time-dependent case that of well-posed Cauchy problem in $(-\infty, \infty)$ (see Section 1.5).

When the Cauchy problem is well posed in both senses of time in some interval I, equality (7.1.8) holds in the square $s, t \in I$ and it follows from it and from (7.1.7) that $S(t, s)S(s, t) = S(t, t) = I = S(s, s) = S(s, t)S(t, s)$, or

$$S(s, t)^{-1} = S(t, s) \quad (s, t \in I). \qquad (7.1.21)$$

Each operator $S(t, s)$ is invertible.

7.1.3 Example. Assume the hypotheses on $A(\cdot)$ are those in Theorem 7.1.1. We look now for a (E)-valued function $\tilde{S}(t, s)$, solution of the integral equation.

$$\tilde{S}(t, s)u = u - \int_s^t \tilde{S}(t, \sigma)A(\sigma)u\, d\sigma \quad (u \in E) \qquad (7.1.22)$$

in the square $0 \leqslant s, t < T$. Existence and uniqueness of $\tilde{S}(t, s)$ are proved by means of an iteration scheme similar to the one used in (7.1.9). The function $\tilde{S}(t, s)u$ is continuously differentiable in s for all u,

$$D_s\tilde{S}(t, s)u = -\tilde{S}(t, s)A(s)u \quad (0 \leqslant s < T), \quad \tilde{S}(t, t)u = u, \qquad (7.1.23)$$

and $\tilde{S}$ satisfies an estimate of the type of (7.1.17). Take $u \in E$ and define

$$\varphi(\sigma) = \tilde{S}(t, \sigma)S(\sigma, s)u.$$

Then

$$\varphi'(\sigma) = D_\sigma\tilde{S}(t, \sigma)S(\sigma, s)u + \tilde{S}(t, \sigma)D_\sigma S(\sigma, s)u$$
$$= -\tilde{S}(t, \sigma)A(\sigma)S(\sigma, s)u + \tilde{S}(t, \sigma)A(\sigma)S(\sigma, s)u = 0$$

in $0 \leqslant \sigma < T$, so that φ is constant; hence $\tilde{S}(t, s)u = \varphi(s) = \varphi(t) = S(t, s)u$, which shows that

$$S(t, s) = \tilde{S}(t, s). \qquad (7.1.24)$$

We obtain in this way nontrivial information on S as a function of s. Similar tricks will work in other situations (see Section 7.2).

7.1.4 Example. It was observed in Section 1.5 that in the time-independent case $A(t) \equiv A \in (E)$, we have

$$\|S(t, s) - I\| = \|S(t - s) - I\| \to 0$$

as $t - s \to 0$. A similar result holds if we place ourselves under the hypotheses of Theorem 7.1.1. In fact, it follows immediately from (7.1.19) that

$$\|S(t, s) - I\| \leqslant Ce^{CT'}|t - s| \qquad (7.1.25)$$

for $0 \leqslant s, t \leqslant T'$. It does not seem to be known whether $S(t - s) \to I$ in (E) as $t - s \to 0$ must necessarily imply that the $A(t)$ are bounded in the general case; when A does not depend on time, the corresponding result is Lemma 2.3.5.

Many variants of Theorem 7.1.1 can be obtained under the basic assumption of boundedness of $A(t)$ by simply modifying the requirements on continuity of $t \to A(t)$ (and modifying accordingly the definition of solution employed). One variant in the direction of greater generality is:

7.1.5 Example. Let $A(t)$ be everywhere defined and bounded for almost all t, $0 \leqslant t < T$. Assume that for each u the function $t \to A(t)u$ is strongly measurable and that

$$\|A(t)\| \leqslant \beta(t) \quad (0 \leqslant t < T), \tag{7.1.26}$$

where β is locally integrable. Then the Cauchy problem for (7.1.1) is properly posed forward and backwards if we weaken the definition of solution thus: a continuous function is said to be a *solution* of (7.1.1) if and only if the integral equation (7.1.9) is satisfied, the integral understood in the sense of Bochner. (Note, incidentally, that $A(\tau)u(\tau, s)$ is strongly measurable in τ if u is continuous; this can be seen approximating u by step functions.) Integrability follows from boundedness of u, since $\|A(\tau)u(\tau, s)\| \leqslant C\beta(\tau)$. The proof of Theorem 7.1.1 can be readily adapted to the present setting; we outline the main steps. The iteration scheme to solve (7.1.9) is the same; inequality (7.1.12) becomes

$$\|u_n(t, s) - u_{n-1}(t, s)\| \leqslant \frac{1}{n!} \left(\int_s^t \beta(\tau) \, d\tau \right)^n \|u_0\| \quad (n = 1, 2, \ldots).$$
$$\tag{7.1.27}$$

Inequality (7.1.15) is now

$$\eta(t) \leqslant \|u_0\| + \int_s^t \beta(s) \eta(s) \, ds,$$

which implies, via Lemma 7.1.2, that

$$\|S(t, s)\| \leqslant \exp\left(\int_s^t \beta(\tau) \, d\tau \right). \tag{7.1.28}$$

Continuity of $S(t, s)$ is proved with two inequalities similar to (7.1.19) and (7.1.20). Examples 7.1.3 and 7.1.4 have obvious generalizations to the present situation.

The following result, on the other hand, strengthens the conclusion of Theorem 7.1.1 under stronger assumptions.

7.1.6 Example. If $t \to A(t)$ is continuous (in the norm of (E)), Theorem 7.1.1 holds with the following refinements: the propagator $S(t, s)$ is continuously differentiable (in the norm of (E)) with respect to both variables in the square $0 \leqslant s, t < T$ and

$$D_t S(t, s) = A(t)S(t, s), \quad D_s S(t, s) = -S(t, s)A(s) \tag{7.1.29}$$

there. In fact, $S(t, s)$ is differentiable in the norm of (E) in $0 \leqslant s, t < T$ also under the hypotheses in Theorem 7.1.1 and both equalities (7.1.29) hold (the derivatives of course are not in general continuous in (E)).

7.1.7   Example. Let $A(\cdot)$, $A_0(\cdot)$ be two operator-valued functions satisfying the assumptions in Example 7.1.5 and let

$$\|A(t) - A_0(t)\| \le \varepsilon \quad (0 \le t \le T').$$

If S is the propagator of (7.1.1), S_0 the propagator of $u'(t) = A_0(t)u(t)$, we have

$$\|S(t,s) - S_0(t,s)\| \le \varepsilon T' \exp\left(2 \int_0^{T'} \beta(\tau)\, d\tau\right). \tag{7.1.30}$$

The result follows from subtracting the integral equation (7.1.9) satisfied by S and the corresponding one for S_0; we obtain

$$\left(S(t,s) - S_0(t,s)\right)u = \int_s^t \left(A(\tau) - A_0(\tau)\right)S(\tau,s)u\, d\tau$$

$$+ \int_s^t A_0(\tau)\left(S(\tau,s) - S_0(\tau,s)\right)u\, d\tau$$

so that, if $\eta(t) = \|(S(t,s) - S_0(t,s))u\|$, we have, making use of (7.1.28),

$$\eta(t) \le \varepsilon T'\left(\exp \int_0^{T'} \beta(\tau)\, d\tau\right)\|u\| + \int_s^t \beta(\tau)\eta(\tau)\, d\tau$$

from which (7.1.30) follows using Lemma 7.1.2.

7.1.8   Example. Under the hypotheses in Example 7.1.6, let $0 \le s \le t \le T$ and choose for each $n = 1, 2, \ldots$ a partition $s = t_{n,0} < t_{n,1} < \cdots < t_{n,m(n)} = t$ in such a way that $\delta_n = \sup_j(t_{n,j} - t_{n,j-1}) \to 0$ as $n \to \infty$. Then

$$S(t,s) = \lim_{n \to \infty} \prod_{j=1}^{m(n)} \exp\left((t_{n,j} - t_{n,j-1})A(t_{n,j})\right) \tag{7.1.31}$$

in the norm of (E), where the product is ordered backwards:

$$\prod_{j=1}^{m(n)} \exp\left((t_{n,j} - t_{n,j-1})A(t_{n,j})\right)$$

$$= \exp\left((t_{n,m(n)} - t_{n,m(n)-1})A(t_{n,m(n)})\right) \cdots \exp\left((t_{n,1} - t_{n,0})A(t_{n,1})\right). \tag{7.1.32}$$

The limit is uniform on compacts in the following sense: if $0 \le s, t \le T' < T$, $\|S(t,s) - S_n(t,s)\|$ can be made small with δ_n, independently of s, t or the particular partition chosen. To see this we proceed as follows: having fixed s, t and the partition $t_{n,0}, \ldots, t_{n,m(n)}$, we define $A_n(\cdot)$ thus:

$$A_n(\tau) = \begin{cases} A(\tau) & (0 \le \tau \le s) \\ A(\tau_{n,j}) & (t_{n,j-1} < \tau \le t_{n,j}, \quad j = 1, 2, \ldots, m(n)) \\ A(\tau) & (t \le \tau \le T'). \end{cases}$$

Then

$$\|A(\tau) - A_n(\tau)\| \leqslant \varepsilon \quad (0 \leqslant \tau \leqslant T')$$

uniformly in $0 \leqslant \tau \leqslant T'$ if δ_n is sufficiently small, independently of the particular partition chosen. It follows then from Example 7.1.7 that

$$\|S(t, s) - S_n(t, s)\| \leqslant C\varepsilon,$$

where S_n is the propagator of $u'(t) = A_n(t)u(t)$. But it is easy to see that $S_n(t, s)$ is the operator in (7.1.32), thus the result follows.

7.1.9 Example. Under the hypotheses of Example 7.1.6, assume that $A(t)$ and $\int_s^t A(\tau)\, d\tau$ commute for each s, t. Then

$$S(t, s) = \exp\left(\int_s^t A(\tau)\, d\tau \right). \tag{7.1.33}$$

This formula is also valid with the hypotheses in Theorem 7.1.1 or Example 7.1.5, the integral $\int A(\tau)\, d\tau$ interpreted "elementwise."

7.1.10 Example. Let $f(\cdot)$ be continuous in $t \geqslant s$, and assume the premises of Theorem 7.1.1 hold. Then, if $u_0 \in E$,

$$u(t, s) = S(t, s)u_0 + \int_s^t S(t, \sigma)f(\sigma)\, d\sigma, \tag{7.1.34}$$

is a genuine solution of the inhomogeneous equation

$$u'(t) = A(t)u(t) + f(t) \quad (s \leqslant t < T). \tag{7.1.35}$$

satisfying the initial condition (7.1.2) (genuine solutions are defined in the same way as for the homogeneous equation). It is the only such solution satisfying (7.1.2). Under the hypotheses of Example 7.1.5, if f is locally summable in $t \geqslant s$, formula (7.1.34) furnishes the only solution (in the sense of Example 7.1.5) of (7.1.35), which equals u_0 for $t = s$.

7.1.11 Example. We assume $A(\cdot)$ continuous in the norm of (E) as in Example 7.1.6. Let $t \to F(t)$ a continuous (E)-valued function, $U_0 \in (E)$. Then

$$U(t, s) = S(t, s)U_0 + \int_s^t S(t, \sigma)F(\sigma)\, d\sigma \tag{7.1.36}$$

is the only solution of the initial-value problem

$$D_t U(t, s) = A(t)U(t, s) + F(t), \quad U(t, t) = U_0, \tag{7.1.37}$$

whereas

$$V(t, s) = V_0 S(t, s) - \int_s^t F(\tau)S(\tau, s)\, d\tau \tag{7.1.38}$$

is the only solution of

$$D_s V(t, s) = -V(t, s)A(s) + F(s), \quad V(t, t) = V_0. \tag{7.1.39}$$

The derivatives here are understood in the norm of (E) and continuous in the same norm.

7.1.12 Example. Let $A(\cdot), P(\cdot)$ be strongly continuous in $0 \leqslant t < T$, as in Theorem 7.1.1, and let S be the solution operator of (7.1.1), $\tilde{S}$ the solution operator of

$$u'(t) = (A(t) + P(t))u(t). \tag{7.1.40}$$

Then $\tilde{S}$ can be obtained as the sum of a perturbation series motivated by formula (7.1.34) (see Theorem 5.1.1) as follows. Define $S_0 = S$,

$$S_n(t,s)u = \int_s^t S(t,\sigma)P(\sigma)S_{n-1}(\sigma,s)u\,d\sigma \quad (0 \leqslant s, t < T, n \geqslant 1). \tag{7.1.41}$$

Then

$$\tilde{S}(t,s) = \sum_{n=0}^{\infty} S_n(t,s) \tag{7.1.42}$$

in the same range of s, t. The convergence is uniform in $0 \leqslant s, t \leqslant T'$ for any $T' < T$. To see this we use (7.1.17) and denote by M a bound for $\|P(t)\|$ in $0 \leqslant t \leqslant T'$; a simple inductive argument shows that

$$\|S_n(t,s)\| \leqslant \frac{M^n}{n!}|t-s|^n e^{C|t-s|} \quad (n \geqslant 1) \tag{7.1.43}$$

in $0 \leqslant s, t \leqslant T'$, C the constant in (7.1.17).

7.2. ABSTRACT PARABOLIC EQUATIONS

Many of the results in this chapter on the equation

$$u'(t) = A(t)u(t) \tag{7.2.1}$$

will be obtained, roughly speaking, on the basis of hypotheses of the following type: (1) the Cauchy problem for the "equation with frozen coefficients" $u'(t) = A(s)u(t)$ (s fixed) is properly posed (that is, $A(s) \in \mathcal{C}_+$, $\mathcal{C}$ or a subclass thereof for all s) and (2) the operator function $t \to A(t)$ depends smoothly on t in one way or other (the assumptions in the previous section are clearly of this type).[2]

The methods used in this section and the next are based on the following considerations, which are for the moment purely formal. Let $S(t,s)$ be the propagator of (7.2.1), and let $S(t; A(s))$ be the propagator of the time-invariant equation.

$$u'(t) = A(s)u(t). \tag{7.2.2}$$

---

[2] Since $A(t)$ is in general unbounded, "smooth dependence" in this section may mean smoothness of the function $t \to A(t)^{-1}$ or of other related functions.

Then

$$D_t S(t - s; A(s)) = A(s) S(t - s; A(s))$$
$$= A(t) S(t - s; A(s)) + (A(s) - A(t)) S(t - s; A(s)) \quad (t \geqslant s)$$
(7.2.3)

with $S(s - s; A(s)) = I$ so that, in view of (7.1.36), we may expect an equality of the type

$$S(t - s; A(s)) = S(t, s) - \int_s^t S(t, \sigma)(A(\sigma) - A(s)) S(\sigma - s; A(s)) \, d\sigma.$$
(7.2.4)

On the other hand,

$$D_s S(t - s; A(t)) = - S(t - s; A(t)) A(t)$$
$$= - S(t - s; A(t)) A(s) + S(t - s; A(t))(A(s) - A(t)) \quad (s \leqslant t)$$
(7.2.5)

with $S(t - t, A(t)) = I$, hence (7.1.38) suggests that

$$S(t - s; A(t)) = S(t, s) - \int_s^t S(t - \tau; A(t))(A(\tau) - A(t)) S(\tau, s) \, d\tau,$$
(7.2.6)

thus we can hope to fabricate $S(t, s)$ as the solution of one of the Volterra integral equations (7.2.4) or (7.2.6). We shall in fact use both, beginning with the last. To solve (7.2.6) directly, however, would need rather strong conditions on $S(t - \tau; A(t))(A(\tau) - A(t))$, thus we shall follow a more roundabout way. Ignoring the dependence on t of the function $(A(\tau) - A(t)) S(\tau, s)$ in (7.2.6), we modify the equation thus:

$$S(t, s) = S(t - s; A(t)) + \int_s^t S(t - \tau; A(t)) R(\tau, s) \, d\tau, \quad (7.2.7)$$

and seek to obtain an integral equation for $R(\tau, s)$. In order to do this, we note that (again formally) (7.2.7) implies

$$D_t S(t, s) = D_t S(t - s; A(t)) + R(t, s) + \int_s^t D_t S(t - \tau; A(t)) R(\tau, s) \, d\tau.$$

Also, applying $A(t)$ on the left to both sides of (7.2.7),

$$A(t) S(t, s) = - D_s S(t - s; A(t)) - \int_s^t D_\tau S(t - \tau; A(t)) R(\tau, s) \, d\tau.$$

Combining the last two equalities, we obtain

$$0 = D_t S(t, s) - A(t) S(t, s)$$
$$= (D_t + D_s) S(t - s; A(t)) + R(t, s)$$
$$+ \int_s^t (D_t + D_\tau) S(t - \tau; A(t)) R(\tau, s) \, d\tau,$$

so that $R(t, s)$ satisfies the integral equation

$$R(t, s) - \int_s^t R_1(t, \tau) R(\tau, s) \, d\tau = R_1(t, s), \qquad (7.2.8)$$

where we have set $R_1(t, s) = -(D_t + D_s) S(t - s; A(t))$. We impose now conditions on the function $t \to A(t)$ that will allow us to justify the formal calculations above. As in Section 4.2, if $0 < \varphi < \pi/2$, we denote by $\Sigma(\varphi)$ the sector $|\arg \lambda| \leqslant \varphi$. The assumptions are (I), (II), (III), and (IV) below, where $0 < T < \infty$.

(I) *For each* t, $0 \leqslant t \leqslant T$, $A(t)$ *is densely defined,* $R(\lambda; A(t))$ *exists in a sector* $\Sigma = \Sigma(\varphi + \pi/2)$ *for some* φ, $0 < \varphi < \pi/2$ *and*

$$\|R(\lambda; A(t))\| \leqslant C/|\lambda| \quad (\lambda \in \Sigma, 0 \leqslant t \leqslant T), \qquad (7.2.9)$$

where neither φ *nor* C *depend on* t.

We recall that, by virtue of Theorem 4.2.1, assumption (I) implies that each $A(t)$ belongs to $\mathcal{C}(\varphi -)$. If Γ is the boundary of the sector Σ oriented clockwise with respect to it, we have

$$S(s; A(t)) = \frac{1}{2\pi i} \int_\Gamma e^{\lambda s} R(\lambda; A(t)) \, d\lambda \quad (s > 0) \qquad (7.2.10)$$

(see (4.2.8)) and we deduce from this formula and (7.2.9) that

$$\|A(t) S(s; A(t))\| \leqslant C/s \quad (s > 0), \qquad (7.2.11)$$

where C is independent of t (that an estimate of this type exists for each t is of course a consequence of the fact that $A \in \mathcal{C}$; see Section 4.2).

Note also that, since $R(\lambda; A(t))$ is assumed to exist for $\lambda = 0$, $A(t)^{-1}$ exists for all t. The next three postulates bear on its dependence on t.

(II) *The function* $t \to A(t)^{-1}$ *is continuously differentiable* (*in the sense of the norm of* (E)) *in* $0 \leqslant t \leqslant T$.

We note that (II) implies that $t \to R(\lambda; A(t))$ is as well continuously differentiable in the norm of (E). To see this, write

$$R(\lambda; A(t)) = A(t)^{-1} (\lambda A(t)^{-1} - I)^{-1} \qquad (7.2.12)$$

and notice that products and inverses of continuously differentiable functions are likewise continuously differentiable. Moreover, making use of the formulas $D(FG) = (DF)G + F(DG)$ and $D(F^{-1}) = -F^{-1}(DF)F^{-1}$, we obtain from (7.2.12) (after some rearrangement of terms) the equality

$$D_t R(\lambda; A(t)) = -A(t) R(\lambda; A(t)) D_t A(t)^{-1} A(t) R(\lambda; A(t)). \qquad (7.2.13)$$

(III) *There exist constants* C *and* ρ, $C > 0$, $0 < \rho < 1$, *such that for every* $\lambda \in \Sigma$ *and every* t, $0 \leqslant t \leqslant T$, *we have*

$$\|D_t R(\lambda; A(t))\| \leqslant C/|\lambda|^{1-\rho}. \qquad (7.2.14)$$

(IV) *The function $DA(t)^{-1}$ is Hölder continuous in (E), that is, there exist constants $C, \alpha > 0$ such that*

$$\|DA(t)^{-1} - DA(s)^{-1}\| \leqslant C|t - s|^{\alpha} \quad (0 \leqslant s, t \leqslant T).$$

$$(7.2.15)$$

We proceed next to show that (under Assumptions (I), (II), and (III) alone) the integral equation (7.2.8) can be solved. The first result is an auxiliary estimate for $R_1(t, s)$. Here and in other inequalities, $C, C', \ldots$ denote constants, not necessarily the same in different expressions.

7.2.1 Lemma. $R_1(t, s)$ *is continuous in $0 \leqslant s < t \leqslant T$ in the norm of (E) and satisfies there the inequality*

$$\|R_1(t, s)\| \leqslant C/(t - s)^{\rho}.$$

$$(7.2.16)$$

Proof. Let Γ be as in the comments following Assumption (I). By virtue of (7.2.10), we have

$$D_t S(t - s; A(t)) = \frac{1}{2\pi i} \int_{\Gamma} \lambda e^{\lambda(t-s)} R(\lambda; A(t)) \, d\lambda$$

$$+ \frac{1}{2\pi i} \int_{\Gamma} e^{\lambda(t-s)} D_t R(\lambda; A(t)) \, d\lambda, \quad (7.2.17)$$

$$D_s S(t - s; A(t)) = -\frac{1}{2\pi i} \int_{\Gamma} \lambda e^{\lambda(t-s)} R(\lambda; A(t)) \, d\lambda \quad (7.2.18)$$

(the differentiations under the integral sign can be easily justified) so that

$$R_1(t, s) = -(D_t + D_s) S(t - s; A(t))$$

$$= -\frac{1}{2\pi i} \int_{\Gamma} e^{\lambda(t-s)} D_t R(\lambda; A(t)) \, d\lambda. \quad (7.2.19)$$

Hence continuity of $R_1(t, s)$ in $0 \leqslant s < t \leqslant T$ follows. Making use of (III) we obtain

$$\|R_1(t, s)\| \leqslant C \int_{\Gamma} e^{\operatorname{Re} \lambda(t-s)} |\lambda|^{\rho-1} |d\lambda| = C'(t - s)^{-\rho}, \quad (7.2.20)$$

thus completing the proof.

Inequality (7.2.16) suffices to show that (7.2.8) has a solution. In fact, we have:

7.2.2 Lemma. *There exists a solution $R(t, s)$ of (7.2.8), which is continuous in $0 \leqslant s < t \leqslant T$ in the norm of (E) and satisfies there the inequality*

$$\|R(t, s)\| \leqslant C'/(t - s)^{\rho}.$$

$$(7.2.21)$$

Proof. Starting with $R_1(t, s)$, we define a sequence of (E)-valued functions recursively by

$$R_n(t, s) = \int_s^t R_1(t, \tau) R_{n-1}(\tau, s) \, d\tau \quad (n = 2, 3, \ldots). \quad (7.2.22)$$

A simple inductive argument based on $(7.2.16)^3$ shows that each $R_n(t, s)$ exists and is continuous in $0 \leqslant s < t \leqslant T$ and

$$\|R_n(t, s)\| \leqslant \frac{C^n \Gamma(1-\rho)^n (t-s)^{(n-1)(1-\rho)-\rho}}{\Gamma(n(1-\rho))}$$

there, where C is the same constant in (7.2.16). It follows that

$$R(t, s) = \sum_{n=1}^{\infty} R_n(t, s) \qquad (7.2.23)$$

converges in the norm of (E) in $0 \leqslant s < t \leqslant T$ and defines a continuous (E)-valued function satisfying (7.2.21), where

$$C' = \Sigma \{ C^n \Gamma(1-\rho)^n / \Gamma(n(1-\rho)) \} T^{(n-1)(1-\rho)}.$$

Since the partial sums of (7.2.23) satisfy the same estimate, an application of the Lebesgue dominated convergence theorem shows, adding up both sides of (7.2.22) from 2 to ∞, that $R(t, s)$ is a solution of the integral equation (7.2.8) as claimed.

We have not used up to now hypothesis (IV). With its help we prove two additional results.

7.2.3   Lemma.   *The function* $R_1(t, s)$ *satisfies*

$$\|R_1(t, s) - R_1(\tau, s)\| \leqslant C \left\{ \frac{t-\tau}{(t-s)(\tau-s)^\rho} + \frac{(t-\tau)^\alpha}{t-s} \right\} \qquad (7.2.24)$$

for $0 \leqslant s < \tau < t \leqslant T$.

Proof.   We make use of formula (7.2.19):

$$R_1(t, s) - R_1(\tau, s) = -\frac{1}{2\pi i} \int_\Gamma e^{\lambda(t-s)} \{ D_t R(\lambda; A(t)) - D_\tau R(\lambda; A(\tau)) \} \, d\lambda$$

$$- \frac{1}{2\pi i} \int_\Gamma (e^{\lambda(t-s)} - e^{\lambda(\tau-s)}) D_\tau R(\lambda; A(\tau)) \, d\lambda$$

$$= I_1 + I_2. \qquad (7.2.25)$$

---

[3] We use here the well known equality involving the beta and gamma functions,

$$\int_s^t (t-\tau)^{\alpha-1} (\tau-s)^{\beta-1} \, d\tau = (t-s)^{\alpha+\beta-1} B(\alpha, \beta)$$

$$= (t-s)^{\alpha+\beta-1} \Gamma(\alpha)\Gamma(\beta)/\Gamma(\alpha+\beta),$$

valid for $\alpha, \beta > 0$. This identity will be utilized many times in the future without explicit mention.

By virtue of (7.2.13) we have

$$D_t R(\lambda; A(t)) - D_\tau R(\lambda; A(\tau))$$

$$= -\{A(t)R(\lambda; A(t)) - A(\tau)R(\lambda; A(\tau))\}D_t A(t)^{-1}A(t)R(\lambda; A(t))$$

$$- A(\tau)R(\lambda; A(\tau))\{D_t A(t)^{-1} - D_\tau A(\tau)^{-1}\}A(t)R(\lambda; A(t))$$

$$- A(\tau)R(\lambda; A(\tau))D_\tau A(\tau)^{-1}\{A(t)R(\lambda; A(t)) - A(\tau)R(\lambda; A(\tau))\}.$$

$$(7.2.26)$$

But

$$A(t)R(\lambda; A(t)) - A(\tau)R(\lambda; A(\tau)) = \lambda\{R(\lambda; A(t)) - R(\lambda; A(\tau))\}$$

$$= \lambda \int_\tau^t D_r R(\lambda; A(r))\, dr$$

so that, in view of (III),

$$\|A(t)R(\lambda; A(t)) - A(\tau)R(\lambda; A(\tau))\| \leq |\lambda| \int_\tau^t C|\lambda|^{\rho-1}\, dr$$

$$= C(t-\tau)|\lambda|^\rho.$$

Note that, by virtue of Assumption (I), $A(t)R(\lambda; A(t)) = \lambda R(\lambda; A(t)) - I$ is uniformly bounded in norm in $\Gamma \times [0, T]$. On the other hand, Assumption (II) implies that $D_t A(t)^{-1}$ is bounded in $[0, T]$. Using these two facts and the preceding inequality, we estimate the first and third terms on the right-hand side of (7.2.26); for the second term we use Assumption (IV). We obtain

$$\|D_t R(\lambda; A(t)) - D_\tau R(\lambda; A(\tau))\| \leq C\{(t-\tau)|\lambda|^\rho + (t-\tau)^\alpha\}$$

$$(7.2.27)$$

and we proceed now to estimate I_1. We have

$$\|I_1\| \leq C \int_\Gamma e^{\operatorname{Re}\lambda(t-s)}\{(t-\tau)|\lambda|^\rho + (t-\tau)^\alpha\}|d\lambda|$$

$$\leq C' \left\{ \frac{t-\tau}{(t-s)^{\rho+1}} + \frac{(t-\tau)^\alpha}{t-s} \right\}$$

$$\leq C' \left\{ \frac{t-\tau}{(t-s)(\tau-s)^\rho} + \frac{(t-\tau)^\alpha}{t-s} \right\}.$$

$$(7.2.28)$$

As for I_2, we have

$$I_2 = -\int_{\tau-s}^{t-s} D_r \left\{ \frac{1}{2\pi i} \int_\Gamma e^{\lambda r} D_\tau R(\lambda; A(\tau))\, d\lambda \right\} dr$$

$$= -\frac{1}{2\pi i} \int_{\tau-s}^{t-s} \left\{ \int_\Gamma \lambda e^{\lambda r} D_\tau R(\lambda; A(\tau))\, d\lambda \right\} dr.$$

Making use of (III) we see that

$$\left\| \int_{\Gamma} \lambda e^{\lambda r} D_{\tau} R(\lambda; A(\tau)) \, d\lambda \right\| \leqslant C \int_{\Gamma} e^{(\operatorname{Re}\lambda)r} |\lambda|^{\rho} |d\lambda| \leqslant C' r^{-(\rho+1)},$$

hence

$$\|I_2\| \leqslant C' \int_{\tau-s}^{t-s} r^{-(\rho+1)} \, dr$$

$$= C'' \left\{ \frac{1}{(\tau-s)^{\rho}} - \frac{1}{(t-s)^{\rho}} \right\}$$

$$= \frac{C''}{(\tau-s)^{\rho}} \left\{ 1 - \left(\frac{\tau-s}{t-s} \right)^{\rho} \right\}$$

$$\leqslant \frac{C''}{(\tau-s)^{\rho}} \left\{ 1 - \frac{\tau-s}{t-s} \right\}$$

$$= C'' \frac{t-\tau}{(t-s)(\tau-s)^{\rho}}.$$

Combining this inequality with (7.2.28), Lemma 7.2.3 follows.

7.2.4 Corollary. Let $\kappa < \min(1-\rho, \alpha)$. Then the function $R(t, s)$ satisfies

$$\|R(t, s) - R(\tau, s)\| \leqslant C \left\{ \frac{t-\tau}{(t-s)(\tau-s)^{\rho}} + \frac{(t-\tau)^{\kappa}}{t-s} \right\} \qquad (7.2.29)$$

for $0 \leqslant s < \tau < t \leqslant T$.

Proof. We obtain from (7.2.8) that

$$R(t, s) - R(\tau, s) = R_1(t, s) - R_1(\tau, s) + \int_{\tau}^{t} R_1(t, r) R(r, s) \, dr$$

$$+ \int_{s}^{\tau} (R_1(t, r) - R_1(\tau, r)) R(r, s) \, dr$$

$$= J_1 + J_2 + J_3.$$

It is obvious from Lemma 7.2.3 that J_1 satisfies the estimate (7.2.29). For J_2 we use Lemma 7.2.1 and Lemma 7.2.2:

$$\|J_2\| \leqslant C \int_{\tau}^{t} \frac{dr}{(t-r)^{\rho}(r-s)^{\rho}}.$$

If $\tau \leqslant r \leqslant t$, we have

$$\frac{r-\tau}{r-s} \leqslant \frac{t-\tau}{t-s},$$

thus

$$\|J_2\| \leqslant C \left(\frac{t-\tau}{t-s} \right)^{\rho} \int_{\tau}^{t} \frac{dr}{(t-r)^{\rho} (r-\tau)^{\rho}}$$

$$= C' \frac{(t-\tau)^{1-\rho}}{(t-s)^{\rho}}, \tag{7.2.30}$$

which implies (7.2.29) for J_2. By virtue of Lemma 7.2.3,

$$\|J_3\| \leqslant C(t-\tau) \int_s^{\tau} \frac{dr}{(t-r)(\tau-r)^{\rho}(r-s)^{\rho}}$$

$$+ C(t-\tau)^{\alpha} \int_s^{\tau} \frac{dr}{(t-r)(r-s)^{\rho}}$$

$$= I_1 + I_2. \tag{7.2.31}$$

We estimate these integrals as follows. Making use of the inequality $a^{\gamma} b^{1-\gamma} \leqslant a + b$ valid whenever $a, b \geqslant 0$, $0 \leqslant \gamma \leqslant 1$, we obtain

$$\frac{(t-\tau)^{1-\gamma}}{t-r} \leqslant \frac{1}{(\tau-r)^{\gamma}}. \tag{7.2.32}$$

Using this for $\gamma = \kappa$ and noting that $\rho + \kappa < 1$,

$$I_1 \leqslant C(t-\tau)^{\kappa} \int_s^{\tau} \frac{dr}{(\tau-r)^{\rho+\kappa}(r-s)^{\rho}}$$

$$\leqslant C' \frac{(t-\tau)^{\kappa}}{(\tau-s)^{2\rho+\kappa-1}} \leqslant C'' \frac{(t-\tau)^{\kappa}}{(\tau-s)^{\rho}}.$$

The estimation of I_2 runs along similar lines; we use now (7.2.32) for $\gamma = 1 - \alpha + \kappa$, obtaining

$$I_2 \leqslant C(t-\tau)^{\kappa} \int_s^{\tau} \frac{dr}{(\tau-r)^{1-\alpha+\kappa}(r-s)^{\rho}}$$

$$< C' \frac{(t-\tau)^{\kappa}}{(\tau-s)^{\rho-\alpha+\kappa}} \leqslant C'' \frac{(t-\tau)^{\kappa}}{(\tau-s)^{\rho}}.$$

We have already gathered the information necessary to handle the Cauchy problem for (7.2.1). The relevant notion of solution is:

(A) $u(\cdot)$ is a solution of (7.2.1) in $s \leqslant t \leqslant T$ if and only if $u(\cdot)$ is continuous in $s \leqslant t \leqslant T$, continuously differentiable in $s < t \leqslant T$, $u(t) \in D(A(t))$, and (7.2.1) is satisfied in $s < t \leqslant T$.

7.2.5 Theorem. Let $\{A(t); 0 \leqslant t \leqslant T\}$ be a family of operators in E satisfying Assumptions (I), (II), (III), and (IV). Then the Cauchy problem for

(7.2.1) *is well posed* (*with* $D = E$) *in* $0 \leqslant t \leqslant T$ *with respect to solutions defined by* (A). *The propagator* $S(t, s)$ *is strongly continuous in* $0 \leqslant s \leqslant t \leqslant T$ *and continuously differentiable* (*in the norm of* (E)) *with respect to* s *and* t *in* $0 \leqslant s < t \leqslant T$. *Moreover,* $S(t, s)E \subseteq D(A(t))$, $A(t)S(t, s)$, *and* $S(t, s)A(s)$ *are bounded,*

$$D_t S(t, s) = A(t)S(t, s), \quad D_s S(t, s) = - \overline{S(t, s)A(s)}, \quad (7.2.33)$$

and

$$\|D_t S(t, s)\| \leqslant C/(t - s), \quad \|D_s S(t, s)\| \leqslant C/(t - s) \quad (7.2.34)$$

in $0 \leqslant s < t \leqslant T$.

The proof of Theorem 7.2.5 depends on the following auxiliary result, which will find further use in the next section.

7.2.6 Lemma. *Let Assumptions* (I), (II), *and* (III) *be satisfied and let*

$$S_h(t, s) = S(t - s; A(t)) + \int_s^{t-h} S(t - \tau; A(t))R(\tau, s)\, d\tau$$

$$(7.2.35)$$

for $0 \leqslant s \leqslant t - h \leqslant T - h$,

$$S_h(t, s) = S(t - s; A(t)) \quad (7.2.36)$$

for $t - h \leqslant s \leqslant t$. *Then*[4] (i)

$$S_h(t, s) \to S(t, s) \quad (7.2.37)$$

in the norm of (E) *as* $h \to 0$, *uniformly in* $0 \leqslant s \leqslant t \leqslant T$. (ii) $D_t S_h(t, s)$ *exists and is continuous in the norm of* (E), $S_h(t, s)E \subseteq D(A(t))$, *and* $A(t)S_h(t, s)$ *is continuous in* (E) *for* $0 \leqslant s < t - h \leqslant T - h$. (iii) *For each* $u \in E$ *and* $k > 0$,

$$D_t S_h(t, s)u - A(t)S_h(t, s)u \to 0 \quad (7.2.38)$$

as $h \to 0$, *uniformly in* $0 < s < t - k \leqslant T - k$. (iv) $S(t, s)$ *is* (E)-*continuous in* $0 \leqslant s < t \leqslant T$, *strongly continuous in* $0 \leqslant s \leqslant t \leqslant T$, *and*

$$S(s, s) = I \quad (0 \leqslant s \leqslant T). \quad (7.2.39)$$

Proof. (i) follows immediately from the fact that, if we set $t_{s,h} = \max(s, t - h)$, then

$$S(t, s) - S_h(t, s) = \int_{t_{s,h}}^t S(t - \tau; A(t))R(\tau, s)\, d\tau$$

everywhere in the triangle $0 \leqslant s \leqslant t \leqslant T$, and from Lemma 7.2.2. We have already observed (see (7.2.17)) that $S(t - s; A(t))$ is continuously differentiable as a (E)-valued function with respect to s and t in $0 \leqslant s < t \leqslant T$. This

---

[4] The function $S(t, s)$ is defined by (7.2.7); existence of the integral is a consequence of the estimate (7.2.21).

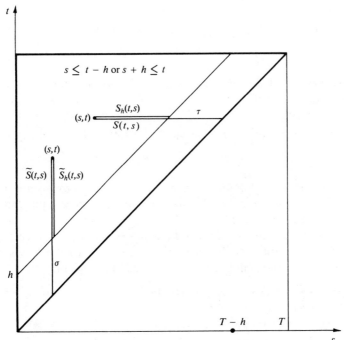

Lines indicate domains of integration in the definition of

$$S, S_h, \tilde{S}, \tilde{S}_h$$

FIGURE 7.2.1

justifies the following calculation in the interval $0 \leqslant s \leqslant t - h$:

$$D_t S_h(t, s) = D_t S(t - s; A(t)) + S(h; A(t)) R(t - h, s)$$
$$+ \int_s^{t-h} D_t S(t - \tau; A(t)) R(\tau, s) \, d\tau$$
$$= (D_t + D_s) S(t - s; A(t)) - D_s S(t - s; A(t))$$
$$+ S(h; A(t)) R(t - h, s)$$
$$+ \int_s^{t-h} (D_t + D_\tau) S(t - \tau; A(t)) R(\tau, s) \, d\tau$$
$$- \int_s^{t-h} D_\tau S(t - \tau; A(t)) R(\tau, s) \, d\tau$$
$$= A(t) S(t - s; A(t)) - R_1(t, s) - \int_s^{t-h} R_1(t, \tau) R(\tau, s) \, d\tau$$
$$+ \int_s^{t-h} A(t) S(t - \tau; A(t)) R(\tau, s) \, d\tau$$
$$+ S(h; A(t)) R(t - h, s). \tag{7.2.40}$$

This equality can be rewritten in the form

$$D_t S_h(t, s) = A(t) S_h(t, s) - R(t, s) + S(h; A(t)) R(t - h; s)$$
$$+ \int_{t-h}^{t} R_1(t, \tau) R(\tau, s) \, d\tau, \qquad (7.2.41)$$

where we have used the integral equation (7.2.8). It follows immediately from these equalities that (ii) holds. To see that (iii) is verified, we write (7.2.41) in the form

$$D_t S_h(t, s) - A(t) S_h(t, s) = S(h; A(t)) R(t - h; s) - R(t, s)$$
$$+ \int_{t-h}^{t} R_1(t, \tau) R(\tau, s) \, d\tau.$$

It is obvious that the combination of the first two terms tends to zero strongly as $h \to 0$ in $s \leqslant t - k$. The integral tends to zero uniformly when $h \to 0$ under the same conditions; this follows easily from the estimates (7.2.16) and (7.2.21) and yields (iii) immediately. The proof of (iv) follows from (i) and from the fact that $S_h(t, s)$ is (E)-continuous in $0 \leqslant s < t \leqslant T$ and strongly continuous in $0 \leqslant s \leqslant t \leqslant T$; (7.2.39) is evident.

Availing ourselves now of hypothesis (IV), we continue with the proof of Theorem 7.2.5. We have

$$A(t) S_h(t, s) = A(t) S(t - s; A(t))$$
$$+ \int_{s}^{t-h} A(t) S(t - \tau; A(t))(R(\tau, s) - R(t, s)) \, d\tau$$
$$- \int_{s}^{t-h} D_\tau S(t - \tau; A(t)) R(t, s) \, d\tau$$
$$= A(t) S(t - s; A(t))$$
$$+ \int_{s}^{t-h} A(t) S(t - \tau; A(t))(R(\tau, s) - R(t, s)) \, d\tau$$
$$+ S(t - s; A(t)) R(t, s) - S(h; A(t)) R(t, s).$$
$$(7.2.42)$$

Observe that, due to the estimate for $R(t, s) - R(\tau, s)$ obtained in Corollary 7.2.4, the integral on the right-hand side of (7.2.42) (which integral is obviously a (E)-continuous function of s, t in $0 \leqslant s \leqslant t - h \leqslant T - h$) converges uniformly on triangles $0 \leqslant s < t - k \leqslant T - k$, $k > 0$ to

$$\int_{s}^{t} A(t) S(t - \tau; A(t))(R(\tau, s) - R(t, s)) \, d\tau, \qquad (7.2.43)$$

which must therefore be itself (E)-continuous in $0 \leqslant s < t \leqslant T$. Since $A(t)$ is closed, we deduce from (7.2.42) that $S(t, s) E \subseteq D(A(t))$ and

$$A(t) S(t, s) = A(t) S(t - s; A(t))$$
$$+ \int_{s}^{t} A(t) S(t - \tau; A(t))(R(\tau, s) - R(t, s)) \, d\tau$$
$$+ S(t - s; A(t)) R(t, s) - R(t, s) \qquad (7.2.44)$$

for $0 \leqslant s < t \leqslant T$, which shows that $A(t)S(t, s)$ is continuous there in view of the comments surrounding (7.2.43). We estimate now the integral on the right-hand side using (7.2.11) and (7.2.29):

$$\left\| \int_s^t A(t)S(t - \tau; A(t))(R(\tau, s) - R(t, s)) \, d\tau \right\|$$

$$\leqslant \frac{C}{t - s} \int_s^t \frac{d\tau}{(\tau - s)^\rho} + \frac{C}{t - s} \int_s^t \frac{d\tau}{(t - \tau)^{1 - \kappa}}$$

$$\leqslant C' \left\{ \frac{1}{(t - s)^\rho} + \frac{1}{(t - s)^{1 - \kappa}} \right\}$$

$$\leqslant \frac{C''}{(t - s)^{1 - \kappa}}.$$

This (and again (7.2.11)) shows that

$$\|A(t)S(t, s)\| \leqslant C/(t - s) \quad (0 \leqslant s < t \leqslant T),$$

which inequality will yield the first estimate (7.2.34) as soon as we prove that $D_t S(t, s) = A(t)S(t, s)$. Note that, as a consequence of the preceding arguments, we have obtained

$$\|A(t)S(t, s) - A(t)S(t - s; A(t))\| \leqslant C/(t - s)^{1 - \kappa} \quad (0 \leqslant s < t \leqslant T), \tag{7.2.45}$$

an inequality that will find application in Section 7.5.

It remains to be shown that $S(t, s)u$ is a solution of (7.2.1) for all $u \in E$. In order to do this, we take $k > 0$ with $s + k < t$; in view of (7.2.38), (7.2.43), and preceding comments,

$$\int_{s+k}^t A(\tau)S(\tau, s)u \, d\tau = \lim_{h \to 0} \int_{s+k}^t A(\tau)S_h(\tau, s)u \, d\tau$$

$$= \lim_{h \to 0} \int_{s+k}^t D_\tau S_h(\tau, s)u \, d\tau$$

$$= S(t, s)u - S(s + k, s)u,$$

thus our claim on $S(t, s)u$ is justified. Actually, much more is true: since $A(\tau)S(\tau, s)$ is (E)-continuous in $s + k \leqslant \tau \leqslant t$, we can write the last inequality thus:

$$\int_{s+k}^t A(\tau)S_h(\tau, s) \, d\tau = S(t, s) - S(s + k, s),$$

hence $D_t S(t, s)$ exists in the norm of (E) and

$$D_t S(t, s) = A(t)S(t, s) \quad (0 \leqslant s < t \leqslant T).$$

We have proved then part (a) in the definition of a properly posed Cauchy problem, since the strong continuity of S has already been shown in Lemma 7.2.6.

Clearly, part (b) will be fulfilled as well if we can prove that any solution (as defined in (A)) of (7.2.1) must perforce be of the form $S(t, s)u$. This will be achieved by essentially the same means as in the time-invariant case (see the end of the proof of Theorem 2.1.1). In the present situation, this would involve differentiation of the function $s \to S(t, s)u(s)$ for an arbitrary solution of (7.2.1). However, we lack information on the behavior of S as a function of s. Taking a hint from Example 7.1.3, we may expect that

$$D_s S(t, s) = - S(t, s) A(s) \quad (0 \leqslant s < t \leqslant T) \qquad (7.2.46)$$

for each t. To prove this directly would not be simple, thus we use a trick similar to that in Example 7.1.3; namely, we construct an operator-valued solution $\tilde{S}$ of equation (7.2.46) and show that $\tilde{S} = S$. The construction of $\tilde{S}$ is based on the integral equation (7.2.4), instead of (7.2.6), which was the starting point for the construction of S. As we did in relation to (7.2.6), we seek an $\tilde{S}$ of the form

$$\tilde{S}(t, s) = S(t - s; A(s)) + \int_s^t \tilde{R}(t, \sigma) S(\sigma - s; A(s)) \, d\sigma. \quad (7.2.47)$$

Formally, we have

$$\begin{aligned} D_s \tilde{S}(t, s) &= D_s S(t - s; A(s)) - \tilde{R}(t, s) \\ &\quad + \int_s^t \tilde{R}(t, \sigma) D_s S(\sigma - s; A(s)) \, d\sigma, \end{aligned}$$

$$\begin{aligned} \tilde{S}(t, s) A(s) &= D_t S(t - s; A(s)) \\ &\quad + \int_s^t \tilde{R}(t, \sigma) D_\sigma S(\sigma - s; A(s)) \, d\sigma. \end{aligned}$$

Therefore, if (7.2.46) is to be satisfied,

$$\begin{aligned} 0 &= D_s \tilde{S}(t, s) + \tilde{S}(t, s) A(s) \\ &= (D_t + D_s) S(t - s; A(s)) - \tilde{R}(t, s) \\ &\quad + \int_s^t \tilde{R}(t, \sigma)(D_\sigma + D_s) S(\sigma - s; A(s)) \, d\sigma. \end{aligned}$$

so that $\tilde{R}$ must be a solution of the integral equation

$$\tilde{R}(t, s) - \int_s^t \tilde{R}(t, \sigma) \tilde{R}_1(\sigma, s) \, d\sigma = \tilde{R}_1(t, s) \qquad (7.2.48)$$

with $\tilde{R}_1(t, s) = (D_t + D_s) S(t - s; A(s))$. The construction of $\tilde{S}$ now proceeds in a way exactly analogous to that of S, thus we limit ourselves to state the main steps and leave the details to the reader.

7.2.7 Lemma. $\tilde{R}_1(t, s)$ is continuous in $0 \leqslant s < t \leqslant T$ in the norm of (E) and satisfies there the inequality

$$\|\tilde{R}_1(t, s)\| \leqslant C/(t - s)^p. \qquad (7.2.49)$$

7.2.8 Lemma. *There exists a solution $\tilde{R}(t,s)$ of (7.2.48), which is continuous in $0 \leqslant s < t \leqslant T$ in the norm of (E) and satisfies there the inequality*

$$\|\tilde{R}(t,s)\| \leqslant C'/(t-s)^{\rho}. \tag{7.2.50}$$

The function $\tilde{R}(t,s)$ is obtained as the sum of the series $\sum_{n=1}^{\infty} \tilde{R}_n(t,s)$, the $\tilde{R}_n$ defined recursively by

$$\tilde{R}_n(t,s) = \int_s^t \tilde{R}_{n-1}(t,\sigma)\tilde{R}_1(\sigma,s)\, d\sigma.$$

We note that the last two results are independent of hypothesis (IV). Using this assumption we obtain:

7.2.9 Lemma. *The function $\tilde{R}_1(t,s)$ satisfies*

$$\|\tilde{R}_1(t,\sigma) - \tilde{R}_1(t,s)\| \leqslant C\left\{ \frac{\sigma - s}{(t-s)(t-\sigma)^{\rho}} + \frac{(\sigma-s)^{\alpha}}{t-s} \right\}$$

in $0 \leqslant s < \sigma < t \leqslant T$.

7.2.10 Corollary. *Let $\kappa < \min(1-\rho, \alpha)$. Then the function $R(t,s)$ satisfies*

$$\|\tilde{R}(t,\sigma) - \tilde{R}(t,s)\| \leqslant C\left\{ \frac{\sigma - s}{(t-s)(t-\sigma)^{\rho}} + \frac{(\sigma-s)^{\kappa}}{t-s} \right\}$$

for $0 \leqslant s < \sigma \leqslant t \leqslant T$.

The analogue of Lemma 7.2.6 is

7.2.11 Lemma. *Let Assumptions (I), (II), and (III) be satisfied, and let*

$$\tilde{S}_h(t,s) = S(t-s; A(s)) + \int_{s+h}^t \tilde{R}(t,\sigma)S(\sigma-s; A(s))\, d\sigma \tag{7.2.51}$$

for $h \leqslant s + h \leqslant t \leqslant T$,

$$\tilde{S}_h(t,s) = S(t-s; A(s))$$

for $s \leqslant t \leqslant s + h$. Then (i)

$$\tilde{S}_h(t,s) \to \tilde{S}(t,s) \tag{7.2.52}$$

in the norm of (E) as $h \to 0$, uniformly in $0 \leqslant s \leqslant t \leqslant T$. (ii) $D_s\tilde{S}_h(t,s)$ exists and is continuous in the norm of (E), $\tilde{S}_h(t,s)A(s)$ is bounded in $D(A(s))$ and its closure $\overline{\tilde{S}_h(t,s)A(s)}$ is continuous in (E) for $h \leqslant s + h \leqslant t \leqslant T$. (iii) For every $u \in E$ and $k > 0$,

$$D_s\tilde{S}_h(t,s)u + \overline{\tilde{S}_h(t,s)A(s)}u \to 0 \tag{7.2.53}$$

as $h \to 0$, uniformly in $h \leqslant s + k \leqslant t \leqslant T$. (iv) $\tilde{S}(t,s)$ is (E)-continuous in

$0 \leqslant s < t \leqslant T$, *strongly continuous in* $0 \leqslant s \leqslant t \leqslant T$ *and*

$$\tilde{S}(t, t) = I \quad (0 \leqslant t \leqslant T). \tag{7.2.54}$$

With the use of Assumption (IV) we prove:

7.2.12 Lemma. *$D_s \tilde{S}(t, s)$ exists in the norm of (E) and is continuous in $0 \leqslant s < t \leqslant T$. For each s and $t \geqslant s$, $\tilde{S}(t, s)A(s)$ is bounded in $D(A(s))$ and its closure satisfies*

$$D_s \tilde{S}(t, s) = - \overline{\tilde{S}(t, s)A(s)}. \tag{7.2.55}$$

Moreover,

$$\|D_s \tilde{S}(t, s)\| \leqslant C/(t - s) \tag{7.2.56}$$

in $0 \leqslant s < t \leqslant T$.

We end now the proof of Theorem 7.2.5. Let $u(\cdot, s)$ be an arbitrary solution of (7.2.1) in $s \leqslant t \leqslant T$ and let t be fixed in that range. Consider the function

$$v(\sigma) = \tilde{S}(t, \sigma)u(\sigma, s)$$

in the interval $s \leqslant \sigma \leqslant t$. On the basis of definition (A) and of the properties of $\tilde{S}$ just obtained, we easily see that $v(\sigma)$ is continuous in $s \leqslant \sigma \leqslant t$ and that $v'(\sigma) = 0$ in $s < \sigma < t$. Hence $v(s) = v(t)$, that is,

$$u(t) = \tilde{S}(t, s)u(s, s). \tag{7.2.57}$$

This clearly implies uniqueness of solutions of (7.2.1); since $t \to S(t, s)u(s, s)$ is a solution with the same initial value, we must have

$$u(t) = S(t, s)u(s, s). \tag{7.2.58}$$

Comparing with (7.2.57), this shows that $\tilde{S}(t, s)$ is none other than $S(t, s)$ in disguise. Collecting all the properties hitherto proved of $S(t, s)$ and $\tilde{S}(t, s)$, we obtain the claims of Theorem 7.2.5 in full.

We note the estimate

$$\overline{\|S(t, s)A(s)} - A(s)S(t - s; A(s))\| \leqslant C/(t - s)^{1 - \kappa} \quad (0 \leqslant s < t \leqslant T), \tag{7.2.59}$$

which is obtained for $\tilde{S}$ in the same way (7.2.45) was deduced for S. It will find use later.

*7.2.13 Example. *A perturbation result* (Kato-Tanabe [1962: 1]). Let $\langle A(t) \rangle$ be a family of operators satisfying Assumptions (I) to (IV), and let $\langle P(t) \rangle$ be another family of operators satisfying:

(V) $D(P(t)) \supseteq D(A(t))$ and

$$\|P(t)R(\lambda; A(t))\| \leqslant C/|\lambda|^{\beta} \quad (0 \leqslant t \leqslant T, \lambda \in \Sigma), \tag{7.2.60}$$

where $C \geqslant 0$, $0 < \beta < 1$ and both constants are independent of λ and t.

(VI) There exist constants $C > 0$, $0 < \gamma \leqslant 1$ independent of s, t such that

$$\|P(t)A(t)^{-1} - P(s)A(s)^{-1}\| \leqslant C|t - s|^\gamma \quad (0 \leqslant s \leqslant t \leqslant T).$$
(7.2.61)

Then the Cauchy problem for

$$u'(t) = (A(t) + P(t))u(t)$$
(7.2.62)

is well posed in the sense of Theorem 7.2.5 (that is, the family $\{A(t) + P(t)\}$, where, by definition, $D(A(t) + P(t)) = D(A(t))$ satisfies all the conclusions in Theorem 7.2.5). The propagator $\hat{S}(t, s)$ of (7.2.59) is obtained by means of the perturbation series

$$\sum_{n=0}^{\infty} S_n(t, s),$$
(7.2.63)

where $S_0 = S$ and

$$S_n(t, s) = \int_s^t S(t, \sigma)P(\sigma)S_{n-1}(\sigma, s)\,d\sigma$$

for $0 < s < t \leqslant T$, $n = 1, 2, \ldots$. (See Example 7.1.12.)

*7.2.14 Example. *A $\mathcal{C}^\infty$ version of Theorem 7.2.5* (Tanabe [1967: 1]). Let Assumptions (I) to (IV) be replaced by the single hypothesis: $R(\lambda; A(t))$ exists for $\lambda \in \Sigma$, $0 \leqslant t \leqslant T$, the function $t \to R(\lambda; A(t))$ is infinitely differentiable in $0 \leqslant t \leqslant T$ for each $\lambda \in \Sigma$ and there exist constants $C_0, C_1, \ldots$ such that

$$\|D_t^n R(\lambda; A(t))\| \leqslant C_n/|\lambda| \quad (\lambda \in \Sigma, 0 \leqslant t \leqslant T)$$

for $n = 0, 1, \ldots$. Then the propagator $S(t, s)$ of (7.2.1) provided by Theorem 7.2.5 is infinitely differentiable in $0 \leqslant s < t \leqslant T$ (in the sense of the norm of (E)). For every $n = 1, 2, \ldots$ there exists a constant C_n' such that

$$\|D_t^n S(t, s)\| \leqslant C_n'/(t - s)^n, \quad \|D_s^n S(t, s)\| \leqslant C_n'/(t - s)^n$$
(7.2.64)

for $0 \leqslant s < t \leqslant T$.

7.3. ABSTRACT PARABOLIC EQUATIONS: WEAK SOLUTIONS

We have already pointed out (see Section 1.3) that strong or genuine solutions of an abstract differential equation are by no means the only suitable ones and that weak solutions of one sort of other may be used to advantage many times. A prime example of this principle will be examined in this section. In fact, we shall show that even if Assumption (IV) in the previous section is discarded, we can conclude that the Cauchy problem for

$$u'(t) = A(t)u(t)$$
(7.3.1)

is still properly posed in $0 \leqslant t \leqslant T$ if we substitute definition (A) of solution by that of *weak solution*, as formulated in the time-independent case in Section 2.4. We shall use, however, a slightly different definition, roughly corresponding to (2.4.10).

(B) $u(\cdot)$ *is a weak solution of* (7.3.1) *in* $s \leqslant t \leqslant T$ *with initial condition*

$$u(s) = u_0 \in E \qquad (7.3.2)$$

if and only if it is continuous in $s \leqslant t \leqslant T$, *satisfies* (7.3.2) *and*

$$\int_s^T \langle u(t), u^{*\prime}(t) + A^*(t) u^*(t) \rangle \, dt = - \langle u_0, u^*(s) \rangle \quad (7.3.3)$$

for every E^*-*valued continuously differentiable function* $u^*(\cdot)$ *defined in* $s \leqslant t \leqslant T$ *such that* $u^*(t) \in D(A(t)^*)$, $A(t)^* u^*(t)$ *is continuous and*

$$u^*(T) = 0. \qquad (7.3.4)$$

7.3.1 Theorem.   *Let* $A(\cdot)$ *satisfy Assumptions* (I), (II), *and* (III) *in the previous section. Then the Cauchy problem for* (7.3.1) *is properly posed in* $0 \leqslant t \leqslant T$ (*with* $D = E$) *with respect to solutions defined by* (B). *The solution operator* $S(t, s)$ *is strongly continuous in* $0 \leqslant s \leqslant t \leqslant T$ *and continuous in the norm of* (E) *in* $0 \leqslant s < t \leqslant T$.

Proof.   We have already pointed out in the previous section that the operators $R_1(t, s)$, $R(t, s)$, $\tilde{R}_1(t, s)$, $\tilde{R}(t, s)$ therein can be constructed without recourse to Assumption (IV) and that some of their properties are retained in this more general setting: in particular, Lemma 7.2.2 holds for $R(t, s)$ as well as its mirror image, Lemma 7.2.8, for $\tilde{R}(t, s)$. We can then construct $S(t, s)$ and $\tilde{S}(t, s)$ by means of the integral formulas (7.2.7) and (7.2.47), respectively, and it is immediate that the continuity properties claimed for S in Theorem 7.2.5 follow, as well as similar properties for $\tilde{S}$. We note, however, that we cannot conclude without further analysis that $S = \tilde{S}$, since the argument in the previous section was based on differentiability of S and $\tilde{S}$ which was a consequence of Assumption (IV).

We begin by proving that

$$u(t) = S(t, s) u \qquad (7.3.5)$$

is a weak solution of (7.3.1) with initial condition $u(s) = u$ in $0 \leqslant t \leqslant T$ for any $u \in E$. To this end, let $u^*(\cdot)$ be a E^*-valued function satisfying the assumptions in Definition (B). We have

$$\int_s^T \langle u(t), u^{*\prime}(t) \rangle \, dt = \int_s^T \langle S(t, s) u, u^{*\prime}(t) \rangle \, dt$$

$$= \lim_{k \to 0+} \lim_{h \to 0+} \int_{s+k}^T \langle S_h(t, s) u, u^{*\prime}(t) \rangle \, dt,$$

$$(7.3.6)$$

where S_h is the operator defined by (7.2.35) and (7.2.36), so that (7.3.6) is justified by Lemma 7.2.6(i). On the other hand,

$$\int_{s+k}^{T} \langle S_h(t,s)u, u^{*\prime}(t)\rangle \, dt = -\langle S_h(s+k,s)u, u^*(s+k)\rangle$$

$$-\int_{s+k}^{T} \langle D_t S_h(t,s)u - A(t)S_h(t,s)u, u^*(t)\rangle \, dt$$

$$-\int_{s+k}^{T} \langle S_h(t,s)u, A(t)^*u^*(t)\rangle \, dt \quad (\text{as } h \to 0+)$$

$$\to -\langle S(s+k,s)u, u^*(s+k)\rangle$$

$$-\int_{s+k}^{T} \langle S(t,s)u, A(t)^*u^*(t)\rangle \, dt \quad (\text{as } k \to 0)$$

$$\to -\langle u, u^*(s)\rangle - \int_{s}^{T} \langle u(t), A(t)^*u^*(t)\rangle \, dt, \quad (7.3.7)$$

which shows that (7.3.3) holds for $u(\cdot)$; note that in the first limit we have used (7.2.38). We must now prove uniqueness of weak solutions and to this end we bring into play the operator $\tilde{S}_h$ defined in (7.2.51). Let $u(\cdot)$ be a weak solution of (7.3.1) in $s \leqslant t \leqslant T$, and let $s < t' < T$, $\Psi(\cdot)$ a continuously differentiable function with values in E^* and support in (s, t'). Since $\tilde{S}_h(t,s)A(s)$ is a bounded operator for $s + h \leqslant t$ (Lemma 7.2.11(ii)) we see (Section 4) that

$$\tilde{S}_h(t,s)^* E^* \subseteq D(A(s)^*)$$

and

$$A(s)^* \tilde{S}_h(t,s)^* = \left(\tilde{S}_h(t,s)A(s)\right)^* = \left(\overline{\tilde{S}_h(t,s)A(s)}\right)^*$$

is continuous in the norm of E^* in the same region. On the other hand, also by Lemma 7.2.11(ii), $D_s\tilde{S}_h(t,s)$ exists and is continuous in the norm of (E) in $s + h \leqslant t$, thus $D_s\tilde{S}_h(t,s)^*$ exists and is continuous in the norm of (E^*) for these values of s and t. It follows that if h is sufficiently small,

$$u_h^*(t) = \tilde{S}_h(t',t)^*\Psi(t)$$

will satisfy the requirements in Definition (B). Accordingly, we have

$$\int_{s}^{t'} \langle \tilde{S}(t',t)u(t), \Psi'(t)\rangle \, dt = \lim_{h \to 0+} \int_{s}^{t'} \langle \tilde{S}_h(t',t)u(t), \Psi'(t)\rangle \, dt$$

$$= \lim_{h \to 0+} \int_{s}^{t'} \langle u(t), \tilde{S}_h(t',t)^*\Psi'(t)\rangle \, dt$$

$$= \lim_{h \to 0+} \int_{s}^{t'} \langle u(t), u_h^{*\prime}(t) + A(t)^*u_h^*(t)\rangle \, dt$$

$$- \lim_{h \to 0+} \int_{s}^{t'} \left\langle \left(D_t\tilde{S}_h(t',t) + \overline{\tilde{S}_h(t',t)A(t)}\right)u(t), \Psi(t)\right\rangle dt.$$

$$(7.3.8)$$

The first term on the right-hand side vanishes since u is a weak solution of (7.3.1) and $u_h^*(s) = 0$; the second limit is seen to be zero taking (7.2.53) into account (note that Ψ vanishes near t'). Taking now $\Psi(s) = \varphi(s)u^*$ in (7.3.8), u^* an arbitrary element of E^* and φ a smooth scalar function with support in (s, t'), we obtain

$$\int_s^{t'} \langle \tilde{S}(t', t)u(t), u^* \rangle \varphi'(t)\, dt = 0, \qquad (7.3.9)$$

thus $\langle \tilde{S}(t', t)u(t), u^* \rangle$ (a fortiori $\tilde{S}(t', t)u(t)$) must be constant. Hence $\tilde{S}(t', s)u(s) = \tilde{S}(t', t')u(t') = u(t')$, and we obtain essentially as in the end of Section 7.2, that weak solutions are unique and that $\tilde{S} = S$. The proof of Theorem 7.3.1 is then complete.

It is natural to ask whether the assumptions in this or in the previous section imply that the intersection (7.1.16) is dense in E. This is in fact not so, as the following example shows.

7.3.2 Example. There exists an operator-valued function $t \to A(t)$ defined in $0 \leqslant t \leqslant 1$ in $E = L^2(0, 1)$ such that hypotheses (I) to (IV) in Section 7.2 are satisfied, but

$$\bigcap_{0 \leqslant t \leqslant 1} D(A(t)) = \{0\}. \qquad (7.3.10)$$

The operators are defined by

$$A(t)u(x) = -\frac{1}{(t - x)^2} u(x), \qquad (7.3.11)$$

$D(A(t))$ consisting of all u such that (7.3.11) belongs to L^2. Clearly each $A(t)$ is self-adjoint and $(A(t)u, u) \leqslant -\|u\|^2$ for $u \in D(A(t))$ so that $\sigma(A(t))$ is contained in $(-\infty, -1]$ for all t; (I) is satisfied for any $\varphi \in (0, \pi/2)$. If $\lambda \notin (-\infty, -1]$, we have

$$R(\lambda; A(t))u(x) = \frac{(t - x)^2}{\lambda(t - x)^2 + 1} u(x) \qquad (7.3.12)$$

and

$$D_t R(\lambda; A(t))u(x) = \frac{2(t - x)}{\left(\lambda(t - x)^2 + 1\right)^2} u(x). \qquad (7.3.13)$$

A simple estimation shows that $2\alpha|\lambda\alpha^2 + 1|^{-2} \leqslant C|\lambda|^{-1/2}$ in any sector $\Sigma(\varphi + \pi/2)$ for $0 \leqslant \alpha \leqslant 1$, so that

$$\|D_t R(\lambda; A(t))\| \leqslant C|\lambda|^{-1/2}$$

and (III) holds. As a particular case of (7.3.12), we see that

$$A(t)^{-1}u(x) = -(t - x)^2 u(x)$$

so that $A(t)^{-1} = -(tI - M)^2$, M the (bounded) operator of multiplication by x and (II) and (IV) are amply satisfied. Let $u \in L^2$ be an element of $\cap D(A(t))$. Then for every t, $0 < t < 1$, we have

$$\frac{1}{2h} \int_{t-h}^{t+h} |u(x)| \, dx$$

$$\leqslant \left(\int_{t-h}^{t+h} \frac{|u(x)|^2}{(t-x)^4} \, dx \right)^{1/2} \left(\frac{1}{4h^2} \int_{t-h}^{t+h} (t-x)^4 \, dx \right)^{1/2} \to 0$$

as $h \to 0$, thus $u(x) = 0$ at every Lebesgue point. It follows that $u(\cdot)$ vanishes almost everywhere.

We note that, in order to prove that the Cauchy problem for (7.3.1) is properly posed for the family $A(\cdot)$ in this example, it is not necessary to use the awesome machinery of Theorem 7.2.5: in a suitable sense, (7.3.1) is in this case equivalent to the ordinary differential equation

$$\frac{du}{dt} = -\frac{1}{(t-x)^2} u \quad (t \geqslant s), \tag{7.3.14}$$

and $S(t, s)$ is given by the formula

$$S(t,s)u(x) = \begin{cases} \exp\left(\dfrac{1}{t-x} - \dfrac{1}{s-x} \right) u(x) & (x < s \text{ or } x > t) \\ 0 & (s \leqslant x \leqslant t). \end{cases} \tag{7.3.15}$$

The example above may be used to illustrate why the obvious generalization of the definition of weak solution used in Section 2.4 for time-independent equations would be inadequate in the present case: in fact, the generalization of (2.4.8) would read as follows:

(B′) *For every* $u^* \in \cap D(A(t)^*)$ *and every test function* φ *with support in* $(-\infty, T]$, *we have*

$$\int_0^T (\langle u(t), u^* \rangle \varphi'(t) + \langle u(t), A(t)^* u^* \rangle \varphi(t)) \, dt$$

$$= -\langle u_0, u^* \rangle \varphi(s). \tag{7.3.16}$$

However, in the case considered in Example 7.3.2 (where $E^* = E$ and $A(t)^* = A(t)$), we have $\cap D(A(t)^*) = \{0\}$. Hence, $u^* = 0$ is the only candidate to appear in (7.3.16).

7.4. ABSTRACT PARABOLIC EQUATIONS: THE ANALYTIC CASE

We prove in this section a complex variable version of Theorem 7.2.5.

7.4.1   Theorem.   *Let* Δ *be an open convex neighborhood of the interval* $[0, T]$ *in the complex plane, and let* $\{A(z); z \in \Delta\}$ *be a family of operators* *satisfying*:

(I′) *For each $z \in \Delta$, $A(z)$ is densely defined, $R(\lambda; A(z))$ exists in a sector $\Sigma = \Sigma(\varphi + \pi/2)$ for some φ independent of z, $0 < \varphi < \pi/2$; if Δ' is a bounded neighborhood of $[0, T]$ with $\overline{\Delta'} \subseteq \Delta$,*

$$\|R(\lambda; A(z))\| \leqslant C/|\lambda| \quad (\lambda \in \Sigma, z \in \Delta'), \qquad (7.4.1)$$

where C may depend only on Δ'.

(II′) *The function*

$$z \to A(z)^{-1} \in (E) \qquad (7.4.2)$$

is analytic in Δ.

Then the Cauchy problem for

$$u'(t) = A(t)u(t) \qquad (7.4.3)$$

is properly posed (with $D = E$) in $0 \leqslant t \leqslant T$ with respect to solutions defined in (A), *Section 7.2. The propagator $S(t, s)$ is strongly continuous in $0 \leqslant s \leqslant t$ and admits an analytic extension $S(\zeta, z)$ (as a function of the two variables) to*

$$\Delta_\varphi = \{(\zeta, z); \zeta, z \in \Delta, \zeta \neq z, |\arg(\zeta - z)| \leqslant \varphi\}. \qquad (7.4.4)$$

Moreover, for these values of ζ, z, $S(\zeta, z)E \subseteq D(A(\zeta))$ (so that $A(\zeta)S(\zeta, z)$ is everywhere defined and bounded), $S(\zeta, z)A(z)$ is bounded in $D(A(z))$ and the equalities

$$D_\zeta S(\zeta, z) = A(\zeta)S(\zeta, z), \quad D_z S(\zeta, z) = \overline{S(\zeta, z)A(z)} \qquad (7.4.5)$$

hold. Finally, for any Δ' as described above,

$$\|D_\zeta S(\zeta, z)\| \leqslant C/|\zeta - z|, \quad \|D_z S(\zeta, z)\| \leqslant C/|\zeta - z| \qquad (7.4.6)$$

(the constant C possibly depending on Δ') for ζ, z in

$$\Delta'_\varphi = \{(\zeta, z); \zeta, z \in \Delta', \zeta \neq z, |\arg(\zeta - z)| \leqslant \varphi\}. \qquad (7.4.7)$$

Proof. We begin by verifying that the assumptions in Theorem 7.2.5 are satisfied for the restriction $\{A(t); 0 \leqslant t \leqslant T\}$ of $A(\cdot)$ to $[0, T]$. This is immediately obvious for (I), (II), and (IV), but not quite so for (III). To establish this assumption we note that (I′) and (II′) imply that $R(\lambda; A(z))$ is analytic in λ, z for $\lambda \in \Sigma$, $z \in \Delta$; to see this we only have to write the complex variable analog of (7.2.12),

$$R(\lambda; A(z)) = A(z)^{-1}(\lambda A(z)^{-1} - I)^{-1},$$

and note that inverses and products of operator-valued analytic functions are analytic. Making use of analyticity of $R(\lambda; A(z))$ with respect to z, we can write

$$D_z R(\lambda; A(z)) = \frac{1}{2\pi i} \int_{|z' - z| = a} R(\lambda; A(z')) \frac{dz'}{(z' - z)^2} \qquad (7.4.8)$$

for $z' \in \Delta'$ (Δ' as in (7.4.1)), where a is so small that all the circles $\{z; |z' - z| \le a\}$ for z in Δ' are contained in another bounded neighborhood Δ'' of $[0, T]$ with $\bar{\Delta}'' \subseteq \Delta$. We note next that, using techniques very similar to those in Lemma 4.2.3 combined with inequality (7.4.1), we can show that $R(\lambda; A(z))$ actually exists in a sector $\Sigma(\varphi'' + \pi/2)$ with $\varphi < \varphi'' < \pi/2$ for every $z \in \Delta'$ (φ'' may of course depend on Δ') and (7.4.1) holds as well in $\Sigma(\varphi'' + \pi/2) \times \Delta'$. We can then estimate the integral (7.4.8) obtaining: if $\varphi' < \varphi''$,

$$\|D_z R(\lambda; A(z))\| \le C/|\lambda| \quad (\lambda \in \Sigma(\varphi' + \pi/2), z \in \Delta'), \quad (7.4.9)$$

where C may depend only on φ' and Δ'. It is clear that this implies (7.2.14) for $\rho \ge 0$. (The substitution of φ by φ' was not necessary in this argument, but will be essential in the proof of Lemma 7.4.2.)

The preceding considerations show that we can apply Theorem 7.2.5 and, in view of the conclusions of that theorem, we only have to establish the statements in Theorem 7.4.1 regarding analytic extension of $S(t, s)$. To do this we reexamine the construction of S in Section 2 and extend all the necessary results to the complex plane. We denote below by Δ' a subset of Δ like the one described in the statement of Theorem 7.4.1; the symbol Δ'_φ indicates the set in (7.4.7).

7.4.2 Lemma. $S(t; A(s))$ *admits an analytic extension* $S(\zeta; A(z))$ *to* $\Sigma(\varphi) \times \Delta$ *such that*

$$\|A(z) S(\zeta; A(z))\| \le C/|\zeta| \quad (7.4.10)$$

for $(\zeta, z) \in \Sigma(\varphi) \times \Delta'$. *Moreover,* $R_1(\zeta, z) = -(D_\zeta + D_z) S(\zeta - z; A(z))$ *satisfies*

$$\|R_1(\zeta, z)\| \le C/|\zeta - z|^\rho \quad (z \in \Delta', |\arg(\zeta - z)| \le \varphi) \quad (7.4.11)$$

for $\rho > 0$ *arbitrary* (*with* C *depending on* ρ *and* Δ').

Proof. As noted in the comments preceding (7.4.9), for every Δ' there exists a $\varphi' > \varphi$ such that $R(\lambda; A(z))$ exists in $\Sigma(\varphi' + \pi/2) \times \Delta'$ and satisfies an estimate of the form (7.4.1) there. The formula

$$S(\zeta; A(z)) = \frac{1}{2\pi i} \int_{\Gamma'} e^{\lambda \zeta} R(\lambda; A(z)) \, d\lambda, \quad (7.4.12)$$

where Γ' is the boundary of $\Sigma(\varphi' + \pi/2)$ (see (7.2.10)), then provides an analytic extension of $S(t; A(z))$ to $\Sigma(\varphi) \times \Delta'$. Using the estimate for $R(\lambda; A(z))$ and proceeding much in the same way as in the second part of Theorem 4.2.1, we obtain the required inequality. The estimate (7.4.11) is obtained working with the analytic continuation of (7.2.19), again using the techniques in the second part of Theorem 4.2.1 and inequality (7.4.9).

We examine next the integral equation (7.2.8) giving birth to $R(t, s)$. In view of Lemma 7.4.2, each member of the sequence $\{R_n(t, s)\}$ defined by

(7.2.22) admits an analytic continuation to Δ_φ given by

$$R_n(\zeta, z) = \int_z^\zeta R_1(\zeta, \tau) R_{n-1}(\tau, z)\, d\tau,$$

(see Figure 7.4.1) where the path of integration is, say, the segment I joining z and ζ (note that if $(z, \zeta) \in \Delta_\varphi$ and $\tau \in I$, then both (ζ, τ) and (τ, z) belong to Δ_φ as well, and a similar statement holds for Δ'_φ). It is easy to deduce inductively from (7.4.11) that if $\rho > 0$, then

$$\|R_n(\zeta, z)\| \leqslant \frac{C^n \Gamma(1-\rho)^n |\zeta - z|^{(n-1)(1-\rho)-\rho}}{\Gamma(n(1-\rho))} \qquad (7.4.13)$$

on each Δ'_φ; thus the series $\Sigma R_n(\zeta, z)$ is uniformly convergent in Δ'_φ and it provides an analytic extension of $R(t, s)$ to Δ_φ, which satisfies an estimate of the form (7.4.11) in each Δ'_φ. Making then again use of Lemma 7.4.2 and of (7.2.7), we see that the desired analytic extension of S is given by

$$S(\zeta, z) = S(\zeta - z; A(z)) + \int_z^\zeta S(\zeta - \tau; A(z)) R(\tau, s)\, d\tau. \quad (7.4.14)$$

It only remains to prove equalities (7.4.5), and the statements preceding them, as well as inequalities (7.4.6). To do this, we write the first equality (7.4.5) for s, t real in the form

$$A(t)^{-1} D_t S(t, s) = S(t, s) \quad (0 \leqslant s < t \leqslant T).$$

Since the two holomorphic functions

$$A(\zeta)^{-1} D_\zeta S(\zeta, z), \quad S(\zeta, z)$$

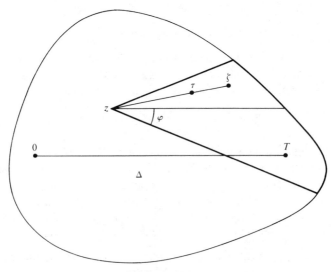

FIGURE 7.4.1

coincide for ζ, z real, they must coincide wherever both are defined, in particular in Δ_φ. The second equality (7.4.5) is equivalent to

$$D_z S(\zeta, z) A(z)^{-1} = - S(\zeta, z)$$

and this is established in an entirely similar way. To prove inequalities (7.4.6), we need to extend the estimate (7.2.29) to complex values of t and s. This is done as follows: we have already observed that for any $\rho > 0$,

$$\|R_1(\zeta, z)\| \leqslant C/|\zeta - z|^\rho \tag{7.4.15}$$

for $(\zeta, z) \in \Delta'_\varphi$, and it follows from (7.4.13) that

$$\|R(\zeta, z)\| \leqslant C/|\zeta - z|^\rho \tag{7.4.16}$$

there as well. The relevant estimates in Section 7.2 can be extended to the complex plane (and at the same time simplified, due to the fact that $\alpha = 1$): (7.2.27) becomes

$$\|D_\zeta R(\lambda; A(\zeta)) - D_z R(\lambda; A(z))\| \leqslant C|\zeta - z| \quad ((\zeta, z) \in \Delta'_\varphi). \tag{7.4.17}$$

With the help of this inequality, (7.2.24) is extended to

$$\|R_1(\zeta, z) - R_1(\tau, z)\| \leqslant C \frac{|\zeta - \tau|}{|\zeta - z| |\tau - z|^\rho} \quad ((\zeta, z) \in \Delta'_\varphi) \tag{7.4.18}$$

valid for any $\rho > 0$ (with C depending on ρ): here τ is any complex number in the segment joining z and ζ. The proof of Corollary 7.2.4 can then be extended to the complex plane: the result is the estimate

$$\|R(\zeta, z) - R(\tau, z)\| \leqslant C \left\{ \frac{|\zeta - \tau|}{|\zeta - z| |\tau - z|^\rho} + \frac{|\zeta - \tau|^\kappa}{|\tau - z|} \right\} \quad ((\zeta, z) \in \Delta'_\varphi), \tag{7.4.19}$$

which holds for any $\rho, \kappa > 0$ with $\rho + \kappa < 1$. Making use of this inequality combined with (7.4.16) in (7.4.14), the first estimate (7.4.6) follows. The second can be obtained working in the same way with the integral equation defining $\tilde{S}(t, s)$. Details are left to the reader.

7.4.3 Example. The "real analytic" version of Theorem 7.4.1 is false, as can be seen with the function $t \to A(t)$ produced in Example 7.3.2; although the real counterparts of the hypotheses of Theorem 7.4.1 are satisfied, the propagator $S(t, s)$ is not real analytic in the neighborhood of any point (s, t), $0 \leqslant s < t \leqslant T$. To see this it suffices to show that the function $u(t, s, x) = S(t, s)u(x)$ provided by (7.3.15) (say, for $u(x) \equiv 1$) is merely $C^{(\infty)}$ (but not real analytic) as a L^2-valued function. Details are omitted; we only point out that the verification is related to the classical counterexample of an infinitely differentiable nonanalytic function $f(\xi) = e^{-1/\xi}$ ($\xi > 0$), $f(\xi) = 0$ ($\xi \leqslant 0$).

Since Assumption (II′) is the same in its real or complex form, the total failure of Theorem 7.4.1 to function in the real domain must be traced to (I′). In fact, even though the family $\{A(t)\}$ can be extended analytically to the entire complex plane by the obvious formula

$$A(z)u(x) = -\frac{1}{(z-x)^2}u(x)$$

(note incidentally that $A(z)$ is bounded off the real axis), we have $\sigma(A(z)) = \{-(z-x)^{-2}; 0 \leqslant x \leqslant 1\}$; Assumption (I′) is not satisfied in any complex neighborhood of $[0,1]$ ($\sigma(A(a+\varepsilon e^{i\varphi}))$ contains the point $-\varepsilon^2 e^{-2i\varphi}$, which surrounds the origin as φ moves from 0 to π).

We note a few additional results that can be obtained by careful examination of the proof of Theorem 7.4.1. In the first place, it can be proved that $S(\zeta, z)$ is strongly continuous in the closure of Δ'_φ (we only have proved strong continuity up to $\zeta = z$ for ζ, z real); then, equalities (7.1.7) and (7.1.8) can be extended to the complex plane:

$$S(z,z) = I, \quad S(\zeta,\tau)S(\tau,z) = S(\zeta,z) \qquad (7.4.20)$$

for $(\zeta,\tau), (\tau,z) \in \Delta_\varphi$ (note that $(\zeta,z) \in \Delta_\varphi$). On the basis of these observations, the following result shows that the conditions in Theorem 7.4.1 are essentially necessary.

7.4.4   Example (Masuda [1972: 1]). Let Δ, φ be as in the statement of Theorem 7.4.1. Assume that $\{S(\zeta, z); (\zeta, z) \in \Delta_\varphi\}$ is a family of operators in (E) such that:

(a) (7.4.20) is satisfied for $(\zeta,\tau), (\tau, z)$ in Δ_φ.
(b) $S(\cdot, \cdot)$ can be extended to a strongly continuous, uniformly bounded function in the closure of Δ_φ in such a way that

$$S(z,z) = I \quad (z \in \Delta). \qquad (7.4.21)$$

(c) $S(\cdot, \cdot)$ is holomorphic in Δ_φ.

For each $z \in \Delta$, define an operator $A(z)$ by

$$A(z)u = \lim_{h \to 0+} h^{-1}(S(z+h, z) - I)u, \qquad (7.4.22)$$

the domain of $A(z)$ consisting precisely of those u for which (7.4.22) exists. Then the family $\{A(z)\}$ has the following properties:

(d) $A(z)$ is densely defined for all $z \in \Delta$.
(e) For any compact subset K of Δ and any $\varepsilon > 0$ there exists constants C, M such that

$$\Sigma = \Sigma(\varphi + \pi/2 - \varepsilon) \cap \{\lambda; |\lambda| \geqslant M\}$$

belongs to $\rho(A(z))$ for $z \in K$,

$$\|R(\lambda; A(z))\| \leqslant C/|\lambda| \quad (\lambda \in \Sigma, z \in K)$$

and for each $\lambda \in \Sigma$, $R(\lambda; A(z))$ is holomorphic in z in some neighborhood of K.

(f) $S(\zeta, z)E \subseteq D(A(\zeta))$, $S(\zeta, z)A(z)$ is bounded in $D(A(z))$ and

$$D_\zeta S(\zeta, z) = A(\zeta)S(\zeta, z), \quad D_z S(\zeta, z) = - \overline{S(\zeta, z)A(z)} .$$

7.5. THE INHOMOGENEOUS EQUATION

We prove in this section two results on the equation

$$u'(t) = A(t)u(t) + f(t), \tag{7.5.1}$$

which are based, respectively, on the hypotheses in Section 7.3 and 7.2. The first one refers to weak solutions of (7.5.1). If f is, say, locally summable in $t \geq 0$, these solutions are defined by generalizing in an obvious way definition (B) in Section 7.3; a function $u(\cdot)$, continuous in $s \leq t \leq T$, is a *weak solution* of (7.5.1) with initial condition

$$u(s) = u_0 \tag{7.5.2}$$

for f locally summable in $t \geq 0$ if and only if

$$\int_s^T \langle u(t), u^{*\prime}(t) + A^*(t)u^*(t) \rangle \, dt = - \langle u_0, u^*(s) \rangle - \int_s^T \langle f(t), u^*(t) \rangle \, dt \tag{7.5.3}$$

for all functions $u^*(\cdot)$ satisfying the assumptions in Definition (B).

As noted in Section 7.1, the analogue of the Lagrange formula (2.4.3) in the time-dependent case is

$$u(t) = S(t, s)u_0 + \int_s^t S(t, \sigma)f(\sigma) \, d\sigma. \tag{7.5.4}$$

It does not seem to be known in general whether every weak solution of (7.5.1) must be representable in the form (7.5.4), at least if we only assume that the Cauchy problem for (7.5.1) is properly posed as defined in Section 7.1; note that the corresponding result for the time-independent case (see Theorem 2.4.6) was proved using the Phillips adjoint of S, whose theory has not been fully extended to the time-dependent case. However, the result holds in the abstract parabolic setting; this will be seen as a consequence of the following result.

7.5.1 Theorem. *Let Assumptions* (I), (II), *and* (III) *in Section* 7.2 *be satisfied. Assume u_0 is an arbitrary element of E and f is continuous in $s \leq t \leq T$ (or, more generally, is locally summable there). Then the function $u(t)$ defined by* (7.5.4) *is the only weak solution of* (7.5.1), (7.5.2) *in $s \leq t \leq T$.*

Proof. The first term on the right-hand side of (7.5.4) has already been shown to be a weak solution of the homogeneous equation (7.3.1) in

Theorem 7.3.1, thus we may assume that $u_0 = 0$. We begin by showing that

$$u(t) = \int_s^t S(t, \sigma) f(\sigma) \, d\sigma$$

is continuous in $s \leqslant t \leqslant T$ (under the sole hypothesis of strong continuity of S). In fact, if $s \leqslant t < t' \leqslant T$, we have

$$\|u(t') - u(t)\| \leqslant \int_t^{t'} \|S(t', \sigma)\| \, \|f(\sigma)\| \, d\sigma$$

$$+ \int_s^t \|(S(t' - s) - S(t - s)) f(s)\| \, ds$$

and continuity follows as in Example 2.4.5. We prove now that $u(\cdot)$ satisfies the inhomogeneous equation in the weak sense. We have

$$\int_s^T \langle u(t), u^{*\prime}(t) \rangle \, dt = \int_s^T \left\langle \int_s^t S(t, \sigma) f(\sigma) \, d\sigma, u^{*\prime}(t) \right\rangle dt$$

$$= \int_s^T \int_\sigma^T \langle S(t, \sigma) f(\sigma), u^{*\prime}(t) \rangle \, dt \, d\sigma$$

$$= \lim_{k \to 0+} \lim_{h \to 0+} \int_s^{T-k} \int_{\sigma+k}^T \langle S_h(t, \sigma) f(\sigma), u^{*\prime}(t) \rangle \, dt \, d\sigma.$$

But

$$\int_s^{T-k} \int_{\sigma+k}^T \langle S_h(t, \sigma) f(\sigma), u^{*\prime}(t) \rangle \, dt \, d\sigma$$

$$= - \int_s^{T-k} \langle S_h(\sigma + k, \sigma) f(\sigma), u^*(\sigma + k) \rangle \, d\sigma$$

$$- \int_s^{T-k} \int_{\sigma+k}^T \langle (D_t S_h(t, \sigma) - A(t) S(t, \sigma)) f(\sigma), u^*(t) \rangle \, dt \, d\sigma$$

$$- \int_s^{T-k} \int_{\sigma+k}^T \langle S_h(t, \sigma) f(\sigma), A(t)^* u^*(t) \rangle \, dt \, d\sigma. \qquad (7.5.5)$$

Making use of the properties of $S_h(t, s)$ established in Lemma 7.2.6, we see that (7.5.5) tends (as $h \to 0+$) to

$$- \int_s^{T-k} \langle S(\sigma + k, \sigma) f(\sigma), u^*(\sigma + k) \rangle \, d\sigma$$

$$- \int_s^{T-k} \int_{\sigma+k}^T \langle S(t, \sigma) f(\sigma), A(t)^* u^*(t) \rangle \, dt \, d\sigma \qquad (7.5.6)$$

On the other hand, this expression tends to

$$- \int_s^T \langle f(t), u^*(t) \rangle \, dt - \int_s^T \int_\sigma^T \langle S(t, \sigma) f(\sigma), A(t)^* u^*(t) \rangle \, dt \, d\sigma$$

$$= - \int_s^T \langle f(t), u^*(t) \rangle \, dt - \int_s^T \langle u(t), A(t)^* u^*(t) \rangle \, dt$$

when $k \to 0+$, which proves (7.5.3) taking into account that $u(s) = u_0 = 0$.

Since uniqueness of weak solutions of (7.5.1) reduces to uniqueness of solutions of the homogeneous equation (7.3.1) (which has been established in Theorem 7.3.1), the proof is complete.

The next result assumes the hypotheses used in Section 7.2 and generalizes to the time-dependent case Example 2.5.7.

7.5.2   Theorem.   *Let Assumptions* (I), (II), (III), *and* (IV) *in Section 7.2 hold. Assume u_0 is an arbitrary element of E and f is Hölder continuous in $s \leqslant t \leqslant T$, that is, for some $\beta > 0$,*

$$\|f(t') - f(t)\| \leqslant C|t' - t|^{\beta} \quad (s \leqslant t, t' \leqslant T). \tag{7.5.7}$$

Then the function $u(t)$ defined by (7.5.4) is the only genuine solution of the inhomogeneous equation (7.5.1) in the sense of Definition (A), *Section 7.2.*

Proof.   Uniqueness again follows from uniqueness for the homogeneous equation, thus we can assume that $u_0 = 0$ in (7.5.4), which we do in the sequel.

In view of (7.2.10), we have

$$S(t - s; A(r)) = \frac{1}{2\pi i} \int_{\Gamma} e^{\lambda(t-s)} R(\lambda; A(r)) \, d\lambda, \tag{7.5.8}$$

where Γ is the contour used in Section 7.2. Hence

$$A(t)S(t - s; A(t)) - A(s)S(t - s; A(s))$$

$$= \frac{1}{2\pi i} \int_{\Gamma} \lambda e^{\lambda(t-s)} \{R(\lambda; A(t)) - R(\lambda; A(s))\} \, d\lambda \tag{7.5.9}$$

in $0 \leqslant s < t \leqslant T$. By virtue of Assumption (III),

$$\|R(\lambda; A(t)) - R(\lambda; A(s))\| \leqslant (t - s) \sup_{s \leqslant \tau \leqslant t} \|D_{\tau} R(\lambda; A(\tau))\|$$

$$\leqslant C(t - s)|\lambda|^{\rho - 1},$$

and we obtain from (7.5.9) that

$$\|A(t)S(t - s; A(t)) - A(s)S(t - s; A(s))\| \leqslant C/(t - s)^{\rho} \quad (0 \leqslant s < t \leqslant T).$$

Combining this inequality with (7.2.45) and (7.2.59), we get

$$\|A(t)S(t, s) - \overline{S(t, s)A(s)}\| \leqslant C/(t - s)^{1-\kappa} \quad (0 \leqslant s < t \leqslant T) \tag{7.5.10}$$

with κ as defined in Corollary 7.2.4. This will be used later.

Write

$$\int_s^t S(t, \sigma) f(\sigma) \, d\sigma = \int_s^t S(t, \sigma)(f(\sigma) - f(t)) \, d\sigma + \int_s^t S(t, \sigma) f(t) \, d\sigma. \tag{7.5.11}$$

Combining (7.5.7) with (7.2.34), we see that $\|A(t)S(t,\sigma)(f(\sigma)-f(t))\|$ is integrable in $s \leqslant \sigma \leqslant t$, so the first integral belongs to $D(A(t))$ and $A(t)$ can be introduced under the integral sign. In relation to the second integrand, we note that

$$A(t)S(t,\sigma)f(t)$$

$$= \left(A(t)S(t,\sigma)-\overline{S(t,\sigma)A(\sigma)}\right)f(t)+\overline{S(t,\sigma)A(\sigma)}f(t)$$

where the first term on the right is integrable in $s \leqslant \sigma \leqslant t$ in view of (7.5.10); for the second we have, making use of (7.2.33):

$$\int_s^{t-h} \overline{S(t,\sigma)A(\sigma)}f(t)\,d\sigma = -\int_s^{t-h} D_\sigma S(t,\sigma)f(t)\,d\sigma$$

$$= S(t,s)f(t) - S(t,t-h)f(t).$$

Hence (7.5.11) belongs to the domain of $A(t)$ for all $t > s$ and

$$A(t)\int_s^t S(t,\sigma)f(\sigma)\,d\sigma = \int_s^t A(t)S(t,\sigma)(f(\sigma)-f(t))\,d\sigma$$

$$+ \int_s^t \left(A(t)S(t,\sigma)-\overline{S(t,\sigma)A(\sigma)}\right)f(t)\,d\sigma$$

$$+ S(t,s)f(t)-f(t). \qquad (7.5.12)$$

An examination of this equality, (7.2.34), (7.5.7), and (7.5.10), shows that

$$A(t)u(t) = A(t)\int_s^t S(t,\sigma)f(\sigma)\,d\sigma$$

is continuous in $s \leqslant t \leqslant T$ and that

$$A(t)\int_s^{t-h} S(t,\sigma)f(\sigma)\,d\sigma \to A(t)u(t) = A(t)\int_s^t S(t,\sigma)f(\sigma)\,d\sigma$$

uniformly there. Hence, if $s \leqslant t' \leqslant T$ we have

$$\int_s^{t'} A(t)u(t)\,dt = \lim_{h\to 0+} \int_{s+h}^{t'} A(t)\int_s^{t-h} S(t,\sigma)f(\sigma)\,d\sigma\,dt,$$

where

$$\int_{s+h}^{t'} A(t)\int_s^{t-h} S(t,\sigma)f(\sigma)\,d\sigma\,dt = \int_s^{t'-h}\int_{\sigma+h}^{t'} A(t)S(t,\sigma)f(\sigma)\,dt\,d\sigma$$

$$= \int_s^{t'-h}\int_{\sigma+h}^{t'} D_t S(t,\sigma)f(\sigma)\,dt\,d\sigma$$

$$= \int_s^{t'-h} S(t',\sigma)f(\sigma)\,d\sigma$$

$$- \int_s^{t'-h} S(\sigma+h,\sigma)f(\sigma)\,d\sigma.$$

Taking limits, we obtain

$$\int_s^{t'} A(t)u(t)\,dt = u(t') - \int_s^{t'} f(t)\,dt,$$

which shows that $u(\cdot)$ is indeed a solution of (7.5.1). This completes the proof of Theorem 7.5.2.

7.6. PARABOLIC EQUATIONS WITH TIME-DEPENDENT COEFFICIENTS

We examine here time-dependent versions of the elliptic operators in Chapter 4. Since the methods are clearly visible in space dimension one and most of the technical complications are absent, we look mostly at this case. In the following preparatory results, β_0 (resp. β_l) is a boundary condition at 0 (resp. at $l > 0$) and $A_p(\beta_0, \beta_l)$ is the operator (written in variational form)

$$A_p(\beta_0, \beta_l)u(x) = (a(x)u'(x))' + b(x)u'(x) + c(x)u(x). \quad (7.6.1)$$

The coefficients a, b are assumed to be continuously differentiable in $0 \leqslant x \leqslant l$ while c is continuous there. As in Section 4.3, $A_p(\beta_0, \beta_l)$ is thought of as an operator in $L^p(0, l)$ with domain consisting of all $u \in W^{2,p}(0, l)$ that satisfy both boundary conditions. Use will also be made of the operator A_p, defined in a similar way, but without including boundary conditions in the definition of the domain (that is, $D(A_p) = W^{2,p}(0, l)$). The admissible range of p will be $1 < p < \infty$.

The first observation concerns solutions of the equation

$$(\lambda I - A_p)u = f \quad (7.6.2)$$

for $f \in L^p$ with nonhomogeneous boundary conditions; to fix ideas, assume both boundary conditions are of type (I), so that one wishes to find $u \in W^{2,p}(0, l)$ satisfying (7.6.2) and

$$u'(0) = \gamma_0 u(0) + d_0, \quad u'(l) = \gamma_l u(l) + d_l. \quad (7.6.3)$$

To solve this problem we use our information on the operator $A_p(\beta_0, \beta_l)$. Let ρ be a real-valued twice continuously differentiable function satisfying the same boundary conditions as u:

$$\rho'(0) = \gamma_0 \rho(0) + d_0, \quad \rho'(l) = \gamma_l \rho(l) + d_l.$$

Then $v = u - \rho$ belongs to $D(A_p(\beta_0, \beta_l))$ and satisfies

$$(\lambda I - A(\beta_0, \beta_l))v = f - (\lambda I - A_p)\rho. \quad (7.6.4)$$

Observe next that

$$\|A_p\rho\| \leqslant C'\|\rho\|_{W^{2,p}} \leqslant C''(|d_0| + |d_l|), \quad (7.6.5)$$

where the constant C'' does not depend on $|d_0|, |d_1|$ if ρ is adequately chosen. We deduce then that if $\lambda \in \rho(A_p(\beta_0, \beta_l))$, the solution of (7.6.2),(7.6.3) given by

$$u = R(\lambda; A_p(\beta_0, \beta_l))(f - (\lambda I - A_p)\rho) + \rho \quad (7.6.6)$$

can be estimated in terms of $\|f\|, |d_0|, |d_1|$. It will be important in what follows to obtain estimates that are "uniform with respect to $A_p(\beta_0, \beta_l)$" in a sense to be precised below. We consider a class $\mathfrak{A}$ of operators $A_p(\beta_0, \beta_l)$ whose coefficients satisfy the inequalities:

$$|a(x)|, |a'(x)|, |b(x)|, |b'(x)|, |c(x)|, |\gamma_0|, |\gamma_l| \leqslant C, \quad a(x) \geqslant \kappa > 0,$$
$$(7.6.7)$$

uniformly for $0 \leqslant x \leqslant l$ and $A_p(\beta_0, \beta_l) \in \mathfrak{A}$.

7.6.1 Lemma. *Let φ_p be the angle in Theorem 7.3.1,*

$$\varphi_p = \operatorname{arc tg}\left\{\left(\frac{p}{p-2}\right)^2 - 1\right\}^{1/2}.$$

Then if $0 < \varphi < \varphi_p$, there exists ω such that the sector $\Sigma = \Sigma(\varphi + \pi/2) = \{\lambda; |\arg \lambda| \leqslant \varphi + \pi/2\}$ belongs to the resolvent set of every $A_p(\beta_0, \beta_l) - \omega I$, $A_p(\beta_0, \beta_l) \in \mathfrak{A}$, and the solution u of (7.6.2),(7.6.3) for $A_p - \omega I$ satisfies

$$\|u\| \leqslant \frac{1}{|\lambda + 1|}\|f\| + C(|d_0| + |d_l|), \tag{7.6.8}$$

$$\|u\|_{W^{1,p}} \leqslant C'\left(\frac{1}{|\lambda + 1|^{1/2}}\|f\| + |\lambda|^{1/2}(|d_0| + |d_l|)\right) \tag{7.6.9}$$

$$\|u\|_{W^{2,p}} \leqslant C''(\|f\| + |\lambda|(|d_0| + |d_l|)) \tag{7.6.10}$$

for $\lambda \in \Sigma$ and $A_p(\beta_0, \beta_l) \in \mathfrak{A}$, where the constants C, C', C'' are independent of $f, d_0, d_1, \lambda, A_p(\beta_0, \beta_l) \in \mathfrak{A}$.

The proof follows from careful observation of the material in Section 4.3. Note first that under the present uniform bounds on the coefficients of $A_p(\beta_0, \beta_l)$, the estimate (4.3.15) can be obtained with ω independent of $A_p(\beta_0, \beta_l)$; it suffices to increase that ω to $\omega + 1$ and (7.6.8) results from (7.6.6). To obtain (7.6.10) we note that the norm of $W^{2,p}$ is equivalent to, say, the graph norm of $A_p(\beta_0, \beta_l) - \omega I$. Finally, to show (7.6.9), we use the fact (Adams [1975: 1, p. 70]) that for every $u \in W^{2,p}(0, l)$ we have

$$\|u\|_{W^{1,p}} \leqslant K\delta\|u\|_{W^{2,p}} + K\delta^{-1}\|u\|_{L^p} \tag{7.6.11}$$

for $0 < \delta \leqslant 1$, where K only depends on p and l. Setting $\delta = |\lambda + 1|^{-1/2}$ and using (7.6.8) and (7.6.10), the desired inequality is obtained.

In the following result we consider time-dependent operators and boundary conditions: here $\beta_0(t), \beta_l(t)$ are, for each t, boundary conditions of type (I),

$$u'(0) = \gamma_0(t)u(0), \quad u'(t) = \gamma_l(t)u(l), \tag{7.6.12}$$

and

$$A_p(t, \beta_0(t), \beta_l(t)) = (a(x,t)u'(x))' + b(x,t)u'(x) + c(x,t)u(x). \tag{7.6.13}$$

We assume that γ_0 and γ_l are defined and continuously differentiable in $0 \leqslant t \leqslant T$ and that $D_t a, a', D_t a', b', D_t b, D_t c$ exist and are continuous in $0 \leqslant x \leqslant l, 0 \leqslant t \leqslant T$ together with a, b, c. Moreover,

$$a(x, t) > 0 \quad (0 \leqslant x \leqslant l, 0 \leqslant t \leqslant T). \tag{7.6.14}$$

In what follows, ω is still the parameter in Lemma 7.6.1 corresponding to the family $\mathfrak{A} = \{ A_p(t; \beta_0(t), \beta_l(t)); 0 \leqslant t \leqslant T \}$.

7.6.2 Theorem. *Under the hypotheses above the family of operators*

$$\{ A_p(\beta_0(t), \beta_l(t)) - \omega I; 0 \leqslant t \leqslant T \} \tag{7.6.15}$$

in $L^p(0, l)$ satisfies the assumptions of Theorem 7.3.1; in particular, the Cauchy problem for the equation

$$u'(t) = A_p(t; \beta_0(t), \beta_l(t)) u(t) \quad (0 \leqslant t \leqslant T) \tag{7.6.16}$$

is properly posed (with respect to the weak solutions defined in (B), Section 7.3).

Proof. We only have to verify assumptions (I), (II), and (III) in Section 7.2. Since the family (7.6.15) fits into Lemma 7.6.1, (I) follows from (7.6.8) with $d_0 = d_l = 0$ and $u = u_\lambda(\cdot, t) = R(\lambda; A(t; \beta_0(t), \beta_l(t)) - \omega I)f$. The second estimate implies that

$$\| u_\lambda(\cdot, t) \|_{W^{1,p}} \leqslant C \| f \| \tag{7.6.17}$$

uniformly for $0 \leqslant t \leqslant T, \lambda \in \Sigma$. This shows uniform boundedness of $u_\lambda(0, t)$, $u_\lambda(l, t)$ since it follows from (4.3.40) that

$$| u_\lambda(0, t) |, | u_\lambda(l, t) | \leqslant C \| u_\lambda(\cdot, t) \|_{W^{1,p}} \tag{7.6.18}$$

for $0 \leqslant t \leqslant T$ and $\lambda \in \Sigma$ fixed (how C depends on λ is unimportant for the moment). The next step is to observe that the function

$$w_\lambda(x, t) = u_\lambda(x, t + h) - u_\lambda(x, t)$$

satisfies the nonhomogeneous boundary value problem

$$\left(\lambda I - \left(A_p(t) - \omega I \right) \right) w_\lambda(x, t) = \left(A_p(t + h) - A_p(t) \right) u_\lambda(x, t + h), \tag{7.6.19}$$

$$w_\lambda'(0, t) = \gamma_0(t) w_\lambda(0, t) + \left(\gamma_0(t + h) - \gamma_0(t) \right) u_\lambda(0, t + h),$$

$$w_\lambda'(l, t) = \gamma_l(t) w_\lambda(l, t) + \left(\gamma_l(t + h) - \gamma_l(t) \right) u_\lambda(l, t + h). \tag{7.6.20}$$

A look at (7.6.10) and (7.6.18) shows that, for $\lambda \in \Sigma$ fixed, the function

$$t \to u_\lambda(\cdot, t) \in W^{2, p} \tag{7.6.21}$$

is continuous in the $W^{2, p}$ norm in $0 \leqslant t \leqslant T$ uniformly for f in $\| f \| \leqslant 1$. We consider next the problem

$$\left(\lambda I-\left(A_p(t)-\omega I\right)\right)v_\lambda(x,t)=\left(D_t a(x,t)u_\lambda'(x,t)\right)'$$
$$+D_t b(x,t)u_\lambda'(x,t)+D_t c(x,t)u(x,t)$$

$$(7.6.22)$$

with boundary conditions

$$v_\lambda'(0,t)=\gamma_0(t)v_\lambda(0,t)+D_t\gamma_0(t)u_\lambda(0,t),$$
$$v_\lambda'(l,t)=\gamma_l(t)v_\lambda(l,t)+D_t\gamma_l(t)u_\lambda(l,t). \qquad (7.6.23)$$

A similar argument, this time applied to $v_\lambda(x,t+h)-v_\lambda(x,t)$, shows that

$$t\to v_\lambda(\cdot,t)\in W^{2,p} \qquad (7.6.24)$$

is continuous for $0\leqslant t\leqslant T$ in the $W^{2,p}$ norm uniformly in $\|f\|\leqslant 1$; here we make use of the continuity of (7.6.21) and of the fact, again consequence of (4.3.40), that $u_\lambda(0,t),u_\lambda(l,t)$ are continuous in $0\leqslant t\leqslant T$ uniformly in $\|f\|\leqslant 1$. It is obvious that (II) will be amply satisfied if we can show that

$$v_\lambda(\cdot,t)=D_t R\left(\lambda;A\left(t;\beta_0(t),\beta_l(t)\right)-\omega I\right)f(\cdot) \qquad (7.6.25)$$

(which is intuitively obvious since the problem (7.6.22),(7.6.23) is obtained differentiating formally with respect to t the boundary value problem satisfied by $u_\lambda(x,t)$). To establish (7.6.25) rigorously, we consider

$$z_\lambda(x,t)=\left(h^{-1}(u_\lambda(x,t+h)-u_\lambda(x,t))-v_\lambda(x,t)\right).$$

An amalgamation of (7.6.19) and (7.6.22) shows that z_λ satisfies

$$\left(\lambda I-\left(A_p(t)-\omega I\right)\right)z_\lambda(x,t)$$
$$=\left(\left(h^{-1}(a(x,t+h)-a(x,t))-D_t a(x,t)\right)u_\lambda'(x,t)\right)'$$
$$+\left(h^{-1}(b(x,t+h)-b(x,t))-D_t b(x,t)\right)u_\lambda'(x,t)$$
$$+\left(h^{-1}(c(x,t+h)-c(x,t))-D_t c(x,t)\right)u_\lambda(x,t)$$
$$+h^{-1}\left(A_p(t+h)-A_p(t)\right)(u_\lambda(x,t+h)-u_\lambda(x,t)),$$

$$z_\lambda'(0,t)=\gamma_0(t)z_\lambda(0,t)+\left(h^{-1}(\gamma_0(t+h)-\gamma_0(t))-D_t\gamma_0(t)\right)u_\lambda(0,t)$$
$$+h^{-1}(\gamma_0(t+h)-\gamma_0(t))(u_\lambda(0,t+h)-u_\lambda(0,t)),$$

$$z_\lambda'(l,t)=\gamma(t)z_\lambda(l,t)+\left(h^{-1}(\gamma_l(t+h)-\gamma_l(t))-D_t\gamma_l(t)\right)u_\lambda(l,t)$$
$$+h^{-1}(\gamma_l(t+h)-\gamma_l(t))(u_\lambda(l,t+h)-u_\lambda(l,t)).$$

thus $z_\lambda\to 0$ as $h\to 0$.

It remains to show that Assumption (III) holds. In possession of (7.6.25) we go back to the boundary value problem defining v_λ and obtain, using (7.6.8), that

$$\|v_\lambda(\cdot, t)\| \leq \frac{C}{|\lambda + 1|} \|u\|_{W^{2,p}} + C(|u_\lambda(0, t)| + |u_\lambda(l, t)|).$$

We use next (7.6.9) and (7.6.10) for u, keeping in mind that u satisfies a homogeneous boundary value problem. In view of (7.6.18), we obtain

$$\|v_\lambda(\cdot, t)\| \leq \frac{C}{|\lambda + 1|^{1/2}} \quad (\lambda \in \Sigma, 0 \leq t \leq T), \qquad (7.6.26)$$

which completes the verification of (III) and thus the proof of Theorem 7.6.2.

7.6.3 Theorem. *Let the hypotheses in Theorem 7.6.2 be satisfied. Assume in addition that $D_t a, D_t a', D_t b, D_t c$, are Hölder continuous in t uniformly with respect to x and that $D_t \gamma_0, D_t \gamma_l$ are Hölder continuous. Then the family (7.6.15) satisfies the assumptions of Theorem 7.2.5; in particular, the Cauchy problem for (7.6.16) is properly posed with respect to the strong solutions defined in (V), Section 7.2).*

Proof. We only have to verify Assumption (IV) in Section 7.2. We can in fact show that if α is the Hölder exponent of $D_t a, D_t a', D_t b, D_t c$, $D_t \gamma_0, D_t \gamma_l$, then

$$\|y_\lambda(\cdot, t, t')\| \leq C \quad (0 \leq t < t' \leq T),$$

where $y_\lambda(x, t, t') = (t' - t)^{-\alpha}(v_\lambda(x, t') - v_\lambda(x, t))$. This is done by examination of the boundary value problem satisfied by y_λ, which is easily deduced from (7.6.22). Details are omitted.

The case where one of the boundary conditions is of type (II) is handled in a similar way. When both boundary conditions are of type (II) the argument becomes in fact much simpler, since all functions employed $(u_\lambda, v_\lambda, \ldots)$ satisfy the homogeneous boundary conditions and no results on the nonhomogeneous problem are necessary. Of course the same is true if β_0, β_l are of type (I) but independent of t.

Finally, we comment briefly on the extension of the results to partial differential operators. Using again the divergence form, we write

$$A(t; \beta(t)) = \sum \sum D^j(a_{jk}(x, t) D^k u) + \sum b_j(x, t) D^j u + c(x, t) u,$$

where the coefficients a_{jk}, b_j, c are defined in $\bar{\Omega} \times [0, T]$, Ω a bounded domain in $\mathbb{R}^m$, and $\beta(t)$ is a time-dependent boundary condition on the boundary Γ. As in dimension one, the case where $\beta(t)$ is the Dirichlet boundary condition is much simpler, and we indicate how to adapt the arguments above. The assumptions are as follows. For each t, the coefficients of the operator and the domain satisfy the smoothness assumptions in Theorem 4.9.3; moreover, $a_{jk}, D^j a_{jk}, b_j$ and c are continuously differentiable with respect to t in $\bar{\Omega} \times [0, T]$ and

$$\sum \sum a_{jk}(x, t)\xi_j\xi_k \geq \kappa|\xi|^2 \quad (\xi \in \mathbb{R}^m, (x, t) \in \bar{\Omega} \times [0, T])$$

for some $\kappa > 0$. The one-dimensional reasoning adapts without changes, the only estimates needed being those for the homogeneous boundary value problem (we note incidentally that the "intermediate" estimate (7.6.9) is unnecessary). Under these conditions the family

$$\{A(t; \beta(t)); 0 \leq t \leq T\} \tag{7.6.27}$$

satisfies the conclusion of Theorem 7.6.3. If one wishes to fit (7.6.27) into Theorem 7.6.4, it is enough to require that $D_t a_{jk}$, $D_t D^j a_{jk}$, $D_t b_j$, and $D_t c$ be Hölder continuous in t uniformly with respect to $x \in \bar{\Omega}$.

The case where the boundary condition is of type (I) must be handled with the help of estimates on the nonhomogeneous boundary value problem (see Example 7.6.6 below).

7.6.4   Example.   State and prove analogues of Theorems 7.6.3 and 7.6.4 in the case $m > 1$ for Dirichlet boundary conditions using the suggestions above and the results on partial differential operators in Chapter 4.

*7.6.5   Example.   *A multidimensional analogue of* (7.6.5). Let Ω be a bounded domain of class $C^{(2)}$, $\gamma(\cdot)$ a function defined and continuously differentiable on the boundary Γ. Consider the space $W^{1,p}(\Omega)$, $1 < p < \infty$. If $d(\cdot)$ is a function in $W^{1,p}(\Omega)$, then there exists a function $\rho(\cdot)$ in $W^{2,p}(\Omega)$ such that

$$D^{\tilde{\nu}}\rho(x) = \gamma(x)\rho(x) + d(x) \quad (x \in \Gamma)$$

in the sense of (4.7.37). Moreover, ρ can be chosen in such a way that there exists a constant C independent of $d(\cdot)$ such that

$$\|\rho\|_{W^{2,p}(\Omega)} \leq C\|d\|_{W^{1,p}(\Omega)}. \tag{7.6.28}$$

The outline of the proof is as follows. By means of boundary patches and their associated maps, we can reduce locally the problem to the case $\Omega = \mathbb{R}^m_+$, $\Gamma = \mathbb{R}^{m-1}$, and then reason as in Example 4.8.19 via an approximation argument.

7.6.6   Example.   Using Example 7.6.5, establish a multi-dimensional analogue of Lemma 7.6.1.

7.6.7   Example.   State and prove extensions of Theorem 7.6.2 and Theorem 7.6.3 in the case $m > 1$ for boundary conditions of type (I).

7.7.   THE GENERAL CASE

We study in this section the abstract differential equation

$$u'(t) = A(t)u(t) \tag{7.7.1}$$

under hypotheses that can be roughly described as follows: each $A(t)$ is assumed to belong to $\mathcal{C}_+$ and relations among the operators $A(t), A(t'), \ldots$ are imposed in order to make possible the construction of $S(t, s)$ under the "product integral" form sketched in Example 7.1.8 in the bounded case.

To simplify future statements, we introduce several ad hoc notations and definitions, several of which were already used in Section 5.6. If E, F are two Banach spaces, we write $F \to E$ to indicate that F is a dense subspace of E and that the identity map from F to E is bounded. Expressions like $A \in \mathcal{C}_+(\omega; E)$, $A \in \mathcal{C}_+(C, \omega, E)$ indicate that A belongs to the indicated classes *in the space E*. Given an operator $A \in \mathcal{C}_+(E)$, we say that F is *A-admissible* if and only if

$$S(t; A)F \subseteq F \quad (t \geqslant 0) \tag{7.7.2}$$

and $t \to S(t; A)$ is strongly continuous in F for $t \geqslant 0$. We denote by A_F the largest restriction of A to F with range in F, that is, the restriction of A with domain $\{u \in D(A) \cap F; Au \in F\}$. The reader is referred to Section 5.6 for examples and relations between these definitions, especially Lemma 5.6.2, which will be useful in the sequel.

A family $\{A(t); 0 \leqslant t \leqslant T\}$ of densely defined operators in the Banach space E is said to be *stable (in E) with stability constants C, ω* if and only if $\cap \rho(A(t)) \supseteq (\omega, \infty)$ and

$$\left\| \prod_{j=1}^{n} R(\lambda; A(t_j)) \right\| \leqslant C(\lambda - \omega)^{-n} \quad (\lambda > \omega) \tag{7.7.3}$$

for every finite set $\{t_j\}$, $0 \leqslant t_1 \leqslant t_2 \leqslant \cdots \leqslant t_n \leqslant T$. Here and in what follows, products like the one in (7.7.3) are arranged in descending order of indices, that is,

$$\prod_{j=1}^{n} B_j = B_n B_{n-1} \cdots B_1. \tag{7.7.4}$$

It is plain (as we see taking $t_1 = t_2 = \cdots = t_n$) that each $A(t)$ in a stable family belongs to $\mathcal{C}_+(C, \omega)$. Two equivalent forms of the definition are proved below.

7.7.1 Lemma. (a) *Inequality (7.7.3) is equivalent to*

$$\left\| \prod_{j=1}^{n} R(\lambda_j; A(t_j)) \right\| \leqslant C \prod_{j=1}^{n} (\lambda_j - \omega)^{-1} \tag{7.7.5}$$

for all $\{t_j\}, \{\lambda_j\}$ such that $0 \leqslant t_1 \leqslant t_2 \leqslant \cdots \leqslant t_n \leqslant T$ and $\lambda_1, \lambda_2, \ldots, \lambda_n > \omega$. (b) *A family $A(\cdot)$ of operators in $\mathcal{C}_+(\omega)$ is stable with stability constants C, ω if and only if*

$$\left\| \prod_{j=1}^{n} S(s_j; A(t_j)) \right\| \leqslant Ce^{\omega(s_1 + \cdots + s_n)} \tag{7.7.6}$$

for all $\{t_j\}$ as above, $s_1, s_2, \ldots, s_n \geqslant 0$, $S(s_j; A(t_j)) = \exp(s_j A(t_j))$.

Proof. Assume that $A(\cdot)$ is stable, and let the $\{s_j\}$ $(s_1, s_2, \ldots, s_n \geqslant 0)$ be rational, $s_j = p_j / q_j$ (p_j, q_j integers). Let n be a positive integer. Consider the operator

$$B_n = \prod_{j=1}^{n} \left\{ \frac{n_j}{s_j} R\left(\frac{n_j}{s_j}; A(t_j) \right) \right\}^{n_j}, \tag{7.7.7}$$

where $n_j = nq_1 \cdots q_n s_j = nq_1 \cdots q_{j-1} p_j q_{j+1} \cdots q_n$. Since n_j/s_j is independent of j, we can use (7.7.3) for $\lambda = n_j/s_j$ and the family

$$\{t_1, \ldots, t_1, t_2, \ldots, t_2, \ldots, t_n, \ldots, t_n\},$$

each t_j repeated n_j times, obtaining

$$\|B_n\| \leqslant C \prod_{j=1}^{n} \left(1 - \frac{\omega s_j}{n_j} \right)^{-n_j}. \tag{7.7.8}$$

Observe next that, in view of (2.1.27) each of the n factors on the right-hand side of (7.7.7) converges strongly to $S(s_j; A(t_j))$, while the product obtained deleting any number of factors from (7.7.7) is uniformly bounded, just as the total product, by virtue of the stability assumption. This is easily seen to imply that

$$B_n \to \prod_{j=1}^{n} S(s_j; A(t_j))$$

strongly, thus we obtain (7.7.6) for s_j rational from (7.7.8). The general case is proved exploiting the strong continuity of each $S(s; A(t_j))$. Inequality (7.7.5) is then obtained noting that

$$\prod_{j=1}^{n} R(\lambda_j; A(t_j)) u = \int_0^{\infty} e^{-\lambda(s_1 + \cdots + s_n)} \prod_{j=1}^{n} S(s_j; A(t_j)) u \, ds_1 \cdots ds_n. \tag{7.7.9}$$

Finally, it is obvious that (7.7.5) implies (7.7.3) if we take all the λ_j equal to λ. This completes the proof of Lemma 7.7.1.

7.7.2　Remark. To verify the stability condition (7.7.3) (in any of its equivalent forms) is usually no easy task, except of course in the trivial case where all the $A(t)$ are m-dissipative, or, more generally, when all the $A(t) - \omega I \in \mathcal{C}_+(1,0)$ for some fixed ω; in fact, in this case, if $\lambda > \omega$ and $0 \leqslant t_1 \leqslant \cdots \leqslant t_n \leqslant T$,

$$\left\| \prod_{j=1}^{n} R(\lambda; A(t_j)) \right\| \leqslant (\lambda - \omega)^{-n}. \tag{7.7.10}$$

The following result generalizes this observation to the case where the above condition is satisfied after a certain (time-dependent) renorming of the space.

7.7.3　Example (Kato [1970; 1]). For each t, $0 \leqslant t \leqslant T$, let $\|\cdot\|_t$ be a norm in E equivalent to the original one. Assume that there exists a constant $c \geqslant 0$

with

$$\|u\|_t \leqslant e^{c|t-s|}\|u\|_s \quad (u \in E, 0 \leqslant s, t \leqslant T), \tag{7.7.11}$$

and let E_t be the space E endowed with the norm $\|\cdot\|_t$. Assume that

$$A(t) \in \mathcal{C}_+(1, \omega, E_t) \quad (0 \leqslant t \leqslant T) \tag{7.7.12}$$

for some ω. Then the family $A(\cdot)$ is stable.

Another method of constructing stable families is by bounded perturbations of other stable families. In this direction we have the following result (which will not be used until next section).

7.7.4 Lemma. *Let $A(\cdot)$ be stable (with constants C, ω) and let $B(\cdot)$ be a family of operators in (E) such that*

$$\|B(t)\| \leqslant M \quad (0 \leqslant t \leqslant T). \tag{7.7.13}$$

Then $\{A(t) + B(t); 0 \leqslant t \leqslant T\}$ is stable (with constants $C, \omega + CM$).

Proof. It was proved in Theorem 5.1.2 that if $A \in \mathcal{C}_+(C, \omega)$ and $B \in (E)$, then $A + B \in \mathcal{C}_+(C, \omega + C\|B\|)$ by using a direct construction (and subsequent estimation) of $R(\lambda; A + B)$ as a power series in $R(\lambda; A)$. The same power series development and a very similar estimation prove the present result. Details are left to the reader.

We are now in condition to demonstrate a very general result to the effect that the Cauchy problem for (7.7.1) is well posed. However, we are forced once again to modify the notion of solution introduced in Section 7.1. In the following definition we assume that $\{A(t); 0 \leqslant t \leqslant T\}$ is a family of densely defined operators in E and F a subspace of E such that $F \subseteq D(A(t)) \ (0 \leqslant t \leqslant T)$.

(C) *The E-valued function $u(t)$ is a solution of (7.7.1) in $s \leqslant t \leqslant T$ if and only if*:
(a) *$u(\cdot)$ is E-continuous and $u(t) \in F$ in $s \leqslant t \leqslant T$.*
(b) *The right-sided derivative*

$$D^+u(t) = \lim_{h \to 0+} h^{-1}(u(t+h) - u(t))$$

exists (in the norm of E) and

$$D^+u(t) = A(t)u(t)$$

in $s \leqslant t < T$.

7.7.5 Theorem. *Let E, F be Banach spaces with $F \to E$ and F reflexive, and let $\{A(t); 0 \leqslant t \leqslant T\}$ be a family of densely defined operators in E. Assume that: (a) $A(\cdot)$ is stable in E (with constants C, ω). (b) F is $A(t)$-admissible for $0 \leqslant t \leqslant T$. (c) The family $\{A(t)_F; 0 \leqslant t \leqslant T\}$ is stable in F (with constants $\tilde{C}, \tilde{\omega}$). (d) $F \subseteq D(A(t))$ and $A(t): F \to E$, belongs to $(F; E)$*

for each t. (e) If $\tilde{A}(t)$ is the restriction of $A(t)$ to F, the map $t \to \tilde{A}(t)$ from $0 \leqslant t \leqslant T$ into $(F; E)$ is continuous. Then the Cauchy problem for (7.7.1) is properly posed with respect to solutions defined in (C) *with $D = F$. If $S(t, s)$ is the evolution operator of* (7.7.1), *then S is E-strongly continuous in the triangle $0 \leqslant s \leqslant t \leqslant T$ and*

$$\|S(t, s)\|_{(E)} \leqslant Ce^{\omega(t-s)} \quad (0 \leqslant s \leqslant t \leqslant T), \tag{7.7.14}$$

where C, ω are the constants in (a). *For each $u \in F$, $S(t, s)u$ is differentiable with respect to s in $0 \leqslant s \leqslant t$ and*

$$D_s S(t, s)u = -S(t, s)A(s)u \quad (0 \leqslant s \leqslant t). \tag{7.7.15}$$

Finally, $S(t, s)F \subseteq F$ and

$$\|S(t, s)\|_{(F)} \leqslant \tilde{C}e^{\tilde{\omega}(t-s)} \quad (0 \leqslant s \leqslant t \leqslant T), \tag{7.7.16}$$

where $\tilde{C}, \tilde{\omega}$ are the constants in (c), *and $S(t, s)$ is F-weakly continuous in $0 \leqslant s \leqslant t \leqslant T$.*

Proof. Let $A_n(t)$ be defined by

$$A_n(t) = A\left(\frac{k-1}{n}T\right) \quad \left(\frac{k-1}{n}T \leqslant t < \frac{k}{n}T\right), \tag{7.7.17}$$

$k = 1, 2, \ldots, n$, and (for the sake of completeness), $A_n(T) = A(T)$. In view of assumption (e),

$$\|\tilde{A}_n(t) - \tilde{A}(t)\|_{(F; E)} \to 0 \tag{7.7.18}$$

uniformly in $0 \leqslant t \leqslant T$, where the tilde, as before, indicates restriction to F. It is obvious that $A_n(\cdot)$ is stable in E and that $A_n(\cdot)_F$ is stable in F, in both cases with the same constants $C, \omega, \tilde{C}, \tilde{\omega}$ postulated in (a) and (c), therefore independent of n.

We examine now the equation

$$u'(t) = A_n(t)u(t). \tag{7.7.19}$$

A moment's reflection shows that if $0 \leqslant s < T$ and $u \in F$, then (7.7.19) has a solution $u_n(\cdot)$ in $s \leqslant t < T$ with initial value

$$u_n(s) = u \in F$$

given by

$$u_n(t) = S_n(t, s)u, \tag{7.7.20}$$

where the operator $S_n(t, s)$ is defined as follows. Let $0 \leqslant s \leqslant t \leqslant T$ and let k, l be two integers, $1 \leqslant k, l \leqslant n$, such that

$$\frac{k-1}{n}T \leqslant s < \frac{k}{n}T, \quad \frac{l-1}{n}T \leqslant t < \frac{l}{n}T,$$

so that $k \leqslant l$ (clearly $k = [ns/T]+1$, $l = [nt/T]+1$, where $[\cdot]$ denotes integral part). Then

$$S_n(t,s) = S\left(t - \frac{l-1}{n}T; A\left(\frac{l-1}{n}T\right)\right)\left\{\prod_{j=k+1}^{l-1} S\left(\frac{1}{n}T; A\left(\frac{j-1}{n}T\right)\right)\right\}$$

$$\times S\left(\frac{k}{n}T - s; A\left(\frac{k-1}{n}T\right)\right) \qquad (7.7.21)$$

if $k < l$ (if $k = l - 1$ the product between curly brackets is taken equal to the identity operator). If $k = l$,

$$S_n(t,s) = S\left(t - s; A\left(\frac{k-1}{n}T\right)\right). \qquad (7.7.22)$$

The word "solution" in the preceding statement is understood in the sense of Definition (C), or more precisely as follows: $u(\cdot)$ is a genuine solution of (7.7.19) in each interval

$$\left[s, \frac{k}{n}T\right), \left[\frac{j-1}{n}T, \frac{j}{n}T\right) \quad (j = k+1, \dots, n),$$

and it is continuous in $s \leqslant t \leqslant T$. Likewise, the function $s \to S(t, s)u$ is continuous in $0 \leqslant s \leqslant t$, continuously differentiable in each interval

$$\left[\frac{j-1}{n}T, \frac{j}{n}T\right) \quad (j = 1, \dots, k-1), \quad \left[\frac{k-1}{n}T, t\right]$$

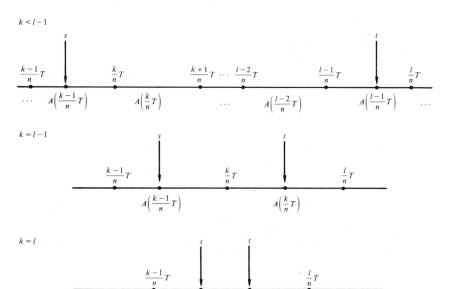

FIGURE 7.7.1

and satisfies there

$$D_s S_n(t,s)u = -S_n(t,s)A(s)u. \tag{7.7.23}$$

As the reader will no doubt suspect by now, the propagator $S(t,s)$ of (7.7.1) will be constructed as the limit of the S_n, in a way quite similar to that employed in Example 7.1.8 (but under totally different hypotheses). Consequently, the next step will be to show that, for each $u \in E$,

$$\lim_{n \to \infty} S_n(t,s)u = S(t,s)u \tag{7.7.24}$$

exists, uniformly in $0 \leqslant s \leqslant t \leqslant T$. In order to do this, we begin by noting that it follows from stability of $A(\cdot)$ in E, stability of $A(\cdot)_F$ in F, $A(t)$-admissibility of F, and (part of) Lemma 7.7.1 that $S_n(t,s)F \subseteq F$ and

$$\|S_n(t,s)\|_{(E)} \leqslant Ce^{\omega(t-s)}, \quad \|S_n(t,s)\|_{(F)} \leqslant \tilde{C}e^{\tilde{\omega}(t-s)} \tag{7.7.25}$$

in $0 \leqslant s \leqslant t \leqslant T$, $n = 1,2,\dots$. Hence (7.7.24) needs to be shown only for u in a dense subset of E, say, for $u \in F$. In order to do this we take profit of our previous observations concerning $S_n(t,s)$. If $u \in F$ and $0 \leqslant s < t < T$, it is easy to see that $\sigma \to S_n(t,\sigma)S_m(\sigma,s)u$ is continuous in $s \leqslant \sigma \leqslant t$ and continuously differentiable there (except perhaps at points of the form kT/n lying in $[s,t]$) and

$$\begin{aligned} D_\sigma S_n(t,\sigma)S_m(\sigma,s)u &= -S_n(t,\sigma)A_n(\sigma)S_m(\sigma,r)u \\ &\quad + S_n(t,\sigma)A_m(\sigma)S_m(\sigma,s)u \\ &= S_n(t,\sigma)(A_m(\sigma)-A_n(\sigma))S_m(\sigma,s)u. \end{aligned} \tag{7.7.26}$$

Integrating in $s \leqslant \sigma \leqslant t$, we obtain

$$(S_m(t,s)-S_n(t,s))u = \int_s^t S_n(t,\sigma)(A_n(\sigma)-A_m(\sigma))S_m(\sigma,s)u\,d\sigma, \tag{7.7.27}$$

hence

$$\|(S_n(t,s)-S_m(t,s))u\|_E \leqslant C\tilde{C}e^{(\omega+\tilde{\omega})(t-s)}\|u\|_F \int_s^t \|\tilde{A}_n(\sigma)-\tilde{A}_m(\sigma)\|_{(F;E)}\,d\sigma, \tag{7.7.28}$$

which estimate, in view of (7.7.18), shows that $S_n(t,s)u$ is convergent for $u \in F$; in view of the first inequality (7.7.25) and the denseness of F, $S_n(t,s)u$ is actually convergent for all $u \in E$ uniformly in the triangle $0 \leqslant s \leqslant t \leqslant T$. Since each S_m is strongly continuous, S is strongly continuous in $0 \leqslant s \leqslant t \leqslant T$. We also have

$$S(s,s) = I.$$

In order to study the differentiability properties of S stated in Theorem 7.7.5, we shall make use of the following result:

7.7.6   Lemma.   *Let $A^+(\cdot)$ be a family of densely defined operators in E satisfying exactly the same assumptions as $A(\cdot)$ in Theorem 7.7.7 (we suppose that F is the same, as well as $C, \omega, \tilde{C}, \tilde{\omega}$). Let $S^+(\cdot, \cdot)$ be the operator function constructed from $A^+(\cdot)$ in the same way as $S(\cdot, \cdot)$ was constructed from $A(\cdot)$. Then*

$$\|(S^+(t, s) - S(t, s))u\|_E \leqslant C\tilde{C}e^{(\omega + \tilde{\omega})(t - s)}\|u\|_F \int_s^t \|\tilde{A}^+(\sigma) - \tilde{A}(\sigma)\|_{(F, E)} \, d\sigma$$

(7.7.29)

in $0 \leqslant s \leqslant t \leqslant T$.

To prove this, we use an argument very similar to that leading to (7.7.28). If $A_n^+(\cdot)$ is defined from $A^+(\cdot)$ as in (7.7.21),(7.7.22), we can prove inequality (7.7.29) for $A_n^+(\cdot)$ and $A_n(\cdot)$ reasoning with the function $s \to S_n^+(t, \sigma)S_n(\sigma, s)$. We make then use of the convergence properties of S_n^+ and S_n proved a moment ago and (7.7.29) results.

Inequality (7.7.29) will be put to work with the following $A^+(\cdot)$: given $0 \leqslant s \leqslant T$, let

$$A^+(t) = \begin{cases} A(t) & (0 \leqslant t \leqslant s) \\ A(s) & (s \leqslant t \leqslant T). \end{cases}$$

(7.7.30)

The assumptions of Lemma 7.7.6 are rather trivially satisfied; therefore (7.7.29) holds. Since the right-hand side is $o(t - s)$ as $t \to s +$ and (as we see easily) $S^+(t, s)u = S(t - s; A(s))u \ (t \geqslant s)$, it follows that $D_t^+ S(t, s)u$ exists at $t = s$ and equals $A(s)u = A(t)u$.

We obtain easily from the definition of S_n that

$$S_n(t, s) = S_n(t, \sigma)S_n(\sigma, s) \quad (0 \leqslant s \leqslant \sigma \leqslant t \leqslant T).$$

(7.7.31)

Applying both sides of this equality to an arbitrary $u \in E$ and taking limits,

$$S(t, s) = S(t, \sigma)S(\sigma, s),$$

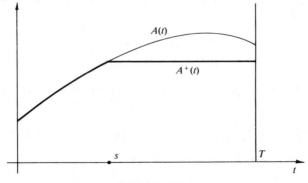

FIGURE 7.7.2

thus

$$S(t + h, s) - S(t, s) = (S(t + h, t) - S(t, t))S(t, s). \quad (7.7.32)$$

Therefore, the required right-sided differentiability of $S(t, s)u$ for $u \in F$ (and the equality $D_t^+ S(t, s)u = A(t)S(t, s)u$) will follow if we show that

$$S(t, s)F \subseteq F$$

for $0 \leqslant s \leqslant t \leqslant T$. We do this next, and the reader will note that reflexivity of F is used only here. If $u \in F$, we have already noted that $S_n(t, s)u \in F$; on account of the uniform boundedness of $\|S_n\|_{(F)}$ (see the second inequality (7.7.25)), $\{S_n(t, s)u\}$ must be bounded in F and must then contain a weakly convergent subsequence (which we denote in the same fashion). Let v be the weak limit of $\{S_n(t, s)u\}$. Since $S_n(t, s)u \to S(t, s)u$ strongly (thus weakly) in E and every continuous linear functional in E is a continuous linear functional in F, it follows that $S(t, s)u = v$.

It is clear that (7.7.16) results from this argument and from the second inequality (7.7.25).

To prove F-weak continuity of S, we argue in a somewhat similar way. Let $(t, s), (t_n, s_n)$, $n = 1, 2, \ldots$ be points in the triangle $0 \leqslant s \leqslant t \leqslant T$ such that $(t_n, s_n) \to (t, s)$, and let $u \in F$. If $S(t_n, s_n)u$ does not converge to $S(t, s)u$ weakly, there exists $u^* \in F^*$, $\varepsilon > 0$ and a subsequence of $\{(t_n, s_n)\}$ (which we design with the same symbol) such that

$$|\langle u^*, S(t_n, s_n)u - S(t, s)u \rangle| \geqslant \varepsilon. \quad (7.7.33)$$

However, in view of the reflexivity of F, we may assume (if necessary after further thinning out of the sequence) that $\{S(t_n, s_n)u\}$ is weakly convergent in F (thus in E) to some $v \in F$. But $S(t_n, s_n)u \to S(t, s)u$ strongly, hence weakly, in E. It follows that $v = S(t, s)u$, which contradicts (7.7.33) and completes our argument.

It only remains to study the s-differentiability of $S(t, s)u$, $u \in F$. The fact that $s \to S(t, s)u$ has a left-sided derivative at $s = t$ can be proved exactly in the same way in which we show that $t \to S(t, s)u$ has a right-sided derivative; we use now

$$A^-(t) = \begin{cases} A(s) & (0 \leqslant t \leqslant s) \\ A(t) & (s \leqslant t \leqslant T) \end{cases} \quad (7.7.34)$$

instead of $A^+(t)$ and we obtain that $D_s^- S(t, s)u = -A(t)u$ for $s = t$. If $0 < s < t$,

$$h^{-1}(S(t, s - h) - S(t, s))u = S(t, s)h^{-1}(S(s, s - h)u - u),$$

so that $D_s^- S(t, s)$ exists in $0 < s \leqslant t$ and

$$D_s^- S(t, s)u = -S(t, s)A(s)u.$$

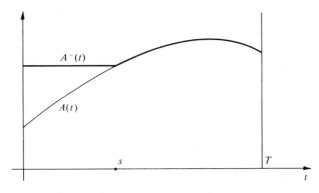

FIGURE 7.7.3

On the other hand, if $0 \leqslant s < t$,

$$h^{-1}(S(t, s + h) - S(t, s))u = S(t, s + h)h^{-1}(u - S(s + h, s)u)$$
$$\to - S(t, s)A(s)u$$

as $h \to 0+$ in view of the right-sided differentiability of $t \to S(t, s)u$ at $t = s$ and the strong continuity and uniform boundedness of S.

Finally, let $u(\cdot)$ be an arbitrary solution of (7.7.1) in $s \leqslant t \leqslant T$ in the sense of Definition (C). Let $t > s$ and write $v(\sigma) = S(t, \sigma)u(\sigma)$, $s \leqslant \sigma \leqslant t$. We have

$$v(\sigma + h) - v(\sigma) = S(t, \sigma + h)u(\sigma + h) - S(t, \sigma)u(\sigma)$$
$$= S(t, \sigma + h)(u(\sigma + h) - u(\sigma))$$
$$+ (S(t, \sigma + h) - S(t, \sigma))u(\sigma)$$

so that the right-sided derivative of $v(\cdot)$ exists in $s < \sigma < t$ and vanishes there. After application of linear functionals, we deduce that $v(\cdot)$ is constant from the following result:

7.7.7 Lemma. *Let η be a function defined and continuous in $a \leqslant t \leqslant b$, and such that $D^+\eta$ exists and equals zero in $a < t < b$. Then η is constant.*

In fact, since $v(\cdot)$ must be constant,

$$u(t) = v(t) = S(t, s)u(s),$$

which ends the uniqueness argument and thus the proof of Theorem 7.7.5.

Lemma 7.7.7 can be proved as follows. Assume η is not constant, and let $a < t_1 < t_2 < b$ with, say, $\eta(t_1) < \eta(t_2)$. Then the graph of η lies below the segment J joining $(t_1, \eta(t_1))$ and $(t_2, \eta(t_2))$ (otherwise, on account of the continuity of η, it is not difficult to prove that there exists t_0, $t_1 < t_0 < t_2$ such that $(t_0, \eta(t_0)) \in J$ and the graph of η lies above J in some interval to the right of t_0, thus $D^+\eta(t_0) > 0$). (See Figure 7.7.4.) Since

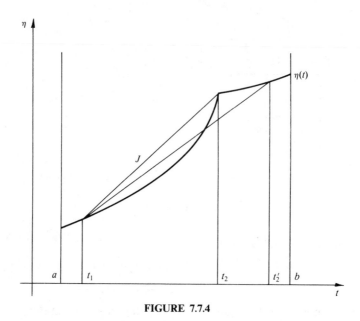

FIGURE 7.7.4

$D^+\eta(t_2) = 0$, there exists $t_2' > t_2$ with $\eta(t_1) \leqslant \eta(t_2')$ and such that $(t_2, \eta(t_2))$ lies above the segment J' joining $(t_1, \eta(t_1))$ and $(t_2', \eta(t_2'))$, which is absurd in view of the preceding argument. We reason in a similar way in the case $\eta(t_1) > \eta(t_2)$.

7.7.8 Remark. We note that uniqueness of solutions of (7.7.1) has been only established under the assumption that $u(t) \in F$ for all t; thus the result does not cover, for instance, the solutions defined in Section 7.3, although these are rather more differentiable than the present ones.

On the other hand, uniqueness was established on the only basis of the behavior of $S(t, s)$ *as a function of s* and it is then valid independently of the hypothesis that F is reflexive.

7.7.9 Remark. There is more than meets the eye in the solutions of (7.7.1) constructed in Theorem 7.7.5. In fact, under the hypotheses there, we can prove the following result:

7.7.10 Lemma. *Let* $u \in F$. *Then* $(t, s) \to A(t)S(t, s)u$ *is* E-*weakly continuous and strongly measurable in* $0 \leqslant s \leqslant t \leqslant T$; *moreover, if* $0 \leqslant s < T$, $t \to A(t)S(t, s)u$ *is* E-*strongly measurable in* $s \leqslant t \leqslant T$ *and*

$$S(t, s)u = u + \int_s^t A(\tau)S(\tau, s)u\, d\tau \quad (s \leqslant t \leqslant T), \qquad (7.7.35)$$

the integral understood in the sense of Lebesgue-Bochner. Therefore, $D_t S(t, s)u$

exists in a set $e = e(s)$ of full measure in $[s, t]$ and

$$D_t S(t, s)u = A(t)S(t, s)u \quad (s \leqslant t \leqslant T)$$

wherever the derivative exists.

Proof. We have proved in Theorem 7.7.5 that $(t, s) \to S(t, s)u$ is F-weakly continuous for $u \in F$. If $u^* \in E^*$, we have

$$\langle u^*, A(t)S(t, s)u \rangle = \langle \tilde{A}(t)^* u^*, S(t, s)u \rangle, \quad (7.7.36)$$

where $\tilde{A}(t)^* : E^* \to F^*$ indicates the adjoint of $\tilde{A}(t)$ (the restriction of $A(t)$ to F) as an operator from F to E. Hypothesis (e) plainly implies that $t \to \tilde{A}(t)^*$ is continuous as a (E^*, F^*)-valued function, thus (7.7.36) yields our claim on E-weak continuity of $A(t)S(t, s)$. The statements on E-strong measurability of $S(t, s)$ as a function of t and s as well as a function of t for each s fixed follow from this well-known result in integration theory of vector-valued functions:

7.7.11 Example (Hille-Phillips [1957; 1, p. 73]). Let f be a function defined (say) in a measurable subset K of n-dimensional space with values in E. Then, if f is weakly continuous, it is strongly measurable in K.

We note that it also follows from Example 7.7.11 that S is F-strongly measurable, either as a function of (t, s) or in each variable separately. To show (7.7.35), we take $u^* \in F^*$, write the (obviously valid) version of this equality for S_n and apply u^* to both sides, obtaining

$$\langle u^*, S_n(t, s)u \rangle = \langle u^*, u \rangle + \int_s^t \langle A(\tau)^* u^*, S_n(\tau, s)u \rangle \, d\tau \quad (0 \leqslant s \leqslant t \leqslant T).$$

$$(7.7.37)$$

We have already observed in the proof of Theorem 7.7.5 that

$$S_n(t, s)u \to S(t, s)u \quad (u \in F) \quad (7.7.38)$$

F-weakly as $n \to \infty$. We let $n \to \infty$ in (7.7.37) and make use of the strong convergence of $S(t, s)u$ in E, uniformly with respect to s and t. The result is (7.7.35), since u^* can be chosen at will. It is a consequence of this equality that $D_t S(t, s)u$ must exist in a set $e = e(s)$ of full measure in (s, T). Since $D_t^+ S(t, s)u$ has been seen to exist everywhere and equals $A(t)S(t, s)u$, the proof of Lemma 7.7.12 is complete.

The restrictions to the t-differentiability of S in Theorem 7.7.7 can be (partly) eliminated if the assumptions in Theorem 7.7.5 are reinforced. A sample result is:

7.7.12 Example (Kato [1970: 1]). Let the hypotheses of Theorem 7.7.5 hold with (c) strengthened as follows:

(c′) $\{A(t)_F; \ 0 \leqslant t \leqslant T\}$ satisfies the assumptions of Example 7.7.3 with respect to a family of norms $\{\|\cdot\|_t; \ 0 \leqslant t \leqslant T\}$ in F such that each of these norms is uniformly convex.

Then the conclusions of Theorem 7.7.7 hold with the following additional statements: if $0 \leqslant s < T$ and $u \in F$, $t \to S(t,s)u$ is F-strongly right continuous in $s \leqslant t \leqslant T$ and strongly continuous there except in a countable set $e = e(s)$; $D_t S(t,s)u$ exists for $t \notin e(s)$, is continuous for $t \notin e(s)$ and equals $A(t)S(t,s)u$.

Finally, we note that the reservations to the t-differentiability of the solutions of (7.7.1) constructed in Theorem 7.7.5 can be totally lifted in a case that covers many of the applications of Theorem 7.7.9.

7.7.13   Theorem.   *Let the assumptions of Theorem 7.7.5 be satisfied both for* $\{A(t);\ 0 \leqslant t \leqslant T\}$ *and for* $\{\hat{A}(t) = -A(T-t);\ 0 \leqslant t \leqslant T\}$. *Then the Cauchy problem for (7.7.1) is well posed in both senses of time (see Section 7.1) in* $0 \leqslant t \leqslant T$ *with respect to the following notion of solution:*

(D) $u(\cdot)$ *is a solution of (7.7.1) in* $0 \leqslant t \leqslant T$ *if and only if:*
(a) $u(\cdot)$ *is continuously differentiable (in the sense of the norm of E) and* $u(t) \in F$ *for* $0 \leqslant t \leqslant T$.
(b) *Equation (7.7.1) is satified in* $0 \leqslant s \leqslant T$.

Proof.   The assumption that both $A(\cdot)$ and $\hat{A}(\cdot)$ are stable in E obviously implies (taking $t_1 = \cdots = t_n$) that each $A(t)$ belongs to $\mathcal{C}(C, \omega)$. (We may and will suppose that the stability constants for $\hat{A}(\cdot)$ are the same as for $A(\cdot)$; a similar assumption will be made for $A(\cdot)_F$ and $\hat{A}(\cdot)_F$.) Accordingly, the solution operator $S_n(t, s)$ of Equation (7.7.19) can now be constructed in the square $0 \leqslant s, t \leqslant T$ using formulas (7.7.21) and (7.7.22) when $s \leqslant t$ and setting

$$S_n(t, s) = S_n(s, t)^{-1} \tag{7.7.39}$$

when $s > t$, and it obeys inequalities (7.7.25) in the extended range, replacing $t - s$ by $|t - s|$. Inequality (7.7.28) becomes

$$\left\| \left(S_n(t, s) - S_m(t, s) \right) u \right\|_E \leqslant C\tilde{C} e^{(\omega + \tilde{\omega})|t - s|} \|u\|_F \left| \int_s^t \|\tilde{A}_n(\sigma) - \tilde{A}_m(\sigma)\|_{(F, E)}\, d\sigma \right| \tag{7.7.40}$$

valid for $0 \leqslant s, t \leqslant T$ and this inequality can be used to establish strong E-convergence of S, uniformly in the whole square, and hence E-strong continuity and the inequality

$$\|S(t, s)\| \leqslant C e^{\omega |t - s|} \quad (0 \leqslant s, t \leqslant T). \tag{7.7.41}$$

Lemma 7.7.8 has an immediate counterpart in the present situation, inequality (7.7.29) becoming

$$\left\| \left(S^+(t, s) - S(t, s) \right) u \right\|_E \leqslant C\tilde{C} e^{(\omega + \tilde{\omega})|t - s|} \|u\|_F \left| \int_s^t \|\tilde{A}^+(\sigma) - \tilde{A}(\sigma)\|_{(F, E)}\, d\sigma \right| \tag{7.7.42}$$

for $0 \leqslant s, t \leqslant T$. Taking advantage of this improved inequality, we can prove, making use of $A^+(t)$ defined by (7.7.30) and of $A^-(t)$ defined by (7.7.34)(see Figures 7.7.1 and 7.7.2) that if $u \in F$, $D_t S(t, s)u$ exists at $s = t$ and equals $A(s)u$. We show next that (7.7.31) holds in $0 \leqslant s, t \leqslant T$ and use (7.7.32), now valid for h of arbitrary sign (as a consequence of (7.7.31) for $0 \leqslant s, \sigma, t \leqslant T$) to show that $D_t S(t, s)u$ exists in $0 \leqslant t \leqslant T$ and satisfies

$$D_t S(t, s)u = A(t)S(t, s)u$$

there. The proof of (7.7.15) in $0 \leqslant s \leqslant T$ follows the same lines and is therefore omitted.

It is interesting to note that we have just fallen short of proving that $(t, s) \to S(t, s)u$ is F-strongly continuous for $u \in F$ in Theorem 7.7.13 (of course, this would imply strong E-continuity of $(t, s) \to A(t)S(t, s)u$, thus continuity of the derivatives of the solutions therein constructed). This additional bit of information can be established under hypotheses of the type used in Example 7.7.12.

7.7.14 Example (Kato [1970: 1]).   Let the assumptions of Example 7.7.12 be satisfied both for $A(\cdot)$ and for the family $\hat{A}(\cdot)$ defined in the statement of Theorem 7.7.13. Then the conclusions of said theorem hold with the additional result that, if $u \in F$,

$$(t, s) \to S(t, s)u$$

is F-continuous in $0 \leqslant s, t \leqslant T$.

7.8. TIME-DEPENDENT SYMMETRIC HYPERBOLIC SYSTEMS IN THE WHOLE SPACE

The results of last section will be applied here to the abstract differential equation

$$u'(t) = \mathcal{Q}(t)u(t) \quad (-\infty < t < \infty). \tag{7.8.1}$$

Here $\mathcal{Q}(t) = (\mathcal{Q}_0'(t))^*$, where $\mathcal{Q}_0(t), \mathcal{Q}_0'(t)$ are defined as in Sections 3.5 and 5.6 (with domain $D(\mathcal{Q}_0) = D(\mathcal{Q}_0') = \mathcal{D}^\nu$) with respect to the time-dependent partial differential operator

$$L(t)u = \sum_{k=1}^{m} A_k(x, t)D^k u + B(x, t)u. \tag{7.8.2}$$

Precise assumptions on the coefficients will be given later; the space E is once again $L^2 = L^2(\mathbb{R}^m)^\nu$, and $F = H^1 = H^1(\mathbb{R}^m)^\nu$. Verification of the various hypotheses in Theorem 7.7.5 involves different degrees of difficulty. Stability of the family $\mathcal{Q}(\cdot)$ in L^2 poses no problem, since each $\mathcal{Q}(t)$ will belong to $\mathcal{C}_+(1, \omega; L^2)$ for sufficiently large ω under assumptions similar to those of Section 5.6 for each t (see Remark 7.7.2). In contrast, $\mathcal{Q}(t)$-admissibility of H^1 and stability of $\mathcal{Q}(\cdot)_{H^1}$ are far from trivial to verify, as a glance at Section 5.6 will show: there we needed a powerful result on singular integrals merely to show that H^1 is $\mathcal{Q}$-admissible. Fortunately, "dynamic" versions of the auxiliary results in Section 5.6 will do the trick also in the time-dependent situation. The first of these extensions is a descendant of Lemma 5.6.3 and takes care of F-admissibility and F-stability in one stroke.

7.8.1   Lemma. *Let $F \to E$, $\{A(t); 0 \leqslant t \leqslant T\}$ a stable family in E. For each $t \in [0, T]$, let $K(t): F \to E$ be a bounded invertible operator ($K(t) \in (F, E), K(t)^{-1} \in (E, F)$) such that the family*

$$\left\{ A_1(t) = K(t)A(t)K(t)^{-1}; 0 \leqslant t \leqslant T \right\} \tag{7.8.3}$$

is stable in E; assume, moreover, that

$$\|K(t)\|_{(F; E)} \leqslant N, \quad \|K(t)^{-1}\|_{(E; F)} \leqslant N \quad (0 \leqslant t \leqslant T) \tag{7.8.4}$$

and that the map $t \to K(t) \in (F; E)$ is of bounded variation. Then F is $A(t)$-admissible for $0 \leqslant t \leqslant T$ and the family $\{A(\cdot)_F\}$ is stable in F.

Proof. The statement on $A(t)$-admissibility of F is a direct consequence of Lemma 5.6.3. Combining (5.6.9) with (5.6.15), we obtain

$$
\begin{aligned}
R(\lambda; A(t)_F) &= R(\lambda; A(t))_F \\
&= \left(K(t)^{-1} R(\lambda; A_1(t)) K(t) \right)_F \\
&= K(t)^{-1} R(\lambda; A_1(t)) K(t), \tag{7.8.5}
\end{aligned}
$$

where λ belongs both to $\rho(A(t))$ and $\rho(A(t_1))$ (note that this will happen if λ is sufficiently large independently of t, as both $A(\cdot)$ and $A_1(\cdot)$ are stable). If $0 \leqslant t_1 \leqslant \cdots \leqslant t_n \leqslant T$, we have

$$
\prod_{j=1}^{n} R(\lambda; A(t_j)_F) = \prod_{j=1}^{n} K(t_j)^{-1} R(\lambda; A_1(t_j)) K(t_j)
$$

$$
= K(t_n)^{-1} \left\{ \prod_{j=2}^{n} R(\lambda; A_1(t_j))(I + L_j) \right\} R(\lambda; A_1(t_1)) K(t_1),
$$

$$\tag{7.8.6}$$

where

$$
\begin{aligned}
L_j &= \left(K(t_j) - K(t_{j-1}) \right) K(t_{j-1})^{-1} \\
&= K(t_j) K(t_{j-1})^{-1} - I. \tag{7.8.7}
\end{aligned}
$$

We estimate now the (F)-norm of (7.8.6). We expand this product and arrange the resulting summands as a polynomial in the L_j; note that, due to lack of commutativity of the factors involved in each summand, the different L_j will appear sandwiched between consecutive $R(\lambda; A(t_j))$, each such L_j corresponding to the choice of L_j rather than I in the parenthesis $(I + L_j)$. If C_1, ω_1 are the stability constants for $A_1(\cdot)$ and N is the constant in (7.8.4), we can then estimate each summand by

$$N^2 C_1^{m+1}(\lambda - \omega_1)^{-n}\|L_{j_1}\| \cdots \|L_{j_m}\|, \qquad (7.8.8)$$

where $L_{j_1}, \ldots, L_{j_m}$ are the L_j appearing in that particular product (note that each such L_j provokes the breaking up of the stability product (7.7.3) into two pieces and thus forces us to take powers of C_1 on the right-hand side, the exponent being equal to the number of L_j in the product plus one). It follows then that (7.8.7) can be estimated by

$$N^2 C_1 \left\{ \prod_{j=2}^{n} \left(1 + C_1\|L_j\|\right) \right\}(\lambda - \omega_1)^{-n}. \qquad (7.8.9)$$

We bring now into play the hypothesis that $K(\cdot)$ is of bounded variation, combined with the first inequality (7.8.4), obtaining

$$\sum_{j=1}^{n} \|L_j\| \leqslant NV,$$

V the total variation of $K(\cdot)$; accordingly, the factor between curly brackets in (7.8.9) is bounded by $\exp(C_1 NV)$ and there exists $\tilde{C} > 0$ such that

$$\left\| \prod_{j=1}^{n} R\big(\lambda; A(t_j)_F\big) \right\|_{(F)} \leqslant \tilde{C}(\lambda - \omega_1)^{-n}. \qquad (7.8.10)$$

This ends the proof of Lemma 7.8.1.

As in the time-independent case, Lemma 7.8.1 will only be used when A_1 is a bounded perturbation of A (see (5.6.16)).

7.8.2 Corollary. *Let $E, F, A(\cdot), K(\cdot)$ be as in Lemma 7.8.1, but assume that*

$$K(t)A(t)K(t)^{-1} = A(t) + B(t), \qquad (7.8.11)$$

where each $B(t)$ belongs to (E) and

$$\|B(t)\|_{(E)} \leqslant M \quad (0 \leqslant t \leqslant T). \qquad (7.8.12)$$

Then the conclusions of Lemma 7.8.1 hold.

Proof. We only need to show that $A(\cdot) + B(\cdot)$ is stable, and this has already been done in Lemma 7.7.4.

It is remarkable that a slight stiffening of the assumptions in Corollary 7.8.2 will totally eliminate the restrictions to the differentiability of $t \rightarrow$

$S(t, s)u$ in Theorem 7.7.5. This is perhaps of marginal interest to us in relation with the application we have in mind, since some of these restrictions can be eliminated anyway on the basis of Theorem 7.7.13; however, it is many times the case in applications that the reinforced assumptions are satisfied without any special effort on our part.

7.8.3   Theorem.   *Let* E, F *be Banach spaces with* $F \to E$, *and let* $\{A(t); \ 0 \leqslant t \leqslant T\}$ *be a stable family in* E *such that* $F \subseteq D(A(t))$ *and* $A(t): F \to E$ *is bounded in* $0 \leqslant t \leqslant T$; *moreover, if* $\tilde{A}(t)$ *is the restriction of* $A(t)$ *to* F, *the map*

$$t \to \tilde{A}(t) \in (F, E) \qquad (7.8.13)$$

is continuous. For each $t \in [0, T]$, *let* $K(t)$ *be a bounded and invertible operator from* F *to* E ($K(t) \in (F, E)$, $K(t)^{-1} \in (E, F)$) *such that*

$$t \to K(t) \in (F, E) \qquad (7.8.14)$$

is strongly continuously differentiable in $0 \leqslant t \leqslant T$ (*i.e.,* $t \to K(t)u$ *is continuously differentiable in* E *for each* $u \in F$). *Assume, finally, that*

$$K(t)A(t)K(t)^{-1} = A(t) + B(t), \qquad (7.8.15)$$

where $B(t) \in (E)$ *for each* t *and* $t \to B(t)$ *is strongly continuous in* $0 \leqslant t \leqslant T$. *Then the Cauchy problem for*

$$u'(t) = A(t)u(t) \qquad (7.8.16)$$

is properly posed in $0 \leqslant t \leqslant T$ *with respect to solutions defined in* (D).[5] *Moreover, if* $u \in F$, $S(t, s)u$ *is continuous in* F *for* $0 \leqslant s \leqslant t \leqslant T$.

Proof.   It follows easily from the uniform boundedness theorem that the family $\{(t' - t)(K(t') - K(t)), \ t' \neq t\}$ is uniformly bounded in $(F; E)$; in other words,

$$\|K(t') - K(t)\|_{(F; E)} \leqslant D|t' - t| \qquad (7.8.17)$$

for $0 \leqslant t, t' \leqslant T$. We can then use the remarks on inverses[6] in Section 1 to deduce that $t \to K(t)^{-1} \in (E; F)$ is as well Lipschitz continuous in $0 \leqslant t \leqslant T$; in particular, $\|K(t)\|_{(F, E)}$ and $\|K(t)^{-1}\|_{(E, F)}$ are bounded in $0 \leqslant t \leqslant T$.

This shows that the assumptions of Corollary 7.8.2 are amply satisfied. It follows that all the hypotheses in Theorem 7.7.5 are themselves fulfilled, except of course that we have not assumed that F is reflexive. Combing the proof we soon realize that the reflexitivity assumption was only used to assure that

$$S(t, s)F \subseteq F \qquad (7.8.18)$$

(see the comments following (7.7.32)); thus, if (7.8.18) can be established by

---

[5]See the statement of Theorem 7.7.13. Strictly speaking, solutions will be only defined for $t \geqslant s$ so that the interval $0 \leqslant t \leqslant T$ in (D) becomes now $s \leqslant t \leqslant T$.

[6]See the comments preceding (1.1).

other means, the conclusions of Theorem 7.7.5 will follow in full. We shall in fact exceed this by proving that $K(t)S(t, s)K(s)^{-1}$ is everywhere defined and bounded in E and

$$(t, s) \to K(t)S(t, s)K(s)^{-1} \tag{7.8.19}$$

is strongly continuous in E for $0 \leqslant s \leqslant t \leqslant T$. In fact, if this is true, it follows that

$$(t, s) \to S(t, s) = K(t)^{-1}\big(K(t)S(t, s)K(s)^{-1}\big)K(s)$$

is strongly continuous in F, hence $t \to A(t)S(t, s)u$ is strongly continuous in E. Since, by virtue of the arguments in Remark 7.7.9, equality (7.7.35) holds everywhere in $s \leqslant t \leqslant T$, all the claims in Theorem 7.8.3 follow. The proof of the strong continuity of (7.8.19) is fairly intricate and will be divided in several steps for the sake of convenience. We bring back into the light the operators $S_n(t, s)$ used in Theorem 7.7.5 to construct the propagator $S(t, s)$ (see (7.7.21) and (7.7.22) and define

$$Q_n(t, s) = K(t)S_n(t, s)K(s)^{-1}. \tag{7.8.20}$$

7.8.4 Lemma. (a) *Under the assumptions of Theorem 7.8.3, we have*

$$\|Q_n(t, s)\| \leqslant C\left(1 + \frac{DNT}{n}\right)^2 e^{(\omega + CM + CDN)(t-s) + CDNT} \quad (0 \leqslant s \leqslant t \leqslant T),$$
$$\tag{7.8.21}$$

where C, ω are the stability constants of $A(\cdot)$, M is a bound for $\|B(t)\|$ in $0 \leqslant t \leqslant T$ as in (7.8.12), D is the constant in (7.8.17), and N is a bound for both $\|K(t)\|_{(F, E)}$ and $\|K(t)^{-1}\|_{(E, F)}$ in $0 \leqslant t \leqslant T$ as in (7.8.4). (b) Q_n is strongly convergent as $n \to \infty$, uniformly in $0 \leqslant s, t \leqslant T$.

Proof. It has been established in Lemma 7.7.2 that if the family $A(\cdot)$ is stable, then

$$\{A(t) + B(t); 0 \leqslant t \leqslant T\}$$

is stable if each $B(t)$ is bounded and (7.7.12) holds; in the present situation, this inequality is a consequence of the strong continuity of $B(\cdot)$ and of the uniform boundedness theorem (see Section 1). (Recall that if C, ω are the stability constants of $A(\cdot)$, $A(\cdot) + B(\cdot)$ possesses stability constants $C, \omega + CM$; since, on the other hand, the hypotheses on $K(\cdot)$ in Lemma 7.8.1 are satisfied, it follows that the conclusions there hold.) We obtain from (7.8.11) that

$$R(\lambda; A(t))^n u = K(t)^{-1} R(\lambda; A(t) + B(t))^n K(t) u \quad (u \in F) \tag{7.8.22}$$

for $0 \leqslant t \leqslant T$, $n \geqslant 1$, and λ sufficiently large; making use of the exponential formula (2.1.27), we deduce that

$$S(s; A(t)) = K(t)^{-1} S(s; A(t) + B(t))K(t)$$

for $s \geq 0$, $0 \leq t \leq T$. Let now $0 \leq s \leq t \leq T$ and let k, l be two integers that stand in the relation to s and t described in the definition of S_n, that is,

$$\frac{k-1}{n} T \leq s < \frac{k}{n} T, \quad \frac{l-1}{n} T \leq t < \frac{l}{n} T \qquad (7.8.23)$$

(see the comments around (7.7.21)). Then we have

$$Q_n(t, s)$$

$$= K(t) K\left(\frac{l-1}{n} T\right)^{-1} S\left(t - \frac{l-1}{n} T; A\left(\frac{l-1}{n} T\right) + B\left(\frac{l-1}{n} T\right)\right) K\left(\frac{l-1}{n} T\right)$$

$$\times \prod_{j=k+1}^{l-1} K\left(\frac{j-1}{n} T\right)^{-1} S\left(\frac{1}{n} T; A\left(\frac{j-1}{n} T\right) + B\left(\frac{j-1}{n} T\right)\right) K\left(\frac{j-1}{n} T\right)$$

$$\times K\left(\frac{k-1}{n} T\right)^{-1} S\left(\frac{k}{n} T - s; A\left(\frac{k-1}{n} T\right) + B\left(\frac{k-1}{n} T\right)\right) K\left(\frac{k-1}{n} T\right) K(s)^{-1}$$

$$(7.8.24)$$

if $k < l$ (the product in the middle is taken equal to I if $k = l - 1$), and

$$Q_n(s, t)$$

$$= K(t) K\left(\frac{k-1}{n} T\right)^{-1} S\left(t - s; A\left(\frac{k-1}{n} T\right) + B\left(\frac{k-1}{n} T\right)\right) K\left(\frac{k-1}{n} T\right) K(s)^{-1}$$

$$(7.8.25)$$

when $k = l$. We define now

$$L(t, s) = K(t) K(s)^{-1} - I$$

$$= (K(t) - K(s)) K(s)^{-1} \quad (0 \leq s, t \leq T) \qquad (7.8.26)$$

(note that the operator L_j in (7.8.7) is none other than $L(t_j, t_{j-1})$) and denote by $\mathfrak{S}_n$ the operator constructed from $A(\cdot) + B(\cdot)$ in the same way S_n was constructed from $A(\cdot)$; clearly

$$\|\mathfrak{S}_n(t, s)\| \leq C e^{(\omega + CM)(t-s)}. \qquad (7.8.27)$$

We note that

$$K\left(\frac{j-1}{n} T\right) K\left(\frac{j-2}{n} T\right)^{-1} = I + L\left(\frac{j-1}{n} T, \frac{j-2}{n} T\right)$$

and replace this expression in the interstices of the product (7.8.24). We develop then in products of the $L((j-1)T/n, (j-2)T/n)$ much in the same way as (7.8.6), but ordering the sum this time by ascending powers.

After doing this we obtain

$$Q_n(t,s) = \left\{ I + L\left(t, \frac{l-1}{n}T\right) \right\} R_n(t,s) \left\{ I + L\left(\frac{k-1}{n}T, s\right) \right\} \quad (7.8.28)$$

with

$R_n(t,s)$

$$= \mathfrak{S}_n(t,s) + \sum_{j=k+1}^{l} \mathfrak{S}_n\left(t, \frac{j-1}{n}T\right) L\left(\frac{j-1}{n}T, \frac{j-2}{2}T\right) \mathfrak{S}_n\left(\frac{j-1}{n}T, s\right)$$

$$+ \sum_{j=k+1}^{l} \mathfrak{S}_n\left(t, \frac{j-1}{n}T\right) L\left(\frac{j-1}{n}T, \frac{j-2}{n}T\right)$$

$$\times \sum_{i=k+1}^{j-1} \mathfrak{S}_n\left(\frac{j-1}{n}T, \frac{i-1}{n}T\right) L\left(\frac{i-1}{n}T, \frac{i-2}{n}T\right) \mathfrak{S}_n\left(\frac{i-1}{n}T, s\right)$$

$$+ \cdots, \quad (7.8.29)$$

where all the terms after $\mathfrak{S}_n$ disappear if $k = l$ (the total number of summands in (7.8.29) is $l - k + 1$). We can make the notation more orderly as follows. Define inductively operators $R_{n,0}(t,s), R_{n,1}(t,s), \ldots$ in the triangle $0 \leqslant s \leqslant t \leqslant T$ by $R_{n,0} = \mathfrak{S}_n$,

$$R_{n,m}(t,s) = \sum_{j=k+1}^{l} \mathfrak{S}_n\left(t, \frac{j-1}{n}T\right) L\left(\frac{j-1}{n}T, \frac{j-2}{n}T\right) R_{n,m-1}\left(\frac{j-1}{n}T, s\right).$$

$$(7.8.30)$$

It is then easy to see that we can write

$$R_n(t,s) = \sum_{m=0}^{l-k} R_{n,m}(t,s). \quad (7.8.31)$$

(Note that if (t,s) is a fixed point in the triangle $0 \leqslant s < t \leqslant T$ and $n \to \infty$, the number of summands in (7.8.31) will also tend to infinity.)

Observe next that, by virtue of (7.8.17),

$$\left\| L\left(\frac{j-1}{n}T, \frac{j-2}{n}T\right) \right\| \leqslant \frac{T}{n} DN, \quad (7.8.32)$$

hence

$$\|R_{n,1}(t,s)\| \leqslant \frac{l-k}{n} TC^2 DN e^{(\omega+CM)(t-s)}$$

$$\leqslant C^2 DN\left(t - s + \frac{T}{n}\right) e^{(\omega+CM)(t-s)}. \quad (7.8.33)$$

Accordingly, each term in the sum (7.8.30) corresponding to $m = 2$ is bounded in norm by

$$C^3(DN)^2\frac{T}{n}\left(\frac{j}{n}T-s\right)e^{(\omega+CM)(t-s)}. \qquad (7.8.34)$$

Hence

$$\|R_{n,2}(t,s)\| \leqslant C^3(DN)^2\frac{T}{n}\sum_{j=k+1}^{l}\left(\frac{j}{n}T-s\right)e^{(\omega+CM)(t-s)}$$

$$\leqslant \frac{C^3(DN)^2}{2!}\left(t-s+\frac{2}{n}T\right)^2 e^{(\omega+CM)(t-s)}, \qquad (7.8.35)$$

noting that

$$\frac{T}{n}\sum_{j=k+1}^{l}\left(\frac{j}{n}T-s\right) \leqslant \int_s^{t+2T/n}(r-s)\,dr.$$

(See Figure 7.8.1.) Replacing the estimate into the expression for $R_{n,3}$ and proceeding inductively in a similar fashion, we obtain

$$\|R_{n,m}(t,s)\| \leqslant \frac{C^{m+1}(DN)^m}{m!}\left(t-s+\frac{m}{n}T\right)^m e^{(\omega+CM)(t-s)} \qquad (7.8.36)$$

in $0 \leqslant s \leqslant t \leqslant T$, $m = 0,1,2,\ldots$. We deduce from (7.8.31) that

$$\|R_{n,m}(t,s)\|, \quad \|R_n(t,s)\| \leqslant Ce^{(\omega+CM+CDN)(t-s)+CDNT}. \qquad (7.8.37)$$

The second inequality implies (7.8.21).

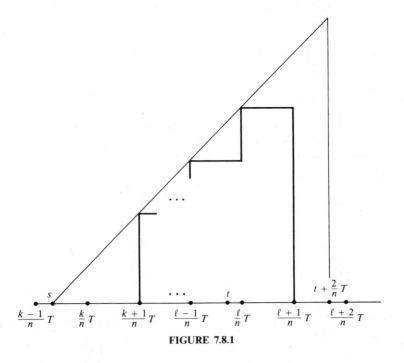

FIGURE 7.8.1

We prove now (b). As a first step, swe shall construct the "evolution operator" $\mathfrak{S}(t, s)$ of the equation $u'(t) = (A(t) + B(t))u(t)$ by a perturbation series like the one used in Example 7.1.12 and prove that $\mathfrak{S}_n \to \mathfrak{S}$. It should be pointed out that convergence of the $\mathfrak{S}_n$ cannot be proved by applying to $A(\cdot) + B(\cdot)$ the machinery of Theorem 7.7.5, since this family does not fulfill the hypotheses therein (in particular, F may not be $A(t) + B(t)$-admissible and, even if it were, $t \to \tilde{A}(t) + \tilde{B}(t)$ may not be $(F; E)$-continuous). Define $\tilde{\mathfrak{S}}_0(t, s) = S(t, s)$,

$$\tilde{\mathfrak{S}}_m(t, s)u = \int_s^t S(t, \sigma)B(\sigma)\tilde{\mathfrak{S}}_{m-1}(\sigma, s)u\, ds \quad (m \geqslant 1).$$

It is rather easy to show inductively (see again Example 7.1.12) that each $\tilde{\mathfrak{S}}_m$ is strongly continuous in $0 \leqslant s \leqslant t \leqslant T$ and that an estimate of the form

$$\|\tilde{\mathfrak{S}}_m(t, s)\| \leqslant \frac{C^{m+1}M^m}{m!}(t-s)^m e^{\omega(t-s)} \quad (m \geqslant 1) \qquad (7.8.38)$$

holds there, the constants C, ω, M with the same meaning as before. It follows that the series

$$\sum_{m=1}^{\infty} \tilde{\mathfrak{S}}_m(t, s) \qquad (7.8.39)$$

is uniformly convergent in $0 \leqslant s \leqslant t \leqslant T$ in the norm of (E), hence its sum $\mathfrak{S}(t, s)$ is strongly continuous there and satisfies

$$\|\mathfrak{S}(t, s)\| \leqslant Ce^{(\omega + MC)(t-s)}. \qquad (7.8.40)$$

The next step is to show that

$$\mathfrak{S}_n(t, s) \to \mathfrak{S}(t, s) \qquad (7.8.41)$$

strongly, uniformly in the triangle $0 \leqslant s \leqslant t \leqslant T$. This is done as follows. We observe first that the operators $\mathfrak{S}_n$ constructed from the "discretized" family $A_n(\cdot) + B_n(\cdot)$ can be obtained from the S_n—similarly constructed from $A_n(\cdot)$—by means of a perturbation series in the same way $\mathfrak{S}$ was obtained from S; precisely, if we set $\mathfrak{S}_{n,0}(t, s) = S_n(t, s)$,

$$\mathfrak{S}_{n, m}(t, s)u = \int_s^t S_n(t, \sigma)B_n(\sigma)\mathfrak{S}_{n, m-1}(\sigma, s)u\, d\sigma \quad (m \geqslant 1),$$

$$\qquad (7.8.42)$$

then

$$\sum_{m=0}^{\infty} \mathfrak{S}_{n, m}(t, s) = \mathfrak{S}_n(t, s) \qquad (7.8.43)$$

strongly, uniformly in $0 \leqslant s \leqslant t \leqslant T$. (The proof of this relation[7] is a some-

---

[7] Equality (7.8.43) can be heuristically justified as follows: $\mathfrak{S}_n(t, s)$ and the function on the left hand side satisfy the same differential equation in $s \leqslant t \leqslant T$ and the same initial condition at $t = s$.

what tedious exercise involving the explicit expression (7.7.21) for S_n, the analogous formula for $\mathfrak{S}_n$, and the perturbation argument used in Theorem 5.1.2 to construct $S(t; A + B)$ from $S(t; A)$ in the particular case where B is bounded; we omit the details.) We compare now the two series (7.8.39) and (7.8.43) term by term, beginning with the observation that the approximations to $\mathfrak{S}_n$ furnished by (7.8.42) satisfy the same inequality (7.8.38) obeyed by the $\tilde{\mathfrak{S}}_m$; given then $\varepsilon > 0$, we can choose an integer m_0 so large that

$$\sum_{m = m_0 + 1}^{\infty} \|\tilde{\mathfrak{S}}_m(t, s) - \mathfrak{S}_{n, m}(t, s)\| \leqslant \varepsilon/2 \qquad (7.8.44)$$

in $0 \leqslant s \leqslant t \leqslant T$, this estimate being independent of n. We note next that, since $\tilde{\mathfrak{S}}_0 = S, \tilde{\mathfrak{S}}_1, \tilde{\mathfrak{S}}_2, \ldots$ are all strongly continuous, given $u \in E$ fixed, each of the sets

$$\{\tilde{\mathfrak{S}}_m(t, s)u; 0 \leqslant s \leqslant t \leqslant T\}$$

is compact in E. We bring into play the obvious estimate

$$\|S_n(t, s)B_n(s)\| \leqslant CMe^{\omega(t - s)} \quad (0 \leqslant s \leqslant t \leqslant T)$$

valid for all n, and the fact that $S_n(t, s)B_n(s) \to S(t, s)B(s)$ strongly, uniformly in the same region: an elementary compactness argument then shows that, given $\delta > 0$, we can pick n_0 so large that

$$\|(S(t, \sigma)B(\sigma) - S_n(t, \sigma)B_n(\sigma))\tilde{\mathfrak{S}}_m(\sigma, s)u\| \leqslant \delta$$

in $0 \leqslant s \leqslant \sigma \leqslant t \leqslant T$ for $n \geqslant n_0$ and $0 \leqslant m \leqslant m_0$, and we can even include the inequality

$$\|S(\sigma, s)u - S_n(\sigma, s)u\| \leqslant \delta$$

in the same range of s, σ, n, by eventual augmentation of n_0 in view of the fact (proved in Theorem 7.7.5) that $S_n \to S$ strongly, uniformly in $0 \leqslant s \leqslant \sigma \leqslant T$. We have now

$$\tilde{\mathfrak{S}}_1(t, s)u - \mathfrak{S}_{n, 1}(t, s)u = \int_s^t (S(t, \sigma)B(\sigma) - S_n(t, \sigma)B_n(\sigma))S(\sigma, s)u \, d\sigma$$

$$+ \int_s^t S_n(t, \sigma)B_n(\sigma)(S(\sigma, s)u - S_n(\sigma, s)u) \, d\sigma,$$

$$(7.8.45)$$

so that, in view of all the previous precautions,

$$\|\tilde{\mathfrak{S}}_1(t, s)u - \tilde{\mathfrak{S}}_{n, 1}(t, s)u\| \leqslant T(1 + CMe^{\omega T})\delta.$$

Replacing this inequality into the inductive successor of (7.8.45) and repeating the trick, we obtain

$$\|\tilde{\mathfrak{S}}_m(t, s)u - \mathfrak{S}_{n, m}(t, s)u\| \leqslant \varepsilon/2m_0 \qquad (7.8.46)$$

in $0 \leqslant s \leqslant t \leqslant T$ for $n \geqslant n_0$, $1 \leqslant m \leqslant m_0$ if δ is chosen adequately small.

Putting this inequality together with (7.8.44) and taking advantage of the arbitrariness of ε, (7.8.41) follows.

With that relation in our bag, we go back to (7.8.30) and (7.8.31). The next step in the proof of part (b) of Lemma 7.8.4 will be to show that each $R_{n,m}$ has a (strong) limit as $n \to \infty$; precisely,

$$R_{n,m}(t,s)u \to \tilde{R}_m(t,s)u \qquad (7.8.47)$$

for each $u \in E$, uniformly in $0 \leqslant s \leqslant t \leqslant T$, the operators $\tilde{R}_m$ defined inductively by the formulas $\tilde{R}_0(t,s) = \mathfrak{S}(t,s)$,

$$\tilde{R}_m(t,s)u = \int_s^t \mathfrak{S}(t,\sigma)K'(\sigma)K(\sigma)^{-1}\tilde{R}_{m-1}(\sigma,s)u\,d\sigma \quad (m \geqslant 1).$$

$$(7.8.48)$$

This is quite obvious when $m = 0$, since $R_{n,0} = \mathfrak{S}_n$. Assume then that (7.8.47) holds for all indices $\leqslant m - 1$; let $u \in E$ and consider the following function in the interval $[0, T]$:

$$g_n(\sigma) = \mathfrak{S}_n\left(t, \frac{j-1}{n}T\right)K'(\sigma)K\left(\frac{j-2}{n}T\right)^{-1}R_{n,m-1}\left(\frac{j-1}{n}T,s\right)u$$

in

$$\frac{j-2}{n}T \leqslant \sigma < \frac{j-1}{n}T \quad \text{for} \quad j = k+1,\ldots,l,$$

$g(\sigma) = 0$ elsewhere. Taking into account that

$$\int_{(j-2)T/n}^{(j-1)T/n} K'(\sigma)K\left(\frac{j-2}{n}\right)^{-1}u\,d\sigma = K\left(\frac{j-1}{n}T\right)K\left(\frac{j-2}{n}T\right)^{-1}u - u$$

$$= L\left(\frac{j-1}{n}T, \frac{j-2}{n}T\right)u,$$

it is obvious that

$$R_{n,m}(t,s)u = \int_0^T g_n(\sigma)\,d\sigma. \qquad (7.8.49)$$

We use now the convergence relation (7.8.41), (7.8.47) (for indices $\leqslant m-1$) and the uniform estimates (7.8.27) and (7.8.37) to deduce that $g_n(\sigma) \to \mathfrak{S}(t,\sigma)K'(\sigma)K(\sigma)^{-1}\tilde{R}_{m-1}(\sigma,s)u$ in $0 \leqslant \sigma \leqslant T$ with $\|g_n(\sigma)\|$ uniformly bounded there, thus we can take limits in (7.8.49) using (a vector-valued analogue) of the dominated convergence theorem and obtain (7.8.47).

Making use of (7.8.40) and of the recursive formula (7.8.48), we easily obtain the estimate

$$\|\tilde{R}_m(t,s)\| \leqslant \frac{C^{m+1}(HN)^m}{m!}(t-s)^m e^{(\omega + MC)(t-s)} \quad (m \geqslant 1) \quad (7.8.50)$$

in $0 \leqslant s \leqslant t \leqslant T$, where H (resp. N) is a bound for $\|K'(\sigma)\|_{(F;E)}$ (resp. for

$\|K(\sigma)^{-1}\|_{(E;\,F)}$ in $0 \leqslant \sigma \leqslant T$. Combining this with (7.8.37) and with the convergence relation (7.8.47) obtained a moment ago, we obtain that

$$R_n(t, s) = \sum_{m=0}^{l-k} R_{n,m}(t, s) \to R(t, s)$$

$$= \sum_{m=0}^{\infty} \tilde{R}_m(t, s) \quad (n \to \infty) \qquad (7.8.51)$$

strongly, uniformly in $0 \leqslant s \leqslant t \leqslant T$. The final step takes advantage of (7.8.28) and of the estimate (7.8.32): the result is

$$Q_n(t, s) \to R(t, s) \quad (n \to \infty) \qquad (7.8.52)$$

in the same sense as in (7.8.51).

We summarize the different steps in their logical order; arrows indicate strong convergence, uniform in $0 \leqslant s \leqslant t \leqslant T$, as $n \to \infty$.

---

(1) $\mathfrak{S}_n(t, s)$ ("evolution operator" of $A_n(t) + B_n(t)$ defined before (7.8.27))	$\to$ (7.8.41) and following comments	$\mathfrak{S}(t, s)$ ("evolution operator" of $A(t) + B(t)$ defined in (7.8.39))
(2) $R_{n,m}(t, s)$ (defined in (7.8.30))	$\to$ (7.8.47) and following comments	$\tilde{R}_m(t, s)$ (defined in (7.8.48))
(3) $R_n(t, s)$ $= \sum_{m=0}^{l-k} R_{n,m}(t, s)$	$\to$ (7.8.51) and preceding comments	$R(t, s) = \sum_{m=0}^{\infty} \tilde{R}_m(t, s)$
(4) $Q_n(t, s)$	$\to$ (3) above and (7.8.28)	$R(t, s)$

End of proof of Theorem 7.8.3. As pointed out at the beginning of the proof (see (7.8.19)), we only have to show that $Q(t, s) = K(t)S(t, s)K(s)^{-1}$ is everywhere defined and (E)-strongly continuous in $0 \leqslant s \leqslant t \leqslant T$. This is done as follows. Pick $u \in E$. Since $Q_n(t, s) = K(t)S_n(t, s)K(s)^{-1}u$ converges in E by Lemma 7.8.4(b), we obtain pre-multiplying by $K(t)^{-1}$ that $S_n(t, s)K(s)^{-1}u$ is convergent in F, a fortiori in E; since $S_n(t, s)K(s)^{-1}u \to S(t, s)K(s)^{-1}u$ in E, it follows that $S_n(t, s)K(s)^{-1}u \to S(t, s)K(s)^{-1}u$ in F, thus $S(t, s)K(s)^{-1}u \in F$ and

$$K(t)S_n(t, s)K(s)^{-1}u \to K(t)S(t, s)K(s)^{-1}u$$

in E, the convergence being uniform in $0 \leqslant s \leqslant t \leqslant T$ by virtue of Lemma 7.8.4(b). Since each $K(t)S_n(t, s)K(s)^{-1}u$ is continuous in $0 \leqslant s \leqslant t \leqslant T$, our claims on $K(t)S(t, s)K(s)^{-1}$ follow and the proof of Theorem 7.8.3 is complete.

The manipulations in the proof of Theorem 7.8.3 are perhaps too intricate to let the basic idea of the argument shine through. An heuristic justification of why we should expect $K(t)S(t, s)K(s)^{-1}$ to be strongly continuous is this. Differentiating formally and using (7.8.15), we obtain

$$D_t\big(K(t)S(t, s)K(s)^{-1}\big) = K(t)A(t)S(t, s)K(s)^{-1}$$
$$+ K'(t)S(t, s)K(s)^{-1}$$
$$= \big(A(t) + B(t) + K'(t)K(t)^{-1}\big)\big(K(t)S(t, s)K(s)^{-1}\big),$$

whereas $K(t)S(t, t)K(t)^{-1} = I$, thus $K(t)S(t, s)K(s)^{-1}$ should be the propagator of

$$u'(t) = \big(A(t) + B(t) + K'(t)K(t)^{-1}\big)u(t)$$

and must then be everywhere defined and strongly continuous. Naturally, it would be hard to try to establish this along the lines of Theorem 7.7.5, since the hypotheses there are far from being satisfied for $A(\cdot) + B(\cdot) + K'(\cdot)K(\cdot)^{-1}$.

We close this section with the announced application of Theorem 7.8.3 to the time-dependent symmetric hyperbolic operator (7.8.2) (and to the associated differential equation (7.8.1)).

7.8.5 Theorem. *Let the matrices $A_1(x, t), \ldots, A_m(x, t), B(x, t)$ be defined and continuously differentiable with respect to $x_1, \ldots, x_m$ in $\mathbb{R}^m \times [0, T]$. Assume all the functions*

$$|A_k(x, t)|, |D^jA_k(x, t)|, |B(x, t)|, |D^jB(x, t)| \quad (1 \leqslant j, k \leqslant m)$$
$$(7.8.53)$$

are bounded in $\mathbb{R}^m \times [0, T]$. Assume, moreover, that $A_k(x, t), D^jA_k(x, t),$ and $B(x, t)$ are continuous in t for $0 \leqslant t \leqslant T$, uniformly with respect to $x \in \mathbb{R}^m$. Then the Cauchy problem for (7.8.1) is properly posed in both senses of time in $0 \leqslant t \leqslant T$ (with respect to solutions defined by (D)), with $F = H^1(\mathbb{R}^m)^\nu$.

Proof. Fortunately, most of the work needed to verify the assumptions of Theorem 7.8.3 in this case was already done in Section 5.6 in for the stationary equation. Continuity of the map (7.8.13) is immediate (we only use here the hypotheses on $A_k(x, t), B(x, t)$, not those on the partial derivatives). The role of the operator function $K(t)$ is played by the constant function

$$\Re = \mathscr{F}^{-1}\big((1 + |\sigma|^2)^{1/2}\mathscr{F}u(\sigma)\big).$$

Using the results of Section 5.6, we see that

$$\Re\mathcal{Q}(t)\Re^{-1} = \mathcal{Q}(t) + Q(t) \quad (0 \leqslant t \leqslant T), \qquad (7.8.54)$$

where $Q(t) = Q_1(t) + Q_2(t)$, both operators being bounded for each t (see their definition in (5.6.36) and (5.6.38). Our only task is then to show that $Q_1(\cdot), Q_2(\cdot)$ are strongly continuous. If $u \in \mathbb{S}^r$,

$$(Q_2(t') - Q_2(t))u = \left(\Re^{-1} - \sum_{k=1}^m D^k \Re^{-1} D^k\right)(B(\cdot, t') - B(\cdot, t))\Re^{-1}u$$
$$- (B(\cdot, t') - B(\cdot, t))u. \qquad (7.8.55)$$

The second term on the right-hand side obviously tends to zero as $t' \to t$ uniformly with respect to u bounded in the L^2 norm. As for the first, taking into account that

$$D^j B(\cdot, t') - D^j B(\cdot, t)$$

tends to zero pointwise in $x \in \mathbb{R}^m$ while remaining uniformly bounded there, we see that it tends to zero also (although of course not uniformly with respect to u). On the other hand, it follows from boundedness of B and $D^j B$ that $\|Q_2(\cdot)\|$ is uniformly bounded in $0 \leqslant t \leqslant T$, thus $Q_2(\cdot)$ is strongly continuous. Now,

$$(Q_1(t') - Q_1(t))u = \sum_{k=1}^m \{\Re, (A_k(\cdot, t') - A_k(\cdot, t))\}D^k \Re^{-1}u.$$

Since $|D^j A_k(x, t') - D^j A_k(x, t)| \to 0$ as $t' - t \to 0$, uniformly for $x \in \mathbb{R}^m$, it follows from Theorem 5.6.6 that Q_1 is actually continuous as a (L^2)-valued function, hence strongly continuous. It follows then that $Q(\cdot) = Q_1(\cdot) + Q_2(\cdot)$ is strongly continuous and all the hypotheses of Theorem 7.8.3 are fulfilled. Since these hypotheses are also satisfied for the family $\{-\mathcal{C}(T - t); 0 \leqslant t \leqslant T\}$ (which obeys the same assumptions as $\mathcal{C}(\cdot)$), the claims of Theorem 7.8.5 follow.

A time-dependent counterpart of Theorem 5.6.7 can be easily proved.

7.8.6 Theorem. *Let the matrices $A_1(x, t), \ldots, A_m(x, t), B(x, t)$ be r times continuously differentiable in $\mathbb{R}^m \times [0, T]$ with respect to $x_1, \ldots, x_m$ and assume all the functions*

$$|D^\alpha A_k(x, t)|, |D^\alpha B(x, t)| \quad (0 \leqslant |\alpha| \leqslant r, 1 \leqslant k \leqslant m) \qquad (7.8.56)$$

are bounded in $\mathbb{R}^m \times [0, T]$. Suppose, moreover, that the $D^\alpha A(x, t)$ ($|\alpha| \leqslant r$) and the $D^\beta B(x, t)$ ($|\beta| < r - 1$) are continuous in t for $0 \leqslant t \leqslant T$, uniformly with respect to $x \in \mathbb{R}^m$. Then the conclusion of Theorem 7.8.5 holds; moreover,

$$S(t, s)H^r \subseteq H^r \quad (0 \leqslant s, t \leqslant T),$$

each $S(t, r)$ is a bounded operator in H^r and $(t, s) \to S(t, r)$ is continuous from $0 \leqslant s, t \leqslant T$ into H^r. In particular, a solution $u(t)$ that starts in H^r ($u(s) \in H^r$) remains in H^r and satisfies

$$\|u(t)\|_{H^r} \leqslant C_r \|u(s)\|_{H^r} \quad (0 \leqslant s, t \leqslant T)$$

for a suitable constant C_r independent of u.

The proof can be carried out modifying that of Theorem 7.8.6 along the lines of Theorem 5.6.7 and is therefore omitted.

7.9. THE CASE WHERE $D(A(t))$ IS INDEPENDENT OF t. THE INHOMOGENEOUS EQUATION

The following result is an important particular case of Theorem 7.8.3.

7.9.1 Theorem. *Let* $\{A(t); 0 \leqslant t \leqslant T\}$ *be a stable family of densely defined operators in* E *with stability constants* C, ω. *Assume that*

$$F = D(A(t)) \qquad (7.9.1)$$

does not depend on t, *and that, for some* $\lambda > \omega$, *the function*

$$t \to A(t)R(\lambda; A(0)) \qquad (7.9.2)$$

is strongly continuously differentiable in $0 \leqslant t \leqslant T$. *Then the conclusions of Theorem 7.8.3 hold for the equation*

$$u'(t) = A(t)u(t). \qquad (7.9.3)$$

Proof. Let $F = D(A(t))$, and let $\|\cdot\|_t$ be the graph norm of $A(t)$ in F or, even better, the equivalent norm

$$\|u\|_t = \|(\lambda I - A(t))u\|. \qquad (7.9.4)$$

Plainly F is a Banach space endowed with any of the $\|\cdot\|_t$. We show that all these norms are equivalent (even if no assumptions are made on the map (7.9.2)). Let $s, t \in [0, T]$ and $\{u_n\}$ a sequence in E with $u_n \to u$, $A(t)R(\lambda; A(s))u_n \to v$. Since $R(\lambda; A(s))u_n \to R(\lambda; A(s))u$ and $A(t)$ is closed, it follows that $A(t)R(\lambda; A(s))u = v$, thus $A(t)R(\lambda; A(s))$ is itself closed; since it is everywhere defined, it results from the closed graph theorem that $A(t)R(\lambda; A(s))$ is bounded. Then, if $u \in F$ and $w = (\lambda I - A(s))u$,

$$\begin{aligned}
\|u\|_t &= \|(\lambda I - A(t))u\| \\
&= \|(\lambda I - A(t))R(\lambda; A(s))w\| \\
&\leqslant C\|w\| = C\|(\lambda I - A(s))u\| = C\|u\|_s,
\end{aligned}$$

which justifies our claim.

We check the other hypotheses. It is obvious that $F \to E$. Using the fact that (7.9.2) is strongly continuously differentiable, we apply the uniform boundedness theorem to the family

$$\{(t' - t)\{A(t')R(\lambda; A(0)) - A(t)R(\lambda; A(0))\}; t \neq t'\}$$

obtaining

$$\|A(t')R(\lambda; A(0)) - A(t)R(\lambda; A(0))\| \leqslant C|t' - t|$$

and this shows that $t \to A(t) \in (F; E)$ is continuous (in fact, Lipschitz continuous). Define

$$K(t) = \lambda I - A(t). \qquad (7.9.5)$$

Then $\|K(t)u\| = \|u\|_t$ for $u \in F$, $\|K(t)^{-1}v\|_t = \|v\|$ for $v \in E$, so that $K(t) \in (F, E)$ and $K(t)^{-1} \in (E, F)$. Finally, it is obvious that (7.8.15) holds with $B(t) \equiv 0$ and that the hypothesis on (7.9.2) implies that $K(t)u$ is continuously differentiable if $u \in F = R(\lambda; A(0))E$. The assumptions of Theorem 7.8.3 are thus verified in full. This completes the proof.

Going back to the general set-up of Sections 7.7 and 7.8 we consider the inhomogeneous equation

$$u'(t) = A(t)u(t) + f(t). \qquad (7.9.6)$$

If $F, A(\cdot)$ satisfy the hypotheses of Theorem 7.7.5 and f is a E-valued continuous function in $0 \leqslant t \leqslant T$, every solution of (7.9.6) in the sense of Definition (D) must perforce be of the form

$$u(t) = S(t, s)u(s) + \int_s^t S(t, s)f(s)\, ds \quad (s \leqslant t \leqslant T).$$

To see this, take $v(\sigma) = S(t, \sigma)u(\sigma)$, $s \leqslant \sigma \leqslant t$ and operate in a way similar to that used to show uniqueness for the homogeneous equation (see the end of the proof of Theorem 7.7.5 and Remark 7.7.8); the result is the equality

$$D^+v(\sigma) = S(t, \sigma)f(\sigma) \quad (s \leqslant \sigma \leqslant t).$$

It is then an easy consequence of Lemma 7.7.7 that

$$u(t) = S(t, s)u(s) + \int_s^t S(t, \sigma)f(\sigma)\, d\sigma; \qquad (7.9.7)$$

(it should be noted that this argument depends only on s-differentiability of $S(t, s)$ and thus holds even if F is not reflexive.) Conversely, if the assumptions on f are strengthened, (7.9.7) will be a solution of (7.9.6). We limit ourselves to proving two results, which can be considered to be "dynamic" counterparts of parts (a) and (b) of Lemma 2.4.2.

7.9.2 Lemma. *Let the assumptions of Theorem 7.8.3 hold, and let $f(t)$ be F-valued and F-continuous in $0 \leqslant t \leqslant T$. Then*

$$u(t) = \int_s^t S(t, \sigma)f(\sigma)\, d\sigma \qquad (7.9.8)$$

is a genuine solution of (7.9.6) *in* $s \leqslant t \leqslant T$ (in the sense of Definition (D).[8]

Proof. Under the present assumptions $S(t, s)$ is F-strongly continuous in $0 \leqslant s \leqslant t \leqslant T$, thus

$$(t, s) \to A(t)S(t, s)f(s)$$

---

[8] See footnote 5, p. 440.

is E-continuous there. This justifies the following computation:

$$\int_s^t A(r)u(r)\,dr = \int_s^t \left\{ \int_s^r A(r)S(r,\sigma)f(\sigma)\,d\sigma \right\}\,dr$$

$$= \int_s^t \left\{ \int_\sigma^t A(r)S(r,\sigma)f(\sigma)\,dr \right\}\,d\sigma$$

$$= \int_s^t \left\{ \int_\sigma^t D_r S(r,\sigma)f(\sigma)\,dr \right\}\,d\sigma$$

$$= \int_s^t S(t,\sigma)f(\sigma)\,d\sigma - \int_s^t f(\sigma)\,d\sigma \qquad (7.9.9)$$

and shows that $u(\cdot)$ is a solution of (7.9.6) in the sense claimed above. This ends the proof.

Other versions of Lemma 7.9.2 can be proved under different hypotheses. We limit ourselves to the following result:

7.9.3 Example.   Let the assumptions of Theorem 7.7.7 be satisfied, and let f be strongly measurable and integrable (as a F-valued function) in $0 \leqslant t \leqslant T$. Then $u(t)$ is continuous in E, $u(t) \in F$ almost everywhere, $t \to A(t)u(t)$ is strongly measurable and integrable in E and

$$u(t) = \int_s^t A(r)u(r)\,dr + \int_s^t f(r)\,dr \quad (s \leqslant t \leqslant T). \qquad (7.9.10)$$

The proof follows that of Lemma 7.9.2 almost verbatim; here we use again the fact that weak continuity implies strong measurability to show that S is F-strongly measurable in $0 \leqslant s \leqslant t \leqslant T$; therefore, $A(t)S(t,\sigma)f(\sigma)$ is E-strongly measurable there (as well as strongly measurable separately in each variable) justifying the calculations.

7.9.4 Lemma.   *Let the assumptions of Theorem 7.9.1 be satisfied, and let f be an E-valued continuously differentiable function in $0 \leqslant t \leqslant T$. Then $u(\cdot)$, given by (7.9.8) is a genuine solution of (7.9.6) in $s \leqslant t \leqslant T$.*

Proof.   Replacing if necessary $A(\cdot)$ by $A(\cdot) - \alpha I$ for α sufficiently large, we may assume the stability constant ω is negative, and thus set $K(t) = A(t)$. We have

$$\left(A(t)^{-1}f(t) \right)' = -A(t)^{-1}A'(t)A(t)^{-1}f(t) + A(t)^{-1}f'(t)$$

$$= A(t)^{-1}g(t), \qquad (7.9.11)$$

where $g(t) = -A'(t)A(t)^{-1}f(t) + f'(t)$ is E-continuous. Hence we have

$$D_\sigma \left(S(t,\sigma)A(\sigma)^{-1}f(\sigma) \right) = -S(t,\sigma)f(\sigma) + S(t,\sigma)A(\sigma)^{-1}g(\sigma),$$

$$(7.9.12)$$

where we have used the properties on σ-differentiability of $S(t,\sigma)$ proved in

Theorem 7.7.5 under much weaker hypotheses. We integrate now (7.9.12) in $s \leqslant \sigma \leqslant t$, obtaining

$$u(t) = -A(t)^{-1}f(t) + S(t,s)A(s)^{-1}f(s) + \int_s^t S(t,\sigma)A(\sigma)^{-1}g(\sigma)\,d\sigma$$

$$= -A(t)^{-1}f(t) + v(t). \qquad (7.9.13)$$

Since $A(s)^{-1}f(s) \in F$ and $g(\cdot)$ is (E)-continuous, $A(\cdot)^{-1}g(\cdot)$ is F-continuous, and by virtue of Lemma 7.9.2, $v(\cdot)$ is a solution of

$$v'(t) = A(t)v(t) + A(t)^{-1}g(t)$$

in $s \leqslant t \leqslant T$. Then

$$u'(t) = -\left(A(t)^{-1}f(t)\right)' + v'(t)$$

$$= A(t)v(t)$$

$$= A(t)u(t) + f(t) \qquad (7.9.14)$$

as claimed.

We close this section with a few results on the homogeneous equation (7.7.1) (the proofs are omitted). In the first one (Example 7.9.4 below) the conclusions of the theorems in Sections 7.7 and 7.8 are essentially unchanged, but the assumptions are considerably weaker. The first task is to modify in a convenient way the notion of stability.

A family $\{A(t); 0 \leqslant t \leqslant T\}$ of densely defined operators in E is called *quasi-stable* if there exists $C > 0$ and a function $\omega(\cdot)$, upper Lebesgue integrable in $0 \leqslant t \leqslant T$, such that $R(\lambda; A(t))$ exists for $\lambda > \omega(t)$ and

$$\left\|\prod_{j=1}^n R(\lambda_j; A(t_j))\right\| \leqslant C \prod_{j=1}^n (\lambda_j - \omega(t_j))^{-1} \qquad (7.9.15)$$

for all $\{t_j\}, \{\lambda_j\}$ such that $0 \leqslant t_1 \leqslant t_2 \leqslant \cdots \leqslant t_n \leqslant T$, and $\lambda_1 > \omega(t_1), \ldots, \lambda_n > \omega(t_n)$. The product on the left-hand side of (7.9.15) is ordered as in the definition of stability.

*7.9.5 Example (Kato [1973: 1]). Let E, F be Banach spaces, $F \to E, \{A(t); 0 \leqslant t \leqslant T\}$ a quasi-stable family of operators in E such that $F \subseteq D(A(t))$ and $A(t): F \to E$ is bounded; moreover, if $\tilde{A}(t)$ is the restriction of $A(t)$ to F, the map

$$t \to \tilde{A}(t) \in (F; E) \qquad (7.9.16)$$

is continuous in $0 \leqslant t \leqslant T$. For each $t \in [0, T]$, let $K(t)$ be a bounded and invertible operator from F onto E such that

$$t \to K(t) \in (F, E) \qquad (7.9.17)$$

is strongly continuous, and there exists another strongly measurable function $t \to \dot{K}(t) \in (F; E)$ such that

$$\|\dot{K}(t)\| \leqslant \beta(t) \quad (0 \leqslant t \leqslant T)$$

(β integrable in $0 \leqslant t \leqslant T$),

$$K(t)u = \int_0^t \dot{K}(s)u\,ds + K(0)u \quad (0 \leqslant t \leqslant T, u \in F) \qquad (7.9.18)$$

and

$$K(t)A(t)K(t)^{-1} = A(t) + B(t) \quad (0 \leqslant t \leqslant T), \qquad (7.9.19)$$

where $B(\cdot)$ is a strongly measurable (E)-valued function such that

$$\|B(t)\| \leqslant \gamma(t) \quad (0 \leqslant t \leqslant T), \qquad (7.9.20)$$

γ integrable in $0 \leqslant t \leqslant T$. Then the conclusions of Theorem 7.8.3 hold in full for the equation

$$u'(t) = A(t)u(t). \qquad (7.9.21)$$

In other words, the Cauchy problem for (7.9.19) is properly posed in $0 \leqslant t \leqslant T$ with respect to solutions defined in (D).

The proof runs along the same lines as that of Theorem 7.8.3, the main difference being that the sequence of partitions $\{t_k = (k-1)T/n\}$ used in the construction of the approximating operator S_n must now be replaced by a suitable sequence of uneven partitions.

Even more general results can be obtained if one is willing to accept less stringent definitions of solution; in fact, we have

7.9.6 Example (Kato [1973: 1]). Let the assumptions in Example 7.9.6 hold, except in that the map (7.9.16) is only assumed to be strongly measurable (in the topology of $(F; E)$) and

$$\int_0^T \|A(t)\|_{(F; E)}\,dt < \infty. \qquad (7.9.22)$$

Then the Cauchy problem for (7.9.22) is properly posed with respect to the following definition of solution:

(E) $u(\cdot)$ *is a solution of* (7.9.21) *in* $s \leqslant t \leqslant T$ *if and only if* $u(t) \in F$ *a.e,* $A(t)u(t)$ *is strongly measurable and integrable in* $s \leqslant t \leqslant T$, *and*

$$u(t) = \int_s^t A(\tau)u(\tau)\,d\tau + u(s) \quad (s \leqslant t \leqslant T).$$

7.10. MISCELLANEOUS COMMENTS

The notion of properly posed abstract Cauchy problem was introduced by Lax (see Section 1.7) including the time-dependent setting treated here. Operators $S(t, s)$ satisfying (7.1.7) and (7.1.8) were called *propagators* by Segal [1963: 1]; other names in use are *evolution operators* or *solution operators*, when arising from a differential equation like (7.1.1). This equation is usually called an *evolution equation*; curiously, the name seems not to be used for its time-invariant counterpart (2.1.1).

The facts on the equation (7.1.1) pertaining to the case $A(t)$ *bounded* are all well known (and rather natural generalizations of even better known results for finite-dimensional vector differential equations). For deep investigations of the equation (7.1.1) in this setting see the treatises of Massera-Schäffer [1966: 1] and Daleckiĭ-Kreĭn [1970: 1]. Of special interest is the result in Example 7.1.8. Some manipulations with formula (7.1.31) and the definition of the exponential function show that the propagator $S(t, s)$ can be obtained in the form

$$S(t, s) = \lim_{n \to \infty} \prod_{j=1}^{m(n)} \left(I + \left(t_{n, j} - t_{n, j-1}\right) A\left(t_{n, j}\right) \right). \qquad (7.10.1)$$

In the finite-dimensional case, this limit was taken by Volterra [1887: 1], [1902: 1] as the definition of the *product integral* of the matrix function $A(\cdot)$ in the interval $s \leqslant r \leqslant t$ (Volterra's name for (7.10.1) is different). The product integral is denoted in various ways, for instance

$$\widehat{\int_s^t}(I + A(r)\, dr), \quad \prod_s^t (I + A(r)\, dr), \quad \prod_s^t e^{A(r)\, dr}, \qquad (7.10.2)$$

and coincides with

$$\exp\left(\int_s^t A(r)\, dr \right)$$

when the different values of $A(\cdot)$ commute. A calculus of matrix-valued functions based on this integral and on a corresponding *product derivative* was developed by Volterra, Schlesinger, Rasch, and others. (We note that (7.10.2) is the *left integral* of Volterra; the *right integral* is defined inverting the order of the factors and is used to construct the propagator or the equation $S' = SA$.) For a detailed exposition of the calculus see Volterra-Hostinský [1938: 1]. During the fifties, a revival of interest took place, motivated on the one hand by quantum mechanical applications (see Feynman [1951: 1]) and on the other hand by the search for generalization of representation results for matrix-valued analytic functions of the type of Nevanlinna's theorem; see, for instance, Ginzburg [1967: 1]. The definition of product integral extends in an obvious way to continuous operator-valued functions in infinite dimensional spaces. See Daleckiĭ-Kreĭn [1970: 1, Ch. III] for a brief account. A thorough exposition with many applications can be found in the recent monograph of Dollard and Friedman [1979: 1], where several generalizations are presented as well.

The material in Sections 7.2, 7.3, and 7.5 is entirely due to Kato and Tanabe [1962: 1]; we have followed closely the original exposition incorporating some simplifications in the estimates in Corollary 7.2.4 due to Kreĭn [1967: 1]. We note, however, that the original estimates of Kato and Tanabe are more precise. Section 7.4 is due to Kato-Tanabe [1967: 1]; some of the results are already in [1962: 1]. The applications in Section 7.6 are rather particular cases of results sketched in Kato-Tanabe [1967: 1]; see also Tanabe [1979: 1]. The material in Sections 7.7, 7.8, and 7.9 is due to Kato [1970: 1]; some of the results in Section 7.9 were discovered earlier by Kato [1953: 1], [1956: 1].

It is of interest to trace the evolution of the results in evolution equations leading to the present exposition. The first theorem on the equation (7.1.1), where the $A(t)$ are permitted to depend on t, is due to Kato [1953: 1]; we note in passing a result of Phillips [1953: 1] corresponding to the case $A(t) = A + P(t)$ with $A \in \mathcal{C}_+$ and $P(t)$ strongly continuous and taking values in (E): the propagator is constructed by means of a perturbation series of the type of (7.1.42). Kato's result, on the other hand, is a forerunner of Theorems 7.7.5 and 7.8.3 in the sense that the propagator $S(t, s)$ is obtained in the same "product integral" form. However, the hypotheses on $A(\cdot)$ are different; the domain $D(A(t))$ is assumed to be independent of t, each $A(t)$ belongs to $\mathcal{C}_+(1, 0)$ and smoothness hypotheses are placed on the $((E)$-valued) function $B(t, s) = (1 - A(t))(I - A(s))^{-1}$. The assumptions in Kato [1956: 1] are of the same type, but more general. Other early works are due to Ladyženskaya [1955: 1], [1956: 1], [1958: 1], Višik [1956: 1], Višik-Ladiženskaya [1956: 1], although these refer to special kinds of operators $A(t)$ such as self-adjoint operators and perturbations thereof and will be commented on in (b). The same observation applies to the works of Sobolevskiĭ [1957: 1], [1957: 2], and [1958: 2]. In this last paper the operators $A(t)$ are self-adjoint and an effort is made to relax the hypothesis, prevalent until then, that $D(A(t))$ must be independent of t. Sobolevskiĭ assumes instead that $D(A(t)^h)$ is constant for certain $h > 0$ and makes use of the integral equations (7.2.4) and (7.2.6) in the construction of the propagator. This method was taken up by Kato [1961: 1] for general $A(t)$ in $\mathcal{C}$; as shown there, the hypothesis of constancy of $D(A(t)^h)$ is (somewhat surprisingly) verified in some important cases where $D(A(t))$ varies with time. These results of Kato generalize earlier theorems of Tanabe [1959: 1], [1960: 1], [1960: 2], [1961: 1], where $D(A(t))$ is assumed constant. Tanabe introduced the construction of the propagator through the integral equations (7.2.8) and (7.2.48). The same method was then carried a big step forward by Kato and Tanabe in [1962: 1], where essentially no relations are postulated between the different $D(A(t))$, as Example 7.3.2 puts in evidence in a rather extreme way. We note that independence on t of the domain of $A(t)$ is truly essential in treating initial-boundary value problems where the boundary conditions depend on t, or even operators like $\mathcal{C}(t)$ in Section 7.8, where $D(\mathcal{C}(t))$ is not easily identified (and may be suspected of changing with time).

As pointed out before, some of the ideas leading to the results in Section 7.7 and following sections can be traced back to Kato [1953: 1]. Other basic ingredients in the proof of Theorems 7.7.5 and 7.8.3 such as the

introduction of the space F, and the requirement of $(F; E)$-smooth dependence can be found in Heyn [1962: 1] and Kisyński [1966: 1]; the use of E-
and F-stable families appears for the first time in Heyn's paper, while
several results in the style of Example 7.7.3 were given by Kisyński. A
similar approach to the equation (7.7.1), where many of the basic ideas in
Kato [1970: 1] are already in evidence, is Goldstein [1969: 4]. We point out
finally that the operators $K(t)$ make their apparition in Kato [1956: 1] and
that the proof of Theorem 7.8.3 follows Yosida [1965/66: 1] (see also [1978:
1]). Certain simplifications to the various arguments have been proposed by
Dorroh [1975: 1]. We note that in the work of Heyn and (partly) in that of
Kisyński mentioned above, the approach to the equation (7.7.1) is through
the Yosida approximation (Example 2.3.13), that is, the solution $u(\cdot)$ is
constructed as $\lim u_n(\cdot)$, where $u_n(\cdot)$ is the solution of

$$u'_n(t) = nA(t)R(n; A(t))u_n(t). \qquad (7.10.3)$$

A similar approach was used by Elliott [1962: 1] and Yosida [1963: 1],
[1965/66: 1] the latter working in locally convex spaces.

     A far-reaching generalization of the ideas in Sections 7.7 and 7.8 is
due to Kato [1973: 1] (see Examples 7.9.5 and 7.9.6) roughly along the lines
of [1970: 1], but relaxing in an essential way many of the hypotheses.

     Numerous other approaches have been used in relation to the
equation (7.7.1), most of them starting with the assumption that each
$A(t)$ belongs to $\mathcal{C}_+$ or to a subclass thereof but making different requirements on t-dependence of $A(t)$; the following is a list of references. See
Baumgärtel [1964: 1], [1964: 2], Bragg [1974: 1], Domslak [1961: 1], [1962:
1], Dyment-Sobolevskiĭ [1970: 1], Enikeeva [1972: 1], Fitzgibbon [1976: 1],
Foias-Gussi-Poenaru [1957: 1], Fujie-Tanabe [1973: 1], Gajewski [1971: 1],
Gibson-Clark [1977: 1], Globevnik [1971: 1], Goldstein [1972: 1], [1974: 1],
[1975/76: 1], Hackman [1968: 1], Inoue [1972: 1], [1974: 1], V. N. Ivanov
[1966: 1], Jakubov [1964: 1], [1964: 2], [1966: 1], [1966: 2], [1966: 3], [1966:
4], [1966: 5], [1967: 4], [1968: 1], [1970: 1], [1970: 2], [1973: 1], Jakubov-
Samedova [1975: 1], Kartsatos-Zigler [1976: 1], Krasnosel'skiĭ-Kreĭn-
Sobolevskiĭ [1956: 1], [1957: 1], Köhnen [1970: 1] [1971: 1], [1972: 1], Krée
[1970: 1], Kreĭn [1957: 1], [1966: 1], [1967: 1], Kreĭn-Daleckiĭ [1970: 1],
Kreĭn-Laptev [1969: 1], Kreĭn-Savčenko [1972: 1], Lopatinskiĭ [1966: 1],
Lovelady [1975: 2], Mamedov [1964: 1], [1964: 2], [1965: 1], [1966: 1],
Massey [1972: 1], Masuda [1967: 1], [1967: 2], [1968: 1], Mlak [1959: 1],
[1960: 1], [1960/61: 1], Murakami [1966: 1], Nemyckiĭ-Vaĭnberg-Gusarova
[1964: 1], Oja [1974: 1], Poulsen [1965: 1], Raskin-Sobolevskiĭ [1967:
1], [1968: 1], [1968: 2], Rota [1963: 1], Sadarly [1969: 1], Sarymsakov-
Murtazaev [1970: 1], Sohr [1973: 1], [1973: 2], [1973: 3], Solomjak
[1960: 1], Strygina-Zabreiko [1971: 1], Suryanarayana [1965: 1], Tanabe-
Watanabe [1966: 1], Tsokos-Rama Mohana Rao [1969: 1], Watanabe
[1967: 1], [1968: 1], Pljuščev [1973: 1], [1975: 1], Yagi [1976: 1], Da Prato-

Iannelli [1976: 1], Datko [1972: 1], Derguzov [1963: 1], [1964: 1], [1964: 2], [1964: 3], [1973: 1], Derguzov-Jakubovič [1963: 1], [1969: 1], Derguzov-Melanina-Timčenko [1975: 1], and Djacenko [1976: 1]. For a thorough treatment of the subject and additional references, see Tanabe [1979: 1].

(a) *Propagators.* It was proved in Chapter 2 that there is a complete equivalence between propagators of time invariant equations and strongly continuous semigroups, that is, strongly continuous operator functions that satisfy the two equations (2.3.1). The equivalence breaks down in the time-varying case: although the propagator of the time-dependent equation (7.1.1) is strongly continuous and satisfies (7.1.7), (7.1.8), not every strongly continuous (E)-valued function satisfying these equations is the propagator of a differential equation. For a trivial counterexample (with dim $E = 1$), it suffices to consider

$$S(t, s) = \alpha(t)/\alpha(s),$$

where α is a continuous, positive, nowhere differentiable function. The same counterexample shows that nothing is gained by requiring (E)-continuity of S. Necessary and sufficient conditions on a (E)-valued function S satisfying (7.1.7) and (7.1.8) in order that S be the propagator of a properly posed Cauchy problem are apparently not known; likewise, the class of operator-valued functions $A(\cdot)$ giving rise to a properly posed Cauchy problem for (7.1.1) has never been fully characterized. For results in these and related areas see Masuda [1972: 1] (Example 7.4.4), Y. Komura [1970: 1], Lovelady [1975: 2], Pljuščev [1975: 1], Herod [1971: 2], and Lezański [1970: 1], [1974: 1]. See also Carroll [1975: 1] and references therein for very general relations between the equation (7.1.1) and the variation-of-constants formula (7.1.34).

(b) *Operational Differential Equations.* The equation (7.1.1) has been extensively investigated using tools on the whole different from semigroup theory. To this category belong some of the early works by Višik-Ladyženskaya, Ladyženskaya, Sobolevskiĭ, and others, quoted at the beginning of this section. A systematic treatment was undertaken by Lions starting with [1958: 1], [1958: 2], [1959: 1] and culminating in this treatise [1961: 1]. Some salient features of Lions' theory can be roughly described as follows. Let V, H be Hilbert spaces with $V \subseteq H, V$ dense in H and the injection $i: V \to H$ continuous. The operators $A(t)$ are defined by time-dependent bounded sesquilinear forms $a(t; u, v)$ in $V \times H$, that is,

$$(A(t)u, v) = a(t; u, v).$$

In more general situations V may depend on t. The solutions under consideration range from strong or genuine solutions in the sense of Section 7.1 to weak solutions as those in Section 7.5 to solutions in the sense of vector-valued distributions. The methods used in the construction of the solutions are, on the one hand, duality arguments combined with a priori estimates (such as versions of the Lax-Milgram lemma) and, on the other hand, approximation methods (for instance, based on finite difference approximations of (7.1.1)) also used in combination with a priori estimates. A notable way to construct approximate solutions of (7.1.1) is the so-called *Faedo-Galerkin method*, an abstract adaptation of the classical Rayleigh-Ritz approximation procedure. It can be roughly exemplified as follows. Choose a linearly independent sequence $w_1, w_2, \ldots$ in V such that finite linear combinations of the $\{w_k\}$ are dense in V; take $u_0 = u(0)$ in V and select coefficients η_{nk} such that

$$u_n = \sum_{k=1}^{n} \eta_{nk} w_k \to u_0.$$

Solve then the ordinary differential system

$$\sum_{k=1}^{n} \eta'_{nk}(t)(w_j, w_k) = \sum_{k=1}^{n} \eta_{nk}(t) a(t, w_j, w_k) \quad (1 \leqslant j \leqslant n)$$

with initial conditions

$$\eta_{jk}(0) = \eta_{jk}$$

and take

$$u_n(t) = \sum_{k=1}^{n} \eta_{nk}(t) w_k$$

as an approximation of $u(t)$.

For an extensive bibliography of early work in this direction, see Lions [1961: 1]; for other applications and additional references, see Lions [1969: 1], Lions-Magenes [1968: 1], [1968: 2], [1970: 1], and Carroll [1969: 2].

Chapter 8

The Cauchy Problem in the Sense of Vector-Valued Distributions

Although the Cauchy problem for certain differential equations and systems is not well posed in the sense of Chapter 1, it fits into the mold of *mildly well posed problems* (Section 8.6) where the solution of the abstract differential equation $u'(t) = Au(t)$ depends continuously on $u(0)$ if the initial condition is measured in the graph norm of a suitable power of A. This concept is closely related to that of properly posed problem in the sense of Hadamard and turns out to be equivalent to a formulation of the Cauchy problem in the sense of distributions for the nonhomogeneous equation $u'(t) = Au(t) + f(t)$ studied in Sections 8.4 and 8.5. The fact that the equation under consideration is a differential equation plays no role here, thus the theory refers to convolution equations.

The rest of the chapter covers preparatory material on vector-valued distributions.

8.1. VECTOR-VALUED DISTRIBUTIONS. SUPPORTS, CONVERGENCE, STRUCTURE RESULTS

Let Ω be an open interval in $\mathbb{R}$, $\mathcal{D}(\Omega)$ the space of test functions in Ω. A sequence $\{\varphi_n\}$ is *convergent to zero* in $\mathcal{D}(\Omega)$ if and only if:

(a) There exists a compact set $K \subset \Omega$ such that

$$\operatorname{supp}(\varphi_n) \subseteq K \quad \text{for all } n. \tag{8.1.1}$$

(b) $\varphi_n^{(m)}(x) \to 0$ uniformly on $\mathbb{R}$ for all $m \geqslant 0$.

461

More generally, $\{\varphi_n\}$ *converges to* $\varphi \in \mathcal{D}(\Omega)$ if and only if $\varphi_n - \varphi \to 0$ according to the previous definition.

Given a Banach space E the space $\mathcal{D}'(\Omega; E)$ of *E-valued distributions defined in* Ω consists of all linear operators $U, V, \dots$ from $\mathcal{D}(\Omega)$ into E that are continuous in the following fashion:

$$\text{if } \varphi_n \to 0 \text{ in } \mathcal{D}(\Omega), \text{ then } U(\varphi_n) \to 0 \text{ in } E, \qquad (8.1.2)$$

where $\{\varphi_n\}$ is a sequence in $\mathcal{D}(\Omega)$.[1] Clearly $\mathcal{D}'(\Omega; E)$ is a linear space. Every piecewise continuous (or, more generally, locally integrable) function f defined in Ω with values in E can be identified with a distribution $U \in \mathcal{D}'(\Omega, E)$ through the formula

$$U(\varphi) = \int \varphi(t) f(t)\, dt \quad (\varphi \in \mathcal{D}(\Omega)) \qquad (8.1.3)$$

and two functions f, g give rise to the same distribution if and only if they coincide almost everywhere (if f and g are piecewise continuous, this means $f = g$ except perhaps at discontinuity points).

We introduce some notational conventions. When $\Omega = \mathbb{R}$, we write $\mathcal{D}(\Omega) = \mathcal{D}$, as agreed in Section 7. We also write $\mathcal{D}'(\mathbb{R}; E) = \mathcal{D}'(E)$, $\mathcal{D}'(\Omega, \mathbb{C}) = \mathcal{D}'(\Omega)$, $\mathcal{D}'(\mathbb{C}) = \mathcal{D}'$. If $f: \Omega \to E$, we indicate the function f (or the distribution in $\mathcal{D}'(\Omega; E)$ it defines) by $f(\hat{t})$; $f(t)$ denotes instead the value of f at t. Distributions will be often written in "functional" notation; for instance, the Dirac measure $\delta \in \mathcal{D}'$ will be denoted by $\delta(\hat{t})$ while, say, $\delta(\hat{t} - 3)$ indicates the Dirac measure centered at $t = 3$, that is, the distribution $\varphi \to \varphi(3)$ in $\mathcal{D}'$. Likewise we shall abuse sometimes the notation writing $U(\varphi)$ as $\int \varphi(t) U(t)\, dt$, where $U \in \mathcal{D}'(\Omega; E)$ and $\varphi \in \mathcal{D}(\Omega)$. Finally, given $U \in \mathcal{D}'(\Omega)$ and $u \in E$, we indicate by $U \otimes u$ the distribution in $\mathcal{D}'(\Omega; E)$ defined by

$$(U \otimes u)(\varphi) = U(\varphi) u. \qquad (8.1.4)$$

As in the scalar case, the derivative of a distribution $U \in \mathcal{D}'(\Omega; E)$ is defined by

$$U'(\varphi) = -U(\varphi') \qquad (8.1.5)$$

---

[1] The definition of the topology of $\mathcal{D}$ is not complete, since it cannot be entirely characterized by sequences; there are nets $\{\varphi_\alpha\}$ converging to zero in $\mathcal{D}$ which do not satisfy (8.1.1). The correct topology of $\mathcal{D}$ is the inductive limit topology of (say) the sequence $\mathcal{D}_n$, where $\mathcal{D}_n$ is the (Fréchet) space of all $\varphi \in \mathcal{D}$ with support in $|t| \leqslant n$ endowed with the topology of uniform convergence of all derivatives; the inductive limit topology is the finest (that is, the largest) one making all injections $i_n: \mathcal{D}_n \to \mathcal{D}$ continuous. A fundamental property of this inductive limit topology is that an operator $U: \mathcal{D} \to E$ is continuous if and only if all its restrictions U_n to $\mathcal{D}_n$ are continuous. It follows then that our definition of continuity (8.1.2) *is* the correct definition of continuity with respect to the topology of $\mathcal{D}$; moreover, since each $\mathcal{D}_n$ is a Fréchet (thus metric) space, we may well use a sequence $\{\varphi_n\}$ in (8.1.1) and (8.1.2). The same observations apply to arbitrary Ω.

and behaves in an entirely similar way. When f is a continuous, piecewise continuously differentiable function with values in E, its ordinary derivative defines the same distribution as the one obtained differentiating f as a distribution.

8.1.1 Lemma.   *Let $U \in \mathcal{D}'(\Omega; E)$ satisfy*

$$U^{(m)} = 0$$

for some $m \geqslant 0$. Then U coincides with a polynomial

$$P(t) = \sum_{j=0}^{m-1} t^j u_j \quad (t \in \Omega), \tag{8.1.6}$$

where $u_0, \ldots, u_{m-1} \in E$.

Proof. We consider first the case $m = 1$. Since $U' = 0$, it follows that $U(\varphi') = 0$ for all $\varphi \in \mathcal{D}(\Omega)$ or, equivalently, that $U(\varphi) = 0$ if $\int \varphi \, dt = 0$. Let φ_0 be any function in $\mathcal{D}(\Omega)$ with nonzero integral $\eta = \int \varphi_0 \, dt$. If $\psi \in \mathcal{D}(\Omega)$, we write $\varphi = \psi - \lambda \varphi_0$, $\lambda = (1/\eta) \int \psi \, dt$. Then $\int \varphi \, dt = 0$ and $U(\psi) = \lambda U(\varphi_0) = (1/\eta) U(\varphi_0) \int \psi \, dt$, showing that $U = u = (1/\eta) U(\varphi_0)$. If $m > 1$, we obtain using the previous result that $U^{(m-1)} = u$ for some $u \in E$ from which (8.1.6) follows by repeated integration.

We define convergence in the space $\mathcal{D}'(\Omega; E)$. In order to do this, we introduce first an auxiliary notion. A set $\mathcal{K} \subseteq \mathcal{D}(\Omega)$ is *bounded* if for every sequence $\{\varphi_n\} \subseteq \mathcal{K}$ and every sequence of real numbers $\{\varepsilon_n\}$ with $\varepsilon_n \to 0$ we have

$$\varepsilon_n \varphi_n \to 0 \quad \text{in } \mathcal{D}(\Omega). \tag{8.1.7}$$

It is immediate to verify that a set $\mathcal{K}$ is bounded if and only if:

(a) There exists a compact set $K \subset \Omega$ such that

$$\operatorname{supp}(\varphi) \subseteq K \quad (\varphi \in \mathcal{K}). \tag{8.1.8}$$

(b) For every $m \geqslant 0$ there exists a constant $C_m < \infty$ with

$$|\varphi^{(m)}(t)| \leqslant C_m \quad (\varphi \in \mathcal{K}, t \in K). \tag{8.1.9}$$

A sequence (or generalized sequence) $\{U_\alpha\}$ of distributions in $\mathcal{D}'(\Omega; E)$ is said to *converge* to a $U \in \mathcal{D}'(\Omega; E)$ if and only if

$$U_\alpha(\varphi) \to U(\varphi)$$

uniformly on bounded subsets of $\mathcal{D}$.

It is easy to see that if $\{f_\alpha\}$ is a generalized sequence of piecewise continuous (or, more generally, locally integrable) functions such that $f_\alpha \to f$ pointwise and $\|f_\alpha(t)\| \leqslant g(t)$, where g is locally integrable, then $f_\alpha \to f$ in the sense of distributions.

A distribution $U \in \mathcal{D}'(\Omega; E)$ is said to *vanish* in an open subset Ω' of Ω (in symbols, $U = 0$ in Ω') if

$$U(\varphi) = 0 \tag{8.1.10}$$

whenever $\mathrm{supp}(\varphi) \subseteq \Omega'$. Denote by $\Omega(U)$ the union of all the open subsets of Ω where U vanishes. We define the *support* of U (in symbols $\mathrm{supp}(U)$) as follows:

$$\mathrm{supp}(U) = \Omega \backslash \Omega(U).$$

It is clear that $\mathrm{supp}(U)$ is closed in Ω. We may easily verify that the present definition of support coincides with the usual one when U is a smooth function. The following result shows that the property that $U = 0$ can be "pieced together."

8.1.2   Lemma. $U = 0$ *in* $\Omega \backslash \mathrm{supp}(U)$.

Proof.   A local compactness argument easily shows that there exist two countable families of open intervals $\{I_n\}, \{J_n\}$ such that (i) $\bar{I}_n \subset J_n$. (ii) $\cup_n I_n = \Omega \backslash \mathrm{supp}(U)$. (iii) $J_n \cap J_{n+1} \neq \varnothing$. (iv) $J_n \cap J_m = \varnothing$ if $n < m - 1$. (v) $U = 0$ in J_n. Let φ_n be a nonnegative function in $\mathcal{D}(\Omega)$ with support in J_n which equals 1 in I_n. The sum $\Phi(t) = \Sigma \varphi_n(t)$ (where all the terms, except for at most two, vanish for each t) is positive in Ω. Then, if $\psi_n = \varphi_n / \Phi$, it is clear that $\psi_n \in \mathcal{D}(\Omega)$, $\mathrm{supp}(\psi_n) \subseteq J_n$, and $\Sigma \psi_n(t) = 1$ $(t \in \Omega(U))$. Hence if $\varphi \in \mathcal{D}(\Omega)$ and $\mathrm{supp}\, \varphi \subseteq \Omega \backslash \mathrm{supp}(U)$,

$$U(\varphi) = U\left(\varphi \sum \psi_n\right) = \sum U(\varphi \psi_n) = 0 \qquad (8.1.11)$$

since $\mathrm{supp}(\varphi \psi_n) \subseteq J_n$ and U vanishes in J_n. Note that the introduction of U into the series is justified by the fact that the $\varphi \psi_n$ vanish identically except for a finite number.

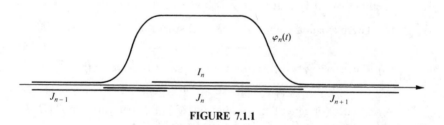

FIGURE 7.1.1

In the case $\Omega = \mathbb{R}$, we shall make use of several subspaces of $\mathcal{D}'(\Omega; E) = \mathcal{D}'(E)$. We denote by $\mathcal{E}'(E)$ (resp. $\mathcal{D}_0'(E)$) the set of all elements of $\mathcal{D}'(E)$ with compact support (resp. with support in $t \geqslant 0$). We shall also write $\mathcal{E}_0'(E) = \mathcal{E}'(E) \cap \mathcal{D}_0'(E)$ and denote by $\mathcal{D}_+'(E)$ the subspace of all $U \in \mathcal{D}'(E)$ with $\mathrm{supp}(U)$ bounded below. Use will be made of $\mathcal{D}_0'(\Omega; E)$ (distributions in $\mathcal{D}'(\Omega; E)$ with support in $t \geqslant 0$) and of the similarly defined $\mathcal{E}_0'(\Omega; E)$.

The results in the remainder of this section will show that, at least locally, a E-valued distribution U is nothing more than the derivative (in the sense of distributions) of a continuous function, which can be chosen to vanish in $t < a$ if U does. Moreover, this correspondence between functions and distributions can be established in such a way that boundedness in $\mathscr{D}'(\Omega; E)$ translates into local boundedness of the corresponding sequence of continuous functions.

8.1.3   Lemma.   *Let $U \in \mathscr{D}'(\Omega; E)$, K a compact subset of Ω. Then there exists an integer $p \geqslant 0$ and a constant $C > 0$ such that*

$$\|U(\varphi)\| \leqslant C\|\varphi\|_p \tag{8.1.12}$$

for every $\varphi \in \mathscr{D}$ with support in K, where

$$\|\varphi\|_p = \sup_{0 \leqslant j \leqslant p} \ \sup_{-\infty < t < \infty} |\varphi^{(j)}(t)|. \tag{8.1.13}$$

Proof.   Assume (8.1.12) does not hold for any p. Then there exists a sequence $\{\varphi_n\} \subseteq \mathscr{D}$ such that

$$\operatorname{supp}(\varphi_n) \subseteq K \tag{8.1.14}$$

and

$$\|U(\varphi_n)\| > n\|\varphi_n\|_n \quad (n \geqslant 1). \tag{8.1.15}$$

Let $\psi_n = \varphi_n / n\|\varphi_n\|_n$. Clearly $\|\psi_n\|_n = n^{-1} \to 0$ as $n \to \infty$, which, together with (8.1.14), implies that $\psi_n \to 0$ in $\mathscr{D}$. But it follows from (8.1.15) that $\|U(\psi_n)\| > 1$, which is a contradiction. This ends the proof.

Let K be, as before, a compact set. We denote by $\mathscr{D}_p(K)$ the space of all p times continuously differentiable functions defined in $\mathbb{R}$ with support in K endowed with the norm (8.1.13). It is clear that $\mathscr{D}_p(K)$ is a Banach space.

8.1.4   Lemma.   *Let K be a compact subset of Ω. Then, given $\varphi \in \mathscr{D}_p(K)$ and $\delta > 0$, there exists $\psi \in \mathscr{D}(\Omega)$ such that*

$$\|\varphi - \psi\|_p \leqslant \delta,$$

where $\|\cdot\|_p$ indicates the norm in $\mathscr{D}_p(K_\delta)$, $K_\delta = \{t; \operatorname{dist}(t, K) \leqslant \delta\}$.

The proof is a standard application of mollifiers (Section 8) and is thus omitted.

8.1.5   Structure Theorem.   *Let $U \in \mathscr{D}'(\Omega; E)$, Ω' an open bounded subset of Ω with $\overline{\Omega}' \subseteq \Omega$. Then there exists a continuous function $f: \mathbb{R} \to E$, and an integer $m \geqslant 0$ such that*

$$U = f^{(m)} \text{ in } \Omega' \tag{8.1.16}$$

(the derivative understood in the sense of distributions).

Proof.   Let $K = \overline{\Omega}'$ and let $\varepsilon > 0$ be such that $K_\varepsilon = \{t; \operatorname{dist}(t, K) \leqslant \varepsilon\}$ is contained in Ω. Applying Lemma 8.1.4 to K_ε (see (8.1.12)), we deduce that U can be extended by continuity to all elements of $\mathscr{D}_p(K_\varepsilon)$ and that the extension (which we denote by the same symbol) satisfies

$$\|U(\varphi)\| \leqslant C\|\varphi\|_p \quad (\varphi \in \mathcal{D}_p(K_\varepsilon)), \tag{8.1.17}$$

where $\|\cdot\|_p$ indicates the norm of $\mathcal{D}_p(K_\varepsilon)$. Loosely speaking, this partial result states that "a distribution can be applied to a sufficiently differentiable function." Define now

$$\eta(t) = \begin{cases} t^{p+1}/(p+1)! & (t \geqslant 0) \\ 0 & (t < 0). \end{cases} \tag{8.1.18}$$

Clearly η is p times continuously differentiable. Then if $\chi \in \mathcal{D}$ and supp $\chi \subseteq K_\varepsilon$, the function $\chi(\hat{s})\eta(t - \hat{s})$ belongs to $\mathcal{D}_p(K_\varepsilon)$ for each t and, moreover, the map $t \to \chi(\hat{s})\eta(t - \hat{s})$ from $\mathbb{R}$ into $\mathcal{D}_p(K_\varepsilon)$ is continuous (as can be easily verified by repeated differentiation). Hence, the function

$$f(t) = U(\chi(\hat{s})\eta(t - \hat{s})) \tag{8.1.19}$$

from $\mathbb{R}$ into E is as well continuous. We observe next that if $\varphi \in \mathcal{D}$,

$$\int \varphi(t) f(t) \, dt = U(\psi),$$

where ψ is the element of $\mathcal{D}$ defined by

$$\psi(s) = \chi(s) \int \eta(t - s) \varphi(t) \, dt$$

$$= \chi(s) \int \eta(t) \varphi(s + t) \, dt$$

(this can be easily proved approximating the integrals involved by Riemann sums). Take now χ with $\chi(t) = 1$ for $t \in K$ and $\varphi \in \mathcal{D}$ with support in Ω'. Let $m = p + 2$. Then

$$f^{(m)}(\varphi) = (-1)^m \int \varphi^{(m)}(t) f(t) \, dt$$

$$= U((-1)^m \psi_m),$$

where

$$\psi_m(s) = \chi(s) \int \eta(t - s) \varphi^{(m)}(t) \, dt$$

$$= (-1)^m \varphi(s) \tag{8.1.20}$$

since $\varphi(s) = 0$ outside of K. This ends the proof.

As the classical example

$$U = \sum_{n=0}^{\infty} \delta^{(n)}(\hat{t} - n)$$

shows, Theorem 8.1.5 and the equivalent Lemma 8.1.3 do not admit a global version even in the scalar case.

8.1.6   Remark.   Note that if $U = 0$ for $t < a$, it results that $f = 0$ for $t < a$. This follows from the fact that $\eta(t - s) = 0$ for $s \geqslant t$. Note also that if U has compact support (that is, if $U \in \mathcal{E}'(E)$), then the function f can be chosen in such a way that (7.1.16) holds in Ω (we only have to take Ω' containing the support of U); the result is *global* in this case.

As an immediate consequence of Theorem 8.1.6, we obtain

8.1.7   Corollary.   *Let $U \in \mathcal{D}'(\Omega; E)$, and assume*

$$\mathrm{supp}(U) = \{0\}.$$

Then

$$U = \delta \otimes u_0 + \delta' \otimes u_1 + \cdots + \delta^{(m)} \otimes u_m \qquad (8.1.21)$$

for some integer m and $u_0, u_1, \ldots, u_m \in E$.

Proof.   Write

$$U = f^{(m)},$$

where $f(t) = 0$ for $t \leqslant 0$. Since $U = 0$ in $t > 0$, it follows from Lemma 8.1.1 that f must coincide there with a polynomial of degree $\leqslant m - 1$. Differentiating, (8.1.21) follows.

We close this section with a "uniform" version of Theorem 8.1.5. Bounded set of distributions are defined in the same fashion as for test functions: a set $\mathcal{U} \subseteq \mathcal{D}'(\Omega; E)$ is *bounded* if and only if, for every sequence $\{U_n\} \subseteq \mathcal{U}$ and every sequence $\{\varepsilon_n\}$ of real numbers with $\varepsilon_n \to 0$, we have

$$\varepsilon_n U_n \to 0 \quad \text{in } \mathcal{D}'(\Omega; E). \qquad (8.1.22)$$

It is easy to see that $\mathcal{U}$ is bounded if and only if for every bounded set $\mathcal{K} \subseteq \mathcal{D}(\Omega)$ there exists a constant C such that

$$\|U(\varphi)\| \leqslant C \quad (\varphi \in \mathcal{K}, U \in \mathcal{U}). \qquad (8.1.23)$$

8.1.8   Theorem.   *Let $\mathcal{U}$ be a bounded set in $\mathcal{D}'(\Omega; E)$, Ω' a bounded open subset of Ω with $\overline{\Omega}' \subseteq \Omega$. Then the integer m in (8.1.16) can be chosen independently of U and the functions f are uniformly bounded on compact subsets of $\mathbb{R}$.*

Theorem 8.1.8 is a consequence of the following uniform version of Lemma 8.1.3.

8.1.9   Lemma.   *Let $\mathcal{U}$ be a bounded set in $\mathcal{D}'(\Omega; E)$, K a compact subset of Ω. Then there exists an integer $p \geqslant 0$ and a constant C independent of*

U such that

$$\|U(\varphi)\| \leqslant C\|\varphi\|_p \tag{8.1.24}$$

for all $U \in \mathcal{U}$ and all $\varphi \in \mathcal{D}$ with support in K.

The proof is much like that of Lemma 8.1.3. If (8.1.24) does not hold for any p, there exists a sequence $\{\varphi_n\}$ in $\mathcal{D}$, each φ_n with support in K, and a sequence $\{U_n\}$ in $\mathcal{U}$ such that

$$\|U_n(\varphi_n)\| > n\|\varphi_n\|_n \quad (n \geqslant 1).$$

We consider this time the sequence $\psi_n = \varphi_n / \|\varphi_n\|_n$. Obviously, $\{\psi_n\}$ is bounded in $\mathcal{D}(\Omega)$, but $\|U_n(\psi_n)\|$ is unbounded, contradicting (8.1.23).

To prove Theorem 8.1.8 we only have to recall that the functions f are given by (8.1.19) and use (8.1.24).

A generalized sequence $\{U_\alpha\}$ in $\mathcal{D}'(\Omega; E)$ is said to be *boundedly convergent* to $U \in \mathcal{D}'(\Omega; E)$ (in symbols, $U_\alpha \rightrightarrows U$) if and only if $U_\alpha \to U$ and $\{U_\alpha\}$ is bounded. It is easy to see that bounded convergence and convergence coincide for sequences, but this is not the case for arbitrary generalized sequences.

8.1.10 Theorem. *Let $\{U_\alpha\}$ be a generalized sequence of distributions converging boundedly to zero in $\mathcal{D}'(\Omega; E)$, Ω' an open bounded subset of Ω with $\bar{\Omega}' \subseteq \Omega$. Then there exists a generalized sequence $\{f_\alpha\}$ of continuous E-valued functions in $\mathbb{R}$, bounded and converging to zero uniformly on compact subsets of R such that*

$$U_\alpha = f_\alpha^{(m)} \quad \text{in } \Omega'.$$

This result is an immediate consequence of Theorem 8.1.8: it suffices to note that the functions f_α are defined by (8.1.19) with $U = U_\alpha$.

8.2. VECTOR-VALUED DISTRIBUTIONS. CONVOLUTION, TEMPERED DISTRIBUTIONS, LAPLACE TRANSFORMS

We shall only define convolution in a very particular case. Throughout this section E, F, G will be three Banach spaces, $(u, v) \to uv$ a bounded bilinear map from $E \times F$ into G, $U \in \mathcal{D}'_+(E)$, $V \in \mathcal{D}'_+(F)$. Assume, for the sake of simplicity, that U and V have support in $t \geqslant 0$ and let $c > 0$. Making use of Theorem 8.1.5 we see that there exist two nonnegative integers m, p, a continuous E-valued function that vanishes for $t < 0$ and a continuous F-valued function g equal to zero in $t < 0$ such that

$$U = f^{(m)} \quad (t < c), \tag{8.2.1}$$

$$V = g^{(p)} \quad (t < c). \tag{8.2.2}$$

We define

$$U * V = (f * g)^{(m+p)} \tag{8.2.3}$$

in $t < c$, that is, for every $\varphi \in \mathcal{D}$ with support in $(-\infty, c)$. Here $f * g$ is the ordinary convolution of f and g,

$$(f * g)(t) = \int f(t - s) g(s) \, ds, \tag{8.2.4}$$

where the interval of integration is actually $0 \leqslant s \leqslant t$. (Note that $f * g$ is continuous.) Since c may be arbitrarily large, it is clear that the definition above works for any $\varphi \in \mathcal{D}$, but it is not immediately apparent that it does not depend on the choice of the representations (8.2.1) and (8.2.2), which are obviously not unique. Assume, say, that we replace (8.2.1) by

$$U = f_1^{(m_1)} \quad (t < c_1).$$

We may assume that $m \leqslant m_1$ and $c \leqslant c_1$ (the other possibilities are dealt with in a similar way). Since $(f - f_1^{(m_1 - m)})^{(m)} = 0$ in $t < c$, it follows from Lemma 8.1.1 that $f - f_1^{(m_1 - m)}$ is a polynomial in $t < c$. But both f and f_1 are zero for $t < 0$, thus $f = f_1^{(m_1 - m)}$ and f_1 is continuously differentiable at least $m_1 - m$ times in $t < c$. Then we have, after a clearly permissible differentiation under the integral sign,

$$(f_1 * g)^{m_1} = (f_1^{(m_1 - m)} * g)^{(m)} = (f * g)^{(m)}$$

in $t < c$. This proves that $U * V$ is defined independently of the particular representations (8.2.1), (8.2.2) chosen. The definition can be easily extended to the case where $U = 0$ for $t < a$, $V = 0$ for $t < b$; we omit the details.

It is a rather simple consequence of the construction of $U * V$ that

$$(U * V)^{(k)} = U^{(k)} * V = U * V^{(k)} \tag{8.2.5}$$

for any U, V and any integer $k \geqslant 0$.

It can also be easily seen that if $U = 0$ in $t < a$ and $V = 0$ in $t < b$, then $U * V = 0$ in $t < a + b$. On the other hand, assume that $U = 0$ for $t > a'$ and $V = 0$ for $t > b'$. Take c so large that the interval $t < c$ contains both the supports of U and V. Then the function f in (8.2.1) must be a polynomial of degree $\leqslant m - 1$ in $t > a'$ and g must likewise be a polynomial of degree $\leqslant p - 1$ in $t > b'$. Hence $f * g$ is a polynomial of degree $\leqslant m + p - 1$ in $t \geqslant a' + b'$, which implies that $U * V = 0$ in $t \geqslant a' + b'$. We can summarize these observations in the formula

$$(\operatorname{supp}(U * V))^c \subseteq (\operatorname{supp} U)^c + (\operatorname{supp}(V))^c, \tag{8.2.6}$$

where $(\ )^c$ indicates convex hull.

It is immediate from its definition that the map $(U, V) \to U * V$ from $\mathcal{D}'_+(E) \times \mathcal{D}'_+(F)$ into $\mathcal{D}'_+(G)$ is bilinear. We also have the following continuity property:

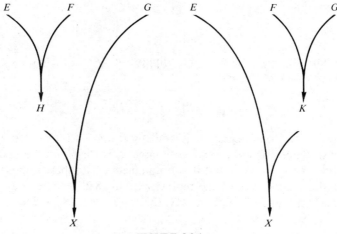

FIGURE 8.2.1

8.2.1   Theorem.   *Let $\{U_\alpha\}$ (resp. $\{V_\alpha\}$) be a generalized sequence in*
$\mathcal{D}'_+(E)$ *(resp. $\mathcal{D}'_+(F)$) such that $U_\alpha \rightrightarrows U$ (resp. $V_\alpha \rightrightarrows V$) and such that $U_\alpha = 0$*
for $t < a$ ($V_\alpha = 0$ for $t < b$). Then $U_\alpha * V_\alpha \rightrightarrows U * V$ in $\mathcal{D}'_+(G)$.[2]

The proof is a direct consequence of Theorem 8.1.8 and will be left
to the reader.

We note, finally, the following associative property of the convolution. Let E, F, G, H, K, X be six Banach spaces and let the two-tailed
arrows in Figure 8.2.1 indicate bounded bilinear maps. Then if the maps
themselves associate (which means the trilinear maps from $E \times F \times G$ into
X obtained from each diagram coincide), we have

$$(U * V) * W = U * (V * W) \tag{8.2.7}$$

for arbitrary $U \in \mathcal{D}'_+(E)$, $V \in \mathcal{D}'_+(F)$, $W \in \mathcal{D}'_+(G)$.

We introduce a subspace of $\mathcal{D}'_+(E)$ consisting of distributions of
"moderate growth at infinity" with a view to the definition of the Laplace
transform. To this end we use instead of $\mathcal{D}$ the Schwartz space $\mathcal{S}$ of *rapidly*
decreasing functions consisting of all infinitely differentiable functions φ
defined in $\mathbb{R}$ and such that $t^j \varphi^{(k)}(t) \to 0$ as $|t| \to \infty$ for all j, k. A sequence
$\{\varphi_n\}$ in $\mathcal{S}$ *converges to zero* if and only if

$$\|\varphi_n\|_{j,k} \to 0 \quad (n \to \infty)$$

---

[2]Note that U (resp. V) must vanish for $t < a$ (resp. for $t < b$). Theorem 3.2.1 is as well
valid for convergence (see Schwartz [1955: 1]), but the present particular case is sufficient here.

for all integers $j, k \geqslant 0$, where

$$\|\varphi\|_{j,k} = \sup_{0 \leqslant l \leqslant k} \sup_{-\infty < t < \infty} (1 + |t|)^j |\varphi^{(l)}(t)|. \qquad (8.2.8)$$

The space $S'(E)$ of *tempered* E-valued distributions consists of all linear operators $U, V, \ldots$ from S into E that are continuous (that is, $S(\varphi_n) \to 0$ in E whenever $\varphi_n \to 0$ in S). As in $\mathcal{D}$, a set $\mathcal{K} \subseteq S$ is *bounded* if $\varepsilon_n \varphi_n \to 0$ in S whenever $\{\varphi_n\}$ is a sequence in $\mathcal{K}$ and the numerical sequence $\{\varepsilon_n\}$ tends to zero. It is easy to see that $\mathcal{K}$ is bounded if and only if for any two integers j, k there exists a constant $C_{j,k} < \infty$ such that

$$\|\varphi\|_{j,k} \leqslant C_{j,k} \quad (\varphi \in \mathcal{K}). \qquad (8.2.9)$$

Convergence of a sequence of distributions $\{U_n\}$ in $S'(E)$ is again defined as uniform convergence on bounded subsets of S.

It is plain that a linear continuous operator from S into E is as well continuous in the topology of $\mathcal{D}$, since $\mathcal{D} \subseteq S$ and the inclusion map is continuous. In other words, $S'(E) \subseteq \mathcal{D}'(E)$.[3] Moreover, a set $\mathcal{K}$ bounded in $\mathcal{D}$ is also bounded in S, thus this inclusion is also continuous. We can apply to distributions in $S'(E)$ all the results in this section and in the previous one. However, most of these results take a considerably simpler form when applied to distributions in $S'(E)$. For instance Lemma 8.1.3 and Theorem 8.1.5 become global in $S'(E)$. In fact, we have

8.2.2 Lemma. *Let $U \in S'(E)$. Then there exists an integer $p \geqslant 0$ and a constant $C > 0$ such that*

$$\|U(\varphi)\| \leqslant C \|\varphi\|_{p,p} \quad (\varphi \in S). \qquad (8.2.10)$$

8.2.3 Structure Theorem. *Let $U \in S'(E)$. Then there exist two integers $m, r \geqslant 0$ and a continuous function $f: \mathbb{R} \to E$ such that*

$$U = f^{(m)} \quad \text{in } \mathbb{R} \qquad (8.2.11)$$

$$|f(t)| = O(|t|^r) \quad (|t| \to \infty). \qquad (8.2.12)$$

The proof of Lemma 8.2.2 is entirely similar to that of Lemma 8.1.3 and will thus be omitted. To prove Theorem 8.2.3 we make use of (8.2.10) to extend U to the space $S_{p,p}$, where $S_{j,k}$ is the Banach space of all k times continuously differentiable functions defined in $-\infty < t < \infty$ such that $\|\varphi\|_{j,k} < \infty$, endowed with the norm $\|\cdot\|_{j,k}$.[4] Assume first that U has support contained in $t \geqslant 0$, and let χ be a $C^{(\infty)}$ function that equals 1 in

---

[3]We are using here implicitly the fact that two distributions in $S'(E)$ that take the same values in $\mathcal{D}$ must coincide. This follows from denseness of $\mathcal{D}$ in S.

[4]An approximation argument of the type of Lemma 8.1.4 is used here.

$t \geq 0$ and vanishes in $t \leq -1$, η the function defined in (8.1.18). Then it is not difficult to see using arguments similar to those in the proof of Theorem 8.1 that the E-valued continuous function

$$f(t) = U(\chi(\hat{s})\eta(t - \hat{s})) \tag{8.2.13}$$

satisfies (8.2.11) for $m = p + 2$. The verification of (8.2.12) follows from the fact that the $\|\cdot\|_{p,p}$-norm of $\chi(\hat{s})\eta(t - \hat{s})$ grows no more than a power of t as $t \to \infty$. Obviously, the argument extends to distributions U with support in $t \geq a$.

In the general case we choose an infinitely differentiable function ψ that equals 1 in, say, $t \geq 0$ and vanishes in $t \leq -1$, write

$$U = U_1 + U_2,$$

where $U_1 = \psi U$, $U_2 = (1 - \psi)U$ and apply the previous particular case to U_1 and to $\tilde{U}_2$ defined by $\tilde{U}_2(\varphi(\hat{s})) = -\tilde{U}_2(\varphi(-\hat{s}))$; if f_1, f_2 are the corresponding continuous functions, then $f(t) = f_1(t) + (-1)^m f_2(-t)$ will be the function required in (8.2.11).

It results clearly from the proof of Theorem 8.2.3 that f may be asked to vanish in $t < a$ if U does. Note also that the following sort of converse holds:

8.2.4   Lemma. *Let $U \in \mathcal{D}'(E)$. Assume that U admits the representation (8.2.11), where f is a E-valued continuous function satisfying (8.2.12). Then $U \in \mathcal{S}'(E)$ (strictly speaking, U can be extended to a distribution in $\mathcal{S}'(E)$).*

The proof is rather simple and left to the reader.

We prove next that the spaces $\mathcal{S}'(E)$ are closed under convolution. We write $\mathcal{S}'_+(E) = \mathcal{D}'_+(E) \cap \mathcal{S}'(E)$; in other words, $\mathcal{S}'_+(E)$ is the set of all $U \in \mathcal{S}'(E)$ with support bounded below.

8.2.5   Theorem. *Let $U \in \mathcal{S}'_+(E)$, $V \in \mathcal{S}'_+(F)$. Then $U * V \in \mathcal{S}'_+(G)$.*

The proof is immediate. Let f, g be two E-valued continuous functions vanishing for large negative t, growing no more than a power of t as $t \to \infty$, and such that

$$U = f^{(m)}, \quad V = g^{(n)} \text{ in } \mathbb{R}.$$

Then, using the definition of convolution

$$U * V = (f * g)^{(m+n)} \text{ in } \mathbb{R},$$

where it is immediate that $f * g$ grows at most like a power of t when $t \to \infty$ as well. Applying Lemma 8.2.4, the result follows.

We define now the Laplace transform of distributions in $\mathcal{S}'_0(E) = \mathcal{D}'_0(E) \cap \mathcal{S}'(E)$. It will be convenient to introduce here some new spaces; we denote by $\mathcal{S}'_\omega(E)$ the space of all distributions U in $\mathcal{D}'(E)$ with support in $t \geq 0$ and such that

$$\exp(-\omega\hat{t})U \in \mathcal{S}'(E), \qquad (8.2.14)$$

where ω is a real number. The *Laplace transform* of $U \in \mathcal{S}'_\omega(E)$ is by definition the function

$$\hat{U}(\lambda) = \mathcal{L}U(\lambda) = U(\exp(-\lambda\hat{t}))$$
$$= (\exp(-\omega\hat{t})U)(\exp(-(\lambda-\omega)\hat{t})) \quad (\mathrm{Re}\,\lambda > \omega). \quad (8.2.15)$$

Formula (8.2.15) requires some explanation. Its right-hand side does not make sense a priori, since $\exp(-(\lambda-\omega)\hat{t})$ does not belong to $\mathcal{S}$ for any λ. We define it to mean $V(\chi(\hat{t})\exp(-(\lambda-\omega)\hat{t}))$ $(V = \exp(-\omega\hat{t})U)$, where χ is any infinitely differentiable function that equals 1 in $t \geq 0$ and vanishes for $t \leq a < 0$. It is easy to show that this definition is independent of the particular χ used. Also, since $\mathcal{S}'_\omega(E) \subseteq \mathcal{S}'_{\omega'}(E)$ for $\omega \leq \omega'$, we may replace ω in (8.2.15) by any $\omega' \geq \omega$; it can be immediately shown that this does not modify the definition.

We begin by observing that $\mathcal{L}U$, which is defined for $\mathrm{Re}\,\lambda > \omega$, is analytic there. In fact, we have

$$h^{-1}(\mathcal{L}U(\lambda+h) - \mathcal{L}U(\lambda)) = (\exp(-\omega\hat{t})U)(\varphi(\hat{t},\lambda)),$$

where

$$\varphi(\hat{t},\lambda) = \chi(\hat{t})h^{-1}\{\exp(-(\lambda+h-\omega)\hat{t}) - \exp(-(\lambda-\omega)\hat{t})\}$$
$$\to -\chi(\hat{t})\hat{t}\exp(-(\lambda-\omega)\hat{t}) \in \mathcal{S}.$$

We obtain as a consequence the familiar formula

$$(\mathcal{L}U)'(\lambda) = -\mathcal{L}(\hat{t}U) \quad (\mathrm{Re}\,\lambda > \omega) \qquad (8.2.16)$$

(note that $\hat{t}U \in \mathcal{S}'_\omega(E)$ if $U \in \mathcal{S}'_\omega(E)$. Observe next that

$$V'(\chi(\hat{t})\exp(-(\lambda-\omega)\hat{t}))$$
$$= -V(\chi'(\hat{t})\exp(-(\lambda-\omega)\hat{t}) - (\lambda-\omega)\chi(\hat{t})\exp(-(\lambda-\omega)\hat{t}))$$
$$= (\lambda-\omega)V(\chi(\hat{t})\exp(-(\lambda-\omega)\hat{t})).$$

Applying this to $V = \exp(-\omega\hat{t})U$, we obtain the no less familiar equality[5]

$$\mathcal{L}U'(\lambda) = \lambda\mathcal{L}U(\lambda) \quad (\mathrm{Re}\,\lambda > \omega), \qquad (8.2.17)$$

where the left-hand side makes sense since $U' \in \mathcal{S}'_\omega(E)$. Note finally that if $U \in \mathcal{S}'_\omega(E)$, then

$$\exp(a\hat{t})U \in \mathcal{S}'_{\omega+a}(E)$$

---

[5] If U coincides with a differentiable function f in $t \geq 0$, we have $U' = f'(t) + \delta \otimes f(0)$, hence (8.2.17) becomes the classical formula $\mathcal{L}f'(\lambda) = \lambda\mathcal{L}f(\lambda) + f(0)$.

and

$$\exp(-\omega\hat{\imath})\big(\exp(a\hat{\imath})U\big)\big(\chi(\hat{\imath})\exp(-(\lambda-\omega)\hat{\imath})\big)$$
$$= \exp(-\omega\hat{\imath})U\big(\chi(\hat{\imath})\exp(-(\lambda-a-\omega)\hat{\imath})\big)$$

for $\mathrm{Re}\,\lambda > \omega + a$ so that

$$\mathcal{L}\big(\exp(a\hat{\imath})U\big) = (\mathcal{L}U)(\lambda - a) \quad (\mathrm{Re}\,\lambda > \omega + a). \tag{8.2.18}$$

Let $U \in \mathcal{S}'_\omega(E)$. Then $\exp(-\omega\hat{\imath})U \in \mathcal{S}'_0(E)$ and, according to Theorem 8.2.3, it admits the representation (8.2.11), where the function f satisfies (8.2.12). Making use of (8.2.17) and (8.2.18), we obtain

$$\mathcal{L}U(\lambda) = \mathcal{L}\big(\exp(\omega\hat{\imath})(\exp(-\omega\hat{\imath})U)\big)$$
$$= \lambda^m(\mathcal{L}f)(\lambda - \omega) \quad (\mathrm{Re}\,\lambda > \omega).$$

But $\|(\mathcal{L}f)(\lambda)\| \leqslant C/|\mathrm{Re}\,\lambda|$ for $\mathrm{Re}\,\lambda \geqslant \delta > 0$. We obtain the following result:

8.2.6   Theorem.   Let $U \in \mathcal{S}'_\omega(E)$. Then there exists an integer m such that

$$\|\mathcal{L}U(\lambda)\| \leqslant C_{\omega'}(1 + |\lambda|)^m \quad (\mathrm{Re}\,\lambda \geqslant \omega') \tag{8.2.19}$$

for any $\omega' > \omega$.

A partial converse of this result is

8.2.7   Theorem.   Let $h(\lambda)$ be a function with values in E, defined and analytic in $\mathrm{Re}\,\lambda \geqslant \omega$. Assume that

$$\|h(\lambda)\| \leqslant C(1 + |\lambda|)^m \quad (\mathrm{Re}\,\lambda \geqslant \omega). \tag{8.2.20}$$

Then there exists a (unique) $U \in \mathcal{S}'_\omega(E)$ with

$$h = \mathcal{L}U. \tag{8.2.21}$$

Proof.   Clearly h must be analytic in an open region containing $\mathrm{Re}\,\lambda \geqslant \omega$. Define

$$f(t) = \frac{1}{2\pi i} \int_{\omega - i\infty}^{\omega + i\infty} e^{\lambda t} \lambda^{-m-2} h(\lambda)\, d\lambda.$$

A deformation-of-contour argument shows that $f(t) = 0$ for $t \leqslant 0$; moreover,

$$\|f(t)\| \leqslant Ce^{\omega t} \quad (t \geqslant 0).$$

An easily justifiable interchange of the order of integration yields

$$(\mathcal{L}f)(\lambda) = \lambda^{-m-2} h(\lambda).$$

Clearly $f \in \mathcal{S}'_\omega(E)$. Then it follows from (8.2.17) that $U = f^{(m+2)}$ belongs to $\mathcal{S}'_\omega(E)$ and satisfies (8.2.21). It only remains to be proved that U is unique. To this end, we observe that if U_1, U_2 are two distributions in $\mathcal{S}'_\omega(E)$ with the same Laplace transform, then we have $\mathcal{L}f_1(\lambda) = \lambda^{-m}\mathcal{L}U_1(\lambda) = \lambda^{-m}\mathcal{L}U_2(\lambda) = \mathcal{L}f_2(\lambda)$, where f_1, f_2 are the functions provided by Theorem 8.2.3; then $f_1 = f_2$ by uniqueness of ordinary Laplace transforms and a fortiori $U_1 = U_2$.

We characterize next the Laplace transforms of certain distributions with compact support:

8.2.8 Theorem. *Let $U \in \mathcal{E}'(E)$ with support in $0 \leqslant t \leqslant a$. Then its Laplace transform $h(\lambda) = \mathcal{L}U(\lambda)$ is a E-valued entire function and satisfies*

$$\|h(\lambda)\| \leqslant C(1 + |\lambda|)^m \quad (\mathrm{Re}\,\lambda \geqslant 0), \tag{8.2.22}$$

$$\|h(\lambda)\| \leqslant C(1 + |\lambda|)^m e^{-a\mathrm{Re}\,\lambda} \quad (\mathrm{Re}\,\lambda < 0), \tag{8.2.23}$$

for some integer m. Conversely, let h be an E-valued entire function satisfying (8.2.22) and (8.2.23). Then there exists a (unique) distribution U in $\mathcal{E}'(E)$ with support in $0 \leqslant t \leqslant a$ such that

$$h(\lambda) = \mathcal{L}U(\lambda). \tag{8.2.24}$$

Proof. Let the support of U be contained in $0 \leqslant t \leqslant a$. By virtue of Remark 8.1.6, there exists an integer m and a continuous function $f: \mathbb{R} \to E$ vanishing in $t \leqslant 0$ and such that

$$U = f^{(m)} \quad \text{in } \mathbb{R}. \tag{8.2.25}$$

Since $U = 0$ for $t \geqslant a$, f must be a polynomial of degree $\leqslant m - 1$ in $t \geqslant a$ (Lemma 8.1.1). Write $f = f_1 + f_2$, where $f_2 = 0$ in $t \leqslant a$, $f_1 = 0$ in $t \geqslant a$. We have

$$\|\mathcal{L}f_1(\lambda)\| \leqslant C \quad (\mathrm{Re}\,\lambda \geqslant 0),$$

$$\|\mathcal{L}f_1(\lambda)\| \leqslant Ce^{-a\mathrm{Re}\,\lambda} \quad (\mathrm{Re}\,\lambda < 0).$$

On the other hand, since f_2 is a polynomial of degree $\leqslant m - 1$ in $t \geqslant a$, $\mathcal{L}f_2(\lambda) = e^{-\lambda a}\Pi(1/\lambda)$, where Π is a polynomial of degree $\leqslant m$. Inequalities (8.2.22) and (8.2.23) then follow immediately from the fact that $\mathcal{L}U(\lambda) = \lambda^m\mathcal{L}f(\lambda) = \lambda^m(\mathcal{L}f_1(\lambda) + \mathcal{L}f_2(\lambda))$, a consequence of (8.2.17).

Conversely, let h be an entire E-valued function satisfying (8.2.22) and (8.2.23). As a consequence of the first inequality and of Theorem 8.2.8, we obtain that (8.2.24) holds for some distribution $U \in S_0'(E)$; consequently, it only remains to be shown that $U = 0$ for $t \geqslant a$. To see this, let f be a continuous E-valued function vanishing for $t \leqslant 0$, growing no more than a power of t at infinity and such that

$$U = f^{(n)}.$$

We may evidently assume that $n \geqslant m + 2$. Clearly $\mathcal{L}f(\lambda) = \lambda^{-n}h(\lambda)$ and by virtue of the classical inversion formula for Laplace transforms,

$$f(t) = \frac{1}{2\pi i}\int_{\omega - i\infty}^{\omega + i\infty} e^{\lambda t}\lambda^{-n}h(\lambda)\,d\lambda,$$

where $\omega > 0$. A deformation of contour then shows that if $t > a$,

$$f(t) = \frac{1}{2\pi i} \int_{-\omega - i\infty}^{-\omega + i\infty} e^{\lambda t} \lambda^{-n} h(\lambda)\, d\lambda$$

$$+ \frac{1}{(n-1)!} \left(\frac{d}{d\lambda}\right)^{n-1}_{\lambda = 0} \left(e^{\lambda t} h(\lambda)\right) \quad (\omega > 0),$$

where the second term on the right-hand side is a polynomial in t of degree $\leqslant n - 1$. As for the first term, observe that it does not depend on $\omega > 0$ and that, by virtue of (8.2.23), if $t > a$ the integrand tends to zero as $\omega \to \infty$ and is uniformly bounded by a constant times $(1 + |\lambda|)^{-2}$. Letting $\omega \to \infty$, we see that the integral vanishes and thus that f coincides in $t > a$ with a polynomial of degree $\leqslant n - 1$; a fortiori $U = 0$ in $t > a$, as we wished to prove.

We observe that, in the proof of the direct part of the theorem, the integers m in (8.2.25) and (8.2.22) may be taken to be equal.

8.2.9   Theorem.   *Let $U \in \mathcal{S}'_\omega(E)$, $V \in \mathcal{S}'_\omega(F)$. Then $U * V \in \mathcal{S}'_\omega(G)$* and

$$\mathcal{L}(U * V)(\lambda) = \mathcal{L}U(\lambda)\mathcal{L}V(\lambda) \quad (\mathrm{Re}\,\lambda > \omega). \tag{8.2.26}$$

Proof.   If $U \in \mathcal{D}'_0(E)$ and $V \in \mathcal{D}'_0(F)$, then the formula

$$\exp(a\hat{t})(U * V) = \exp(a\hat{t})U * \exp(a\hat{t})V, \tag{8.2.27}$$

valid for any a, can be easily verified. Making use of (8.2.27) for $a = -\omega$ and of (8.2.18), we see that we may limit ourselves to prove Theorem 8.2.8 in the case $\omega = 0$, where the fact that $U * V \in \mathcal{S}'_0(G)$ follows from Theorem 8.2.5. Let $U \in \mathcal{S}'_0(E)$, $V \in \mathcal{S}'_0(F)$, and let f (resp. g) be a E-valued (resp. F-valued) continuous function growing at most as a power of t at infinity and such that $U = f^{(m)}$ (resp. $V = g^{(n)}$) for two integers m, n. We use (8.2.17) and obtain

$$\mathcal{L}(U * V) = \mathcal{L}\left((f * g)^{(m+n)}\right)$$

$$= \lambda^{m+n}\mathcal{L}(f * g)$$

$$= \lambda^m(\mathcal{L}f)(\lambda)\lambda^n(\mathcal{L}g)(\lambda) = (\mathcal{L}U)(\lambda)(\mathcal{L}V)(\lambda).$$

We have made use here of formula (8.2.26) for f and g; this can be proved essentially as in the classical theory of Laplace transforms and is left to the reader.

We close this section with a definition. For $\alpha > 0$, let

$$Y_\alpha(t) = \begin{cases} t^{\alpha-1}/\Gamma(\alpha) & (t \geqslant 0) \\ 0 & (t < 0). \end{cases} \tag{8.2.28}$$

Then $Y_\alpha \in S_0'$ and the following properties can be easily verified:

$$Y_1' = \delta \tag{8.2.29}$$

(note that Y_1 is just the Heaviside function $Y_1(t) = \frac{1}{2}(1 + \operatorname{sign} t)$),

$$Y_\alpha * Y_\beta = Y_{\alpha+\beta} \quad (\alpha, \beta > 0), \tag{8.2.30}$$

$$Y_\alpha' = Y_{\alpha-1} \quad (\alpha > 1). \tag{8.2.31}$$

The distribution Y_α can actually be constructed for any complex α (Schwartz [1966: 1]), but this shall not be used here. Y_α is used to define derivatives of fractional order. If, say $U \in \mathcal{D}_0'(E)$ and $\beta \geqslant 0$, we set

$$D^\beta U = D^{n+1} Y_{n+1-\beta} * U,$$

where n is an integer such that $n \geqslant \beta$. It is not difficult to see that this definition does not depend on n and coincides with the usual one when β is an integer.

8.3. CONVOLUTION AND TRANSLATION INVARIANT OPERATORS. SYSTEMS: THE STATE EQUATION

The operator τ_h of translation by h in $\mathcal{D}'(E)$ or $\mathcal{D}_+'(E)$ is defined by the formula

$$(\tau_h U)(\varphi(\hat{t})) = U(\varphi(\hat{t} + h)) = \delta(\hat{t} - h) * U$$

for any real h. It is not difficult to verify directly from the definition that convolution commutes with translation operators, that is,

$$U * (\tau_h V) = (\tau_h U) * V = \tau_h(U * V). \tag{8.3.1}$$

Conversely, any linear continuous map that commutes with translation operators must be given by a convolution. Precisely, we have

8.3.1   Theorem. *Let E, X be two arbitrary Banach spaces, $\mathfrak{M} : \mathcal{D}_+'(E) \to \mathcal{D}_+'(X)$ a linear operator, continuous in the following sense: if F_α is a generalized sequence in $\mathcal{D}_+'(E)$ such that $F_\alpha \rightrightarrows 0$ and $F_\alpha = 0$ for $t < a$ for all α, then $\mathfrak{M}F_\alpha \to 0$ in $\mathcal{D}'(X)$. Assume, moreover, that $\mathfrak{M}$ commutes with translation operators, that is,*

$$\mathfrak{M}(\tau_h U) = \tau_h(\mathfrak{M}U) \tag{8.3.2}$$

for any real h, and that $\mathfrak{M}F = 0$ in $t \leqslant a$ whenever $F = 0$ in $t \leqslant a$. Then there exists a unique distribution $S \in \mathcal{D}_+'((E, X))$ with support in $t \geqslant 0$ such that

$$\mathfrak{M}F = S * F. \tag{8.3.3}$$

Proof.   We define S by the formula

$$S(\varphi)u = \mathfrak{M}(\delta \otimes u)(\varphi) \quad (\varphi \in \mathcal{D}, u \in E). \tag{8.3.4}$$

For any fixed φ, $S(\varphi)$ is a linear operator from E into X. Continuity (or boundedness) of $S(\varphi)$ follows from the fact that if $\{u_n\}$ is a sequence in E converging to zero, then $\delta \otimes u_n \rightrightarrows 0$ in $\mathscr{D}'(E)$, whereas $\delta \otimes u_n = 0$ for $t < 0$; hence $\mathfrak{M}(\delta \otimes u_n) \to 0$ in $\mathscr{D}'(X)$, and then so does $S(\varphi)u_n$ in X. Since the map $\varphi \to S(\varphi) \in (E, X)$ is linear, in order to show that $S \in \mathscr{D}'((E, F))$, we only have to prove that $S(\varphi_n) \to 0$ whenever $\varphi_n \to 0$, $\{\varphi_n\}$ a sequence in $\mathscr{D}$. In order to do this, we need some more information on the topology of distribution spaces. Let E be, for the moment, an arbitrary Banach space, Ω an open interval in $\mathbb{R}$.

8.3.2 Lemma. *Let $\mathfrak{F} \subseteq \mathscr{D}'(\Omega; E)$ be a bounded set, and let $\{\varphi_n\}$ be a sequence converging to zero in $\mathscr{D}(\Omega)$. Then*

$$\lim F(\varphi_n) = 0$$

uniformly for F in $\mathfrak{F}$.

Proof. By virtue of Lemma 8.1.9, we know that given a compact subset K of Ω, there exists an integer $p \geq 0$ and a constant C such that

$$\|F(\varphi)\| \leq C\|\varphi\|_p \quad (F \in \mathcal{K}) \tag{8.3.5}$$

for any $\varphi \in \mathscr{D}$ with $\mathrm{supp}(\varphi) \subseteq K$, where $\|\cdot\|_p$ is the semi-norm defined in (8.1.13). In view of the definition of convergence in $\mathscr{D}$, Lemma 8.3.2 follows.

End of proof of Theorem 8.3.1. Consider the set $\mathfrak{U} = \{\delta \otimes u; \|u\| \leq 1\}$ $\subseteq \mathscr{D}'(E)$. Direct application of the definition shows that $\mathfrak{U}$ is bounded. Then $\mathfrak{U}' = \mathfrak{M}(\mathfrak{U})$, being the image of a bounded set through the operator $\mathfrak{M}$, must be as well bounded. According to Lemma 8.3.2, $\mathfrak{M}(\delta \otimes u)(\varphi_n) = S(\varphi_n)u$ tends to zero uniformly for $\|u\| \leq 1$, which shows that $S: \mathscr{D} \to (E, X)$ is a continuous operator, that is, $S \in \mathscr{D}'((E, X))$. It results from the definition of S and from the hypotheses that $S = 0$ in $t \leq 0$, so that to complete the proof of Theorem 8.3.1, we only have to show (8.3.3). In order to do this we begin by observing that $\mathfrak{M}$ commutes with differentiation, that is,

$$(\mathfrak{M}F)^{(m)} = \mathfrak{M}F^{(m)} \tag{8.3.6}$$

for any integer $m \geq 0$. This results for $m = 1$ observing that if $F \in \mathscr{D}'_+(E)$, $-h^{-1}(\tau_h F - F) \rightrightarrows F'$ in $\mathscr{D}'(E)$ and for $m > 1$ by induction. We note next that we only need to verify (8.3.3) "locally," that is, in all intervals $(-\infty, c)$, $c > 0$. Making then use of Theorem 8.1.5 we conclude that it is only necessary to show that (8.3.3) holds when f is a continuous function with values in E whose support is bounded below. Let then $c > 0$, m a nonnegative integer, and g a continuous function with values in (E, X) such that

$$S = g^{(m)} \quad \text{in } t \leq c \tag{8.3.7}$$

and $g(t) = 0$ in $t < a$. Then we have

$$S * f = (g * f)^{(m)}$$

$$= D_t^m \int g(\hat{t} - s) f(s)\, ds$$

$$= D_t^m \lim_{n \to \infty} \frac{1}{n} \sum_{k=-\infty}^{\infty} g\left(\hat{t} - \frac{k}{n}\right) f\left(\frac{k}{n}\right)$$

$$= \lim_{n \to \infty} \frac{1}{n} \sum_{k=-\infty}^{\infty} g^{(m)}\left(\hat{t} - \frac{k}{n}\right) f\left(\frac{k}{n}\right)$$

$$= \lim_{n \to \infty} \frac{1}{n} \sum_{k=-\infty}^{\infty} \tau_{k/n} Sf\left(\frac{k}{n}\right)$$

$$= \lim_{n \to \infty} \frac{1}{n} \sum_{k=-\infty}^{\infty} \tau_{k/n} \mathfrak{M}\left(\delta \otimes f\left(\frac{k}{n}\right)\right)$$

$$= \mathfrak{M}\left(\lim_{n \to \infty} \frac{1}{n} \sum_{k=-\infty}^{\infty} \tau_{k/n} \delta \otimes f\left(\frac{k}{n}\right)\right)$$

$$= \mathfrak{M}f \quad (t < c). \tag{8.3.8}$$

Justification of the different steps in (8.3.8) runs as follows. The fourth equality results from the fact that the Riemann sums for $g * f$ approximate it uniformly for t on compact subsets of $\mathbb{R}$, thus in the sense of distributions, and from continuity of the differentiation operator in $\mathcal{D}'(E)$. The fifth is evident, and the sixth follows from the definition of S. Finally, the seventh is a consequence of the easily verified fact that $n^{-1} \Sigma \tau_{k/n} \delta \otimes f(k/n)$ converges boundedly to f, its support remaining bounded below.

The uniqueness of S results from the fact that any $\tilde{S} \in \mathcal{D}'((E, X))$ that satisfies (8.3.3) must also satisfy (8.3.4); in fact, we must have $\tilde{S}u = \tilde{S} * ((\delta \otimes u)) = \mathfrak{M}(\delta \otimes u) = Su$ for all $u \in E$.

Theorem 8.3.1 bears on (a version of) system theory. Generally speaking, a *system* is a map from a set $\mathcal{I}$ of *inputs* into a set $\mathcal{O}$ of *outputs* or *states*. In the continuous time version, inputs and outputs are functions of t (or, more generally distributions), the input describing the exterior influence applied to the system at time t while the output represents the state of the system at the same time t. A very general version of this kind of situation can be formalized as follows.

A *linear system* is a linear map $\mathfrak{M}: \mathcal{D}'_+(E) \to \mathcal{D}'_+(X)$, which enjoys the two following properties:

(a) *Causality. If $F = 0$ for $t < a$, then $\mathfrak{M}F = 0$ for $t < a$.*

(b) *Continuity. If $\{F_\alpha\}$ is a generalized sequence in $\mathcal{D}'_+(E)$ such that $F_\alpha \rightrightarrows$ in $\mathcal{D}'_+(E)$ and $F_\alpha = 0$ for $t < a$ (a independent of α), then $\mathfrak{M}F_\alpha \to 0$ in $\mathcal{D}'_+(X)$.*

Condition (a) is rather natural in that it simply prescribes that cause precedes effect; in other words, the output $\mathfrak{M} F$ cannot make itself felt before the input F that originates it begins to act. That some continuity condition like (b) must be imposed it is obvious, if only on mathematical grounds; in fact, if inputs that differ by little could produce outputs that lie wide apart, the usefulness of our system as a model for a physical situation would be small, since any kind of prediction based on the model would be impossible (or, at least, extremely sensible to observation errors.) However, it is not necessarily true that the notion of continuity that we are using is the adequate one in each application. The same comment can be made about the choice of input and output spaces; while to use distributions as inputs seems to be natural in view of, say, the interpretation of $\delta(\hat{t} - t_0) \otimes u$ as an "impulsive excitation" at time t_0, so useful in applications, it is not necessarily true, however, that distributional outputs will have any physical interpretation.

Leaving aside the question of motivation we go back now to the linear system $\mathfrak{M}: \mathcal{D}'_+(E) \to \mathcal{D}'_+(X)$. We call the system *time-invariant* if $\mathfrak{M}$ commutes with translation operators, that is, if (8.3.2) holds for all h. Roughly speaking, this means that "the result of the experiment does not depend on the starting time" or, that "the system itself does not change in time." It follows from Theorem 8.3.1 that there exists in this case a unique $S \in \mathcal{D}'((E, X))$ with support in $t \geqslant 0$ such that

$$U = \mathfrak{M} F = S * F \quad (F \in \mathcal{D}'_+(E)). \tag{8.3.9}$$

We call S the *Green distribution* of the system. Observe that, since $Su = \mathfrak{M}(\delta \otimes u)$, the distribution S can be computed "experimentally" by feeding inputs of the form $\delta \otimes u$ to the system and observing the corresponding outputs.

We call the system *invertible* if $\mathfrak{N} = \mathfrak{M}^{-1}: \mathcal{D}'_+(X) \to \mathcal{D}'_+(E)$ exists and is continuous in the same sense as $\mathfrak{M}$, and $\mathfrak{N} U = 0$ in $t < a$ whenever $U = 0$ in $t < a$. It follows from Theorem 8.3.1 that there exists a distribution $P \in \mathcal{D}'((X; E))$ with support in $t \geqslant 0$ that relates input and output through the equation

$$P * U = F \quad (U \in \mathcal{D}'_+(X)). \tag{8.3.10}$$

We call (8.3.7) the *state equation* of the system. It is immediate that P is the convolution inverse of S; in other words,

$$P * S = \delta \otimes I, \quad S * P = \delta \otimes J, \tag{8.3.11}$$

where I is the identity operator in E, J the identity operator in X. In practice, it is usually the state equation that is given, and our task is to reconstruct the system $\mathfrak{M}$ from it, which of course amounts to finding the Green distribution S by solving the equations (8.3.11), that is, finding a convolution inverse of P with support in $t \geqslant 0$.

We deal in the sequel with an invertible system. Up to now we have exclusively considered the case where the system is initially at rest (which simply means that the state is zero before the input is applied). However, the following situation often arises in practice: we know the history of the system up to some time $t = t_0$ (we may assume that $t_0 = 0$), that is, we are given an *initial state* U_0, which is an element of $\mathcal{D}_+(X)$ with support in $t \leqslant 0$. For the sake of simplicity we assume this initial state to be a smooth function. At time $t = 0$ the state equation (8.3.10) takes over and determines from then on the evolution of the system. If $U \in \mathcal{D}_0'(X)$ represents the state of the system for $t > 0$, we must then have

$$P * (U_0 + U) = F \quad (t \geqslant 0), \tag{8.3.12}$$

where F is the input applied after $t = 0$. In other words, the state of the system after $t = 0$ and its past history must stand in the following relation:

$$P * U = - P * U_0 + F \quad (t \geqslant 0). \tag{8.3.13}$$

Since $U = 0$ for $t < 0$, the only solution of (8.3.13) (if any) must be

$$U = - S * \{ P * U_0 + F \}_+, \tag{8.3.14}$$

where $\{\cdot\}_+$ indicates "truncation at zero";

$$\{V\}_+ = V \text{ for } t \geqslant 0, \quad \{V\}_+ = 0 \text{ for } t < 0. \tag{8.3.15}$$

However, a distribution satisfying (8.3.15) may not exist at all.

8.3.3   Example.   Let

$$V(\varphi) = \sum_{n=1}^{\infty} \frac{1}{n} \left\{ \varphi \left(\frac{1}{n} \right) - \varphi \left(-\frac{1}{n} \right) \right\} \quad (\varphi \in \mathcal{D}).$$

Then $V \in \mathcal{D}'$. However, there is no $W \in \mathcal{D}'$ such that $W = \{V\}_+$.

Consequently, (8.3.13) will only make sense for certain inputs F.

We look at (8.3.14) more closely. It is clear that, in general, the state of the system at $t \geqslant 0$ will depend on all of its past history (the system has "infinitely long memory"). The system has *finite memory* if P has compact support, say $[0, a]$; in that case only the history of the system for $t \geqslant - a$ has any relevance on its behavior for $t > 0$. Finally, the system is *memoryless* if supp$(P) = 0$; here the past of the system is unimportant and only its present or "initial state" matters for its future evolution. We know from Corollary 8.1.7 that supp$(P) = \{0\}$ if and only if

$$P = \delta \otimes P_0 + \delta' \otimes P_1 + \cdots + \delta^{(m)} \otimes P_m,$$

where $P_0, P_1, \ldots, P_m \in (X; E)$, so that we immediately obtain the following result:

8.3.4   Theorem.   *The time-invariant, invertible linear system $\mathfrak{M}$ is memoryless if and only if its state equation is purely differential:*

$$P_0 U + P_1 U' + \cdots + P_m U^{(m)} = F, \tag{8.3.16}$$

where $P_0, P_1, \ldots, P_m \in (X; E)$.

We go back now to the equation (8.3.12). Even in the case where the system has infinitely long memory, some restrictions on P are reasonable. To justify this we invoke the *principle of fading memory* of rational mechanics (Truesdell [1960: 1]). Roughly speaking, the principle asserts that "the effects of what happened in the remote past will be felt but little in the present." This can be of course formalized in many ways. We choose the following precise statement: *if K is a compact set in $\mathbb{R}$ and $\mathcal{U}$ is a bounded set of distributions in $\mathcal{D}'(X)$ such that $\text{supp}(U) \subseteq K$ for all $U \in \mathcal{U}$, then*

$$\lim_{h \to -\infty} P * \tau_h U = 0 \quad \text{in } \mathcal{D}'(E) \tag{8.3.17}$$

uniformly for U in $\mathcal{U}$.

It is not difficult to see that, if P is the derivative (of arbitrarily high order) of a $(X; E)$-valued function tending to zero at infinity, (8.3.17) holds with all the stipulated conditions. We shall take this comment as a justification for taking P in $\mathcal{S}_0'(X; E)$ in the next section.

We note the following notational convention to be used in the sequel. Given a set $e \in \mathbb{R}$ bounded below, we define

$$(e)^d = [m, \infty), \tag{8.3.18}$$

where $m = \inf e$. In this way, the causality relation (a) can be written

$$(\text{supp}(\mathfrak{M}F))^d \subseteq (\text{supp}(F))^d. \tag{8.3.19}$$

8.4. THE CAUCHY PROBLEM IN THE SENSE OF DISTRIBUTIONS

We examine in this section the equation

$$P * U = F, \tag{8.4.1}$$

where $P \in \mathcal{D}_0'((X; E))$, X and E two arbitrary Banach spaces.

We say that the *Cauchy problem* for (8.4.1) is *well posed in the sense of distributions* if:

(a) *For every $F \in \mathcal{D}_+'(E)$ there exists a solution $U \in \mathcal{D}_+'(X)$ of (8.4.1) with*

$$(\text{supp}(U))^d \subseteq (\text{supp}(F))^d \tag{8.4.2}$$

(the symbol $(\cdot)^d$ was defined in (8.3.18)).

(b) *If $\{F_\alpha\}$ is a generalized sequence in $\mathcal{D}_+'(E)$ with $\text{supp}(F_\alpha) \subseteq [a, \infty)$ for all α and some $a > -\infty$, $F_\alpha \overset{\rightarrow}{\to} 0$ in $\mathcal{D}_+'(E)$, and $\{U_\alpha\}$ is a generalized sequence in $\mathcal{D}_+'(X)$ with $P * U_\alpha = F_\alpha$, then $U_\alpha \to 0$ in $\mathcal{D}_+'(X)$.*

Clearly (b) implies uniqueness of the solutions of (8.4.1) in the following sense: if $P * U = 0$ and $U \in \mathcal{D}_+'(X)$, then $U = 0$ (but there might well be nonzero distributions in $\mathcal{D}'(X)$ that are annihilated by P.) In fact, the following more general uniqueness result holds: if $F \in \mathcal{D}'_+(E)$ and $U \in \mathcal{D}'_+(X)$ is a solution of (8.4.1), then (8.4.2) must hold (if this were not true, let $\tilde{U}$ the solution of (8.4.1), (8.4.2) provided by (a); then $P * (U - \tilde{U})$ $= 0$, which implies $U = \tilde{U}$, absurd).

8.4.1 Theorem. *Let $P \in \mathcal{D}'_0((X; E))$. The Cauchy problem for (8.4.1) is well posed in the sense of distributions if and only if P has a convolution inverse with support in $t \geqslant 0$, that is, if there exists $S \in \mathcal{D}'_0((E; X))$ satisfying*

$$P * S = \delta \otimes I, \quad S * P = \delta \otimes J, \tag{8.4.3}$$

where I (resp. J) is the identity operator in E (resp. X).

Proof. If P has a convolution inverse $S \in \mathcal{D}'_0((E; X))$, then it is clear (making use of the associativity property (8.2.7) of the convolution) that $U = S * F$ satisfies (8.4.1) and (8.4.2) (this last relation follows from (8.2.6).) Moreover, U is the only solution of (8.4.1) in $\mathcal{D}'_+(X)$, since $U = S *(P * U)$ for any $U \in \mathcal{D}'_+(X)$. The continuity condition (b) follows then from continuity of the convolution (Theorem 8.2.1). Conversely, assume that conditions (a) and (b) are satisfied for equation (8.4.1). Then the map $\mathcal{M}: \mathcal{D}'_+(E) \to \mathcal{D}'_+(X)$ defined by the equation satisfies all the conditions in Theorem 8.3.1; accordingly, there exists a $S \in \mathcal{D}'_0((E; X))$ such that

$$\mathcal{M}F = S * F \quad \left(F \in \mathcal{D}'_+(E) \right).$$

It remains to be shown that S is the convolution inverse of P. To see this, take $F = \delta \otimes u$ ($u \in E$); then $\mathcal{M}F = S * F = Su$ and $P * Su = \delta \otimes u$, which shows the first equality (8.4.3). Taking now $U = \delta \otimes u$ with $u \in X$ and observing that $P * U = Pu$, we must have $\delta \otimes u = \mathcal{M}(Pu) = S * Pu$. This proves the second equality.

It is not hard to see that the Cauchy problem in the sense of distributions generalizes the Cauchy problem defined in Chapter 1. To see this, take $A \in \mathcal{C}_+$ and consider $S(\cdot)$, the propagator of $u'(t) = Au(t)$. Let $X = D(A)$, endowed with its graph norm, and define

$$S(\varphi)u = \int_0^\infty \varphi(t)S(t)u\,dt \quad (u \in E, \varphi \in \mathcal{D}). \tag{8.4.4}$$

It is a direct consequence of Lemma 3.4 that if $u \in D(A)$, then $S(\varphi)u \in D(A)$ with

$$AS(\varphi)u = -\varphi(0)u - \int_0^\infty \varphi'(t)S(t)u\,dt, \tag{8.4.5}$$

Since the right-hand side of (8.4.5) is a continuous function of u, and A is closed, it follows that $S(\varphi)u \in D(A)$ and (8.4.5) holds for any $u \in E$. This clearly shows that $S(\varphi_n) \to 0$ in $(E; X)$ if $\varphi_n \to 0$ in $\mathcal{D}$ and therefore that $S \in \mathcal{D}'((E; X))$. It is evident that supp(S) is contained in $t \geqslant 0$.

Consider now the distribution $P = \delta' \otimes I - \delta \otimes A \in \mathcal{D}'((X; E))$. It follows immediately from (8.4.5) that $P * S = \delta \otimes I$; on the other hand, since A and $S(t)$ commute for all t, we must also have $S * P = \delta \otimes J$. It results then from Theorem 8.4.1 that the Cauchy problem for the equation

$$(D - A)U = (\delta' \otimes I - \delta \otimes A) * U = F \qquad (8.4.6)$$

is well posed in the sense of distributions, the only solution of (8.4.6) in $\mathcal{D}'_+(X)$ given by

$$U = S * F.$$

This reduces to the familiar formula

$$U(t) = \int_0^t S(t - s) F(s)\, ds$$

of Section 2.4 when F, say, coincides with a $(E; E)$-valued continuous function in $t \geq 0$.

We note that, in Chapter 1, the definition of well-posed Cauchy problem involved only the homogeneous equation $u' = Au$ with arbitrary initial condition; the inhomogeneous equation $u' = Au + f$ was an afterthought and the results for its solutions in Section 2.4 and elsewhere a consequence of the postulates for the homogeneous case. The emphasis here is, on the contrary, on the inhomogeneous equation (8.4.1), while (8.4.2) plays the role of the initial condition $u(t_0) = 0$ for t_0 to the left of the support of F. This shift in orientation is unavoidable in the present context since distributions cannot be forced to take values at individual points; this precludes the consideration of arbitrary initial conditions. Moreover, it is rather unclear what "initial condition" means since P may be, say, a differential operator of fractional order or an even more general entity.

In relation with the comments at the end of the previous section, we note that conditions (a) and (b) are exactly what is needed in order that the equation (8.4.1) be the state equation of a linear, time-invariant invertible system.

It follows from general properties of inverses that there can be only one solution to (8.4.3) in $\mathcal{D}'_+((E; X))$. This uniqueness result also holds locally, as the following result shows:

8.4.2 Lemma. Let $S_1, S_2 \in \mathcal{D}'_0((-\infty, a); (E, X))$ be two solutions of (8.4.3) in $t < a$. Then $S_1 = S_2$ in $t < a$.

In fact, $S_1 = S_1 * (\delta \otimes I) = S_1 * (P * S_2) = (S_1 * P) * S_2 = (\delta \otimes J) * S_2$ $= S_2$ in $t < a$. We use here the easily verified fact that a convolution $U * V$ in $t < a$ depends only on the values of U in $t < a$ if $V = 0$ for $t < 0$. This is an immediate consequence of the definition.

This local uniqueness result has a local existence counterpart, namely,

8.4.3 Lemma. Assume that, for each $a > 0$ we can construct $S_a \in \mathcal{D}'_0((-\infty, a); (E, X))$ satisfying (8.4.3) in $t < a$. Then there exists $S \in \mathcal{D}'_0((E, X))$ satisfying (8.4.3) in $-\infty < t < \infty$.

To prove Lemma 8.4.3 we simply define $S = S_a$ in $t < a$ and apply Lemma 8.4.2 to show that $S_a = S_{a'}$ in $t < a$ if $a \leqslant a'$, which gives sense to the definition.

We introduce some notational conventions. Distributions $P \in \mathcal{D}'_0((X; E))$ and $S \in \mathcal{D}'_0((E; X))$ related by (8.4.4) will be naturally denoted $S = P^{-1}, P = S^{-1}$. This and the following notations will only be used in the spaces $\mathcal{D}'_0$, that is, if both P and S have support in $t \geqslant 0$. We write $P \in \mathcal{D}'_0((X; E)) \cap \mathcal{D}'_0((E; X))^{-1}$, or simply $P \in \mathcal{D}'_0((E; X))^{-1}$ to indicate that P^{-1} exists, that is, that there is a solution of (8.4.3) in $\mathcal{D}'_0((E; X))$. When no danger of confusion could arise we simply write $P \in \mathcal{D}_0 \cap \mathcal{D}'^{-1}_0$ or $P \in \mathcal{D}'^{-1}_0$. Similar interpretations will be attached to expressions like $P \in \mathbb{S}'_0 \cap \mathcal{D}'^{-1}_0, P \in \mathbb{S}'_0 \cap \mathbb{S}'^{-1}_\omega$, and so on.

Our main goal is to characterize those distributions P that belong to the class $\mathcal{D}'_0((E; X))^{-1}$. We assume here that $P \in \mathbb{S}'_0((X; E))$ (see the comments at the end of the previous section) although a result for the case $P \in \mathcal{D}'_+((X; E))$ is also available (Corollary 8.4.9).

We define the *characteristic function* $\mathfrak{P}$ of P by

$$\mathfrak{P}(\lambda) = \mathfrak{P}(\lambda; P) = \mathcal{L}P(\lambda), \tag{8.4.7}$$

where $\mathcal{L}$ indicates Laplace transform. It is a consequence of the results in Section 8.2 that $\mathfrak{P}(\lambda)$ exists and is a $(X; E)$-valued analytic function in a region containing the half plane $\operatorname{Re}\lambda > 0$; we shall denote by $\mathfrak{p}(P)$ the largest open connected region (containing $\operatorname{Re}\lambda > 0$) where $\mathfrak{P}$ is analytic. It follows from standard arguments on operator inverses (Section 1) that $\mathfrak{r}(P)$, the subset of $\mathfrak{p}(P)$ where $\mathfrak{P}(\lambda)$ is invertible, is open (although it may be empty), and that $\mathfrak{P}(\lambda)^{-1} = \mathfrak{R}(\lambda)$ is a $(E; X)$-valued analytic function in $\mathfrak{r}(P)$. By analogy with the operator case we shall call $\mathfrak{r}(P)$ the *resolvent set* of P and $\mathfrak{R}(\lambda) = \mathfrak{R}(\lambda; P)$ the *resolvent* (or *resolvent operator*) of P. The complement $\mathfrak{s}(P)$ of $\mathfrak{r}(P)$ is the *spectrum* of P.

Our first result is a characterization of those P in $\mathbb{S}'$ that possess an inverse in $\mathbb{S}'_\omega$.

8.4.4 Lemma. *Assume that* $P \in \mathbb{S}'_0((X; E)) \cap \mathbb{S}'_\omega((E; X))^{-1}$ *for some* $\omega \in \mathbb{R}$. *Then* $\mathfrak{R}(\lambda) = \mathfrak{P}(\lambda)^{-1}$ *exists for* $\operatorname{Re}\lambda > \omega$ *and there exists an integer m such that*

$$\|\mathfrak{R}(\lambda)\|_{(E, X)} \leqslant C_{\omega'}(1 + |\lambda|)^m \quad (\operatorname{Re}\lambda \geqslant \omega') \tag{8.4.8}$$

for every $\omega' > \omega$. *Conversely, assume that* $\mathfrak{R}(\lambda)$ *exists in* $\operatorname{Re}\lambda \geqslant \omega$ *and*

$$\|\mathfrak{R}(\lambda)\|_{(E, X)} \leqslant C(1 + |\lambda|)^m \quad (\operatorname{Re}\lambda \geqslant \omega). \tag{8.4.9}$$

Then $P \in \mathbb{S}'_\omega((E; X))^{-1}$.

The proof is an immediate consequence of Theorems 8.2.6 and 8.2.7. In fact, if $P \in \mathbb{S}'^{-1}_\omega$, we obtain from (8.4.3) and Theorem 8.2.9 that

$$\mathfrak{P}(\lambda)\mathcal{L}S(\lambda) = I, \quad \mathcal{L}S(\lambda)\mathfrak{P}(\lambda) = J \quad (\operatorname{Re}\lambda > \omega)$$

so that $\Re(\lambda) = \mathcal{L}S(\lambda)$ and Theorem 8.2.6 applies. On the other hand, assume that (8.4.9) holds. Then it follows from Theorem 8.1.7 that

$$\Re(\lambda) = \mathcal{L}S(\lambda) \quad (\operatorname{Re}\lambda > \omega),$$

where S is a distribution in $\mathcal{S}'_\omega((E; X))$. Then $\mathcal{L}(P * S)(\lambda) = \mathfrak{P}(\lambda)\mathcal{L}S(\lambda) = I$, $\mathcal{L}(S * P)(\lambda) = \mathcal{L}S(\lambda)\mathfrak{P}(\lambda) = J$, thus both equalities 8.4.3 follow from uniqueness of the Laplace transform (see the end of the proof of Theorem 8.2.7).

8.4.5 Example. Let A be a closed operator in E with domain $D(A)$. We consider the equation

$$(D - A)U = F. \tag{8.4.10}$$

Here $X = D(A)$ (endowed with the graph norm $\|u\|_X = \|u\| + \|Au\|$) and $P = \delta' \otimes I - \delta \otimes A$. Clearly $\mathfrak{P}(\lambda) = \lambda I - A$ (so that $\mathfrak{p}(P) = \mathbb{C}$). According to our general definition $\mathfrak{r}(R)$ consists of all those λ such that $\lambda I - A$ (as an operator from $X = D(A)$ into E) has a bounded inverse $\Re(\lambda)$. Since $\|u\| \leqslant \|u\|_X$ ($u \in X$), it follows that $\Re(\lambda) \in (E, E)$ and

$$\|\Re(\lambda)\|_{(E, E)} \leqslant \|\Re(\lambda)\|_{(E, X)}. \tag{8.4.11}$$

On the other hand, assume that $\Re(\lambda) = (\lambda I - A)^{-1} \in (E, E)$. Then, since $A\Re(\lambda) = \lambda\Re(\lambda) - I$, it is clear that $\Re(\lambda) \in (E; X)$ as well and

$$\|\Re(\lambda)\|_{(E, X)} \leqslant 1 + (1 + |\lambda|)\|\Re(\lambda)\|_{(E, E)}. \tag{8.4.12}$$

It follows that $\mathfrak{r}(P) = \rho(A)$ and that $\Re(\lambda; P) = R(\lambda; A)$; moreover, inequalities (8.4.8) or (8.4.9) will hold in the norm of (E, X) if and only if they hold (with a different integer m) in the norm of (E, E).

8.4.6 Example. We take now two closed operators A, B with domains $D(A)$ and $D(B)$ and $X = D(A) \cap D(B)$ endowed with the "joint graph norm"

$$\|u\|_X = \|u\| + \|Au\| + \|Bu\|.$$

It is easy to see that $\|\cdot\|_X$ makes X a Banach space. We consider the equation

$$(D^2 - BD - A)U = F. \tag{8.4.13}$$

Here $P = \delta'' \otimes I - \delta' \otimes B - \delta \otimes A$, $\mathfrak{p}(P) = \mathbb{C}$, $\mathfrak{P}(\lambda) = \lambda^2 I - \lambda B - A$, $\mathfrak{r}(P)$ is the set of all λ for which $\lambda^2 I - \lambda B - A$ has an inverse $\Re(\lambda)$ bounded as an operator from E into X. Clearly (8.4.11) holds as well now; on the other hand, if $\Re(\lambda) = (\lambda^2 I - \lambda B - A)^{-1}$ exists and belongs to (E, E) together with $A\Re(\lambda)$, we obtain from the equality $B\Re(\lambda) = \lambda\Re(\lambda) - \lambda^{-1}A\Re(\lambda) - \lambda^{-1}I$ that

$$\|\Re(\lambda)\|_{(E, X)} \leqslant |\lambda|^{-1} + (1 + |\lambda|)\|\Re(\lambda)\|_{(E, E)} + (1 + |\lambda|^{-1})\|A\Re(\lambda)\|_{(E, E)}. \tag{8.4.14}$$

Consequently inequalities (8.4.8) or (8.4.9) hold in the norm of (E, X) if and only if they hold (with a different m) for $\Re(\lambda)$ and $A\Re(\lambda)$ in the norm of $(E; E)$. A similar statement holds if we change $A\Re(\lambda)$ by $B\Re(\lambda)$; if either A or B are bounded, the estimate on $\Re(\lambda)$ in the $(E; E)$ norm suffices.

We note that if $(\lambda^2 I - \lambda B - A)^{-1}$ exists and belongs to (E, E), it is also bounded as an operator from E into X (we only have to apply the closed graph theorem to the operators $A\Re(\lambda)$ and $B\Re(\lambda)$).

8.4.7 Example. Let

$$P = \delta' \otimes I - \mu, \tag{8.4.15}$$

where $X \subseteq E$ and μ is a measure with values in (X, E); assume in addition that

$$\int_0^\infty \|\mu(dt)\| < \infty.$$

Here $\mathfrak{p}(P)$ contains the half plane $\operatorname{Re}\lambda \geqslant 0$ and

$$\mathfrak{P}(\lambda) = \lambda I - \int_0^\infty e^{-\lambda t} \mu(dt)$$

(the precise extent of $\mathfrak{p}(P)$ depends of course on μ). Equation (8.4.1) is the *integro-differential equation*

$$DU(\hat{t}) = \mu * U(\hat{t}) + F(\hat{t})$$
$$= \int_0^t U(\hat{t} - s)\mu(ds) + F(\hat{t}). \tag{8.4.16}$$

A particular instance is that where $X = D(A) \cap D(B)$ (A and B as in the previous example) and $\mu = \delta(\hat{t}) \otimes A + \delta(\hat{t} - h) \otimes B$ for some $h > 0$; here $\mathfrak{p}(P) = \mathbb{C}$, $\mathfrak{P}(\lambda) = \lambda I - A - e^{-\lambda h} B$, and (8.4.16) is the *difference-differential equation* (or *delay differential equation*)

$$U'(\hat{t}) = AU(\hat{t}) + BU(\hat{t} - h) + F(\hat{t}).$$

The next theorem is a characterization of the class $\mathcal{S}_0' \cap \mathcal{D}_0'^{-1}$. In order to state the result, we introduce the following definition. A *logarithmic region* in the complex plane is a set Λ of the form

$$\Lambda = \Lambda(\alpha, \beta) = \{\lambda; \operatorname{Re}\lambda \geqslant \beta + \alpha \log|\lambda|\}, \tag{8.4.17}$$

where α, β are two nonnegative numbers. (See Figure 8.4.1.)

8.4.8 Theorem. *A distribution* $P \in \mathcal{S}_0'((X; E))$ *is an element of* $\mathcal{D}_0'((E; X))^{-1}$ *if and only if* $\Re(\lambda) = \mathfrak{P}(\lambda)^{-1}$ *exists in a logarithmic region* Λ *and*

$$\|\Re(\lambda)\|_{(E, X)} \leqslant C(1 + |\lambda|)^m \quad (\lambda \in \Lambda) \tag{8.4.18}$$

for some $C \geqslant 0$ *and some nonnegative integer* m.

Proof. Denote by $\mathcal{K}_a$ the set of all functions in $\mathcal{D}$ that equal 1 in $0 \leqslant t \leqslant a$ and vanish in $t \geqslant 2a$. Assume that $P \in \mathcal{D}'_0((E; X))^{-1}$ and let $S = P^{-1}$. If $\varphi \in \mathcal{K}_a$, it follows from (8.4.3) that

$$P * \varphi S = \delta \otimes I - \Phi, \quad \varphi S * P = \delta \otimes J - \Psi, \qquad (8.4.19)$$

where $\Phi = P * (1 - \varphi)S$, $\Psi = (1 - \varphi)S * P$. Since $\Phi = \delta \otimes I - P * \varphi S$ and φS belongs to $\in \mathcal{E}'_0((E; X))$, a fortiori to $\mathcal{S}'_0((E; X))$, it follows from Theorem 8.2.5 that $P * \varphi S$, thus also Φ, belongs to $\mathcal{S}'_0((E; E))$; on the other hand, the support of $(1 - \varphi)S$ is contained in $t \geqslant a$ thus we obtain from (8.2.6) that $\Phi = 0$ for $t < a$. We use then Theorem 8.2.3 to deduce the existence of a $(E; E)$-valued continuous function f vanishing for $t \leqslant a$ and growing at most like a power of t at infinity, such that

$$\Phi = f^{(p)}.$$

Since f vanishes for $t \leqslant a$, we have

$$\|\mathcal{L}\Phi(\lambda)\|_{(E, E)} \leqslant C|\lambda|^p e^{-a \operatorname{Re} \lambda} \quad (\operatorname{Re} \lambda \geqslant 1). \qquad (8.4.20)$$

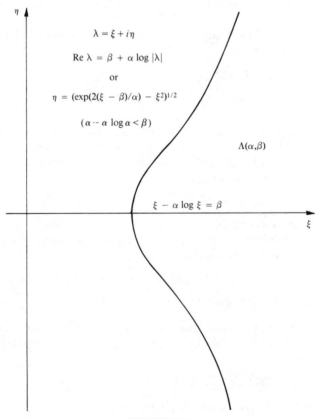

FIGURE 8.4.1

We choose now γ, $0 < \gamma < 1$ and take λ in a logarithmic region $\Lambda(\alpha, \beta)$ with $\alpha = p/a$ and $\beta \geqslant a^{-1}\log(C\gamma^{-1})$ so large that $\Lambda(\alpha, \beta)$ is contained in the half plane $\operatorname{Re}\lambda \geqslant 1$ (see Figure 8.4.1). Then we deduce from (8.4.20) that

$$\|\mathfrak{L}\Phi(\lambda)\|_{(E, E)} \leqslant \gamma \quad (\lambda \in \Lambda(\alpha, \beta)). \tag{8.4.21}$$

On the other hand, since $\varphi S \in \mathcal{S}'_0((E; X))$, we obtain from Theorem 8.2.6 that, if m is large enough,

$$\|\mathfrak{L}(\varphi S)(\lambda)\|_{(E; X)} \leqslant C'(1 + |\lambda|)^m \quad (\operatorname{Re}\lambda \geqslant 1). \tag{8.4.22}$$

Taking the Laplace transform of the first equation (8.4.19) it follows that

$$\mathfrak{P}(\lambda)\mathfrak{L}(\varphi S)(\lambda) = I - \mathfrak{L}\Phi(\lambda). \tag{8.4.23}$$

We use (8.4.21) to deduce that $(I - \mathfrak{L}\Phi(\lambda))^{-1} = \sum_{n=0}^{\infty}(\mathfrak{L}\Phi(\lambda))^n$ exists in $\Lambda(\alpha, \beta)$ and

$$\|(I - \mathfrak{L}\Phi(\lambda))^{-1}\|_{(E, E)} \leqslant (1 - \gamma)^{-1} \tag{8.4.24}$$

there. Postmultiplying (8.4.23) by $(I - \mathfrak{L}\Phi(\lambda))^{-1}$, we see that $\mathfrak{R}_r(\lambda) = \mathfrak{L}(\varphi S)(\lambda)(I - \mathfrak{L}\Phi(\lambda))^{-1}$ is a right inverse for $\mathfrak{P}(\lambda)$ in $\Lambda(\alpha, \beta)$ and, in view of (8.4.22) and (8.4.24), we have

$$\|\mathfrak{R}_r(\lambda)\|_{(E, X)} \leqslant C(1 + |\lambda|)^m \quad (\lambda \in \Lambda(\alpha, \beta)). \tag{8.4.25}$$

We apply identical arguments to the second equation (8.4.19); we prove, modifying if necessary the parameters α, β, that $(I - \mathfrak{L}\Psi(\lambda))^{-1}$ exists and belongs to (X, X) in $\Lambda(\alpha, \beta)$. Taking the Laplace transform of the second equation and premultiplying by $(I - \mathfrak{L}\Psi(\lambda))^{-1}$, we show that $\mathfrak{R}_l(\lambda) = (I - \mathfrak{L}\Psi(\lambda))^{-1}\mathfrak{L}(\varphi S)$ is a left inverse for $\mathfrak{P}(\lambda)$. It results that $\mathfrak{R}(\lambda)$ $(= \mathfrak{R}_r(\lambda) = \mathfrak{R}_l(\lambda))$ exists and satisfies (8.4.18) in $\Lambda(\alpha, \beta)$.

We prove now the converse. Assume the conditions of Theorem 8.4.8 are satisfied, let Γ be the boundary of the logarithmic region $\Lambda(\alpha, \beta)$ oriented clockwise with respect to Λ, and let $p > m + 1$ be a positive integer. Define

$$T_p(t) = \frac{1}{2\pi i}\int_\Gamma \lambda^{-p}e^{\lambda t}\mathfrak{R}(\lambda)\, d\lambda. \tag{8.4.26}$$

The integrand grows no more than $|\lambda|^{\alpha t - p + m}$ as $|\lambda| \to \infty$; thus T_p is a continuous $(E; X)$-valued function in

$$-\infty < t < (p - m - 1)/\alpha \tag{8.4.27}$$

and a standard deformation of contour in the integral shows that $T_p(t) = 0$ for $t \leqslant 0$.

Since $P \in \mathcal{S}'_0((X; E))$, there exists an integer q and a continuous $(X; E)$-valued function f vanishing in $t \leqslant 0$, growing no more than a power of t at infinity and such that

$$P = f^{(q)}. \tag{8.4.28}$$

We have

$$(f * T_p)(t) = \frac{1}{2\pi i} \int_\Gamma \lambda^{-p} \left(e^{\lambda t} \int_0^t e^{-\lambda s} f(s)\, ds \right) \Re(\lambda)\, d\lambda$$

$$(0 \leqslant t < (p - m - 1)/\alpha), \quad (8.4.29)$$

where the interchanging of order of integration is justified by the fact that the integrand (of the double integral) grows at most like $|\lambda|^{\alpha(t-s)-p+m}$ when $|\lambda| \to \infty$. Let now

$$\eta(\lambda) = e^{\lambda t} \int_t^\infty e^{-\lambda s} f(s)\, ds.$$

We have

$$|\eta(\lambda)| = O(|\mathrm{Re}\,\lambda|^{-1}) \quad (|\mathrm{Re}\,\lambda| \to \infty).$$

It follows then from yet another deformation-or-contour argument that

$$\frac{1}{2\pi i} \int_\Gamma \lambda^{-p} \eta(\lambda) \Re(\lambda)\, d\lambda = 0,$$

therefore the upper limit in the inner integral in (8.4.29) can be replaced by ∞. Doing this and making use of the equality $\mathfrak{P}(\lambda) = \lambda^q (\mathfrak{L}f)(\lambda)$ that results from (8.4.28), we obtain

$$(f * T_p)(t) = \left(\frac{1}{2\pi i} \int_\Gamma \lambda^{-(p+q)} e^{\lambda t}\, d\lambda \right) I$$

$$= Y_{p+q}(t) I \quad (0 \leqslant t < (p - m - 1)/\alpha), \quad (8.4.30)$$

where Y_m is the distribution defined at the end of Section 8.2. If we define $S_p = D^p T_p$, the following equality results:

$$P * S_p = \delta \otimes I \quad (t < (p - m - 1)/\alpha). \quad (8.4.31)$$

The convolution $S_p * f$ can be computed using essentially the same devices; we obtain

$$S_p * P = \delta \otimes J \quad (t < (p - m - 1)/\alpha). \quad (8.4.32)$$

Making use of (8.4.31), (8.4.32) of the arbitrariness of p and of the "construction-by-pieces" Lemma 8.4.3, Theorem 8.4.8 follows.

Theorem 8.4.8 can be extended—if in a somewhat awkward form—to the case $P \in \mathcal{D}_0'$. In fact, let $\mathcal{K} = \cup_{a > 0} \mathcal{K}_a$. Then we have

8.4.9   Corollary.   *Let* $P \in \mathcal{D}_0'((X; E))$. *Then* $P \in \mathcal{D}_0'((E; X))^{-1}$ *if and only if* $\varphi P \in \mathcal{D}_0'((E; X))^{-1}$ *for all* $\varphi \in \mathcal{K}$, *that is, if* φP *satisfies the hypotheses of Theorem 8.4.8 for all* $\varphi \in \mathcal{K}$.

Proof.   Let $\varphi \in \mathcal{K}_a$, $a > 0$. If $\varphi P \in \mathcal{D}_0'^{-1}$, then there exists $S_\varphi \in \mathcal{D}_0'$ with $\varphi P * S_\varphi = \delta \otimes I$, $S_\varphi * \varphi P = \delta \otimes J$. But $\varphi P = P$ in $t < a$, so that $P * S_\varphi =$

$\delta \otimes I$, $S_\varphi * P = \delta \otimes J$ in $t < a$ and Lemma 8.4.3 applies. Conversely, let $P \in \mathcal{D}_0'^{-1}$ and let $S = P^{-1}$. Then, if $\varphi \in \mathcal{K}_a$, we have $\varphi P * \varphi S = \delta \otimes I - \Phi$, $\varphi S * \varphi P = \delta \otimes I - \Psi$, where

$$\Phi = \delta \otimes I - \varphi P * \varphi S$$
$$= -(1 - \varphi) P * (1 - \varphi) S + (1 - \varphi) P * S + P * (1 - \varphi) S$$

belongs to $\mathcal{S}_0'((E; E))$ and vanishes for $t \leq a$; as for Ψ, which admits a similar expression, it belongs to $\mathcal{S}_0'((X, X))$ and vanishes in $t \leq a$. We can then proceed just as in the proof of the first half of Theorem 8.4.8 to show that $\Re(\lambda; \varphi P) = \mathfrak{P}(\lambda; \varphi P)^{-1}$ satisfies the assumptions there; hence $\varphi P \in \mathcal{D}_0'^{-1}$, as claimed.

8.4.10 Example. Assume $E = X$, $B \in (E, E)$.

(a) Let $P(\hat{t}) = \delta'(\hat{t}) \otimes I - \delta(\hat{t} - h) \otimes B$. Then $P \in \mathcal{E}_0' \cap \mathcal{S}_0'^{-1}$ and

$$S(\hat{t}) = P^{-1}(\hat{t})$$
$$= Y_1(\hat{t}) \otimes I + Y_2(\hat{t} - h) \otimes B + Y_3(\hat{t} - 2h) \otimes B^2 + \cdots,$$

the series convergent in $\mathcal{S}_0'((E; E))$. Here $\mathfrak{p}(P) = \mathbb{C}$, $\mathfrak{P}(\lambda) = \lambda I - e^{-\lambda h} B$, $\mathfrak{r}(P)$ is the set of all λ such that $\lambda e^{\lambda h} \in \rho(B)$ and $\Re(\lambda; P) = e^{\lambda h} R(\lambda e^{\lambda h}; B)$; in particular, $\mathfrak{r}(P)$ includes the set $\{\lambda; |\lambda| e^{h \operatorname{Re} \lambda} \geq r\}$, where r is the norm of B (or, more generally, its spectral radius). (See Figure 8.4.2.)

(b) Let $P(\hat{t}) = \delta(\hat{t}) \otimes I + \delta'(\hat{t} - h) \otimes B$. Then $P \in \mathcal{E}_0' \cap \mathcal{D}_0'^{-1}$ (but not to $\mathcal{S}_\omega'^{-1}$ for any ω) and

$$S(\hat{t}) = P^{-1}(\hat{t})$$
$$= \delta(\hat{t}) \otimes I + \delta'(\hat{t} - h) \otimes B + \delta''(\hat{t} - 2h) \otimes B^2 + \cdots,$$

the series being convergent in $\mathcal{D}_0'((E; E))$. We have $\mathfrak{p}(P) = \mathbb{C}$, $\mathfrak{P}(\lambda) = I - \lambda e^{-\lambda h} B$, $\mathfrak{r}(P) = \{\lambda; \lambda^{-1} e^{\lambda h} \in \rho(B)\}$, and $\Re(\lambda) = \lambda^{-1} e^{\lambda h} R(\lambda^{-1} e^{\lambda h}; B)$. The resolvent set $\mathfrak{r}(P)$ includes the set $\{\lambda; |\lambda|^{-1} e^{h \operatorname{Re} \lambda} \geq r\}$, r the spectral radius of B. (see Figure 8.4.3.)

(c) Let $P(\hat{t}) = \delta(\hat{t}) \otimes I - \delta(\hat{t} - h) \otimes B$. Then $P \in \mathcal{E}_0' \cap \mathcal{S}_0'^{-1}$ and

$$S(\hat{t}) = \delta(\hat{t}) \otimes I + \delta(\hat{t} - h) \otimes B + \delta(\hat{t} - 2h) \otimes B^2 + \cdots,$$

the series convergent in $\mathcal{S}_0'$. Again $\mathfrak{p}(P) = \mathbb{C}$. Now we have $\mathfrak{P}(\lambda) = I - e^{-\lambda h} B$, $\mathfrak{r}(P)$ is the set of all λ such that $e^{\lambda h} \in \rho(B)$ and $\Re(\lambda; P) = e^{\lambda h} R(e^{\lambda h}; B)$. The resolvent set contains the half plane $\operatorname{Re} \lambda > h^{-1} \log r$.

(d) Let $P = \delta' \otimes I - \mu$, μ a measure with values in $(E; E)$ with

$$\int_0^\infty \|\mu(dt)\| = \omega < \infty.$$

Then $P \in \mathcal{S}_\omega'^{-1}$ and

$$S = \sum_{n=0}^\infty Y_{n+1} * \mu^{*n}$$

where $*n$ indicates convolution powers. Here $\mathfrak{p}(P)$ contains the half plane

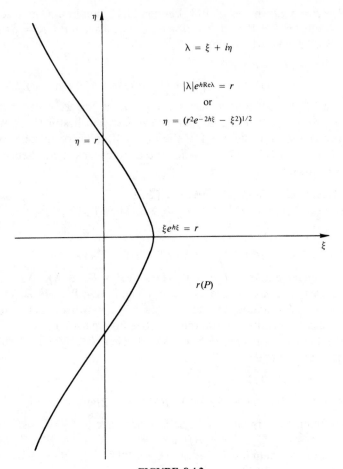

$$\lambda = \xi + i\eta$$

$$|\lambda|e^{h\operatorname{Re}\lambda} = r$$

or

$$\eta = (r^2 e^{-2h\xi} - \xi^2)^{1/2}$$

$\eta = r$

$\xi e^{h\xi} = r$

$r(P)$

FIGURE 8.4.2

$\operatorname{Re}\lambda \geq 0$ and

$$\mathfrak{P}(\lambda) = \lambda I - \int_0^\infty e^{-\lambda t}\mu(dt)$$

$$= \lambda\left(I - \frac{1}{\lambda}\int_0^\infty e^{-\lambda t}\mu(dt)\right).$$

Accordingly $r(P)$ contains the set of all λ for which

$$\frac{1}{|\lambda|}\int_0^\infty e^{-(\operatorname{Re}\lambda)t}\|\mu(dt)\| < 1$$

and

$$\mathfrak{R}(\lambda) = \frac{1}{\lambda}\sum_{n=0}^\infty \left(\frac{1}{\lambda}\int_0^\infty e^{-\lambda t}\mu(dt)\right)^n$$

for those λ.

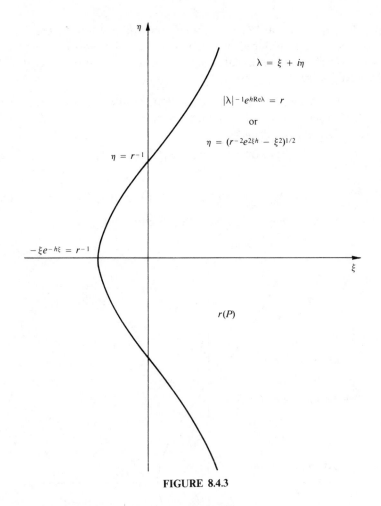

$$\lambda = \xi + i\eta$$

$$|\lambda|^{-1}e^{h\,\text{Re}\lambda} = r$$

or

$$\eta = (r^{-2}e^{2\xi h} - \xi^2)^{1/2}$$

$$\eta = r^{-1}$$

$$-\xi e^{-h\xi} = r^{-1}$$

$$r(P)$$

FIGURE 8.4.3

8.4.11 Example. Let A be a closed operator in E, $X = D(A)$ with the graph norm, $P = \delta' \otimes I - \delta \otimes A$. Then $P \in \mathfrak{D}_0'^{-1}$ if and only if $\rho(A)$ contains some logarithmic region Λ and

$$\|R(\lambda)\| \leqslant C(1 + |\lambda|)^m \quad (\lambda \in \Lambda),$$

where $\|\cdot\|$ indicates the $(E; E)$-norm (see Example 8.4.5).

8.5. THE ABSTRACT PARABOLIC CASE

In accordance with the notation introduced in the previous section, a distribution $P \in \mathcal{S}_0'((X; E)) \cap \mathfrak{D}_0'((E; X))^{-1}$ is said to be an element of $C^{(\infty)}((0, \infty); (E; X))^{-1}$ (or simply of $(C^{(\infty)})^{-1}$) if the solution of (8.4.3) is an infinitely differentiable $(E; X)$-valued function in $t > 0$. In keeping with the nomenclature introduced in Chapter 4, we call P *abstract parabolic* in this case. We characterize the abstract parabolic distributions in $\mathcal{E}_0'((X; E))$ in the next result.

8.5.1   Theorem.   *Let $P \in \mathcal{E}_0'((X; E))$. Then P is abstract parabolic if and only if for every $\alpha > 0$ there exists $\beta = \beta(\alpha) > 0$ such that $\mathfrak{r}(P)$ contains the reverse logarithmic region*

$$\Omega(\alpha, \beta) = \{\lambda;\ \mathrm{Re}\,\lambda \geqslant \beta - \alpha \log|\lambda|\} \tag{8.5.1}$$

and

$$\|\mathfrak{R}(\lambda)\|_{(E, X)} \leqslant C(1 + |\lambda|)^m \quad (\lambda \in \Omega(\alpha, \beta)), \tag{8.5.2}$$

where C (but not m) may depend on α. (See Figure 8.5.1.)

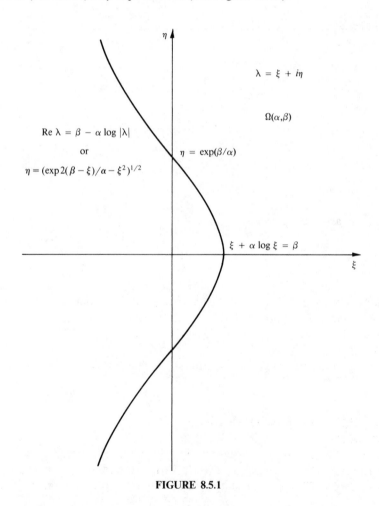

FIGURE 8.5.1

Proof. Assume the hypotheses are satisfied. Then so are those of Lemma 8.4.4 and $P \in \mathcal{S}'_\omega((E; X))^{-1}$ for some $\omega > 0$. Let $S = P^{-1}$. We may deform the contour of integration in (8.4.26) to $\Sigma(\alpha)$, boundary of $\Omega(\alpha, \beta(\alpha))$ oriented in the same way as Γ. The integrand in the resulting integral grows no more that $|\lambda|^{-\alpha t - p + m}$ as $|\lambda| \to \infty$ so that if $t > (m+1)/\alpha$ we can differentiate p times under the integral sign obtaining

$$S(t) = \frac{1}{2\pi i} \int_{\Sigma(\alpha)} e^{\lambda t} \Re(\lambda) \, d\lambda \quad (t > (m+1)/\alpha). \tag{8.5.3}$$

Moreover, the same type of argument shows that (8.5.3) can be differentiated n times under the integral sign if $t > (m+n+1)/\alpha$. Since α is arbitrary this establishes that $S(\cdot)$ is infinitely differentiable in $t > 0$.[6]

We prove the converse. Assume that $S = P^{-1} \in C^{(\infty)}((0, \infty); (E; X))$ and that $P \in \mathcal{S}'_0((X; E))$. Let $\mathrm{supp}(P) \subseteq [0, b]$ and, given $\alpha > 0$, choose $a > 0$ and an integer p in such a way that

$$p \geqslant \alpha(1+b), \quad a \leqslant \tfrac{1}{2}, \quad a \leqslant (1+b)/2p. \tag{8.5.4}$$

Let $\varphi \in \mathcal{K}_a$. As in Theorem 2.5 we have

$$P * \varphi S = \delta \otimes I - \Phi, \quad \varphi S * P = \delta \otimes I - \Psi, \tag{8.5.5}$$

but now $\Phi = P * (1 - \varphi) S = \delta \otimes I - P * \varphi S$ is an infinitely differentiable $(E; E)$-valued function with support in $a \leqslant t \leqslant 2a + b$, whereas Ψ is a $(X; X)$-valued function enjoying the same privileges. After repeated integration by parts, we obtain

$$\mathfrak{L}\Phi(\lambda) = \lambda^{-p} \int_a^{2a+b} e^{-\lambda t} \Phi^{(p)}(t) \, dt,$$

whence

$$\|\mathfrak{L}\Phi(\lambda)\|_{(E, E)} \leqslant C|\lambda|^{-p} \quad (\mathrm{Re}\,\lambda \geqslant 0) \tag{8.5.6}$$

and

$$\|\mathfrak{L}\Phi(\lambda)\|_{(E, E)} \leqslant C|\lambda|^{-p} e^{-(2a+b)\mathrm{Re}\,\lambda} \quad (\mathrm{Re}\,\lambda < 0). \tag{8.5.7}$$

Let $0 < \gamma < 1$. Choose β in such a way that every $\lambda \in \Omega(\alpha, \beta)$ satisfies the inequality

$$|\lambda| \geqslant \max(\omega, 1) \tag{8.5.8}$$

with $\omega = (C/\gamma)^{1/p}$; then it follows from (8.5.6) that

$$\|\mathfrak{L}\Phi(\lambda)\|_{(E, E)} \leqslant \gamma \quad (\lambda \in \Omega(\alpha, \beta)) \tag{8.5.9}$$

---

[6]This half of Theorem 8.5.1 does not make use of the hypothesis that $P \in \mathcal{S}'_0$; it is sufficient to assume that $P \in \mathcal{S}'_0$.

if $\text{Re}\,\lambda \geqslant 0$. Taking β larger, if necessary, we can assume that

$$Ce^{-\beta(2a+b)} \leqslant \gamma.$$

Also, by virtue of the first two conditions (8.5.4), we have

$$p \geqslant \alpha(2a+b),$$

so that it follows from the last two estimates, from (8.5.7), and from (8.5.8) that inequality (8.5.9) actually holds for all $\lambda \in \Omega(\alpha, \beta)$.

We write next $\varphi S = S + (\varphi - 1)S$ and deduce that there is an integer q *independent of* φ such that for every $\varphi \in \mathcal{H}$ there exists a continuous $(E; X)$-valued function f_φ vanishing for $t \leqslant 0$ and such that

$$\varphi S = f_\varphi^{(q)} \text{ in } R. \tag{8.5.10}$$

Since $\varphi S = 0$ for $t \geqslant 2a$, f_φ must be there a polynomial of degree $\leqslant q - 1$. Proceeding now exactly as in the proof of Theorem 8.2.8, we obtain

$$\|\mathcal{L}(\varphi S)(\lambda)\|_{(E; X)} \leqslant C|\lambda|^q \quad (\text{Re}\,\lambda \geqslant 0, |\lambda| \geqslant 1) \tag{8.5.11}$$

and

$$\|\mathcal{L}(\varphi S)(\lambda)\|_{(E; X)} \leqslant C|\lambda|^q e^{-2a\text{Re}\,\lambda} \quad (\text{Re}\,\lambda < 0, |\lambda| \geqslant 1). \tag{8.5.12}$$

Let $\lambda \in \Omega(\alpha, \beta)$. We deduce from the first and third inequalities (8.5.4) that $-2a\,\text{Re}\,\lambda \leqslant 2a\alpha \log |\lambda| - 2a\beta \leqslant (\alpha(1+b)/p)\log|\lambda| - 2a\beta \leqslant \log|\lambda| - 2a\beta$ so that (8.5.11) and (8.5.12) combine to yield

$$\|\mathcal{L}(\varphi S)(\lambda)\|_{(E; X)} \leqslant C|\lambda|^{q+1} \quad (\lambda \in \Omega(\alpha, \beta)). \tag{8.5.13}$$

We make use now of (8.5.13) and of (8.5.9) in the style of Theorem 8.4.8 to show that $\mathfrak{R}_r(\lambda) = \mathcal{L}(\varphi S)(\lambda)(I - \mathcal{L}\Phi(\lambda))^{-1}$ is a right inverse for $\mathfrak{P}(\lambda)$ in $\Omega(\alpha, \beta)$ satisfying (8.5.2) for $m = q + 1$. The fact that $\mathfrak{R}_r(\lambda)$ is a two-sided inverse follows, again as in the proof of Theorem 8.4.8, from arguments of the same type applied to the function Ψ in the second equality (8.5.5).

8.5.2 Remark. As a byproduct of the proof of Theorem 8.5.1, we have obtained that a distribution P in $(C^{(\infty)})^{-1}$ must be in $\mathcal{S}_\omega'^{-1}$ for some $\omega > 0$. But more is true: in fact, it follows from the representation (8.5.3) and a simple estimation that there exists a $\omega > 0$ such that for every $\varepsilon > 0$ there exists a constant $C_\varepsilon > 0$ with

$$\|S(t)\|_{(E; X)} \leqslant C_\varepsilon e^{\omega t} \quad (t \geqslant 0).$$

8.5.3 Remark. Theorem 8.5.1 implies Theorem 4.1.3 on the characterization of operators A in the class $\mathcal{C}^\infty$ (see the remarks on the relation between the Cauchy problem in the strong sense and in the sense of distributions after Theorem 8.4.1). To see that the estimate (8.5.2) can be strengthened to (4.1.6), it suffices to notice that the integer q in (8.5.10) can be chosen equal to zero in this case; although $f_\varphi = \varphi S$ is merely strongly continuous, the argument works. The estimation must of course be carried out in the $(E; E)$-norm.

8.6. APPLICATIONS; EXTENSIONS OF THE NOTION OF PROPERLY POSED CAUCHY PROBLEM

There exist abstract differential equations for which the Cauchy problem, although perfectly natural, does not fit into the mold of properly posed problems introduced in Chapter 1. We examine below an instance of this situation.

8.6.1 Example (Beals [1972: 2]). Consider the following system of partial differential equations for two functions $u(x, t), v(x, t)$:

$$D_t u = D_x u + D_x v, \quad D_t v = D_x v \quad (x \in \mathbb{R}, t \geqslant 0) \tag{8.6.1}$$

with initial conditions

$$u(x,0) = u_0(x), \quad v(x,0) = v_0(x). \tag{8.6.2}$$

We can write (8.6.1) as a differential equation

$$\mathfrak{u}'(t) = \mathfrak{A}\mathfrak{u}(t) \tag{8.6.3}$$

in the space $L^2(\mathbb{R})^2$; in matrix notation the operator $\mathfrak{A}$ is

$$\mathfrak{A} = \begin{bmatrix} D_x & D_x \\ 0 & D_x \end{bmatrix} \tag{8.6.4}$$

whose domain $D(\mathfrak{A})$ consists of all $\mathfrak{u} = (u, v) \in L^2(\mathbb{R})^2$ such that $u', v' \in L^2(\mathbb{R})$ (that is, $D(A) = H^1(\mathbb{R})^2 = H^1(\mathbb{R}) \times H^1(\mathbb{R})$). We analyze (8.6.3) using Fourier transforms as in the examples in Chapter 1. The operator $\tilde{\mathfrak{A}} = \mathfrak{F}\mathfrak{A}\mathfrak{F}^{-1}$ in $L^2(\mathbb{R}_\sigma)^2$ is the operator of multiplication by the matrix

$$\tilde{\mathfrak{A}}(\sigma) = -i\sigma \begin{bmatrix} 1 & 1 \\ 0 & 1 \end{bmatrix} \tag{8.6.5}$$

with domain $D(\tilde{\mathfrak{A}})$ consisting of all $(\tilde{u}, \tilde{v}) \in L^2(\mathbb{R}_\sigma)^2$ with $\sigma\tilde{u}, \sigma\tilde{v} \in L^2(\mathbb{R}_\sigma)$. In view of the isometric character of $\mathfrak{F}$, we may examine

$$\tilde{\mathfrak{u}}'(t) = \tilde{\mathfrak{A}}\tilde{\mathfrak{u}}(t) \tag{8.6.6}$$

instead of (8.6.3). We have

$$\exp(t\tilde{\mathfrak{A}}(\sigma)) = \sum_{k=0}^{\infty} \frac{(-i\sigma t)^k}{k!} \begin{bmatrix} 1 & k \\ 0 & 1 \end{bmatrix}$$

$$= \begin{bmatrix} e^{-i\sigma t} & -i\sigma t e^{-i\sigma t} \\ 0 & e^{-i\sigma t} \end{bmatrix}. \tag{8.6.7}$$

If $\tilde{\mathfrak{u}} = (\tilde{u}, \tilde{v}) \in D(\tilde{\mathfrak{A}}^2) = \{(\tilde{u}, \tilde{v}) \in L^2(\mathbb{R}_\sigma)^2, \sigma^2\tilde{u}, \sigma^2\tilde{v} \in L^2(\mathbb{R}_\sigma)\}$, then $\tilde{\mathfrak{u}}(t) = \mathfrak{S}(t)(\tilde{u}, \tilde{v}) = \exp(t\tilde{\mathfrak{A}}(\sigma))(\tilde{u}, \tilde{v})$ is a genuine solution of (8.6.6) in $t \geqslant 0$ (in fact, in $-\infty < t < \infty$), although the operator $\mathfrak{S}(t)$ is not bounded in $L^2(\mathbb{R}_\sigma)^2$ if $t \neq 0$. However, it is plain that there exists a constant C such that

$$\|\tilde{\mathfrak{u}}(t)\| \leqslant C(1 + |t|)(\|\tilde{\mathfrak{u}}\| + \|\tilde{\mathfrak{A}}\tilde{\mathfrak{u}}\|) \quad (-\infty < t < \infty). \tag{8.6.8}$$

Obviously, the same conclusion holds for the original equation (8.6.3): the assumption on the initial condition $u = (u, v)$ is that $u, v \in H^2(\mathbb{R})$.

We take this example as motivation for the following definition, where additional generality is attempted.

Let A be a densely defined operator in an arbitrary Banach space E. The Cauchy problem for the equation

$$u'(t) = Au(t) \tag{8.6.9}$$

is *mildly well posed in* $t \geq 0$ if and only if

(a) *The subspace* $D = D(A^\infty) = \cap_{n=1}^\infty D(A^n)$ *is dense in* E *and for each* $u_0 \in D$ *there exists a solution[7]* $u(\cdot)$ *of (8.6.9) in* $t \geq 0$ *with*

$$u(0) = u_0. \tag{8.6.10}$$

(b) *For every* $T > 0$ *there exists an integer* $p = p(T)$ *and a constant* $C = C(T)$ *such that*

$$\|u(t)\| \leq C(\|u(0)\| + \|A^p u(0)\|) \tag{8.6.11}$$

for any solution $u(\cdot)$ *of (8.6.9) defined in* $0 \leq t \leq T$ *and such that* $u(0) \in D(A^p)$.

We note that the first assumption coincides with (a) in the definition of well posed Cauchy problem in Section 1.2 (except for identification of D). Note also that (b), unlike its counterpart in Section 1.2 applies to solutions not necessarily defined in $t \geq 0$ but only in a finite subinterval $0 \leq t \leq T$. Obviously, it implies uniqueness of any such solution $u(\cdot)$, at least in $0 \leq t \leq T$, if $u(0) \in D(A^p)$, p the integer corresponding to T. If we assume that $\rho(A) \neq \varnothing$, uniqueness follows in general: if $u(\cdot)$ is an arbitrary solution of (8.6.9) in $0 \leq t \leq T$, then $v(\cdot) = R(\lambda)^p u(\cdot)$ is a solution to which uniqueness applies, and $v(0) = 0$ if $u(0) = 0$.

The mildly well posed Cauchy problem in $-\infty < t < \infty$ is correspondingly defined.

8.6.2 Theorem. *Let* A *be a densely defined operator with* $\rho(A) \neq \varnothing$ *and assume the Cauchy problem for (8.6.9) is mildly well posed in* $t \geq 0$. *Then the Cauchy problem for* $(D - A)U = F$ *is well posed in the sense of distributions, that is, there exists* $S \in \mathcal{D}_0'((E; D(A)))$ *such that*

$$(\delta' \otimes I - \delta \otimes A) * S = \delta \otimes I, \quad S * (\delta' \otimes I - \delta \otimes A) = \delta \otimes J, \tag{8.6.12}$$

where I *(resp.* J*) is the identity operator in* E *(resp.* $D(A)$*). Conversely, let the Cauchy problem for a closed, densely defined operator* A *be well posed in the sense of distributions. Then the Cauchy problem for (8.6.9) is mildly well posed in* $t \geq 0$.

---

[7]Solutions are defined as in Chapter 1.

Proof. Assume the Cauchy problem for (8.6.9) is mildly well posed in $t \geq 0$ with $\rho(A) \neq \varnothing$. As in Section 1.2 we define the propagator $S(\cdot)$ of (8.6.9) for $u \in D$ by

$$S(t)u = u(t),$$

where $u(\cdot)$ is the solution of (8.6.9) with $u(0) = u$ provided by assumption (a). Unlike in Section 1.2, however, $S(t)$ is in general unbounded and not defined in all of E.

We begin by observing that if $\lambda \in \rho(A)$, then the norms $\|u\| + \|A^p u\|$ and $\|(\lambda I - A)^p u\|$ are equivalent in the space $D(A^p)$. Accordingly, it follows from (b) that $\|S(t)u\| \leq C\|(\lambda I - A)^p u\|$ ($0 \leq t \leq T$, $u \in D$). Hence $S(t)R(\lambda)^p$ is bounded in D, uniformly in $0 \leq t \leq T$. We note next that

$$R(\lambda)^p S(t)u = S(t)R(\lambda)^p u \tag{8.6.13}$$

for $0 \leq t \leq T$, $u \in D$. To prove this equality we observe that the right-hand side of (8.6.13) is a solution of (8.6.9) in $t \geq 0$, and so is $R(\lambda)^p S(\cdot)u$ (note that we obviously have $(\lambda I - A)^p D = D$). Since both solutions have the same initial value, they must coincide everywhere and (8.6.13) results. Using the uniform boundedness of the right-hand side and the continuity of $S(\cdot)u$ for $u \in D$, we deduce that $R(\lambda)^p S(t)$ has an extension $\overline{R(\lambda)^p S(t)}$, uniformly bounded and strongly continuous in $0 \leq t \leq T$. We shall use this fact to show that

$$(Y_p * Su)(t) = \frac{1}{(p-1)!} \int_0^t (t-s)^{p-1} S(s)u \, ds \tag{8.6.14}$$

(in principle defined only for $u \in D$) admits as well a uniformly bounded and strongly continuous extension $\overline{(Y_p * S)}(t)$ in $0 \leq t \leq T$. To see this we write equation (8.6.9) for $u(s) = S(s)u$ ($u \in D$), subtract $\lambda S(s)u$ from both sides and premultiply by $R(\lambda)$, obtaining

$$(Y_1 * Su)(t) = \int_0^t S(s)u \, ds$$

$$= \lambda \int_0^t R(\lambda)S(s)u \, ds - R(\lambda)S(t)u + R(\lambda)u, \tag{8.6.15}$$

which proves our claims about $Y_p * S$ in the event that $p = 1$. An obvious iteration argument takes care of the case $p \geq 1$.

The distribution $S \in \mathcal{D}_0'((E; X))$ promised in the statement of Theorem 8.6.2 is defined as follows: if φ is a test function with support in $t < T$ and $u \in E$, we set

$$S(\varphi)u = (-1)^{p+1} \int_{-\infty}^T \overline{(Y_{p+1} * S)}(t)u\varphi^{(p+1)}(t) \, dt. \tag{8.6.16}$$

If $u \in D$, we have

$$
\begin{aligned}
A(Y_{p+1} * S)(t)u &= (Y_{p+1} * AS)(t)u \\
&= (Y_{p+1} * S')(t)u \\
&= (Y_p * S)(t)u - Y_{p+1} \otimes u.
\end{aligned}
$$

An approximation argument then shows that

$$
\overline{(Y_{p+1} * S)(t)} E \subseteq D(A)
$$

with

$$
A \, \overline{(Y_{p+1} * S)(t)} E = \overline{(Y_p * S)(t)} u - Y_{p+1} \otimes I
$$

so that (8.6.16), used for arbitrary $T > 0$ (and p depending on T) defines a distribution in $\mathcal{D}'((E; X))$; that the definition is consistent for different values of T (and p) is a simple matter of integration by parts.

Finally, we verify both equations (8.6.12). To show the first we check that $(\delta' \otimes I - \delta \otimes A) * Su = \delta \otimes u$ for $u \in D$ and extend to $u \in E$ by denseness of D. The second equation reduces in an entirely similar fashion to $S * (\delta' \otimes u - \delta \otimes Au) = \delta \otimes u$ for $u \in D$, and this can be shown proving that $AS(t)u = S(t)Au$ for $u \in D$. This is done arguing as in the proof of (8.6.13) and we omit the details.

We prove now the converse. Assume that $\delta' \otimes I - \delta \otimes A \in \mathcal{D}_0((E; X))^{-1}$ and let $S \in \mathcal{D}_0((E; X))$ be the solution of (8.6.12). We begin by checking assumption (a). To show denseness of D, we consider S—as we may—as a distribution in $\mathcal{D}_0'((E; E))$ and define the *adjoint* distribution $S^* \in \mathcal{D}_0'((E^*; E^*))$ by

$$
S^*(\varphi) = S(\varphi)^* \quad (\varphi \in \mathcal{D}). \tag{8.6.17}
$$

Since $AS(\varphi) = S(\varphi)A$, it follows that $S(\varphi)A$ is a bounded operator (with domain $D(A)$) for each $\varphi \in \mathcal{D}$, thus the second equality (4.8) implies that $A^*S(\varphi)^* = (S(\varphi)A)^*$ is bounded. Accordingly $S(\varphi)^* \in (E^*; D(A^*))$, $D(A^*)$ endowed with the graph norm. Also, it follows that $S(\varphi)^*$ and $A^*S(\varphi)^*$ depend continuously on φ as $(E^*; E^*)$-valued functions, thus $S^* \in \mathcal{D}_0'((E^*; D(A^*))$. Taking adjoints in (8.6.12), we obtain the corresponding adjoint versions:

$$
(\delta' \otimes I^* - \delta \otimes A^*) * S^* = \delta \otimes I^*, \quad S^* * (\delta' \otimes I^* - \delta \otimes A^*) = \delta \otimes J^*,
$$

$$
\tag{8.6.18}
$$

where I^* (resp. J^*) is the identity operator in E^* (resp. in $D(A^*)$). We note, incidentally, that $D(A^*)$ may not be dense in E^*.

Let $\mathfrak{E}$ be the subspace of E consisting of all elements of the form $S(\varphi)u$, $\varphi \in \mathcal{D}((0, \infty))$, $u \in E$. Since $AS(\varphi)u = -S(\varphi')u$, it is clear that $\mathfrak{E} \subseteq D(A^\infty)$. Assume $\mathfrak{E}$ is not dense in E. Then there exists $u^* \neq 0$ in E^*

such that $\langle S^*(\varphi)u^*, u \rangle = \langle u^*, S(\varphi)u \rangle = 0$ for all $\varphi \in \mathcal{D}((0, \infty))$, $u \in E$, so that $S^*(\varphi)u^* = 0$ for all such φ. This means that $S^*u^* = 0$ in $t > 0$; since S^* itself has support in $t \geqslant 0$, it follows that $\operatorname{supp} S^*u^* = \{0\}$. Applying Corollary 8.1.7, we deduce that

$$S^*u^* = \delta \otimes u_0^* + \cdots + \delta^{(m)} \otimes u_m^*,$$

where $u_0^*, \ldots, u_m^* \in D(A^*)$ $(m \geqslant 0)$. Replacing this expression in the first equality (8.6.18), we deduce that

$$(\delta' \otimes I^* - \delta \otimes A^*) * S^*u^* = \delta' \otimes u_0^* + \cdots + \delta^{(m+1)} \otimes u_m^*$$

$$- \delta \otimes A^*u_0 - \cdots - \delta^{(m)} \otimes A^*u_m^* = \delta \otimes u^*.$$

Equating coefficients we obtain $A^*u_0 = -u^*$, $A^*u_1^* = u_0^*, \ldots, A^*u_m^* = u_{m-1}^*$, $u_m^* = 0$, which implies that $u_{m-1}^* = \cdots = u_1^* = u^* = 0$, whence a contradiction results. This proves that $\mathfrak{C}$, thus D, is dense in E.

To construct the solutions postulated in (a) we use Theorem 8.4.8. Let $\Lambda = \Lambda(\alpha, \beta)$ be the logarithmic region where $R(\lambda)$ exists and satisfies (8.4.18), Γ the boundary of Λ oriented clockwise with respect to Λ. For $p > m + 1$, we define

$$u_p(t) = \sum_{j=0}^{p-1} \frac{t^j}{j!} A^j u_0 + \frac{1}{2\pi i} \int_\Gamma \lambda^{-p} e^{\lambda t} R(\lambda) A^p u_0 \qquad (8.6.19)$$

in $0 \leqslant t < t_p = (p - m - 2)/\alpha$. (See Example 2.5.1). We check that $u_p(\cdot)$ is a solution of (8.6.9) in $0 \leqslant t < t_p$ with $u(0) = u_0$. To prove (as we must) that $u_p(t) = u_{p'}(t)$ if $t < t_p$ when $p' > p$, it is obviously enough to show that any solution $u(\cdot)$ of (8.6.9) in a finite interval $0 \leqslant t \leqslant T$ with $u(0) = 0$ must vanish identically. This is done observing that the function $u(\cdot)$ satisfies $(\delta' \otimes I - \delta \otimes A) * u(\hat{t}) = \delta(\hat{t}) \otimes u(0)$ and convolving both sides on the left by S; the formula

$$u(\hat{t}) = S(\hat{t})u(0) \qquad (8.6.20)$$

results.

We shall use (8.6.20) as well to prove (b). Let $T > 0$, p an integer, such that $T < t_p = (p - m - 2)/\alpha$, $u(\hat{t})$ an arbitrary solution of (8.6.9) in $0 \leqslant t \leqslant T$ with $u(0) \in D(A^p)$, $u_p(\hat{t})$ the solution provided by (8.6.19). By the uniqueness result (8.6.20), $u(\hat{t}) = u_p(\hat{t})$ and the estimate (8.6.11) can be instantly obtained from (8.6.19). We omit the remaining details.

We note the following particular case of Theorem 8.6.1, which is proved using essentially the same arguments.

8.6.3   Theorem.   Let A be a closed, densely defined operator such that $\delta' \otimes I - \delta \otimes A \in \mathring{S}_\omega'((E; X))^{-1}$. Then the Cauchy problem for (8.6.9) is mildly well posed in $t \geqslant 0$. Moreover, for each $\omega' > \omega$ there exists a constant C and an *integer p such that any solution of (8.6.9) in $t \geqslant 0$ with $u(0) \in D(A^p)$ satisfies*

$$\|u(t)\| \leqslant Ce^{\omega't}(\|u\| + \|A^p u\|) \quad (t \geqslant 0). \tag{8.6.21}$$

In the following result another relation between the Cauchy problem in the sense of distributions and the ordinary Cauchy problem is established. Although the latter will be properly posed, we need to use here linear topological spaces.

Let A be an arbitrary operator in E such that $\rho(A) \neq \varnothing$. We introduce a translation invariant metric $d(\cdot, \cdot)$ in $D(A^\infty)$ as follows:

$$d(u, v) = \sum_{n=0}^{\infty} 2^{-n} \|A^n u - A^n v\| (1 + \|A^n u - A^n v\|)^{-1}.$$

It can be easily shown that existence of $R(\lambda)$ implies that all powers A^n are closed, thus d makes $D(A^\infty)$ a Fréchet space (Dunford-Schwartz [1957: 1]). We note that a sequence $\{u_m\}$ in $D(A^\infty)$ converges to $u \in D(A^\infty)$ if and only if $A^n u_m \to A^n u$ as $m \to \infty$ for $n = 0, 1, \ldots$. Note also that the operator A is continuous in $D(A^\infty)$: in fact, $d(Au, Av) \leqslant 2d(u, v)$.

8.6.4   Theorem.   *Let A be a closed, densely defined operator in E such that $\delta' \otimes I - \delta \otimes A \in \mathcal{D}_0'((E; D(A))^{-1}$. Then the Cauchy problem for*

$$u'(t) = \hat{A}u(t) \quad (t \geqslant 0) \tag{8.6.22}$$

($\hat{A}$ the restriction of A to $D(A^\infty)$) is well posed; more precisely, for every $u \in D(A^\infty)$ there exists a solution of (8.6.22) and solutions depend continuously on their initial data (equivalently, $\hat{A}$ generates a uniformly continuous semigroup in $D(A^\infty)$). Conversely, let A be an operator with $\rho(A) \neq \varnothing$ and $D(A^\infty)$ dense in E, and assume the Cauchy problem for (8.6.22) is well posed with every $u \in D(A^\infty)$ as initial datum of a solution. Then $\delta' \otimes I - \delta \otimes A \in \mathcal{D}_0'((E; D(A))^{-1}$.

This result is little more than a restatement of Theorem 8.6.2. Assume that $\delta' \otimes I - \delta \otimes A \in \mathcal{D}_0'((E; X))$, and let S be the solution of (8.6.12). Let $u_0 \in D(A^\infty)$ and let $u(\hat{t})$ be the solution of (8.6.9) constructed "by pieces" using formula (8.6.19). An examination of this formula shows that $u(t) \in D(A^\infty)$ for $t \geqslant 0$; moreover, given an arbitrary integer q we can take p large enough in (8.6.19) and show that $\lim A^j(h^{-1}(u(t+h) - u(t))$ exists and $A^j u'(t)$ is continuous in $0 \leqslant t \leqslant T$ for $j = 0, 1, \ldots, q$. This covers the existence statement (a) in the definition of properly posed Cauchy problems. The continuous dependence assumption (b) follows essentially as in the proof of Theorem 8.6.2.

We must finally show that the propagator $\hat{S}(t)$ of (8.6.22) is continuous in the topology of $(D(A^\infty))$ (that is, that $\hat{S}(t)u \to u$ in $D(A^\infty)$ uniformly for u in bounded sets of $D(A^\infty)$). This is a rather obvious consequence of the equality

$$\hat{S}(t)u - u = \int_0^t \hat{S}(s)\hat{A}u \, ds$$

and of the fact that a set $\mathcal{K}$ is bounded in the space $D(A^\infty)$ when and only when $\{A^n u; u \in \mathcal{K}\}$ is bounded in n for each $n = 0, 1, 2, \ldots$.

To prove the converse, let $\hat{S}(t)$ be the semigroup generated by $\hat{A}$ in $D(A^\infty)$. It follows from the definition of the topology of $D(A^\infty)$ that, given $T > 0$, there exists an integer p and a constant C such that

$$\|\hat{S}(t)u\| \leqslant C\|(\lambda I - \hat{A})^p u\| \quad (u \in D(A^\infty), 0 \leqslant t \leqslant T). \qquad (8.6.23)$$

We check easily that $\hat{A}$ commutes with $\hat{S}(t)$; thus if $\lambda \in \rho(A)$, $R(\lambda)$ commutes with $\hat{A}$, and we deduce from (8.6.23) that $R(\lambda)^p \hat{S}(t)$—a priori only defined in $D(A^\infty)$—possesses an extension $\overline{R(\lambda)^p \hat{S}(t)}$, uniformly bounded and strongly continuous in $0 \leqslant t \leqslant T$. After this is established the proof proceeds in essentially the same way as that of the corresponding portion of Theorem 8.6.2 (see the comments following (8.6.13)). We omit the details.

8.6.5   Remark.   In some situations where A is a differential operator (such as the one examined in Example 8.6.1) the norm of the space $D(A^p)$ is equivalent, say, to a Sobolev norm involving norms of partial derivatives of order less than or equal to a multiple of p. In these instances, the mildly well-posed Cauchy problem is related to the Cauchy problem in the sense of Hadamard (see Section 1.7 for a thorough discussion). However, the equivalence is not complete in that the use of Banach spaces imposes global rather than local convergence (see again Section 1.7). In other cases, the norm of $D(A^p)$ may not be equivalent to a Sobolev norm.

We close this section with an example where the framework of mildly well-posed Cauchy problems proves too restrictive to handle the initial value problem for a partial differential equation.

8.6.6   Example (Beals (1972: 2]).   We consider the following lower order perturbation of the system (8.6.1):

$$D_t u = D_x u + D_x v, \quad D_t v = D_x v + u \quad (x \in \mathbb{R}, t \geqslant 0) \qquad (8.6.24)$$

with initial conditions (8.6.2) and write it as an abstract differential equation

$$\mathfrak{u}'(t) = \mathfrak{B} \mathfrak{u}(t) \qquad (8.6.25)$$

in $L^2(\mathbb{R})^2$, with

$$\mathfrak{B} = \begin{bmatrix} D_x & D_x \\ 1 & D_x \end{bmatrix}.$$

The domain of $\mathfrak{B}$ is again $H^1(\mathbb{R})^2$. The Fourier analysis of (8.6.25) is essentially the same as that of (8.6.3): this time the equivalence is with

$$\tilde{\mathfrak{u}}'(t) = \tilde{\mathfrak{B}} \tilde{\mathfrak{u}}(t), \qquad (8.6.26)$$

where $\tilde{\mathfrak{B}}$ is the operator of multiplication by

$$\tilde{\mathfrak{B}}(\sigma) = \begin{bmatrix} -i\sigma & -i\sigma \\ 1 & -i\sigma \end{bmatrix}$$

with domain $D(\tilde{\mathfrak{B}})$ consisting of all $\tilde{u} = (\tilde{u}, \tilde{v})$ with $\sigma\tilde{u}, \sigma\tilde{v} \in L^2(\mathbb{R}_\sigma)$. We have

$$\exp(t\tilde{\mathfrak{B}}(\sigma)) = e^{-it\sigma} \begin{bmatrix} \cosh(-i\sigma)^{1/2}t & (-i\sigma)^{1/2}\sinh(-i\sigma)^{1/2}t \\ (-i\sigma)^{-1/2}\sinh(-i\sigma)^{1/2}t & \cosh(-i\sigma)^{1/2}t \end{bmatrix}.$$

If $\tilde{u} = (\tilde{u}, \tilde{v})$ has, say, compact support in $L^2(\mathbb{R}_\sigma)^2$, then $\mathfrak{S}(t)u = \exp(t\tilde{\mathfrak{B}}(\sigma))u$ is a solution of (8.6.26). However, $\mathfrak{S}(t)$ is an unbounded operator if $t > 0$. Unlike in Example 8.6.1, the unboundedness of $\mathfrak{S}(t)$ cannot be remedied by measuring the initial condition u in the norm of some $D(\mathfrak{B}^p)$. However, a sort of extension of the notion of mildly well-posed problem works here: in fact, it is obvious that if $\alpha > 1$, solutions of (8.6.25) satisfy an inequality of the form

$$\|u(t)\| \leqslant C \sum_{n=0}^\infty \frac{K^n}{(n!)^\alpha} \|\mathfrak{B}^n u(0)\|. \tag{8.6.27}$$

This motivates the introduction of the *abstract Gevrey spaces* $G(\alpha; A)$ associated with an operator A in an arbitrary Banach space E. These are subspaces of $D(A^\infty)$ consisting of all u such that $\Sigma(n!)^{-\alpha}K^n\|A^n u\| < \infty$ for some $K > 0$ (equivalently, $\sup(n!)^{-\alpha}K^n\|A^n u\| < \infty$ for some $K > 0$). Theorem 8.6.2 has a generalization to this situation, the notion of Cauchy problem well posed in the sense of vector-valued distributions being extended to the realm of *ultradistributions*, linear continuous operators in spaces of test functions restricted by growth conditions on the derivatives. See Beals [loc. cit.], Chazarain [1971: 1], and Cioränescu [1977: 1] (where more general types of ultradistributions are considered).

8.6.7 Example. We have seen in Example 1.4.1 that the Cauchy problem for the Schrödinger equation

$$u'(t) = \mathcal{Q}u(t) \tag{8.6.28}$$

is not well posed in $L^r(\mathbb{R}^m)$ if $r \neq 2$. However, the Cauchy problem for (8.6.28) is mildly well posed in $L^r(\mathbb{R}^m)$ for $1 \leqslant r < \infty$; in fact, an estimate of the type of (8.6.21) holds in the L^r norm (with $\omega = 0$) if p is large enough. This is a consequence of the fact that the operator

$$Bu = \mathcal{F}^{-1}\left((1 + |\sigma|^p)^{-1} e^{i\kappa|\sigma|^2}\mathcal{F}u\right)$$

is bounded in $L^r(R^m)$ $(1 \leqslant r < \infty)$ independently of κ. The required properties of B follow from Theorem 8.1.

An entirely similar analysis applies to the symmetric hyperbolic equation (1.6.21).

8.7. MISCELLANEOUS COMMENTS

The theory of distributions with values in a linear topological space was developed (and nearly preempted) by L. Schwartz in [1957: 1] and [1958: 1].

Almost all the definitions and results here can be found there in enormously general versions. Theorem 8.3.1 is due to Lions [1960: 2], who also introduced the notion of the Cauchy problem in the sense of distributions for the equation

$$U' - AU = F. \tag{8.7.1}$$

The generalization to the convolution equation (8.4.1) is due to the author [1976: 1]. In [1960: 2] Lions defines a *distribution semigroup* as an operator-valued distribution $S \in \mathcal{D}'((0, \infty); (E; E))$ satisfying the *generalized semigroup equation*

$$S(\varphi * \psi) = S(\varphi)S(\psi) \tag{8.7.2}$$

for arbitrary test functions $\varphi, \psi \in \mathcal{D}'((0, \infty))$. This is the analogue of the second equation (2.3.1); semigroup distributions are thus natural generalizations of the semigroups in Section 2.5(a), where no specific behavior near $t = 0$ is postulated. If certain assumptions regarding that behavior are introduced the distribution semigroup in question is called *regular* by Lions and an *infinitesimal generator* A can be defined (roughly, as the closure of $\lim S(\varphi_n')$, where $\{\varphi_n\} \subset \mathcal{D}'((0, \infty))$ and $\varphi_n \to \delta$). The distribution S can be shown to belong to $\mathcal{D}'((E; D(A))$ and to satisfy

$$S' - AS = \delta \otimes I, \quad S' - SA = \delta \otimes J. \tag{8.7.3}$$

Conversely, any solution of (8.7.3) is a regular distribution semigroup with infinitesimal generator A, thus establishing a relation between the abstract differential equation (8.7.1) and distribution semigroups very similar to that between the equation (2.1.1) and strongly continuous semigroups. The central problem of the theory is of course, the identification of the generators A or equivalently of those operators A such that $\delta' \otimes I - \delta \otimes A$ possesses a convolution inverse. This was done by Lions [1960: 2] only in the case where S grows exponentially at infinity. At the same time, Foias [1960: 1] coinsidered semigroup distributions of *normal* operators in Hilbert space, identified in general their infinitesimal generators and gave growth conditions on S based on the location of the spectrum of A in subregions of logarithmic regions. The general case was handled by Chazarain [1968: 1], [1971: 1], who also extended Lions' theory to the case of *ultradistribution semigroups* (where S belongs to a space of vector valued Roumieu-Gevrey ultradistributions) in [1968: 2], [1971: 1]; the corresponding results for Beurling ultradistributions and for Sato hyperfunctions are respectively due to Emami-Rad [1973: 1] and to Ouchi [1971: 1]. Semigroup distributions that are smooth in $t > 0$ have been studied by Da Prato-Mosco [1965: 1], [1965: 2] in the case where S coincides with a vector-valued analytic function in a sector containing the half axis $t > 0$ and by Barbu [1968: 2] [1969: 1] when S is infinitely differentiable in $t > 0$; we note that his result generalizes the result of Pazy (Theorem 4.1.3) commented on in Section 4.11. Barbu also gives necessary and sufficient conditions on A in order that

S possess different "degrees of smoothness" for $t > 0$ related to a priori bounds on its derivatives; in particular, a characterization is obtained for those A that make S real analytic in $t > 0$. These theorems are new even in the semigroup case.

Among other results we mention an "exponential formula" of the type of the Yosida approximation due to the author [1970: 1] (for additional material on this score see Ciorănescu [1972: 1], who provided as well a perturbation result in [1973: 2]). The connection between distribution semigroups and ordinary semigroups in the Fréchet space $D(A^\infty)$ (Theorem 8.6.3) was established by Fujiwara [1966: 1] for distribution semigroups of exponential growth and by Ujishima [1972: 1] in the general case (see also Oharu [1973: 1] and Guillement-Lai [1975: 1]). The extension to ultradistribution semigroups is due to Ciorănescu [1977: 1]. An earlier and somewhat different (but essentially equivalent) treatment is due to Beals [1972: 1], [1972: 2] (see Example 8.6.6). This is an abstract version of a method due to Ohya [1964: 1] and Leray-Ohya [1964: 1] for the treatment of hyperbolic systems with multiple characteristics of which those in Examples 8.6.5 and 8.6.6 are particular cases. For additional results on distribution semigroups and on the equation (8.7.1) in the sense of distributions see Ciorănescu [1974: 1], Emami-Rad [1975: 1], Da Prato [1966: 3], Krabbe [1975: 1], Ujishima [1969: 1], [1970: 1], Malik [1971: 1], [1972: 1], [1975: 2], Larsson [1967: 1], Shirasai-Hirata [1964: 1], Mosco [1965: 1], [1967: 1], Yoshinaga [1963: 1], [1964: 1], [1965: 1], [1971: 1]. Several portions of the theory have been generalized to distribution semigroups with values operators in a linear topological space: we mention the results of Vuvunikjan [1971: 3] and [1972: 1] (see also the reviews in MR) where generation theorems are given using "resolvent sequences" in the style of T. Komura, Dembart, and Okikiolu (see Section 2.5(d)); moreover, extensions of the results of Barbu on smoothness of S can be found as well. See also Vuvunikjan-Ivanov [1974: 1].

Many of the results above on distribution semigroups do not involve in any essential way the difference between the equation (8.4.6) and its generalization (8.4.1); for instance, Chazarain's results on characterization of generators are in fact stated for a P involving combinations of (possibly fractional) derivatives. It was shown by Ciorănescu [1974: 2] that Chazarain's theorems extend to an arbitrary P having compact support, both in the distribution and ultradistribution case; in fact, Ciorănescu coinsiders also the case of several time variables. A somewhat different proof (Theorem 8.4.8) was given by the author [1976: 1], where the restriction that P have compact support is removed (see also Corollary 8.4.9). Theorem 8.5.1 is a generalization (due to the author) of Barbu's result mentioned above for infinitely differentiable distribution semigroups. The rest of Barbu's results on quasianalytic and analytic classes was also generalized by the author ([1980: 1]). Other results for the equation (8.4.6) (such as the Trotter-Kato theorem and the Lax equivalence theorem) have been extended to the general equation (8.4.1) (see the author [1982: 2]).

Hereditary Differential Equations. The systems described shortly after Theorem 8.3.1 do not in general satisfy Huygens' principle; the state of the system at a single time $s < t$ does not suffice to determine the state at time t (compare with Section 2.5(b)). That this type of situation is common in physical phenomena was well known more than a century ago, although its mathematical modeling had to wait for some time. Hereditary equations were introduced (and named) by Picard [1907: 1]. Their importance in physical, and especially biological phenomena was recognized by Volterra [1909: 1], [1928: 1], [1931: 1], who undertook a systematic study of what would be called now ordinary integro-differential equations. We quote from [1931: 1, p. 142]: "L'état d'un système biologique à un moment donné semble donc bien devoir dépendre des rencontres ayant eu lieu pendant une période plus on moins longue précédant ce moment; et dans les chapitres qu'on vient de développer, on a, somme toute, négligé la durée de cette periode. Il convient de tenir compte maintenant de l'influence du passé."

"On rencontre, *en physique*, dans l'étude de l'élasticité, du magnétisme, de l'électricité, bien des phénomènes analogues de retard, traînage ou histérésis. On peut dire que dans le monde inorganique il existe aussi une mémoire du passé, comme la mémoire du fil de torsion dont la déformation actuelle dépend des états antérieurs."

A few lines below, Volterra remarks: "Mais lorsqu'en physique l'hérédité entre en jeu, les équations différentielles et aux dérivées partielles ne peuvent pas suffire; sinon les données initiales détermineraient l'avenir. Pour faire jouer un rôle a la *suite continue* des états antérieurs (infinité de paramètres ayant la puissance du continu) il a fallu recourir à des équations intégrales et intégro-différentielles où figurent des intégrales sous lesquelles entrent les paramètres caractéristiques du système fonctions du temps pendant une période antérieure a l'instant considéré; on a même introduit des types plus généraux d'équations aux dérivées fonctionnelles."

After these investigations, interest in hereditary differential equations faded for almost two decades, except for work such as that of Minorski [1942: 1], where *delay differential equations* $x'(t) = f(t, x(t), x(t - h))$ are examined. Beginning in the late forties, an intense revival occurred, stimulated in part by control theory; the objects of study were *functional differential equations*, in general of the form

$$x'(t) = f(t, x_t), \qquad (8.7.4)$$

where for each t, f is a vector-valued functional defined in a suitable function space of vector-valued "histories" $x_t = \{x(s), -\infty < s \leqslant t\}$. Assuming the histories are continuous functions, the functional is continuous, linear and time independent, and the system has finite memory, one obtains the *linear autonomous* equation

$$x'(t) = \int_{-r}^{0} \mu(ds)x(t+s), \tag{8.7.5}$$

where μ is a matrix-valued measure of appropriate dimension. For an excellent treatment of functional differential equations, we refer the reader to the recent monograph of Hale [1977: 1], where extensive references and many examples can be found. We point out in passing that there exists an interesting and important relation between the equation (8.7.5) and the theory of the abstract Cauchy problem in Chapter 2, which can be roughly described as follows. Let $x(\cdot)$ be, say, a continuously differentiable solution of (8.7.5) with

$$x(t) = x_0(t) \quad (-r \leqslant t \leqslant 0). \tag{8.7.6}$$

Consider the function

$$t \to u(t) = \{x(t+s); -r \leqslant s \leqslant 0\}$$

defined in $t \geqslant 0$ with values in the "space of histories" $\mathcal{H}_r$ of all (vector-valued) continuous functions defined in $-r \leqslant t \leqslant 0$. Then $u(\cdot)$ satisfies the abstract differential equation

$$u'(t) = Au(t) \quad (t \geqslant 0) \tag{8.7.7}$$

in the space $E = \mathcal{H}_r$, where A is the differentiation operator

$$Ax(s) = x'(s) \tag{8.7.8}$$

with domain consisting of all continuously differentiable functions $x(\cdot)$ satisfying the "nonlocal boundary condition"

$$x'(0) - \int_{-r}^{0} \mu(ds)x(s)\, ds = 0. \tag{8.7.9}$$

Conversely, if $u(\cdot)$ is a $\mathcal{H}_s$-valued function satisfying (8.7.7), we must have $u(t)(s) = x(t+s)$ $(-r \leqslant s \leqslant 0)$ because of the definition of A; thus condition (8.7.9) guarantees the functional differential equation (8.7.5). This equivalence was used by Krasovskiĭ, Hale and Shimanov (see Hale [1977: 1, Ch. 7]) to derive many deep results for the linear autonomous equation, as well as for their nonlinear counterparts.

The study of functional differential equations in infinite-dimensional spaces is of very recent data. Some (but not all) of the literature below makes use of variants of the "reduction to differential equations" outlined above for the equation (8.7.5). See Artola [1967: 1], [1969: 1], [1969: 2], [1970: 1], Datko [1976: 1], [1977: 1], Dickerson-Gibson [1976: 1], Hlaváček [1971: 1], Korenevskiĭ-Fesčenko [1975: 1], R. K. Miller [1975: 1], Monari [1971: 1], Pachpatte [1976: 1], Slemrod [1976: 1], Travis-Webb [1974: 1], Webb [1974: 1], [1976: 1], Zamanov [1967: 1], [1968: 1], [1968: 3], [1969: 1], and Zaplitnaja [1973: 1].

The theory in this chapter formally contains the theory of functional differential equations of the form (8.7.5), but only in reference to the Cauchy problem in the sense of distributions. A theory of the Cauchy problem for (8.4.1) in the strong sense (as that in Chapter 2 for differential equations) seems to have been developed only in particular instances (some of which one references quoted above). In most of them $P = \delta' \otimes I - \delta \otimes A - P_1$, where P_1 is in some sense subordinate to A. Also, nonlinear equations are studied in many cases by application of the theory of nonlinear semigroups. Spaces of histories have been studied, among others, by Coleman-Mizel [1966: 1].

References

References

This section lists papers and monographs quoted in the text or closely related to the contents of the book. Additional references can be found in Amerio-Prouse [1971: 1], Butzer-Berens [1967: 1], Carroll [1969: 2] and [1979: 2], Carroll-Showalter [1976: 1], Chernoff [1974: 1], Daleckiĭ-Kreĭn [1970: 1], Hille-Phillips [1957: 1], Kreĭn [1967: 1], Tanabe [1979: 1], and Zaidman [1979: 1].

L. Š. Abdulkerimov
[1966: 1]   Solvability of the Cauchy problem in a Banach space with a generating operator of a semigroup of class (A). (Russian.) *Izv. Akad. Nauk Azerbaĭdžan SSR Ser. Fiz.-Tehn. Mat. Nauk* (1966), no. 1, 48–55. (MR 33, #6081).
[1969: 1]   The solvability of the Cauchy problem for differential equations with constant operators in a Banach space. (Russian.) *Izv. Akad. Nauk Azerbaĭdžan SSR Ser. Fiz.-Tehn. Mat. Nauk* (1969), no. 3, 57–60. (MR 43, #5188).

L. Š. Abdulkerimov and S. Ja. Jakubov
[1966: 1]   Investigation of the Cauchy problem for quasilinear differential equations of the parabolic type in a Banach space. (Russian.) *Izv. Akad. Nauk Azerbaĭdžan SSR Ser. Fiz.-Tehn. Mat. Nauk* (1966), no. 2, 35–42. (MR 34, #1734).

[1969: 1]   The solvability of the Cauchy problem for higher-order nonlinear parabolic equations in a Banach space. (Russian.) *Izv. Akad. Nauk Azerbaĭdžan SSR Ser. Fiz.-Tehn. Mat. Nauk* (1969), no. 3, 35–40. (MR 43, #749).

M. A. Ackoglu and J. Cunsolo
[1970: 1]   An ergodic theorem for semigroups. *Proc. Amer. Math. Soc.* **24** (1970), 161–170. (MR 40, #1578).

V. M. Adamjan and D. Z. Arov
[1965: 1]   On scattering operators and contraction semigroups in Hilbert spaces. (Russian.) *Dokl. Akad. Nauk SSSR* **165** (1965), 9–12. (MR 32, #6240).

R. A. Adams
[1975: 1]   *Sobolev Spaces.* Academic Press, New York, 1975.

G. N. Agaev
[1969: 1]   The Cauchy problem for nonlinear parabolic equations. (Russian.) *Studia Math.* **32** (1969), 153–179. (MR 40, #580).

S. Agmon
[1962: 1]   Unique continuation and lower bounds for solutions of abstract differential equations. *Proc. Internat. Congr. Mathematicians (Stockholm, 1962)*, 301–305; Inst. Mittag-Leffler, Djursholm, 1953. (MR 31, #469).
[1962: 2]   On the eigenfunctions and the eigenvalues of general elliptic boundary value problems. *Comm. Pure Appl. Math.* **15** (1962), 119–147. (MR 26 # 5288).
[1965: 1]   *Lectures on Elliptic Boundary Value Problems.* Van Nostrand Mathematical Studies No. 2, Van Nostrand, Princeton, 1965.
[1966: 1]   *Unicité et Convexité dans les Problèmes Différentiels.* Séminaire de Mathématiques Supérieures No. 13 (Été, 1965). Les Presses de l'Université de Montréal, Montréal, Quebec, 1966.
[1975: 1]   Spectral properties of Schrödinger operators and scattering theory. *Ann. Scuola Norm. Sup. Pisa Cl. Sci.* (4), 2 (1975), 151–218. (MR 53, #1053).

S. Agmon, A. Douglis and L. Nirenberg
[1959: 1]   Estimates near the boundary for solutions of elliptic partial differential equations satisfying general boundary conditions I. *Comm. Pure Appl. Math.* **12** (1959), 623–727. (MR 23, #A2610).
[1964: 1]   Estimates near the boundary for solutions of elliptic partial differential equations satisfying general boundary conditions II. *Comm. Pure Appl. Math.* **17** (1964), 35–92. (MR 28, #5252).

S. Agmon and L. Nirenberg
[1963: 1]   Properties of solutions of ordinary differential equations in Banach space. *Comm. Pure Appl. Math.* **16** (1963), 121–239. (MR 27, #5142).
[1967: 1]   Lower bounds and uniqueness theorems for solutions of differential equations in a Hilbert space. *Comm. Pure Appl. Math.* **20** (1967), 207–229. (MR 34, #4665).

512   References

D. A. Ahundov
[1972: 1]   Wave operators for an abstract quasilinear second-order hyperbolic
             equation. (Russian.) *Izv. Akad. Nauk. Azerbaĭdžan. SSR Ser. Fiz.-Tehn.
             Mat. Nauk* (1972), no. 3, 63–68. (MR 50, #8114).

D. A. Ahundov and S. Ja. Jakubov
[1969: 1]   The uniformly correct Cauchy problem for abstract first-order quasi-
             linear differential equations in a Banach space. (Russian.) *Akad. Nauk
             Azerbaĭdžan SSR Dokl.* **25** (1969), 3–7. (MR 42, #897).
[1970: 1]   A scattering problem for a first-order quasilinear equation and its
             applications (Russian.) *Izv. Akad. Nauk Azerbaĭdžan SSR Ser. Fiz.-
             Tehn. Mat. Nauk* (1970), no. 3, 23–30. (MR 44, #3040).

O. Akinyele
[1976: 1]   Stability of motions of non-linear abstract Cauchy problems. *Boll Un.
             Mat. Ital.* A(5) **13** (1976), 322–331. (MR 55, #794).

Ja. A. Al'ber
[1968: 1]   A continuous regularization of linear operator equations in Hilbert
             space. (Russian.) *Mat. Zametki* **4** (1968), 503–509. (MR 39, #590).

R. G. Aliev
[1974: 1]   Differential equations in a Banach space whose solutions decrease
             faster than an exponential. (Russian, English summary.) *Vestnik
             Moskov. Univ. Ser. I Mat. Meh.* **29** (1974), 3–7. (MR 51, #10784).
[1975: 1]   Almost periodic solutions of linear differential equations in a Banach
             space. (Russian.) *Dokl. Akad. Nauk Azerbaĭdžan. SSR Ser. Fiz.-Tehn.
             Mat. Nauk* **25** (1975), No. 2, 29–34. (MR 52, #11233).
[1976: 1]   New integral inequalities, and their applications. (Russian, English
             summary.) *Izv. Akad. Nauk Azerbaĭdžan. SSSR Ser. Fiz.-Tehn. Mat.
             Nauk* **2** (1976), 133–139. (MR 55, #6236).

L. Amerio
[1960: 1]   Sulle equazioni differenziali quasi-periodiche astratte. *Ricerche Mat.* **9**
             (1960), 255–274. (MR 24, #A896).
[1961: 1]   Sulle equazioni lineari quasi-periodiche negli spazi Hilbertiani. *Atti.
             Accad. Naz. Lincei Rend. Cl. Sci. Fis. Mat. Nat.* (8) **31** (1961),
             110–117. (MR 25, #3389).
[1961: 2]   Sulle equazioni lineari quasi-periodiche negli spazi Hilbertiani, II. *Atti.
             Accad. Naz. Lincei Rend. Cl. Sci. Fis. Mat. Nat.* (8) **31** (1961),
             197–205. (MR 26, #2879).
[1963: 1]   Soluzioni quasi-periodiche di equazioni quasi-periodiche negli spazi
             hilbertiani. *Ann. Mat. Pura Appl.* (4) **61** (1963), 259–277. (MR 27,
             #6160).
[1963: 2]   Su un teorema di minimax per le equazioni differenziali astratte. *Atti
             Accad. Naz. Lincei Rend. Cl. Sci. Fis. Mat. Nat.* (8) **35** (1963),
             409–416. (MR 29, #5126).
[1965: 1]   Abstract almost-periodic functions and functional equations, *Boll. Un.
             Mat. Ital.* (3) **20** (1965), 287–334. (MR 32, #8039).

[1967: 1]   Almost-periodic solutions of the equation of Schrödinger type. *Atti Accad. Naz. Lincei Cl. Sci. Fis. Mat. Natur.* (8) **43** (1967), 147–153. (MR 39, #1784).

[1967: 2]   Almost-periodic solutions of the equation of Schrödinger type, II. *Atti Accad. Naz. Lincei Rend. Cl. Sci. Fis. Mat. Nat.* (8) **43** (1967), 265–270. (MR 38, #3566).

L. Amerio and G. Prouse
[1971: 1]   *Almost-Periodic Functions and Functional Equations.* Van Nostrand-Reinhold Co., New York, 1971.

N. Angelescu, G. Nenciu, and M. Bundaru
[1975: 1]   On the perturbation of Gibbs semigroups. *Comm. Math. Phys.* **42** (1975), 29–30. (MR 55, #6236).

V. N. Anosov and P. E. Sobolevskiĭ
[1971: 1]   Coercive solvability of a boundary value problem for a second order elliptic equation in a Banach space, I, II. (Russian.) *Differencial'nye Uravnenija* **7** (1971), 2030–2044; ibid. **7** (1971), 2191–2198. (MR 44, #7166).

[1972: 1]   The coercive solvability of parabolic equations. (Russian.) *Mat. Zametki* **11** (1972), 409–419. (MR 46, #558).

B. G. Ararkcjan
[1973: 1]   Behavior as $t \to \infty$ of the solutions of certain operator equations. (Russian.) *Izv. Akad. Nauk Armjan. SSR Ser. Mat.* **8** (1973), 226–234, 276. (MR 48, #9018).

A. Ardito and P. Ricciardi
[1974: 1]   An a priori bound for the Cauchy problem in Banach space *Atti Accad. Naz. Lincei Rend. Cl. Sci. Fis. Mat. Nat.* (8) **56** (1974), 473–481. (MR 53, #6043).

P. Arminjon
[1970: 1]   Existence et régularité de solutions élémentaires d'un opérateur différentiel abstrait. *Math. Ann.* **187** (1970), 117–126. (MR 41, #6012).

[1970: 2]   Un théorème de régularité pour une équation différentielle abstraite. *Canad. J. Math.* **22** (1970), 288–296. (MR 54, #7983).

[1973: 1]   Sur un probleme de existence de Lions pour une équation différentielle opérationnelle. *Aequationes Math.* **9** (1973), 91–98. (MR 47, #5392).

M. Artola
[1964: 1]   Sur les solutions de certaines équations différentielles opérationnelles du second ordre dans un espace de Hilbert. *C. R. Acad. Sci. Paris* **259** (1964), 4463–4465. (MR 30, #2210).

[1967: 1]   Équations paraboliques à retardement. *C. R. Acad. Sci. Paris* **264** (1967), A668–A671. (MR 35, #3297).

[1969: 1]   Sur les perturbations des équations d'évolution: Applications à des problèmes de retard. *Ann. Sci. École Norm. Sup.* (4) **2** (1969), 137–253. (MR 39, #5917).

[1969: 2]   Sur une équation d'évolution de premier ordre à argument retardé.
              C. R. Acad. Sci. Paris 268 (1969), A1540–A1543. (MR 39, #5972).
[1970: 1]   Dérivées intérmediaires dans les espaces de Hilbert pondérés; applica-
              tion au comportement a l'infini des solutions des ēquation d'évolution.
              Rend. Sem. Mat. Univ. Padova 43 (1970), 177–202. (MR 44, #3120).

S. Aširov
[1972: 1]   A remark on the Čapligin method for an abstract parabolic equation.
              (Russian.) Izv. Akad. Nauk Turkmen SSR Ser. Fiz.-Tehn. Him. Geol.
              Nauk 2 (1972) 3–7. (MR 45, #2291).

M. A. Atahodžaev
[1976: 1]   A fourth order linear differential equation in Hilbert space. (Russian.)
              Dokl. Akad. Nauk SSSR 230 (1976), 1265–1266. (MR 54, #5573).

V. A. Babalola
[1974: 1]   Semigroups of operators on locally convex spaces. Trans. Amer. Math.
              Soc. 199 (1974), 163–179. (MR 52, #4023).

D. G. Babbitt
[1969/70: 1]   Wiener integral representations for certain semigroups which have
              infinitesimal generators with matrix coefficients. J. Math. Mech. 19
              (1969/70), 1051–1067. (MR 42, #5097).

A. L. Badoev
[1968: 1]   Semigroups generated by linear systems of neutral type. (Russian.)
              Sakharth. SSSR Mecn. Akad. Moambe 51 (1968), 15–18. (MR 41,
              #9047).

C. Baiocchi
[1965: 1]   Regolarità e unicità della soluzione di una equazione differenziale
              astratta. Rend. Sem. Mat. Univ. Padova 35 (1965), 380–417. (MR 33,
              #446).
[1966: 1]   Teoremi di esistenza e regolarità per certe classi di equazioni differen-
              ziali astratte. Ann. Mat. Pura Appl. (4) 72 (1966), 395–417. (MR 34,
              #7927).
[1966: 2]   Un criterio di risolubilità di equazioni lineari tra spazi di Banach. Ist.
              Lombardo Accad. Sci. Lett. Rend. A100 (1966) 936–950. (MR 35,
              #4745).
[1967: 1]   Sulle equazioni differenziali astratte lineari del primo e del secondo
              ordine negli spazi di Hilbert. Ann. Mat. Pura Appl. (4) 76 (1967),
              233–304. (MR 36, #6745).
[1967: 2]   Soluzioni ordinarie e generalizzate del problema di Cauchy per
              equazioni differenziali astratte lineari del secondo ordine in spazi di
              Hilbert. Ricerche Mat. 16 (1967), 27–95. (MR 37, #1819).
[1970: 1]   Teoremi di regolarità per le soluzioni di un'equazione differenziale
              astratta. Symposia Mathematica, Vol. VI: (Convegno sulle Problemi di
              Evoluzione, INDAM, Roma, Maggio 1970), 269–323. Academic Press,
              London, 1971. (MR 49, #9652).

C. Baiocchi and M. S. Baouendi
[1977: 1]   Singular evolution equations. *J. Functional Analysis* **25** (1977), no. 2, 103–120. (MR 55, #10850).

A. B. Bakusinskiĭ
[1972: 1]   The solution by difference methods of an ill-posed Cauchy problem for a second order abstract differential equation. (Russian.) *Differencial'nye Uravnenija* **8** (1972), 881–890. (MR 46, #5782).
[1972: 2]   Difference methods of solving ill-posed Cauchy problems for evolution equations in a complex Banach space. (Russian.) *Differencial'nye Uravnenija* **8** (1972), 1661–1668, 1716. (MR 47, #3784).

M. Balabane
[1976: 1]   Puissances fractionnaires d'un opérateur générateur d'un semi-groupe distribution régulier. (English summary.) *Ann. Inst. Fourier* (Grenoble) **26** (1976), 157–203. (MR 53, #6353).

M. K. Balaev
[1976: 1]   Correctness of the Cauchy problem for arbitrary order operator-differential equations of parabolic type. (Russian.) *Izv. Akad. Nauk Azerbaĭdžan. SSR Ser. Fiz.-Tehn. Mat. Nauk* (1976), no. 5, 25–31. (MR 56, #3424).

A. V. Balakrishnan
[1958: 1]   Abstract Cauchy problems of the elliptic type. *Bull. Amer. Math. Soc.* **64** (1958), 290–291. (MR 21, #4294).
[1959: 1]   An operational calculus for infinitesimal generators of semigroups. *Trans. Amer. Math. Soc.* **91** (1959), 330–353. (MR 21, #5904).
[1960: 1]   Fractional powers of closed operators and the semigroups generated by them. *Pacific J. Math.* **10** (1960), 419–437. (MR 22, #5899).

J. M. Ball
[1977: 1]   Strongly continuous semigroups, weak solutions and the variation of constants formula. *Proc. Amer. Math. Soc.* **63** (1977), 372–373. (MR 56, #1128).

S. Banach
[1932: 1]   *Théorie des Opérations Linéaires.* Monografje Matematyczne, Warsaw, 1932.

M. S. Baouendi and P. Grisvard
[1968: 1]   Sur une équation d'évolution changeant de type. *J. Functional Analysis* **2** (1968), 352—367. (MR 40, #6034).

P. Baras, J.-C. Hassan, and L. Veron
[1977: 1]   Compacité de l'opérateur définissant la solution d'une équation d'évolution non homogéne. (English summary.) *C. R. Acad. Sci. Paris* **284** (1977), A799–A802. (MR 55, #3869).

V. Barbu

[1968: 1]   On regularity of weak solutions of abstract differential equations in
            Hilbert spaces. *Atti. Accad. Naz. Lincei Rend. Cl. Sci. Fis. Mat. Natur.*
            (8) **45** (1968), 129–134. (MR 41, #5749).

[1968: 2]   Les semi-groupes distributions différentiables. *C. R. Acad. Sci. Paris*
            267 (1968), A875–A878. (MR 39, #4703).

[1969: 1]   Differentiable distribution semi-groups. *Ann. Scuola Norm. Sup. Pisa*
            (3) **23** (1969), 413–419. (MR 41, #2456).

[1974: 1]   *Semigrupuri de contractii neliniare in spatii Banach.* (Romanian.) Editura
            Academiei Republicii Socialiste Romania, Bucharest, 1974. English
            translation: [1976: 1] *Nonlinear semigroups and differential equations in
            Banach spaces.* Editura Academiei Republicii Socialiste Romania,
            Bucharest; Noordhoff International Publishing, Leiden, 1976.

J. Barros-Neto

[1966: 1]   Remarks on non-existence of solutions for an abstract differential
            equation. *An. Acad. Sci. Brasil Ci.* **38** (1966), 1–4. (MR 34, #1647).

J. Bartak

[1976: 1]   Lyapunov stability at constantly acting disturbances of an abstract
            differential equation of the second order. *Czechoslovak Math. J.* **26**
            (1976), 411–437. (MR 54, #3109).

H. Baumgärtel

[1964: 1]   Einige Bemerkungen zur Differentialgleichung $X'(t) = P(t)X(t)$ für
            Operatorfunktionen. *Abh. Deutsch Akad. Wiss. Berlin Kl. Math. Phys.
            Tech.* (1964), no. 5, 19 pp. (MR 30, #319).

[1964: 2]   Einige Bemerkungen zur Differentialgleichung $X'(t) = P(t)X(t)$ für
            Operatorfunktionen. *Wiss. Z. Humboldt-Univ. Berlin Math.-Natur.
            Reihe* **13** (1964), 181. (MR 29, #6140).

R. Beals

[1972: 1]   Semigroups and abstract Gevrey spaces. *J. Functional Analysis* **10**
            (1972), 300–308. (MR 50, #14355).

[1972: 2]   On the abstract Cauchy problem. *J. Functional Analysis* **10** (1972),
            281–299. (MR 51, #8859).

M. Becker

[1976: 1]   Linear approximation processes in locally convex spaces I. Semigroups
            of operators and saturation. *Aequationes Math.* **14** (1976), 73–81. (MR
            53, #8756).

[1976: 2]   *Über den Satz von Trotter mit Anwendungen auf die Approx-
            imationstheorie.* Forschungbericht des Landes Nordhein-Westphalen
            No. 2577, Westdeutscher Verlag, Munich, 1976.

M. Becker, P. L. Butzer, and R. J. Nessel

[1976/77: 1]   Saturation for Favard operators in weighted function spaces. *Studia
               Math.* **59** (1976/77), 73–81. (MR 55, #10927).

M. Becker and R. J. Nessel
[1975: 1]   Iteration von Operatoren und Saturation in lokalkonvexen Raümen.
              Forschungbericht des Landes Nordhein-Westphalen No. 2470, Westde-
              utscher Verlag, Opladen (1975), 27–49.

A. Belleni-Morante and C. Vitocolonna
[1974: 1]   Discretization of the time variable in evolution equations. *Inst. Math.
              Appl.* **14** (1974), 105–112. (MR 49, #11312).

R. Bellman
[1970: 1]   *Introduction to Matrix Analysis.* 2nd ed., McGraw-Hill, New York,
              1970.

Ju. Ja. Belov
[1976: 1]   Some systems of linear differential equations with a small parameter.
              (Russian.) *Dokl. Akad. Nauk SSSR* **231** (1976), 1037–1040. (MR 55,
              #3455).
[1976: 2]   The approximation of a certain linear system of second order differen-
              tial equations. (Russian.) *Mat. Zametki* **20** (1976), no. 5, 693–702. (MR
              56, #6039).

A. G. Belyĭ and Yu. A. Semenov
[1975: 1]   Kato's inequality and semigroup product-formulas. (Russian.) *Func-
              tional Anal. i Prilozen.* **9** (1975), 59–60. (MR 52, #15100).

H. Berens
[1964: 1]   Approximationssätze für Halbgruppenoperatoren in intermediären
              Räumen. *Schr. Math. Inst. Univ. Münster* no. 32 (1964). (MR 30,
              #2351).
[1965: 1]   Equivalent representations for the infinitesimal generator of higher
              orders in semi-group theory. *Nederl Akad. Wetensch. Proc. Ser. A
              68 = Indag. Math.* **27** (1965), 497–512. (MR 35, #2173).

H. Berens and P. L. Butzer
[1964: 1]   Approximation theorems for semi-group operators in intermediate
              spaces. *Bull. Amer. Math. Soc.* **70** (1964), 689–692. (MR 29, #5110).
[1966: 1]   Über die Stetigkeit von Halbgruppen von Operatoren in Intermediären
              Raumen. *Math. Ann.* **163** (1966), 204–211. (MR 33, #6419).

H. Berens, P. L. Butzer, and U. Westphal
[1968: 1]   Representation of fractional powers of infinitesimal generators of
              semigroups. *Bull. Amer. Math. Soc.* **74** (1968), 191–196. (MR 36,
              #3163).

H. Berens and U. Westphal
[1968: 1]   A Cauchy problem for a generalized wave equation. *Acta Sci. Mat.
              Szeged* **29** (1968), 93–106. (MR 39, #819).

C. Berg
[1974: 1]   On the potential operators associated with a semigroup. *Studia Math.*
              51 (1974), 111–113. (MR 50, #3000).

[1975: 1]   Sur les semigroupes de convolution. *Lecture Notes in Mathematics* **404**
             (1975), 1–26, Springer, Berlin, 1975. (MR 52, #778).

C. A. Berger and L. A. Coburn
[1970: 1]   One parameter semigroups of isometries. *Bull. Amer. Math. Soc.* **76**
             (1970), 1125–1129. (MR 42, #893).

S. Bergman and M. Schiffer
[1953: 1]   *Kernel Functions and Elliptic Differential Equations in Mathematical
             Physics.* Academic Press, New York, 1953.

E. Berkson
[1966: 1]   Semi-groups of scalar type operators and a theorem of Stone. *Illinois J.
             Math.* **10** (1966), 345–352. (MR 33, #583).

E. Berkson, R. J. Fleming and J. Jamison
[1976: 1]   Groups of isometries on certain ideals of Hilbert space operators.
             Math. Ann. **220** (1976), 151–156. (MR 53, #14211).

E. Berkson, R. Kaufman, and H. Porta
[1974: 1]   Möbius transformations of the disc and one-parameter groups of
             isometries of H^p. *Trans. Amer. Math. Soc.* **199** (1974), 223–239. (MR
             50, #14365).

E. Berkson and H. Porta
[1974: 1]   Hermitian operator and one-parameter groups of isometries in Hardy
             spaces. *Trans. Amer. Math. Soc.* **185** (1974), 331–344. (MR 49, #3597).
[1976: 1]   One-parameter groups of isometries on Hardy spaces of the torus.
             Trans. Amer. Math. Soc. **220** (1976), 373–391. (MR 54, #590).
[1977: 1]   One-parameter groups of isometries on Hardy spaces of the
             torus: spectral theory. *Trans. Amer. Math. Soc.* **227** (1977), 357–370.
             (MR 56, #9322).

M. L. Bernardi
[1974: 1]   Sulla regolarità delle soluzioni di equazioni differenziali lineari astratte
             del primo ordine in domini variabili. (English summary.) *Boll. Un.
             Mat. Ital.* **10** (1974), 182–201. (MR 51, #10875).
[1976: 1]   Su alcune equazioni d'evoluzione singolari. (English summary.) *Boll.
             Un. Mat. Ital. B* (5) **13** (1976), 498–517. (MR 56, #767).

L. Bers, F. John, and M. Schechter
[1964: 1]   *Partial Differential Equations.* Lectures in Applied Mathematics, vol. 3,
             American Mathematical Society; Wiley, New York, 1964.

A. Beurling
[1970: 1]   On analytic extension of semigroups of operators. *J. Funct. Analysis* **6**
             (1970), 387–400. (MR 54, #7960).

V. A. Bezverhniĭ
[1975: 1]   The self-adjointness of an operator of Dirac type in a space of
             vector-valued functions. (Russian.) *Mat. Zametki* **18** (1975), 3–7. (MR
             52, #11649).

A. T. Bharucha-Reid

[1958: 1]   Ergodic projections for semi-groups of periodic operators. *Studia Math.*
             17 (1958), 189–197. (MR 20, #6664).

[1960: 1]   *Elements of the Theory of Markoff Processes and their Applications.*
             McGraw-Hill, New York, 1960.

[1965: 1]   Markov branching processes and semigroups of operators. *J. Math.*
             Anal. Appl. **12** (1965), 513–536. (MR 32, #8402).

G. Birkhoff

[1964: 1]   Well-set Cauchy problems and C_0- semigroups. *J. Math. Anal. Appl.* **8**
             (1964), 303–324. (MR 28, #2343).

[1967: 1]   *Lattice Theory*, 3rd. ed. Amer. Math. Soc. Colloquium Publications,
             vol. 25, Providence, 1967.

G. Birkhoff and T. Mullikin

[1958: 1]   Regular partial differential equations. *Proc. Amer. Math. Soc.* **9** (1958),
             18–25. (MR 20, #2528).

M. Biroli

[1971: 1]   Solutions presque periodiques des inéquations d'évolution parabo-
             liques. (Italian summary.) *Ann. Mat. Pura Appl.* (4) **88** (1971), 51–70
             (MR 48, #4836).

[1972: 1]   Sur les solutions bornées et presque periodiques des équation et
             inéquations d'evolution. (English summary.) *Ann. Mat. Pura Appl.* (4)
             93 (1972), 1–79. (MR 47, #9380).

[1974: 1]   Sur l'unicité et la presque periodicité de la solution bornée d'une
             equation non linéaire d'évolution du type elliptique. (Italian summary.)
             Boll. Un. Mat. Ital. (4) **9** (1974), 767–774. (MR 51, #3983).

[1974: 2]   Sur l'équation non linéaire de Schrödinger. (Italian summary.) *Atti*
             Accad. Naz. Lincei. Rend. Cl. Sci. Fis. Mat. Natur. (8) **54** (1973),
             854–859, (1974). (MR 51, #8615).

E. Bishop and R. R. Phelps

[1961: 1]   A proof that every Banach space is subreflexive. *Bull. Amer. Math.*
             Soc. **67** (1961), 97–98. (MR 23, #A503)

F. Bloom

[1977: 1]   Continuous dependence on initial geometry for a class of abstract
             equations in Hilbert space. *J. Math. Anal. Appl.* **58** (1977), 293–297.
             (MR 56, #783).

L. E. Bobisud

[1975: 1]   Some asymptotic properties of solutions of nonlinear abstract wave
             equations. *J. Math. Anal. Appl.* **49** (1975), 680–691. (MR 50, #14399).

L. E. Bobisud, L. F. Boron, and J. Calvert

[1972: 1]   On the degeneration of solutions of the abstract differential equation
             $\varepsilon u'' + Au = 0$. *Rocky Mountain J. Math.* **2** (1972), 619–625. (MR 47,
             #2210).

L. E. Bobisud and J. Calvert

[1970: 1]   Singularly perturbed differential equations in a Hilbert space. *J. Math. Anal. Appl.* **30** (1970), 113–127. (MR 40, #6031).

[1972: 1]   On the degeneration of solutions of the abstract differential equation $\varepsilon u'' + Au = 0$. *Rocky Mountain J. Math.* **2** (1972), 619–625. (MR 47, #2210).

[1973: 1]   Energy bounds and virial theorem for abstract wave equations. *Pacific J. Math.* **47** (1973), 27–37. (MR 48, #4545).

[1974: 1]   Singular perturbation of a time-dependent Cauchy problem in a Hilbert space. *Pacific J. Math.* **54** (1974), 45–53. (MR 50, #10474).

L. Bobisud and R. Hersh

[1972: 1]   Perturbation and approximation theory for higher-order abstract Cauchy problems. *Rocky Mountain J. Math.* **2** (1972), 57–73. (MR 45, #3981).

S. Bochner

[1949: 1]   Diffusion equation and stochastic processes. *Proc. Nat. Acad. Sci. U.S.A.* **35** (1949), 368–370. (MR 10, 720).

S. Bochner and J. von Neumann

[1935: 1]   On compact solutions of operational differential equations. *Ann. Math.* **36** (1935), 255–291.

P. Bolis Basit and L. Zend

[1972: 1]   A generalized Bohr-Neugebauer theorem. *Differencial'nye Uravnenija* **8** (1972), 1343–1348.

D. A. Bondy

[1976: 1]   An application of functional operator models to dissipative scattering theory. *Trans. Amer. Math. Soc.* **223** (1976), 1–43. (MR 55, #1106).

D. G. Bourgin and R. J. Duffin

[1939: 1]   The Dirichlet problem for the vibrating string equation. *Bull. Amer. Math. Soc.* **45** (1939), 851–858.

L. R. Bragg

[1969: 1]   Hypergeometric operator series and related partial differential equations. *Trans. Amer. Math. Soc.* **143** (1969), 319–336. (MR 40, #1691).

[1974: 1]   Linear evolution equations that involve products of commutative operators. *SIAM J. Math. Anal.* **5** (1974), 327–335. (MR 50, #8157).

[1974/75: 1]   Singular non-homogeneous abstract Cauchy and Dirichlet type problems related by a generalized Stieltjes transform. *Indiana Univ. Math. J.* **24** (1974/75), 183–195. (MR 50, #712).

[1974/75: 2]   Related non-homogeneous partial differential equations. *Applicable Anal.* **4** (1974/75), 161–189. (MR 52, #14562).

[1977: 1]   Some abstract Cauchy problems in exceptional cases. *Proc. Amer. Math. Soc.* **65** (1977), 105–112. (MR 58, #11738)

L. R. Bragg and J. W. Dettman

[1968: 1]   Related problems in partial differential equations. *Bull. Amer. Math. Soc.* **74** (1968), 375–378. (MR 37, #560).

[1968: 2]   An operator calculus for related partial differential equations. *J. Math. Anal. Appl.* **22** (1968), 459–467. (MR 37, #561).
[1968: 3]   Related partial differential equations and their applications. *SIAM J. Appl. Math.* **16** (1968), 459–467. (*MR* 37, #3170).

P. Brenner
[1966: 1]   The Cauchy problem for symmetric hyperbolic systems in L_p. *Math. Scand.* **19** (1966), 27–37. (MR 35, #3299).
[1972: 1]   Initial value problems in L_p for systems with variable coefficients. *Math. Scand.* **30** (1972), 141–149. (MR 48, #2602).
[1973: 1]   The Cauchy problem for systems in L_p and $L_{p,\alpha}$. *Ark. Mat.* **11** (1973), 75–101. (MR 48, #2601).
[1975: 1]   On $L_p - L_p$ estimates for the wave equation. *Math. Z.* **145** (1975), 251–254. (MR 52, #8658).
[1977: 1] $L_p - L_p$ estimates for Fourier integral operators related to hyperbolic equations. *Math. Z.* **152** (1977), 273–286. (MR 55, #3877).

P. Brenner, V. Thomée, and L. B. Wahlbin
[1975: 1]   *Besov Spaces and Applications to Difference Methods for Initial Value Problems.* Lecture Notes in Mathematics, vol. 434, Springer, New York, 1975.

H. Brézis
[1973: 1]   *Operateurs Maximaux Monotones et Semi-Groupes de Contractions dans les Espaces de Hilbert.* Notas de Matemática, vol. 50. North-Holland–Elsevier, Amsterdam-London-New York, 1973.

H. Brézis, W. Rosenkrantz, and B. Singer
[1971: 1]   On a degenerate elliptic-parabolic equation occurring in the theory of probability. *Comm. Pure Appl. Math.* **24** (1971), 395–416. (MR 44, #1941).

H. Brézis and W. A. Strauss
[1973: 1]   Semi-linear second-order elliptic equations in L^1. *J. Math. Soc. Japan* **25** (1973), 565–590. (MR 49, #826).

N. I. Briš and N. I. Jurčuk
[1971: 1]   The Goursat problem for abstract second order linear differential equations. (Russian.) *Differencial'nye Uravnenija* **7** (1971), 1017–1030, 1139. (MR 40, #7538).

A. R. Brodsky
[1967: 1]   On the asymptotic behavior of solutions of the wave equation. *Proc. Amer. Math. Soc.* **18** (1967), 207–208. (MR 35, #3289).

L. J. F. Broer and J. De Graaf
[1971: 1]   Linear dynamical systems in Hilbert space. Instability of continuous sysems (IUTAM Sympos., Herrenalb, 1969), 376–378. Springer, Berlin, 1971. (MR 51, #1049).

F. Browder

[1962: 1]   On nonlinear wave equations. *Math. Z.* **80** (1962), 249–264. (MR 26, #5283).

[1976: 1]   Nonlinear operators and nonlinear equations of evolution in Banach spaces. *Nonlinear functional analysis* (*Proc. Sympos. Pure Math.* XVIII, *Part.* 2, *Chicago*, 1968), 1–308. Amer. Math Soc., Providence, 1976. (MR 53, #807).

A. B. Buche

[1968: 1]   Approximation of semigroups of operators in Frèchet spaces. *Proc. Japan Acad.* **44** (1968), 816–819. (MR 39, #820).

[1971: 1]   On the cosine-sine operator functional equations. *Aequationes Math.* **6** (1971), 231–234. (MR 45, #961).

[1973: 1]   Perturbation theory for the generalized semigroup. *J. Mathematical and Physical Sci.* **7** (1973), 233–244. (MR 48, #12142).

[1975: 1]   On an exponential-cosine operator-valued functional equation. *Aequationes Math.* **13** (1975), 233–241. (MR 53, #3789).

A. B. Buche and A. T. Bharucha-Reid

[1968: 1]   On some functional equations associated with semigroups of operators. *Proc. Nat. Acad. Sci. U.S.A.* **60** (1968), 1170–1174. (MR 37, #5724).

[1971: 1]   On the generalized semigroup relation in the strong operator topology. *Nederl. Akad. Wetensch. Proc. Ser. A 74 = Indag. Math.* **33** (1971), 26–31. (MR 43, #6771).

T. Burak

[1973: 1]   Two-point problems and analiticity of solutions of abstract parabolic equations. *Israel J. Math.* **16** (1973), 404–417. (MR 52, #6493).

[1973: 2]   Two-point problems and analyticity of solutions of abstract parabolic equations. *Israel J. Math.* **16** (1973), 418–445. (MR 52, #6494).

J. A. Burns and T. L. Herdman

[1976: 1]   Adjoint semigroup theory for a class of functional differential equations. *SIAM J. Math. Anal.* **7** (1976), 729–745. (MR 54, #5578).

G. H. Butcher and J. A. Donaldson

[1975: 1]   Regular and singular perturbation problems for a singular abstract Cauchy problem. *Duke Math. J.* **42** (1975), 435–445. (MR 51, #8887).

P. L. Butzer

[1956: 1]   Sur la théorie des demi-groupes et classes de saturation de certaines intégrales singulières. *C. R. Acad. Sci. Paris* **243** (1956), 1473–1475. (MR 18, 585).

[1957: 1]   Halbgruppen von Linearen Operatoren und eine Anwendung in der Approximationstheorie. *J. Reine Angew. Math.* **197** (1957), 112–120. (MR 18, 585).

[1957: 2]   Über den Grad der Approximations des Identitätsoperators durch Halbgruppen von Linearen Operatoren und Anwendungen auf die Theorie der Singulären Integrale, *Math. Ann.* **133** (1957), 410–425. (MR 20, #1232).

[1958: 1]   Zur Frage der Saturationsklassen singulärer Integraloperatoren. *Math. Z.* **70** (1958), 93–112. (MR 20, #6039).

P. L. Butzer and H. Berens
[1967: 1]   *Semi-groups of Operators and Approximation.* Springer, New York, 1967. (MR 37, #5558).

P. L. Butzer and W. Köhnen
[1961: 1]   Approximation invariance of semi-group operators under perturbations. *J. Approximation Theory* **2** (1969), 389–393. (MR 40, #6299).

P. L. Butzer and R. J. Nessel
[1971: 1]   *Fourier-Analysis and Approximation* I. Birkhäuser-Verlag, Basel, 1971.

P. L. Butzer and S. Pawelke
[1968: 1]   Semi-groups and resolvent operators. *Arch Rat. Mech. Anal.* **30** (1968), 127–147. (MR 37, #3397).

P. L. Butzer and H. G. Tillmann
[1960: 1]   An approximation theorem for semi-groups of operators, *Bull. Amer. Math. Soc.* **66** (1960), 191–193. (MR 22, #3986).
[1960: 2]   Approximation theorems for semi-groups of bounded linear transformations, *Math. Ann.* **140** (1960), 256–262. (MR 22, #8348).

P. L. Butzer and U. Westphal
[1972: 1]   On the Cayley transform and semigroup operators. Hilbert space operators and operator algebras (*Proc. Internat. Conf., Tihany, 1970*), 89–97. Colloq. Math. Soc. János Bolyai, 5. North-Holland, Amsterdam, 1972. (MR 50, 14356).

Ja. V. Bykov and N. I. Fomina
[1973: 1]   The singular solutions of nonlinear differential equations. I. (Russian.) *Differencial'nye Uravnenija* **9** (1973), 983–994. (MR 48, #655).
[1976: 1]   The singular solutions of nonlinear differential equations. II. (Russian.) *Differencial'nye Uravnenija* **12** (1976), 1161–1173, 1339. (MR 55, #3457).

A. P. Calderón
[1965: 1]   Commutators of singular integral operators. *Proc. Nat. Acad. Sci. U.S.A.* **53** (1965), 1092–1099. (MR 31, #1575).

B. Calvert
[1976: 1]   The equation $A(t, u(t))' + B(t, u(t)) = 0$. *Math. Proc. Cambridge Phylos. Soc.* **79** (1976), no. 3, 545–561. (MR 53, #8983).

J. T. Cannon
[1975: 1]   Convergence criteria for a sequence of semi-groups. *Applicable Anal.* **5** (1975), 23–31. (MR 54, #1016).

A. Carasso
[1971: 1]   The abstract backwards beam equation. *SIAM J. Math. Anal.* **2** (1971), 193–212. (MR 44, #5636).

R. W. Carroll

[1963: 1]   On the singular Cauchy problem. *J. Math. Mech.* **12** (1963), 69–102. (MR 26, #5284).

[1963: 2]   Problems in linked operators, I. *Math. Ann.*, **151** (1963), 272–282. (MR 29, #3902).

[1963: 3]   Some degenerate Cauchy problems with operator coefficients. *Pacific J. Math.* **13** (1963), 471–485. (MR 29, #367).

[1964: 1]   On the structure of the Green's operator. *Proc. Amer. Math. Soc.* **15** (1964), 225–230. (MR 32, #2757).

[1965: 1]   Some differential problems related to spectral theory in several operators. *Atti Accad. Naz. Lincei Rend. Cl. Sci. Fis. Mat. Natur.* (8) **39** (1965), 170–174. (MR 33, #3142).

[1965: 2]   Problems in linked operators. II. *Math. Ann.* **160** (1965), 233–256. (MR 36, #6981).

[1966: 1]   On the structure of some abstract differential problems, I. *Ann. Mat. Pura Appl.* (4) **72** (1966), 305–318. (MR 35, #7140).

[1966: 2]   On the spectral determination of the Green's operator. *J. Math. Mech.* **15** (1966), 1003–1018. (MR 34, #3141).

[1967: 1]   On the propagator equation. *Illinois J. Math.* **11** (1967), 506–527. (MR 35, #4547).

[1967: 2]   Some growth and convexity theorems for second-order equations. *J. Math. Anal. Appl.* **17** (1967), 508–518. (MR 35, #510).

[1969: 1]   On the structure of some abstract differential problems, II. *Ann. Mat. Pura Appl.* (4) **81** (1969), 93–109. (MR 41, #870).

[1969: 2]   *Abstract Methods in Partial Differential Equations.* Harper and Row, New York, 1969.

[1973: 1]   On some hyperbolic equations with operator coefficients. *Proc. Japan Acad.* **49** (1973), 233–238. (MR 51, #6458).

[1975: 1]   Some remarks on kernels, recovery formulas and evolution equations *Rocky Mountain J. Math.* **5** (1975), 61–74. (MR 51, #3936).

[1976: 1]   Singular Cauchy problems in symmetric spaces. *J. Math. Anal. Appl.* **56** (1976), 41–54.

[1978: 1]   The group theoretic nature of certain recursion relations for singular Cauchy problems. *J. Math. Anal. Appl.* **63** (1978), 23–41.

[1979: 1]   Transmutation and separation of variables. *Appl. Analysis* **8** (1979), 253–263.

[1979: 2]   *Operational Calculus and Abstract Differential Equations.* Notas de Matemática, vol. 67, North-Holland, Amsterdam, 1979.

R. W. Carroll and J. Neuwirth

[1964: 1]   Some uniqueness theorems for differential equations with operator coefficients. *Trans. Amer. Math. Soc.* **110** (1964), 459–472. (MR 31, #465).

R. W. Carroll and R. Showalter

[1976: 1]   *Singular and Degenerate Cauchy Problems.* Academic Press, New York, 1976.

R. W. Carroll and E. State
[1971: 1]   Existence theorems for some weak abstract variable domain hyperbolic problems. *Canad. J. Math.* **23** (1971), 611–626. (MR 44, #1933).

R. W. Carroll and C. L. Wang
[1965: 1]   On the degenerate Cauchy problem. *Canad. J. Math.* **17** (1965), 245–256. (MR 36, #489).

A. Čaryev
[1966: 1]   Criteria for the boundedness of solutions of differential equations of order two in Hilbert space. (Russian.) *Izv. Akad. Nauk Turkmen. SSR Ser. Fiz.-Tehn. Him. Geol. Nauk* (1966), no. 2, 8–15. (MR 33, #4429).

A. L. Cauchy
[1821: 1]   *Cours d'Analyse de l'École Royale Politechnique*, 1re. *Partie: Analyse Algebrique.* Imprimerie Royale, Paris, 1821. Reprinted as: *Oeuvres Complètes*, IIe. serie, tome III, Gauthier-Villars, Paris, 1897.

M. W. Certain
[1974: 1]   One-parameter semigroups holomorphic away from zero. *Trans. Amer. Math. Soc.* **187** (1974), 377–389. (MR 49, #1216).

M. W. Certain and T. G. Kurtz
[1977: 1]   Landau-Kolomogorov inequalities for semigroups and groups. *Proc. Amer. Math. Soc.* **63** (1977), 226–230. (MR 56, #16445).

N. Chafee
[1977: 1]   Behavior of solutions leaving the neigborhood of a saddle point for a nonlinear evolution equation. *J. Math. Anal. Appl.* **58** (1977) 312–325 (MR 56, #342)

J. Chazarain
[1968: 1]   Problèmes de Cauchy au sens des distributions vectorielles et applications. *C. R. Acad. Sci. Paris.* **266** (1968), A10–A13. (MR 37, #5691).
[1968: 2]   Problèmes de Cauchy dans des espaces d'ultra-distributions. *C. R. Acad. Sci. Paris* **267** (1968), A564–A566. (MR 38, #531).
[1968: 3]   Un résultat de perturbation pour les générateurs de semi-groupes et applications. *C. R. Acad. Sci. Paris 267* (1968), A13–A15. (MR 38, #571).
[1971: 1]   Problèmes de Cauchy abstraits et applications a quelques problèmes mixtes. *J. Functional Analysis* **7** (1971), 386–446. (MR 43, #2570).
[1971: 2]   Le problème de Cauchy pour les opérateurs hyperboliques, non necessairement stricts, qui satisfont à la condition de Lévi. *C. R. Acad. Sci. Paris* **273** (1971), A1218–A1221. (MR 45, #2308).

H. Chebli
[1975: 1]   Opérateurs de translation généralisée et semi-groupes de convolution. *Lecture Notes in Math.*, vol. 404, 35–39, Springer, Berlin, 1975. (MR 51, #10745).

P. R. Chernoff

[1968: 1]   Note on product formulas for operator semigroups. *J. Functional Analysis* **2** (1968), 238–242. (MR 37, #6793).

[1970: 1]   Semigroup product formulas and addition of unbounded operators. *Bull. Amer. Math. Soc.* **76** (1970), 395–398. (MR 41, #2457).

[1972: 1]   Some remarks on quasi-analytic vectors. *Trans. Amer. Math. Soc.* **167** (1972), 105–113. (MR 45, #4193).

[1972: 2]   Perturbations of dissipative operators with relative bound one. *Proc. Amer. Math. Soc.* **33** (1972), 72–74. (MR 45, #5804).

[1972: 3]   Nonassociative addition of unbounded operators and a problem of Brézis and Pazy. *Bull. Amer. Math. Soc.* **78** (1972), 562–563. (MR 46, #2480).

[1973: 1]   Essential self-adjointness of powers of generators of hyperbolic equations. *J. Functional Analysis* **12** (1973), 401–414. (MR 51, #6119).

[1974: 1]   *Product Formulas, Nonlinear Semigroups, and Addition of Unbounded Operators*. Memoirs of the American Mathematical Society, Providence, 1974 (MR 54, #5899).

[1976: 1]   On the converse of the Trotter product formula. *Illinois J. Math.* **20** (1976), 348–353. (MR 52, #15101).

[1976: 2]   Two counterexamples in semigroup theory in Hilbert space. *Proc. Amer. Math. Soc.* **56** (1976), 253–255. (MR 53, #3790).

P. R. Chernoff and J. A. Goldstein

[1972: 1]   Admissible subspaces and the denseness of the intersection of the domains of semigroup generators. *J. Functional Analysis* **9** (1972), 460–468. (MR 45, #7517).

J. Chevalier

[1969: 1]   Sur la continuité simple des semi-groupes d'opérateurs (English summary.) *Bull. Soc. Roy. Sci. Liège* **38** (1969), 459–461. (MR 41, #9048).

[1970: 1]   Semi-groupes d'opérateurs et problèmes aux limites pur les systèmes d'équations aux dérivées partielles d'évolution. *Bull. Soc. Roy. Sci. Liège* **39** (1970), 349–393. (MR 43, #3841).

[1975: 1]   Sur la régularité de la solution du problème de Cauchy abstrait faible. (English summary.) *Bull. Roy. Soc. Liege* **44** (1975), 551–554. (MR 53, #11428).

K. L. Chung

[1962: 1]   On the exponential formulas of semi-group theory. *Math. Scand* **10** (1962), 153–162. (MR 26, #2886).

I. Ciorănescu

[1971: 1]   La caractérisation spectrale d'opérateurs générateurs des semi-groupes distributions d'ordre fini de croissance. *J. Math. Anal. Appl.* **34** (1971), 34–41. (MR 43, #3842).

[1972: 1]   Sur les semi-groupes distributions d'ordre fini de croissance. *Rev. Roumaine Math. Pures Appl.* **17** (1972), 845–849. (MR 47, #7513).

[1973: 1]   Accretive operators in Banach spaces and semigroups of nonlinear contractions I, II. (Romanian. French summary.) *Stud. Cerc. Mat.* **25** (1973), 791–826; ibid. **25** (1973), 959–991. (MR 50, #8174).

[1973: 2]   Sur la somme des genérateurs de certains semigroupes distributions continus. (Italian summary.) *Rend. Mat.* **6** (1973), 785–801. (MR 50, #14357).

[1974: 1]   An example of a distribution semigroup. (Romanian.) *Stud. Cerc. Mat.* **26** (1974), 357–365. (MR 50, #5542).

[1974: 2]   Sur les solutions fondamentales d'ordre fini de croissance. *Math. Ann.* **211** (1974), 37–46. (MR 50, #8059).

[1977: 1]   Espaces abstraits de classe (M_p) de Beurling et les semi-groupes ultra-distributions. *C. R. Acad. Sci. Paris* **284** (1977), 1439–1441. (MR 56, #3631).

I. Ciorănescu and L. Zsidó
[1976: 1]   On spectral subspaces of some unbounded groups of operators. *Rev. Roumaine Math. Pures Appl.* **21** (1976), 817–850. (MR 54, #5904).

[1976: 2]   Analytic generators for one-parameter groups. *Tôhoku Math. J.* **28** (1976), 327–362. (MR 55, #3872).

E. A. Coddington
[1961: 1]   *An Introduction to Ordinary Differential Equations.* Prentice-Hall, Englewood Cliffs, N.J., 1961.

E. A. Coddington and N. Levinson
[1955: 1]   *Theory of Ordinary Differential Equations.* McGraw-Hill, New York, 1955.

P. J. Cohen and M. Lees
[1961: 1]   Asymptotic decay of solutions of differential inequalities. *Pacific J. Math.* **11** (1961), 1235–1249. (MR 23, #A3427).

P. M. Cohn
[1961: 3]   *Lie Groups.* Cambridge University Press, Cambridge, 1961.

B. D. Coleman and V. Mizel
[1966: 1]   Norms and semi-groups in the theory of fading memory. *Arch. Rat. Mech. Anal.* **23** (1966), 87–123. (MR 23, #1966).

I. Colojoara and C. Foias
[1968: 1]   *Theory of Generalized Spectral Operators.* Gordon and Breach, New York, 1968.

J. B. Cooper
[1975: 1] L^1-sequential convergence on Stone spaces and Stone's theorem on unitary semigroups. *Canad. Math. Bull.* **18** (1975), 191–193. (MR 53, #3792).

J. L. B. Cooper
[1947: 1]   One-parameter semigroups of isometric operators in Hilbert space. *Ann. of Math.* **48** (1947), 827–842. (MR 10, 257).

[1948: 1]   Symmetric operators in Hilbert space. *Proc. London Math. Soc.* (2) **50** (1948), 11–55. (MR 9, 446).

J. M. Cooper
[1971: 1]   Two-point problems for abstract evolution equations. *J. Differential Equations* **9** (1971), 453–495. (MR 45, #7322).

C. Corduneanu
[1961: 1]   *Functii Aproape-Periodice.* Editura Academiei Republicii Populare Romine, Bucharest, 1961. English translation: [1968: 1] *Almost Periodic Functions.* Academic Press, New York, 1968.

R. Courant and D. Hilbert
[1953: 1]   *Methods of Mathematical Physics*, vol. I. Interscience, New York, 1953.
[1962: 1]   *Methods of Mathematical Physics*, vol. II. Interscience-Wiley, New York, 1962.

M. G. Crandall
[1974: 1]   Semigroups of nonlinear transformations and evolution equations. Proceedings of the International Congress of Mathematicians (Vancouver, B. C. 1974), Vol. 2, 257–262. Canad. Math. Congress, Montreal, 1975. (MR 54, #8367).

M. G. Crandall and R. S. Phillips
[1968: 1]   On the extension problem for dissipative operators. *J. Functional Analysis* **2** (1968), 147–176. (MR 36, #6775).

M. G. Crandall and A. Pazy
[1968/69: 1]   On the differentiability of weak solutions of a differential equation in Banach space. *J. Math. Mech.* **18** (1968/69), 1007–1016. (MR 39, #3349).

M. Crouzeix and P. A. Raviart
[1976: 1]   Approximation des équations d'évolution linéaires par des méthodes à pas multiples (English summary.) *C. R. Acad. Sci. Paris* **283** (1976), A367–A370. (MR 54, #14377).

J. R. Cuthbert
[1971: 1]   On semi-groups such that $T_t - I$ is compact for some $t > 0$. *Z. Wahrscheinlichkeitstheorie und Verw. Gebiete* **18** (1971), 9–16. (MR 43, #5346).

Ju. L. Daleckiĭ
[1959: 1]   A linear equation relative to elements of a normed ring. (Russian.) *Uspehi Mat. Nauk* **14** (1959), no. 1 (85), 165–168. (MR 21, #2926).
[1960: 1]   Representability of solutions of operator equations in the form of continual integrals. (Russian.) *Dokl. Akad. Nauk SSSR* **134** (1960), 1013–1016. (MR 30, #1411).
[1961: 1]   Spatio-temporal integrals related to some differential equations and systems. (Russian.) *Dokl. Akad. Nauk SSSR* **137** (1961), 268–271. (MR 28, #284).

[1961: 2]   Continual integrals and characteristics related to a group of operators. (Russian.) *Dokl. Akad. Nauk SSSR* **141** (1961), 1290–1293. (MR 28, #1515).

[1961: 3]   Fundamental solutions of an operator equation and continual integrals. (Russian.) *Izv. Vysš. Učebn. Zaved. Matematika* (1961), no. 3 (22), 27–48. (MR 28, #474).

[1962: 1]   An asymptotic method for certain differential equations with oscilllating coefficients. (Russian.) *Dokl. Akad. Nauk SSSR* **143** (1962), 1026–1029. (MR 24, #A2726).

[1962: 2]   Functional integrals associated with operator evolution equations. (Russian.) *Uspehi Mat. Nauk 17* (1962), no. 5 (107), 3–115. (MR 28, #4389).

[1966: 1]   Elliptic operators in functional derivatives, and diffusion equations connected with them. (Russian.) *Dokl. Akad. Nauk SSSR* **171** (1966), 21–24. (MR 35, #3749).

[1967: 1]   Infinite-dimensional elliptic operators and the corresponding parabolic equations. (Russian.) *Uspehi Mat. Nauk* **22** (1967), no. 4 (136), 3–54. (MR 36, #6821).

[1976: 1]   The selfadjointness and maximal dissipativity of differential operators for functions of infinite-dimensional argument. (Russian.) *Dokl. Akad. Nauk SSSR* **227** (1976), 784–787. (MR 53, #3795).

[1976: 2]   An estimate of the solutions of the Cauchy problem for second-order parabolic equations that does not depend on the dimension. (Russian.) *Ukrain. Mat. Ž.* **28** (1976), 807–812, 862. (MR 55, #8552).

Ju. L. Daleckiĭ and E. A. Fadeeva
[1972: 1]   Hyperbolic equations with operator coefficients, and ultraparabolic systems. (Russian.) *Ukrain. Mat. Ž.* **24** (1792), 92–95. (MR 45, #3982).

Ju. L. Daleckiĭ and S. G. Kreĭn
[1950: 1]   On differential Equations in Hilbert space. *Ukrain. Mat. Ž.* **2** (1950), no. 4, 71–79 (Russian.) (MR 13, 954).

[1970: 1]   *Stability of Solutions of Differential Equations in Banach Space.* (Russian.) Izdat. "Nauka," Moscow, 1970. English translation: [1974: 1], Amer. Math. Soc., Providence, 1974.

G. Da Prato
[1965: 1]   Regolarizzazione di alcuni semigruppi distribuzioni. Atti del Convegno su le Equazioni alle Derivate Parziali (Nervi, 1965), 52–54, Edizioni Cremonese, Rome, 1966. (MR 34, #6540).

[1966: 1]   Semigruppi di crescenza *n*. *Ann. Scuola Norm. Sup. Pisa* (3) **20** (1966), 753–682. (MR 36, #5760).

[1966: 2]   Nouveau type de semi-groupes. *C. R. Acad. Sci. Paris* **262** (1966), A998–A998. (MR 34, #623).

[1966: 3]   Semigruppi regolarizzabili. *Ricerche Mat.* **15** (1966), 223–248. (MR 37, #793).

[1968: 1]   Semigruppi periodici. *Ann. Mat. Pura Appl.* (4) **78** (1968), 55–67. (MR 39, #2016).

[1968: 2]   Somma di generatori infinitesimali di semigruppi di contrazione e
             equazioni di evoluzione in spazi di Banach. (English summary.) *Ann.*
             Mat. Pura Appl. (4) **78** (1968), 131–157. (MR 39, #2017).
[1968: 3]   Somme de générateurs infinitésimaux de classe C_0. (Italian summary.)
             Atti Accad. Naz. Lincei Rend. Cl. Sci. Fis. Mat. Natur. (8) **45** (1968),
             14–21. (MR 40, #1811).
[1968: 4]   Somma di generatori infinitesimali di semigruppi di contrazione e
             equazioni di evoluzione in spazi di Banach riflessivi. *Boll. Un. Mat.*
             Ital. (4) **1** (1968), 138–141. (MR 38, #6256).
[1968: 5]   Équations opérationnelles dans les espaces de Banach et applications.
             C. R. Acad. Sci. Paris **266** (1968), A60–A62. (MR 40, #1801).
[1968: 6]   Équations opérationnelles dans les espaces de Banach (cas analytique).
             C. R. Acad. Sci. Paris **266** (1968), A277–A279. (MR 39, #6117).
[1968: 7]   Somma di generatori infinitesimali di semigruppi analitici. *Rend. Sem.*
             Mat. Univ. Padova **40** (1968), 151–161. (MR 37, #6794).
[1969: 1]   Somme d'applications non linéaires et solutions globales d'équations
             quasi-linéaires dans des espaces de Banach. *Boll. Un. Mat Ital.* (4) **2**
             (1969), 229–240. (MR 40, #1683).
[1970: 1]   Weak solutions for linear abstract differential equations in Banach
             spaces. *Advances in Math.* **5** (1970), 181–245. (MR 43, #6772).
[1974: 1]   Sums of linear operators. Linear operators and approximation II (Proc.
             Conf. Oberwolfach Math. Res. Inst., Oberwolfach, 1974), 461–472.
             Internat. Ser. Numer Math. 25, Birkhäuser, Basel, 1974. (MR 55,
             #3823).
[1975: 1]   Somme d'opérateurs linéaires et équations différentielles opération-
             nalles. *J. Math. Pures Appl.* **54** (1975), 305–387. (MR 56, #6794).

G. Da Prato and E. Giusti
[1967: 1]   Equazioni di Schrödinger e delle onde per l'operatore di Laplace
             iterato in $L^p(R^n)$. (English summary.) *Ann. Mat. Pura Appl.* (4) **76**
             (1967), 377–397. (MR 37, #570).
[1967: 2]   Equazioni di evoluzione in L^p. *Ann. Scuola Norm. Sup. Pisa* (3) **21**
             (1967), 485–505. (MR 37, #5552).
[1967: 3]   Una caratterizzazione dei generatori di funzioni coseno astratte. *Boll.*
             Un. Mat Ital. (3) **22** (1967), 357–362. (MR 39, #2018).

G. Da Prato and P. Grisvard
[1975: 1]   Sommes d'opérateurs linéaires et équations différentielles opération-
             nelles. *J. Math. Pures. Appl.* **54** (1975), 305–387. (MR 56, #1129).

G. Da Prato and M. Iannelli
[1976: 1]   On a method for studying abstract evolution equations in the hyper-
             bolic case. *Comm. Partial Diff. Equations* **1** (1976), 585–608. (MR 56,
             #1130).

G. Da Prato and U. Mosco
[1965: 1]   Semigruppi distribuzioni analitici. *Ann. Scuola Norm. Sup. Pisa* (3) **19**
             (1965), 367–396. (MR 32, #2899).

[1965: 2]   Regolarizzazione dei semigruppi distribuzioni analitici. *Ann. Scuola Norm. Sup. Pisa* (3) **19** (1965), 563–576. (MR 34, #3342).

R. Datko

[1968: 1]   An extension of a theorem of A. M. Lyapunov to semi-groups of operators. *J. Math. Anal. Appl.* **24** (1968), 290–295. (MR 37, #6795).

[1970: 1]   Extending a theorem of A. M. Liapunov to Hilbert space. *J. Math. Anal. Appl.* **32** (1970), 610–616. (MR 42, #3614).

[1970: 2]   The variable gradient as a tool in the study of stability. Seminar on Differential Equations and Dynamical Systems, II (Univ. of Maryland, College Park, Maryland, 1969), 34–43. *Lecture Notes in Math.*, vol. 144, Springer, Berlin, 1970. (MR 52, #8594).

[1972: 1]   Uniform asymptotic stability of evolutionary processes in a Banach space. *SIAM J. Math. Anal.* **3** (1972), 428–445. (MR 47, #9004).

[1976: 1]   Linear differential-difference equations in a Banach space. *Proceedings of the IEEE Conference on Decision and Control and the 15th Symposium on Adaptive Processes* (Clearwater, Fla., 1967), 735–742. Inst. Electr. Electron. Engrs., New York, 1976. (MR 56, #16087).

[1977: 1]   Linear autonomous neutral differential equations in a Banach space. *J. Diff. Equations* **25** (1977), 258–274. (MR 56, #6053).

[1977: 2]   The uniform asymptotic stability of certain neutral differential-difference equations. *J. Math. Anal. Appl.* **58** (1977), 510–526. (MR 55, #13047).

[1978: 1]   Representation of solutions and stability of linear differential-difference equations in a Banach space. *J. Diff. Equations* **29** (1978), 105–166. (MR 58, #6619).

E. B. Davies

[1977: 1]   Asymptotic analysis of some abstract evolution equations. *J. Functional Analysis* **25** (1977), 81–101. (MR 56, #9318).

F. S. De Blasi and J. Myjak

[1977: 1]   On generic asymptotic stability of differential equations in Banach space. *Proc. Amer Math. Soc.* **65** (1977), 47–51. (MR 56, #6040).

J. De Graaf

[1971: 1]   A constructive approach to one-parameter semigroups of operators in Hilbert space. *Arch. Rat. Mech. Anal.* **43** (1971), 125–153. (MR 48, #12144).

[1972: 1]   On a class of lateral boundary value problems associated with infinitesimal generators. *Arch. Rat. Mech. Anal.* **47** (1972), 330–342. (MR 49, #1217).

M. V. Dehspande and N. E. Joshi

[1971: 1]   On compactness of tensor products of semi-groups of linear operators. *J. Mathematical and Physical Sci.* **5** (1971), 271–274. (MR 46, #716).

K. De Leeuw

[1960: 1]   On the adjoint semigroup and some problems in the theory of approximation. *Math. Z.* **73** (1960), 219–234. (MR 22, #2909).

J. Delsarte

[1938: 1]   Sur certaines transformations fonctionnelles relatives aux équations
             linéaires aux derivées partielles de second ordre. *C. R. Acad. Sci. Paris*
             206 (1938). 1780–1782.

J. Delsarte and J. L. Lions

[1957: 1]   Transmutations d'opérateurs différentiels dans le domaine complexe.
             Comm. Math. Helv. **32** (1957), 113–128. (MR 19, 959).

B. Dembart

[1973: 1]   Perturbation of semigroups on locally convex spaces. *Bull. Amer.*
             Math. Soc. **79** (1973), 986–991. (MR 48, #949).

[1974: 1]   On the theory of semigroups of operators on locally convex spaces. *J.*
             Functional Analysis **16** (1974), 123–160. (MR 49, #9679).

J. D. DePree and H. S. Klein

[1974: 1]   Semi-groups and collectively compact sets of linear operators. *Pacific*
             J. Math. **55** (1974). 55–63. (MR 51, #11188).

V. I. Derguzov

[1963: 1]   On the stability of the solutions of Hamiltonian equations in Hilbert
             space with unbounded periodic operator coefficients. (Russian.) *Dokl.*
             Akad. Nauk SSSR **152** (1963), 1294–1296. (MR 28, #394).

[1964: 1]   On the stability of the solutions of the Hamilton equations with
             unbounded periodic operator coefficients. (Russian.) *Mat Sb.* (*N. S.*)
             63 (105) (1964), 591–619. (MR 29, #325).

[1964: 2]   Sufficient conditions for the stability of Hamiltonian equations with
             unbounded periodic operator coeficients. (Russian) *Mat Sb.* (*N. S.*) **64**
             (106) (1964), 419–435. (MR 29, #4965).

[1964: 3]   Necessary conditions for the strong stability of Hamiltonian equations
             with unbounded periodic operator coefficients. (Russian. English
             summary.) *Vestnik Leningrad. Univ. Ser. Mat. Meh. Astronom.* **19**
             (1964), no. 4, 18–30. (MR 30, #2211).

[1966: 1]   Continuous dependence of the maximal exponent of the exponential
             growth of solutions of a linear Hamiltonian equation with periodic
             operator coefficients. (Russian.) *Problems Math. Anal. Boundary Value*
             Problems Integr. Equations (Russian), 120–134.

[1969: 1]   The existence of periodic motions and motions of a special form for
             Hamiltonian systems with small parameter. (Russian.) *Dokl. Akad.*
             Nauk SSSR (1969), 757–760. (MR 40, #457).

[1969: 2]   Regions of strong stability of linear Hamiltonian equations with peri-
             odic operator coefficients. (Russian. English summary.) *Vestnik*
             Leningrad. Univ. **24** (1969), no. 13, 20–30. (MR 40, #7567).

[1970: 1]   Solution of an anticanonical equation with periodic coefficients. (Rus-
             sian.) *Dokl. Akad. Nauk SSSR* **193** (1970), 17–20. (MR 42, #3390).

[1973: 1]   Operator Hamiltonian equations and anticanonical equations with
             periodic coefficients. (Russian.) *Problems of Mathematical Analysis No.*
             4: Integral and Differential Operators. Differential Equations. 9–36.
             Izdat., Leningrad. Univ., Leningrad, 1973. (MR 49, #7547).

V. I. Derguzov and V. Ja. Jakubovič

[1963: 1]   Existence of solutions of linear Hamiltonian equations with un-
            bounded operator coefficients. (Russian.) *Dokl. Akad. Nauk SSSR* **151**
            (1963), 1264–1267. (MR 28, #313).

[1969: 1]   The existence of solutions of linear Hamiltonian equations with un-
            bounded operators. (Russian.) *Problems of Math. Anal. No. 2: Linear
            Operators and Operator Equations*, 3–28. Izdat., Leningrad. Univ.,
            Leningrad, 1969. (MR 42, #2136).

V. I. Derguzov, T. M. Melanina, and N. A. Timčenko

[1975: 1]   Application of the methods of perturbation theory to linear differential
            equations with periodic coefficients. (Russian.) *Problems of Mathemati-
            cal Analysis No. 5: Linear and Nonlinear Differential Equations, Dif-
            ferential Operators*, 47–66. Izdat., Leningrad. Univ., Leningrad, 1975.
            (MR 54, #3117).

L. De Simon

[1964: 1]   Un'applicazione della teoria degli integrali singolari allo studio delle
            equazioni differenziali lineare astratte del primo ordine. *Rend. Sem.
            Mat. Univ. Padova* **34** (1964), 205–223. (MR 31, #467).

J. W. Dettman

[1969: 1]   Initial-boundary value problems related through the Stieltjes trans-
            form. *J. Math. Anal. Appl.* **25** (1969), 341–349. (MR 38, #4816).

[1973: 1]   Perturbation techniques in related differential equations. *J. Diff. Equa-
            tions* **14** (1973), 547–558. (MR 48, #9089).

A. Devinatz

[1954: 1]   A note on semi-groups of unbounded self-adjoint operators. *Proc.
            Amer. Math. Soc.* **5** (1954), 101–102. (MR 15, 632).

A. A. Dezin

[1965: 1]   On the theory of operators of the type $d/dt - A$. (Russian.) *Dokl.
            Akad. Nauk SSSR* **164** (1965), 963–966. (MR 32, #6085).

[1967: 1]   Operators with first derivative with respect to "time" and non-local
            boundary conditions. (Russian.) *Izv. Akad. Nauk SSSR Ser. Mat.* **31**
            (1967), 61–86. (MR 35, #4619).

G. Di Blasio

[1973: 1]   Un metodo per lo studio dell'equazione di Boltzmann linearizzata.
            (English summary.) *Rend. Mat.* **6** (1973), 619–628. (MR 51, #3956).

J. Dickerson and J. Gibson

[1976: 1]   Stability of linear functional differential equations in Banach spaces. *J.
            Math. Anal. Appl.* **55** (1976), 150–155. (MR 53, #13786).

J. Dieudonné

[1950: 1]   Deux exemples singuliers d'équations différentielles. *Acta Sci. Math.
            Szeged* **12** (1950), 38–40. (MR 11, 729).

Z. Ditzian

[1969: 1]   On Hille's first exponential formula. *Proc. Amer. Math. Soc.* **22** (1969), 351–355. (MR 39, #6118).

[1970: 1]   Note on Hille's exponential formula. *Proc. Amer. Math. Soc.* **25** (1970), 351–352. (MR 41, #866).

[1971: 1]   Exponential formulae for semi-groups of operators in terms of the resolvent. *Israel J. Math.* **9** (1971), 541–553. (MR 43, #3843).

[1971: 2]   On the rate of convergence of exponential formulae. *Bull. Inst. Politehn. Iasi (N.S.)* **17** (1971), 75–78. (MR 45, #7518).

[1975: 1]   Some remarks on inequalities of Landau and Kolmogorov. *Aequationes Math.* **12** (1975), 145–151. (MR 52, #1403).

Z. Ditzian and C. P. May

[1973: 1]   Saturation classes for exponential formulae of semi-groups of operators. Spline Functions and Approximation Theory (Proc. Sympos., Univ. Alberta, Edmonton, Alberta, 1972), 83–99. Internat. Ser. Numer. Math. Vol. 21, Birkhäuser, Basel, 1973. (MR 52, #9001).

J. Dixmier

[1950: 1]   Les moyennes invariantes dans les semi-groupes et leurs applications. *Acta Sci. Math. Szeged* **12** (1950), 213–227. (MR 12, 267).

S. V. Djačenko

[1976: 1]   Semigroups of almost negative type, and their applications. (Russian.) *Dokl. Akad. Nauk SSSR* **229** (1976), 1306–1309. (MR 54, #1017).

M. Djedour

[1972: 1]   Existence and approximation of weak solutions of abstract differential equations. *Ann. Scuola Norm. Sup. Pisa* **26** (1972), 463–478. (MR 51, #10786).

J. D. Dollard and C. N. Friedman

[1979: 1]   *Product Integration, with Applications to Partial Differential Equations.* Encyclopedia of Mathematics and Its Applications, vol. 10, Addison-Wesley, Reading, Mass., 1979.

Ju. I. Domšlak

[1961: 1]   Behavior at infinity of solutions of the evolution equation with unbounded operator. (Russian.) *Izv. Akad. Nauk Azerbaĭdžan SSR Ser. Fiz.-Mat. Tehn. Nauk* (1961), no. 3, 9–22. (MR 25, #3253a).

[1962: 1]   Behavior at infinity of solutions of the evolution equation with unbounded operator and non-linear perturbation. (Russian.) *Izv. Akad. Nauk Azerbaĭdžan Ser. Fiz.-Mat. Tehn. Nauk* (1962), no. 1, 3–14. (MR 25, #3253a).

[1965: 1]   On the behavior of solutions of quasilinear parabolic equations in a Banach space. (Russian.) *Izv. Akad. Nauk Azerbaĭdžan SSR Ser. Fiz.-Tehn. Mat. Nauk* (1965), no. 4, 10–19. (MR 33, #6177).

[1970: 1]   The oscillability of the solutions of vectorial differential equations. (Russian.) *Dokl. Akad. Nauk SSSR* **193** (1970), 21–23. (MR 43, #626).

[1971: 1]   The oscillatoriness of the solutions of second-order differential equations. (Russian.) *Differencial'nye Uravnenija* **7** (1971), 205–214. (MR 43, #7708).

[1972: 1]   The oscillatoriness and non-oscillatoriness of the solutions of vector differential equations. (Russian.) *Differencial'nye Uravnenija* **7** (1971), 961–969. (MR 44, #4335).

[1975: 1]   The oscillatory properties of the solutions of quasilinear and second-order differential equations in Hilbert space and in Euclidean space. (Russian.) *Differencial'nye Uravnenija* **11** (1975), 255–261, 394–395. (MR 51, #10787).

Ju. I. Domslak and A. Čaryev

[1968: 1]   Certain conditions for the unboundedness of solutions of differential equations in Hilbert space. (Russian.) *Izv. Akad. Nauk Turkmen. SSR Ser. Fiz.-Tehn. Him. Geol. Nauk* (1968), no. 2, 3–8. (MR 37, #4379).

[1968: 2]   Certain conditions for unboundedness of the solutions of second order differential equations in Hilbert space. (Russian.) *Izv. Akad. Nauk Azerbaĭdžan SSR Ser. Fiz.-Tehn. Mat. Nauk* (1968), no. 3, 124–128. (MR 40, #7604).

J. A. Donaldson

[1970: 1]   A singular abstract Cauchy problem. *Proc. Nat. Acad. Sci. U.S.A.* **66** (1970), 269–274. (MR 42, #706).

[1971: 1]   New integral representations for solutions of Cauchy's problem for abstract parabolic equations. *Proc. Nat. Acad. Sci. U.S.A.* **68** (1971), 2025–2027. (MR 44, #4414).

[1972: 1]   An operational calculus for a class of abstract operator equations. *J. Math. Anal. Appl.* **37** (1972), 167–184. (MR 45, #753).

[1975: 1]   The Cauchy problem for a first-order system of abstract operator equations. *Bull. Amer. Math. Soc.* **81** (1975), 576–578. (MR 54, #7985).

[1977: 1]   The abstract Cauchy problem. *J. Differential Equations* **25** (1977), 400–409. (MR 56, #12470).

J. A. Donaldson, S. G. Gibson, and R. Hersh

[1973: 1]   On the invariance principle of scattering theory. *J. Functional Analysis* **14** (1973), 131–145. (MR 50, #2949).

J. A. Donaldson and J. A. Goldstein

[1976: 1]   Some remarks on uniqueness for a class of singular Cauchy problems. *Proc. Amer. Math. Soc.* **54** (1976), 149–153. (MR 52, #11234)

J. A. Donaldson and R. Hersh

[1970: 1]   A perturbation series for Cauchy's problem for higher-order abstract parabolic equations. *Proc. Nat. Acad. Sci. U.S.A.* **67** (1970), 41–44. (MR 42, #2321).

C. C. Y. Dorea

[1976: 1]   Differentiability preserving properties of a class of semigroups. *Z. Warsch. Verw. Geb.* **36** (1976), 13–26. (MR 54, #3868).

J. R. Dorroh

[1966: 1]   Contraction semigroups in a function space. *Pac. J. Math.* **19** (1966), 35–38. (MR 34, #1860).

[1967: 1]   Some properties of a partial differential operator. *Ill. J. Math.* **11** (1967), 177–188. (MR 36, #2014).

[1975: 1]   A simplified proof of a theorem of Kato on linear evolution equations. *J. Math. Soc. Japan* **27** (1975), 474–478. (MR 52, #1404).

J. R. Dorroh and T. F. Lin

[1973: 1]   Markov processes with quasi-linear first-order forward differential equations. *J. Math. Anal. Appl.* **41** (1973), 205–225. (MR 47, #5970).

[1976: 1]   Markov processes with quasi-linear parabolic forward differential equations. *Adv. in Math.* **19** (1976), 19–47. (MR 52, #15703).

Ju. A. Dubinskiĭ

[1968: 1]   The Cauchy problem for operator-differential equations. (Russian.) *Dokl. Akad. Nauk SSSR* **181** (1968), 1046–1049. (MR 38, #3635).

[1969: 1]   A certain class of differential operator equations of higher order. (Russian.) *Dokl. Akad. Nauk SSSR* **187** (1969), 982–985. (MR 42, #3621).

[1971: 1]   Boundary value problems for certain classes of differential-operator equations of higher order. (Russian.) *Dokl. Akad. Nauk SSSR* **196** (1971), 32–34. (MR 43, #745).

N. Dunford

[1938: 1]   On one parameter groups of linear transformations. *Ann. of Math.* **39** (1938), 569–573.

N. Dunford and J. T. Schwartz

[1958: 1]   *Linear Operators*, part I. Interscience, New York, 1958.

[1963: 1]   *Linear Operators*, part II. Interscience-Wiley, New York, 1963.

N. Dunford and I. Segal

[1946: 1]   Semi-groups of operators and the Weierstrass theorem. *Bull. Amer. Math. Soc.* **52** (1946), 911–914. (MR8, 386)

D. R. Dunninger and H. A. Levine

[1976: 1]   Uniqueness criteria for solutions of singular boundary value problems. *Trans. Amer. Math. Soc.* **221** (1976), 289–301. (MR 53, #8596).

[1976: 2]   Uniqueness criteria for solutions to abstract boundary value problems. *J. Differential Equations* **22** (1976), 368–378. (MR 54, #10781).

R. Durrett

[1977: 1]   On Lie generators for one-parameter semigroups. *Houston J. Math.* **3** (1977), 163–164. (MR 56, #13005).

D. A. Dyment and P. E. Sobolevskiĭ

[1970: 1]   Coercive solvability of the Cauchy problem for differential equations in a Banach space that has a variable domain of definition. (Russian.) *Voronež. Gos. Univ. Trudy Mat. Fak. Vyp.* **4** (1970), 131–136. (MR 53, #8597).

[1970: 2]   The coercive solvability of the Cauchy problem for parabolic differential equations in a Banach space. (Russian.) *Voronež. Gos. Univ. Trudy*

Naučn.-Issled. Inst. Mat. VGU 1 Sb. Statei po Teorii Operatornyh Uravnenii (1970), 60–72. (MR 54, #13242).

[1971: 1]   Coercive solvability of abstract parabolic equations in Bochner spaces. (Russian.) *Voronež Gos. Univ. Trudi Naučn.-Issled. Inst. Mat. VGU Vyp.* **3** (1971), 27–34. (MR 47, #879).

E. B. Dynkin

[1965: 1]   *Markov Processes*, vol. I. Springer, Berlin, 1965.
[1965: 2]   *Markov Processes*, vol. II. Springer, Berlin, 1965.

A. A. Èfendieva

[1965: 1]   Behavior of solutions of the Cauchy problem for a differential equation of the second order in a Banach space in the case of unlimited time increase. (Russian.) *Akad. Nauk Azerbaïdžan SSSR Dokl.* **21** (1965), no. 11, 3–8. (MR 33, #1566).

M. Eguchi and Y. Kijima

[1975/75: 1]   A note on semigroups of Markov operators on $C(X)$. *Kodai Mat. Sem. Rep.* **26** (1974/75), 109–112. (MR 52, #1405).

D. M. Eĭdus

[1964: 1]   On the limiting amplitude principle. (Russian.) *Dokl. Akad. Nauk SSSR* **158** (1964), 794–797. (MR 29, #4982).
[1968: 1]   The behavior of the solutions of certain linear differential equations in Hilbert space for large values of the argument. (Russian.) *Izv. Vysš. Učebn. Zaved. Matematika*, no. 10 (77) (1968), 100–107. (MR 40, #1684).

J. Eisenfeld

[1970: 1]   Regularity conditions in linear hydrodynamic stability. *J. Math. Anal. Appl.* **31** (1970), 167–181. (MR 41, #5724).

I. Ekeland

[1979: 1]   Nonconvex minimization problems. *Bull. Amer. Math. Soc.* (N.S.) **1** (1979), 443–474.

J. Elliott

[1962: 1]   The equation of evolution in a Banach space. *Trans. Amer. Math. Soc.* **103** (1962), 470–483. (MR 25, #4368).
[1967: 1]   Lateral conditions for semi-groups in partially ordered spaces. *J. Math. Mech.* **16** (1967), 1071–1093. (MR 36, #734).
[1969: 1]   Lateral conditions for semigroups involving mappings in L^p, I. *J. Math. Anal. Appl.* **25** (1969), 388–410. (MR 39, #3350).

L. E. Elsgolts and S. B. Norkin

[1971: 1]   *Introduction to the Theory and Applications of Differential Equations with Deviating Arguments*. Izdat. "Nauka," Moscow, 1971. English translation: [1973: 1], Academic Press, London, 1973.

H. A. Emami-Rad

[1973: 1]   Les semi-groupes distributions de Beurling. *C. R. Acad. Sci. Paris* **276** (1972), A117–A119. (MR 56, #6465).

[1975: 1]   Semi-groupe distribution engendré par $-A^\alpha$. (English summary.) *C. R. Acad. Sci. Paris* (1975), A337–A339. (MR 55, #13293).

M. R. Embry and A. Lambert
[1977: 1]   Weighted translation semigroups. *Rocky Mountain J. Math.* **7** (1977), 333–344. (MR 56, #9319).
[1977: 2]   Subnormal weighted translation semigroups. *J. Functional Analysis* **24** (1977), 268–275. (MR 55, #6227).

M. R. Embry, A. L. Lambert, and L. J. Wallen
[1975: 1]   A simplified treatment of the structure of semigroups of partial isometries. *Michigan Math. J.* **22** (1975), 175–179. (MR 52, #11655).

T. H. Enikeeva
[1968: 1]   The Cauchy problem for a certain class of second-order nonlinear differential equations in Hilbert space. (Russian.) *Dokl. Akad. Nauk SSSR* **182** (1968), 1264–1267. (MR 38, #1376).
[1969: 1]   On the Cauchy problem for a class of second-order nonlinear differential equations in a Banach space. (Russian.) *Uspehi Mat. Nauk* **24** (1969), no. 1 (145), 195–196. (MR 39, #591).
[1969: 2]   Certain properties of solutions of the Cauchy problem for nonlinear second order equations in Banach and Hilbert spaces. (Russian. English summary.) *Vestnik Moskov. Univ. Ser. I Mat. Meh.* **24** (1969), no. 2, 40–49. (MR 40, #7605).
[1972: 1]   The Cauchy problem for evolution equations. (Russian.) *Uspehi Mat. Nauk* **27** (1972), no. 2 (164), 171–172. (MR 52, #11235).

S. N. Ethier
[1976: 1]   A class of degenerate diffusion processes occurring in population genetics. *Comm. Pure Appl. Math.* **29** (1976), 483–492. (MR 55, #1509).

M. A. Evgrafov
[1961: 1]   Structure of solutions of exponential growth for some operator equations. *Trudy Mat. Inst. Steklov* **60** (1961), 145–180 (Russian). (MR 25, #252).

I. D. Evzerov
[1977: 1]   Domains of definition of fractional powers of ordinary differential operators in L_p spaces. (Russian.) *Mat. Zametki* **21** (1977), 509–518. (MR 56, #12987).

I. D. Evzerov and P. E. Sobolevskiĭ
[1973: 1]   Fractional powers of ordinary differential operators. (Russian.) *Differencial'nye Uravnenija* **9** (1973), 228–240, 393. (MR 47, #5657).
[1976: 1]   The resolvent and fractional powers of ordinary differential operators in spaces of smooth functions. (Russian.) *Differencial'nye Uravnenija* **12** (1976), 227–233, 378. (MR 53, #8968).

J. Faraut
[1970: 1]   Semi-groupes de mesures complexes et calcul simbolique sur les
            générateurs infinitesimaux de semi-groupes d'opérateurs. *Ann Inst.
            Fourier Grenoble* **20** (1970), 235–301. (MR 54, #8348).

W. G. Faris
[1967: 1]   Product formulas for perturbations of linear propagators. *J. Functional
            Analysis* **1** (1967), 93–108. (MR 36, #5761).
[1967: 2]   The product formula for semigroups defined by Friedrichs extensions.
            Pacific J. Math. **22** (1967), 47–70 (MR 35, #5975).

H. O. Fattorini
[1969: 1]   Sur quelques équations différentielles pour les distributions vectorielles.
            C. R. Acad. Sci. Paris **268** (1969), A707–A709. (MR 39, #4666).
[1969: 2]   Ordinary differential equations in linear topological spaces, I. *J. Diff.
            Equations* **5** (1969), 72–105. (MR 43, #3593).
[1969: 3]   Ordinary differential equations in linear topological spaces, II. *J. Diff.
            Equations* **6** (1969), 50–70. (MR 43, #3594).
[1970: 1]   On a class of differential equations for vector-valued distributions.
            Pacific J. Math. **32** (1970), 79–104. (MR 45, #2309).
[1970: 2]   Extension and behavior at infinity of solutions of certain linear opera-
            tional differential equations. *Pacific J. Math.* **33** (1970), 583–615. (MR
            41, #8789).
[1970: 3]   A representation theorem for distribution semigroups. *J. Functional
            Analysis* **6** (1970), 83–96. (MR 54, #5901).
[1970: 4]   Uniformly bounded cosine functions in Hilbert space. *Indiana Univ.
            Math. J.* **20** (1970), 411–425. (MR 42, #2319).
[1971: 1]   The abstract Goursat problem. *Pacific J. Math.* **37** (1971), 51–83. (MR
            46, #9793).
[1971: 2]   Un teorema de perturbación para generadores de funciones coseno.
            Rev. Unión Matemática Argentina **25** (1971), 200–211. (MR 48, 4825)
[1973: 1]   The underdetermined Cauchy problem in Banach spaces. *Math. Ann.*
            200 (1973), 103–112. (MR 48, #659).
[1974: 1]   Two-point boundary value problems for operational differential equa-
            tions. *Ann. Scuola Norm. Sup. Pisa* I (1974), 63–79. (MR 54, #10782).
[1976: 1]   Some remarks on convolution equations for vector-valued distribu-
            tions. *Pacific J. Math.* **66** (1976), 347–371. (MR 57, #13471).
[1980: 1]   Vector-valued distributions having a smooth convolution inverse.
            Pacific J. Math. **90** (1980), 347–372.
[1980: 2]   Structure theorems for vector-valued ultradistributions. *J. Functional
            Analysis* **39** (1980), 381–407.
[1981: 1]   Some remarks on second-order abstract Cauchy problems. Funkcialaj
            Ekvacioj **24** (1981) 331–344.
[1983: 1]   On the angle of dissipativity of ordinary and partial differential
            operators. To appear.
[1983: 2]   Convergence and approximation theorems for vector-valued distribu-
            tions. To appear.

[1983: 3]   A note on fractional derivatives of semigroups and cosine functions.
             To appear.
[1983: 4]   Singular perturbation and boundary layer for an abstract Cauchy
             problem. To appear.

H. O. Fattorini and A. Radnitz
[1971: 1]   The Cauchy problem with incomplete initial data in Banach spaces.
             Michigan Math. Jour. **18** (1971), 291–320. (MR 44, #4415).

J. Favard
[1933: 1]   *Leçons sur les Fonctions Presque-Périodiques.* Gauthier-Villars, Paris,
             1933.

A. Favini
[1974: 1]   Sulle equazioni differenziali astratte degeneri. (English summary.) *Rend.
             Sem. Mat. Univ. Padova* **52** (1974), 243–263 (1975). (MR 53, #961).
[1974: 2]   Su certe equazioni astratte ultraparaboliche. *Rend. Sem. Mat. Univ.
             Padova* **51** (1974), 221–256 (1975). (MR 54, #8349).
[1974: 3]   Su certe equazioni astratte a coefficienti operatoriali. (English
             summary.) *Boll. Un. Mat. Ital.* **9** (1974), 463–475. (MR 54, #3449).
[1975: 1]   Su una equazione astratta di tipo Tricomi. *Rend. Sem. Mat. Univ.
             Padova* **53** (1975), 257–267. (MR 54, #1078).
[1975: 1]   Su un problema ai limiti per certe equazioni astratte del secondo
             ordine. (English summary.) *Rend. Sem. Mat. Univ. Padova* **53** (1975),
             211–230. (MR 56, #12472).
[1977: 1]   Su una equazione astratta alle derivate parziali. (English summary.)
             Rend. Mat. **9** (1976), 665–700 (1977). (MR 55, #6043).

W. Feller
[1952: 1]   The parabolic partial differential equation and the associated semi-
             groups of transformations. *Ann. of Math.* (2) **55** (1952), 468–519. (MR
             13, 948).
[1952: 2]   On a generalization of Marcel Riesz' potentials and the semi-groups
             generated by them. *Comm. Sém. Math. Univ. Lund* (Medd. Lunds
             Univ. Mat. Sem.), Tome Supplémentaire (1952), 72–81. (MR 14, 561).
[1953: 1]   Semi-groups of transformations in general weak topologies. *Ann. of
             Math.* (2) **57** (1953), 287–308. (MR 14, 881).
[1953: 2]   On the generation of unbounded semi-groups of bounded linear opera-
             tors. *Ann. of Math.* (2) **58** (1953), 166–174. (MR 14, 1093).
[1953: 3]   On positivity preserving semigroups of transformations on $C[r_1, r_2]$.
             Ann. Soc. Polon. Math. **25** (1953), 85–94. (MR 14, 1094).
[1954: 1]   The general diffusion operator and positivity preserving semi-groups in
             one dimension. *Ann. of Math.* (2) **60** (1954), 417–436. (MR 16, 488).
[1954: 2]   Diffusion processes in one dimension. *Trans. Amer. Math. Soc.* **77**
             (1954), 1–31. (MR 16, 150).
[1955: 1]   On second-order differential operators. *Ann. of Math.* (2) **61** (1955),
             90–105. (MR 16, 824).

[1955: 2]   On differential operators and boundary conditions. *Comm. Pure Appl. Math.* **8** (1955), 203–216. (MR 16, 927).

[1956: 1]   On generalized Sturm-Liouville operators. *Proceedings of the Conference on Differential Equations* (dedicated to A. Weinstein), 251–270. University of Maryland Press, College Park, Maryland, 1956. (MR 18, 575).

[1956: 2]   Boundaries induced by non-negative matrices. *Trans. Amer. Math. Soc.* **83** (1956), 19–54. (MR 19, 892).

[1957: 2]   On boundaries defined by stochastic matrices. *Applied probability. Proceeding of Symposia in Applied Mathematics*, vol. VII, 35–40. McGraw-Hill Book Co., New York, 1957. (MR 19, 988).

[1957: 3]   On boundaries and lateral conditions for the Kolmogorov differential equations. *Ann. of Math.* (2) **65** (1957), 527–270. (MR 19, 892).

[1957: 4]   Generalized second order differential operators and their lateral conditions. *Illinois J. Math.* **1** (1957), 459–504. (MR 19, 1054).

[1957: 5]   Sur une forme intrinsèque pour les opérateurs différentiels du second ordre. *Publ. Inst. Statist. Univ. Paris* **6** (1957), 291–301. (MR 20, #7126).

[1958: 1]   On the intrinsic form for second order differential operators. *Illinois J. Math.* **2** (1958), 1–18. (MR 19, #1052).

[1959: 2]   Differential operators with the positive maximum property. *Illinois J. Math.* **3** (1959), 182–186.

[1968: 1]   *An Introduction to Probability Theory and Its Applications*, vol. I, 3rd ed., Wiley, New York, 1968.

[1971: 2]   *An Introduction to Probability Theory and Its Applications*, vol. II, 2nd ed., Wiley, New York, 1971.

R. P. Feynman
[1951: 1]   An operator calculus having applications in quantum electrodynamics. *Phys. Rev.* **84** (1951), 108–128. (MR 13, 410).

G. Fichera
[1960: 1]   On a unified theory of boundary value problems for elliptic-parabolic equations of the second order. *Boundary Value Problems in Differential Equations* (R. Langer, ed.), University of Wisconsin Press, Madison, 1960.

J. P. Fink and W. S. Hall
[1973: 1]   Entrainment of frequency in evolution equations. *J. Diff. Equations* **14** (1973), 9–41. (MR 49, #10986).

W. E. Fitzgibbon
[1976: 1]   Nonlinear perturbation of linear evolution equations in a Banach space. *Ann. Mat. Pura Appl.* **110** (1976), 279–293. (MR 55, #3890).

H. Flaschka and G. Strang
[1971: 1]   The correctness of the Cauchy problem. *Advances in Math.* **6** (1971), 347–379. (MR 43, #5147).

R. J. Fleming, J. A. Goldstein, and J. E. Jamison
[1976: 1]   One parameter groups of isometries on certain Banach spaces. *Pacific
              J. Math.* **64** (1976), 145–151. (MR 54, #3372).

V. A. Fock
[1974: 1]   *Fundamentals of Quantum Mechanics.* (Russian.) Izdateltsvo "Nauka,"
              Moscow, 1976. English transl.: [1978: 1] Mir Publishers, Moscow,
              1978.

S. R. Foguel
[1964: 1]   A counterexample to a problem of Sz.-Nagy. *Proc. Amer. Math. Soc.*
              15 (1964), 788–790. (MR 29, #2646).

C. Foias
[1958: 1]   On strongly continuous semigroups of spectral operators in Hilbert
              spaces. *Acta Sci. Math. Szeged* **19** (1958), 188–191. (MR 23, #A2057).
[1960: 1]   Remarques sur le semi-groupes distributions d'opérateurs normaux.
              Portugaliae Math. **19** (1960), 227–242. (MR 26, #610).
[1961: 1]   Sur une propriété de non-analiticité des semi-groupes d'opérateurs
              isométriques. *Rev. Mat. Pures Appl.* **6** (1961), 551–552. (MR 26,
              #2888).
[1975: 1]   On the Lax-Phillips nonconservative scattering theory. *J. Functional
              Analysis* **19** (1975), 273–301. (MR 52, #6463).

C. Foias, G. Gussi, and V. Poenaru
[1957: 1]   L'étude de l'équation $du/dt = A(t)u$ pour certaines classes d'opérateurs
              non bornés de l'espace de Hilbert. *Trans. Amer. Math. Soc.* **86** (1957),
              335–347. (MR 19, #1185).
[1957: 2]   On the basic approximation theorem for semigroups of linear opera-
              tors. *Proc. Nat. Acad. Sci. U.S.A.* **43** (1957), 616–618. (MR 19, #757).
[1958: 1]   Generalized solutions of a quasilinear differential equation in Banach
              space. *Dokl. Akad. Nauk. SSSR (N.S.)* **119** (1958), 884–887. (MR 20,
              #4963).
[1958: 2]   Sur les solutions généralisées de certaines équations linéaires et quasi
              linéaires dans l'espace de Banach. *Rev. Math. Pures Appl.* **3** (1958),
              283–304. (MR 21, #6465).

C. Foias and B. Sz.-Nagy
[1967: 1]   *Analyse Harmonique des Opérateurs de l'Espace de Hilbert.* Masson,
              Paris–Akadémiai Kiadó, Budapest, 1967. English transl. (re-
              vised): [1970: 1] *Harmonic Analysis of Operators in Hilbert Space.*
              North-Holland, Amsterdam–American Elsevier, New York–Akadémiai
              Kiadó, Budapest, 1970.

H. Fong
[1976: 1]   Weak convergence for semigroups implies strong convergence of
              weighted averages. *Proc. Amer. Math. Soc.* **56** (1976), 157–161, **61**
              (1977), 186. (MR 53, #8928).

G. Forsythe and W. Wasow
[1960: 1]   *Finite-Difference Methods for Partial Differential Equations.* Wiley, New
              York, 1960.

J. M. Freeman
[1969/70: 1]   The tensor product of semigroups and the operator equation
                $SX - XT = A$. *J. Math. Mech.* **19** (1969/70), 819–828. (MR 43,
                #7961).

A. Friedman
[1963: 1]   *Generalized Functions and Partial Differential Equations.* Prentice-Hall,
             Englewood Cliffs, N.J., 1963.
[1964: 2]   Uniqueness of solutions of ordinary differential inequalities in Hilbert
             space. *Arch. Rat. Mech. Anal.* **17** (1964), 353–357. (MR 30, #1412).
[1966: 1]   Differentiability of solutions of ordinary differential equations in Hil-
             bert space. *Pacific J. Math.* **16** (1966), 267–271. (MR 32, #7906).
[1969: 1]   *Partial Differential Equations.* Holt, Rinehart and Winston, New York,
             1969.
[1969: 2]   Singular perturbation for the Cauchy problem and for boundary value
             problems. *J. Diff. Equations* **5** (1969), 226–261. (MR 38, #1355).

A. Friedman and Z. Schuss
[1971: 1]   Degenerate evolution equations in Hilbert space. *Trans. Amer. Math.
             Soc.* **161** (1971), 401–427. (MR 44, #853).

C. N. Friedman
[1972: 2]   Semigroup product formulas, compressions, and continual observation
             in quantum mechanics. *Indiana U. Math. J.* **21** (1972), 1001–1011.
             (MR 45, #6327).

K. O. Friedrichs
[1934: 1]   Spektraltheorie halbbeschränkter Operatoren I. *Math. Ann.* **109** (1934),
             465–487.
[1944: 1]   The identity of weak and strong extensions of differential operators.
             Trans. Amer. Math. Soc. **55** (1944), 132–151. (MR 5, 188).
[1954: 1]   Symmetric hyperbolic linear differential equations. *Comm. Pure Appl.
             Math.* **7** (1954), 345–392. (MR 16, 44).
[1958: 1]   Symmetric positive linear differential equations. *Comm. Pure Appl.
             Math.* **11** (1958), 333–418. (MR 20, #7147).

Y. Fujie and H. Tanabe
[1973: 1]   On some parabolic equations of evolution in Hilbert space. *Osaka J.
             Math.* **10** (1973), 115–130. (MR 48, #12143).

D. Fujiwara
[1966: 1]   A characterisation of exponential distribution semi-groups. *J. Math.
             Soc. Japan* **18** (1966), 267–274. (MR 33, #6420).
[1967: 1]   Concrete characterization of the domains of fractional powers of some
             elliptic differential operators of the second order. *Proc. Japan Acad.* **43**
             (1967), 82–86. (MR 35, #7170).
[1968: 1] L^p-theory for characterizing the domain of the fractional powers of
             $-\Delta$ in the half space. *J. Fac. Sci. Univ. Tokyo Sect. I* **15** (1968),
             169–177. (MR 39, #823).

M. Fukamiya
[1940: 1]   On one-parameter groups of operators. *Proc. Imp. Acad. Tokyo* **16**
             (1940), 262–265. (MR 2, 105).

H. Gajewski
[1971: 1]   Zur Approximation nichtlinearer Evolutionsgleichungen durch Ab-
             strakte Differentialgleichungen mit Lipschitzestetigen Operatoren.
             Math. Nachr. **48** (1971), 377–385. (MR 52, #11235).
[1972: 1]   Über eine Klasse nichtlinearer abstrakter Wellengleichungen im
             Hilbert-Raum. *Math. Nachr.* **52** (1972), 371–383. (MR 47, #2159).

D. Gaspar
[1971: 1]   On a certain class of semi-groups of operators in Hilbert spaces.
             (Russian summary.) *Bull. Acad. Polon. Sci. Sér. Sci. Math. Astronom.
             Phys.* **19** (1971), 463–468. (MR 46, #717).

A. T. Gašymov
[1975: 1]   A difference method of solution of the Cauchy problem for abstract
             equations (Russian.) *Izv. Akad. Nauk Azerbaĭdžan SSR Ser. Fiz.-Tehn.
             Mat. Nauk* (1975), 17–23. (MR 52, #8595).

M. G. Gašymov
[1971: 1]   On the theory of evolution equations of regular type. (Russian.) *Dokl.
             Akad. Nauk SSSR* **200** (1971), 13–16. (MR 45, #7497).
[1972: 1]   On the theory of higher order evolution equations. (Russian.) *Dokl.
             Akad. Nauk SSSR* **206** (1972), 780–783. (MR 48, #906).

M. Gaultier
[1972/73: 1] Solutions "faibles periodiques" d'équations d'évolution linéaires
             du premier ordre pérturbées. Publ. Math. Univ. Bordeaux Année I
             (1972/73) Exp. no. 1, 20 pp. (MR 54, #13243).

Z. G. Gegečkori and G. V. Demidov
[1973: 1]   The convergence of the weak approximation method. (Russian.) *Dokl.
             Akad. Nauk SSSR* **213** (1973), 264–266. (MR 49, #5502).

I. M. Gelfand
[1939: 1]   On one-parametrical groups of operators in a normed space. *Dokl.
             Akad. Nauk SSSR* **25** (1939), 713–718.

I. M. Gelfand and G. E. Šilov
[1958: 1]   *Generalized Functions*, vol. I: *Properties and Operations.* Goztekhizdat,
             Moscow, 1958. English trans.: [1964: 1], Academic Press, New York,
             1964.
[1958: 2]   *Generalized Functions*, vol. II: *Spaces of Fundamental and Generalized
             Functions.* Gostekhizdat, Moscow, 1958. English transl.: [1968: 1],
             Academic Press, New York, 1968.
[1958: 3]   *Generalized Functions*, vol. III: *Theory of Differential Equations.*
             Gostekhizdat, Moscow, 1958. English transl.: [1967: 1], Academic Press,
             New York, 1967.

A. E. Gel'man
[1963: 1]   Analytic solutions of a class of operator equations. (Russian.) *Dokl. Akad. Nauk SSSR* **153** (1963), 1234–1237. (MR 29, #2657).
[1968: 1]   Properties of bounded solutions of differential equations in *B*-spaces. (Russian.) *Dokl. Akad. Nauk SSSR* **178** (1968), 515–517. (MR 37, #4436).

A. E. Gel'man and E. N. Gerst
[1968: 1]   A generalization of Ljapunov's first stability theorem for certain classes of differential equations in Banach spaces. (Russian.) *Dokl. Akad. Nauk SSSR* **178** (1968), 1241–1244. (MR 36, #6748).

M. G. Gendler
[1956: 1]   On one-parameter semigroups of functional transformations. (Russian.) *Dokl. Akad. Nauk SSSR (N.S.)* **111** (1956), 524–527. (MR 20, #6662).

E. N. Geršt
[1969: 1]   A generalization of Ljapunov's first theorem on stability to certain classes of differential equations in Banach space. (Russian.) *Differencial'nye Uravnenija* **5** (1969), 833–847. (MR 40, #3023).

L. M. Gerštein and P. E. Sobolevskiĭ
[1974: 1]   Coercive solvability of general boundary value problems for second order elliptic differential equations in a Banach space. (Russian.) *Differencial'nye Uravnenija* **10** (1974), 2059–2061, 2086. (MR 50, #10476).
[1975: 1]   Coercive solvability of general boundary value problems for second order elliptic differential equations in a Banach space II. (Russian.) *Differencial'nye Uravnenija* **11** (1975), 1335–1337, 1351. (MR 51, #10788).

G. Geymonat
[1972: 1]   Quelques remarques sur l'utilisation des méthodes itératives dans l'approximation des solutions des équations paraboliques linéaires. *Symposia Mathematica, vol. X (Convegno di Analisi Numerica, INDAM, Roma, Gennaio 1972)* 381–402. Academic Press, London, 1972. (MR 55, #5991).

A. G. Gibson
[1972: 1]   A discrete Hille-Yosida-Phillips theorem. *J. Math. Anal. Appl.* **39** (1972), 761–770. (MR 47, #881).

J. S. Gibson and L. G. Clark
[1977: 1]   Sensitivity analysis for a class of evolution equations. *J. Math. Anal. Appl.* **58** (1977), 22–31. (MR 56, #3426).

I. I. Gihman and A. V. Skorohod
[1975: 1]   *The Theory of Stochastic Processes* I. Springer, Berlin, 1975.
[1975: 2]   *The Theory of Stochastic Processes* II. Springer, Berlin, 1975.
[1979: 1]   *The Theory of Stochastic Processes* III. Springer, Berlin, 1979.

D. Gilbarg and N. S. Trudinger
[1977: 1]   *Elliptic Partial Differential Equations of the Second Order.* Springer, New York, 1977.

T. A. Gillespie and T. T. West
[1971/72: 1]   Operators generating weakly compact groups. *Indiana Univ. Math. J.* **21** (1971/72), 671–688. (MR 47, #4040).

H. A. Gindler and J. A. Goldstein
[1975: 1]   Dissipative operator versions of some classical inequalities. *J. Analyse Math.* **28** (1975), 213–238. (MR 58, #2434).

Ju. P. Ginzburg
[1967: 1]   On the multiplicative representation of $\mathcal{J}$-contracting operator functions, *I. Mat. Issled.* **2** (1967), n. 2, 52–83. (MR 38, #1551a).

E. Giusti
[1967: 1]   Funzioni coseno periodiche. *Boll. Un. Mat. Ital.* (3) **22** (1967), 478–485. (MR 38, #1559).

J. Glimm and A. Jaffe
[1969: 1]   Singular perturbations of self-adjoint operators. *Comm. Pure Appl. Math.* **22** (1969), 401–414. (MR 43, #7955).

J. Globevnik
[1971: 1]   On fractional powers of linear positive operators acting in Banach spaces. *Glasnik Mat. Ser. III* **6** (26) (1971), 79–96. (MR 46, #718).
[1971: 2]   A note on $A^{1/2}$ where $-A$ generates a bounded group. *Glasnik Mat. Ser. III* **6** (26) (1971), 301–306. (MR 46, #4274).

V. P. Gluško
[1971: 1]   The smoothness of the solutions of degenerate differential equations in a Banach space. (Russian.) *Dokl. Akad. Nauk SSSR* **198** (1971), 20–22. (MR 43, #7747).

V. P. Gluško and S. G. Kreĭn
[1968: 1]   Degenerate linear differential equations in a Banach space. (Russian.) *Dokl. Akad. Nauk SSSR* **181** (1968), 784–789. (MR 38, #393).

A. N. Godunov
[1974: 1]   Peano's theorem in an infinite-dimensional Hilbert space is false, even in a weakened formulation. (Russian.) *Mat. Zametki* **15** (1974), 467–477. (MR 50, #5127).
[1974: 2]   Linear differential equations in locally convex spaces. (Russian. English summary.) *Vestnik Moskov. Univ. Ser. I Mat. Meh.* **29** (1974), 31–39. (MR 52, #933).

L. Ju. Gofman
[1976: 1]   The uniqueness of the solution of the Cauchy problem for a convolution equation. (Russian. English summary.) *Vestnik Moskov. Univ. Ser. I Mat. Meh.* **31** (1976), 3–11. (MR 53, #6361).

J. A. Goldstein

[1969: 1]   Semigroups and second-order differential equations. *J. Functional Analysis* **4** (1969), 50–70. (MR 40, #7875).

[1969: 2]   Some remarks on infinitesimal generators of analytic semigroups. *Proc. Amer. Math. Soc.* **22** (1969), 91–93. (MR 39, #4706).

[1969: 3]   An asymptotic property of solutions of wave equations. *Proc. Amer. Math. Soc.* **23** (1969), 359–363. (MR 40, #3365).

[1969: 4]   Abstract evolution equations. *Trans. Amer. Math. Soc.* **141** (1969), 159–185. (MR 40, #789).

[1970: 1]   Corrigendum on: Time-dependent hyperbolic equations. *J. Functional Analysis* **6** (1970), 347. (MR 44, #665).

[1970: 2]   An asymptotic property of solutions of wave equations, II. *J. Math. Anal. Appl.* **32** (1970), 392–399. (MR 42, #2183).

[1970: 3]   A Lie product formula for one-parameter groups of isometries on Banach spaces. *Math. Ann.* **186** (1970), 299–306. (MR 41, #9049).

[1970: 4]   On a connection between first and second order differential equations in Banach spaces. *J. Math. Anal. Appl.* **30** (1970), 246–251. (MR 41, #2182).

[1970: 5]   *Semigroups of Operators and Abstract Cauchy Problems.* Tulane University Lecture Notes, New Orleans, 1970.

[1971: 1]   On bounded solutions of abstract differential equations. *Ann. Univ. Ferrara* **16** (1971), 39–43. (MR 47, #9346).

[1972: 1]   On the growth of solutions of inhomogeneous abstract wave equations. *J. Math. Anal. Appl.* **37** (1972), 650–654. (MR 46, #2226).

[1972: 2]   Some counterexamples involving selfadjoint operators. *Rocky Mountain J. Math.* **2** (1972), 143–149. (MR 44, #7348).

[1972: 3]   Approximation of nonlinear semigroups and evolution equations. *J. Math. Soc. Japan* **24** (1972), 558–573. (MR 46, #6115).

[1973: 1]   Groups of isometries in Orlicz spaces. *Pacific J. Math.* **48** (1973), 387–393. (MR 52, #11564).

[1973: 2]   On the domain of the square of the infinitesimal generator of a group. *Ann. Acad. Brasil. Cien.* **45** (1973), 1–3. (MR 49, #5932).

[1974: 1]   A perturbation theorem for evolution equations and some applications. *Illinois J. Math.* **18** (1974), 196–207. (MR 50, #3001).

[1974: 2]   On the convergence and approximation of cosine functions. *Aeq. Math.* **10** (1974), 201–205. (MR 50, #10901).

[1975/76: 1]   Variable domain second-order evolution equations. *Applicable Anal.* **5** (1976), 283–291. (MR 54, #3450).

[1976: 1]   Semigroup-theoretic proofs of the central limit theorem and other theorems of analysis. *Semigroup Forum* **12** (1976), 189–206, 388. (MR 54, #3496a).

[1976: 2]   Semigroup-theoretic proofs of the central limit theorem and other theorems of analysis. *Semigroup Forum* **12** (1976), 189–206. Corrigendum. *Semigroup Forum* **12** (1976), 388. (MR 54, #3496).

J. Goldstein and A. Lubin

[1974: 1]   On bounded solutions of nonlinear differential equations in Hilbert space. *SIAM J. Math. Anal.* **5** (1974), 837–840. (MR 50, #7714).

J. A. Goldstein and C. A. Monlezun
[1976: 1]   Temporally inhomogeneous scattering theory II. Approximation theory
             and second-order equations. *SIAM J. Math. Anal.* **7** (1976), 276–290.
             (MR 54, #14173).

J. A. Goldstein and J. T. Sandefur Jr.
[1976: 1]   Asymptotic equipartition of energy for differential equations in Hilbert
             space. *Trans. Amer. Math. Soc.* **219** (1976), 397–406. (MR 53, #13767).

I. I. Golicev
[1974: 1]   Certain differential equations in Hilbert space. (Russian.) *Dokl. Akad.
             Nauk SSSR* **214** (1974), 989–992. (MR 49, #3282).

K. K. Golovkin
[1969: 1]   *Parametrically Normed Spaces and Normed Massives.* Izdat. "Nauka,"
             Moscow, 1969. English transl.: [1971: 1], Amer. Math. Soc., Provi-
             dence, 1971.

D. P. Goodall and I. M. Michael
[1973: 1]   The domains of self-adjoint extensions of a Schrödinger operator. *J.
             London Math. Soc.* **7** (1973), 265–271. (MR 48, #7015).

R. W. Goodman
[1964: 1]   One-sided invariant subspaces and domains of uniqueness for hyper-
             bolic equations. *Proc. Amer Math. Soc.* **15** (1964), 653–660. (MR 30,
             #360).
[1967: 1]   On localization and domains of uniqueness. *Trans. Amer. Math. Soc.*
             127 (1967), 98–106. (MR 34, #6266).
[1970: 1]   One-parameter groups generated by operators in an enveloping alge-
             bra. *J. Functional Analysis* **6** (1970), 218–236. (MR 42, #3229).

V. R. Gopala Rao and T. W. Ting
[1977: 1]   Pointwise solutions of pseudoparabolic equations in the whole space. *J.
             Differential Equations* **23** (1977), 125–161. (MR 55, #3514).

V. I. Gorbačuk and M. L. Gorbačuk
[1976: 1]   Some questions on the spectral theory of differential equations of
             elliptic type in a space of vector-valued functions. (Russian.) *Ukrain.
             Mat. Ž.* **28** (1976), 313–324, 428. (MR 55, #3458).

M. L. Gorbačuk and A. N. Kočubeĭ
[1971: 1]   Self-adjoint boundary value problems for a second-order differential
             equation with unbounded operator coefficients. (Russian.) *Funkcional
             Anal. i Prilozen.* **5** (1971), no. 4, 67–68. (MR 44, #5820).

J. de Graaf
[1971: 1]   A constructive approach to one-parameter semi-groups of operators in
             Hilbert space. *Arch. Rat. Mech. Anal.* **43** (1971), 125–153. (MR 48,
             #12144).
[1971/72: 1] Propagative linear dynamical systems. *Arch. Rat. Mech. Anal.* **44**
             (1971/72), 157–164. (MR 49, #765).

J. De Graaf and L. A. Peletier
[1969: 1]   On the solution of Schrödinger-like wave equations. *Rend. Sem. Mat. Univ. Padova* **42** (1969), 329–340. (MR 40, #6079).

H. Grabmüller
[1972: 1]   Approximation von Evolutionsgleichungen. *Z. Angew. Math. Mech.* **53** (1973), T159–T160. (MR 51, #3957).
[1975: 1]   Relativ Akkretive Operatoren und Approximation von Evolutionsgleichungen 2. Art. I, II. *Math. Nachr.* **66** (1975), 67–87, 89–100. (MR 51, #11143).

N. Grădinaru
[1973: 1]   Approximating difference schemes of Cauchy problems for evolution systems and approximation of two-parameter semigroups of linear operators. (Romanian summary.) *An. Sti. Univ. Al. I. Cuza Iasi Sect. I a Mat.* (*N.S.*) **19** (1973), 161–173. (MR 49, #3594).

I. S. Gradstein and I. M. Ryžik
[1963: 1]   *Tables of Integrals, Sums, Series and Products.* (Russian.) Gostekhizdat, Moscow, 1963.

R. J. Griego and R. Hersh
[1969: 1]   Random evolutions, Markov chains, and systems of partial differential equations. *Proc. Nat. Acad. Sci. U.S.A.* **62** (1959), 305–308. (MR 42, #5099).
[1971: 1]   Theory of random evolutions with applications to partial differential equations. *Trans. Amer. Math. Soc.* **156** (1971), 405–418. (MR 43, #1261).

P. Grisvard
[1965: 1]   Espaces d'interpolation et équations opérationnelles. *C. R. Acad. Sci. Paris* **260** (1965), 1536–1538. (MR 30, #4138).
[1966: 1]   Problèmes aux limites résolus par le calcul opérationnel. *C. R. Acad. Sci. Paris* **262** (1966), A1306–A1308. (MR 33, #7582).
[1967: 1]   Équations opérationnelles abstraites dans les espaces de Banach et problèmes aux limites dans des ouverts cylindriques. *Ann. Scuola Norm. Sup. Pisa* (3) **21** (1967), 307–347. (MR 37, #5675).
[1969: 1]   Équations différentielles abstraites. *Ann. Sci. École Norm. Sup.* (4) (1969), 311–395. (MR 42, #5101).

K. Gröger
[1976: 1]   An iteration method for nonlinear second-order evolution equations. *Comm. Univ. Math. Univ. Carolinae* **17** (1976), 575–592. (MR 54, #1044).

L. Gross
[1966: 1]   The Cauchy problem for the coupled Maxwell and Dirac equations. *Comm. Pure Appl. Math.* **19** (1966), 1–15. (MR 32, #7932).

L. A. Groza

[1970: 1]   The asymptotic solutions of certain linear differential equations in a Banach space that contain an unbounded operator. (Russian.) *Differencial'nye Uravnenija* **6** (1970), 289–297. (MR 42, #635).

[1973: 1]   Asymptotic solutions of second-order linear homogeneous equations in a Banach space with a small parameter multiplying the highest derivative. (Russian.) *Differencial'nye Uravnenija* **9** (1973), 1393–1402, 1547. (MR 48, #6596).

I. S. Gudovič

[1966: 1]   An abstract scheme for finite-difference methods (Russian.) *Z. Vycisl. Mat. i Mat. Fiz.* **6** (1966), 916–921. (MR 34, #6585).

A. N. Gudovič and N. N. Gudovič

[1970: 1]   Convergence of a method of fractional step type for evolution equations. (Russian.) *Voronež. Gos. Trudy Naučn.-Issled. Inst. Mat. VGU Vyp. Sb. Statei po Teorii Operatornyh Uravnenii* (1970), 35–43. (MR 54, #13244).

I. S. Gudovič and S. G. Kreĭn

[1971: 1]   Certain boundary value problems that are elliptic in a subspace. (Russian.) *Mat. Sb. (N.S.)* **84** (126) (1971), 595–606. (MR 43, #7764).

J. P. Guillement and Pham Te Lai

[1975: 1]   Sur la caractérisation des semi-groupes distributions. *J. Fac. Sci. Univ. Tokyo Sect. IA Math.* **22** (1975), 299–318. (MR 56, #3676).

H. Günzler and S. Zaidman

[1969/70: 1]   Almost-periodic solutions of a certain abstract differential equation. *J. Math. Mech.* **19** (1969/70), 155–158. (MR 40, #7607).

K. Gustafson

[1966: 1]   A perturbation lemma. *Bull. Amer. Math. Soc.* **72** (1966), 334–338. (MR 32, #4555).

[1968: 1]   The angle of an operator and positive operator products. *Bull. Amer. Math. Soc.* **74** (1968), 488–492. (MR 36, #5718).

[1968: 2]   Positive (noncommuting) operator products and semigroups. *Math. Z.* **105** (1968), 160–172. (MR 37, #3398).

[1968: 3]   A note on left multiplication of semigroup generators. *Pacific J. Math.* **24** (1968), 463–465. (MR 36, #6979).

[1969: 1]   Doubling perturbation sizes and preservation of operator indices in normed linear spaces. *Proc. Cambridge Philos. Soc.* **66** (1969), 281–294. (MR 40, #4808).

K. Gustafson and G. Lumer

[1972: 1]   Multiplicative perturbation of semigroup generators. *Pacific J. Math.* **41** (1972), 731–742. (MR 47, #5652).

K. E. Gustafson and P. A. Rejto

[1973: 1]   Some essentially self-adjoint Dirac operators with spherically symmetric potentials. *Israel J. Math.* **14** (1973), 63–75. (MR 50, #8127).

K. Gustafson and D. Sather

[1972: 1]   Large nonlinearities and monotonicity. *Arch. Rat. Mech. Anal.* **48** (1972), 109–122. (MR 49, #1239).

[1973: 2]   Large nonlinearities and closed linear operators. *Arch. Rat. Mech. Anal.* **52** (1973), 10–19. (MR 49, #1251).

K. Gustafson and K. Sato

[1969: 1]   Some perturbation theorems for nonnegative contraction semigroups. *J. Math. Soc. Japan* **21** (1969), 200–204. (MR 38, #6397).

K. Gustafson and B. Zwahlen

[1969: 1]   On the cosine of unbounded operators. *Acta Sci. Math. Szeged* **30** (1969), 33–34. (MR 39, #6099).

A. Guzmán

[1976: 1]   Growth properties of semigroups generated by fractional powers of certain linear operators. *J. Functional Analysis* **23** (1976), 331–352. (MR 55, #1133).

K. S. Ha

[1972: 1]   On convergence of semigroups of operators in Banach spaces. *J. Korean Math. Soc.* **9** (1972), 91–99. (MR 48, #12145).

[1973: 1]   Une note sur la convergence des semi-groupes des opérateurs dans les espaces de Banach. *Kyungpook Math.* **13** (1973), 87–95. (MR 48, #4826).

M. Hackman

[1968: 1]   The abstract time-dependent Cauchy problem. *Trans. Amer. Mat. Soc.* **133** (1968), 1–50. (MR 40, #3355).

J. Hadamard

[1923: 1]   *Lectures on Cauchy's Problem in Linear Partial Differential Equations.* Yale University Press, New Haven, 1923. Reprinted by Dover, New York, 1952.

[1924: 1]   Principe de Huygens et prolongement analytique. *Bull. Soc. Math. France* **52** (1924), 241–278.

[1924: 2]   Le principe de Huygens. *Bull. Soc. Math. France* **52** (1924), 241–278.

J. Hale

[1977: 1]   *Theory of Functional Differential Equations.* Springer, New York, 1977.

P. R. Halmos

[1964: 1]   On Foguel's answer to Nagy's question. *Proc. Amer. Math. Soc.* **15** (1964), 791–793.

G. H. Hardy and E. M. Wright

[1960: 1]   *An Introduction to the Theory of Numbers*, 4th ed. Oxford University Press, Oxford, 1960.

M. Hasegawa

[1964: 1]   A note on the convergence of semi-groups of operators. *Proc. Japan. Acad.* **40** (1964), 262–266. (MR 30, #3376).

[1966: 1]   On contraction semi-groups and (di)-operators. *J. Math. Soc. Japan* **18** (1966), 290–302. (MR 34, #1861).

[1967: 1]   Some remarks on the generation of semi-groups of linear operators. *J. Math. Anal. Appl.* **20** (1967), 454–463. (MR 36, #735).

[1967: 2]   On the convergence of resolvents of operators. *Pacific J. Math.* **21** (1967), 35–47. (MR 35, #797).

[1968: 1]   Correction to "On the convergence of resolvents of operators." *Pacific J. Math.* **27** (1968), 641. (MR 38, #6380).

[1969: 1]   On a property of the infinitesimal generators of semi-groups of linear operators. *Math. Ann.* **182** (1969). 121–126. (MR 40, #3356).

[1971: 1]   On quasi-analytic vectors for dissipative operators. *Proc. Amer. Math. Soc.* **29** (1971), 81–84. (MR 43, #981).

M. Hasegawa and R. Sato
[1976: 1]   A general ratio ergodic theorem for semigroups. *Pacific J. Math.* **62** (1976), 435–437.

H. Heinz and W. Von Wahl
[1975: 1]   Zu einem Satz von F. E. Browder über nichtlineare Wellengleichungen. *Math. Z.* **141** (1975), 33–45. (MR 51, #1510).

J. Hejtmanek
[1976]   Stark stetige Halbgruppen und Gruppen als Lösungen von Cauchy-Problemen in der Mathematischen Physik. (English summary.) *Monatsh. Math.* **82** (1976), 151–162. (MR 55, #3870).

L. L. Helms
[1961: 1]   A representation of the infinitesimal generator of a diffusion process. *Bull. Amer. Math. Soc.* **67** (1961), 479–482. (MR 24, #A2256).

J. W. Helton
[1974: 1]   Discrete time systems, operator models and scattering theory. *J. Functional Analysis* **16** (1974), 15–38. (MR 56, #3652).

D. Henry
[1981: 1]   *Geometric Theory of Semilinear Parabolic Equations.* Lecture Notes in Mathematics, vol. 840, Springer-Verlag, Berlin–New York, 1981.

J. V. Herod
[1971: 1]   A product integral representation for an evolution system. *Proc. Amer. Math. Soc.* **27** (1971), 549–556. (MR 43, #1001).

[1971: 2]   A pairing of a class of evolution systems with a class of generators. *Trans. Amer. Math. Soc.* **157** (1971), 247–260. (MR 43, #6778).

R. Hersh
[1964: 1]   Boundary conditions for equations of evolution. *Arch. Rational Mech. Anal.* **16** (1964), 243–264. (MR 34, #7943).

[1969: 1]   Direct solution of a general one-dimensional linear parabolic equation via an abstract Plancherel formula. *Proc. Nat. Acad. Sci. U.S.A.* **63** (1969), 648–654. (MR 41, #7476).

[1970: 1]   Explicit solution of a class of higher-order abstract Cauchy problems. *J. Diff. Equations* **8** (1970), 570–579. (MR 42, #5102).

[1974: 1]   Random evolutions: a survey of results and problems. *Rocky Mountain J. Math.* **4** (1974), 443–477. (MR 52, #15676).

H. Hess, R. Schrader, and D. A. Uhlenbrock

[1977: 1]   Domination of semigroups and generalization of Kato's inequality. *Duke Math. J.* **44** (1977), 892–904. (MR 56, #16441).

E. Heyn

[1962: 1]   Die Differentialgleichung $dT/dt = P(t)T$ für Operator-funktionen. *Math. Nachr.* **24** (1962), 281–330. (MR 28, #308).

E. Hille

[1936: 1]   Notes on linear transformations. I. *Trans. Amer. Math. Soc.* **39** (1936), 131–153.

[1938: 1]   On semi-groups of linear transformations in Hilbert space. *Proc. Nat. Acad. Sci.* **24** (1938), 159–161.

[1938: 2]   Analytic semi-groups in the theory of linear transformations. 9ième Congrès Math. Scand., Helsingfors, 1938, 135–145.

[1939: 1]   Notes on linear transformations, II. Analyticity of semi-groups. *Ann. Math.* (2) **40** (1939), 1–47.

[1942: 1]   Representation of one-parameter semi-groups of linear transformations. *Proc. Nat. Acad. Sci. U.S.A.* **28** (1942), 175–178. (MR 4, 13).

[1942: 2]   On the analytical theory of semi-groups. *Proc. Nat. Acad. Sci. U.S.A.* **28** (1942), 421–424.

[1947: 1]   Sur les semi-groupes analytiques. *C. R. Acad. Sci. Paris* **225** (1947), 445–447. (MR 9, 193).

[1948: 1]   *Functional Analysis and Semi-Groups.* Amer. Math. Soc. Colloquium Publications, vol. 31, New York, 1948.

[1950: 1]   On the differentiability of semi-group operators. *Acta Sci. Math. Szeged* **12**, Leopoldo Fejér et Frederico Riesz LXX annos natis dedicatus, Pars B (1950), 19–24. (MR 12, 110).

[1951: 1]   On the generation of semi-groups and the theory of conjugate functions. *Proc. R. Physiogr. Soc. Lund*, **21**:14 (1951), 130–142. (MR 14, 882).

[1952: 1]   On the integration problem for Fokker-Planck's equation in the theory of stochastic processes. *Den 11te Skandinaviske Matematikerkongress, Trondheim*, 1949, 183–194, Johan Grundt Tanums Forlag, Oslo, 1952. (MR 14, 758).

[1952: 2]   On the generation of semi-groups and the theory of conjugate functions. *Comm. Sém. Math. Univ. Lund* (Medd. Lunds Univ. Mat. Sem.), Tome Supplémentaire (1952), 122–134. (MR 14, 882).

[1952: 3]   A note on Cauchy's problem. *Ann. Soc. Polon. Math.* **25** (1952), 56–68. (MR 15, 39).

[1953: 1]   Quelques remarques sur les équations de Kolmogoroff. *Bull. Soc. Math. Phys. Serbie* **5** (1953), 3–14. (MR 15, 880).

[1953: 2]   Sur le problème abstrait de Cauchy. *C. R. Acad. Sci. Paris* **236** (1953), 1466–1467. (MR 15, 325).

[1953: 3]   Le problème abstrait de Cauchy. *Univ. e Politecnico Torino Rend. Sem. Mat.* **12** (1953), 95–103. (MR 15, 71).

[1954: 1]   Une généralisation du problème de Cauchy. *Ann. Inst. Fourier Grenoble* **4** (1954), 31–48. (MR 15, 718).

[1954: 2]   The abstract Cauchy problem and Cauchy's problem for parabolic differential equations. *J. Analyse Math.* **3** (1954), 81–196. (MR 16, 45).

[1954: 3]   On the integration of Kolmogoroff's differential equations. *Proc. Nat. Acad. Sci. U.S.A.* **40** (1954), 20–25. (MR 15, 706).

[1954: 4]   Sur un théorème de perturbation. *Univ. e Politecnico Torino Rend. Sem. Mat.* **13** (1954), 169–184. (MR 16, 833).

[1956: 1]   Perturbation methods in the study of Kolmogoroff's equations. *Proceedings of the International Congress of Mathematicians, 1954, Amsterdam*, vol. III, 365–376, Erven P. Noordhoff N. V., Groningen; North-Holland Publishing Co., Amsterdam, 1956. (MR 19, 327).

[1957: 1]   Problème de Cauchy: existence et unicité des solutions. *Bull. Math. Soc. Sci. Math. Phys. R. P. Roumaine (N.S.)* **1** (1957), 141–143. (MR 20, #6042).

[1959: 1]   *Analytic Function Theory*, vol. I. Ginn and Co., New York, 1959.

[1962: 1]   *Analytic Function Theory*, vol. II. Ginn and Co., New York, 1962.

[1972: 1]   *Methods in Classical and Functional Analysis*. Addison-Wesley, Reading, Mass., 1972.

[1972: 2]   On the Landau-Kallman-Rota inequality. *J. Approximation Theory* **6** (1972), 117–122. (MR 49, #767).

[1972: 3]   Generalizations of Landau's inequality to linear operators. *Linear operators and approximation* (Proc. Conf. Oberwolfach, 1971), 20–32. *Internat. Ser. Numer. Math.*, vol. 20, Birkhäuser, Basel, 1972. (MR 53, #6354).

E. Hille and R. S. Phillips
[1957: 1]   *Functional Analysis and Semi-Groups*. Amer. Math. Soc. Colloquium Publications, vol. 31, Providence, R.I., 1957.

F. Hirsch
[1972: 1]   Intégrales de résolvantes. *C. R. Acad. Sci. Paris* **274** (1972), A303–A306. (MR 45, #4208).

[1972: 2]   Intégrales de résolvantes et calcul symbolique. (English summary.) *Ann. Inst. Fourier Grenoble* **22** (1972), 239–264. (MR 51, #3958).

[1976: 1]   Domaines d'opérateurs répresentés comme intégrales de résolvantes. *J. Functional Analysis* **23** (1976), 199–217. (MR 55, #1134).

I. Hlaváček
[1971: 1]   On the existence and uniqueness of solution of the Cauchy problem for linear integro-differential equations with operator coefficients. *Apl. Mat.* **16** (1971), 64–80. (MR 45, #9206).

J. A. R. Holbrook
[1974: 1]   A Kallman-Rota inequality for nearly Euclidean spaces. *Advances in Math.* **14** (1974), 335–345. (MR 56, #12980).

C. S. Hönig

[1973: 1]   The Green function of a linear differential equation with a lateral condition. *Bull. Amer. Math. Soc.* **79** (1973), 587–593. (MR 47, #581).

E. Hopf

[1937: 1]   *Ergodentheorie.* Springer, Berlin, 1937. Reprinted by Chelsea, New York, 1948.

L. Hörmander

[1955: 1]   On the theory of general partial differential operators. *Acta Math.* **94** (1955), 161–248. (MR 17, 853).

[1960: 1]   Estimates for translation invariant operators in L^p spaces. *Acta Math.* **104** (1960), 93–140. (MR 22, #12389).

[1961: 1]   Hypoelliptic convolution equations. *Math. Scand.* **9** (1961), 178–184. (MR 25, #3265).

[1967/68: 1]   Convolution equations in convex domains. *Invent. Math.* **4** (1967/68), 306–317. (MR 37, #1978).

[1969: 1]   *Linear Partial Differential Operators.* 3rd revised printing. Springer, New York, 1969.

H. W. Hövel and U. Westphal

[1972: 1]   Fractional powers of closed operators. *Studia Math.* **42** (1972), 177–194. (MR 46, #2481).

L. C. Hsu

[1960: 1]   An estimation for the first exponential formula in the theory of semi-groups of linear operators. *Czechoslovak Math. J.* **10** (85) (1960), 323–328. (MR 22, #9868).

T. J. Hughes, T. Kato, and J. E. Marsden

[1976: 1]   Well-posed quasi-linear second-order hyperbolic systems with applications to nonlinear elastodynamics and general relativity. *Arch. Rat. Mech. Anal.* **63** (1976), 273–294. (MR 54, #8041).

M. Iannelli

[1976: 1]   On a certain class of semilinear evolution equations. *J. Math. Anal. Appl.* **56** (1976), 351–367. (MR 54, #13648).

Š. I. Ibragimov

[1969: 1]   An analog of the method of lines for a first order equation in a Banach space. (Russian.) *Differencial'nye Uravnenija* **5** (1969), 772–774. (MR 40, #507).

[1972: 1]   The method of lines for a first-order differential equation in a Banach space. (Russian.) *Izv. Akad. Nauk Kazah. SSSR Ser. Fiz. Mat.* (1972), 33–40. (MR 46, #9477).

Š. I. Ibragimov and D. A. Ismailov

[1970: 1]   An analogue of the method of straight lines for a first-order nonlinear differential equation in a Banach space. (Russian.) *Azerbaĭdžan. Gos. Univ. Učen. Zap. Ser Fiz.-Mat. Nauk* **1** (1970), 57–69. (MR 45, #7323).

K. Igari
[1976: 1]   A necessary condition for well-posed Cauchy problems. *J. Math. Kyoto Univ.* **16** (1976), 531–543. (MR 55, #3512).

A. Inoue
[1972: 1]   L'opérateur d'onde et de diffusion pour un système évolutif $(d/dt)+iA(t)$. *C. R. Acad. Sci. Paris* **275** (1972), A1323–A1325. (MR 46, #9794).
[1974: 1]   Wave and scattering operators for an evolving system $d/dt - iA(t)$. *J. Math. Soc. Japan* **26** (1974), 606–624. (MR 50, #10571).

G. Iooss
[1972: 1]   Stabilité de la solution périodique secondaire intervenant dans certaines problèmes d'évolution. *C. R. Acad. Sci. Paris* **274** (1972), A108–A111. (MR 45, #5833).

D. A. Ismailov and Ja. D. Mamedov
[1975: 1]   Approximate methods for the solution of abstract differential equations in Banach space. (Russian.) *Differencial'nye Uravnenija* **11** (1975), 921–923, 950. (MR 52, #3684).

V. I. Istratescu and A. I. Istratescu
[1973: 1]   Some remarks on a class of semi-groups of operators. I. *Z. Warsch. verw. Geb.* **26** (1973), 241–243. (MR 48, #12146).
[1974: 1]   Some remarks on a class of semi-groups of operators II. (Italian summary.) *Atti. Accad. Naz. Lincei Rend. Cl. Sci. Fis. Mat. Natur.* **56** (1974), 52–54. (MR 51, #11189).

K. Itô and H. P. McKean Jr.
[1974: 1]   *Diffusion Processes and Their Sample Paths.* Springer, Berlin, 1974.

L. A. Ivanov
[1968: 1]   An operator that realizes the evolution of subspaces. (Russian.) *Voronež. Gos. Univ. Trudy Sem. Funkcional. Anal. Vyp.* **10** (1968), 44–50. (MR 54, #1013).

V. K. Ivanov
[1974: 1]   The value of the regularization parameter in ill-posed control problems. (Russian.) *Differencial'nye Uravnenija* **10** (1974), 2279–2285, 2311–2312. (MR 51, #8860).

V. N. Ivanov
[1966: 1]   *The Abstract Cauchy Problem and Perturbation Theory of Closed Operators.* Works of the Saratov Inst. of Mechanization of Agriculture No. 39, Part III. Privolžsk. Knižn. Isdat., Saratov, 1966. (MR 37, #3166).

V. V. Ivanov
[1973: 1]   A resolvent sequence in questions on the generation of summable semigroups of operators. (Russian.) *Dokl. Akad. Nauk SSSR* **213** (1973), 282–285. (MR 49, #5929).
[1974: 1]   Conjugate semigroups of operators in a locally convex space. (Russian.) *Optimizacija Vyp.* **15** (1974), 149–153, 186. (MR 53, #1143).

[1974: 2]   The absolute *n*-resolvent, and the generation of certain classes of nonsummable semigroups. (Russian.) *Optimizacija Vyp.* **15** (1974), 133–148, 186. (MR 53, #11429).

O. A. Ivanova

[1970: 1]   The adjoint *n*-parameter semi-group of operators. (Ukrainian. Russian summary.) *Visnik L'viv Politehn. Inst. No. 44* (1970), 114–118, 214. (MR 46, #2482).

K. Jacobs

[1979: 1]   Fastperiodizitätseigenschaften allgemeiner Halbguppen in Banach-Räumen. *Math. Z.* **67** (1957), 83–92. (MR 19, #295).

S. Ja. Jakubov

[1964: 1]   Investigation of the Cauchy problem for evolution equations of hyperbolic type. (Russian.) *Akad. Nauk Azerbaĭdžan SSR Dokl.* **20** (1964), no. 4, 3–6. (MR 29, #3758).

[1964: 2]   Solvability of the Cauchy problem for evolution equations. (Russian.) *Dokl. Akad. Nauk SSSR* **156** (1964), 1041–1044. (MR 33, #2981).

[1966: 1]   Solvability of the Cauchy problem for differential equations of higher orders in a Banach space with variable operators having non-constant domain of definition. (Russian.) *Akad. Nauk Azerbaĭdžan SSR Dokl.* **22** (1966), no. 2, 6–10. (MR 33, #6084).

[1966: 2]   Investigation of the Cauchy problem for differential equations of hyperbolic type in a Banach space. (Russian.) *Akad. Nauk Azerbaĭdžan SSR Dokl.* **22** (1966), no. 4, 3–7. (MR 34, #465).

[1966: 3]   Quasi-linear differential equations in abstract spaces. (Russian.) *Akad. Nauk Azerbaĭdžan SSR Dokl.* **22** (1966), no. 8, 8–12. (MR 34, #6267).

[1966: 4]   The Cauchy problem for differential equations of the second order in a Banach space. (Russian.) *Dokl. Akad. Nauk SSSR* **168** (1966), 759–762. (MR 33, #6178).

[1966: 5]   Differential equations of higher orders with variable unbounded operators in a Banach space. (Russian.) *Izv. Akad. Nauk Azerbaĭdžan SSR Ser Fiz.-Tehn. Mat. Nauk* (1966), no. 1, 20–27. (MR 34, #464).

[1966: 6]   The Cauchy problem for differential equations of second order of parabolic type in a Banach space. (Russian.) *Izv. Akad. Nauk Azerbaĭdžan SSR Ser Fis.-Tehn Mat. Nauk* (1966), no. 4, 3–8. (MR 34, #6268).

[1967: 1]   Solvability of the Cauchy problem for differential equations in a Banach space (Russian.) *Functional Anal. Certain Problems Theory of Differential Equations and Theory of Functions* (Russian.), 187–206. Izdat. Akad. Nauk Azerbaĭdžan SSR, Baku, 1967. (MR 35, #4549).

[1967: 2]   Nonlinear abstract hyperbolic equations. (Russian.) *Dokl. Akad. Nauk SSSR* **176** (1967), 46–49. (MR 36, #4178).

[1967: 3]   Non-local solvability of boundary value problems for quasilinear partial differential equations of hyperbolic type. (Russian.) *Dokl. Akad Nauk SSSR* **176** (1967), 279–282. (MR 36, #5533).

[1967: 4]   The solvability of the Cauchy problem for abstract parabolic equations with variable operators. (Russian.) *Dokl. Akad. Nauk SSSR* **176** (1967), 545–548. (MR 36, #4179).

[1968: 1]   A linear abstract parabolic equation. (Russian.) *Special Problems of Functional Analysis and Their Applications to the Theory of Differential Equations and the Theory of Functions* (Russian.), 164–187. Izdat. Akad. Nauk Azerbaĭdžan SSR, Baku, 1968. (MR 40, #4575).

[1970: 1]   Uniform correctness of the Cauchy problem for equations of evolution and applications. (Russian.) *Funkcional Anal. i Priložen.* **4** (1970), no. 3, 86–94. (MR 42, #3453).

[1970: 2]   A uniformly well-posed Cauchy problem for abstract hyperbolic equations. (Russian.) *Izv. Vyss. Ucebn. Zaved. Matematika* (1970), no. 12 (103), 108–113. (MR 43, #7970).

[1970: 3]   The Cauchy problem for non-linear second-order hyperbolic equations I, II. (Russian.) *Izv. Akad. Nauk Azerbaĭdžan. SSR Fiz.-Tehn. Mat. Nauk* (1970), no. 3, 16–22, no. 5, 16–21. (MR 46, #479).

[1970: 4]   Solvability of the Cauchy problem for abstract quasilinear second-order hyperbolic equations, and their applications. (Russian.) *Trudy Moskov Mat. Obšč.* **23** (1970), 37–60.

[1971: 1]   Solution of mixed problems for partial differential equations by the oeprator method. (Russian.) *Dokl. Akad. Nauk SSSR* **196** (1971), 545–548. (MR 43, #2572).

[1971: 2]   Solvability "in the large" of the Cauchy problem for a first-order quasilinear hyperbolic system. (Russian.) *Dokl. Akad. Nauk SSSR* **196** (1971), 793–796. (MR 43, #716).

[1973: 1]   Correctness of a boundary value problem for second order linear evolution equations. (Russian.) *Izv. Akad. Nauk Azerbaĭdžan. SSR Ser. Fiz.-Tehn. Mat. Nauk* (1973), 37–42. (MR 52, #8596).

S. Ja. Jakubov and A. B. Aliev
[1976: 1]   Solvability "in the large" of the Cauchy problem for quasilinear equations. (Russian.) *Dokl. Akad. Nauk SSSR* **226** (1976), 1025–1028. (MR 54, #6690).

S. Ja. Jakubov and M. K. Balaev
[1976: 1]   Correct solvability of operator equations on the whole line. (Russian.) *Dokl. Akad. Nauk SSSR* **229** (1976), 562–565. (MR 54, #3119).

S. Ja. Jakubov and U. L. Ismailova
[1974: 1]   "Global" solvability of a mixed problem for a quasilinear equation of Schrödinger type. (Russian.) *Izv. Akad. Nauk Azerbaĭdžan. SSR Ser. Fiz-Tehn. Mat. Nauk* (1974), no. 6, 67–71. (MR 51, #13974).

S. Ja. Jakubov and S. G. Samedova
[1975: 1]   The correctness of the Cauchy problem for parabolic equations. (Russian.) *Akad. Nauk Azerbaĭdžan SSR Dokl.* **31** (1975), no. 2, 17–20. (MR 52, #936).

L. I. Jakut
[1963: 1]   On the convergence of finite-difference methods for the solution of evolution equations. (Russian.) *Voronez. Gos. Univ. Trudy Sem. Funkcional Anal.*, no. 7 (1963), 160–177.

[1963: 2]   On the basic problem of convergence of difference schemes. (Russian.)
            Dokl. Akad. Nauk SSSR **151** (1963), 76–79. (MR 25, #5375)
[1964: 1]   Theorems of Lax for non-linear evolutionary equations. (Russian.)
            Dokl. Akad. Nauk SSSR **156** (1964), 1304–1307. (MR 29, #4984).
[1969: 1]   Finite difference methods for equations of parabolic type in a Banach
            space. (Russian.) *Differencial'nye Uravnenija* **5** (1969), 1225–1235.
            (MR 40, #6020).
[1969: 2]   Finite difference methods for nonlinear evolutionary equations. (Rus-
            sian.) *Differencial'nye Uravnenija* **5** (1969), 1415–1425. (MR 40,
            #6021).

R. C. James
[1963: 1]   Characterizations of reflexivity. *Studia Math.* **23** (1963), 205–216. (MR
            30, #431).

M. Janusz
[1973: 1]   Singular perturbation methods for differential equations in Banach
            spaces. *Math. Balkanica* **3** (1973), 321–334. (MR 50, #10477).

F. John
[1955: 1]   Numerical solution of the heat equation for preceding times. *Ann. Mat.
            Pura Appl.* **40** (1955), 129–142. (MR 19, 323).
[1955: 2]   A note on "improper" problems for partial differential equations.
            Comm. Pure Appl. Math. **8** (1955), 591–594. (MR 17, 746).
[1959: 1]   Numerical solution of problems which are not well-posed in the sense
            of Hadamard. *Proc. Rome Symp. Prov. Int. Comp. Center* (1959),
            103–116. (MR 21, #6704).
[1960: 1]   Continuous dependence on data for solutions of partial differential
            equations with a prescribed bound. *Comm. Pure Appl. Math.* **13**
            (1960), 551–585. (MR 23, #A317).
[1978: 1]   *Partial Differential Equations.* Springer, New York, 1978.

K. Jörgens
[1958: 1]   An asymptotic expansion in the theory of neutron transport. *Comm.
            Pure Appl. Math.* **11** (1958), 219–242. (MR 21, #2189). (MR 21,
            #2189).
[1967: 1]   Zur Spektraltheorie der Schrödinger-Operatoren. *Math. Zeitschr.* **96**
            (1967), 355–372. (MR 34, #8003).

K. Jörgens and J. Weidmann
[1973: 1]   *Spectral Properties of Hamiltonian Operators.* Lecture Notes in Math,
            vol. 313, Springer, New York, 1973.

H. D. Junghenn and C. T. Taam
[1974: 1]   Arcs defined by one-parameter semigroups of operators. *Proc. Amer.
            Math. Soc.* **44** (1974), 113–120. (MR 48, #9448).

N. I. Jurčuk

[1974: 1]   Boundary value problems for equations that contain in their principal part operators of the form $d^{2m+1}/dt^{2m+1} + A$. (Russian.) *Diferencial'nye Uravnenija* **10** (1974), 759–762, 768. (MR 49, #934).

[1976: 1]   A priori estimates of the solutions of boundary value problems for certain operator-differential equations. (Russian.) *Differencial'nye Uravnenija* **12** (1976), 729–739, 774. (MR 54, #766).

[1976: 2]   Boundary value problems for differential equations with operator coefficients that depend on a parameter I. A priori estimates. (Russian.) *Differencial'nye Uravnenija* **12** (1976), 1645–1661, 1726. (MR 55, #10794).

T. Kakita

[1974/75: 1]   On the existence of time-periodic solutions of some non-linear evolution equations. *Applicable Anal.* **4** (1974/75), 63–76. (MR 56, #12529).

S. Kakutani

[1938: 1]   Iteration of linear operators in complex Banach spaces. *Proc. Imp. Acad. Tokyo* **14** (1938), 295–300.

R. R. Kallman and G. C. Rota

[1970: 1]   On the inequality $\|f'\|^2 \leqslant 4\|f\| \, \|f''\|$. *Inequalities* II. (O. Shisha, ed.). Academic Press, New York, 1970, 187–192. (MR 43, #3791).

T. F. Kalugina

[1977: 1]   Functions of a nonnegative operator in a Banach space. (Russian. English summary.) *Vestnik Moskov. Univ. Ser. I Mat. Meh.* (1977), no. 2, 32–40. (MR 56, #12939).

S. Kantorovitz

[1970: 1]   On the operational calculus for groups of operators. *Proc. Amer. Math. Soc.* **26** (1970), 603–608. (MR 42, #6649).

B. G. Karasik

[1976: 1]   Regularization of an ill-posed Cauchy problem for operator-differential equations of arbitrary order. (Russian.) *Izv. Akad. Nauk Azerbaĭdžan. SSR Ser. Fiz.-Tehn. Mat. Nauk* (1976), 9–14. (MR 56, #9020).

A. G. Kartsatos

[1969: 1]   Boundedness of solutions to nonlinear equations in Hilbert space. *Proc. Japan Acad.* **15** (1969), 339–341. (MR 40, #4576).

A. G. Kartsatos and W. R. Zigler

[1976: 1]   Rothe's method and weak solutions of perturbed evolution equations in reflexive Banach spaces. *Math. Ann.* **219** (1976), 159–166. (MR 52, #11679).

T. Kato

[1951: 1]   Fundamental properties of Hamiltonian operators of Schrödinger type. *Trans. Amer. Math. Soc.* **70** (1951), 195–211. (MR 12, 781).

[1951: 2]   On the existence of solutions of the helium wave equation. *Trans. Amer. Math. Soc.* **70** (1951), 212–218. (MR 12, 781).

[1952: 1]   Notes on some inequalities for linear operators. *Math. Ann.* **125** (1952), 208–212. (MR 14, 766).

[1953: 1]   Integration of the equation of evolution in a Banach space. *J. Math. Soc. Japan* **5** (1953), 208–234. (MR 15, 437).

[1954: 1]   On the semi-groups generated by Kolmogoroff's differential equations. *J. Math. Soc. Japan* **6** (1954), 1–15. (MR 19, 279).

[1956: 1]   On linear differential equations in Banach spaces. *Comm. Pure Appl. Math.* **9** (1956), 479–486. (MR 19, 279).

[1959: 1]   Remarks on pseudo-resolvents and infinitesimal generators of semi-groups. *Proc. Japan Acad.* **35** (1959), 467–478. (MR 22, #8347).

[1960: 1]   Note on fractional powers of linear operators. *Proc. Japan Acad.* **36** (1960), 94–96. (MR 22, #12400).

[1961: 1]   Abstract evolution equations of parabolic type in Banach and Hilbert spaces. *Nagoya Math. J.* **19** (1961), 93–125. (MR 26, #631).

[1961: 2]   Fractional powers of dissipative operators. *J. Math. Soc. Japan* **13** (1961), 246–274. (MR 25, #1453).

[1961: 3]   A generalization of the Heinz inequality. *Proc. Japan Acad.* **37** (1961), 305–308. (MR 26, #2876).

[1962: 1]   Fractional powers of dissipative operators II. *J. Math. Soc. Japan* **14** (1962), 242–248. (MR 27, #1851).

[1966: 1]   Wave operators and similarity for some non-selfadjoint operators. *Math. Ann.* **162** (1966), 258–279. (MR 32, #8211).

[1967: 1]   On classical solutions of the two-dimensional non-stationary Euler equation. *Arch. Rat. Mech. Anal.* **25** (1967), 188–200. (MR 35, #1939).

[1967: 2]   Some mathematical problems in quantum mechanics. *Progress Theor. Phys.* (*Supplement*) **40** (1967), 3–19. (MR 36, #1169).

[1967: 3]   Scattering theory with two Hilbert spaces. *J. Functional Analysis* **1** (1967), 342–369. (MR 36, #3164).

[1970: 1]   Linear evolution equations of "hyperbolic" type. *J. Fac. Sci. Univ. Tokyo Sect. I* **17** (1970), 241–258. (MR 43, #5347).

[1970: 2]   A characterization of holomorphic semigroups. *Proc. Amer. Math. Soc.* **25** (1970), 495–498. (MR 41, #9050).

[1971: 1]   On an inequality of Hardy, Littlewood and Pólya. *Advances in Math.* **7** (1971), 217–218. (MR 45, #2531).

[1972: 1]   Schrödinger operators with singular potentials. Proceedings of the International Symposium on Partial Differential Equations and the Geometry of Normed Spaces (Jerusalem, 1972). *Israel J. Math.* **13** (1972), 135–148. (MR 48, #12155).

[1973: 1]   Linear evolution equations of "hyperbolic" type II. *J. Math. Soc. Japan* **25** (1973), 646–666. (MR 48, #4827).

[1974: 1]   On the Trotter-Lie product formula. *Proc. Japan Acad.* **50** (1974), 694–698. (MR 54, #5900).

[1975: 1]   Quasi-linear equations of evolution, with applications to partial differential equations. *Lecture Notes in Math.*, vol. 448, Springer, Berlin, 1975, 25–70. (MR 53, #11252).

[1975: 2]   Singular Perturbation and Semigroup Theory. Turbulence and Navier-Stokes Equations (Proc. Conf. Univ. Paris-Sud, Orsay, 1975), 104–112. *Lecture Notes in Math.*, vol. 565, Springer, Berlin, 1976. (MR 56, #16447).

[1976: 1]   *Perturbation Theory for Linear Operators*, 2nd ed. Springer, New York, 1976.

T. Kato and H. Fujita
[1962: 1]   On the nonstationary Navier-Stokes system. *Rend. Sem. Mat. Univ. Padova* **32** (1962), 243–260. (MR 26, #495).

T. Kato and S. T. Kuroda
[1970: 1]   Theory of simple scattering and eigenfunction expansions. *Functional Analysis and Related Fields* (Proc. Conf. for M. Stone, Univ. Chicago, Chicago, Ill. 1968), 99–131. Springer, New York, 1970. (MR 52, #6464a).

[1971: 1]   The abstract theory of scattering. *Rocky Mountain J. Math.* **1** (1971), 127–171. (MR 52, #6464b).

T. Kato and H. Tanabe
[1962: 1]   On the abstract evolution equation. *Osaka Math. J.* **14** (1962), 107–133. (MR 25, #4367).

[1967: 1]   On the analyticity of solutions of evolution equations. *Osaka J. Math.* **4** (1967), 1–4. (MR 36, #5482).

R. M. Kauffman
[1971: 1]   Unitary groups and differential operators. *Proc. Amer. Math. Soc.* **30** (1971), 102–106. (MR 43, #7966).

J. L. Kelley
[1955: 1]   *General Topology*, Van Nostrand, New York, 1955.

D. G. Kendall
[1954: 1]   Bernstein polynomials and semi-groups of operators. *Math. Scand.* **2** (1954), 185–186. (MR 16, 717).

[1959: 1]   Sur quelques critères classiques de compacité dans certains espaces fonctionnels et la théorie des semi-groupes de transformations. *J. Math. Pures Appl.* (9) **38** (1959), 235–244. (MR 21, #6544).

H. Kielhöfer
[1973: 1]   Stabilität bei fastlinearen Evolutionsgleichungen. Methoden und Verfahren der Mathematischen Physik, Band 9, 161–170. Bibliographisches Inst., Mannheim, 1973. (MR 51, #3636).

[1975: 1]   Stability and semilinear evolution equations in Hilbert space. *Arch. Rat. Mech. Anal.* **57** (1975), 150–165. (MR 56, #787).

A. I. Kirjanen
[1974: 1]   On the question of the stability of a nonlinear differential equation in Hilbert space. (Russian.) *Methods and Models of Control*, no. 7, 65–69,

121. Redakcionno-Izdat. Otdel. Rizsk. Politehn. Inst., Riga, 1974. (MR 52, #11238).

N. V. Kislov

[1972: 1] Certain operator-differential equations of mixed type. (Russian.) *Trudy Moskov. Orden. Lenin. Ènerget. Inst. Vyp. Vysš. Mat.* (1972), 60–69. (MR 49, #764).

[1975: 1] A boundary value problem for a second-order operator-differential equation of the type of the Tricomi equation. (Russian.) *Differencial'nye Uravnenija* 11 (1975), 718–726, 766. (MR 52, #938).

J. Kisyński

[1963: 1] Sur les équations hyperboliques avec petit paramètre. *Colloq. Math.* 10 (1963), 331–343. (MR 27, #6036).

[1963/64: 1] Sur les opérateurs de Green des problèmes de Cauchy abstraits. *Studia Math.* (1963/64), 285–328. (MR 28, #4393).

[1966: 1] On temporally inhomogeneous equations of evolution in Banach spaces. (Russian summary.) *Bull. Acad. Polon. Sci. Sér. Sci. Math. Astronom. Phys.* 14 (1966), 623–626. (MR 34, #7924).

[1967: 1] A proof of the Trotter-Kato theorem on approximation of semi-groups. *Colloq. Math.* 18 (1967), 181–184. (MR 36, #3165).

[1968: 1] A remark on continuity of one-parameter semi-groups of operators. *Colloq. Math.* 19 (1968), 277–284. (MR 37, #3399).

[1970: 1] On second-order Cauchy's problem in a Banach space. *Bull. Acad. Polon. Sci. Sér Sci. Math. Astronom. Phys.* 18 (1970), 371–374. (MR 42, #4851).

[1971: 1] On operator-valued solutions of D'Alembert's functional equation, I. *Colloq. Math.* 23 (1971), 107–114. (MR 47, #2428).

[1972: 1] On operator-valued solutions of D'Alembert's functional equation, II. *Studia Math.* 42 (1972), 43–66. (MR 47, #2428).

[1972: 2] On cosine operator functions and one-parameter semigroups of operators. *Studia Math.* 44 (1972), 93–105. (MR 47, #890).

M. Kline and I. W. Kay

[1965: 1] *Electromagnetic Theory and Geometrical Optics.* Interscience-Wiley, New York, 1965.

R. Kluge and G. Bruckner

[1974: 1] Über einige Klassen nichtlinearer Differentialgleichungen und ungleichungen im Hilbert-Raum. *Math. Nachr.* 64 (1974), 5–32. (MR 51, #1511).

R. J. Knops and L. E. Payne

[1971: 1] Growth estimates for solutions of evolutionary equations in Hilbert space with applications in elastodynamics. *Arch. Rational Mech. Anal.* 41 (1971), 363–398. (MR 48, #9068).

W. Köhnen

[1970: 1] Das Anfangswertverhalten von Evolutionsgleichungen in Banachräumen. Teil I: Approximationseigenschaften von Evolutionsoperatoren. *Tohoku Math. J.* (2) **22** (1970), 566–596. (MR 43, #3844).

[1971: 1]   Das Anfangswertverhalten von Evolutionsgleichungen in Banachräumen. Teil II: Anwendungen. *Tohoku Math. J.* (2) **23** (1971), 621–639. (MR 46, #2483).

[1972: 1]   Approximationsprobleme bei der Störung von Halbgruppenoperatoren. Constructive theory of functions (*Proc. Internat. Conf., Varna, 1970*) 201–207. Izdat. Bolgar. Akad. Nauk, Sofia, 1972. (MR 52, #1406).

L. V. Koledov

[1976: 1]   Fractional powers of elliptic operators and approximation-moment scales. (Russian.) *Dokl. Akad. Nauk SSSR* **227** (1976), 26–28. (MR 53, #6357).

G. A. Kolupanova

[1972: 1]   A certain differential equation of hyperbolic type in a Banach space. (Russian.) *Dokl. Akad. Nauk SSSR* **207** (1972), 1044–1047. (MR 48, #4546).

H. Komatsu

[1961: 1]   Abstract analyticity in time and unique continuation property of solutions of a parabolic equation. *J. Fac. Sci. Univ. Tokyo Sect. I* **9**, 1–11. (MR 26, #6193).

[1964: 1]   Semi-groups of operators in locally convex spaces. *J. Math. Soc. Japan* **16** (1964), 230–262. (MR 32, #368).

[1966: 1]   Fractional powers of operators. *Pacific J. Math.* **19** (1966), 285–346. (MR 34, #1862).

[1967: 1]   Fractional powers of operators, II. Interpolation spaces. *Pacific J. Math.* **21** (1967), 89–111. (MR 34, #6533).

[1969: 1]   Fractional powers of operators, III. Negative powers. *J. Math. Soc. Japan* **21** (1969), 205–220. (MR 39, #3340).

[1969: 2]   Fractional powers of operators, IV. Potential operators. *J. Math. Soc. Japan* **21** (1969), 221–228. (MR 39, #3341).

[1970: 1]   Fractional powers of operators, V. Dual operators. *J. Fac. Sci. Univ. Tokyo Sect. I* **17** (1970), 373–396. (MR 42, #6650).

[1972: 1]   Fractional powers of operators, VI. Interpolation of nonnegative operators and imbedding theorems. *J. Fac. Sci. Univ. Tokyo Sect. IA Math.* **19** (1972), 1–63. (MR 47, #7510).

T. Komura

[1968: 1]   Semigroups of operators in locally convex spaces. *J. Functional Analysis* **2** (1968), 258–296. (MR 38, #2634).

Y. Komura

[1970: 1]   On linear evolution operators in reflexive Banach spaces. *J. Fac. Sci. Univ. Tokyo Sect. IA Math.* **17** (1970), 529–542. (MR 44, #5467).

Y. Konishi
[1971/72: 1]   Cosine functions of operators in locally convex spaces. *J. Fac. Sci. Univ. Tokyo Sect IA Math.* **18** (1971/72), 443–463. (MR 48, #2846).
[1976: 1]   Une remarque sur des problems de perturbations singulieres et d'évolution. *C. R. Acad. Sci. Paris* **282** (1976), A775–A777. (MR 54, #3138).
[1976: 2]   Convergence des solutions d'équations elliptiques semi-linéaires dans L^1. *C. R. Acad. Sci. Paris* **283** (1976), A489–A490. (MR 55, #854).

V. I. Kononenko
[1974: 1]   A nth-order ordinary differential equation in a linear topological space. (Russian.) *Differencial'nye Uravnenija* **10** (1974), 1727–1729, 1736. (MR 50, #10478).

A. A. Kononova
[1974: 1]   The Cauchy problem for the inverse heat equation in an infinite strip. (Russian.) *Ural. Gos. Univ. Mat. Zap. 8 tetrad' 4 58-63,* **134** (1974). (MR 51, #3693).

D. G. Korenevskiĭ and T. S. Fesčenko
[1975: 1]   On the abstract Cauchy problem for differential equations with deviating argument. Stability of the solutions. (Russian.) *Differencial'nye Uravnenija* **11** (1975), 1895–1898, 1910. (MR 52, #1120).

G. Köthe
[1966: 1]   *Linear Topological Spaces I.* Springer, Berlin, 1966.
[1979: 1]   *Linear Topological Spaces II.* Springer, Berlin, 1979.

O. M. Kozlov
[1958: 1]   Fractional powers of self-adjoint extensions of operators and some boundary problems. (Russian.) *Dokl. Akad. Nauk SSSR* **123** (1958), 971–974. (MR 23, #A3904).

O. N. Kozlova
[1972: 1]   Analyticity of the semigroup of the Cauchy problem for an equation with a hypoelliptic operator. (Russian.) Trudy Rižsk. Inst. Inž. Graždan. Aviacii Vyp. 195 Kraevye Zadači Differencial'nyh Uravneniĭ (1972), 52–67. (MR 56, #12606).

G. Krabbe
[1975: 1]   A new algebra of distributions; initial-value problems involving Schwartz distributions. I. *Publ. Math. Debrecen* **22** (1975), no. 3–4, 279–292. (MR 52, #14992).

H. Kraljević and S. Kurepa
[1970: 1]   Semigroups in Banach spaces. *Glasnik Mat. Ser. III* **5** (25) (1970), 109–117. (MR 45, #9190).

M. A. Krasnosel'skiĭ and S. G. Kreĭn
[1957: 1]   On differential equations in Banach space. (Russian.) *Proc. Third All-Union Math. Congr.*, vol. III. Izdat. Akad. Nauk SSSR, Moscow, 1957, 73–80.

[1964: 1]   On operator equations in functional spaces. (Russian.) *Proc. Fourth All-Union Math. Congr.*, vol. II. Izdateltsvo "Nauka," Leningrad, 1965, 292–299 (Russian).

M. A. Krasnosel'skiĭ, S. G. Kreĭn, and P. E. Sobolevskiĭ
[1956: 1]   On differential equations with unbounded operators in Banach spaces. (Russian.) *Dokl. Akad. Nauk SSSR* (N.S.) **111** (1956), 19–22. (MR 19, #550).
[1957: 1]   On differential equations with unbounded operators in Hilbert space. (Russian.) *Dokl. Akad. Nauk SSSR* **112** (1957), 990–993. (MR 19, #747).

M. A. Krasnosel'skiĭ and E. A. Lifšic
[1968: 1]   Relatedness principles for differential equations with unbounded operators in a Banach space. (Russian.) *Funkcional Anal. i Priložen.* **2** (1968), no. 4, 58–62. (MR 39, #7255).

M. A. Krasnosel'skiĭ, E. I. Pustyl'nik, and P. P. Zabreĭko
[1965: 1]   On fractional powers of elliptic operators. (Russian.) *Dokl. Akad. Nauk SSSR* **165** (1965), 990–993. (MR 33, #406).
[1965: 2]   A problem on fractional powers of operators. (Russian.) *Uspehi Mat. Nauk* **20** (1965), no. 6 (126), 87–89. (MR 33, #7867).

M. A. Krasnosel'skiĭ and P. E. Sobolevskiĭ
[1959: 1]   Fractional powers of operators acting in Banach spaces. (Russian.) *Dokl. Akad. Nauk SSSR* **129** (1959), 499–502. (MR 21, #7447).
[1962: 1]   The structure of the set of solutions of equations of parabolic type. (Russian.) *Dokl. Akad. Nauk SSSR* **146** (1962), 26–29. (MR 27, #2730).
[1964: 1]   The structure of the set of solutions of an equation of parabolic type. (Russian.) *Ukrain. Mat. Ž.* **16** (1964), 319–333. (MR 29, #3763).

K. Kraus, L. Polley, and G. Reents
[1977: 1]   Generators of infinite direct products of unitary groups. *J. Mathematical Phys.* **18** (1977), 2166–2171. (MR 56, #12893).

P. Krée
[1970: 1]   Équations différentielles dissipatives. *C. R. Acad. Sci. Paris* **270** (1970), A1583–A1585. (MR 42, #3454).

M. G. Kreĭn and Ju. L. Daleckiĭ
[1970: 1]   Certain results and problems in the theory of stability of solutions of differential equation in a Banach space. (Russian. English summary.) *Proc. Fifth. Internat. Conf. on Nonlinear Oscillations (Kiev, 1969)*, vol. 1: *Analytic methods in the theory of nonlinear oscillations*, 332–347. Izdanie Inst. Mat. Akad. Nauk Ukrain. SSR, Kiev, 1970. (MR 47, #7168).

S. G. Kreĭn
[1957: 1]   Differential equations in a Banach space and their application in

hydromechanics. (Russian.) *Uspehi Mat. Nauk* **12** (1957), no. 1 (73), 208–211 (MR 19, 36).

[1966: 1]   Correctness of the Cauchy problem and the analyticity of solutions of the evolution equation. (Russian.) *Dokl. Akad. Nauk SSSR* **171** (1966), 1033–1036. (MR 34, #7925).

[1967: 1]   *Linear Differential Equations in Banach Spaces*. (Russian.) Izdat. "Nauka," Moscow, 1967. English translation: [1971: 1], Amer. Math. Soc., Providence, 1971.

S. G. Kreĭn and G. I. Laptev

[1962: 1]   Boundary-value problems for an equation in Hilbert space. (Russian.) *Dokl. Akad. Nauk SSSR* **146** (1962), 535–538. (MR 27, #6000).

[1966: 1]   Boundary value problems for second order differential equations in a Banach space, I. (Russian.) *Differencial'nye Uravnenija* **2** (1966), 382–390. (MR 33, #7662).

[1966: 2]   Correctness of boundary value problems for a differential equation of the second order in a Banach space, II. (Russian.) *Differencial'nye Uravnenija* (1966), 919–926. (MR 34, #3409).

[1969: 1]   An abstract scheme for the examination of parabolic problems in noncylindric regions. (Russian.) *Differencial'nye Uravnenija* **5** (1969), 1458–1469. (MR 41, #649).

S. G. Kreĭn, G. I. Laptev, and G. A. Cvetkova

[1970: 1]   The Hadamard correctness of the Cauchy problem for an evolution equation. (Russian.) *Dokl. Akad. Nauk SSSR* **192** (1970), 980–983. (MR 42, #637).

S. G. Kreĭn and S. Ja. L'vin

[1973: 1]   A general initial problem for a differential equation in a Banach space. (Russian.) *Dokl. Akad. Nauk SSSR* **211** (1973), 530–533. (MR 49, #7549).

S. G. Kreĭn and O.I. Prozorovskaya

[1957: 1]   An analogue of Seidel's method for operator equations. (Russian.) *Voronež. Gos. Univ. Trudy Sem. Funkcional Anal.* **5** (1957), 35–38. (MR 20, #2081).

[1960: 1]   Analytic semigroups and incorrect problems for evolutionary equations. (Russian.) *Dokl. Akad. Nauk SSSR* **133** (1960), 277–280 (MR 27, #1845).

[1963: 1]   Approximate methods of solving ill-posed problems (Russian.) *Ž. Vyčisl. Mat. i Mat. Fiz.* **3** (1963), 120–130. (MR 27, #3094).

S. G. Kreĭn and L. N. Šablickaja

[1966: 1]   Stability of difference schemes for the Cauchy problem. (Russian.) *Ž. Vyčisl. Mat. i Mat. Fiz.* **6** (1966), 648–664. (MR 33, #8118).

S. G. Kreĭn and Ju. B. Savčenko

[1972: 1]   Exponential dichotomy for partial differential equations. (Russian.) *Differencial'nye Uravnenija* **8** (1972), 835–844. (MR 46, #3941).

S. G. Kreĭn and P. E. Sobolevskiĭ
[1958: 1]   A differential equation with an abstract elliptical operator in Hilbert
            space. (Russian.) *Dokl. Akad. Nauk SSSR* **118** (1958), 233–236. (MR
            20, #6043).

H. O. Kreiss
[1959: 1]   Über Matrizen die Beschränkte Halbgruppen erzeugen. *Mat. Scand.* **7**
            (1959), 71–80. (MR 22, #1820).
[1962: 1]   Über die Stabilitätsdefinition für Differenzengleichungen die partielle
            Differentialgleichungen approximieren. *BIT* **2** (1962), 153–181. (MR
            29, #2992).
[1963: 1]   Über Sachgemässe Cauchyprobleme. *Math. Scand.* **13** (1963), 109–128.
            (MR 29, #6177).

K. Kubota and T. Shirota
[1966: 1]   On certain condition for the principle of limiting amplitude. *Proc.
            Japan Acad.* **42** (1966), 1155–1160. (MR 38, #1397).
[1967: 1]   The principle of limiting amplitude. *J. Fac. Sci. Hokkaido Univ. Ser. I*
            (20) (1967), 31–52. (MR 38, #3568).
[1967: 2]   On certain condition for the principle of limiting amplitude, II. *Proc.
            Japan Acad.* **43** (1967), 458–463. (MR 38, #1398).

S. Kurepa
[1955: 1]   Semigroups of unbounded self-adjoint transformations in Hilbert space.
            Hrvatsko Prirod. Drustvo Glasnik Mat.-Fiz. Astr. Ser. II **10** (1955),
            233–238. (MR 18, 139).
[1958: 1]   On the continuity of semi-groups of normal transformations in Hilbert
            space. *Glasnik Mat. Fiz. Astr. Drustvo Mat. Fiz. Hrvatske Ser. II* **13**
            (1958), 81–87. (MR 21, #1539).
[1958: 2]   Semigroups of normal transformations in Hilbert space. *Glasnik Mat.-
            Fiz. Astr. Drustvo Mat. Fiz. Hrvatske Ser. II* **13** (1958), 257–266. (MR
            21, #4361).
[1962: 1]   A cosine functional equation in Banach algebras. *Acta Sci. Math.
            Szeged* **23** (1962), 255–267. (MR 26, #2901).
[1972: 1]   Uniformly bounded cosine functions in a Banach space. *Math. Bal-
            kanica* **2** (1972), 109–115. (MR 47, #9352).
[1973: 1]   A weakly measurable selfadjoint cosine function. *Glasnik Mat. Ser. III*
            8 (28) (1973), 73–79. (MR 48, #12147).
[1976: 1]   Decomposition of weakly measurable semigroups and cosine operator
            functions. *Glasnik Mat. Ser III* **11** (31), (1976) 91–95. (MR 53,
            #1140).

S. T. Kuroda
[1970: 1]   Some remarks on scattering for Schrödinger operators. *J. Fac. Sci.
            Univ. Tokyo Sect. I* **17** (1970), 315–329. (MR 54, #112).

T. G. Kurtz
[1969: 1]   Extensions of Trotter's operator semigroup approximation theorems. *J.
            Functional Analysis* **3** (1969), 354–375. (MR 39, #3351).

[1970: 1]   A general theorem on the convergence of operator semigroups. *Trans. Amer. Math. Soc.* **148** (1970), 23–32. (MR 41, #867).

[1971: 1]   Comparison of semi-Markov and Markov processes. *Ann. Math. Statist.* **42** (1971), 991–1002. (MR 43, #4112).

[1972: 1]   A random Trotter product formula. *Proc. Amer. Math. Soc.* **35** (1972), 147–154. (MR 46, #2484).

[1973: 1]   Convergence of sequences of semigroups of nonlinear operators with an application to gas kinetics. *Trans. Amer. Math. Soc.* **186** (1973), 259–272. (MR 51, #11222).

[1973: 2]   A limit theorem for perturbed operator semigroups with applications to random evolutions. *J. Functional Analysis* **12** (1973), 55–67. (MR 51, #1477).

[1975: 1]   Semigroups of conditioned shifts and approximation of Markoff processes. *Ann. of Prob.* **3** (1975), 618–642. (MR 52, #4425).

[1976: 1]   An abstract averaging theorem. *J. Functional Anal.* **23** (1976), 135–144. (MR 54, #13628).

[1977: 1]   Applications of an abstract perturbation theorem to ordinary differential equations. *Houston J. Math.* **3** (1977), 67–82. (MR 54, #13245).

V. O. Kutovoĭ

[1976: 1]   The spectrum of a Sturm-Liouville equation with an unbounded operator-valued coefficient. (Russian.) *Ukrain. Mat. Z* **28** (1976), 473–482, 574. (MR 54, #13633).

B. Kvedaras and I. Macionis

[1975: 1]   The Cauchy problem for a degenerate differential equation. (Russian.) *Litovsk. Math. Sb.* **15** (1975), 121–131. (MR 54, #7987).

J.-P. Labrousse

[1972: 1]   Une caractérisation topologique des générateurs infinitésimaux de semi-groupes analytiques et de contractions sur un espace de Hilbert. *Atti Accad. Naz. Lincei Rend. Cl. Sci. Fis. Mat. Natur.* (8) **52**(1972), 631–636. (MR 48, #4828).

C. Lacomblez

[1974: 1]   Sur une équation d'évolution du second ordre en t à coefficients dégénerés ou singuliers. *C. R. Acad. Sci. Paris* **278** (1974), 241–244. (MR 54, #8350).

G. Ladas and V. Lakshmikantham

[1971: 1]   Lower bounds and uniqueness for solutions of evolution inequalities in Hilbert space. *Arch. Rat. Mech. Anal.* **43** (1971), 293–303. (MR 49, #678).

[1972: 1]   Lower bounds and uniqueness for solutions of evolution inequalities in a Hilbert space. Ordinary differential equations (*Proc. Conf. Math. Res. Center, Naval Res. Lab., Washington, D.C., 1971*), 473–487, Academic Press, New York, 1972. (MR 54, #3121).

[1972: 2]   *Differential Equations in Abstract Spaces.* Academic Press, New York, 1972.

O. A. Ladyženskaya
[1955: 1]   On the solution of non-stationary operator equations of various types.
              (Russian.) *Dokl. Akad. Nauk SSSR* **102** (1955), 207–210. (MR 17,
              161).
[1956: 1]   On the solutions of non-stationary operator equations. (Russian.) *Mat.
              Sb. N.S.* **39** (81), (1956), 491–594. (MR 19, 279).
[1958: 1]   On non-stationary operator equations and their applications to linear
              problems of mathematical physics. (Russian.) *Mat. Sb. (N.S.)* **45** (87)
              (1958), 123–158. (MR 22, #12290).

J. Lagnese
[1969: 1]   Note on a theorem of C. T. Taam concerning bounded solutions of
              nonlinear differential equations. *Proc. Amer. Math. Soc.* **20** (1969),
              351–356. (MR 38, #2418).
[1970: 1]   On equations of evolution and parabolic equations of higher order in *t*.
              J. Math. Anal. Appl. **32** (1970), 15–37. (MR 42, #2193).
[1972: 1]   Exponential stability of solutions of differential equations of Sobolev
              type. *SIAM J. Math. Anal.* **3** (1972), 625–636. (MR 47, #609).
[1973: 1]   Approximation of solutions of differential equations in Hilbert space.
              J. Math. Soc. Japan **25** (1973), 132–143. (MR 46, #9479).
[1973: 2]   Singular differential equations in Hilbert space. *SIAM J. Math. Anal.*
              4 (1973), 623–637. (MR 48, #4443).
[1974: 1]   Existence, uniqueness and limiting behavior of solutions of a class of
              differential equations in Banach space. *Pacific J. Math.* **53** (1974),
              473–485. (MR 50, #13781).
[1976: 1]   The final value problem for Sobolev equations. *Proc. Amer. Math. Soc.*
              56 (1976), 247–252. (MR 54, #7988).

V. Lakshmikantham
[1963: 1]   Differential equations in Banach spaces and extensions of Lyapunov's
              method. *Proc. Cambridge Phylos. Soc.* **59** (1963), 373–381.
[1964: 1]   Properties of solutions of abstract differential inequalities. *Proc. London
              Math. Soc.* **14** (1964), 74–82.

V. Lakshmikantham and S. Leela
[1969: 1]   *Differential and Integral Inequalities, Theory and Applications*, vol. I.
              Academic Press, New York, 1969.
[1969: 2]   *Differential and Integral Inequalities. Theory and Applications*, vol. II.
              Academic Press, New York, 1969.

A. Lambert
[1974/75: 1] Equivalence for groups and semigroups of operators. *Indiana Univ.
              Math. J.* **24** (1974–75), 879–885. (MR 51, #1478).

E. Lanconelli
[1968: 1]   Valutazioni in $L_p(R^n)$ della soluzione del problema di Cauchy per
              l'equazione di Schrödinger. *Boll. Un. Mat. Ital.* (4) **1** (1968), 591–607.
              (MR 38, #2452).

[1968: 2]   Valutazioni in B_p della soluzione del problema di Cauchy per l'equazione delle onde. *Boll. Un. Mat. Ital.* (4) **1** (1968), 780–790. (MR 39, #4461).

[1970: 1]   Sui moltiplicatori in L_p e in B_p^0 e applicazioni al problema di Cauchy. (English summary.) *Rend. Mat.* (6) **3** (1970), 33–88. (MR 41, #7489).

C. E. Langenhop

[1967: 1]   Note on a stability condition of C. T. Taam. *J. Math. Mech.* **16** (1967), 1287–1289. (MR 37, #5513).

H. Langer

[1962: 1]   Über die Wurzeln eines Maximalen Dissipativen Operatoren. *Acta Math. Acad. Sci. Hungary* **13** (1962), 415–424. (MR 26, #1757).

H. Langer and V. Nollau

[1966: 1]   Einige Bemerkungen über dissipative Operatoren im Hilbertraum. *Wiss. Z. Techn. Univ. Dresden* **15** (1966), 669–673. (MR 35, #774).

G. I. Laptev

[1966: 1]   Eigenvalue problems for differential equations of the second order in Banach and Hilbert spaces. (Russian.) *Differencial'nye Uravnenija* **2** (1966), 1151–1160. (MR 35, #1900).

[1968: 1]   Boundary value problems for second order differential equations in Hilbert space with a variable selfadjoint operator. (Russian.) *Dokl. Akad. Nauk SSSR* **179** (1968), 283–286. (MR 37, #3223).

[1968: 2]   Strongly elliptic second order equations in Hilbert space. (Russian.) *Litovsk. Mat. Sb.* **8** (1968), 87–99. (MR 39, #1829).

[1969: 1]   On the theory of the operational calculus of linear unbounded operators. (Russian.) *Dokl. Akad. Nauk SSSR* **185** (1969), 760–763. (MR 39, #6105).

[1970: 1]   An operational calculus for linear unbounded operators, and semigroups. (Russian.) *Funkcional. Anal. i Prilozen.* **4** (1970), no. 4, 31–40. (MR 43, #3845).

[1971: 1]   Exponential solutions of the Cauchy problem with a constant operator. (Russian.) *Differencial'nye Uravnenija* **7** (1971), 232–243. (MR 44, #3008).

E. W. Larsen

[1975: 1]   Solution of neutron transport problems in L^1. *Comm. Pure Appl. Math.* **28** (1975), 729–746. (MR 56, #10656).

E. Larsson

[1967: 1]   Generalized distribution semi-groups of bounded linear operators. *Ann. Scuola Norm. Sup. Pisa* **3** (1967), 137–159. (MR 38, #572).

A. Lasota and J. A. Yorke

[1971: 1]   Bounds for periodic solutions of differential equations in Banach spaces. *J. Diff. Equations* **10** (1971), 83–91. (MR 43, #5133).

B. Latil

[1968: 1]   Singular perturbations of Cauchy's problem. *J. Math. Anal. Appl.* **23**
             (1968), 683–698. (MR 38, #1378).

R. Lattès and J. L. Lions

[1967: 1]   *Méthode de Quasi-Réversibilité et Applications*, Dunod, Paris, 1967.
             (MR 38, #874). English translation: [1969: 1] *The Method of Quasi-
             Reversibility. Applications to Partial Differential Equations.* Translated
             and edited by R. Bellman. American Elsevier Publishing Co., Inc.,
             New York, 1969.

M. M. Lavrentiev

[1956: 1]   On the Cauchy problem for the Laplace equation. (Russian.) *Izvest.
             Akad. Nauk SSSR Ser. Mat.* **120** (1956), 819–842. (MR 19, 426).

[1957: 1]   On the problem of Cauchy for linear elliptic equations of second order.
             (Russian.) *Dokl. Akad. Nauk SSSR* **112** (1957), 195–197. (MR 19,
             553).

[1967: 1]   *Some Improperly Posed Problems in Mathematical Physics.* Springer,
             New York, 1967.

M. M. Lavrentiev and A. H. Amirov

[1977: 1]   Conditional correctness of the Cauchy problem. (Russian.) *Funkcional
             Anal. i Prilozen.* **11** (1977), 80–82. (MR 56, #788).

M. M. Lavrentiev and M. A. Atahodzaev

[1976: 1]   The Cauchy problem for a complete evolution equation of second
             order. (Russian.) *Dokl. Akad. Nauk SSSR* **230** (1976), 1033–1034.
             (MR 58, #28954b).

M. M. Lavrentiev, V. G. Romanov, and V. G. Vasiliev

[1970: 1]   *Multi-dimensional Inverse Problems for Differential Equations.* Lecture
             Notes in Mathematics, vol. 167, Springer, Berlin, 1970.

A. Lax

[1956: 1]   On Cauchy's problem for partial differential equations with multiple
             characteristics. *Comm. Pure Appl. Math.* **9** (1956), 135–169. (MR 18,
             397).

P. D. Lax

[1952: 1]   Operator theoretic treatment of hyperbolic equations. *Bull. Amer.
             Math. Soc.* **58** (1952), 182.

[1956: 1]   A stability theorem for solutions of abstract differential equations, and
             its application to the study of local behavior of solutions of elliptic
             equations. *Comm. Pure Appl. Math.* **9** (1956), 747–766. (MR 19,
             #281).

[1957: 1]   A Phragmén-Lindelöf theorem in harmonic analysis and its application
             to some questions in the theory of elliptic equations. *Comm. Pure Appl.
             Math.* **10** (1957), 361–389. (MR 20, #229).

[1957: 2]   Asymptotic solutions of oscillatory initial value problems. *Duke Math.
             J.* **24** (1957), 627–646. (MR 20, #4096).

[1958: 1]   Differential equations, difference equations and matrix theory. *Comm. Pure Appl. Math.* **11** (1958), 175–194. (MR 20, #4572).

[1961: 1]   On the stability of difference approximations to solutions of hyperbolic equations with variable coefficients. *Comm. Pure Appl. Math.* **14** (1961), 497–520. (MR 26, #3215).

[1961: 2]   Survey of stability of difference schemes for solving initial value problems for hyperbolic equations. *Proc. Sympos. Appl. Math.*, vol. XV, 251–258. Amer. Math. Soc., Providence, R.I., 1963. (MR 28, #3549).

P. D. Lax and A. N. Milgram

[1954: 1]   Parabolic equations. *Contributions to the theory of partial differential equations.* 167–190. *Annals of Mathematics Studies No. 33*, Princeton University Press, Princeton, N.J., 1954. (MR 16, 709).

P. D. Lax and R. S. Phillips

[1960: 1]   Local boundary conditions for dissipative symmetric linear differential operators. *Comm. Pure Appl. Math.* **13** (1960), 427–255. (MR 22, #9718).

[1964: 1]   Scattering theory. *Bull. Amer. Math. Soc.* **70** (1964), 130–142. (MR 29, #5133).

[1966: 1]   Scattering theory for transport phenomena. *Functional Analysis (Proc. Conf., Irvine, Calif. 1966)*, 119–130, Academic Press, London, 1967. (MR 36, #3166).

[1967: 1]   *Scattering Theory.* Academic Press, New York–London, 1967.

[1971: 1]   Scattering theory, *Rocky Mount. J. Math.* **1** (1971), 173–223. (MR 54, #758).

[1972: 1]   Scattering theory for the acoustic equation in an even number of space dimensions. *Indiana U. Math. J.* **22** (1972), 101–134. (MR 46, #4014).

[1972: 2]   On the scattering frequencies of the Laplace operator for exterior domains. *Comm. Pure Appl. Math.* **25** (1972), 85–101. (MR 45, #5531).

[1973: 1]   Scattering theory for dissipative hyperbolic systems. *J. Functional Anal.* **14** (1973), 172–235. (MR 50, #5502).

[1976: 1]   *Scattering Theory for Automorphic Functions.* Princeton Univ. Press, Princeton, N.J., 1976.

P. D. Lax and R. D. Richtmyer

[1956: 1]   Survey of the stability of linear finite difference equations. *Comm. Pure Appl. Math.* **9** (1956), 267–293. (MR 18, 48).

S. Leader

[1954: 1]   On the infinitesimal generator of a semigroup of positive transformations with local character condition. *Proc. Amer. Math. Soc.* **5** (1954), 401–406. (MR 16, 145).

M. Lees

[1962: 1]   Asymptotic behaviour of solutions of parabolic differential inequalities. *Canadian J. Math.* **14** (1962), 626–631. (MR 28, #354).

J. Lehner and G. M. Wing
[1956: 1]   Solution of the linearized Boltzmann transport equation for the slab
             geometry. *Duke Math. J.* **23** (1956), 125–142. (MR 18, 50).

J. Lelong-Ferrand
[1958: 1]   Sur les groupes à un paramètre de transformations des variétés diffé-
             rentiables. *J. Math. Pures et Appl.* **37** (1958), 269–278. (MR 20,
             #4874).

A. Lenard
[1971: 1]   Probabilistic version of Trotter's exponential product formula in
             Banach algebras. *Acta Sci. Math.* (*Szeged*) **32** (1971), 101–107. (MR
             49, #3595).

J. Leray and Y. Ohya
[1964: 1]   Systèmes linéaires hyperboliques non stricts. *Deuxième Colloq. d'Anal.*
             Fonct., Centre Belge de Recherches Math., Louvain, 1964. (MR 32,
             #7956).

H. A. Levine
[1970: 1]   Logarithmic convexity and the Cauchy problem for some abstract
             second-order differential inequalities. *J. Diff. Equations* **8** (1970), 34–55.
             (MR 41, #3945).
[1970: 2]   Logarithmic convexity, first order differential inequalities and some
             applications. *Trans. Amer. Math. Soc.* **152** (1970), 299–320. (MR 43,
             #746).
[1972: 1]   Some uniqueness and growth theorems in the Cauchy problem for
             $Pu_{tt} + Mu_t + Nu = 0$ in Hilbert space. *Math. Z.* **126** (1972), 345–360.
             (MR 47, #3865).
[1973: 1]   On the uniqueness of bounded solutions to $u'(t) = A(t)u(t)$ and
             $u''(t) = A(t)u(t)$ in Hilbert space. *SIAM. J. Math. Anal.* **4** (1973),
             250–259. (MR 48, #9021).
[1973: 2]   Some nonexistence and instability theorems to formally parabolic
             equations of the form $Pu_t = -Au + F(u)$. *Arch. Rat. Mech. Anal.* **51**
             (1973), 371–386. (MR 50, #714).
[1974: 1]   Instability and nonexistence of global solutions to nonlinear wave
             equations of the form $Pu_{tt} = -Au + F(u)$. *Trans. Amer. Math. Soc.*
             192 (1974), 1–21. (MR 49, #9436).
[1975: 1]   Uniqueness and growth of weak solutions to certain linear differential
             equations in Hilbert space. *J. Diff. Equations* **8** (1975), 73–81. (MR 50,
             #10479).
[1977: 1]   An equipartition of energy theorem for weak solutions of evolutionary
             equations in Hilbert space: The Lagrange identity method. *J. Diff.*
             Equations **24** (1977), 197–210. (MR 55, #10795).

H. A. Levine and L. E. Payne
[1975: 1]   On the nonexistence of global solutions to some abstract Cauchy
             problems of standard and nonstandard types. *Rend. Mat.* (6) **8** (1976),
             413–428. (MR 52, #11680).

B. M. Levitan
[1966: 1] On the integration of almost-periodic functions with values in a Banach space. (Russian.) *Izv. Akad. Nauk SSSR Ser. Mat.* **30** (1966), 1101–1110. (MR 34, #561).

T. Lezański
[1970: 1] Sur l'intégration directe des équations d'évolution. *Studia Math.* **34** (1970), 149–163. (MR 56, #6042).
[1974: 1] Sur l'intégration directe des équations d'évolution, 2. *Studia Math.* **52** (1974), 81–108. (MR 58, #6584).

H. M. Lieberstein
[1972: 1] *Theory of Partial Differential Equations.* Academic Press, New York, 1972.

J. H. Lightbourne, III.
[1976: 1] Periodic solutions and perturbed semigroups of linear operators. Nonlinear systems and applications (*Proc. Internat. Conf., Univ. Texas, Arlington, Texas 1976*), 591–602. Academic Press, New York, 1977. (MR 56, #16466).

J.-L. Lions
[1956: 1] Opérateurs de Delsarte en problèmes mixtes. *Bull. Soc. Math. France* **81** (1956), 9–95. (MR 19, 556).
[1957: 1] Une remarque sur les applicatione du théorème de Hille-Yosida. *J. Math. Soc. Japan* **9** (1957), 62–70. (MR 19, #424).
[1958: 1] Équations différentielles à coefficients opérateurs non bornés. *Bull. Soc. Math. France* **86** (1958), 321–330. (MR 23, #A2624).
[1958: 2] Sur certaines équations aux derivées partielles à coefficients opérateurs non bornés. *J. Analyse Math.* **6** (1958), 333–355. (MR 23, #A2623).
[1959: 1] Équations différentielles du premier ordre dans un espace de Hilbert. *C. R. Acad. Sci. Paris* **248** (1959), 1099–1102. (MR 21, #757).
[1960: 1] Problèmes mixtes abstraits. *Proc. Internat. Congress Math., 1958*, 389–397. Cambridge University Press, Cambridge, 1960. (MR 23, #A3925).
[1960: 2] Les semi-groupes distributions. *Portugal. Math.* **19** (1960), 141–164. (MR 26, #611).
[1961: 1] *Équations Différentielles Opérationnelles et Problèmes aux Limites.* Springer, Berlin, 1961.
[1962: 1] Espaces d'interpolation et domaines de puissances fractionnaires d'opérateurs. *J. Math. Soc. Japan* **14** (1962), 233–241. (MR 27, #2850).
[1963: 1] Remarques sur les équations différentielles opérationnelles. *Osaka Math. J.* **15** (1963), 131–142. (MR 27, #422).
[1963: 2] Quelques remarques sur les équations différentielles opérationnelles du 1er ordre. *Rend. Sem. Mat. Univ. Padova* **33** (1963), 213–225. (MR 27, #6155).
[1966: 1] Sur la stabilisation de certains problèmes mal posés. (English summary.) *Rend. Sem. Mat. Fis. Milano* **36** (1966), 80–87. (MR 35, #6965).

[1969: 1]   *Quelques Méthodes de Résolution de Certains Problèmes aux Limites Non Linéaires*. Dunod-Gauthier-Villars, Paris, 1969.
[1969: 2]   Some linear and non-linear boundary value problems for evolution equations. Lectures in Differential Equations, vol. I, 97–121, Van Nostrand, New York, 1969. (MR 40, #3089).

J.-L. Lions and E. Magenes
[1968: 1]   *Problèmes aux Limites Non Homogènes at Applications*, vol. I. Dunod, Paris, 1968.
[1968: 2]   *Problèmes aux Limites Non Homogènes et Applications*, vol. II. Dunod, Paris, 1968.
[1970: 1]   *Problèmes aux Limites Non Homogènes et Applications*, vol. III. Dunod, Paris, 1970.

J. L. Lions and P. Raviart
[1966: 1]   Remarques sur la résolution et l'approximation d'équations d'évolutions couplées. *ICC Bull.* **5** (1966), 1–21. (MR 34, #4650).

J. L. Lions and W. A. Strauss
[1965: 1]   Some non-linear evolution equations. *Bull. Soc. Math. France* **93** (1965), 43–96. (MR 33, #7663).

W. Littman
[1963: 1]   The wave operator and L^p norms. *J. Math. Mech.* **12** (1963), 55–68. (MR 26, #4043).

Ju. I. Ljubič
[1963: 1]   Completeness conditions for a system of eigenvectors of a correct operator. (Russian.) *Uspehi Mat. Nauk* **18** (1963), no. 1 (109), 165–171. (MR 27, #4081).
[1965: 1]   Conservative operators. (Russian.) *Uspehi Mat. Nauk* **20** (1965), 211–225. (MR 34, #4925).
[1966: 1]   Investigating the deficiency of the abstract Cauchy problem. (Russian.) *Dokl. Akad. Nauk SSSR* **166** (1966), 783–786. (MR 33, #6086).
[1966: 2]   Construction of a semigroup of operators with prescribed norm. (Russian.) *Teor. Funkcii funkcional. Anal. i Prilozen. Vyp.* **3** (1966), 21–25. (MR 34, #6541).
[1966: 3]   The classical and local Laplace transform in the abstract Cauchy problem. (Russian.) *Uspehi Mat. Nauk* **21** (1966), no. 3 (129), 3–51. (MR 33, #7889).

Ju. I. Ljubič and V. I. Macaev
[1960: 1]   On the spectral theory of linear operators in Banach space. (Russian.) *Dokl. Akad. Nauk SSSR* **131** (1960), 21–23. (MR 22, #3980).

Ju. I. Ljubič and V. A. Tkačenko
[1965: 1]   Uniqueness and approximation thorems for a local Laplace transformation. (Russian.) *Dokl. Akad. Nauk SSSR* **164** (1965), 273–276. (MR 32, #8051).
[1966: 1]   Theory and certain applications of the local Laplace transform. (Russian.) *Mat. Sb. (N.S.)* **70** (112), (1966), 416–437. (MR 33, #7890).

J. Löfstrom
[1965: 1]   On certain interpolation spaces related to generalized semi-groups. *Math. Scand.* **16** (1965), 41–54. (MR 32, #2921).

F. E. Lomovcev and N. I. Jurčuk
[1976: 1]   The Cauchy problem for second order hyperbolic operator differential equations. (Russian.) *Differencial'nye Uravnenija* **12** (1976), 2242–2250, 2301. (MR 55, #10796).

Ja. B. Lopatins'kiĭ
[1966: 1]   A class of evolution equations. (Ukrainian.) *Dopovĭdĭ Akad. Nauk. Ukrain. RSR* (1966), 711–714. (MR 33, #4431).

D. L. Lovelady
[1973/74: 1]   A Hammerstein-Volterra integral equation with a linear semigroup convolution kernel. *Indiana Univ. Math. J.* **23** (1973/74), 615–622. (MR 48, #7044).
[1974/75: 1]   Semigroups of linear operators on locally convex spaces with Banach subspaces. *Indiana Univ. Math. J.* **24** (1974/75), 1191–1198. (MR 52, #1407).
[1975: 1]   Un criterio per la continuità dell'addizione di Lie-Chernoff. (English summary.) *Ann. Univ. Ferrara Sez. VII (N.S.)* **20** (1975), 65–68. (MR 51, #11190).
[1975: 2]   On the generation of linear evolution operators. *Duke Math. J.* **42** (1975), 57–69. (MR 53, #1336).
[1975: 3]   Ergodic methods for closed linear operator equations. *J. Math. Anal. Appl.* **51** (1975), 151–157. (MR 54, #1018).

V. Lovicar
[1975: 1]   Almost periodicity of solutions of the equation $x'(t) = A(t)x(t)$ with unbounded commuting operators. *Časopis Pěst. Mat.* **100** (1975), 36–45. (MR 54, #3122).

G. Lumer
[1961: 1]   Semi-inner product spaces. *Trans. Amer. Math. Soc.* **100** (1961), 29–43. (MR 24, #A2860).
[1964: 1]   Spectral operators, Hermitian operators and bounded groups. *Acta Sci. Math. Szeged* **25** (1964), 75–85. (MR 29, #6329).
[1973: 1]   Potential-like operators and extensions of Hunt's theorem for σ-compact spaces. *J. Functional Analysis* **13** (1973), 410–416. (MR 50, #1046).
[1974: 1]   Perturbations de générateurs infinitésimaux du type "changement de temps." *Ann. Inst. Fourier* **23** (1974), 271–279. (MR 49, #3588).
[1975: 1]   Problème de Cauchy pour opérateurs locaux. *C. R. Acad. Sci. Paris* **281** (1975), 763–765. (MR 52, #1509).
[1975: 2]   Problème de Cauchy pour opérateurs locaux et "changement de temps." *Ann. Inst. Fourier* **23** (1974), 409–466. (MR 54, #8056).
[1975: 3]   Problème de Cauchy avec valeurs au bord continues. *C. R. Acad. Sci. Paris* **281** (1975), 805–807. (MR 52, #15098).

[1976: 1]   Problème de Cauchy avec valeurs au bord continues, comportement
               asymptotyque, et applications. *Lecture Notes in Mathematics*, no. 563,
               193–201. Springer, Berlin, 1976. (MR 58, #28957)

[1976: 2]   Problème de Cauchy et fonctions sous-harmoniques, *Lecture Notes in
               Mathematics*, No. 563, 202–218. Springer, Berlin, 1976. (MR 58,
               #28957).

[1977: 1]   Équations d'évolution pour opérateurs locaux non localement fermés.
               C. R. Acad. Sci. Paris **284** (1977), 1361–1363 (MR 56, #1104).

[1977: 2]   Équations d'évolution en norme uniforme pour opérateurs elliptiques.
               Régularité des solutions. *C. R. Acad. Sci. Paris* **284** (1977), 1435–1437.
               (MR 55, #13078).

G. Lumer and L. Paquet
[1977: 1]   Semi-groups holomorphes et équations d'évolution. *C. R. Acad. Sci.
               Paris* **284** (1977), 237–240. (MR 55, #1135).

G. Lumer and R. S. Phillips
[1961: 1]   Dissipative operators in a Banach space. *Pac. J. Math.* **11** (1961),
               679–698. (MR 24, #A2248).

V. Ju. Lunin
[1973: 1]   Estimates of the solutions of abstract parabolic equations with nonlin-
               ear perturbations. (Russian.) *Vestnik. Moskov. Univ. Ser. I Mat. Meh.*
               28 (1972), 56–63. (MR 49, #738).

[1974: 1]   Estimates of the solutions of abstract parabolic equations. (Russian.)
               Uspehi Mat. Nauk **29** (1974), 213–214. (MR 53, #966).

D. Lutz
[1977: 1]   Gruppen und Halbgruppen verallgemeinerter Skalaroperatoren. (En-
               glish summary.) *Manuscripta Math.* **20** (1977), 105–117. (MR 55,
               #8881).

M. A. Lyapunov
[1892: 1]   Dissertation, Kharkov, 1892. Published as *Problème Général de la
               Stabilité du Mouvement* in 1983; reprinted in Ann. Math. Studies 17,
               Princeton, 1949. English transl.: *Stability of Motion*, Academic Press,
               New York, 1966.

O. B. Lykova
[1975: 1]   The reduction principle for a differential equation with an unbounded
               operator coefficient. (Russian.) *Ukrain. Mat. Ž.* **27** (1975), 240–243,
               286. (MR 52, #3687).

V. I. Macaev and Ju. A. Palant
[1962: 1]   On the powers of a bounded dissipative operator. (Russian.) *Ukrain.
               Mat. Z.* **14** (1962), 329–337 (MR 26, #4184).

W. Magnus
[1954: 1]   On the exponential solution of differential equations for a linear
               operator. *Comm. Pure Appl. Math.* **7** (1954), 649–673. (MR 16, 790).

V. V. Maĭorov

[1974: 1] A study of the stability of the solutions of abstract parabolic equations with nearly constant almost periodic coefficients. (Russian.) *Vestnik Jaroslav. Univ. Vyp.* **7** (1974), 124–133. (MR 56, #16090).

M. A. Malik

[1967: 1] Weak solutions of abstract differential equations. *Proc. Nat. Acad. Sci. U.S.A.* **57** (1967), 883–884. (MR 35, #512).

[1971: 1] Regularity of elementary solutions of abstract differential equations. *Boll. Un. Mat. Ital.* (4) **4** (1971), 78–84. (MR 44, #7168).

[1972: 1] Weak generalized solutions of abstract differential equations. *J. Math. Anal. Appl.* **40** (1972). (MR 47, #2399).

[1973: 1] Existence of solutions of abstract differential equations in a local space. *Canad. Math. Bull.* **16** (1973), 239–244. (MR 48, #2832).

[1975: 1] Uniqueness of generalized solutions of abstract differential equations. *Canad. Math. Bull.* **18** (1975), 379–382. (MR 52, #1502).

[1975: 2] On elementary solutions of abstract differential operators. (Italian summary.) *Boll. Un. Mat. Ital.* (4) **12** (1975), 330–332. (MR 54, #671).

Ja. D. Mamedov

[1960: 1] Asymptotic stability of differential equations in a Banach space with an unbounded operator. (Russian.) *Azerbaĭdžan Gos. Učen. Zap. Ser. Fiz.-Mat. i Him. Nauk* (1960), no. 1, 3–7. (MR 36, #2928).

[1964: 1] One-sided conditional estimates for investigating the solutions of differential equations of parabolic type in Banach spaces. (Russian.) *Azerbaĭdžan Gos. Univ. Ucen. Zap. Ser. Fiz.-Mat. i Him. Nauk* (1964), no. 5, 17–25. (MR 36, #6750).

[1964: 2] Some properties of the solutions of non-linear equations of hyperbolic type in Hilbert space. (Russian.) *Dokl. Akad. Nauk SSSR* **158** (1964), 45–48. (MR 29, #3759).

[1965: 1] Certain properties of solutions of non-linear equations of parabolic type in Hilbert space. (Russian.) *Dokl. Akad. Nauk SSSR* **161** (1965), 1011–1014. (MR 31, #3723).

[1966: 1] One-sided estimates in conditions for asymptotic stability of solutions of differential equations involving unbounded operators. (Russian.) *Dokl. Akad. Nauk SSSR* **166** (1966), 533–535. (MR 33, #386).

[1967: 1] One-sided estimates in the conditions for investigating solutions of differential equations with unbounded operators and retarded argument in a Banach space. (Russian.) *Problems Comput. Math.*, 173–181. Izdat. Akad. Nauk Azerbaĭdžan SSR, Baku (1967). (MR 36, #6738).

[1968: 1] One-sided estimates of solutions of differential equations with retarded argument in a Banach space. (Russian. English summary.) *Trudy Sem. Teor. Differencial. Uravneniĭ s Otklon. Argumentom Univ. Družby Narodov Patrisa Lumumby* **6** (1968), 135–146. (MR 38, #4785).

Ja. D. Mamedov and P. E. Sobolevskiĭ

[1966: 1] On a theorem of E. Hille. (Russian.) *Akad. Nauk Azerbaĭdžan SSR Dokl.* **22** (1966), no. 7, 7–9. (MR 34, #4927).

K. S. Mamiĭ

[1965: 1]   On the boundedness of solutions of a linear homogeneous second-order equation in Hilbert space. (Russian.) *Litovsk. Mat. Sb.* **5** (1965), 593–604. (MR 33, #6087).

[1965: 2]   On the boundedness of solutions of a linear homogeneous second order equation in Hilbert and Banach spaces. (Russian.) *Ukrain. Mat. Ž.* **17** (1965), no. 6, 31–41. (MR 33, #1570).

[1966: 1]   Boundedness of solutions of nonlinear differential equations of the second order in a Banach space. (Ukrainian. Russian and English summaries.) *Dopovidi Akad. Nauk Ukraïn RSR* (1966), 715–719. (MR 33, #2919).

[1967: 1]   Boundedness and stability of solutions of certain non-linear differential equations of the second order in a Hilbert space, I. (Russian.) *Vestnik Moskov. Univ. Ser. I Mat. Meh.* **22** (1967), no. 1, 25–34. (MR 34, #6271).

[1967: 2]   Boundedness of solutions of certain non-linear differential equations of the second order in a Hilbert space, II. (Russian.) *Vestnik Moskov. Univ. Ser. I Mat. Meh.* **22** (1967), no. 2, 3–9. (MR 35, #513).

K. S. Mamiĭ and D. D. Mirzov

[1971: 1]   Behavior of the solution on the half-axis of a certain second-order nonlinear differential equation. (Russian.) *Differencial'nye Uravnenija* **7** (1971), 1330–1332, 1334. (MR 44, #5547).

P. Mandl

[1968: 1]   *Analytical Treatment of One-Dimensional Markov Processes.* Springer, Berlin, 1968.

J. E. Marsden

[1968: 1]   Hamiltonian one parameter groups. *Arch. Rat. Mech. Anal.* **28** (1968), 362–396. (MR 37, #1735).

[1968: 2]   A Banach space of analytic functions for constant coefficient equations of evolution. *Canad. Math. Bull.* **11** (1968), 599–601. (MR 39, #3130).

B. Marshall, W. Strauss, and S. Wainger

[1980: 1] $L^p - L^q$ estimates for the Klein-Gordon equation. *J. Math. Pures Appl.* **59** (1980), 417–440.

J. T. Marti

[1965: 1]   Existence d'une solution unique de l'équation linéarisée de Boltzmann et développement de cette solution suivant des fonctions propres. *C. R. Acad. Sci. Paris* **260** (1965), 2692–2694. (MR 31, #5115).

[1967: 1]   Über Anfangs-und Eigenwertprobleme aus der Neutronentransporttheorie. (English summary.) *Z. Angew. Math. Phys.* **18** (1967), 247–259. (MR 37, #1142).

R. H. Martin Jr.

[1975: 1]   Invariant sets for perturbed semigroups of linear operators. *Ann. Mat. Pura Appl.* (4) **105** (1975), 221–239. (MR 52, #11240).

[1976: 1]   *Nonlinear Operators and Differential Equations in Banach Spaces.* Wiley-Interscience, New York, 1976.

[1977: 1]   Nonlinear perturbations of linear evolution systems. *J. Math. Soc. Japan* **29** (1977), 233–252. (MR 56, #6045).

A. A. Martinjuk

[1972: 1]   The approximation of the solutions of linear equations in a Banach space. (Russian.) *Differencial'nye Uravnenija* **8** (1972), 2988–2993, 2111. (MR 47, #582).

[1975: 1]   Some direct versions of the Galerkin-Krylov method. (Russian.) *Ukrain Mat Z.* **27** (1975), 599–605, 715. (MR 52, #6467).

K. Maruo

[1974: 1]   Integral equation associated with some non-linear evolution equation. *J. Math. Soc. Japan* **26** (1974), 433–439. (MR 51, #3986).

J. L. Massera and J. J. Schäffer

[1966: 1]   *Linear Differential Equations and Function Spaces.* Academic Press, New York, 1966.

F. J. Massey

[1972: 1]   Abstract evolution equations and the mixed problem for symmetric hyperbolic systems. *Trans. Amer. Math. Soc.* **168** (1972), 165–188. (MR 45, #7283).

[1976: 1]   Analyticity of solutions of nonlinear evolution equations. *J. Diff. Equations* **22** (1976), 416–127. (MR 55, #8499). (MR 55, #8499).

[1977: 1]   Semilinear parabolic equations with L^1 initial data. *Indiana Univ. Math. J.* **26** (1977), 399–412. (MR 58, #17526).

K. Masuda

[1967: 1]   Asymptotic behavior in time of solutions for evolution equations. *J. Functional Analysis* **1** (1967), 84–92. (MR 36, #1817).

[1967: 2]   A note on the analyticity in time and the unique continuation property for solutions of diffusion equations. *Proc. Japan Acad.* **43** (1967), 420–422. (MR 36, #5529).

[1968: 1]   A unique continuation theorem for solutions of wave equations with variable coefficients. *J. Math. Anal. Appl.* **21** (1968), 369–376. (MR 36, #2965).

[1972: 1]   On the holomorphic evolution operators. *J. Math. Anal. Appl.* **39** (1972), 706–711. (MR 46, #7963).

[1972: 2]   Anti-locality of the one-half power of elliptic differential operators. *Publ. Res. Inst. Math. Sci.* **8** (1972), 207–210. (MR 48, #9454).

L. Máté

[1962: 1]   Extension of a semigroup of operators. (Hungarian.) *Magyar Tud. Akad. Mat. Fiz. Oszt. Közl.* **12** (1962), 217–222. (MR 27, #5134).

[1962: 2]   On the semigroup of operators in a Frechet space. (Russian.) *Dokl. Akad. Nauk SSSR* **142** (1962), 1247–1250. (MR 25, #2458).

A. Matsumura
[1976/77: 1]   On the asymptotic behavior of solutions of semilinear wave equations. *Publ. RIMS, Kyoto Univ.* **12** (1976/77), 169–189. (MR 54, #8048).

T. Matzuzawa
[1970: 1]   Sur les équations $u'' - t^a A u = f$. *Proc. Japan Acad.* **46** (1970), 609–613. (MR 33, #747).

C. Mayer
[1969: 1]   Opérateurs pseudo-négatifs et semi-groupes de contractions. *C. R. Acad. Sci. Paris* **268** (1969), A586–A589. (MR 39, #4707).

V. G. Maz'ja and B. A. Plamenevskiĭ
[1971: 1]   The asymptotics of the solutions of differential equations with operator coefficients. (Russian.) *Dokl. Akad. Nauk SSSR* **196** (1971), 512–151. (MR 43, #2565).

V. G. Maz'ja and P. E. Sobolevskiĭ
[1962: 1]   On generating operators of semi-groups. (Russian.) *Uspehi Mat. Nauk* **17** (1962), no. 6 (108), 151–154. (MR 27, #1838).

T. Mazumdar
[1974: 1]   Generalized projection theorem and weak noncoercive evolution problems in Hilbert space. *J. Math. Anal. Appl.* **46** (1974), 143–168. (MR 52, #6468).
[1975: 1]   Regularity of solutions of linear coercive evolution equations with variable domain I. *J. Math. Anal. Appl.* **52** (1975), 625–647. (MR 55, #1109).

C. A. McCarthy and J. G. Stampfli
[1964: 1]   On one-parameter groups and semigroups of operators in Hilbert space. *Acta Sci. Mat. Szeged* **25** (1964), 6–11. (MR 30, #1402).

A. McNabb
[1972: 1]   Initial value method for linear boundary value problems. *J. Math. Anal. Appl.* **39** (1972), 495–526. (MR 46, #7654).

A. McNabb and A. Schumitzky
[1973: 1]   Factorization and embedding for general linear boundary value problems. *J. Diff. Equations* **14** (1973), 518–546. (MR 49, #771).
[1974: 1]   An initial value problem for linear causal boundary value problems. *J. Diff. Equations* **15** (1974), 322–349. (MR 49, #7558).

L. A. Medeiros
[1967: 1]   An application of semi-groups of class C_0. *Portugal. Math.* **26** (1967), 71–77. (MR 40, #3367).
[1969: 1]   On nonlinear differential equations in Hilbert spaces. *Amer. Math. Monthly* **76** (1969), 1024–1027. (MR 40, #1685).
[1969: 2]   The initial value problem for nonlinear wave equations in Hilbert space. *Trans. Amer. Math. Soc.* **136** (1969), 305–327. (MR 42, #2184).

E. Ja. Melamed
[1964: 1]   The application of semigroups to the study of boundedness of the solutions of a certain differential equation in a Banach space. (Russian.) *Izv. Vyss. Učebn. Zaved. Matematika* (1964), no. 6 (43), 123–133. (MR 30, #2212).
[1965: 1]   Application of the method of semigroups to the investigation of the boundedness of the solutions of a linear partial differential equation. (Russian.) *Mat. Sb. (N.S.)* **68** (110), (1965), 228–241. (MR 34, #3142).
[1969: 1]   The exponential growth of the solution of a second order abstract Cauchy problem. (Ukrainian. English and Russian summary.) *Dopovīdī Akad. Nauk Ukraïn. RSR Ser. A* (1969), 797–801, 861. (MR 42, #2138).
[1971: 1]   Apropos a certain theorem of E. Hille. (Russian.) *Izv. Vysš. Učebn. Zaved. Matematika* (1971) (111), 69–75. (MR 45, #2296).

I. V. Melnikova
[1975: 1]   The method of quasi-reversibility for abstract parabolic equations in a Banach space. (Russian.) *Trudy Inst. Mat. i Meh. Ural. Naucn. Centr. Akad. Nauk SSSR Vyp. 17 Metody Resenija Uslovno-korrekt Zadac.* (1975), 27–56, 141. (MR 56, #16091).

P. R. Meyers
[1970: 1]   On contractive semigroups and uniform asymptotic stability. *J. Res. Nat. Bur. Standards Sect. B* **74B** (1970), 115–120. (MR 42, #2315).
[1970: 2]   Contractifiable semigroups. *J. Res. Nat. Bureau Standards B* **74B** (1970), 315–322. (MR 43, #5367).

V. A. Mihaĭlec
[1974: 1]   Solvable and sectorial boundary value problems for the operator Sturm-Liouville equation. (Russian.) *Ukrain. Mat. Ž.* **26** (1974), 450–459, 573. (MR 50, #13782).
[1975: 1]   Boundary value problems for the operator Sturm-Liouville equation with a complete system of eigen- and associated functions. (Russian.) *Differencial'nye Uravnenija* **11** (1975), 1595–1600, 1715. (MR 52, #3688).

J. Mika
[1977: 1]   Singularly perturbed evolution equations in Banach spaces. *J. Math. Anal. Appl.* **58** (1977), 189–201. (MR 56, #6046)

J. J. H. Miller
[1967: 1]   On power-bounded operators and operators satisfying a resolvent condition. *Numer. Math.* **10** (1967), 389–396. (MR 36, #3147).
[1968: 1]   On the resolvent of a linear operator associated with a well-posed Cauchy problem. *Math. Comp.* **22** (1968), 541–548. (MR 38, #1543).

J. J. H. Miller and G. Strang
[1966: 1]   Matrix theorems for partial differential and difference equations. *Math. Scand.* **18** (1966), 113–133. (MR 35, #206).

K. Miller
[1975: 1]   Logarithmic convexity results for holomorphic semigroups. *Pacific J. Math.* **58** (1975), 549–551. (MR 51, #1375).

R. K. Miller
[1975: 1]   Volterra integral equations in a Banach space. *Funkcial. Ekvac.* **18** (1975), 163–193. (MR 53, #14062).

A. I. Miloslavskiĭ
[1976: 1]   The decay of the solutions of an abstract parabolic equation with a periodic operator coefficient. (Russian.) *Izv. Severo-Kavkaz. Naučn. Centra Vysš. Školy Ser. Estestv. Nauk.* (1976) (116), 12–15. (MR 55, #8500).

G. N. Milštein
[1975: 1]   Exponential stability of positive semigroups in a linear topological space. (Russian.) *Izv. Vysš. Učebn. Zaved. Matematika* (1975) (160), 35–42. (MR 54, #3498).

N. Minorski
[1942: 1]   Self-excited oscillations in dynamical systems possessing retarded actions. *J. Appl. Mech.* **9** (1942), 65–71.

P. A. Mišnaevskiĭ
[1971: 1]   The attainment of almost periodic conditions, and the almost periodicity of solutions of differential equations in a Banach space. (Russian.) *Vestnik Moskov. Univ. Ser. I Mat. Meh.* **26** (1971), no. 3, 69–76. (MR 44, #582).
[1972: 1]   Almost periodic solutions of equations of elliptic type. (Russian.) *Differencial'nye Uravnenija* **8** (1972), 921–924. (MR 46, #5786).
[1976: 1]   On the spectral theory for the Sturm-Liouville equation with an operator-valued coefficient. (Russian.) *Izv. Akad. Nauk SSSR Ser. Mat.* **40** (1976), 152–189, 222. (MR 54, #10790).

I. Miyadera
[1951: 1]   On one-parameter semi-groups of operators. *J. Math. Tokyo* **1** (1951), 23–26. (MR 14, 564).
[1952: 1]   Generation of a strongly continuous semi-group of operators. *Tôhoku Math. J.* (2) (1952), 109–114. (MR 14, #564).
[1954: 1]   On the generation of a strongly ergodic semi-group of operators. *Proc. Japan Acad.* **30** (1954), 335–340. (MR 16, 374).
[1954: 2]   A note on strongly (C, α)-ergodic semi-groups of operators. *Proc. Japan Acad.* **30** (1954), 797–800. (MR 16, 1031).
[1954: 3]   On the generation of a strongly ergodic semi-group of operators. *Tôhoku Math. J.* (2) **6** (1954), 38–52. (MR 16, 374).
[1954: 4]   On the generation of strongly ergodic semi-groups of operators, II. *Tôhoku Math. J.* (2) **6** (1954), 231–242. (MR 16, 1031).
[1955: 1]   A note on strongly ergodic semi-groups of operators. *Kōdai Math. Sem. Rep.* **7** (1955), 55–57. (MR 17, 988).

[1956: 1]   On the representation theorem by the Laplace transformation of vector-valued functions. *Tôhoku Math. J.* (2) **8** (1956), 170–180. (MR 18, 748).

[1959: 1]   A note on contraction semi-groups of operators. *Tôhoku Math. J.* (2) **11** (1959), 98–105. (MR 21, #5903).

[1959: 2]   Semi-groups of operators in Fréchet space and applications to partial differential equations. *Tôhoku Math. J.* (2) **11** (1959), 162–183. (MR 21, #7745).

[1966: 1]   A perturbation theorem for contraction semi-groups. *Proc. Japan Acad.* **42** (1966), 755–758. (MR 34, #8204).

[1966: 2]   On perturbation theory for semi-groups of operators. *Tôhoku Math. J.* (2) **18** (1966), 299–210. (MR 35, #795).

[1972: 1]   On the generation of semi-groups of linear operators. *Tôhoku Math. J.* (2) **24** (1972), 251–261. (MR 48, #953).

I. Miyadera, S. Oharu, and N. Okazawa
[1972/73: 1]   Generation theorems of semi-groups of linear operators. *Publ. Res. Inst. Math. Sci.* **8** (1972–73), 509–555. (MR 49, #11315).

W. Mlak
[1959: 1]   Limitations and dependence on parameter of solutions of non-stationary differential operator equations. *Ann Pol. Math.* **6** (1959), 305–322. (MR 21, #7345).

[1960: 1]   Integration of differential equations with unbounded operators in abstract *L*-spaces. *Bull. Acad. Polon. Sci. Sér Sci. Math. Astr. Phys.* **8** (1960), 163–168. (MR 22, #8178).

[1960/61: 1]   Differential equations with unbounded operators in Banach spaces. *Ann. Pol. Math.* **9** (1960/61), 101–111. (MR 23, #A2875).

[1961: 1]   A note on approximation of solutions of abstract differential equations. *Ann. Pol. Math.* **10** (1961), 273–268. (MR 23, #A3388).

[1966: 1]   On semi-groups of contractions in Hilbert spaces. *Studia Math.* **26** (1966), 263–272. (MR 33, #7871).

C. Monari
[1971: 1]   Su una equazione parabolica con termine di ritardo non lineare, discontinuo rispetto all'incognita: soluzioni periodiche. *Richerche Mat.* **20** (1971), 118–142. (MR 49, #11073).

Ch. J. Monlezun
[1974: 1]   Temporally inhomogeneous scattering theory. *J. Math. Anal. Appl.* **47** (1974), 133–152. (MR 51, #8856).

R. T. Moore
[1967: 1]   Duality methods and perturbation of semigroups. *Bull. Amer. Math. Soc.* **73** (1967), 548–553. (MR 36, #5759).

[1968: 1]   *Measurable, Continuous and Smooth Vectors for Semigroups and Group Representations.* Memoirs Amer. Math. Soc., No. 78, Providence, R.I., 1968.

[1971: 1]   Generation of equicontinuous semigroups by hermitian and sectorial
             operators, I. *Bull. Amer. Math. Soc.* **77** (1971), 224–229. (MR 43,
             #3846a).
[1971: 2]   Generation of equicontinuous semigroups by hermitian and sectorial
             operators, II. *Bull. Amer. Math. Soc.* **77** (1971), 368–373. (MR 43,
             #3846b).

V. A. Morozov
[1973: 1]   Linear and nonlinear ill-posed problems. (Russian.) *Mathematical
             Analysis*, vol. 11, 129–278, 180. Akad. Nauk. SSSR Vsesojuz. Inst.
             Naučn. i Tehn. Informacii, Moscow, 1973. (MR 51, #13799).

U. Mosco
[1965: 1]   Semigruppi distribuzioni prolungabili analiticamente. *Atti del Con-
             vegno su le Equazioni alle Derivate Parziali* (Nervi, 1965), 86–89.
             Edizioni Cremonese, Roma, 1966. (MR 34, #6525).
[1967: 1]   On the duality for linear operators and distribution semigroups. (Italian
             summary.) *Ann. Mat. Pura Appl.* (4) **75** (1967), 121–141. (MR 40,
             #3359).

V. B. Moseenkov
[1977: 1]   Almost periodic solutions of nonlinear wave equations. (Russian.)
             Ukrain. Mat Ž. **29** (1977), no. 1, 116–122, 143, (MR 56, #822).

T. W. Mullikin
[1959: 1]   Semi-groups of class C_0 in L_p determined by parabolic differential
             equations. *Pacific J. Math.* **9** (1959), 791–804. (MR 22, #189).

H. Murakami
[1966: 1]   On non-linear ordinary and evolution equations. *Funkcial. Ekvac.* **9**
             (1966), 151–162. (MR 35, #514).

M. Murata
[1973: 1]   Anti-locality of certain functions of the Laplace operator. *J. Math.
             Soc. Japan* **25** (1973), 556–564. (MR 49, #5555).

M. K. V. Murthy
[1971: 1]   A survey of some recent developments in the theory of the Cauchy
             problem. *Boll. Un. Mat. Ital.* **4** (1971), 473–562 (MR 45, #2310).

M. M. Mustafaev
[1975:]   Asymptotic behavior with respect to a small parameter of the solution
             of a certain boundary value problem for an evolution equation of
             elliptic type. (Russian.) *Izv. Akad. Nauk Azerbaĭdžan. SSR Ser. Fiz.-
             Tehn. Mat. Nauk* (1975) no. 1, 82–87. (MR 56, #12478).

A. D. Myškis
[1972: 1]   *Linear Differential Equations with Retarded Argument*, 2nd ed. (Rus-
             sian.) Izdat. "Nauka," Moscow, 1972.

M. Nagumo
[1936: 1]   Einige analytische Untersuchungen in linearen metrischen Ringen.
             Japan J. Math. **13** (1936), 61–80.
[1956: 1]   On linear hyperbolic systems of partial differential equations in the
             whole space. *Proc. Japan Acad.* **32** (1956), 703–706. (MR 21, #3664).

B. Nagy
[1974: 1]   On cosine operator functions in Banach spaces. *Acta Sci. Math. Szeged*
             36 (1974), 281–289. (MR 51, #11191).
[1976: 1]   Cosine operator functions and the abstract Cauchy problem. *Period.
             Math. Hungar.* **7** (1976), 213–217. (MR 56, #9023).

H. Nakano
[1941: 1]   Über den Beweis des Stoneschen Satzes. *Ann. of Math.* **42** (1941),
             665–667. (MR 3, 51).

D. S. Nathan
[1935: 1]   One-parameter groups of transformations in abstract vector spaces.
             Duke Math. J. **1** (1935), 518–526.

J. Naumann
[1976: 1]   Periodic solutions to evolution inequalities in Hilbert spaces (Italian
             summary.) *Boll. Un. Mat. Ital.* **13** (1976), 686–711. (MR 55, #3460).

J. Nečas
[1974: 1]   Application of Rothe's method to abstract parabolic equations.
             Czechoslovak Math. J. **24** (99), (1974), 496–500. (MR 50, #1069).

E. Nelson
[1964: 1]   Feynman integrals and the Schrödinger equation. *J. Math. Phys.* **5**
             (1964), 332–343. (MR 28, #4397).

S. Nelson and R. Triggiani
[1979: 1]   Analytic properties of cosine operators. *Proc. Amer. Math. Soc.* **74**
             (1979), 101–104.

V. V. Nemyckiĭ, M. M. Vaĭnberg, and R. S. Gusarova
[1964: 1]   Operator differential equations. (Russian.) *Math. Analysis* (1964)
             165–235. Akad. Nauk SSSR Inst. Naučn. Informacii, Moscow, 1966.
             (MR 33, #4500).

J. W. Neuberger
[1964: 1]   A quasi-analyticity condition in terms of finite differences. *Proc.
             London Math. Soc.* **14** (1964), 245–269. (MR 28, #3130).
[1970: 1]   Analyticity and quasi-analyticity for one-parameter semigroups. *Proc.
             Amer. Math. Soc.* **25** (1970), 488–494. (MR 41, #4296).
[1971/72: 1] Lie generators for strongly continuous equi-uniformly continuous
             one parameter semigroups on a metric space. *Indiana Univ. Math. J.*
             21 (1971/72), 961–971. (MR 45, #964).
[1973: 1]   Quasi-analyticity and semigroups. *Bull. Amer. Math. Soc.* **78** (1972),
             909–922. (MR 47, #4061).

[1973: 2]   Lie generators for one parameter semigroups of transformations. *J. Reine Angew. Math.* **258** (1973), 133–136. (MR 46, #9796).

[1973: 3]   Quasi-analytic semigroups of bounded linear transformations. *J. London Math. Soc.* **7** (1973), 259–264. (MR 49, #1218).

J. von Neumann

[1932: 1]   Über einen Satz von Herrn M. H. Stone. *Ann. Math.* (2) **33** (1932), 567–573.

[1932: 2]   Proof of the quasi-ergodic hypothesis. *Proc. Nat. Acad. Sci. U.S.A.* **18** (1932), 70–82.

R. Nevanlinna

[1936: 1]   *Eindeutige Analytische Funktionen.* Springer-Verlag, Berlin, 1936.

J. Neveu

[1955: 1]   Sur une hypotèse de Feller à propos de l'équation de Kolmogoroff. *C. R. Acad. Sci. Paris* **240** (1955), 590–591. (MR 16, 716).

[1958: 1]   Théorie des semi-groupes de Markov. *Univ. California Publ. Statistics* **2** (1958), 319–394. (MR 21, #1538).

N. V. Nikloenko

[1973: 1]   Conditional stability of the family of periodic solutions of differential equations in Banach spaces. (Russian.) *Izv. Vyss. Učebn. Zaved. Matematika* (1973), no. 9, 54–60. (MR 49, #5504).

L. Nirenberg

[1962: 1]   Comportement a l'infini pour des équations différentielles ordinaires dans un espace de Banach. *Les Équations aux Dérivées Partielles* (Paris, 1962), 163–173. Éditions du Centre National de la Recherche Scientifique, Paris, 1963. (MR 28, #2320).

L. Nirenberg and F. Trèves

[1971: 1]   Remarks on the solvability of linear equations of evolution. *Symposia Mathematica*, vol. VII (Convegno sui Problemi di Evoluzione, INDAM, Roma, Maggio 1970, 325–338). Academic Press, London, 1971. (MR 48, #1210).

W. Nollau

[1975: 1]   Über gebrochene Potenzen infinitesimaler Generatoren Markovscher Übergangswahrscheinlichkeiten I. *Math. Nachr.* **65** (1975), 235–246. (MR 51, #1479).

K. Nowak

[1973: 1]   On approximation of the solution of linear differential equations in Banach spaces by a difference process. *Comment. Math. Prace Mat.* **17** (1973), 187–199. (MR 49, #772).

H. S. Nur
[1971: 1]   Singular perturbations of differential equations in abstract spaces.
              Pacific. J. Math. **36** (1971), 775–780. (MR 43, #7750).

A. E. Nussbaum
[1959: 1]   Integral representation of semigroups of unbounded self-adjoint opera-
              tors. *Ann. Math.* **69** (1959), 133–141. (MR 21, #305).
[1970: 1]   Spectral representation of certain one-parametric families of symmetric
              operators in Hilbert space. *Trans. Amer. Math. Soc.* **99** (1970), 419–429.
              (MR 42, #3616).

E. Obrecht
[1975: 1]   Sul problema di Cauchy per le equazioni paraboliche astratte di ordine
              n. (English summary.) *Rend. Sem. Mat. Univ. Padova* **53** (1975),
              231–256. (MR 54, #7989).

J. Odhnoff
[1964: 1]   Un example de non-unicité d'une équation différentielle opération-
              nelle. *C. R. Acad. Sci. Paris* **258** (1964), 1689–1691. (MR 28, #4339).

H. Ogawa
[1965: 1]   Lower bounds for solutions of differential inequalities in Hilbert space.
              Proc. Amer. Math. Soc. **16** (1965), 1241–1243. (MR 32, #2759).
[1967: 1]   Lower bounds for solutions of parabolic differential inequalities.
              Canad. J. Math. **19** (1967), 667–672. (MR 41, #627).
[1970: 1]   On the maximum rate of decay of solutions of parabolic differential
              inequalities. *Arch. Rational Mech. Anal.* **38** (1970), 173–177. (MR 41,
              #8800).

S. Oharu
[1966: 1]   On the convergence of semi-groups of operators. *Proc. Japan Acad.* **42**
              (1966), 880–884. (MR 35, #3481).
[1971/72: 1] Semigroups of linear operators in a Banach space. *Publ. Res. Inst.*
              Math. Sci. **7** (1971/72), 205–260. (MR 47, #885).
[1973: 1]   Eine Bemerkung zur Charakterisierung der Distributionenhalbgruppen.
              Math. Ann. **204** (1973), 189–198. (MR 52, #15102).
[1976: 1]   The embedding problem for operator groups. *Proc. Japan Acad.* **52**
              (1976), 106–108. (MR 54, #3501).

S. Oharu and H. Sunouchi
[1970: 1]   On the convergence of semigroups of linear operators. *J. Functional*
              Analysis **6** (1970), 293–304. (MR 42, #2317).

Y. Ohya
[1964: 1]   Le problème de Cauchy pour les équations hyperboliques à caractéris-
              tiques multiples. *J. Math. Soc. Japan* **16** (1964), 268–286. (MR 31,
              #3693).

P. Oja
[1974: 1]   The solution of evolution equations by Galerkin's method. (Russian.)
             Tartu Riikl. Ul. Toimetised Vih. **342** (1974), 237–248. (MR 52, #939).

T. Okazawa
[1969: 1]   Two perturbation theorems for contraction semigroups in a Hilbert
             space. *Proc. Japan Acad.* **45** (1969), 850–853. (MR 42, #2312).
[1971: 1]   A perturbation theorem for linear contraction semigroups on reflexive
             Banach spaces. *Proc. Japan Acad.* **47** (1971), suppl. II, 947–949. (MR
             47, #2429).
[1973: 1]   Operator semigroups of class (D_n). *Math. Japon.* **18** (1973), 33–51.
             (MR 50, #14359).
[1973: 2]   Perturbations of linear m-accretive operators. *Proc. Amer. Math. Soc.*
             37 (1973), 169–174. (MR 47, #2403).
[1974: 1]   A generation theorem for semi-groups of growth order α. *Tôhoku
             Math. J.* **26** (1974), 39–51. (MR 49, #1219).
[1975: 1]   Remarks on linear m-accretive operators in a Hilbert space. *J. Math.
             Soc. Japan* **27** (1975), 160–165. (MR 51, #3950).

G. O. Okikiolu
[1966: 1]   On the infinitesimal generator of the Poisson operator. *Proc. Cam-
             bridge Philos. Soc.* **62** (1966), 713–718. (MR 34, #1836).
[1967: 1]   On the infinitesimal generator of the Poisson operator, II. The n-
             dimensional case. *Proc. Cambridge Philos. Soc.* **63** (1967), 1021–1025.
             (MR 36, #736).
[1969: 1]   On semi-groups of operators and the resolvents of their generators.
             Proc. Cambridge Philos. Soc. **66** (1969), 533–540. (MR 40, #4812).
[1974: 1]   Uniformly bounded projections and semi-groups of operators. *Proc.
             Cambridge Philos. Soc.* **75** (1974), 199–217. (MR 48, #12148).

O. A. Oleinik and E. V. Radkevič
[1971: 1]   Second-order equations with nonnegative characteristic form. (Russian.)
             Mathematical Analysis (1969), 7–252. Akad. Nauk. SSSR Vsesojuzn.
             Inst. Naucn. i Tehn. Informacii, Moscow, 1971. English trans.: [1973:
             1], Plenum Press, New York-London, 1973.

A. Olobummo
[1963: 1]   Monotone semi-groups of bounded operators. *J. Math. Mech.* **12**
             (1963), 385–390. (MR 26, #5433).
[1964: 1]   A note on perturbation theory for semi-groups of operators. *Proc.
             Amer. Math. Soc.* **15** (1964), 818–822. (MR 29, #6320).
[1967: 1]   Dissipative ordinary differential operators of even order. *Trans. Amer.
             Math. Soc.* **129** (1967), 130–139. (MR 35, #4768).

A. Olobummo and R. S. Phillips
[1965: 1]   Dissipative ordinary differential operators. *J. Math. Mech.* **14** (1965),
             929–949. (MR 32, #8216).

M. I. Orlov
[1973: 1]   An abstract Gregory formula on semigroups of class $(0, A)$. (Russian.)
             Funkcional. Anal. i Priložen. **7** (1973), no. 4, **90**. (MR 48, #9449).

[1977: 1]   Degenerate differential operators in weighted Hölder spaces. (Russian.)
             Mat. Zametki **21** (1977), 759–568. (MR 56, #12988).

V. P. Orlov and P. E. Sobolevskiĭ
[1975: 1]   The resolvent of a degenerate operator in a Banach space. (Russian.)
             Dokl. Akad. Nauk SSSR **221** (1975), 1035–1037. (MR 52, #941).
[1975: 2]   The resolvent of a degenerate operator on a Banach space. (Russian.)
             Differencial'nye Uravnenija **11** (1975), 858–868, 948. (MR 52, #940).

G. D. Orudžev
[1976: 1]   A description of self-adjoint extensions of higher order operator-dif-
             ferential expressions. (Russian.) *Akad. Nauk Azerbaĭdžan. SSR Dokl.*
             32 (1976), 9–12. (MR 56, #12480).

V. B. Osipov
[1968: 1]   Asymptotic integration and construction of a boundary layer for an
             equation in a Banach space. (Russian.) *Litovsk. Mat. Sb.* **8** (1968),
             581–589. (MR 40, #509).
[1970: 1]   Asymptotic integration of a second order differential equation in a
             Banach space. (Russian.) *Izv. Vysš. Učebn. Zaved. Matematika* (1970)
             no. 6 (97), 76–82. (MR 43, #5134).
[1970:2]   The evolution equation with degeneracy in a Banach space. (Russian.)
             Dal'nevostočn. Mat. Sb. **1** (1970), 81–85. (MR 52, #3993).

S. Ouchi
[1971: 1]   Hyperfunctions solutions of the abstract Cauchy problem. *Proc. Japan
             Acad.* **47** (1971), 541–544. (MR 45, #7465).
[1973: 1]   Semi-groups of operators in locally convex spaces. *J. Math. Soc. Japan*
             25 (1973), 265–276. (MR 51, #1480).

B. G. Pachpatte
[1976: 1]   On the stability and asymptotic behavior of solutions of integro-dif-
             ferential equations in Banach spaces. *J. Math. Anal. Appl.* **53** (1976),
             604–617. (MR 53, #6265).

E. W. Packel
[1969: 1]   A semigroup analogue of Foguel's counterexample. *Proc. Amer. Math.
             Soc.* **21** (1969), 240–244. (MR 38, #6400).
[1972: 1]   A simplification of Gibson's theorem on discrete operator semigroups.
             J. Math. Anal. Appl. **39** (1972), 586–589. (MR 47, #882).

T. V. Panchapagesan
[1969: 1]   Semi-groups of scalar type operators in Banach spaces. *Pacific J. Math.*
             30 (1969), 489–517. (MR 55, #6237).

C. V. Pao
[1969: 1]   The existence and stability of solutions of nonlinear operator differen-
             tial equations. *Arch. Rational Mech. Anal.* **35** (1969), 16–29. (MR 39,
             #7256).
[1972: 1]   Semigroups and asymptotic stability of nonlinear differential equa-
             tions. *SIAM J. Math. Anal.* **3** (1972), 371–379. (MR 46, #976).

[1973: 1]   On the asymptotic stability of differential equations in Banach spaces. *Math. Systems Theory* **7** (1973), 25–31. (MR 48, #2518).

[1974/75: 1]   On the uniqueness problem of differential equations in normed spaces. *Applicable Anal.* **4** (1974/75), 269–282. (MR 53, #967).

C. V. Pao and W. G. Vogt

[1969: 1]   On the stability of nonlinear operator differential equations, and applications. *Arch. Rat. Mech. Anal.* **35** (1969), 30–46. (MR 39, #7257).

P. L. Papini

[1969: 1]   Un'osservazione sui semi-gruppi di isometrie in certi spazi di Banach. *Boll. Un. Mat. Ital.* (4) **2** (1969), 682–685. (MR 41, #2461).

[1969: 2]   Un'osservazione sui prodotti semi-scalari negli spazi di Banach. *Boll. Un. Mat. Ital.* (4) **2** (1969), 686–689. (MR 41, #5945).

[1971: 1]   Semi-gruppi di isometrie. (English summary.) *Boll. Un. Mat. Ital.* **4** (1971), 950–961. (MR 46, #7964).

R. Pavec

[1971: 1]   Équation d'évolution avec condition sur une combinaison linéaire des valeurs de la solution en plusieurs points. *C. R. Acad. Sci. Paris* **272** (1971), 1242–1244.

N. Pavel

[1972: 1]   Sur certaines équations différentielles abstraites. (Italian summary.) *Boll. Un. Mat. Ital.* **6** (1972), 397–409. (MR 49, #720).

[1974: 1]   Approximate solutions of Cauchy problems for some differential equations in Banach spaces. *Funkcial. Ekvac.* **17** (1974), 85–94. (MR 50, #10840).

[1974: 2]   On an integral equation. *Rev. Roumaine Math. Pures Appl.* **19** (1974), 237–244. (MR 52, #1433).

L. E. Payne

[1975: 1]   *Improperly Posed Problems in Partial Differential Equations.* SIAM Regional Conf. Series in Applied Math., no. 22. SIAM, Philadelphia, Penn., 1975.

A. Pazy

[1967: 1]   Asymptotic expansions of solutions of differential equations in Hilbert space. *Arch. Rat. Mech. Anal.* **24** (1967), 193–218. (MR 35, #515).

[1968: 1]   Asymptotic behavior of the solution of an abstract evolution equation and some applications. *J. Diff. Equations* **4** (1968), 493–509. (MR 38, #1418).

[1968: 2]   On the differentiability and compactness of semigroups of linear operators. *J. Math. Mech.* **17** (1968), 1131–1141. (MR 37, #6797).

[1970: 1]   Semi-groups of nonlinear contractions in Hilbert space. *Problems in Non-linear Analysis* (CIME, IV Ciclo, Varenna, 1970), 343–430. Edizioni Cremonese, Rome, 1971. (MR 45, #965).

[1971: 1]   Asymptotic behavior of contractions in Hilbert space. *Israel J. Math.* **9** (1971), 235–240. (MR 43, #7988).

[1971: 2]   Approximation of the identity operator by semigroups of linear operators. *Proc. Amer. Math. Soc.* **30** (1971), 147–150. (MR 44, #4568).

[1972: 1]   On the applicability of Lyapunov's theorem in Hilbert space. (*SIAM J. Math. Anal.* **3** (1972), 291–294. (MR 47, #5653).

[1974: 1]   *Semi-groups of Linear Operators and Applications to Partial Differential Equations.* Univ. of Maryland Lecture Notes No. 10, College Park, Md., 1974.

[1975: 1]   A class of semi-linear equations of evolution. *Israel J. Math.* **20** (1975), 23–36. (MR 51, #11192).

H. Pecher
[1975: 1]   Globale Klassische Lösungen des Cauchyproblems semilinearer parabolischer Differentialgleichungen. *J. Functional Anal.* **20** (1975), 286–303. (MR 57, #13197).

I. G. Petrovskiĭ
[1938: 1]   Über das Cauchysche Problem für ein System linearer partieller Differentialgleichungen im Gebiete der nichtanalytischen Funktionen. *Bull. Univ. Moscou Sér. Int.* **1**, no. 7 (1938), 1–74.

R. S. Phillips
[1951: 1]   On one-parameter semi-groups of linear transformations. *Proc. Amer. Math. Soc.* **2** (1951), 234–237. (MR 12, 617).

[1951: 2]   A note on ergodic theory. *Proc. Amer. Math. Soc.* **2** (1951), 662–669. (MR 13, 138).

[1951: 3]   Spectral theory for semi-groups of linear operators. *Trans. Amer. Math. Soc.* **71** (1951), 393–415. (MR 13, 469).

[1952: 1]   On the generation of semigroups of linear operators. *Pacific J. Math.* **2** (1952), 343–369. (MR 14, 383).

[1953: 1]   Perturbation theory for semi-groups of linear operators. *Trans. Amer. Math. Soc.* **74** (1953), 199–221. (MR 18, 882).

[1954: 1]   An inversion formula for Laplace transforms and semi-groups of linear operators. *Ann. of Math.* (2) **59** (1954), 325–356. (MR 15, 718).

[1954: 2]   A note on the abstract Cauchy problem. *Proc. Nat. Acad. Sci. U.S.A.* **40** (1954), 244–248. (MR 15, 880).

[1955: 1]   Semi-groups of operators. *Bull. Amer. Math. Soc.* **61** (1955), 16–33. (MR 16, 833).

[1955: 2]   The adjoint semi-group. *Pacific J. Math.* **5** (1955), 269–283. (MR 17, 64).

[1957: 1]   Dissipative hyperbolic systems. *Trans. Amer. Math. Soc.* **86** (1957), 109–173. (MR 19, 863).

[1959: 1]   Dissipative operators and parabolic partial differential equations. *Comm. Pure Appl. Math.* **12** (1959), 249–276. (MR 21, #5810).

[1959: 2]   On a theorem due to Sz.-Nagy. *Pacific J. Math.* **9** (1959), 169–173. (MR 21, #4360).

[1959: 3]   Dissipative operators and hyperbolic systems of partial differential equations. *Trans. Amer. Math. Soc.* **90** (1959), 193–254. (MR 21, #3669).

[1961: 1]   Semigroup methods in the theory of partial differential equations.
             Modern Mathematics for the Engineer: 2nd *Series*, 100–132, McGraw-
             Hill, New York, 1961. (MR 23, #B2195).
[1961: 2]   On the integration of the diffusion equation with boundary conditions.
             Trans. Amer. Math. Soc. **98** (1961), 62–84. (MR 23, #A3921).
[1962: 1]   Semi-groups of positive contraction operators. *Czechoslovak Math. J.*
             12 (87), (1962), 294–313. (MR 26, #4195).

É. Picard
[1907: 1]   La mécanique classique et ses approximations succesives. *Riv. di
             Scienza* **I** (1907), 1–67.

M. A. Piech
[1969: 1]   A fundamental solution of the parabolic equation in Hilbert space. *J.
             Functional Anal.* **3** (1969), 85–114. (MR 40, #4815).
[1970: 1]   A fundamental solution of the parabolic equation in Hilbert space
             II: The semigroup property. *Trans. Amer. Math. Soc.* **150** (1970),
             257–286. (MR 43, #3847).
[1972: 1]   Differential equations on abstract Wiener space. *Pacific J. Math.* **43**
             (1972), 465–473. (MR 47, #9096).
[1975: 1]   The Ornstein-Uhlenbeck semigroup in an infinite dimensional L_2
             setting. *J. Functional Anal.* **18** (1975), 213–222. (MR 52, #1911).

S. Piskarev
[1979: 1]   Discretization of an abstract hyperbolic equation. (Russian. English
             summary.) *Tartu Riikl. Uel. Toimetised No. 500 Trudy Mat. i Meh.*,
             No. 25 (1979), 3–23.
[1979: 2]   Solution of an inhomogeneous abstract linear hyperbolic equation.
             (Russian. English summary.) *Tartu Riikl. Uel. Toimetised No. 500
             Trudy Mat. i Meh.*, No. 25 (1979), 24–32.
[1979: 3]   On the approximation of holomorphic semigroups (Russian.) *Tartu
             Riikl. Uel. Toimetised No. 500 Trudy Mat. i Meh.*, No. 24 (1979), 3–14.

L. Pitt
[1969: 1]   Products of Markovian semi-groups of operators. *Z. Warsch. Verw.
             Geb.* **12** (1969), 246–254. (MR 41, #1119).

B. A. Plamenevskiĭ
[1972: 1]   The asymptotic behavior of the solutions of differential equations in a
             Banach space. (Russian.) *Dokl. Akad. Nauk SSSR* **202** (1972), 34–37.
             (MR 44, #7098).

Ju. V. Pljuščev
[1973: 1]   Integration of an evolution equation in a Banach space. (Russian.)
             Differencial'nye Uravnenija **9** (1973), 761–763, 787. (MR 54, #4113).
[1975: 1]   Nonhomogeneous semigroups of commuting operators. (Russian.) *Mat.
             Zametki* **17** (1975), 57–65. (MR 50, #14360).
[1975: 2]   Inhomogeneous semigroups of linear operators and the Cauchy prob-
             lem for evolution equations of "hyperbolic" type. (Russian.) *Collection*

of articles on applications of functional analysis (Russian), 122–135. *Voronež. Technolog. Inst., Voronezh*, 1975. (MR 56, #12982).

V. A. Pogorelenko and P. E. Sobolevskiĭ

[1967: 1]   Hyperbolic equations in a Hilbert space. (Russian.) *Sibirsk. Mat. Ž.* **8** (1967), 123–145. (MR 35, #516).

[1967: 2]   Non-local existence theorem of solutions of non-linear hyperbolic equations in a Hilbert space. (Russian.) *Ukrain. Mat. Ž.* **19** (1967), no. 1, 113–15. (MR 34, #7928).

[1967: 3]   Hyperbolic equations in a Banach space. (Russian.) *Uspehi Mat. Nauk* **22** (1967), no. 1 (133), 170–172. (MR 34, #8014).

[1970: 1]   The solvability of mixed problems for one-dimensional quasilinear hyperbolic equations. (Russian.) *Ukrain. Mat. Ž.* **22** (1970), 114–121. (MR 41, #4002).

[1972: 1]   Periodic solutions of quasilinear hyperbolic equations. (Russian.) *Voronez. Gos. Univ. Trudy Mat. Fak.* (1972), 85–89. (MR 55, #3462).

A. Ē. Polička and P. E. Sobolevskiĭ

[1976: 1]   Correct solvability of a difference boundary value problem in a Bochner space. (Russian.) *Ukrain. Mat. Ž.* **28** (1976), 511–523, 574. (MR 55, #795).

G. Pólya

[1928: 1]   Über die Funktionalgleichung der Exponentialfunktion im Matrizenkalkül. *Sitz. Acad. Wiss. Berlin* (1928), 96–99.

G. Pólya and G. Szego

[1954: 1]   *Aufgaben und Lehrsätze aus der Analysis I.* Zweite Auflage. Springer, Berlin, 1954.

S. M. Ponomarev

[1972: 1]   The convergence of semigroups. (Russian.) *Dokl. Akad. Nauk SSSR* **204** (1972), 42–44. (MR 46, #7965).

[1972: 2]   Semigroups of the classes $A^{n,\alpha}$ and the abstract Cauchy problem. (Russian.) *Differencial'nye Uravnenija* **8** (1972), 2038–2047. (MR 47, #2430).

E. L. Post

[1930: 1]   Generalized differentiation. *Trans. Amer. Math. Soc.* **32** (1930), 723–781.

E. T. Poulsen

[1965: 1]   Evolutionsgleichungen in Banach-Räumen. *Math Z.* **90** (1965), 286–309. (MR 32, #4407).

A. I. Povolockiĭ and V. V. Masjagin

[1971: 1]   Differential equations in locally convex spaces. (Russian.) *Leningrad. Gos. Ped. Inst. Ucen. Zap.* **404** (1971), 406–414. (MR 55, #13030).

A. Povzner

[1964: 1]   A global existence theorem for a nonlinear system and the defect index
             of a linear operator. (Russian.) *Sibirsk. Mat. Ž.* **5** (1964), 377–386.
             (MR 29, #336)

R. T. Powers and Ch. Radin

[1976: 1]   Average boundary conditions in Cauchy problems. *J. Functional Analy-
             sis* **23** (1976), 23–32. (MR 56, #9025).

G. A. Pozzi

[1968: 1]   Problemi di Cauchy e problemi ai limiti per equazioni di evoluzioni del
             tipo Schrödinger lineari e non lineari, I. L'equazione lineare astratta.
             Ann. Mat. Pura Appl. (4) **78** (1968), 197–258. (MR 37, #5553).

[1969: 1]   Problemi di Cauchy e problemi ai limiti per equazioni di evoluzione del
             tipo de Schrödinger lineari e non lineari, II. Applicazioni alle equazioni
             a derivate parziali. L'equazione non lineare. *Ann. Mat. Pura Appl.* (4)
             81 (1969), 205–248. (MR 40, #3024).

L. M. Prokopenko

[1963: 1]   The uniqueness of the solution of the Cauchy problem for differential-
             operator equations. (Russian.) *Dokl. Akad. Nauk SSSR* **148** (1963),
             1030–1033. (MR 26, #6587).

[1971: 1]   Integral estimates for analytic semigroups and generalized solutions of
             the Cauchy problem. (Ukrainian. English and Russian summaries.)
             Dopovidi Akad. Nauk Ukrain. RSR Ser. A (1971), 209–212, 284. (MR
             44, #4569).

O. I. Prozorovskaja

[1962: 1]   On the rate of decrease of solutions of evolution equations. (Russian.)
             Sibirsk. Mat. Ž (1962), 391–408. (MR 27, #598).

[1970: 1]   The rate of convergence of a semigroup operator to the identity.
             (Russian.) *Voronež. Gos. Univ. Trudy Naučn.-Issled. Inst. Mat. VGU
             Vyp. l Sb. Stateĭ po Teorii Operatornyh Uravneniĭ* (1970), 11–119. (MR
             54, #13625).

C. Pucci

[1955: 1]   Sui problemi di Cauchy non "bene posti." *Rend. Accad. Naz. Lincei* **18**
             (1955), 473–477. (MR 19, 426).

[1958: 1]   Discussione del problema di Cauchy per le equazioni di tipo ellittico.
             Ann. Mat. Pura Appl. **46** (1958), 131–153. (MR 29, #1440).

V. T. Purmonen

[1975: 1]   Ein Abstraktes Cauchysches Problem. *Ann. Acad. Sci. Fenn. Ser. AI
             Math.* **1** (1975), 125–142. (MR 52, #14520).

B. K. Quinn

[1971: 1]   Solutions with shocks: An example of an L_1-contractive semigroup.
             Comm. Pure Appl. Math. **24** (1971), 125–132. (MR 42, #6428).

A. Radnitz
[1970: 1]   The incomplete Cauchy problem in Banach spaces. Ph. D. dissertation, University of California, Los Angeles, Calif., 1970.

Ja. V. Radyno and N. I. Jurčuk
[1976: 1]   The Cauchy problem for certain abstract hyperbolic equations of even order. (Russian.) *Differencial'nye Uravnenija* **12** (1976), 331–342. (MR 53, #756).

M. B. Ragimov and G. I. Novruzov
[1976: 1]   Almost periodic solutions of differential equations in a Banach space. (Russian.) *Akad. Nauk Azerbaĭdžan SSR Dokl.* **32** (1976), no. 12, 3–4. (MR 56, #6049).

A. S. Rao
[1973/74: 1]   On the Stepanov-almost periodic solution of an abstract differential equation. *Indiana Univ. Math. J.* **23** (1973/74), 205–208. (MR 48, #664).
[1974/75: 1]   On the Stepanov-almost periodic solution of a second-order operator differential equation. *Proc. Edinburgh Math. Soc.* **19** (1974/75), 261–263. (MR 53, #132).
[1975: 1]   Strong almost periodicity of certain solutions of operator differential equations. (Italian summary.) *Ist. Lombardo Accad. Sci. Lett. Rend. A* **109** (1975), 243–248. (MR 54, #5574).
[1975: 2]   On second-order differential operators with Bohr-Neugebauer type property. *Canad. Math. Bull.* **18** (1975), 393–396. (MR 55, #796).
[1977: 1]   On the Stepanov-bounded solutions of certain abstract differential equations. *Indian J. Pure Appl. Math.* **8** (1977), 120–123. (MR 56, #12484).

A. S. Rao and W. Hengartner
[1974: 1]   On the spectrum of almost periodic solutions of an abstract differential equation. *J. Austral. Math. Soc.* **18** (1974), 385–387. (MR 51, #3637).
[1974: 2]   On the existence of a unique almost periodic solution of an abstract differential equation. *J. London Math. Soc.* **8** (1974), 577–581. (MR 50, #13785).

D. K. Rao
[1973: 1]   Symmetric perturbations of a self-adjoint operator. *Rev. Colombiana Mat.* **7** (1973), 81–84. (MR 49, #3572).

V. G. Raskin
[1973: 1]   Solutions bounded on the halfaxis of a second order nonlinear elliptic equation in a Banach space. (Russian.) *Differencial'nye Uravnenija* **9** (1973), 768–770, 788. (MR 48, #665).
[1976: 1]   The solvability of operator-differential equations of arbitrary order. (Russian.) *Differencial'nye Uravnenija* **12** (1976), 362–364. (MR 53, #13770).

V. G. Raskin and A. I. Jasakov
[1970: 1]   On the question of exponential stability of the solutions of differential
             equations in a Banach space. (Russian.) *Voronež. Gos. Univ. Trudy
             Naučn.-Issled. Inst. Mat. VGU Vyp. l Sb. Stateĭ po Teorii Operatornyh
             Uravnenii* (1970), 120–127. (MR 54, #13247).

V. G. Raskin and P. E. Sobolevskiĭ
[1967: 1]   The Cauchy problem for differential equations of second order in
             Banach spaces. (Russian.) *Sibirsk. Mat. Ž.* **8** (1967), 70–90. (MR 34,
             #7929).
[1968: 1]   A priori estimates of the solutions of a certain class of nonlinear
             nonstationary equations. (Russian.) *Ukrain. Mat. Ž.* **20** (1968),
             547–552. (MR 37, #6588).
[1968: 2]   The stabilization of the solutions of a certain class of second-order
             nonlinear equations in Hilbert space. (Russian.) *Voronež. S.-h. Inst.
             Zap.* **35** (1968), 420–423. (MR 44, #4337).
[1969: 1]   Expansion of the solutions of second-order differential equations in
             Banach spaces in series of exponential solutions. (Russian.) *Differen-
             cial'nye Uravnenija* **5** (1969), 543–545. (MR 40, #3025).

V. A. Rastrenin
[1976: 1]   Application of the difference method to boundary value problems for
             abstract elliptic equations. (Russian.) *Differencial'nye Uravnenija* **12**
             (1976), 2014–2018, 2109. (MR 55, #10797).

R. Rautmann
[1973: 1]   Differentialgleichungen in Banachräumen mit einer zweiten Norm.
             Methoden und Verfahren der Mathematischen Physik, Band 8, 183–222,
             Bibliographisches Institut, Mannheim, 1973. (MR 52, #11244).

P. A. Raviart
[1967: 1]   Sur l'approximation de certaines équations d'évolution linéaires et non
             linéaires. *J. de Math. Pures Appl.* **46** (1967), 11–107. (MR 38, #3636a).
[1967: 2]   Sur l'approximation de certaines équations d'évolution linéaires et
             non-linéaires (suite). *J. de Math. Pures Appl.* **46** (1967), 109–183. (MR
             38, #3636b).

K. W. Reed, Jr.
[1965: 1]   On a problem in the neutron transport equation. *J. Math. Anal. Appl.*
             10 (1965), 161–165. (MR 30, #5770).
[1966: 1]   On weak and strong solutions for the neutron transport equation. *J.
             d'Analyse Math.* **17** (1966), 347–368. (MR 37, #4541).

M. Reed and B. Simon
[1973: 1]   Tensor products of closed operators in Banach spaces. *J. Functional
             Anal.* **13** (1973), 107–124. (MR 50, #1036).

F. Rellich
[1939: 1]   Störungstheorie der Spektralzerlegung III. *Math. Ann.* **116** (1939),
             555–570.

G. E. H. Reuter
[1955: 1]   A note on contraction semigroups. *Math. Scand.* **3** (1955), 275–280.
(MR 17, #988).

G. Restrepo
[1970: 1]   A remark on asymptotic stability. *Bol. Soc. Mat. Mexicana* (2) **15**
(1970), 37–39. (MR 51, #11194).

P. Ricciardi and L. Tubaro
[1973: 1]   Local existence for differential equations in Banach space. (Italian
summary.) *Boll. Un. Mat. Ital.* **8** (1973), 306–316. (MR 48, #11712).

R. D. Richtmyer
[1966: 1]   The abstract theory of the linear inhomogeneous Cauchy problem.
(Russian.) *Certain Problems Numer. Appl. Math.* (Russian), 84–91.
Izdat. "Nauka," Sibirsk. Otdel., Novosibirsk, 1966. (MR 35, #6951).

R. D. Richtmyer and K. W. Morton
[1967: 1]   *Difference Methods for Initial-Value Problems.* 2nd ed. Interscience-
Wiley, New York, 1967.

F. Riesz
[1938: 1]   Some mean ergodic theorems. *J. London Math. Soc.* **13** (1938), 274–278.

V. K. Romanko
[1967: 1]   On the theory of operators of the form $d^m/dt^m - A$. (Russian.) *Differencial'nye Uravnenija* **3** (1967), 1957–1970. (MR 36, #5548).
[1969: 1]   Systems with first-order derivative in the "time." (Russian.) *Differencial'nye Uravnenija* **5** (1969), 174–185. (MR 39, #5974).
[1976: 1]   Boundary value problems for certain operator-differential equations.
(Russian.) *Dokl. Akad. Nauk SSSR* **227** (1976), 812–815. (MR 53,
#3452).
[1977: 1]   Boundary value problems for operator-differential equations that are
not solved with respect to the highest derivative. (Russian.) *Dokl.
Akad. Nauk SSSR* (1977), 1030–1033. (MR 56, #12485).

N. P. Romanoff
[1947: 1]   On one-parameter groups of linear transformations, I. *Ann. of Math.*
48 (1947), 216–233. (MR 8, 520).

W. A. Rosenkrantz
[1974: 1]   An application of the Hille-Yosida theorem to the construction of
martingales. *Indiana Univ. Math. J.* **24** (1975), 613–625. (MR 50,
#8722).
[1976: 1]   A strong continuity theorem for a class of semigroups of type Γ with
an application to martingales. *Indiana Univ. Math. J.* **25** (1976),
171–178. (MR 52, #9390).

G. C. Rota
[1962: 1]   An "alternierende Verfahren" for general positive operators. *Bull.
Amer. Math. Soc.* **68** (1962), 95–102. (MR 24, #A3671).

[1963: 1]   A limit theorem for the time-dependent evolution equation. *Equazioni Differenziali Astratte*. C.I.M.E., Roma, 1963.

J. P. Roth
[1973: 1]   Sur les semi-groupes à contraction invariants sur un espace homogène. *C. R. Acad. Sci. Paris* **277** (1973), A1091–A1094. (MR 51, #3905).
[1974: 1]   Restriction a un ouvert d'un générateur infinitésimal local. *C. R. Acad. Sci. Paris* **279** (1974), 363–366. (MR 51, #3906).
[1976: 1]   Opérateurs dissipatifs et semigroupes dans les espaces de fonctions continues. (English summary.) *Ann. Inst. Fourier (Grenoble)* **26** (1976), 1–97. (MR 56, #6480).
[1977: 1]   Opérateurs elliptiques comme générateurs infinitésimaux de semi-groupes de Feller. *C. R. Acad. Sci. Paris* **284** (1977), 755–757. (MR 55, #6011).

W. A. Roth
[1972/73]   Goursat problems for $u_{rs} = Lu$. *Indiana Univ. Math. J.* **22** (1972/73), 779–788. (MR 46, #9568).

H. L. Royden
[1968: 1]   *Real Analysis*, Macmillan, New York, 1968.

D. L. Russell
[1975: 1]   Decay rates for weakly damped systems in Hilbert space obtained with control-theoretic methods. *J. Diff. Equations* **19** (1975), 344–370. (MR 54, #13248).

T. L. Saaty
[1958/59: 1]   An application of semi-group theory to the solution of a mixed type partial differential equation. *Univ. e Politec. Torino Rend. Sem. Mat.* **18** (1958/59), 53–76. (MR 22, #9748).

S. M. Sadarly
[1969: 1]   On the theory of the Cauchy limit problem for differential equations with unbounded operators in a Banach space. (Russian.) *Differencial'nye Uravnenija* **5** (1969), 1431–1437.

L. A. Sahnovič
[1970: 1]   Generalized wave operators. (Russian) *Mat. Sb. (N.S.)* **81** (123) (1970), 209–227. (MR 54, #3446).

Y. Saitō
[1979: 1]   *Spectral Representation for Schrödinger Operators with Long-Range Potentials*. Lecture Notes in Mathematics, vol. 727, Springer, Berlin, 1979.

H. Salehi
[1972: 1]   Stone's theorem for a group of unitary operators over a Hilbert space. *Proc. Amer. Math. Soc.* **31** (1972), 480–484. (MR 45, #9176).

J. T. Sandefur
[1977: 1]   Higher order abstract Cauchy problems. *J. Math. Anal. Appl.* **60**
             (1977), 728–742. (MR 56, #9027).

G. P. Samosjuk
[1961: 1]   Solution of a one-dimensional diffusion problem with integro-differen-
             tial boundary conditions. (Russian.) *Vestnik Leningrad. Univ.* **16** (1961),
             5–12. (MR 24, #A2146).

A. Š. Šapatava
[1972: 1]   The convergence of double-layer regulated difference schemes in com-
             plex Hilbert space. (Russian.) *Sakharth. SSR Mech. Akad. Moambe* **67**
             (1972), 285–288. (MR 49, #5869).

T. A. Sarymsakov and E. E. Murtazaev
[1970: 1]   Semigroups of linear operators and evolution equations in weakly
             normed spaces. (Russian.) *Dokl. Akad. Nauk Uz. SSR* (1970), no. 8,
             3–5. (MR 51, #13760).

K. Sato
[1968: 1]   On the generators of non-negative contraction semi-groups in Banach
             lattices. *J. Math. Soc. Japan* **20** (1968), 423–426. (MR 37, #6798).
[1970: 1]   On dispersive operators in Banach lattices. *Pacific J. Math.* **33** (1970),
             429–443. (MR 42, #2318).
[1970: 2]   Positive pseudo-resolvents in Banach lattices. *J. Fac. Sci. Univ. Tokyo
             Sect. I* **17** (1970), 305–313. (MR 53, #8966).
[1972: 1]   A note on infinitesimal generators and potential operators of contrac-
             tion semigroups. *Proc. Japan Acad.* **48** (1972), 450–453. (MR 49,
             #1220).
[1972: 2]   Potential operators for Markov processes. *Proc. Sixth Berkeley Symp.
             Math. Stat. Prob.* **3** (1972), 193–211. (MR 53, #11768).

R. Sato
[1974: 1]   A note on operator convergence for semigroups. *Comment. Math.
             Univ. Carolinae* **15** (1974), 127–129. (MR 49, #5890).
[1976: 1]   Invariant measures for bounded amenable semigroups of operators.
             Proc. Japan Acad. **52** (1976), 215–218. (MR 56, #1132).

S. A. Sawyer
[1970: 1]   A formula for semigroups, with an application to branching diffusion
             processes. *Trans. Amer. Math. Soc.* **152** (1970), 1–38. (MR 42, #1226).

B. Scarpellini
[1974: 1]   On the spectra of certain semi-groups. *Math. Ann.* **211** (1974), 323–336.
             (MR 52, #9003).

H. H. Schaefer
[1974: 1]   *Banach Lattices and Positive Operators.* Springer, New York, 1974.

M. Schechter
[1971: 1]   *Spectra of Partial Differential Operators.* North-Holland-Elsevier, New
             York, 1971.

[1974: 1]   A unified approach to scattering. *J. Math. Pures Appl.* **53** (1974), 373–396. (MR 51, #1436).

L. I. Schiff
[1955: 1]   *Quantum Mechanics*, 2nd ed. McGraw-Hill, New York, 1955.

E. J. P. G. Schmidt
[1974/75:1]   On scattering by time-dependent perturbations. *Indiana Univ. Math. J.* **24** (1974/75), 925–935. (MR 52, #1360).

W. R. Schneider
[1975: 1]   Resolvents, semigroups and Gibbs states for infinite coupling constants. *Comm. Math. Phys.* **43** (1975), 121–129. (MR 56, #3652).

A. Schoene
[1970: 1]   Semigroups and a class of singular perturbation problems. *Indiana Univ. Math. J.* **20** (1970), 247–264. (MR 44, #852).

Z. Schuss
[1972: 1]   Regularity theorems for solutions of a degenerate evolution equation. *Arch. Rat. Mech. Anal.* **46** (1972), 200–211. (MR 48, #11714).

L. Schwartz
[1950: 1]   Les équations d'évolution liées au produit de composition. *Ann. Inst. Fourier Grenoble* **2** (1950), 19–49. (MR 13, 242).
[1957: 1]   Théorie des distributions a valeurs vectorielles, I. *Ann. Inst. Fourier* **7** (1957), 1–141. (MR 21, #6534).
[1958: 1]   Théorie des distributions a valeurs vectorielles, II. *Ann. Inst. Fourier* **8** (1958), 1–209. (MR 22, #8322).
[1958: 2]   *Lectures on Mixed Problems in Partial Differential Equations and the Representation of Semi-Groups.* Tata Institute for Fundamental Research, Bombay, 1958.
[1966: 1]   *Théorie des distributions.* Nouvelle édition, Hermann, Paris, 1966.

I. E. Segal
[1963: 1]   Non-linear semi-groups. *Ann. of Math.* (2) **78** (1963), 339–364. (MR 27, #2879).

I. E. Segal and R. W. Goodman
[1965: 1]   Anti-locality of certain Lorentz-invariant operators. *J. Math. Mech.* **14** (1965), 629–638. (MR 31, #4411).

T. I. Seidman
[1970: 1]   Approximation of operator semi-groups. *J. Functional Analysis* **5** (1970), 160–166. (MR 40, #7866).

Z. B. Seidov
[1962: 1]   Investigation of the solution of second-order nonlinear differential equations in Hilbert space. (Russian.) *Azerbaĭdžan Gos. Univ. Ucen. Zap. Ser. Fiz.-Mat i Him. Nauk* (1962), no. 4, 49–52. (MR 36, #2929).

Yu. A. Semenov
[1977: 1]   Schrödinger operators with L_{loc}^p potentials. *Comm. Math Phys.* **53**
            (1977) 277–284 (MR 58, #211).
[1977: 2]   On the Lie-Trotter theorems in L^p spaces. *Lett. Math. Phys.* **1** (1977),
            379–385. (MR 58, #12514).

F. D. Sentilles
[1970: 1]   Semigroups of operators in $C(S)$. *Canad. J. Math.* **22** (1970), 47–54.
            (MR 54, #1019).

M. Shinbrot
[1968: 1]   Asymptotic behavior of solutions of abstract wave equations. *Proc.*
            Amer. Math. Soc. **19** (1968), 1403–1406. (MR 37, #6801).

R. Shirasai and Y. Hirata
[1964: 1]   Convolution maps and semi-group distributions. *J. Sci. Hiroshima*
            Univ. Ser. A-I Math **28** (1964), 71–88. (MR 30, #1396).

R. E. Showalter
[1969: 1]   Partial differential equations of Sobolev-Galpern type. *Pacific J. Math.*
            31 (1969), 787–793. (MR 40, #6085).
[1970: 1]   Local regularity of solutions of Sobolev-Galpern partial differential
            equations. *Pacific J. Math.* **34** (1970), 781–787. (MR 42, #2153).
[1970: 2]   Well-posed problems for a partial differential equation of order $2m + 1$.
            SIAM J. Math. Anal. **1** (1970), 214–231. (MR 41, #5787).
[1972: 1]   Weak solutions of nonlinear evolution equtions of Sobolev-Galpern
            type. *J. Diff. Equations* **11** (1972), 252–265. (MR 45, #2342).
[1972: 2]   Existence and representation theorems for a semilinear Sobolev equa-
            tion in Banach space. *SIAM J. Math. Anal.* **3** (1972), 527–543. (MR
            47, #3788).
[1973: 1]   Equations with operators forming a right angle. *Pacific J. Math.* **45**
            (1973), 357–362. (MR 47, #7517).
[1973/74: 1]   Degenerate evolution equations and applications. *Indiana Univ.*
            Math. J. **23** (1973/74), 655–677. (MR 48, #12157).
[1974: 1]   The final value problem for evolution equations. *J. Math. Anal. Appl.*
            47 (1974), 563–572. (MR 50, #5131).
[1975: 1]   Nonlinear degenerate evolution equations and partial differential equa-
            tions of mixed type. *SIAM J. Math. Anal.* **6** (1975), 25–42. (MR 52,
            #15154a). Errata. *SIAM J. Math. Anal.* **6** (1975), 358. (MR 52,
            #15154b).
[1976: 1]   Regularization and approximation of second-order evolution equa-
            tions. *SIAM J. Math. Anal.* **7** (1976), 461–472. (MR 54, #10794).
[1977: 1]   *Hilbert Space Methods for Partial Differential Equations.* Pitman,
            London, 1977.

R. E. Showalter and T. W. Ting
[1970: 1]   Pseudoparabolic partial differential equations. *SIAM J. Math. Anal.* **1**
            (1970), 1–26. (MR 55, #10857).

[1971: 1]   Asymptotic behavior of solutions of pseudoparabolic partial differential equations. *Ann. Mat. Pura Appl.* (4) **90** (1971), 241–258. (MR 47, #5414).

N. A. Sidorov

[1972: 1]   The Cauchy problem for a certain class of differential equations. (Russian.) *Differencial'nye Uravnenija* **8** (1972), 1521–1524. (MR 46, #5787).

B. Simon

[1971: 2]   *Quantum Mechanics for Hamiltonians Defined as Quadratic Forms.* Princeton University Press, Princeton, N.J., 1971.

[1973: 1]   Ergodic semigroups of positivity preserving self-adjoint operators. *J. Functional. Analysis* **12** (1973), 335–339. (MR 50, #10900).

[1973: 2]   Essential self-adjointness of Schrödinger operators with positive potentials. *Math. Ann.* **201** (1973), 211–220. (MR 49, #1987).

[1973: 3]   Essential self-adjointness of Schrödinger operators with singular potentials. *Arch. Rat. Mech. Anal.* **52** (1973), 44–48. (MR 49, #3312).

B. Simon and R. Hoegh-Krohn

[1972: 1]   Hypercontractive semigroups and two-dimensional self-coupled Bose fields. *J. Functional Anal.* **9** (1972), 121–180. (MR 45, #2528).

J. Simon

[1972: 1]   Existence et unicité des solutions d'équations et d'inéquations d'évolution sur]−∞,∞[. *C. R. Acad. Sci. Paris* **274** (1972), 1045–1047. (MR 45, #2532).

I. B. Simonenko

[1970: 1]   A justification of the method of averaging for abstract parabolic equations. (Russian.) *Dokl. Akad. Nauk SSSR* **191** (1970), 33–34. (MR 41, #5750).

[1973: 1]   Higher approximations of the averaging method for abstract parabolic equations. (Russian.) *Mat. Sb.* **92** (134) (1973), 541–549, 647. (MR 49, #5506).

E. Sinestrari

[1976: 1]   Accretive differential operators. *Boll. Un. Mat. Ital. B* (5) **13** (1976), 19–31. (MR 54, #13636).

K. Singbal-Vedak

[1965: 1]   A note on semigroups of operators on a locally convex space. *Proc. Amer. Math. Soc.* **16** (1965), 696–702. (MR 32, #370).

[1972: 1]   Semigroups of operators on a locally convex space. *Comment. Math. Prace Mat.* **16** (1972), 53–74. (MR 48, #12141).

S. P. Šišatskiĭ

[1968: 1]   The correctness of boundary value problems for a second-order differential operator with a self-adjoint negative operator in a Hilbert space. (Russian.) *Funkcional Anal. i Priložen.* **2** (1968), no. 2, 81–85. (MR 38, #395).

S. Sjöstrand

[1970: 1] On the Riesz means of the solutions of the Schrödinger equation. *Ann. Scuola Norm. Sup. Pisa* (3) **24** (1970), 331–348. (MR 42, #5110).

M. Slemrod

[1972: 1] The linear stabilization problem in Hilbert space. *J. Functional Analysis* **11** (1972), 334–345. (MR 50, #7724).

[1976: 1] Asymptotic behavior of C_0 semi-groups as determined by the spectrum of the generator. *Indiana Univ. Math. J.* **25** (1976), 783–892. (MR 56, #9321).

[1976: 2] Existence, uniqueness and stability for a simple fluid with fading memory. *Bull. Amer. Math. Soc.* **82** (1976), 581–583. (MR 53, #15075).

A. D. Sloan

[1975: 1] Analytic domination by fractional powers with linear estimates. *Proc. Amer. Math. Soc.* **51** (1975), 94–96. (MR 51, #3938).

M. Slociński

[1974: 1] Unitary dilation of two-parameter semi-groups of contractions. *Bull. Acad, Polon. Sci. Sér. Sci. Math. Astronom. Phys.* **22** (1974), 1011–1014. (MR 52, #1409).

V. V. Smagin and P. E. Sobolevskiĭ

[1970: 1] Comparison theorems for the norms of the solutions of linear homogeneous differential equations in Hilbert spaces. (Russian.) *Differencial'nye Uravnenija* **6** (1970), 2005–2010. (MR 43, #3848).

[1970: 1] Theorems on the comparison of semigroups. (Russian.) *Voronez. Gos. Univ. Trudy Naučn.-Issled. Inst. Mat. VGU Vyp. l. Sb. Stateĭ po Teorii Operatornyh Uravneniĭ* (1970), 145–148. (MR 54, #13631).

J. A. Smoller

[1965: 1] Singular perturbations and a theorem of Kisyński. *J. Math. Anal. Appl.* **12** (1965), 105–114. (MR 31, #6040).

[1965: 2] Singular perturbations of Cauchy's problem. *Comm. Pure Appl. Math.* **18** (1965), 665–677. (MR 32, #2709).

P. E. Sobolevskiĭ

[1957: 1] Approximate methods of solving differential equations in Banach spaces. (Russian.) *Dokl. Akad. Nauk SSSR* **115** (1957), 240–243. (MR 20, #1050).

[1957: 2] On equations with operators forming an acute angle. (Russian.) *Dokl. Akad. Nauk SSSR* **116** (1957), 754–757. (MR 20, #4194).

[1958: 1] Generalized solutions of first order differential equations in Hilbert space. (Russian.) *Dokl. Akad. Nauk SSSR* **122** (1958), 994–996. (MR 21, #1436).

[1958: 2] First-order differential equations in Hilbert space with a variable positive definite self-adjoint operator a fractional power of which has a constant domain of definition. (Russian.) *Dokl. Akad. Nauk SSSR* **123** (1958), 984–987. (MR 23, #A3905).

[1958: 3]   Non-stationary equations of viscous fluid dynamics. (Russian.) *Dokl. Akad. Nauk SSSR* **128** (1959), 45–48. (MR 22, #1763).

[1960: 1]   On the use of fractional powers of self-adjoint operators for the investigation of some nonlinear differential equations in Hilbert space. (Russian.) *Dokl. Akad. Nauk SSSR* **130** (1960), 272–275. (MR 23, #A1118).

[1960: 2]   On the smoothness of generalized solutions of the Navier-Stokes equations. (Russian.) *Dokl. Akad. Nauk SSSR* **131** (1960), 758–760. (MR 25, #3282).

[1961: 1]   Local and nonlocal existence theorems for nonlinear second-order parabolic equations. (Russian.) *Dokl. Akad. Nauk SSSR* **136** (1961), 292–295. (MR 25, #5291).

[1961: 2]   Parabolic equations in a Banach space with an unbounded variable operator, a fractional power of which has a constant domain of definition. (Russian.) *Dokl. Akad. Nauk SSSR* **138** (1961), 59–62. (MR 28, #2360).

[1961: 3]   Equations of parabolic type in a Banach space. (Russian.) *Trudy Moskov. Mat. Obšč.* **10** (1961), 297–350. (MR 25, #5297).

[1962: 1]   On second-order differential equations in a Banach space. (Russian.) *Dokl. Akad. Nauk SSSR* **146** (1962), 774–777. (MR 27, #6004).

[1964: 1]   Applications on the method of fractional powers of operators to the study of the Navier-Stokes equations. (Russian.) *Dokl. Akad. Nauk SSSR* **155** (1964), 50–53. (MR 28, #4258).

[1964: 2]   Investigation of partial differential equations using fractional powers of operators. (Russian.) *Proc. Fourth All-Union Mat. Congr.*, Izdat. "Nauka," Leningrad, 1964, 519–525.

[1964: 3]   Coerciveness inequalities for abstract parabolic equations. (Russian.) *Dokl. Akad. Nauk SSSR* **157** (1964), 52–55. (MR 29, #3762).

[1964: 4]   The structure of the set of solutions of an equation of parabolic type. (Russian.) *Ukrainian Mat. Ž.* **16** (1964), 319–333. (MR 29, #3763).

[1964: 5]   An investigation of the Navier-Stokes equations by methods of the theory of parabolic equations in Banach spaces. (Russian.) *Dokl. Akad. Nauk SSSR* **156** (1964), 745–748. (MR 29, #1462).

[1965: 1]   Generalized solutions of first-order differential equations in Banach space. (Russian.) *Dokl. Akad. Nauk SSSR* **165** (1965), 486–489. (MR 32, #1909).

[1966: 1]   Equations $v'(t) + A(t)v(t) = f(t)$ in a Hilbert space, with $A(t)$ a positive self-adjoint operator depending measurably on t. (Russian.) *Akad. Nauk Azerbaĭdžan SSR Dokl.* **22** (1966), no. 8, 13–16. (MR 34, #6273).

[1966: 2]   On fractional powers of weakly positive operators. (Russian.) *Dokl. Akad. Nauk SSSR* **166** (1966), 1296–1299. (MR 33, #3137).

[1966: 3]   Investigation of general boundary value problems for parabolic equations by the method of differential equations in a Banach space. (Russian.) *Dokl. Akad. Nauk SSSR* **170** (1966), 1024–1027. (MR 34, #6274).

[1967: 1]   Comparison theorems for fractional powers of operators. (Russian.) *Dokl. Akad. Nauk SSSR* **174** (1967), 294–297. (MR 37, #2034).

[1967: 2] Parabolic equations in spaces of means. (Russian.) *Dokl. Akad. Nauk SSSR* (1967), 281–284. (MR 36, #6752).

[1967: 3] The application of fractional powers of operators to the investigation of quasilinear parabolic and elliptic equations and systems. (Russian.) *Ukrain. Mat. Ž.* **19** (1967), no. 3, 127–130. (MR 35, #4565).

[1968: 1] Elliptic equations in a Banach space. (Russian.) *Differencial'nye Uravnenija* **4** (1968), 1346–1348. (MR 37, #6627).

[1968: 2] A certain type of differential equations in a Banach space. (Russian.) *Differencial'nye Uravnenija* **4** (1968), 2278–2280. (MR 39, #593).

[1968: 3] The Bubnov-Galerkin method for parabolic equations in Hilbert space. (Russian.) *Dokl. Akad. Nauk SSSR* **178** (1968), 548–551. (MR 36, #6753).

[1968: 4] Differential equations with unbounded operators which generate non-analytic semigroups. (Russian.) *Dokl. Akad. Nauk SSSR* **183** (1968), 292–295. (MR 40, #3026).

[1968: 5] Quasilinear equations in a Banach space. (Russian.) *Dokl. Akad. Nauk SSSR* **183** (1968), 1020–1023. (MR 39, #5907).

[1968: 6] Fractional integration by parts. (Russian.) *Dokl. Akad. Nauk SSSR* **183** (1968), 1269–1272. (MR 38, #6203).

[1970: 1] The s-numbers of positive operators. (Russian.) *Dokl. Akad. Nauk SSSR* **193** (1970), 1245–1247. (MR 42, #5087).

[1971: 1] Degenerate parabolic operators. (Russian.) *Dokl. Akad. Nauk SSSR* **196** (1971), 302–304. (MR 43, #3849).

[1971: 2] Semigroups of growth α. (Russian.) *Dokl. Akad. Nauk SSSR* **196** (1971), 535–537. (MR 43, #3850).

[1971: 3] The coercive solvability of difference equations. (Russian.) *Dokl. Akad. Nauk SSSR* **201** (1971), 1063–1066. (MR 44, #7375).

[1972: 1] Fractional powers of elliptic operators in the ideals $\mathfrak{S}_p$. (Russian.) *Voronež. Gos. Univ. Trudy Naučn.-Issled. Inst. Mat. VGU Vyp. 5 Sb. Statei Teor. Operator i Differencial. Uravnenii* (1972), 114–117. (MR 56, #12989).

[1977: 1] Fractional powers of coercive-positive sums of operators. (Russian.) *Sibirsk. Mat. Z.* **18** (1977), 636–657. (MR 56, #12492).

P. E. Sobolevskiĭ and M. N. Titenskiĭ
[1969: 1] An investigation of the solvability of prescribed parabolic equations in a Banach space and a generalized Bubnov-Galerkin method for their approximate solution. (Russian.) *Differencial'nye Uravnenija* **5** (1969), 1495–1502.

H. Sohr
[1973: 1] Schwache Lösungen von Evolutionsgleichungen mit Anwendungen auf Partielle Differentialgleichungen. *Methoden und Verfahren der Mathematischen Physik, Band 11*, 117–142. Bibliographisches Institut, Mannheim, 1973. (MR 50, #13787).

[1973: 2] Verallgemeinerung eines Satzes von O. Ladyzenskaja über Schrödingergleichungen. *Methoden und Verfahren der Mathematischen Physik, Band. 11*, 143–162. (MR 50, #13788).

[1973: 3]   Über Evolutionsgleichungen. *Vorträge der Wissenschaftlichen Jahrestagung der Gesselschaft für Angewandte Mathematik und Mechanik* (Ljubljana, 1972). *Z. Angew Math. Mech.* **53** (1973), T169–T170. (MR 52, #6136).

M. Z. Solomjak

[1958: 1]   Application of semigroup theory to the study of differential equations in Banach spaces. (Russian.) *Dokl. Akad. Nauk SSSR* **122** (1958), 766–769. (MR 21, #3775).

[1959: 1]   Analyticity of a semigroup generated by an elliptic operator in L^p spaces. (Russian.) *Dokl. Akad. Nauk SSSR* **127** (1959), 37–39. (MR 22, #12398).

[1960: 1]   On differential equations in Banach spaces. (Russian.) *Izv. Vyss. Ucebn. Zaved. Matematika* (1960), no. 1 (14), 198–209. (MR 23, #A312).

A. R. Sourur

[1974: 1]   Semigroups of scalar-type operators on Banach spaces. *Trans. Amer. Math. Soc.* **200** (1974), 207–232. (MR 51, #1481).

[1974: 2]   On groups and semigroups of spectral operators on a Banach space. *Acta Sci. Math. (Szeged)* **36** (1974), 291–294. (MR 55, #6239).

M. Sova

[1966: 1]   Continuity of semigroups of operators in general operator topologies. (Russian.) *Czechoslovak Math. J.* **16** (91) (1966), 315–338. (MR 34, #3346).

[1966: 2]   Cosine operator functions. *Rozprawy Mat.* **49** (1966), 47 pp. (MR 33, #1745).

[1968: 1]   Problème de Cauchy pour équations hyperboliques opérationnelles à coefficients constants non-bornés. *Ann. Scuola Norm. Sup. Pisa* (3) **22** (1968), 67–100. (MR 38, #437).

[1968: 2]   Unicité des solutions exponentielles des équations différentielles opérationnelles linéaires. *Boll. Un. Mat. Ital.* (4) **1** (1968), 629–699. (MR 39, #594).

[1968: 3]   Solutions périodiques des équations différentielles opérationnelles: Le méthode des développements de Fourier. *Časopis Pěst. Mat.* **93** (1968), 386–421. (MR 40, #1686).

[1968: 4]   Semigroups and cosine functions of normal operators in Hilbert spaces. *Časopis Pěst. Mat.* **93** (1968), 437–458, 480. (MR 40, #3361).

[1969: 1]   Problèmes de Cauchy paraboliques abstraits de classes supérieures et les semi-groupes distributions. *Richerche Mat.* **18** (1969), 215–228. (MR 54, #3499).

[1970: 1]   Équations différentielles opérationnelles linéaires du second ordre à coefficients constants. *Rozprawy Československé Akad. Věd Řada Mat. Přírod. Věd* **80** (1970), sešit 7, 69 pp. (MR 43, #7804).

[1970: 2]   Équations hyperboliques avec petit paramètre dans les espaces de Banach généraux. *Colloq. Math.* **21** (1970), 303–320. (MR 54, #10796).

[1970: 3]   Régularité de l'évolution linéaire isochrone. *Czechloslovak Math. J.* **20** (1970), 251–302. (MR 56, #16094).

[1970: 4]   Perturbations numériques des evolutions paraboliques et hyperboliques. *Časopis Pešt. Mat.* **96** (1971), 406–407.

[1972: 1]   Encore sur les équations hyperboliques avec petit paramètre dans les espaces de Banach généraux. *Colloq. Math.* **25** (1972), 135–161. (MR 54, #10797).

[1977: 1]   On Hadamard's concepts of correctness. *Časopis Pešt. Mat.* **102** (1977), 234–269. (MR 56, #12977).

J. W. Spellmann

[1969: 1]   Concerning the infinite differentiability of semigroup motions. *Pacific J. Math.* **30** (1969), 519–523. (MR 40, #3360).

[1970: 1]   Concerning the domains of generators of linear semigroups. *Pacific J. Math.* **35** (1970), 503–509. (MR 43, #3851).

J. D. Staffney

[1976: 1]   Integral representations of fractional powers of infinitesimal generators. *Illinois J. Math.* **20** (1976), 124–133. (MR 53, #14217).

E. M. Stein

[1970: 1]   *Singular Integrals and Differentiability Properties of Functions.* Princeton University Press, Princeton, N.J., 1970.

E. M. Stein and G. Weiss

[1971: 1]   *Introduction to Fourier Analysis on Euclidean Spaces.* Princeton University Press, Princeton, N.J., 1971.

H. B. Stewart

[1974: 1]   Generation of analytic semigroups by strongly elliptic operators. *Trans. Amer. Math. Soc.* **199** (1974), 141–162. (MR 50, #10532).

M. H. Stone

[1930: 1]   Linear transformations in Hilbert space III. *Proc. Nat. Acad. Sci. U.S.A.* **16** (1930), 172–175.

[1932: 1]   On one parameter unitary groups in Hilbert space. *Ann. Math.* **33** (1932), 643–648.

G. Strang

[1966: 1]   Necessary and inufficient conditions for well-posed Cauchy problems. *J. Diff. Equations* **2** (1966), 107–114. (MR 33, #2928).

[1967: 1]   On strong hyperbolicity. *J. Math. Kyoto Univ.* **6** (1967), 397–417. (MR 36, #529).

[1969: 1]   On multiple characteristics and the Levi-Lax conditions for hyperbolicity. *Arch. Rational Mech. Anal.* **33** (1969), 358–373. (MR 39, #4509).

[1969: 2]   On numerical ranges and holomorphic semigroups. *J. Analyse Math.* **22** (1969), 299–318. (MR 41, #2463).

[1969: 3]   Hyperbolic initial-boundary value problems in two unknowns. *J. Diff. Equations* **6** (1969), 161–171. (MR 39, #633).

[1969: 4]   Approximating semigroups and the consistency of difference schemes. *Proc. Amer. Math. Soc.* **20** (1969), 1–7. (MR 38, #1561).

B. Straughan
[1976: 1]   Growth and uniqueness theorems for an abstract nonstandard wave
             equation. *SIAM J. Math. Anal.* **7** (1976), 519–528. (MR 53, #13770).

W. A. Strauss
[1966: 1]   Evolution equations non-linear in the time derivative. *J. Math. Mech.*
             15 (1966), 49–82. (MR 32, #8217).
[1967: 1]   The initial-value problem for certain non-linear evolution equations.
             Amer. J. Math. **89** (1967), 249–259. (MR 36, #553).
[1967: 2]   On the solutions of abstract nonlinear equations. *Proc. Amer. Math.
             Soc.* **18** (1967), 116–119. (MR 36, #746).
[1970: 1]   Further applications of monotone methods to partial differential equa-
             tions. *Nonlinear Functional Analysis* (*Proc. Sympos. Pure Math.*, vol.
             XVIII, part 1, Chicago, Ill., 1968), 282–288. Amer. Math. Soc., Provi-
             dence, R.I., 1970. (MR 42, #8113).
[1970: 2]   Local exponential decay of a group of conservative nonlinear opera-
             tors. *J. Functional Analysis* **6** (1970), 152–156. (MR 43, #728).

D. W. Stroock and S. R. S. Varadhan
[1979: 1]   *Multidimensional Diffusion Processes.* Springer, New York, 1979.

S. O. Strygina and P. P. Zabreĭko
[1971: 1]   The periodic solutions of evolution equations. (Russian.) *Mat. Zametki*
             9 (1971), 651–662. (MR 44, #4295).

Ju. M. Suhov
[1970: 1]   A one-parameter semigroup of operators that is generated by the
             Gibbs distribution. (Russian.) *Uspehi Mat. Nauk* **25** (1970), no. 1 (151),
             199–200. (MR 41, #7480).

H. Sunouchi
[1970: 1]   Convergence of semi-discrete differential schemes for abstract Cauchy
             problems. *Tôhoku Math. J.* **22** (1970), 394–408. (MR 43, #983).

C. Sunyach
[1975: 1]   Semi-groupes sous-markoviens faiblement continus. *C. R. Acad. Sci.
             Paris.* **280** (1975), A1201–A1203. (MR 51, #11198).

P. Suryanarayana
[1965: 1]   The higher-order differentiability of solutions of abstract differential
             equations. *Pacific J. Math.* **22** (1967), 543–561. (MR 37, #554).

G. A. Suvorčenkova
[1976: 1]   The solutions of the Cauchy problem for a first-order linear differential
             equation with operator coefficients. (Russian.) *Uspehi Mat. Nauk* **31**
             (1976), 263–264. (MR 54, #13634).

M. Suzuki
[1976: 1]   Generalized Trotter's formula and systematic approximants of ex-
             ponential operators and inner derivations with applications to many-
             body problems. *Comm. Math. Phys.* **51** (1976), 183–190. (MR 54,
             #13632).

P. Swerling
[1957: 1] Families of transformations in the function spaces H^p. *Pacific J. Math.* **7** (1957), 1015–1029. (MR 19, #434).

B. Sz.-Nagy
[1938: 1] On semi-groups of self-adjoint transformations in Hilbert space. *Proc. Nat. Acad. Sci. U.S.A.* **24** (1938), 559–560.
[1947: 1] On uniformly bounded linear transformations in Hilbert space. *Acta Sci. Math. Szeged* **11** (1947), 152–157. (MR 9, 191).
[1953: 1] Sur les contractions de l'espace de Hilbert. *Acta Sci. Math. Szeged* **15** (1953), 87–92.
[1954: 1] Transformations de l'espace de Hilbert, fonctions de type positif sur un groupe. *Acta Sci. Math. Szeged* **15** (1954), 104–114. (MR 15, 719).

B. Sz.-Nagy and C. Foias
[1966: 1] *Analyse Harmonique des Opérateurs de l'Espace de Hilbert*. Masson et Cie., Paris-Akadémiai Kiadó, Budapest, 1966. Revised and augmented edition: [1969: 1] *Harmonic Analysis of Operators in Hilbert Space*. North-Holland, Amsterdam–American Elsevier, New York-Akadémiai Kiadó, Budapest, 1969.

C. T. Taam
[1966: 1] Stability, periodicity and almost periodicity of solutions of nonlinear differential equations in Banach spaces. *J. Math. Mech.* **15** (1966), 849–876. (MR 33, #4433).
[1967: 1] On nonlinear diffusion equations. *J. Diff. Equations* **3** (1967), 482–499. (MR 36, #501).
[1968: 1] Nonlinear differential equations in Banach spaces, and applications. *Michigan Math. J.* **15** (1968), 177–186. (MR 37, #3224).

C. T. Taam and J. N. Welch
[1966: 1] Compact solutions of nonlinear differential equations in Banach spaces. *Michigan Math. J.* **13** (1966), 271–284. (MR 33, #7665).

T. Takahashi and S. Oharu
[1972: 1] Approximation of operator semigroups in a Banach space. *Tôhoku Math. J.* **24** (1972), 505–528. (MR 50, #7722).

H. Tanabe
[1959: 1] On a class of the equations of evolution in a Banach space. *Osaka Math. J.* **11** (1961), 121–145. (MR 22, #4964).
[1960: 1] Remarks on the equations of evolution in a Banach space. *Osaka Math. J.* **12** (1960), 145–166. (MR 23, #A2756a).
[1960: 2] On the equations of evolution in a Banach space. *Osaka Math. J.* **12** (1960), 363–376. (MR 23, #A2756b).
[1961: 1] Convergence to a stationary state of the solution of some kind of differential equations in a Banach space. *Proc. Japan Acad.* **37** (1961), 127–130. (MR 24, #A895).
[1961: 2] Evolutional equations of parabolic type. *Proc. Japan Acad.* **37** (1961), 610–613. (MR 25, #4366).

[1964: 1]   Note on singular perturbation for abstract differential equations. *Osaka J. Math.* **1** (1964), 239–252. (MR 30, #2368).

[1964: 2]   On differentiability in time of solutions of some type of boundary value problems. *Proc. Japan Acad.* **40** (1964), 649–653. (MR 32, #7944).

[1965: 1]   On differentiability and analyticity of solutions of weighted elliptic boundary value problems. *Osaka J. Math.* **2** (1965), 163–190. (MR 32, #289).

[1965: 2]   Note on uniqueness of solutions of differential inequalities of parabolic type. *Osaka J. Math.* **2** (1965), 191–204. (MR 32, #290).

[1966: 1]   On estimates for derivatives of solutions of weighted elliptic boundary value problems. *Osaka J. Math.* **3** (1966), 163–182. (MR 34, #480).

[1967: 1]   On the regularity of solutions of abstract differential equations of parabolic type in Banach space. *J. Math. Soc. Japan* **19** (1967), 521–542. (MR 36, #2931).

[1967: 2]   On the regularity of solutions of abstract differential equations in Banach space. *Proc. Japan Acad.* **43** (1967), 305–307. (MR 36, #1802).

[1979: 1]   *Equations of Evolution.* Pitman, London, 1979.

H. Tanabe and M. Watanabe

[1966: 1]   Note on perturbation and degeneration of abstract differential equations in Banach space. *Funkcial. Ekvac.* **9** (1966), 163–170. (MR 35, #517).

A. Taylor

[1958: 1]   *Introduction to Functional Analysis.* Wiley, New York, 1958.

A. P. Terehin

[1974: 1]   Semigroups of operators and mixed properties of elements of a Banach space. (Russian.) *Mat. Zametki* **16** (1974), 107–115. (MR 51, #3961).

[1975: 1]   A bounded group of operators, and best approximation. (Russian.) *Differencial'nye Uravnenija i Vyčisl. Mat. Vyp.* **2** (1975), 3–28, 177. (MR 55, #3650).

F. Terkelsen

[1969: 1]   On semigroups of operators in locally convex spaces. *Proc. Amer. Math. Soc.* **22** (1969), 340–343. (MR 40, #1813).

[1973: 1]   Boundedness properties for semigroups of operators. *Proc. Amer. Math. Soc.* **40** (1973), 107–111. (MR 48, #4883).

A. N. Tikhonov

[1944: 1]   On the stability of inverse problems. *Dokl. Akad. Nauk SSSR* **39** (1944), 195–198. (MR 5, 184).

A. N. Tikhonov and V. Y. Arsenin

[1977: 1]   *Solutions of Ill-Posed Problems.* V. H. Winston and Sons-Wiley, New York, 1977.

H. G. Tillmann

[1960: 1]   Approximationssätze für Halbgruppen von Operatoren in Topologischen Vektorräumen. *Arch. Mat.* **11** (1960), 194–199. (MR 22, #8439).

D. B. Topoljanskiĭ and Ja. M. Zaprudskiĭ
[1974: 1]   A study of the stability of the Bubnov-Galerkin method for nonsta-
            tionary operator equations with variable coefficients. (Russian.) *Ukrain.
            Mat. Ž.* **26** (1974), 621–633, 716. (MR 50, #6160).

M. Tougeron-Sablé
[1973: 1]   Régularité pour des équations d'évolution. Application a la théorie
            spectrale. (Italian summary.) *Boll. Un. Mat. Ital.* **7** (1973), 1–11. (MR
            47, #5654).

C. C. Travis
[1976: 1]   On the uniqueness of solutions to hyperbolic boundary value prob-
            lems. *Trans Amer. Math. Soc.* **216** (1976), 327–336. (MR 52, #11337).

C. C. Travis and G. F. Webb
[1974: 1]   Existence and stability for partial functional differential equations.
            Trans. Amer. Math. Soc. **200** (1974), 395–418. (MR 52, #3690).
[1977: 1]   Compactness, regularity and uniform continuity properties of strongly
            continuous cosine families. *Houston J. Math.* **3** (1977), 555–567. (MR
            58, #17957).
[1978: 1]   Cosine families and abstract nonlinear second order differential equa-
            tions. *Acta Math. Sci. Hung.* **32** (1978), 75–96. (MR 58, #17404).
[1978: 2]   Existence, stability and compactness in the α-norm for partial func-
            tional differential equations. *Trans. Amer. Math. Soc.* **240** (1978),
            129–143. (MR 58, #17405).

W. Trebels and U. Westphal
[1972: 1]   A note on the Landau-Kallman-Rota-Hille inequality. Linear opera-
            tors and approximation (*Proc. Conf. Oberwolfach, 1971*), 115–119.
            Internat. Ser. Numer. Math., vol. 20, Birkhäuser, Basel, 1972. (MR 53,
            #6355).

V. A. Trenogin
[1963: 1]   Existence and asymptotic behavior of the solution of the Cauchy
            problem for a first-order differential equation with small parameter in
            a Banach space. (Russian.) *Dokl. Akad. Nauk SSSR* **152** (1963), 63–66.
            (MR 29, #1409).
[1964: 1]   Existence and asymptotic behavior of solutions of "solitary wave" type
            for differential equations in a Banach space. (Russian.) *Dokl. Akad
            Nauk SSSR* **156** (1964), 1033–1036. (MR 29, #6348).
[1966: 1]   Boundary value problems for abstract elliptic equations. (Russian.)
            Dokl. Akad. Nauk SSSR **170** (1966), 1028–1031. (MR 34, #6275).
[1967: 1]   Linear equations in a Banach space with a small parameter. (Russian.)
            *Proc. Sixth Interuniv. Sci. Conf. of the Far East on Physics and Mathe-
            matics*, vol. 3: *Differential and Integral Equations* (Russian), 211–216,
            Khabarovsk. Gos. Ped. Inst., Khabarovsk, 1967. (MR 41, #885).

F. Trèves
[1961: 1]   Differential equations in Hilbert spaces. *Proc. Sympos. Pure Math.*,
            vol. IV, 83–99. American Mathematical Society, Providence, R.I.,
            1961. (MR 25, #3387).

[1970: 1] An abstract nonlinear Cauchy-Kovalewska theorem. *Trans. Amer. Math. Soc.* **150** (1970), 77–92. (MR 43, #669).

[1973: 1] Concatenation of second-order evolution equations applied to local solvability and hypoellipticity. *Comm. Pure Appl. Math.* **26** (1973), 201–250. (MR 49, #5554).

H. Triebel

[1978: 1] *Interpolation Theory, Function Spaces, Differential Operators.* VEB Deutscher Verlag der Wissenschaften, Berlin, 1978.

H. F. Trotter

[1958: 1] Approximation of semi-groups of operators. *Pacific J. Math.* **8** (1958), 887–919. (MR 21, #2190).

[1959: 1] On the product of semi-groups of operators. *Proc. Amer. Math. Soc.* **10** (1959), 545–551. (MR 21, #7446).

[1959: 2] An elementary proof of the central limit theorem. *Archiv Math.* **10** (1959), 226–234. (MR 21, #7559).

[1974: 1] Approximation and perturbation of semigroups. *Linear Operators and Approximation Theory II* (P. L. Butzer and B. Sz.-Nagy, editors). Birkhäuser, Basel, 1974, 3–21. (MR 52, #4024).

C. Truesdell and R. Toupin

[1960: 1] *The Classical Field Theories.* Handbuch der Physik, Bd. III/1, Springer, Berlin, 1960, 226–793.

C. P. Tsokos and M. Rama Mohana Rao

[1969: 1] On the stability and boundedness of differential systems in Banach spaces. *Proc. Cambridge Philos. Soc.* **65** (1969), 507–512. (MR 38, #3560).

M. Tsutsumi

[1971: 1] Some nonlinear evolution equations of second order. *Proc. Japan Acad.* **47** (1971), suppl. II, 950–955. (MR 47, #585).

[1972: 1] On solutions of semilinear differential equations in a Hilbert space. *Math. Japon.* **17** (1972), 173–193. (MR 50, #7723).

D. A. Uhlenbrock

[1971: 1] Perturbation of statistical semigroups in quantum statistical mechanics. *J. Math. Phys.* **12** (1971), 2503–2512. (MR 44, #5063).

T. Ujishima

[1969: 1] Some properties of regular distribution semigroups. *Proc. Japan Acad.* **45** (1969), 224–227. (MR 40, #7867).

[1970: 1] On the strong continuity of distribution semigroups. *J. Fac. Sci. Univ. Tokyo Sect. I* **17** (1970), 363–372. (MR 43, #5348).

[1971: 1] On the abstract Cauchy problem and semi-groups of linear operators in locally convex spaces. *Sci. Papers College Gen. Ed. Univ. Tokyo* **21** (1971), 116–122. (MR 47, #886).

[1972: 1] On the generation and smoothness of semigroups of operators. *J. Fac. Sci. Univ. Tokyo Sect. 1A Math.* **19** (1972), 65–127. (MR 46, #7966).

Correction. *J. Fac. Sci. Univ. Tokyo Sect. 1A Math.* (1973), 187–189. (MR 48, #952).

[1975/76: 1]   Approximation theory for semi-groups of linear operators and its application to the approximation of wave equations. *Japan J. Math.* (*N.S.*) **1** (1975/76), 185–224. (MR 54, #8354).

L. I. Vaĭnerman

[1974: 1]   Dissipative boundary value problems for a second-order differential equation with an unbounded variable operator coefficient. (Russian.) *Ukrain. Mat. Ž.* **26** (1974), 530–534, 575. (MR 50, #13790).

[1974: 2]   Selfadjoint boundary value problems for strongly elliptic and hyperbolic second-order equations in Hilbert space. (Russian.) *Dokl. Akad. Nauk SSSR* **218** (1974), 745–748. (MR 52, #3994).

[1976: 1]   The selfadjointness of abstract differential operators of hyperbolic type. (Russian.) *Mat. Zametki* **20** (1976), 703–708. (MR 56, #3429).

[1977: 1]   A hyperbolic equation with degeneracy in Hilbert space. (Russian.) *Sibirsk. Mat. Z.* **18** (1977), 736–746, 955. (MR 56, #1608).

L. I. Vaĭnerman and M. L. Gorbačuk

[1975: 1]   Boundary value problems for a second-order differential equation of hyperbolic type in Hilbert space. (Russian.) *Dokl. Akad. Nauk SSSR* **221** (1975), 763–766. (MR 52, #942).

V. I. Vaĭnerman and Ju. M. Vuvunikjan

[1974: 1]   Fundamental functions for differential operators in a locally convex space. (Russian.) *Dokl. Akad. Nauk SSSR* **214** (1974), 15–18. (MR 49, #1221).

G. Vainikko and P. E. Oja

[1975: 1]   The convergence and rate of convergence of the Galerkin method for abstract evolution equations. (Russian.) *Differencial'nye Uravnenija* **11** (1975), 1269–1277, 1350. (MR 56, #12936).

G. Vaĭnikko and M. Šlapikiene

[1971: 1]   On a certain theorem of S. G. Kreĭn on perturbation of operators that generate analytic semigroups. (Russian.) *Tartu Riikl. Ül.Toimetised Vih.* **220** (1968), 190–204. (MR 41, #2453).

K. V. Valikov

[1964: 1]   Characteristic indices of solutions of differential equations in a Banach space. (Russian.) *Dokl. Akad. Nauk SSSR* **158** (1964), 1010–1013. (MR 29, #6349).

[1964: 2]   Some criteria for stability of motion in Hilbert space. (Russian.) *Uspehi Mat. Nauk* **19** (1964), no. 4 (118), 179–184. (MR 29, #6142).

[1965: 1]   The asymptotic behavior of solutions of a nonlinear parabolic equation. (Russian.) *Dokl. Akad. Nauk SSSR* **160** (1965), 743–745. (MR 35, #592).

[1967: 1]   Nonlinear differential equations in a Banach space, close to linear ones. (Russian.) *Differencial'nye Uravnenija* **3** (1967), 1692–1707. (MR 36, #5484).

A. Van Daele
[1975: 1]   On the spectrum of the analytic generator. *Math. Scand.* **37** (1975), 307–318. (MR 53, #3793).

O. Vejvoda and I. Straškraba
[1972: 1]   Periodic solutions to abstract differential equations. *Proceedings of Equadiff III, Brno, 1972*, 199–203. Folia Fac. Sci. Natur. Univ. Purkynianae Brunensis, Ser. Monograph., Tomus 1, Purkyně Univ., Brno, 1973. (MR 50, #5132).

M. A. Veliev
[1972: 1]   Certain conditions for the stability of the Bubnov-Galerkin method for a second order equation in Hilbert space. (Russian.) *Azerbaĭdažan. Gos. Univ. Ucen. Zap. Ser. Fiz.-Mat. Nauk* (1972), 17–24. (MR 55, #8504).
[1973: 1]   The stability of the solutions of differential equations with a nonlinear potential operator in Hilbert space. (Russian.) *Dokl. Akad. Nauk SSSR* **208** (1973), 21–24. (MR 47, #5393).
[1973: 2]   Stability of the Bubnov-Galerkin method for certain linear boundary value problems in Hilbert space. (Russian.) *Akad. Nauk Azerbaĭdžan. SSR Dokl.* **29** (1973), 7–10. (MR 49, #767).
[1975: 1]   A method for the regularization of singularly perturbed differential equations in Hilbert space. (Russian.) *Dokl. Akad. Nauk SSSR* **220** (1975), 1008–1011. (MR 52, #3691).

M. A. Veliev and Ja. D. Mamedov
[1973: 1]   Stability of the solutions of abstract differential equations in Hilbert space I. (Russian.) *Demonstratio Math.* **5** (1973), 311–329. (MR 54, #672a).
[1974: 1]   Stability of the solutions of abstract differential equations in Hilbert space II. (Russian.) *Demonstratio Math.* **7** (1974), 73–84. (MR 54, #627b).

A. Venni
[1975: 1]   Un problema ai limiti per un'equazione astratta del secondo ordine. (English summary.) *Rend. Sem. Mat. Univ. Padova* **53** (1975), 291–314. (MR 55, #8505).

I. Vidav
[1968: 1]   Existence and uniqueness of negative eigenfunctions of the Boltzmann operator. *J. Math. Anal. Appl.* **22** (1968), 144–155.
[1970: 1]   Spectra of perturbed semigroups with applications to transport theory. *J. Math. Anal. Appl.* **30** (1970), 264–279.

R. Vilella Bressan
[1974: 1]   On an evolution equation in Banach spaces. (Italian summary.) *Rend. Sem. Mat. Univ. Padova* **51** (1974), 281–291 (MR 51, #13801).

A. I. Virozub
[1975: 1]   The energy completeness of a system of elementary solutions of a differential equation in Hilbert space. (Russian.) *Funkcional. Anal. i Priložen.* **9** (1975), vyp. 1, 52–53. (MR 51, #3940).

M. I. Višik

[1956: 1]   The Cauchy problem for equations with operator coefficients; mixed
              boundary problem for systems of differential equations and approxi-
              mation methods for their solution. (Russian.) *Mat. Sb.* **39** (81) (1956),
              51–148. (MR 18, #215).

M. I. Višik and O. A. Ladyženskaya

[1956: 1]   Boundary value problems for partial differential equations and certain
              classes of operator equations. (Russian.) *Uspehi Mat. Nauk* **11** (1956), 6
              (72), 41–97. (MR 20, #1092).

M. I. Višik and L. A. Ljusternik

[1957: 1]   Regular degeneration and boundary layer for linear differential equa-
              tion with small parameter. (Russian.) *Uspehi Mat. Nauk* **12** (1957), **5**
              (77), 3–122. (MR 25, #322).

C. Visser

[1938: 1]   On the iteration of linear operations in Hilbert space. *Nederl. Akad.
              Wetensch.* **41** (1938), 487–495.

J. Voigt

[1977: 1]   On the perturbation theory for strongly continuous semigroups. *Math.
              Ann.* **229** (1977), 163–171. (MR 56, #3677).

Ju. I. Volkov

[1973: 1]   Certain relations that are connected with semigroups of operators.
              (Russian.) *Izv. Vysš. Ucebn. Zaved. Matematika* (1973), No. 10 (137),
              23–28. (MR 49, #5930).

V. Volterra

[1887: 1]   Sui fondamenti della teoria delle equazioni differenziali lineari, I.
              Mem. Soc. Ital. Sci. (3), **6** (1877), 1–104.

[1902: 1]   Sui fondamenti della teoria delle equazioni differenziali lineari, II.
              Mem. Soc. Ital. Sci. (3), **12** (1902), 3–68.

[1909: 1]   Sulle equazioni integrodifferenziali della teoria dell'elasticità. *Atti Reale
              Accad. Lincei* **18** (1909), 295–314.

[1928: 1]   Sur la théorie mathématique des phénomènes héréditaires. *J. Math.
              Pures Appl.* **7** (1928), 249–298.

[1931: 1]   *Théorie Mathématique de la Lutte pour la Vie*, Gauthier-Villars, Paris,
              1931.

V. Volterra and B. Hostinský

[1938: 1]   *Opérations Infinitésimales Linéaires*, Gauthier-Villars, Paris, 1931.

G. A. Voropaeva

[1970: 1]   A Cauchy type problem for pseudo-differential equations with opera-
              tor coefficients. (Russian.) *Izv. Vysš. Učebn. Zaved. Matematika*, no. 7
              (98) (1970), 35–39. (MR 45, #8968).

[1976: 1]   On the question of the uniqueness of the weak solution of an equation
              with operator coefficients. (Russian.) *Izv. Vysš. Učebn. Zaved. Mate-
              matika*, no. 7 (170) (1976), 11–21. (MR 56, #12490).

D. Voslamber

[1972: 1]   On exponential approximations for the evolution operator. *J. Math. Anal. Appl.* **37** (1972), 403–411. (MR 45, #967).

Ju. M. Vuvunikjan

[1971: 1]   Quasiexponential semigroups of endomorphisms of a locally convex space. (Russian.) *Sibirski Mat. Ž.* **12** (1971), 284–294. (MR 47, #887).

[1971: 2]   Γ⟨m_k⟩-semigroups of endomorphisms of locally convex spaces. (Russian.) *Voronez Gos. Univ. Trudy Naučn. Issled. Inst. Mat. VGU Vyp.* **3** (1971), 11–18. (MR 47, #888).

[1971: 3]   Generation of semigroup-valued generalized functions of endomorphisms of a locally convex space. (Russian.) *Dokl. Akad. Nauk SSSR* **198** (1971), 269–272. (MR 47, #9439a).

[1972: 1]   Criteria of smoothness and analyticity of a semigroup generalized function in a locally convex space. (Russian.) *Dokl. Akad. Nauk SSSR* **203** (1972), 270–73. (MR 47, #9349b).

[1974: 1]   Evolutionary representations of the algebra of summable functions in a locally convex space. (Russian.) *Dokl. Akad. Nauk SSSR* **216** (1974), 724–727. (MR 50, #14362).

Ju. M. Vuvunikjan and V. V. Ivanov

[1974: 1]   Generation of evolutionary representations, and the method of the resolvent sequence. (Russian.) *Sibirski Mat. Ž.* **15** (1974), 1422–1424. (MR 50, #14363).

M. von Waldenfels

[1964: 1]   Positive Halbgruppen auf einem *n*-dimensionaler Torus. *Arch. Math.* **15** (1964), 191–203.

L. Waelbroeck

[1964: 1]   Les semi-groupes différentiables. *Deuxième Colloq. sur l'Analyse Fonct.*, 97–103, Centre Belge de Recherches Math., Librairie Universitaire, Louvain, 1964. (MR 32, #1569).

[1969: 1]   Differentiability of Hölder-continuous semigroups. *Proc. Amer. Math. Soc.* **21** (1969), 451–454. (MR 39, #2021).

W. von Wahl

[1971: 1] L^p-decay rates for homogeneous wave-equations. *Math. Z.* **120** (1971), 93–106. (MR 43, #6604).

L. J. Wallen

[1969: 1]   Semigroups of partial isometries. *Bull. Amer. Math. Soc.* **75** (1969), 763–764. (MR 40, #1790).

[1969/70: 1]   Semi-groups of partial isometries. *J. Math. Mech.* **19** (1969/70), 745–750. (MR 41, #868).

[1975: 1]   Translational representations of one-parameter semigroups. *Duke Math. J.* **42** (1975), 111–119. (MR 51, #11196).

W. L. Walter

[1969: 1]   Gewöhnliche Differential-Ungleichungen im Banachräum. *Arch. Math. (Basel)* **20** (1969), 36–47. (MR 39, #5908).

J. A. Walker
[1976: 1]   On the application of Liapunov's direct method to linear dynamical systems. *J. Math. Anal. Appl.* **53** (1976), 187–220. (MR 53, #13774).

C. L. Wang
[1975: 1]   A uniqueness theorem on the degenerate Cauchy problem. *Canad. Math. Bull.* **18** (1975), 417–421. (MR 53, #1065).

P. K. C. Wang
[1967: 1]   Équivalence asymptotique d'équations d'évolution. *C. R. Acad. Sci. Paris.* **264** (1967), A629–A632. (MR 35, #6956).

J. R. Ward
[1976: 1]   Existence of solutions to quasilinear differential equations in a Banach space. *Bull. Austral. Math. Soc.* **15** (1976), 421–430. (MR 55, #3464).

M. Watanabe
[1967: 1]   On the abstract semi-linear differential equation. *Sci. Rep. Niigata Univ. Ser. A*, no. 5 (1967), 9–16. (MR 38, #1379).
[1968: 1]   Quasi-linear equations of evolution in a Banach space. *Sci. Rep. Niigata Univ. Ser. A*, no. 6 (1968), 37–49. (MR 39, #7258).
[1969: 1]   On the abstract quasi-linear differential equation. *Sci. Rep. Niigata Univ. Ser. A*, no. 7 (1969), 1–16. (MR 42, #639).
[1972: 1]   On the differentiability of semi-groups of linear operators in locally convex spaces. *Sci. Rep. Niigata Univ. Ser. A*, no. 9 (1972), 23–34. (MR 46, #719).
[1973: 1]   On the characterization of semi-groups of linear operators. *Sci. Rep. Niigata Univ. Ser. A*, no. 10 (1973), 43–50. (MR 48, #4830).
[1977: 1]   A remark on fractional powers of linear operators in Banach spaces. *Proc. Japan Acad.* **53** (1977), no. 1, 4–7. (MR 56, #12943).

G. F. Webb
[1972: 1]   Continuous nonlinear perturbations of linear accretive operators in Banach spaces. *J. Functional Analysis* **10** (1972), 191–203. (MR 50, #14407).
[1972: 2]   Nonlinear perturbations of linear accretive operators in Banach spaces. *Israel J. Math.* **12** (1972), 237–248. (MR 47, #9350)
[1974: 1]   Autonomous nonlinear functional differential equations and nonlinear semigroups. *J. Math. Anal. Appl.* **46** (1974), 1–12. (MR 50, #722).
[1976: 1]   Linear functional differential equations with L^2 initial functions. *Funkcial. Ekvac.* **19** (1976), 65–77. (MR 54, #13250).
[1977: 1]   Regularity of solutions to an abstract inhomogeneous linear differential equation. *Proc. Amer. Math. Soc.* **62** (1977), 271–277. (MR 55, #5975).
[1977: 2]   Exponential representation of solutions to an abstract semi-linear differential equation. *Pacific J. Math.* **70** (1977), 269–279. (MR 58, #11768).
[1977: 3]   Regularity of solutions to an abstract inhomogeneous linear differential equation. *Proc. Amer. Math. Soc.* **62** (1977), 271–277. (MR 55, #5975).

B. Weiss

[1967: 1]   Abstract vibrating systems. *J. Math. Mech* **17** (1967), 241–255. (MR
             36, #5485).

F. B. Weissler

[1979: 1]   Semilinear evolution equations in Banach spaces. *J. Functional Anal.*
             32 (1979), 277–296.

J. N. Welch

[1971: 1]   Spectral properties of a linear operator connected with solutions of
             linear differential equations in Banach spaces. *Math. Ann.* **192** (1971),
             253–256. (MR 45, #4212).

U. Westphal

[1968: 1]   Über Potenzen von Erzeugern von Halbgruppenoperatoren. *Abstract
             Spaces and Approximation* (*Proc. Conf., Oberwolfach, 1968*), 82–91.
             Bikhäuser, Basel, 1969. (MR 41, #2464).

[1970: 1]   Ein Kalkül für gebrochene Potenzen infinitesimaler Erzeuger von
             Halbgruppen und Gruppen von Operatoren. Halbgruppenerzeuger.
             Compositio Math. **22** (1970), 67–103. (MR 41, #6010a).

[1970: 2]   Ein Kalkül für gebrochene Potenzen infinitesimaler Erzeuger von
             Halbgruppen und Gruppen von Operatoren. Gruppenerzeuger. *Com-
             positio Math.* **22** (1970), 104–136. (MR 41, #6010b).

[1974: 1]   Gebrochene Potenzen abgeschlossener Operatoren, definiert mit Hilfe
             gebrochener Differenzen. *Linear Operators and Approximation, II*
             (*Proc. Conf., Oberwolfach Math. Res. Inst., Oberwolfach, 1974*), 23–27.
             Internat. Ser. Numer. Math. 25, Birkhäuser, Basel, 1974. (MR 53,
             #3791).

[1974: 2]   An approach to fractional powers of operators via fractional dif-
             ferences. *Proc. London Math. Soc.* (3) (1974), 557–576. (MR 50,
             #14364).

H. Weyl

[1909: 1]   Über gewöhnliche lineare Differentialgleichungen mit singulären Stel-
             len und ihre Eigenfunktionen. *Nachr. Akad. Wiss. Göttingen. Math.-
             Phys. Kl.* (1909), 37–64.

[1910: 1]   Über gewöhnliche Differentialgleichungen mit singulären Stellen und
             ihre Eigenfunktionen. *Nachr. Akad. Wiss. Gottingen. Math.-Phys. Kl.*
             (1910), 442–467.

[1910: 2]   Über gewöhnliche Differentialgleichungen mit Singularitäten und die
             zugehörigen Entwicklungen willkürlicher Funktionen. *Math. Ann.* **68**
             (1910), 220–269.

D. V. Widder

[1931: 1]   Necessary and sufficient conditions for the representation of a func-
             tion as a Laplace integral. *Trans. Amer. Math. Soc.* **33** (1931), 851–892.

[1944: 1]   Positive temperatures in an infinite rod. *Trans. Amer. Math. Soc.* **55**
             (1944), 85–95. (MR 5, 203).

[1946: 1]   *The Laplace Transform*. Princeton University Press, Princeton, N.J.,
             1946.

[1971: 1]   *An Introduction to Transform Theory*. Academic Press, New York, 1971.
[1975: 1]   *The Heat Equation*. Academic Press, New York, 1975.

N. Wiener
[1939: 1]   The ergodic theorem. *Duke Math. J.* **5** (1939), 1–18.

C. H. Wilcox
[1966: 1]   Wave operators and asymptotic solutions of wave propagation prob-
             lems of classical physics. *Arch. Rat. Mech. Anal.* **22** (1966), 37–78.
             (MR 33, #7675).
[1973: 1]   Scattering states and wave operators in the abstract theory of scatter-
             ing. *J. Functional Analysis* **12** (1973), 257–274. (MR 49, #9651).

C. Wild
[1972: 1]   Domaines de croissance de semi-groupes d'opérateurs sur un espace de
             Banach: définition et propriétés. *C. R. Acad. Sci. Paris* **275** (1972),
             A643–A645.
[1972: 2]   Domaines de croissance de semi-groupes d'opérateurs sur un espace de
             Banach: géneration et théorie spectrale. *C. R. Acad. Sci. Paris* **275**
             (1972), A697–A700. (MR 46, #937).
[1974: 1]   Domaines de croissance d'un semi-groupe localement équicontinue sur
             un espace localement convexe et applications. *C. R. Acad. Sci. Paris*
             279 (1972), 933–935. (MR 51, #3998).
[1975: 1]   Semi-groupes de classe *p*. *C. R. Acad. Sci. Paris* **280** (1975),
             A1595–A1598. (MR 52, #11656).
[1976: 1]   Semi-groupes de classe *p* avec poids. *C. R. Acad. Sci. Paris* **282** (1976),
             A1087–A1090. (MR 54, #3500).
[1976: 2]   Semi-groupes distributions de classe *p*. *C. R. Acad. Sci. Paris* **282**
             (1976), A1147–A1149. (MR 55, #3871).
[1977: 1]   Semi-groupes de croissance $\alpha < 1$ holomorphes. *C. R. Acad. Sci. Paris*
             285 (1977), A437–A440. (MR 56, #6468).

A. Wintner
[1945: 1]   The non-local existence problem of ordinary differential equations.
             Amer. J. Math. **67** (1945), 277–284. (MR 6, 225).

K. T. Wong
[1974: 1]   On extended structures of a closed operator related to semigroup
             theory and the abstract Cauchy problem. *Bull. Amer. Math. Soc.* **80**
             (1974), 350–354. (MR 49, #5931).

R. Wüst
[1971: 1]   Generalizations of Rellich's theorem on the perturbation of (essen-
             tially) self-adjoint operators. *Math. Z.* **119** (1971), 276–280. (MR 43,
             #7959).
[1973: 1]   A convergence theorem for self-adjoint operators applicable to Dirac
             operators with cutoff potentials. *Math. Z.* **131** (1973), 339–349.
             (MR 50, #8129).

A. Yagi

[1976: 1]   On the abstract linear evolution equation in Banach spaces. *J. Math. Soc. Japan* **28** (1976), 290–303. (MR 53, #1337).

A. Yoshikawa

[1971: 1]   Fractional powers of operators, interpolation theory and imbedding theorems. *J. Fac. Sci. Univ. Tokyo Sect. IA Math.* **18** (1971), 335–362. (MR 47, #5655).

[1972: 1]   Note on singular perturbation of linear operators. *Proc. Japan Acad.* **48** (1972), 595–598. (MR 48, #4831).

K. Yoshinaga

[1963: 1]   Ultra-distributions and semi-group distributions. *Bull. Kyushu Inst. Tech. Math. Natur. Sci.*, no. 10 (1963), 1–24. (MR 28, #466).

[1964: 1]   On Liouville's differentiation. *Bull. Kyushu Inst. Tech. Math. Natur. Sci.*, no. 11 (1964), 1–17. (MR 30, #5071).

[1965: 1]   Values of vector-valued distributions and smoothness of semi-group distributions. *Bull. Kyushu Inst. Tech. Math. Natur. Sci.*, no. 12 (1965), 1–27. (MR 32, #6215).

[1971: 1]   Fractional powers of an infinitesimal generator in the theory of semi-group distributions. *Bull. Kyushu Inst. Techn. Math. Natur. Sci.* **18** (1971), 1–15. (MR 54, #5902).

K. Yosida

[1936: 1]   On the group imbedded in the metrical complete ring. *Japan J. Math.* **13** (1936), 7–26.

[1938: 1]   Mean ergodic theorems in Banach spaces. *Proc. Imp. Acad. Sci. Tokyo* **14** (1938), 292–294.

[1948: 1]   On the differentiability and the representation of one-parameter semi-groups of linear operators. *J. Math. Soc. Japan* **1** (1948), 15–21. (MR 10, #462).

[1949: 1]   Integration of Fokker-Planck's equation in a compact Riemannian space. *Ark. Mat.* **1** (1949), 71–75. (MR 11, 443).

[1949: 2]   An operator-theoretical treatment of temporally homogeneous Markoff process. *J. Math. Soc. Japan* **1** (1949), 244–253. (MR 12, 190).

[1951: 1]   Integration of Fokker-Planck's equation with a boundary condition. *J. Math. Soc. Japan* **3** (1951), 69–73. (MR 13, 560).

[1951: 2]   Integrability of the backward diffusion equation in a compact Riemannian space. *Nagoya Math. J.* **3** (1951), 1–4. (MR 13, 560).

[1952: 1]   On the integration of diffusion equations in Riemannian spaces. *Proc. Amer. Math. Soc.* **3** (1952), 864–873. (MR 14, 560).

[1952: 2]   On Cauchy's problem in the large for wave equations. *Proc. Japan Acad.* **28** (1952), 396–406. (MR 14, 757).

[1953: 1]   On the fundamental solution of the parabolic equation in a Riemannian space. *Osaka Math. J.* **5** (1953), 65–74. (MR 15, 36).

[1954: 1]   On the integration of the temporally inhomogeneous diffusion equation in a Riemannian space. *Proc. Japan Acad.* **30** (1954), 19–23. (MR 16, 270).

[1954: 2]   On the integration of the temporally inhomogeneous diffusion equation in a Riemannian space, II. *Proc. Japan Acad.* **30** (1954), 273–275. (MR 16, 370).

[1955: 1]   A characterization of the second order elliptic differential operators. *Proc. Japan Acad.* **31** (1955), 406–409. (MR 17, 494).

[1955: 2]   On the generating parametrix of the stochastic processes. *Proc. Nat. Acad. Sci. U.S.A.* **41** (1955), 240–244. (MR 17, 167).

[1956: 1]   An operator-theoretical integration of the wave equation. *J. Math. Soc. Japan* **8** (1956), 79–92. (MR 19, 424).

[1957: 1]   An operator-theoretical integration of the temporally inhomogeneous wave equation. *J. Fac. Sci. Univ. Tokyo Sect. I.7* (1957), 463–466. (MR 19, #423).

[1958: 1]   On the differentiability of semi-groups of linear operators. *Proc. Japan Acad.* **34** (1958), 337–340. (MR 20, #5435).

[1959: 1]   An abstract analyticity in time for solutions of a diffusion equation. *Proc. Japan Acad.* **35** (1959), 109–113. (MR 21, #4298).

[1960: 1]   Fractional powers of infinitesimal generators and the analyticity of the semi-groups generated by them. *Proc. Japan Acad.* **36** (1960), 86–89. (MR 22, #12399).

[1961: 1]   Ergodic theorems for pseudo-resolvents. *Proc. Japan Acad.* **37** (1961), 422–425. (MR 25, #3375).

[1963: 1]   On the integration of the equation of evolution. *J. Fac. Sci. Univ. Tokyo Sect. I* **9** (1963), 397–402.

[1963: 2]   Holomorphic semi-groups in a locally convex linear topological space. *Osaka Math. J.* **15** (1963), 51–57. (MR 27, #2871).

[1965: 1]   A perturbation theorem for semi-groups of linear operators. *Proc. Japan Acad.* **41** (1965), 645–647. (MR 33, #6422).

[1965/66: 1]   Time dependent evolution equations in a locally convex space. *Math. Ann.* **162** (1965/66), 83–86. (MR 32, #7910).

[1968: 1]   The existence of the potential operarator associated with an equicontinuous semigroup of class (C_0). *Studia Math.* **31** (1968), 531–533. (MR 38, #2841).

[1971: 1]   On the existence and characterization of abstract potential operators. *Troisieme Colloque sur l'Analyse Fonctionnelle (Liege, 1970)*, 129–136. Vander, Louvain, 1971. (MR 53, #11393).

[1972: 1]   Abstract potential operators on Hilbert space. *Publ. Res. Inst. Math. Sci.* **8** (1972), 201–205. (MR 47, #4062).

[1978: 1]   *Functional Analysis*, 5th ed. Springer, Berlin, 1978.

J. Zabczyk

[1975: 1]   A note on C_0-semigroups. *Bull. Acad. Polon. Sci. Sér Sci. Math. Astronom. Phys.* **23** (1975), 895–898. (MR 52, #4025).

P. P. Zabreĭko and A. V. Zafievskiĭ

[1969: 1]   A certain class of semigroups. (Russian.) *Dokl. Akad. Nauk SSSR* **189** (1969), 934–937. (MR 41, #9053).

A. V. Zafievskiĭ

[1970: 1]   Semigroups that have singularities summable with a power weight at zero. (Russian.) *Dokl. Akad. Nauk SSSR* **195** (1970), 24–27. (MR 43, #3852).

S. D. Zaidman

[1957: 1]   Sur la perturbation presque-périodique des groupes et semi-groups de transformations d'un espace de Banach. *Rend. Mat. e Appl.* (5) (1957), 197–206. (MR 20, #1918).

[1963: 1]   Un teorema di esistenza per un problema non bene posto. *Atti. Accad. Naz. Lincei Rend. Cl. Sci. Fis. Mat. Natur.* (8), **35** (1963), 17–22. (MR 30, #357).

[1963: 2]   Quasi-periodicità per una equazione operazionale del primo ordine. *Atti Accad. Naz. Lincei Rend. Sem. Cl. Sci. Fis. Mat. Natur.* (8), **35** (1963), 152–157. (MR 29, #5134).

[1964: 1]   Un teorema de esistenza globale per alcune equazioni differenziali astratte. *Richerche Mat.* **13** (1964). (MR 30, #358).

[1964: 2]   Soluzioni quasi-periodiche per alcune equazioni differenziali in spazi hilbertiani. *Ricerche Mat.* **13** (1964), 118–134. (MR 29, #3910).

[1965: 1]   Convexity properties for weak solutions of some differential equations in Hilbert spaces. *Canad. J. Math.* **17** (1965), 802–807. (MR 34, #4643).

[1969: 1]   On elementary solutions of abstract differential equations. *Boll. Un. Mat. Ital.* (4) **2** (1969), 487–490. (MR 41, #612).

[1969: 2]   Uniqueness of bounded solutions for some abstract differential equations. *Ann. Univ. Ferrara Sez. VII* (*N.S.*) **14** (1969), 101–104. (MR 41, #7246).

[1969: 3]   Spectrum of almost-periodic solutions for some abstract differential equations. *J. Math. Anal. Appl.* **28** (1969), 336–338. (MR 39, #7303).

[1969: 4]   Bounded solutions of some abstract differential equations. *Proc. Amer. Math. Soc.* **23** (1969), 340–342. (MR 39, #7259).

[1971: 1]   Some results on almost-periodic differential equations. *Rend. Sem. Mat. Fis. Milano* (1971), 137–142. (MR 46, #5789).

[1971: 2]   On a certain approximaton property for first-order abstract differential equations. *Rend. Sem. Mat. Univ. Padova* **46** (1971), 191–198. (MR 46, #6097).

[1972: 1]   A remark on harmonic analysis of strongly almost-periodic groups of linear operators. *Ann. Scuola Norm. Sup. Pisa* **26** (1972), 645–648. (MR 52, #4026).

[1972: 2]   Remarks on boundedness of solutions of abstract differential equations. *Ann. Univ. Ferrara Sez. VII* (*N.S.*) **17** (1972), 43–45. (MR 48, #1144).

[1974: 1]   Remarks on weak solutions of differential equations in Banach spaces. *Boll. Un. Mat. Ital.* (4), **9** (1974), 638–643. (MR 52, #3692).

[1976: 1]   Lower bounds for solutions of weak evolution inequalities. *Boll. Un. Mat. Ital.* (4), **13** (1976), 92–106. (MR 53, #13775).

[1979: 1]   *Abstract Differential Equations.* Pitman, London, 1979.

M. Zaki

[1974: 1]   Almost automorphic solutions of certain abstract differential equations. *Ann. Mat. Pura Appl.* (4), **101** (1974), 91–114. (MR 51, #1059).

T. A. Zamanov

[1963: 1]   Dynamical systems and one-parameter groups of operators in Fréchet spaces. (Russian.) *Akad. Nauk Azerbaĭdžan. SSR Trudy Inst. Mat. Meh.* **2** (10), (1963), 104–128. (MR 27, #5999).

[1967: 1]   Differential equations with retarded argument in abstract spaces. (Russian.) *Izv. Akad. Nauk Azerbaĭdžan SSR Ser. Fiz.-Tehn Mat. Nauk 1967*, no. 1, 3–7. (MR 36, #1803).

[1968: 1]   The stability of the solutions of evolutionary equations with retarded argument in a Banach space. (Russian.) *Izv. Akad. Nauk Azerbaĭdžan SSR Ser Fiz-Tehn. Mat. Nauk* (1968), no. 3, 129–135. (MR 40, #6017).

[1968: 2]   A certain operator equation in Hilbert space. (Russian. English summary.) *Vestnik Moskov. Univ. Ser. I Mat. Meh.* **23** (1968), no. 1, 35–40. (MR 36, #5486).

[1968: 3]   On the theory of the solutions of evolution equations with retarded argument. (Russian.) *Vestnik Moskov. Univ. Ser. I Mat. Meh.* **23** (1968), no. 4, 37–44. (MR 39, #1775).

[1969: 1]   The stability of the solutions of differential equations with retarded argument under perturbations which are bounded in the mean. (Russian.) *Differencial'nye Uravnenija* **5** (1969), 488–498. (MR 40, #503).

A. T. Zaplitnaja

[1973: 1]   A certain operator equation in Hilbert space. (Russian.) *Ukrain Mat. Ž.* **25** (1973), 796–797. (MR 48, #9026).

A. G. Zarubin

[1970: 1]   The Cauchy problem for a certain class of nonstationary quasilinear operator equations. (Russian.) *Voronež. Gos. Univ. Trudy Naučn.-Issled. Inst. Mat. VGU Vyp. l Sb. stati po teorii operatornyh uravnenii* (1970), 82–90. (MR 55, #798).

[1970: 2]   The Bubnov-Galerkin method for a certain class of quasilinear equations. (Russian.) *Voronež. Gos. Univ. Trudy Naučn.-Issled. Inst. Mat. VGU Vyp. l Sb. statei po teorii operatornyh uravnenii* (1970), 91–99. (MR 55, #799).

A. G. Zarubin and M. F. Tiunčik

[1973: 1]   Approximate solution of a certain class of nonlinear nonstationary equations. (Russian.) *Differencial'nye Uravnenija* **9** (1973), 1966–1974, 2114. (MR 49, #5508).

V. V. Žikov

[1965: 1]   Abstract equations with almost-periodic coefficients. (Russian.) *Dokl. Akad. Nauk SSSR* **163** (1965), 555–558. (MR 32, #6014).

[1965: 2]   Almost periodic solutions of differential equations in Hilbert space. (Russian.) *Dokl. Akad. Nauk SSSR* **165** (1965), 1227–1230. (MR 33, #355).

[1966: 1]   On the question of harmonic analysis of bounded solutions of operator equations. (Russian.) *Dokl. Akad. Nauk SSSR* **169** (1966), 1254–1257. (MR 33, #6090).

[1967: 1]   Almost periodic solutions of differential equations in Banach spaces.
              (Russian.) *Teor. Funkciĭ Funkcional. Anal. i Priložen. Vyp.* **4** (1967),
              176–188. (MR 36, #6754).
[1970: 1]   On the problem of the existence of almost periodic solutions of
              differential and operator equations. (Russian.) *Akad. Nauk Azerbaĭdžan
              SSR Dokl.* **26** (1970), no. 7, 3–7. (MR 44, #5944).
[1970: 2]   Almost periodic solutions of linear and nonlinear equations in a
              Banach space. (Russian.) *Dokl. Akad. Nauk SSSR* **195** (1970), 278–281.
              (MR 43, #667).
[1970: 3]   A certain complement to the classical theory of Favard. (Russian.)
              Mat. Zametki **7** (1970), 239–246. (MR 41, #2134).
[1971: 1]   The existence of solutions, almost periodic in the sense of Levitan, of
              linear systems (second supplement to the classical theory of Favard).
              (Russian.) *Mat. Zametki* **9** (1971), 409–414. (MR 43, #7720).
[1972: 1]   On the theory of admissibility of pairs of function spaces. (Russian.)
              Dokl. Akad. Nauk SSSR **205** (1972), 1281–1283. (MR 47, #5394).
[1975: 1]   Some new results in the abstract Favard theory. (Russian.) *Mat.
              Zametki* **17** (1975), 33–40. (MR 51, #1060).

V. V. Žikov and B. M. Levitan
[1977: 1]   Favard theory. (Russian.) *Uspehi Mat. Nauk* **32** (1977), 123–171. (MR
              57, #10159).

L. A. Zitokovskiĭ
[1971: 1]   Existence theorems and uniqueness classes of the solutions of func-
              tional equations with heredity. (Russian.) *Differencial'nye Uravnenija* **7**
              (1971), 1377–1384. (MR 44, #7099).

A. Zygmund
[1959: 1]   *Trigonometric Series*, Cambridge University Press, Cambridge, 1959.

Notation and Subject Index

Notation and Subject Index